Microwave Engineering

Theory and Techniques

INTERNATIONAL ADAPTATION

Microwave Engineering

Theory and Techniques

Fourth Edition

INTERNATIONAL ADAPTATION

David M. Pozar

University of Massachusetts at Amherst

WILEY

Microwave Engineering

Theory and Techniques

Fourth Edition

INTERNATIONAL ADAPTATION

Contributing Subject Matter Experts: Prabu K, NIT Karnataka, Surathkal, India; Santosh A. Janawade, RV College of Engineering, Bengaluru, India

Founded in 1807, John Wiley & Sons, Inc. has been a valued source of knowledge and understanding for more than 200 years, helping people around the world meet their needs and fulfill their aspirations. Our company is built on a foundation of principles that include responsibility to the communities we serve and where we live and work. In 2008, we launched a Corporate Citizenship Initiative, a global effort to address the environmental, social, economic, and ethical challenges we face in our business. Among the issues we are addressing are carbon impact, paper specifications and procurement, ethical conduct within our business and among our vendors, and community and charitable support. For more information, please visit our website: www.wiley.com/go/citizenship.

ISBN: 978-1-119-77061-9

ISBN: 978-1-119-77062-6 (ePub)

ISBN: 978-1-119-77063-3 (ePdf)

Printed and bound by CPI Group (UK) Ltd, Croydon, CR0 4YY

C9781119770619_181023

Contents

Preface

The continuing popularity of *Microwave Engineering* is gratifying. I have received many letters and emails from students and teachers from around the world with positive comments and suggestions. I think one reason for its success is the emphasis on the fundamentals of electromagnetics, wave propagation, network analysis, and design principles as applied to modern RF and microwave engineering. As I have stated in earlier editions, I have tried to avoid the handbook approach in which a large amount of information is presented with little or no explanation or context, but a considerable amount of material in this book is related to the design of specific microwave circuits and components, for both practical and motivational value. I have tried to base the analysis and logic behind these designs on first principles, so the reader can see and understand the process of applying fundamental concepts to arrive at useful results. The engineer who has a firm grasp of the basic concepts and principles of microwave engineering and knows how these can be applied toward practical problems is the engineer who is the most likely to be rewarded with a creative and productive career.

For this International Adaptation I again solicited detailed feedback from teachers and readers for their thoughts about how the book should be revised. The most common requests were for more material on active circuits, noise, nonlinear effects, and wireless systems. This edition, therefore, now has separate chapters on noise and nonlinear distortion, and active devices. In Chapter 1, the section on applications of microwaves has been rewritten to include applications in different domains, and new material on electromagnetic compatibility versus electromagnetic interference has been added. For Chapter 7, the section on dividers and couplers has been updated to include categorization of three-port networks as T-junctions, power dividers, and circulators. The discussion on T-junction power dividers and Wilkinson power dividers is followed by a table listing the differences between the two networks. For Chapter 8, on microwave filters, a table showing the comparison between Richards' transformation and stepped-impedance method has been added. In Chapter 10, the coverage of noise has been expanded, along with more material on intermodulation distortion and related nonlinear effects. In Chapter 11, on active devices, I have added updated material on bipolar junction and field effect transistors, including data for a number of commercial devices (Schottky and PIN diodes, and Si, GaAs, GaN, and SiGe transistors), and these sections have been reorganized and rewritten. New material on Ridley–Watkins–Hilsum (RWH) theory and two-valley model theory has also been added in the chapter. The coverage of klystrons has been updated to include working principles of different types of klystrons. Chapters 12 and 13 treat active circuit design, and discussions of differential amplifiers, inductive degeneration for nMOS amplifiers, and differential FET and Gilbert cell mixers have been added. In Chapter 14, on RF and microwave systems, I have updated and added new material on wireless communications systems, including link budget, link margin, digital modulation methods, and bit error rates. The section on radiation hazards has been updated and rewritten. Description of additional types of antennas has been added, and the material on wireless communication standards and IEEE safety standards has been updated to include new standards. Other new material includes a section on transients on transmission lines (material that was originally in the first edition, cut from later editions, and now brought back by popular demand), the theory of power waves, a discussion of higher order modes and frequency effects for microstrip line, and a discussion of how to determine unloaded Q from resonator measurements. This edition also has numerous new or revised problems and examples, including several questions of the "open-ended" variety, as well as multiple choice questions. Material that has been cut from this edition includes the quasi-static numerical analysis of microstrip line, excitation of waveguides using aperture coupling, Bode Fano criterion, cavity perturbations, filters using coupled resonators, and some material related to microwave tubes. Finally, working from the original source files, I have made hundreds of corrections and rewrites of the original text.

Today, microwave and RF technology is more pervasive than ever. This is especially true in the commercial sector, where modern applications include cellular telephones, smartphones, 3G and WiFi wireless networking, millimeter wave collision sensors for vehicles, direct broadcast satellites for radio, television, and networking, global positioning systems, radio frequency identification tagging, ultra wideband radio and radar systems, and microwave remote sensing systems for the environment. Defense systems continue to rely heavily on microwave technology for passive and active sensing, communications, and weapons control systems. There should be no shortage of challenging problems in RF and microwave engineering in the foreseeable future, and there will be a clear need for engineers having both an understanding of the fundamentals of microwave engineering and the creativity to apply this knowledge to problems of practical interest.

Modern RF and microwave engineering predominantly involves distributed circuit analysis and design, in contrast to the waveguide and field theory orientation of earlier generations. The majority of microwave engineers today design planar components and integrated circuits without direct recourse to electromagnetic analysis. Microwave computer-aided design (CAD) software and network analyzers are the essential tools of today's microwave engineer, and microwave engineering education must respond to this shift in emphasis to network analysis, planar circuits and components, and active circuit design. Microwave engineering will always involve electromagnetics (many of the more sophisticated microwave CAD packages implement rigorous field theory solutions), and students will still benefit from an exposure to subjects such as waveguide modes and coupling through apertures, but the change in emphasis to microwave circuit analysis and design is clear.

This text is written for a two-semester course in RF and microwave engineering for seniors or first-year graduate students. It is possible to use *Microwave Engineering* with or without an electromagnetics emphasis. Many instructors today prefer to focus on circuit analysis and design, and there is more than enough material in Chapters 2, 4–8, and 10–14 for such a program with minimal or no field theory requirement. Some instructors may wish to begin their course with Chapter 14 on systems in order to provide some motivational context for the study of microwave circuit theory and components. This can be done, but some basic material on noise from Chapter 10 may be required.

Two important items that should be included in a successful course on microwave engineering are the use of CAD simulation software and a microwave laboratory experience. Providing students with access to CAD software allows them to verify results of the design-oriented problems in the text, giving immediate feedback that builds confidence and makes the effort more rewarding. Because the drudgery of repetitive calculation is eliminated, students can easily try alternative approaches and explore problems in more detail. The effect of line losses, for example, is explored in several examples and problems; this would be effectively impossible without the use of modern CAD tools. In addition, classroom exposure to CAD tools provides useful experience upon graduation. Most of the commercially available microwave CAD tools are very expensive, but several manufacturers provide academic discounts or free "student versions" of their products. Feedback from reviewers was almost unanimous, however, that the text should not emphasize a particular software product in the text or in supplementary materials.

A hands-on microwave instructional laboratory is expensive to equip but provides the best way for students to develop an intuition and physical feeling for microwave phenomena. A laboratory with the first semester of the course might cover the measurement of microwave power, frequency, standing wave ratio, impedance, and scattering parameters, as well as the characterization of basic microwave components such as tuners, couplers, resonators, loads, circulators, and filters. Important practical knowledge about connectors, waveguides, and microwave test equipment will be acquired in this way. A more advanced laboratory session can consider topics such as noise figure, intermodulation distortion, and mixing. Naturally, the type of experiments that can be offered is heavily dependent on the test equipment that is available.

INSTRUCTOR RESOURCES

Additional resources for instructors are available on the Wiley website (www.wiley.com/college/pozar). These include PowerPoint slides, a suggested laboratory manual, and solutions manual for all problems in the text.

ACKNOWLEDGMENTS

It is a pleasure to acknowledge the many students, readers, and teachers who have used the first three editions of *Microwave Engineering*, and have written with comments, praise, and suggestions. I would also like to thank my colleagues in the microwave engineering group at the University of Massachusetts at Amherst for their support and collegiality over many years. In addition I would like to thank Bob Jackson (University of Massachusetts) for suggestions on MOSFET amplifiers and related material; Juraj Bartolic (University of Zagreb) for the simplified derivation of the μ-parameter stability criteria; and Jussi Rahola (Nokia Research Center) for his discussions of power waves. I am also grateful to the following people for providing new photographs for this edition: Kent Whitney and Chris Koh of Millitech Inc., Tom Linnenbrink and Chris Hay of Hittite Microwave Corp., Phil Beucler and Lamberto Raffaelli of LNX Corp., Michael Adlerstein of Raytheon Company, Bill Wallace of Agilent Technologies Inc., Jim Mead of ProSensing Inc., Bob Jackson and B. Hou of the University of Massachusetts, J. Wendler of M/A-COM Inc., Mohamed Abouzahra of Lincoln Laboratory, and Dev Gupta, Abbie Mathew, and Salvador Rivera of Newlans Inc. I would also like to thank Sherrill Redd, Philip Koplin, and the staff at Aptara, Inc. for their professional efforts during production of this book. Also, thanks to Ben for help with PhotoShop.

David M. Pozar
Amherst

Review of Electromagnetic Theory

Learning Objectives

After completing this chapter, you will be able to

- Understand the theoretical principles and concepts of electromagnetics, Maxwell's equations and its applications
- Learn about the importance of EMI/EMC, and applications of microwaves in various disciplines
- Learn about the concepts of electrostatics, magnetostatics, and their applications

- Apply basic knowledge of science, mathematics, and engineering to the analysis and design of electromagnetic components involving magnetic fields and electric fields as well as electromagnetic waves
- Describe different types of medium, wave propagation, and their incidence on materials

We begin our study of microwave engineering with a brief overview of the history and major applications of microwave technology, followed by a review of some of the fundamental topics in electromagnetic theory that we will need throughout the book. Further discussion of these topics may be found in references [1–8].

1.1 INTRODUCTION TO MICROWAVE ENGINEERING

The field of radio frequency (RF) and microwave engineering generally covers the behavior of alternating current signals with frequencies in the range of 100 MHz (1 MHz = 10^6 Hz) to 1000 GHz (1 GHz = 10^9 Hz). RF frequencies range from very high frequency (VHF) (30–300 MHz) to ultra high frequency (UHF) (300–3000 MHz), while the term *microwave* is typically used for frequencies between 3 and 300 GHz, with a corresponding electrical wavelength between $\lambda = c/f = 10$ cm and $\lambda = 1$ mm, respectively. Signals with wavelengths on the order of millimeters are often referred to as *millimeter waves*. Figure 1.1 shows the location of the RF and microwave frequency bands in the electromagnetic spectrum. Because of the high frequencies (and short wavelengths), standard circuit theory often cannot be used directly to solve microwave network problems. In a sense, standard circuit theory is an approximation, or special case, of the broader theory of electromagnetics as described by Maxwell's equations. This is due to the fact that, in general, the lumped circuit element approximations of circuit theory may not be valid at high RF and microwave frequencies. Microwave components often act as *distributed elements*, where the phase of the voltage or current changes significantly over the physical extent of the device because the device dimensions are on the order of the electrical wavelength. At much lower frequencies the wavelength is large enough that there is insignificant phase variation across the dimensions of a component. The other extreme of frequency can be identified as optical engineering, in which the wavelength is much shorter than the dimensions of the component. In this case Maxwell's equations can be simplified to the geometrical optics regime, and optical systems can be designed with the theory of geometrical optics. Such techniques are sometimes applicable to millimeter wave systems, where they are referred to as *quasi-optical*.

In RF and microwave engineering, then, one must often work with Maxwell's equations and their solutions. It is in the nature of these equations that mathematical complexity arises since Maxwell's equations involve vector differential or integral operations on vector field quantities, and these fields are functions of spatial coordinates. One of the goals of this book is to try to reduce the complexity of a field theory solution to a result that can be expressed in terms of simpler circuit theory, perhaps extended to include distributed elements (such as transmission lines) and concepts (such as reflection coefficients and scattering

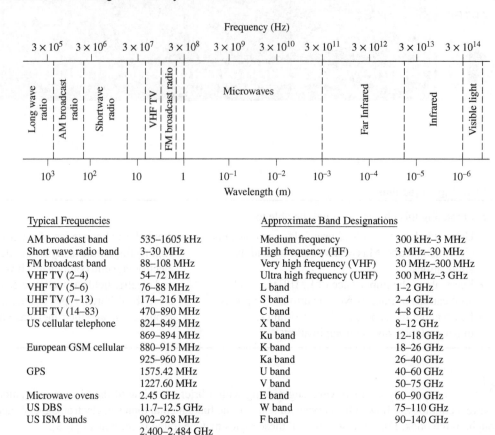

Typical Frequencies		Approximate Band Designations	
AM broadcast band	535–1605 kHz	Medium frequency	300 kHz–3 MHz
Short wave radio band	3–30 MHz	High frequency (HF)	3 MHz–30 MHz
FM broadcast band	88–108 MHz	Very high frequency (VHF)	30 MHz–300 MHz
VHF TV (2–4)	54–72 MHz	Ultra high frequency (UHF)	300 MHz–3 GHz
VHF TV (5–6)	76–88 MHz	L band	1–2 GHz
UHF TV (7–13)	174–216 MHz	S band	2–4 GHz
UHF TV (14–83)	470–890 MHz	C band	4–8 GHz
US cellular telephone	824–849 MHz	X band	8–12 GHz
	869–894 MHz	Ku band	12–18 GHz
European GSM cellular	880–915 MHz	K band	18–26 GHz
	925–960 MHz	Ka band	26–40 GHz
GPS	1575.42 MHz	U band	40–60 GHz
	1227.60 MHz	V band	50–75 GHz
Microwave ovens	2.45 GHz	E band	60–90 GHz
US DBS	11.7–12.5 GHz	W band	75–110 GHz
US ISM bands	902–928 MHz	F band	90–140 GHz
	2.400–2.484 GHz		
	5.725–5.850 GHz		
US UWB radio	3.1–10.6 GHz		

FIGURE 1.1 | The electromagnetic spectrum.

parameters). A field theory solution generally provides a complete description of the electromagnetic field at every point in space, which is usually much more information than we need for most practical purposes. We are typically more interested in terminal quantities such as power, impedance, voltage, and current, which can often be expressed in terms of these extended circuit theory concepts. It is this complexity that adds to the challenge, as well as the rewards, of microwave engineering.

Applications of Microwave Engineering

Just as the high frequencies and short wavelengths of microwave energy make for difficulties in the analysis and design of microwave devices and systems, these same aspects provide unique opportunities for the application of microwave systems. The following considerations can be useful in practice:

- Antenna gain is proportional to the electrical size of the antenna. At higher frequencies, more antenna gain can be obtained for a given physical antenna size, and this has important consequences when implementing microwave systems.
- More bandwidth (directly related to data rate) can be realized at higher frequencies. A 1% bandwidth at 600 MHz is 6 MHz, which (with binary phase shift keying modulation) can provide a data rate of about 6 Mbps (megabits per second), while at 60 GHz a 1% bandwidth is 600 MHz, allowing a 600 Mbps data rate.
- Microwave signals travel by line of sight and are not bent by the ionosphere as are lower frequency signals. Satellite and terrestrial communication links with very high capacities are therefore possible, with frequency reuse at minimally distant locations.
- The effective reflection area (radar cross section) of a radar target is usually proportional to the target's electrical size. This fact, coupled with the frequency characteristics of antenna gain, generally makes microwave frequencies preferred for radar systems.

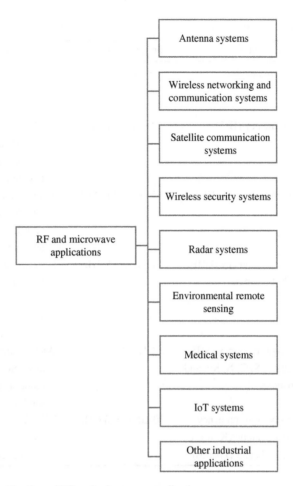

FIGURE 1.2 | Broad classification of RF and microwave applications.

- Various molecular, atomic, and nuclear resonances occur at microwave frequencies, creating a variety of unique applications in the areas of basic science, remote sensing, medical diagnostics and treatment, and heating methods.

The majority of today's applications of RF and microwave technology are to antennas, wireless networking and communication systems, wireless security systems, radar systems, environmental remote sensing, navigation, microwave radiometry, medical systems, Internet of Things (IoT) systems, and other applications. As the frequency allocations listed in Figure 1.1 show, RF and microwave communication systems are pervasive, especially today when wireless connectivity promises to provide voice and data access to "anyone, anywhere, at any time." The applications of RF and microwaves are shown in Figure 1.2, and their typical uses are listed in Table 1.1.

Antenna systems
- A microwave antenna is used to broadcast information between two or more locations. It is also used in radio astronomy, radar, and electronic warfare.
- Recent advances in 3D printing and additive manufacturing have enabled complex RF structures to be realized. Metamaterial and fractal-based antennas have revolutionized the antenna technology in the recent years. These antennas with their new shape, direction, and applications have helped in overcoming the challenges associated with traditional antenna technology.

Wireless communication
- Modern wireless telephony is based on the concept of cellular frequency reuse, a technique first proposed by Bell Labs in 1947 but not practically implemented until the 1970s. By this time advances in miniaturization, as well as increasing demand for wireless communication, drove the introduction

TABLE 1.1 | Microwave Frequency Bands and Their Typical Applications

Frequency Band	Frequency Range	Typical Uses
Medium frequency	300 kHz–3 MHz	AM medium wave broadcasting, coast-to-sea communication, snow avalanches, amateur radio, marine patrol
High frequency (HF)	3 MHz–30 MHz	RFID, short wave broadcasting, flight communication beyond the horizon, amateur and civic radio, radar communication beyond the horizon, mobile marine communication, and sky waves
Very high frequency (VHF)	30 MHz–300 MHz	Television broadcasting, FM, line-of-vision communication between aircrafts or ground to aircraft, mobile land, weather radio, marine communication, and amateur radio
Ultra high frequency (UHF) (includes L band)	300 MHz–3 GHz	Networks and communication devices: 2G, 3G, 4G, and LTE, microwave ovens, TV broadcasting, microwave instruments and communication, wireless LAN, radio astronomy, GPS, cellular devices, Bluetooth, GMRS radio, two-way FRS, and amateur radio
Super high frequency (SHF) (includes S, C, X, Ku, and K bands)	3 GHz–30 GHz	Wi-Fi, Wireless LAN, Wi-Max, modern communication technologies, microwave devices, TV broadcasting, modern radars and communication, broadcasting satellites, radio astronomy, and amateur radio
Extremely high frequency (EHF) (includes K, Ka, U, V, E, W, F, and D bands)	30 GHz–300 GHz	High frequency microwave relays, radio astronomy, amateur radio, remote sensing at microwave frequency, millimeter wave scanner, and guided energy weapons

of several early cellular telephone systems in Europe, the United States, and Japan. The Nordic Mobile Telephone (NMT) system was deployed in 1981 in the Nordic countries, the Advanced Mobile Phone System (AMPS) was introduced in the United States in 1983 by AT&T, and NTT in Japan introduced its first mobile phone service in 1988. All these early systems used analog FM modulation, with their allocated frequency bands divided into several hundred narrowband voice channels. These early systems are usually referred to now as *first-generation* cellular systems, or 1G.

- *Second-generation* (2G) cellular systems achieved improved performance by using various digital modulation schemes, with systems such as GSM, CDMA, DAMPS, PCS, and PHS being some of the major standards introduced in the 1990s in the United States, Europe, and Japan. These systems can handle digitized voice, as well as some limited data, with data rates typically in the 8 to 14 kbps range. In recent years there has been a wide variety of new and modified standards to transition to handheld services that include voice, texting, data networking, positioning, and Internet access. These standards are variously known as 2.5G, 3G, 3.5G, 3.75G, 4G, 5G, 6G, and beyond, with current plans to provide data rates up to at least 1 Gbps. The number of subscribers to wireless services seems to be keeping pace with the growing power and access provided by modern handheld wireless devices; as of 2020 there were more than seven billion cell phone users worldwide.

- Satellite systems also depend on RF and microwave technology, and satellites have been developed to provide cellular (voice), video, and data connections worldwide. Two large satellite constellations, Iridium and Globalstar, were deployed in the late 1990s to provide worldwide telephony service. Unfortunately, these systems suffered from both technical drawbacks and weak business models and have led to multibillion dollar financial failures. However, smaller satellite systems, such as the Global Positioning Satellite (GPS) system and the Direct Broadcast Satellite (DBS) system, have been extremely successful.

- Wireless local area networks (WLANs) provide high-speed networking between computers over short distances, and the demand for this capability is expected to remain strong. One of the newer examples of wireless communication technology is ultra wide band (UWB) radio, where the broadcast signal occupies a very wide frequency band but with a very low power level (typically below the ambient radio noise level) to avoid interference with other systems.

- Metropolitan area network (MAN) protocols, such as Worldwide Interoperability for Microwave Access (WiMAX) based on the IEEE 8.2.16 specification, operate between 2 and 11 GHz. The commercial executions are in the range of 2.3 GHz, 2.5 GHz, 3.5 GHz, and 5.8 GHz.
- Wide area mobile broadband wireless access protocols are based on the IEEE 8.2.20 specification, operating between 1.6 and 2.3 GHz. It offers great mobility and building penetration characteristics with better spectral efficiency.

Satellite communication systems
- Microwaves are mostly used for geostationary satellites while radio waves are used for low-flying earth satellites. Satellite communication uses microwaves because of their transparency, less fading, and high directivity properties.

Wireless security systems
- Audio and video surveillance has become an essential component of security systems. RF and microwaves are used in wireless security cameras to transmit and receive audio and video signals and are highly useful in microwave home security systems.

Radar systems
- Radar is used for detecting and locating air, ground, and seagoing targets. Radar systems find application in military, commercial, and scientific fields. It is also useful in surveillance, navigation, guidance of weapons, electronic warfare, and fire control. In the commercial sector, radar technology is used for air traffic control, motion detectors (door openers and security alarms), vehicle collision avoidance, maritime safety, and distance measurement. Scientific applications of radar include weather prediction, remote sensing of the atmosphere, the oceans, and the ground, as well as medical diagnostics and therapy.
- Frequency selective surfaces (FSS) are essentially periodic surfaces and are textured electromagnetic surfaces which can be designed for different types of spatial electromagnetic filtering properties. FSS-based radar dome (RADOME), being an integral part of stealth technology, is classified and constitutes a frontier area of microwave research and development. The design and development is of highly interdisciplinary nature and involves applied electromagnetics.

Navigation
- The Global Positioning System (GPS), Global Navigation Satellite System (GNSS), and the Russian GLONASS applications are using the frequency between 1.2 GHz and 1.6 GHz for navigation applications.

Environmental remote sensing
- Microwave radiometry (MWR), which is the passive sensing of microwave energy emitted by an object, is used for remote sensing of the atmosphere and the earth. The sensitive receivers of MWRs measure thermal electromagnetic radiation released by atmospheric gases. They are usually equipped with multiple receiving channels in order to derive the characteristic emission spectrum of the atmosphere or extraterrestrial objects. MWRs are used in a variety of remote sensing applications such as climate monitoring and weather forecasting.

Medical applications
- Microwaves are highly useful in biomedical applications such as microwave imaging, thermotherapy, photoacoustic imaging in biomedicine, tumor and lung water detection.
- Microwave thermotherapy uses microwaves to destroy excess tissue and it is mainly used for certain bladder or intestinal problems.
- Radio waves are important in nuclear magnetic resonance imaging, i.e. MRI.

IoT applications
- Wireless connectivity is the enabler for IoT applications. Long term evolution (LTE) technology has evolved over time with more efficient use of RF spectrum for IoT connectivity.
- IoT-enabled microwave ovens are getting a lot of attention because of their fully automated control over voice, mobile, and web. The interesting feature of an IoT-enabled microwave oven is that it scans the barcode of the product and cooks it automatically by fetching the instructions from the "online database." Apart from that it can also "tweet," once the food is ready.

Industrial applications

- Microwave technology is used in food, semiconductor, and various other industries. For example, in food industry, microwaves are used in various food processes such as cooking and reheating, roasting food grains, drying, moisture leveling, and absorbing water molecules. In semiconductor and other industries, microwaves are used in reactive ion etching, chemical vapor processing, and drying and reaction processes.

A Short History of Microwave Engineering

Microwave engineering is often considered a fairly mature discipline because the fundamental concepts were developed more than 50 years ago, and probably because radar, the first major application of microwave technology, was intensively developed as far back as World War II. However, recent years have brought substantial and continuing developments in high-frequency solid-state devices, microwave integrated circuits, and computer-aided design techniques, and the ever-widening applications of RF and microwave technology to wireless communication, networking, sensing, and security have kept the field active and vibrant.

The foundations of modern electromagnetic theory were formulated in 1873 by James Clerk Maxwell, who hypothesized, solely from mathematical considerations, electromagnetic wave propagation and the idea that light was a form of electromagnetic energy. Maxwell's formulation was cast in its modern form by Oliver Heaviside during the period from 1885 to 1887. Heaviside was a reclusive genius whose efforts removed many of the mathematical complexities of Maxwell's theory, introduced vector notation, and provided a foundation for practical applications of guided waves and transmission lines. Heinrich Hertz, a German professor of physics and a gifted experimentalist who understood the theory published by Maxwell, carried out a set of experiments during the period 1887–1891 that validated Maxwell's theory of electromagnetic waves. It is interesting to observe that this is an instance of a discovery occurring after a prediction has been made on theoretical grounds—a characteristic of many of the major discoveries throughout the history of science. All of the practical applications of electromagnetic theory—radio, television, radar, cellular telephones, and wireless networking—owe their existence to the theoretical work of Maxwell.

Because of the lack of reliable microwave sources and other components, the rapid growth of radio technology in the early 1900s occurred primarily in the HF to VHF range. It was not until the 1940s and the advent of radar development during World War II that microwave theory and technology received substantial interest. In the United States, the Radiation Laboratory was established at the Massachusetts Institute of Technology to develop radar theory and practice. A number of talented scientists, including N. Marcuvitz, I. I. Rabi, J. S. Schwinger, H. A. Bethe, E. M. Purcell, C. G. Montgomery, and R. H. Dicke, among others, gathered for a very intensive period of development in the microwave field. Their work included the theoretical and experimental treatment of waveguide components, microwave antennas, small-aperture coupling theory, and the beginnings of microwave network theory. Many of these researchers were physicists who returned to physics research after the war, but their microwave work is summarized in the classic 28-volume Radiation Laboratory Series of books that still finds application today.

Communication systems using microwave technology began to be developed soon after the birth of radar, benefiting from much of the work that was originally done for radar systems. The advantages offered by microwave systems, including wide bandwidths and line-of-sight propagation, have proved to be critical for both terrestrial and satellite communication systems and have thus provided an impetus for the continuing development of low-cost miniaturized microwave components. We refer the interested reader to references [1] and [2] for further historical perspectives on the fields of wireless communication and microwave engineering.

Before we begin the study of Maxwell's equations, let us briefly review the concepts of EMC and EMI.

Electromagnetic Compatibility and Electromagnetic Interference

Electromagnetic compatibility (EMC) is the capability of electrical systems and equipment to operate reasonably in their electromagnetic atmosphere, by controlling the unintended generation, reception, and propagation of electromagnetic energy which can cause undesirable effects such as electromagnetic interference (EMI) or physical damage in functioning equipment. The main objective of EMC is the precise function of different equipment in a common electromagnetic environment. The sources of EMI can be natural events such as solar radiation or electrical storms and other electrical systems or electronic devices. The various EMC tests such as conducted immunity, radiated immunity, and conducted and radiated emission are used to evaluate how a device will react to the exposed electromagnetic energy and the amount of EMI generated by the devices. Both characteristics of EMC are vital in design and engineering considerations of any system. If a device fails to satisfy the EMC test, it undergoes undesirable changes such as product failure, safety risks, and data loss. The EMI/EMC control test and measurement offers EMI/EMC evaluation and certification testing of flight, crew, and ground support equipment such as instrumentation, communication, computation, biomedicine, robotics, guidance, and navigation. Consumer goods such as cellular phones, laptops, microwave ovens, and satellite TV dishes all must undergo EMI/EMC testing to ensure they do not cause harmful interference and accept interference without causing undesired effects in real-world conditions.

1.2 MAXWELL'S EQUATIONS

Electric and magnetic phenomena at the macroscopic level are described by Maxwell's equations, as published by Maxwell in 1873. This work summarized the state of electromagnetic science at that time and hypothesized from theoretical considerations the existence of the electrical displacement current, which led to the experimental discovery by Hertz of electromagnetic wave propagation. Maxwell's work was based on a large body of empirical and theoretical knowledge developed by Gauss, Ampere, Faraday, and others. A first course in electromagnetics usually follows this historical (or deductive) approach, and it is assumed that the reader has had such a course as a prerequisite to the present material. Several references are available [3–7] that provide a good treatment of electromagnetic theory at the undergraduate or graduate level.

This chapter will outline the fundamental concepts of electromagnetic theory that we will require later in the book. Maxwell's equations will be presented, and boundary conditions and the effect of dielectric and magnetic materials will be discussed. Wave phenomena are of essential importance in microwave engineering, and thus much of the chapter is spent on topics related to plane waves. Plane waves are the simplest form of electromagnetic waves and so serve to illustrate a number of basic properties associated with wave propagation. Although it is assumed that the reader has studied plane waves before, the present material should help to reinforce the basic principles in the reader's mind and perhaps to introduce some concepts that the reader has not seen previously. This material will also serve as a useful reference for later chapters.

With an awareness of the historical perspective, it is usually advantageous from a pedagogical point of view to present electromagnetic theory from the "inductive," or axiomatic, approach by beginning with Maxwell's equations. The general form of time-varying Maxwell equations, then, can be written in "point," or differential, form as

$$\nabla \times \bar{\mathcal{E}} = \frac{-\partial \bar{B}}{\partial t} - \bar{\mathcal{M}}, \tag{1.1a}$$

$$\nabla \times \bar{\mathcal{H}} = \frac{\partial \bar{D}}{\partial t} + \bar{J}, \tag{1.1b}$$

$$\nabla \cdot \bar{D} = \rho, \tag{1.1c}$$

$$\nabla \cdot \bar{B} = 0. \tag{1.1d}$$

The MKS system of units is used throughout this book. The script quantities represent time-varying vector fields and are real functions of spatial coordinates x, y, z, and the time variable t. These quantities are defined as follows:

$\bar{\mathcal{E}}$ is the electric field, in volts per meter (V/m).[1]

$\bar{\mathcal{H}}$ is the magnetic field, in amperes per meter (A/m).

---

[1] As recommended by the *IEEE Standard Definitions of Terms for Radio Wave Propagation, IEEE Standard 211-1997*, the terms "electric field" and "magnetic field" are used in place of the older terminology of "electric field intensity" and "magnetic field intensity."

$\bar{D}$ is the electric flux density, in coulombs per meter squared (Coul/m^2).

$\bar{B}$ is the magnetic flux density, in webers per meter squared (Wb/m^2).

$\bar{\mathcal{M}}$ is the (fictitious) magnetic current density, in volts per meter squared (V/m^2).

$\bar{J}$ is the electric current density, in amperes per meter squared (A/m^2).

ρ is the electric charge density, in coulombs per meter cubed (Coul/m^3).

The sources of the electromagnetic field are the currents $\bar{\mathcal{M}}$ and $\bar{J}$ and the electric charge density ρ. The magnetic current $\bar{\mathcal{M}}$ is a fictitious source in the sense that it is only a mathematical convenience: the real source of a magnetic current is always a loop of electric current or some similar type of magnetic dipole, as opposed to the flow of an actual magnetic charge (magnetic monopole charges are not known to exist). Since electric current is really the flow of charge, it can be said that the electric charge density ρ is the ultimate source of the electromagnetic field.

In free-space, the following simple relations hold between the electric and magnetic field intensities and flux densities:

$$\bar{B} = \mu_0 \bar{H}, \tag{1.2a}$$

$$\bar{D} = \epsilon_0 \bar{\mathcal{E}}, \tag{1.2b}$$

where $\mu_0 = 4\pi \times 10^{-7}$ henry/m is the permeability of free-space, and $\epsilon_0 = 8.854 \times 10^{-12}$ farad/m is the permittivity of free-space. We will see in the next section how media other than free-space affect these constitutive relations.

Equations (1.1a)–(1.1d) are linear but are not independent of each other. For instance, consider the divergence of (1.1a). Since the divergence of the curl of any vector is zero [vector identity (B.12), from Appendix B], we have

$$\nabla \cdot \nabla \times \bar{\mathcal{E}} = 0 = -\frac{\partial}{\partial t}(\nabla \cdot \bar{B}) - \nabla \cdot \bar{\mathcal{M}}.$$

Since there is no free magnetic charge, $\nabla \cdot \bar{\mathcal{M}} = 0$, which leads to $\nabla \cdot \bar{B} = 0$, or (1.1d). The *continuity equation* can be similarly derived by taking the divergence of (1.1b), giving

$$\nabla \cdot \bar{J} + \frac{\partial \rho}{\partial t} = 0, \tag{1.3}$$

where (1.1c) was used. This equation states that charge is conserved, or that current is continuous, since $\nabla \cdot \bar{J}$ represents the outflow of current at a point, and $\partial \rho / \partial t$ represents the charge buildup with time at the same point. It is this result that led Maxwell to the conclusion that the displacement current density $\partial \bar{D}/\partial t$ was necessary in (1.1b), which can be seen by taking the divergence of this equation.

The above differential equations can be converted to integral form through the use of various vector integral theorems. Thus, applying the divergence theorem (B.15) to (1.1c) and (1.1d) yields

$$\oint_S \bar{D} \cdot d\bar{s} = \int_V \rho \, dv = Q, \tag{1.4}$$

$$\oint_S \bar{B} \cdot d\bar{s} = 0, \tag{1.5}$$

where Q in (1.4) represents the total charge contained in the closed volume V (enclosed by a closed surface S). Applying Stokes' theorem (B.16) to (1.1a) gives

$$\oint_C \bar{\mathcal{E}} \cdot d\bar{l} = -\frac{\partial}{\partial t} \int_S \bar{B} \cdot d\bar{s} - \int_S \bar{\mathcal{M}} \cdot d\bar{s}, \tag{1.6}$$

which, without the $\bar{\mathcal{M}}$ term, is the usual form of *Faraday's law* and forms the basis for *Kirchhoff's voltage law*. In (1.6), C represents a closed contour around the surface S, as shown in Figure 1.3. *Ampere's law* can

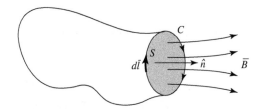

FIGURE 1.3 | The closed contour C and surface S associated with Faraday's law.

be derived by applying Stokes' theorem to (1.1b):

$$\oint_C \bar{H} \cdot d\bar{l} = \frac{\partial}{\partial t} \int_S \bar{D} \cdot d\bar{s} + \int_S \bar{J} \cdot d\bar{s} = \frac{\partial}{\partial t} \int_S \bar{D} \cdot d\bar{s} + \mathcal{I}, \tag{1.7}$$

where $\mathcal{I} = \int_S \bar{J} \cdot d\bar{s}$ is the total electric current flow through the surface S. Equations (1.4)–(1.7) constitute the integral forms of Maxwell's equations.

The above equations are valid for arbitrary time dependence, but most of our work will be involved with fields having a sinusoidal, or harmonic, time dependence, with steady-state conditions assumed. In this case phasor notation is very convenient, and so all field quantities will be assumed to be complex vectors with an implied $e^{j\omega t}$ time dependence and written with roman (rather than script) letters. Thus, a sinusoidal electric field polarized in the $\hat{x}$ direction of the form

$$\bar{\mathcal{E}}(x, y, z, t) = \hat{x} A(x, y, z) \cos(\omega t + \phi), \tag{1.8}$$

where A is the (real) amplitude, ω is the radian frequency, and ϕ is the phase reference of the wave at $t = 0$, has the phasor for

$$\bar{E}(x, y, z) = \hat{x} A(x, y, z) e^{j\phi}. \tag{1.9}$$

We will assume cosine-based phasors in this book, so the conversion from phasor quantities to real time-varying quantities is accomplished by multiplying the phasor by $e^{j\omega t}$ and taking the real part:

$$\bar{\mathcal{E}}(x, y, z, t) = \text{Re}\{\bar{E}(x, y, z) e^{j\omega t}\}, \tag{1.10}$$

as substituting (1.9) into (1.10) to obtain (1.8) demonstrates. When working in phasor notation, it is customary to suppress the factor $e^{j\omega t}$ that is common to all terms.

When dealing with power and energy, we will often be interested in the time average of a quadratic quantity. This can be found very easily for time harmonic fields. For example, the average of the square of the magnitude of an electric field, given as

$$\bar{\mathcal{E}} = \hat{x} E_1 \cos(\omega t + \phi_1) + \hat{y} E_2 \cos(\omega t + \phi_2) + \hat{z} E_3 \cos(\omega t + \phi_3), \tag{1.11}$$

has the phasor form

$$\bar{E} = \hat{x} E_1 e^{j\phi_1} + \hat{y} E_2 e^{j\phi_2} + \hat{z} E_3 e^{j\phi_3}, \tag{1.12}$$

can be calculated as

$$\begin{aligned}
|\bar{\mathcal{E}}|^2_{\text{avg}} &= \frac{1}{T} \int_0^T \bar{\mathcal{E}} \cdot \bar{\mathcal{E}} \, dt \\
&= \frac{1}{T} \int_0^T \left[E_1^2 \cos^2(\omega t + \phi_1) + E_2^2 \cos^2(\omega t + \phi_2) + E_3^2 \cos^2(\omega t + \phi_3) \right] dt \\
&= \frac{1}{2}\left(E_1^2 + E_2^2 + E_3^2 \right) = \frac{1}{2}|\bar{E}|^2 = \frac{1}{2}\bar{E} \cdot \bar{E}^*.
\end{aligned} \tag{1.13}$$

Then the root-mean-square (rms) value is $|\bar{E}|_{\text{rms}} = |\bar{E}|/\sqrt{2}$.

Assuming an $e^{j\omega t}$ time dependence, we can replace the time derivatives in (1.1a)–(1.1d) with $j\omega$. Maxwell's equations in phasor form then become

$$\nabla \times \bar{E} = -j\omega\bar{B} - \bar{M}, \tag{1.14a}$$

$$\nabla \times \bar{H} = j\omega\bar{D} + \bar{J}, \tag{1.14b}$$

$$\nabla \cdot \bar{D} = \rho, \tag{1.14c}$$

$$\nabla \cdot \bar{B} = 0. \tag{1.14d}$$

The Fourier transform can be used to convert a solution to Maxwell's equations for an arbitrary frequency ω into a solution for arbitrary time dependence.

The electric and magnetic current sources, $\bar{J}$ and $\bar{M}$, in (1.14) are volume current densities with units A/m^2 and V/m^2. In many cases, however, the actual currents will be in the form of a current sheet, a line current, or an infinitesimal dipole current. These special types of current distributions can always be written as volume current densities through the use of delta functions. Figure 1.4 shows examples of this procedure for electric and magnetic currents.

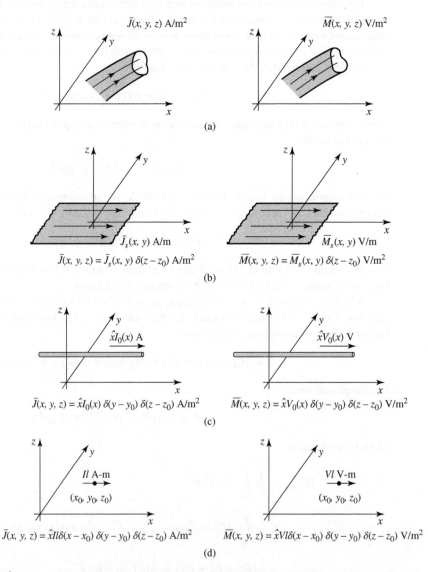

FIGURE 1.4 | Arbitrary volume, surface, and line currents. (a) Arbitrary electric and magnetic volume current densities. (b) Arbitrary electric and magnetic surface current densities in the $z = z_0$ plane. (c) Arbitrary electric and magnetic line currents. (d) Infinitesimal electric and magnetic dipoles parallel to the x-axis.

1.3   FIELDS IN MEDIA AND BOUNDARY CONDITIONS

In the preceding section it was assumed that the electric and magnetic fields were in free-space, with no material bodies present. In practice, material bodies are often present; this complicates the analysis but also allows the useful application of material properties to microwave components. When electromagnetic fields exist in material media, the field vectors are related to each other by the constitutive relations.

For a dielectric material, an applied electric field $\bar{E}$ causes the polarization of the atoms or molecules of the material to create electric dipole moments that augment the total displacement flux, $\bar{D}$. This additional polarization vector is called $\bar{P}_e$, the *electric polarization*, where

$$\bar{D} = \epsilon_0 \bar{E} + \bar{P}_e. \tag{1.15}$$

In a linear medium the electric polarization is linearly related to the applied electric field as

$$\bar{P}_e = \epsilon_0 \chi_e \bar{E}, \tag{1.16}$$

where χ_e, which may be complex, is called the *electric susceptibility*. Then,

$$\bar{D} = \epsilon_0 \bar{E} + \bar{P}_e = \epsilon_0 (1 + \chi_e) \bar{E} = \epsilon \bar{E}, \tag{1.17}$$

where

$$\epsilon = \epsilon' - j\epsilon'' = \epsilon_0 (1 + \chi_e) \tag{1.18}$$

is the complex permittivity of the medium. The imaginary part of ϵ accounts for loss in the medium (heat) due to damping of the vibrating dipole moments. (Free-space, having a real ϵ, is lossless.) Due to energy conservation, as we will see in Section 1.6, the imaginary part of ϵ must be negative (ϵ'' positive). The loss of a dielectric material may also be considered as an equivalent conductor loss. In a material with conductivity σ, a conduction current density will exist:

$$\bar{J} = \sigma \bar{E}, \tag{1.19}$$

which is *Ohm's law* from an electromagnetic field point of view. Maxwell's curl equation for $\bar{H}$ in (1.14b) then becomes

$$\begin{aligned} \nabla \times \bar{H} &= j\omega \bar{D} + \bar{J} \\ &= j\omega \epsilon \bar{E} + \sigma \bar{E} \\ &= j\omega \epsilon' \bar{E} + (\omega \epsilon'' + \sigma) \bar{E} \\ &= j\omega \left(\epsilon' - j\epsilon'' - j\frac{\sigma}{\omega} \right) \bar{E}, \end{aligned} \tag{1.20}$$

where it is seen that loss due to dielectric damping ($\omega \epsilon''$) is indistinguishable from conductivity loss (σ). The term $\omega \epsilon'' + \sigma$ can then be considered as the total effective conductivity. A related quantity of interest is the *loss tangent*, defined as

$$\tan \delta = \frac{\omega \epsilon'' + \sigma}{\omega \epsilon'}, \tag{1.21}$$

which is seen to be the ratio of the real to the imaginary part of the total displacement current. Microwave materials are usually characterized by specifying the real relative permittivity (the *dielectric constant*),[2] ϵ_r, with $\epsilon' = \epsilon_r \epsilon_0$, and the loss tangent at a certain frequency. These properties are listed in Appendix H for several types of materials. It is useful to note that, after a problem has been solved assuming a lossless dielectric, loss can easily be introduced by replacing the real ϵ with a complex $\epsilon = \epsilon' - j\epsilon'' = \epsilon'(1 - j\tan \delta) = \epsilon_0 \epsilon_r (1 - j\tan \delta)$.

---

[2] The *IEEE Standard Definitions of Terms for Radio Wave Propagation, IEEE Standard 211-1997,* suggests that the term "relative permittivity" be used instead of "dielectric constant." The *IEEE Standard Definitions of Terms for Antennas, IEEE Standard 145-1993,* however, still recognizes "dielectric constant." Since this term is commonly used in microwave engineering work, it will occasionally be used in this book.

In the preceding discussion it was assumed that $\bar{P}_e$ was a vector in the same direction as $\bar{E}$. Such materials are called *isotropic* materials, but not all materials have this property. Some materials are *anisotropic* and are characterized by a more complicated relation between $\bar{P}_e$ and $\bar{E}$, or $\bar{D}$ and $\bar{E}$. The most general linear relation between these vectors takes the form of a tensor of rank two (a dyad), which can be written in matrix form as

$$\begin{bmatrix} D_x \\ D_y \\ D_z \end{bmatrix} = \begin{bmatrix} \epsilon_{xx} & \epsilon_{xy} & \epsilon_{xz} \\ \epsilon_{yx} & \epsilon_{yy} & \epsilon_{yz} \\ \epsilon_{zx} & \epsilon_{zy} & \epsilon_{zz} \end{bmatrix} \begin{bmatrix} E_x \\ E_y \\ E_z \end{bmatrix} = [\epsilon] \begin{bmatrix} E_x \\ E_y \\ E_z \end{bmatrix}. \tag{1.22}$$

It is thus seen that a given vector component of $\bar{E}$ gives rise, in general, to three components of $\bar{D}$. Crystal structures and ionized gases are examples of anisotropic dielectrics. For a linear isotropic material, the matrix of (1.22) reduces to a diagonal matrix with elements ϵ.

An analogous situation occurs for magnetic materials. An applied magnetic field may align magnetic dipole moments in a magnetic material to produce a *magnetic polarization* (or magnetization) $\bar{P}_m$. Then,

$$\bar{B} = \mu_0(\bar{H} + \bar{P}_m). \tag{1.23}$$

For a linear magnetic material, $\bar{P}_m$ is linearly related to $\bar{H}$ as

$$\bar{P}_m = \chi_m \bar{H}, \tag{1.24}$$

where χ_m is a complex *magnetic susceptibility*. From (1.23) and (1.24),

$$\bar{B} = \mu_0(1 + \chi_m)\bar{H} = \mu\bar{H}, \tag{1.25}$$

where $\mu = \mu_0(1 + \chi_m) = \mu' - j\mu''$ is the complex permeability of the medium. Again, the imaginary part of χ_m or μ accounts for loss due to damping forces; there is no magnetic conductivity because there is no real magnetic current. As in the electric case, magnetic materials may be anisotropic, in which case a tensor permeability can be written as

$$\begin{bmatrix} B_x \\ B_y \\ B_z \end{bmatrix} = \begin{bmatrix} \mu_{xx} & \mu_{xy} & \mu_{xz} \\ \mu_{yx} & \mu_{yy} & \mu_{yz} \\ \mu_{zx} & \mu_{zy} & \mu_{zz} \end{bmatrix} \begin{bmatrix} H_x \\ H_y \\ H_z \end{bmatrix} = [\mu] \begin{bmatrix} H_x \\ H_y \\ H_z \end{bmatrix}. \tag{1.26}$$

An important example of anisotropic magnetic materials in microwave engineering is the class of ferrimagnetic materials known as *ferrites*; these materials and their applications will be discussed further in Chapter 9.

If linear media are assumed (ϵ, μ not depending on $\bar{E}$ or $\bar{H}$), then Maxwell's equations can be written in phasor form as

$$\nabla \times \bar{E} = -j\omega\mu\bar{H} - \bar{M}, \tag{1.27a}$$

$$\nabla \times \bar{H} = j\omega\epsilon\bar{E} + \bar{J}, \tag{1.27b}$$

$$\nabla \cdot \bar{D} = \rho, \tag{1.27c}$$

$$\nabla \cdot \bar{B} = 0. \tag{1.27d}$$

The constitutive relations are

$$\bar{D} = \epsilon\bar{E}, \tag{1.28a}$$

$$\bar{B} = \mu\bar{H}, \tag{1.28b}$$

where ϵ and μ may be complex and may be tensors. Note that relations like (1.28a) and (1.28b) generally cannot be written in time domain form, even for linear media, because of the possible phase shift between $\bar{D}$ and $\bar{E}$, or $\bar{B}$ and $\bar{H}$. The phasor representation accounts for this phase shift by the complex form of ϵ and μ.

Maxwell's equations (1.27a)–(1.27d) in differential form require known boundary values for a complete and unique solution. A general method used throughout this book is to solve the source-free Maxwell equations in a certain region to obtain solutions with unknown coefficients and then apply boundary conditions to solve for these coefficients. A number of specific cases of boundary conditions arise, as discussed in what follows.

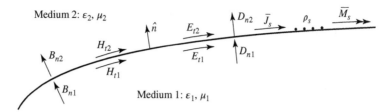

FIGURE 1.5 | Fields, currents, and surface charge at a general interface between two media.

Fields at a General Material Interface

Consider a plane interface between two media, as shown in Figure 1.5. Maxwell's equations in integral form can be used to deduce conditions involving the normal and tangential fields at this interface. The time-harmonic version of (1.4), where S is the closed "pillbox"-shaped surface shown in Figure 1.6, can be written as

$$\oint_S \bar{D} \cdot d\bar{s} = \int_V \rho \, dv. \tag{1.29}$$

In the limit as $h \to 0$, the contribution of D_{tan} through the sidewalls goes to zero, so (1.29) reduces to

$$\Delta S D_{2n} - \Delta S D_{1n} = \Delta S \rho_s,$$

or

$$D_{2n} - D_{1n} = \rho_s, \tag{1.30}$$

where ρ_s is the surface charge density on the interface. In vector form, we can write

$$\hat{n} \cdot (\bar{D}_2 - \bar{D}_1) = \rho_s. \tag{1.31}$$

A similar argument for $\bar{B}$ leads to the result that

$$\hat{n} \cdot \bar{B}_2 = \hat{n} \cdot \bar{B}_1, \tag{1.32}$$

because there is no free magnetic charge.

For the tangential components of the electric field we use the phasor form of (1.6),

$$\oint_C \bar{E} \cdot d\bar{l} = -j\omega \int_S \bar{B} \cdot d\bar{s} - \int_S \bar{M} \cdot d\bar{s}, \tag{1.33}$$

in connection with the closed contour C shown in Figure 1.7. In the limit as $h \to 0$, the surface integral of $\bar{B}$ vanishes (because $S = h\Delta\ell$ vanishes). The contribution from the surface integral of $\bar{M}$, however, may be

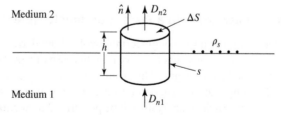

FIGURE 1.6 | Closed surface S for equation (1.29).

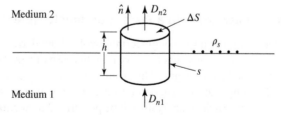

FIGURE 1.7 | Closed contour C for equation (1.33).

nonzero if a magnetic surface current density $\bar{M}_s$ exists on the surface. The Dirac delta function can then be used to write

$$\bar{M} = \bar{M}_s \delta(h), \tag{1.34}$$

where h is a coordinate measured normal from the interface. Equation (1.33) then gives

$$\Delta \ell E_{t1} - \Delta \ell E_{t2} = -\Delta \ell M_s,$$

or

$$E_{t1} - E_{t2} = -M_s, \tag{1.35}$$

which can be generalized in vector form as

$$(\bar{E}_2 - \bar{E}_1) \times \hat{n} = \bar{M}_s. \tag{1.36}$$

A similar argument for the magnetic field leads to

$$\hat{n} \times (\bar{H}_2 - \bar{H}_1) = \bar{J}_s, \tag{1.37}$$

where $\bar{J}_s$ is an electric surface current density that may exist at the interface. Equations (1.31), (1.32), (1.36), and (1.37) are the most general expressions for the boundary conditions at an arbitrary interface of materials and/or surface currents.

Fields at a Dielectric Interface

At an interface between two lossless dielectric materials, no charge or surface current densities will ordinarily exist. Equations (1.31), (1.32), (1.36), and (1.37) then reduce to

$$\hat{n} \cdot \bar{D}_1 = \hat{n} \cdot \bar{D}_2, \tag{1.38a}$$

$$\hat{n} \cdot \bar{B}_1 = \hat{n} \cdot \bar{B}_2, \tag{1.38b}$$

$$\hat{n} \times \bar{E}_1 = \hat{n} \times \bar{E}_2, \tag{1.38c}$$

$$\hat{n} \times \bar{H}_1 = \hat{n} \times \bar{H}_2. \tag{1.38d}$$

In words, these equations state that the normal components of $\bar{D}$ and $\bar{B}$ are continuous across the interface, and the tangential components of $\bar{E}$ and $\bar{H}$ are continuous across the interface. Because Maxwell's equations are not all linearly independent, the six boundary conditions contained in the above equations are not all linearly independent. Thus, the enforcement of (1.38c) and (1.38d) for the four tangential field components, for example, will automatically force the satisfaction of the equations for the continuity of the normal components.

Fields at the Interface with a Perfect Conductor (Electric Wall)

Many problems in microwave engineering involve boundaries with good conductors (e.g., metals), which can often be assumed as lossless ($\sigma \to \infty$). In this case of a perfect conductor, all field components must be zero inside the conducting region. This result can be seen by considering a conductor with finite conductivity ($\sigma < \infty$) and noting that the skin depth (the depth to which most of the microwave power penetrates) goes to zero as $\sigma \to \infty$. (Such an analysis will be performed in Section 1.7.) If we also assume here that $\bar{M}_s = 0$, which would be the case if the perfect conductor filled all the space on one side of the boundary, then (1.31), (1.32), (1.36), and (1.37) reduce to the following:

$$\hat{n} \cdot \bar{D} = \rho_s, \tag{1.39a}$$

$$\hat{n} \cdot \bar{B} = 0, \tag{1.39b}$$

$$\hat{n} \times \bar{E} = 0, \tag{1.39c}$$

$$\hat{n} \times \bar{H} = \bar{J}_s, \tag{1.39d}$$

where ρ_s and $\bar{J}_s$ are the electric surface charge density and current density, respectively, on the interface, and $\hat{n}$ is the normal unit vector pointing out of the perfect conductor. Such a boundary is also known as an *electric wall* because the tangential components of $\bar{E}$ are "shorted out," as seen from (1.39c), and must vanish at the surface of the conductor.

The Magnetic Wall Boundary Condition

Dual to the preceding boundary condition is the *magnetic wall* boundary condition, where the tangential components of $\bar{H}$ must vanish. Such a boundary does not really exist in practice but may be approximated by a corrugated surface or in certain planar transmission line problems. In addition, the idealization that $\hat{n} \times \bar{H} = 0$ at an interface is often a convenient simplification, as we will see in later chapters. We will also see that the magnetic wall boundary condition is analogous to the relations between the voltage and current at the end of an open-circuited transmission line, while the electric wall boundary condition is analogous to the voltage and current at the end of a short-circuited transmission line. The magnetic wall condition, then, provides a degree of completeness in our formulation of boundary conditions and is a useful approximation in several cases of practical interest.

The fields at a magnetic wall satisfy the following conditions:

$$\hat{n} \cdot \bar{D} = 0, \tag{1.40a}$$

$$\hat{n} \cdot \bar{B} = 0, \tag{1.40b}$$

$$\hat{n} \times \bar{E} = -\bar{M}_s, \tag{1.40c}$$

$$\hat{n} \times \bar{H} = 0, \tag{1.40d}$$

where $\hat{n}$ is the normal unit vector pointing out of the magnetic wall region.

The Radiation Condition

When dealing with problems that have one or more infinite boundaries, such as plane waves in an infinite medium, or infinitely long transmission lines, a condition on the fields at infinity must be enforced. This boundary condition is known as the *radiation condition* and is essentially a statement of energy conservation. It states that, at an infinite distance from a source, the fields must either be vanishingly small (i.e., zero) or propagating in an outward direction. This result can easily be seen by allowing the infinite medium to contain a small loss factor (as any physical medium would have). Incoming waves (from infinity) of finite amplitude would then require an infinite source at infinity and so are disallowed.

1.4   THE WAVE EQUATION AND BASIC PLANE WAVE SOLUTIONS

The Helmholtz Equation

In a source-free, linear, isotropic, homogeneous region, Maxwell's curl equations in phasor form are

$$\nabla \times \bar{E} = -j\omega\mu\bar{H}, \tag{1.41a}$$

$$\nabla \times \bar{H} = j\omega\epsilon\bar{E}, \tag{1.41b}$$

and constitute two equations for the two unknowns, $\bar{E}$ and $\bar{H}$. As such, they can be solved for either $\bar{E}$ or $\bar{H}$. Taking the curl of (1.41a) and using (1.41b) gives

$$\nabla \times \nabla \times \bar{E} = -j\omega\mu\nabla \times \bar{H} = \omega^2\mu\epsilon\bar{E},$$

which is an equation for $\bar{E}$. This result can be simplified through the use of vector identity (B.14), $\nabla \times \nabla \times \bar{A} = \nabla(\nabla \cdot \bar{A}) - \nabla^2\bar{A}$, which is valid for the rectangular components of an arbitrary vector $\bar{A}$. Then,

$$\nabla^2\bar{E} + \omega^2\mu\epsilon\bar{E} = 0, \tag{1.42}$$

because $\nabla \cdot \bar{E} = 0$ in a source-free region. Equation (1.42) is the *wave equation*, or *Helmholtz equation*, for $\bar{E}$. An identical equation for $\bar{H}$ can be derived in the same manner:

$$\nabla^2 \bar{H} + \omega^2 \mu \epsilon \bar{H} = 0. \tag{1.43}$$

A constant $k = \omega \sqrt{\mu \epsilon}$ is defined and called the *propagation constant* (also known as the *phase constant*, or *wave number*), of the medium; its units are 1/m.

As a way of introducing wave behavior, we will next study the solutions to the above wave equations in their simplest forms, first for a lossless medium and then for a lossy (conducting) medium.

Plane Waves in a Lossless Medium

In a lossless medium, ϵ and μ are real numbers, and so k is real. A basic plane wave solution to the above wave equations can be found by considering an electric field with only an $\hat{x}$ component and uniform (no variation) in the x and y directions. Then, $\partial/\partial x = \partial/\partial y = 0$, and the Helmholtz equation of (1.42) reduces to

$$\frac{\partial^2 E_x}{\partial z^2} + k^2 E_x = 0. \tag{1.44}$$

The two independent solutions to this equation are easily seen, by substitution, to be of the form

$$E_x(z) = E^+ e^{-jkz} + E^- e^{jkz}, \tag{1.45}$$

where E^+ and E^- are arbitrary amplitude constants.

The above solution is for the time harmonic case at frequency ω. In the time domain, this result is written as

$$\mathcal{E}_x(z, t) = E^+ \cos(\omega t - kz) + E^- \cos(\omega t + kz), \tag{1.46}$$

where we have assumed that E^+ and E^- are real constants. Consider the first term in (1.46). This term represents a wave traveling in the $+z$ direction because, to maintain a fixed point on the wave ($\omega t - kz =$ constant), one must move in the $+z$ direction as time increases. Similarly, the second term in (1.46) represents a wave traveling in the negative z direction—hence the notation E^+ and E^- for these wave amplitudes. The velocity of the wave in this sense is called the *phase velocity* because it is the velocity at which a fixed phase point on the wave travels, and it is given by

$$v_p = \frac{dz}{dt} = \frac{d}{dt}\left(\frac{\omega t - \text{constant}}{k}\right) = \frac{\omega}{k} = \frac{1}{\sqrt{\mu \epsilon}} \tag{1.47}$$

In free-space, we have $v_p = 1/\sqrt{\mu_0 \epsilon_0} = c = 2.998 \times 10^8$ m/sec, which is the speed of light.

The *wavelength*, λ, is defined as the distance between two successive maxima (or minima, or any other reference points) on the wave at a fixed instant of time. Thus,

$$(\omega t - kz) - [\omega t - k(z + \lambda)] = 2\pi,$$

so

$$\lambda = \frac{2\pi}{k} = \frac{2\pi v_p}{\omega} = \frac{v_p}{f}. \tag{1.48}$$

A complete specification of the plane wave electromagnetic field should include the magnetic field. In general, whenever $\bar{E}$ or $\bar{H}$ is known, the other field vector can be readily found by using one of Maxwell's curl equations. Thus, applying (1.41a) to the electric field of (1.45) gives $H_x = H_z = 0$, and

$$H_y = \frac{j}{\omega \mu} \frac{\partial E_x}{\partial z} = \frac{1}{\eta}(E^+ e^{-jkz} - E^- e^{jkz}), \tag{1.49}$$

where $\eta = \omega \mu / k = \sqrt{\mu/\epsilon}$ is known as the *intrinsic impedance* of the medium. The ratio of the $\bar{E}$ and $\bar{H}$ field components is seen to have units of impedance, known as the *wave impedance*; for plane waves the wave impedance is equal to the intrinsic impedance of the medium. In free-space the intrinsic impedance is $\eta_0 = \sqrt{\mu_0/\epsilon_0} = 377\,\Omega$. Note that the $\bar{E}$ and $\bar{H}$ vectors are orthogonal to each other and orthogonal to the direction of propagation ($\pm\hat{z}$); this is a characteristic of transverse electromagnetic (TEM) waves.

EXAMPLE 1.1   BASIC PLANE WAVE PARAMETERS

A plane wave propagating in a lossless dielectric medium has an electric field given as $E_x = E_0\cos(\omega t - \beta z)$ with a frequency of 2.5 GHz and a wavelength in the material of 4.0 cm. Determine the propagation constant, the phase velocity, the relative permittivity of the medium, and the wave impedance.

Solution
From (1.48) the propagation constant is $k = \dfrac{2\pi}{\lambda} = \dfrac{2\pi}{0.04} = 157.08\text{ m}^{-1}$, and from (1.47) the phase velocity is

$$v_p = \frac{\omega}{k} = \frac{2\pi f}{k} = \lambda f = 0.04 \times 2.5 \times 10^9 = 1.0 \times 10^8 \text{ m/sec.}$$

This is slower than the speed of light by a factor of 3.0. The relative permittivity of the medium can be found from (1.47) as

$$\epsilon_r = \left(\frac{c}{v_p}\right)^2 = \left(\frac{3.0 \times 10^8}{1.0 \times 10^8}\right)^2 = 9.0$$

The wave impedance is

$$\eta = \eta_0/\sqrt{\epsilon_r} = \frac{377}{\sqrt{9.0}} = 125.67\,\Omega$$

Plane Waves in a General Lossy Medium

Now consider the effect of a lossy medium. If the medium is conductive, with a conductivity σ, Maxwell's curl equations can be written, from (1.41a) and (1.20) as

$$\nabla \times \bar{E} = -j\omega\mu\bar{H}, \tag{1.50a}$$

$$\nabla \times \bar{H} = j\omega\epsilon\bar{E} + \sigma\bar{E}. \tag{1.50b}$$

The resulting wave equation for $\bar{E}$ then becomes

$$\nabla^2\bar{E} + \omega^2\mu\epsilon\left(1 - j\frac{\sigma}{\omega\epsilon}\right)\bar{E} = 0, \tag{1.51}$$

where we see a similarity with (1.42), the wave equation for $\bar{E}$ in the lossless case. The difference is that the quantity $k^2 = \omega^2\mu\epsilon$ of (1.42) is replaced by $\omega^2\mu\epsilon[1 - j(\sigma/\omega\epsilon)]$ in (1.51). We then define a *complex propagation constant* for the medium as

$$\gamma = \alpha + j\beta = j\omega\sqrt{\mu\epsilon}\sqrt{1 - j\frac{\sigma}{\omega\epsilon}} \tag{1.52}$$

where α is the *attenuation constant* and β is the *phase constant*. If we again assume an electric field with only an $\hat{x}$ component and uniform in x and y, the wave equation of (1.51) reduces to

$$\frac{\partial^2 E_x}{\partial z^2} - \gamma^2 E_x = 0, \tag{1.53}$$

which has solutions

$$E_x(z) = E^+ e^{-\gamma z} + E^- e^{\gamma z}. \tag{1.54}$$

The positive traveling wave then has a propagation factor of the form

$$e^{-\gamma z} = e^{-\alpha z}e^{-j\beta z},$$

which in the time domain is of the form

$$e^{-\alpha z} \cos(\omega t - \beta z).$$

We see that this represents a wave traveling in the $+z$ direction with a phase velocity $v_p = \omega/\beta$, a wavelength $\lambda = 2\pi/\beta$, and an exponential damping factor. The rate of decay with distance is given by the attenuation constant, α. The negative traveling wave term of (1.54) is similarly damped along the $-z$ axis. If the loss is removed, $\sigma = 0$, and we have $\gamma = jk$ and $\alpha = 0$, $\beta = k$.

As discussed in Section 1.3, loss can also be treated through the use of a complex permittivity. From (1.52) and (1.20) with $\sigma = 0$ but $\epsilon = \epsilon' - j\epsilon''$ complex, we have that

$$\gamma = j\omega\sqrt{\mu\epsilon} = jk = j\omega\sqrt{\mu\epsilon'(1 - j\tan\delta)}, \tag{1.55}$$

where $\tan\delta = \epsilon''/\epsilon'$ is the loss tangent of the material.

The associated magnetic field can be calculated as

$$H_y = \frac{j}{\omega\mu}\frac{\partial E_x}{\partial z} = \frac{-j\gamma}{\omega\mu}(E^+ e^{-\gamma z} - E^- e^{\gamma z}). \tag{1.56}$$

The intrinsic impedance of the conducting medium is now complex,

$$\eta = \frac{j\omega\mu}{\gamma}, \tag{1.57}$$

but is still identified as the wave impedance, which expresses the ratio of electric to magnetic field components. This allows (1.56) to be rewritten as

$$H_y = \frac{1}{\eta}(E^+ e^{-\gamma z} - E^- e^{\gamma z}). \tag{1.58}$$

Note that although η of (1.57) is, in general, complex, it reduces to the lossless case of $\eta = \sqrt{\mu/\epsilon}$ when $\gamma = jk = j\omega\sqrt{\mu\epsilon}$.

Plane Waves in a Good Conductor

Many problems of practical interest involve loss or attenuation due to good (but not perfect) conductors. A good conductor is a special case of the preceding analysis, where the conductive current is much greater than the displacement current, which means that $\sigma \gg \omega\epsilon$. Most metals can be categorized as good conductors. In terms of a complex ϵ, rather than conductivity, this condition is equivalent to $\epsilon'' \gg \epsilon'$. The propagation constant of (1.52) can then be adequately approximated by ignoring the displacement current term, to give

$$\gamma = \alpha + j\beta \simeq j\omega\sqrt{\mu\epsilon}\sqrt{\frac{\sigma}{j\omega\epsilon}} = (1+j)\sqrt{\frac{\omega\mu\sigma}{2}}. \tag{1.59}$$

The *skin depth*, or characteristic depth of penetration, is defined as

$$\delta_s = \frac{1}{\alpha} = \sqrt{\frac{2}{\omega\mu\sigma}}. \tag{1.60}$$

Thus the amplitude of the fields in the conductor will decay by an amount $1/e$, or 36.8%, after traveling a distance of one skin depth, because $e^{-\alpha z} = e^{-\alpha\delta_s} = e^{-1}$. At microwave frequencies, for a good conductor, this distance is very small. The practical importance of this result is that only a thin plating of a good conductor (e.g., silver or gold) is necessary for low-loss microwave components.

EXAMPLE 1.2   SKIN DEPTH AT MICROWAVE FREQUENCIES

Compute the skin depth of aluminum, copper, gold, and silver at a frequency of 10 GHz.

Solution

The conductivities for these metals are listed in Appendix G. Equation (1.60) gives the skin depths as

$$\delta_s = \sqrt{\frac{2}{\omega \mu \sigma}} = \sqrt{\frac{1}{\pi f \mu_0 \sigma}} = \sqrt{\frac{1}{\pi (10^{10})(4\pi \times 10^{-7})}} \sqrt{\frac{1}{\sigma}}$$

$$= 5.03 \times 10^{-3} \sqrt{\frac{1}{\sigma}}.$$

$$\text{For aluminum: } \delta_s = 5.03 \times 10^{-3} \sqrt{\frac{1}{3.816 \times 10^7}} = 8.14 \times 10^{-7} \text{m}.$$

$$\text{For copper: } \quad \delta_s = 5.03 \times 10^{-3} \sqrt{\frac{1}{5.813 \times 10^7}} = 6.60 \times 10^{-7} \text{m}.$$

$$\text{For gold: } \quad \delta_s = 5.03 \times 10^{-3} \sqrt{\frac{1}{4.098 \times 10^7}} = 7.86 \times 10^{-7} \text{m}.$$

$$\text{For silver: } \quad \delta_s = 5.03 \times 10^{-3} \sqrt{\frac{1}{6.173 \times 10^7}} = 6.40 \times 10^{-7} \text{m}.$$

These results show that most of the current flow in a good conductor occurs in an extremely thin region near the surface of the conductor.

The intrinsic impedance inside a good conductor can be obtained from (1.57) and (1.59). The result is

$$\eta = \frac{j\omega \mu}{\gamma} \simeq (1+j)\sqrt{\frac{\omega \mu}{2\sigma}} = (1+j)\frac{1}{\sigma \delta_s}. \tag{1.61}$$

Notice that the phase angle of this impedance is 45°, a characteristic of good conductors. The phase angle of the impedance for a lossless material is 0°, and the phase angle of the impedance of an arbitrary lossy medium is somewhere between 0° and 45°.

Table 1.2 summarizes the results for plane wave propagation in lossless and lossy homogeneous media.

TABLE 1.2 | Summary of Results for Plane Wave Propagation in Various Media

	Type of Medium		
Quantity	Lossless $(\epsilon'' = \sigma = 0)$	General Lossy	Good Conductor $(\epsilon'' \gg \epsilon'$ or $\sigma \gg \omega \epsilon')$
Complex propagation constant	$\gamma = j\omega\sqrt{\mu \epsilon}$	$\gamma = j\omega\sqrt{\mu \epsilon}$ $= j\omega\sqrt{\mu \epsilon'}\sqrt{1 - j\dfrac{\sigma}{\omega \epsilon'}}$	$\gamma = (1+j)\sqrt{\omega \mu \sigma /2}$
Phase constant (wave number)	$\beta = k = \omega\sqrt{\mu \epsilon}$	$\beta = \text{Im}\{\gamma\}$	$\beta = \text{Im}\{\gamma\} = \sqrt{\omega \mu \sigma /2}$
Attenuation constant	$\alpha = 0$	$\alpha = \text{Re}\{\gamma\}$	$\alpha = \text{Re}\{\gamma\} = \sqrt{\omega \mu \sigma /2}$
Impedance	$\eta = \sqrt{\mu / \epsilon} = \omega \mu / k$	$\eta = j\omega \mu / \gamma$	$\eta = (1+j)\sqrt{\omega \mu / 2\sigma}$
Skin depth	$\delta_s = \infty$	$\delta_s = 1/\alpha$	$\delta_s = \sqrt{2/\omega \mu \sigma}$
Wavelength	$\lambda = 2\pi /\beta$	$\lambda = 2\pi /\beta$	$\lambda = 2\pi /\beta$
Phase velocity	$v_p = \omega /\beta$	$v_p = \omega /\beta$	$v_p = \omega /\beta$

1.5 GENERAL PLANE WAVE SOLUTIONS

Some specific features of plane waves were discussed in Section 1.4, but we will now look at plane waves from a more general point of view and solve the wave equation by the method of separation of variables. This technique will find application in succeeding chapters. We will also discuss circularly polarized plane waves, which will be important for the discussion of ferrites in Chapter 9.

In free-space, the Helmholtz equation for $\bar{E}$ can be written as

$$\nabla^2 \bar{E} + k_0^2 \bar{E} = \frac{\partial^2 \bar{E}}{\partial x^2} + \frac{\partial^2 \bar{E}}{\partial y^2} + \frac{\partial^2 \bar{E}}{\partial z^2} + k_0^2 \bar{E} = 0, \tag{1.62}$$

and this vector wave equation holds for each rectangular component of $\bar{E}$:

$$\frac{\partial^2 E_i}{\partial x^2} + \frac{\partial^2 E_i}{\partial y^2} + \frac{\partial^2 E_i}{\partial z^2} + k_0^2 E_i = 0, \tag{1.63}$$

where the index $i = x, y$, or z. This equation can be solved by the method of *separation of variables*, a standard technique for treating such partial differential equations. The method begins by assuming that the solution to (1.63) for, say, E_x, can be written as a product of three functions for each of the three coordinates:

$$E_x(x, y, z) = f(x)g(y)h(z). \tag{1.64}$$

Substituting this form into (1.63) and dividing by fgh gives

$$\frac{f''}{f} + \frac{g''}{g} + \frac{h''}{h} + k_0^2 = 0, \tag{1.65}$$

where the double primes denote the second derivative. The key step in the argument is to recognize that each of the terms in (1.65) must be equal to a constant because they are independent of each other. That is, f''/f is only a function of x, and the remaining terms in (1.65) do not depend on x, so f''/f must be a constant, and similarly for the other terms in (1.65). Thus, we define three separation constants, k_x, k_y, and k_z, such that

$$f''/f = -k_x^2; \quad g''/g = -k_y^2; \quad h''/h = -k_z^2;$$

or

$$\frac{d^2 f}{dx^2} + k_x^2 f = 0; \quad \frac{d^2 g}{dy^2} + k_y^2 g = 0; \quad \frac{d^2 h}{dz^2} + k_z^2 h = 0. \tag{1.66}$$

Combining (1.65) and (1.66) shows that

$$k_x^2 + k_y^2 + k_z^2 = k_0^2. \tag{1.67}$$

The partial differential equation of (1.63) has now been reduced to three separate ordinary differential equations in (1.66). Solutions to these equations have the forms $e^{\pm jk_x x}$, $e^{\pm jk_y y}$, and $e^{\pm jk_z z}$, respectively. As we saw in the previous section, the terms with + signs result in waves traveling in the negative x, y, or z direction, while the terms with − signs result in waves traveling in the positive direction. Both solutions are possible and are valid; the amount to which these various terms are excited is dependent on the source of the fields and the boundary conditions. For our present discussion we will select a plane wave traveling in the positive direction for each coordinate and write the complete solution for E_x as

$$E_x(x, y, z) = A e^{-j(k_x x + k_y y + k_z z)}, \tag{1.68}$$

where A is an arbitrary amplitude constant. Now define a wave number vector $\bar{k}$ as

$$\bar{k} = k_x \hat{x} + k_y \hat{y} + k_z \hat{z} = k_0 \hat{n}. \tag{1.69}$$

Then from (1.67), $|\bar{k}| = k_0$, and so $\hat{n}$ is a unit vector in the direction of propagation. Also define a position vector as

$$\bar{r} = x\hat{x} + y\hat{y} + z\hat{z}; \tag{1.70}$$

then (1.68) can be written as

$$E_x(x, y, z) = Ae^{-j\bar{k}\cdot\bar{r}}. \tag{1.71}$$

Solutions to (1.63) for E_y and E_z are, of course, similar in form to E_x of (1.71), but with different amplitude constants:

$$E_y(x, y, z) = Be^{-j\bar{k}\cdot\bar{r}}, \tag{1.72}$$

$$E_z(x, y, z) = Ce^{-j\bar{k}\cdot\bar{r}}. \tag{1.73}$$

The x, y, and z dependences of the three components of $\bar{E}$ in (1.71)–(1.73) must be the same (same k_x, k_y, k_z), because the divergence condition that

$$\nabla \cdot \bar{E} = \frac{\partial E_x}{\partial x} + \frac{\partial E_y}{\partial y} + \frac{\partial E_z}{\partial z} = 0$$

must also be applied in order to satisfy Maxwell's equations, and this implies that E_x, E_y, and E_z must each have the same variation in x, y, and z. (Note that the solutions in the preceding section automatically satisfied the divergence condition because E_x was the only component of $\bar{E}$, and E_x did not vary with x.) This condition also imposes a constraint on the amplitudes A, B, and C because if

$$\bar{E}_0 = A\hat{x} + B\hat{y} + C\hat{z},$$

we have

$$\bar{E} = \bar{E}_0 e^{-j\bar{k}\cdot\bar{r}},$$

and

$$\nabla \cdot \bar{E} = \nabla \cdot (\bar{E}_0 e^{-j\bar{k}\cdot\bar{r}}) = \bar{E}_0 \cdot \nabla e^{-j\bar{k}\cdot\bar{r}} = -j\bar{k} \cdot \bar{E}_0 e^{-j\bar{k}\cdot\bar{r}} = 0,$$

where vector identity (B.7) was used. Thus, we must have

$$\bar{k} \cdot \bar{E}_0 = 0, \tag{1.74}$$

which means that the electric field amplitude vector $\bar{E}_0$ must be perpendicular to the direction of propagation, $\bar{k}$. This condition is a general result for plane waves and implies that only two of the three amplitude constants, $A, B,$ and C, can be chosen independently.

The magnetic field can be found from Maxwell's equation,

$$\nabla \times \bar{E} = -j\omega\mu_0\bar{H}, \tag{1.75}$$

to give

$$\begin{aligned}
\bar{H} &= \frac{j}{\omega\mu_0}\nabla \times \bar{E} = \frac{j}{\omega\mu_0}\nabla \times (\bar{E}_0 e^{-j\bar{k}\cdot\bar{r}}) \\
&= \frac{-j}{\omega\mu_0}\bar{E}_0 \times \nabla e^{-j\bar{k}\cdot\bar{r}} \\
&= \frac{-j}{\omega\mu_0}\bar{E}_0 \times (-j\bar{k})e^{-j\bar{k}\cdot\bar{r}} \\
&= \frac{k_0}{\omega\mu_0}\hat{n} \times \bar{E}_0 e^{-j\bar{k}\cdot\bar{r}} \\
&= \frac{1}{\eta_0}\hat{n} \times \bar{E}_0 e^{-j\bar{k}\cdot\bar{r}} \\
&= \frac{1}{\eta_0}\hat{n} \times \bar{E}, \tag{1.76}
\end{aligned}$$

where vector identity (B.9) was used in obtaining the second line. This result shows that the magnetic field vector $\bar{H}$ lies in a plane normal to $\bar{k}$, the direction of propagation, and that $\bar{H}$ is perpendicular to $\bar{E}$. See

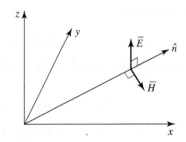

FIGURE 1.8 | Orientation of the $\bar{E}$, $\bar{H}$, and $\bar{k} = k_0\hat{n}$ vectors for a general plane wave.

Figure 1.8 for an illustration of these vector relations. The quantity $\eta_0 = \sqrt{\mu_0/\epsilon_0} = 377\ \Omega$ in (1.76) is the intrinsic impedance of free-space.

The time domain expression for the electric field can be found as

$$\bar{\mathcal{E}}(x, y, z, t) = \text{Re}\left\{\bar{E}(x, y, z)e^{j\omega t}\right\}$$
$$= \text{Re}\left\{\bar{E}_0 e^{-j\bar{k}\cdot\bar{r}} e^{j\omega t}\right\}$$
$$= \bar{E}_0 \cos(\bar{k}\cdot\bar{r} - \omega t), \tag{1.77}$$

assuming that the amplitude constants A, B, and C contained in $\bar{E}_0$ are real. If these constants are not real, their phases should be included inside the cosine term of (1.77). It is easy to show that the wavelength and phase velocity for this solution are the same as obtained in Section 1.4.

EXAMPLE 1.3   CURRENT SHEETS AS SOURCES OF PLANE WAVES

An infinite sheet of surface current can be considered as a source for plane waves. If an electric surface current density $\bar{J}_s = J_0\hat{x}$ exists on the $z = 0$ plane in free-space, find the resulting fields by assuming plane waves on either side of the current sheet and enforcing boundary conditions.

Solution
Since the source does not vary with x or y, the fields will not vary with x or y but will propagate away from the source in the $\pm z$ direction. The boundary conditions to be satisfied at $z = 0$ are

$$\hat{n}\times(\bar{E}_2 - \bar{E}_1) = \hat{z}\times(\bar{E}_2 - \bar{E}_1) = 0,$$
$$\hat{n}\times(\bar{H}_2 - \bar{H}_1) = \hat{z}\times(\bar{H}_2 - \bar{H}_1) = J_0\hat{x},$$

where $\bar{E}_1$, $\bar{H}_1$ are the fields for $z < 0$, and $\bar{E}_2$, $\bar{H}_2$ are the fields for $z > 0$. To satisfy the second condition, $\bar{H}$ must have a $\hat{y}$ component. Then for $\bar{E}$ to be orthogonal to $\bar{H}$ and $\hat{z}$, $\bar{E}$ must have an $\hat{x}$ component. Thus the fields will have the following form:

$$\text{for } z < 0, \qquad \bar{E}_1 = \hat{x}A\eta_0 e^{jk_0 z},$$
$$\bar{H}_1 = -\hat{y}A e^{jk_0 z},$$
$$\text{for } z > 0, \qquad \bar{E}_2 = \hat{x}B\eta_0 e^{-jk_0 z},$$
$$\bar{H}_2 = \hat{y}B e^{-jk_0 z},$$

where A and B are arbitrary amplitude constants. The first boundary condition, that E_x is continuous at $z = 0$, yields $A = B$, while the boundary condition for $\bar{H}$ yields the equation

$$-B - A = J_0.$$

Solving for A, B gives

$$A = B = -J_0/2,$$

which completes the solution.

Circularly Polarized Plane Waves

The plane waves discussed previously all had their electric field vector pointing in a fixed direction and so are called *linearly polarized* waves. In general, the *polarization* of a plane wave refers to the orientation of the electric field vector, which may be in a fixed direction or may change with time.

Consider the superposition of an $\hat{x}$ linearly polarized wave with amplitude E_1 and a $\hat{y}$ linearly polarized wave with amplitude E_2, both traveling in the positive $\hat{z}$ direction. The total electric field can be written as

$$\bar{E} = (E_1\hat{x} + E_2\hat{y})e^{-jk_0z}. \tag{1.78}$$

A number of possibilities now arise. If $E_1 \neq 0$ and $E_2 = 0$, we have a plane wave linearly polarized in the $\hat{x}$ direction. Similarly, if $E_1 = 0$ and $E_2 \neq 0$, we have a plane wave linearly polarized in the $\hat{y}$ direction. If E_1 and E_2 are both real and nonzero, we have a plane wave linearly polarized at the angle

$$\phi = \tan^{-1}\frac{E_2}{E_1}.$$

For example, if $E_1 = E_2 = E_0$, we have

$$\bar{E} = E_0(\hat{x} + \hat{y})e^{-jk_0z},$$

which represents an electric field vector at a 45° angle from the x-axis.

Now consider the case in which $E_1 = jE_2 = E_0$, where E_0 is real, so that

$$\bar{E} = E_0(\hat{x} - j\hat{y})e^{-jk_0z}. \tag{1.79}$$

The time domain form of this field is

$$\bar{\mathcal{E}}(z, t) = E_0[\hat{x}\cos(\omega t - k_0z) + \hat{y}\cos(\omega t - k_0z - \pi/2)]. \tag{1.80}$$

This expression shows that the electric field vector changes with time or, equivalently, with distance along the z-axis. To see this, pick a fixed position, say $z = 0$. Equation (1.80) then reduces to

$$\bar{\mathcal{E}}(0, t) = E_0[\hat{x}\cos\omega t + \hat{y}\sin\omega t], \tag{1.81}$$

so as ωt increases from zero, the electric field vector rotates counterclockwise from the x-axis. The resulting angle from the x-axis of the electric field vector at time t, at $z = 0$, is then

$$\phi = \tan^{-1}\left(\frac{\sin\omega t}{\cos\omega t}\right) = \omega t,$$

which shows that the polarization rotates at the uniform angular velocity ω. Since the fingers of the right hand point in the direction of rotation of the electric field vector when the thumb points in the direction of propagation, this type of wave is referred to as a *right-hand circularly polarized* (RHCP) wave. Similarly, a field of the form

$$\bar{E} = E_0(\hat{x} + j\hat{y})e^{-jk_0z} \tag{1.82}$$

constitutes a *left-hand circularly polarized* (LHCP) wave, where the electric field vector rotates in the opposite direction. See Figure 1.9 for a sketch of the polarization vectors for RHCP and LHCP plane waves.

The magnetic field associated with a circularly polarized wave may be found from Maxwell's equations or by using the wave impedance applied to each component of the electric field. For example, applying (1.76) to the electric field of an RHCP wave as given in (1.79) yields

$$\bar{H} = \frac{E_0}{\eta_0}\hat{z} \times (\hat{x} - j\hat{y})e^{-jk_0z} = \frac{E_0}{\eta_0}(\hat{y} + j\hat{x})e^{-jk_0z} = \frac{jE_0}{\eta_0}(\hat{x} - j\hat{y})e^{-jk_0z},$$

which is also seen to represent a vector rotating in the RHCP sense.

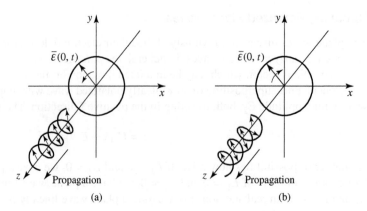

FIGURE 1.9 | Electric field polarization for (a) RHCP and (b) LHCP plane waves.

1.6 ENERGY AND POWER

In general, a source of electromagnetic energy sets up fields that store electric and magnetic energy and carry power that may be transmitted or dissipated as loss. In the sinusoidal steady-state case, the time-average stored electric energy in a volume V is given by

$$W_e = \frac{1}{4}\text{Re} \int_V \bar{E} \cdot \bar{D}^* \, dv, \tag{1.83}$$

which in the case of simple lossless isotropic, homogeneous, linear media, where ϵ is a real scalar constant, reduces to

$$W_e = \frac{\epsilon}{4} \int_V \bar{E} \cdot \bar{E}^* \, dv. \tag{1.84}$$

Similarly, the time-average magnetic energy stored in the volume V is

$$W_m = \frac{1}{4}\text{Re} \int_V \bar{H} \cdot \bar{B}^* \, dv, \tag{1.85}$$

which becomes

$$W_m = \frac{\mu}{4} \int_V \bar{H} \cdot \bar{H}^* \, dv, \tag{1.86}$$

for a real, constant, scalar μ.

We can now derive Poynting's theorem, which leads to energy conservation for electromagnetic fields and sources. If we have an electric source current $\bar{J}_s$ and a conduction current $\sigma\bar{E}$ as defined in (1.19), then the total electric current density is $\bar{J} = \bar{J}_s + \sigma\bar{E}$. Multiplying (1.27a) by $\bar{H}^*$ and multiplying the conjugate of (1.27b) by $\bar{E}$ yields

$$\bar{H}^* \cdot (\nabla \times \bar{E}) = -j\omega\mu|\bar{H}|^2 - \bar{H}^* \cdot \bar{M}_s,$$
$$\bar{E} \cdot (\nabla \times \bar{H}^*) = \bar{E} \cdot \bar{J}^* - j\omega\epsilon^*|\bar{E}|^2 = \bar{E} \cdot \bar{J}_s^* + \sigma|\bar{E}|^2 - j\omega\epsilon^*|\bar{E}|^2,$$

where $\bar{M}_s$ is the magnetic source current. Using these two results in vector identity (B.8) gives

$$\nabla \cdot (\bar{E} \times \bar{H}^*) = \bar{H}^* \cdot (\nabla \times \bar{E}) - \bar{E} \cdot (\nabla \times \bar{H}^*)$$
$$= -\sigma|\bar{E}|^2 + j\omega(\epsilon^*|\bar{E}|^2 - \mu|\bar{H}|^2) - (\bar{E} \cdot \bar{J}_s^* + \bar{H}^* \cdot \bar{M}_s).$$

Now integrate over a volume V and use the divergence theorem:

$$\int_V \nabla \cdot (\bar{E} \times \bar{H}^*) \, dv = \oint_S \bar{E} \times \bar{H}^* \cdot d\bar{s}$$
$$= -\sigma \int_V |\bar{E}|^2 \, dv + j\omega \int_V (\epsilon^*|\bar{E}|^2 - \mu|\bar{H}|^2) \, dv - \int_V (\bar{E} \cdot \bar{J}_s^* + \bar{H}^* \cdot \bar{M}_s) \, dv, \tag{1.87}$$

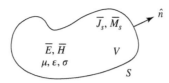

FIGURE 1.10 | A volume V, enclosed by the closed surface S, containing fields $\bar{E}$, $\bar{H}$, and current sources $\bar{J}_s$, $\bar{M}_s$.

where S is a closed surface enclosing the volume V, as shown in Figure 1.10. Allowing $\epsilon = \epsilon' - j\epsilon''$ and $\mu = \mu' - j\mu''$ to be complex to allow for loss, and rewriting (1.87), gives

$$-\frac{1}{2} \int_V (\bar{E} \cdot \bar{J}_s^* + \bar{H}^* \cdot \bar{M}_s)\, dv = \frac{1}{2} \oint_S \bar{E} \times \bar{H}^* \cdot d\bar{s} + \frac{\sigma}{2} \int_V |\bar{E}|^2\, dv$$

$$+ \frac{\omega}{2} \int_V (\epsilon'' |\bar{E}|^2 + \mu'' |\bar{H}|^2)\, dv + j\frac{\omega}{2} \int_V (\mu' |\bar{H}|^2 - \epsilon' |\bar{E}|^2)\, dv. \tag{1.88}$$

This result is known as *Poynting's theorem*, after the physicist J. H. Poynting (1852–1914), and is basically a power balance equation. Thus, the integral on the left-hand side represents the complex power P_s delivered by the sources $\bar{J}_s$ and $\bar{M}_s$ inside S:

$$P_s = -\frac{1}{2} \int_V (\bar{E} \cdot \bar{J}_s^* + \bar{H}^* \cdot \bar{M}_s)\, dv. \tag{1.89}$$

The first integral on the right-hand side of (1.88) represents complex power flow out of the closed surface S. If we define a quantity $\bar{S}$, called the *Poynting vector*, as

$$\bar{S} = \bar{E} \times \bar{H}^*, \tag{1.90}$$

then this power can be expressed as

$$P_o = \frac{1}{2} \oint_S \bar{E} \times \bar{H}^* \cdot d\bar{s} = \frac{1}{2} \oint_S \bar{S} \cdot d\bar{s}. \tag{1.91}$$

The surface S in (1.91) must be a closed surface for this interpretation to be valid. The real parts of P_s and P_o in (1.89) and (1.91) represent time-average powers.

The second and third integrals in (1.88) are real quantities representing the time-average power dissipated in the volume V due to conductivity, dielectric, and magnetic losses. If we define this power as P_ℓ we have

$$P_\ell = \frac{\sigma}{2} \int_V |\bar{E}|^2\, dv + \frac{\omega}{2} \int_V (\epsilon'' |\bar{E}|^2 + \mu'' |\bar{H}|^2)\, dv, \tag{1.92}$$

which is sometimes referred to as *Joule's law*. The last integral in (1.88) can be seen to be related to the stored electric and magnetic energies, as defined in (1.84) and (1.86).

With the above definitions, Poynting's theorem can be rewritten as

$$P_s = P_o + P_\ell + 2j\omega(W_m - W_e). \tag{1.93}$$

In words, this complex power balance equation states that the power delivered by the sources (P_s) is equal to the sum of the power transmitted through the surface (P_o), the power lost to heat in the volume (P_ℓ), and 2ω times the net reactive energy stored in the volume.

Power Absorbed by a Good Conductor

Practical transmission lines involve imperfect conductors, leading to attenuation and power losses, as well as the generation of noise. To calculate loss and attenuation due to an imperfect conductor we must find the power dissipated in the conductor. We will show that this can be accomplished using only the fields at the surface of the conductor, which is a very helpful simplification when calculating attenuation.

Consider the geometry of Figure 1.11, which shows the interface between a lossless medium and a good conductor. A field is incident from $z < 0$, and the field penetrates into the conducting region,

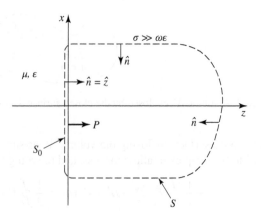

FIGURE 1.11 | An interface between a lossless medium and a good conductor with a closed surface $S_0 + S$ for computing the power dissipated in the conductor.

$z > 0$. The real average power entering the conductor volume defined by the cross-sectional area S_0 at the interface and the surface S is given from (1.91) as

$$P_{\text{avg}} = \frac{1}{2}\text{Re} \int_{S_0+S} \bar{E} \times \bar{H}^* \cdot \hat{n}\, ds, \tag{1.94}$$

where $\hat{n}$ is a unit normal vector pointing into the closed surface $S_0 + S$, and $\bar{E}$, $\bar{H}$ are the fields over this surface. The contribution to the integral in (1.94) from the surface S can be made zero by proper selection of this surface. For example, if the field is a normally incident plane wave, the Poynting vector $\bar{S} = \bar{E} \times \bar{H}^*$ will be in the $\hat{z}$ direction, and so tangential to the top, bottom, front, and back of S, if these walls are made parallel to the z-axis. If the wave is obliquely incident, these walls can be slanted to obtain the same result. If the conductor is good, the decay of the fields away from the interface at $z = 0$ will be very rapid, so the right-hand end of S can be made far enough away from $z = 0$ such that there is negligible contribution to the integral from this part of the surface S. The time-average power entering the conductor through S_0 can then be written as

$$P_{\text{avg}} = \frac{1}{2}\text{Re} \int_{S_0} \bar{E} \times \bar{H}^* \cdot \hat{z}\, ds. \tag{1.95}$$

From vector identity (B.3) we have

$$\hat{z} \cdot (\bar{E} \times \bar{H}^*) = (\hat{z} \times \bar{E}) \cdot \bar{H}^* = \eta \bar{H} \cdot \bar{H}^*, \tag{1.96}$$

since $\bar{H} = \hat{n} \times \bar{E}/\eta$, as generalized from (1.76) for conductive media, where η is the intrinsic impedance (complex) of the conductor. Equation (1.95) can then be written as

$$P_{\text{avg}} = \frac{R_s}{2} \int_{S_0} |\bar{H}|^2\, ds, \tag{1.97}$$

where

$$R_s = \text{Re}\{\eta\} = \text{Re}\left\{ (1+j)\sqrt{\frac{\omega\mu}{2\sigma}} \right\} = \sqrt{\frac{\omega\mu}{2\sigma}} = \frac{1}{\sigma\delta_s} \tag{1.98}$$

is defined as the *surface resistance* of the conductor. The magnetic field $\bar{H}$ in (1.97) is tangential to the conductor surface and needs only to be evaluated at the surface of the conductor; since H_t is continuous at $z = 0$, it does not matter whether this field is evaluated just outside the conductor or just inside the conductor. In the next section we will show how (1.97) can be evaluated in terms of a surface current density flowing on the surface of the conductor, where the conductor can be approximated as perfect.

EXAMPLE 1.4   ENERGY AND POWER

In a nonmagnetic medium $\bar{E} = 8 \sin\left(2\pi \times 10^7 t - 0.6x\right) a_z$ V/m. Find

  i. ϵ_r
 ii. η
iii. Time-averaged power

Solution

Since $\alpha = 0$ and $\beta = 0.6 \neq \dfrac{\omega}{c}$ where $\omega = 2\pi \times 10^7 t$ and $\mu = \mu_0$ and $\epsilon = \epsilon_0 \epsilon_r$

  i. Hence $\beta = 0.6 = \omega\sqrt{\mu\epsilon} \implies \epsilon_r = 8.21$

 ii. $\eta = \sqrt{\dfrac{\mu}{\epsilon}} = 131.57\ \Omega$

iii. $P = E \times H = \dfrac{E_0^2}{n} \sin^2(\omega t - \beta x)a_x$

Time-averaged power, $P_{\text{avg}} = \displaystyle\int\limits_0^T P\,dt = \dfrac{E_0^2}{2n}a_x = \dfrac{64}{2 \times 10\pi^2}a_x = 324.23 a_x \dfrac{\text{mW}}{\text{m}^2}$

1.7   PLANE WAVE REFLECTION FROM A MEDIA INTERFACE

A number of problems to be considered in later chapters involve the behavior of electromagnetic fields at the interface of various types of media, including lossless media, lossy media, a good conductor, or a perfect conductor, and so it is beneficial at this time to study the reflection of a plane wave normally incident from free-space onto a half-space of an arbitrary material. The geometry is shown in Figure 1.12, where the material half-space $z > 0$ is characterized by the parameters ϵ, μ, and σ.

General Medium

With no loss of generality we can assume that the incident plane wave has an electric field vector oriented along the x-axis and is propagating along the positive z-axis. The incident fields can then be written, for $z < 0$, as

$$\bar{E}_i = \hat{x}E_0 e^{-jk_0 z}, \tag{1.99a}$$

$$\bar{H}_i = \hat{y}\frac{1}{\eta_0}E_0 e^{-jk_0 z}, \tag{1.99b}$$

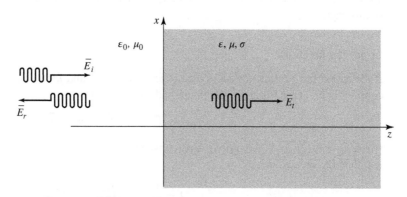

FIGURE 1.12 | Plane wave reflection from an arbitrary medium; normal incidence.

where η_0 is the impedance of free-space and E_0 is an arbitrary amplitude. Also in the region $z < 0$, a reflected wave may exist with the form

$$\bar{E}_r = \hat{x}\Gamma E_0 e^{+jk_0 z}, \tag{1.100a}$$

$$\bar{H}_r = -\hat{y}\frac{\Gamma}{\eta_0}E_0 e^{+jk_0 z}, \tag{1.100b}$$

where Γ is the unknown *reflection coefficient* of the reflected electric field. Note that in (1.100), the sign in the exponential terms has been chosen as positive, to represent waves traveling in the $-\hat{z}$ direction of propagation, as derived in (1.46). This is also consistent with the Poynting vector $\bar{S}_r = \bar{E}_r \times \bar{H}_r^* = -|\Gamma|^2|E_0|^2\hat{z}/\eta_0$, which shows power to be traveling in the $-\hat{z}$ direction for the reflected wave.

As shown in Section 1.4, from equations (1.54) and (1.58), the transmitted fields for $z > 0$ in the lossy medium can be written as

$$\bar{E}_t = \hat{x}TE_0 e^{-\gamma z}, \tag{1.101a}$$

$$\bar{H}_t = \frac{\hat{y}TE_0}{\eta}e^{-\gamma z}, \tag{1.101b}$$

where T is the *transmission coefficient* of the transmitted electric field and η is the intrinsic impedance (complex) of the lossy medium in the region $z > 0$. From (1.57) and (1.52) the intrinsic impedance is

$$\eta = \frac{j\omega\mu}{\gamma}, \tag{1.102}$$

and the propagation constant is

$$\gamma = \alpha + j\beta = j\omega\sqrt{\mu\epsilon}\sqrt{1 - j\sigma/\omega\epsilon}. \tag{1.103}$$

We now have a boundary value problem where the general form of the fields are known via (1.99)–(1.101) on either side of the material discontinuity at $z = 0$. The two unknown constants Γ and T are found by applying boundary conditions for E_x and H_y at $z = 0$. Since these tangential field components must be continuous at $z = 0$, we arrive at the following two equations:

$$1 + \Gamma = T, \tag{1.104a}$$

$$\frac{1 - \Gamma}{\eta_0} = \frac{T}{\eta}. \tag{1.104b}$$

Solving these equations for the reflection and transmission coefficients gives

$$\Gamma = \frac{\eta - \eta_0}{\eta + \eta_0}, \tag{1.105a}$$

$$T = 1 + \Gamma = \frac{2\eta}{\eta + \eta_0}. \tag{1.105b}$$

This is a general solution for reflection and transmission of a normally incident wave at the interface of an arbitrary material, where η is the intrinsic impedance of the material. We now consider three special cases of this result.

Lossless Medium

If the region for $z > 0$ is a lossless dielectric, then $\sigma = 0$, and μ and ϵ are real quantities. The propagation constant in this case is purely imaginary and can be written as

$$\gamma = j\beta = j\omega\sqrt{\mu\epsilon} = jk_0\sqrt{\mu_r\epsilon_r}, \tag{1.106}$$

where $k_0 = \omega\sqrt{\mu_0\epsilon_0}$ is the propagation constant (wave number) of a plane wave in free-space. The wavelength in the dielectric is

$$\lambda = \frac{2\pi}{\beta} = \frac{2\pi}{\omega\sqrt{\mu\epsilon}} = \frac{\lambda_0}{\sqrt{\mu_r\epsilon_r}}, \tag{1.107}$$

the phase velocity is

$$v_p = \frac{\omega}{\beta} = \frac{1}{\sqrt{\mu\epsilon}} = \frac{c}{\sqrt{\mu_r\epsilon_r}}, \tag{1.108}$$

(slower than the speed of light in free-space) and the intrinsic impedance of the dielectric is

$$\eta = \frac{j\omega\mu}{\gamma} = \sqrt{\frac{\mu}{\epsilon}} = \eta_0\sqrt{\frac{\mu_r}{\epsilon_r}}. \tag{1.109}$$

For this lossless case, η is real, so both Γ and T from (1.105) are real, and $\bar{E}$ and $\bar{H}$ are in phase with each other in both regions.

Power conservation for the incident, reflected, and transmitted waves can be demonstrated by computing the Poynting vectors in the two regions. Thus, for $z < 0$, the complex Poynting vector is found from the total fields in this region as

$$\begin{aligned}
\bar{S}^- &= \bar{E} \times \bar{H}^* = (\bar{E}_i + \bar{E}_r) \times (\bar{H}_i + \bar{H}_r)^* \\
&= \hat{z}|E_0|^2\frac{1}{\eta_0}(e^{-jk_0z} + \Gamma e^{jk_0z})(e^{-jk_0z} - \Gamma e^{jk_0z})^* \\
&= \hat{z}|E_0|^2\frac{1}{\eta_0}(1 - |\Gamma|^2 + \Gamma e^{2jk_0z} - \Gamma^* e^{-2jk_0z}) \\
&= \hat{z}|E_0|^2\frac{1}{\eta_0}(1 - |\Gamma|^2 + 2j\Gamma \sin 2k_0z),
\end{aligned} \tag{1.110a}$$

since Γ is real. For $z > 0$ the complex Poynting vector is

$$\bar{S}^+ = \bar{E}_t \times \bar{H}_t^* = \hat{z}\frac{|E_0|^2|T|^2}{\eta},$$

which can be rewritten, using (1.105), as

$$\bar{S}^+ = \hat{z}|E_0|^2\frac{4\eta}{(\eta + \eta_0)^2} = \hat{z}|E_0|^2\frac{1}{\eta_0}(1 - |\Gamma|^2). \tag{1.110b}$$

Now observe that at $z = 0$, $\bar{S}^- = \bar{S}^+$, so that complex power flow is conserved across the interface. Next consider the time-average power flow in the two regions. For $z < 0$ the time-average power flow through a 1 m^2 cross section is

$$P^- = \frac{1}{2}\text{Re}\left\{\bar{S}^- \cdot \hat{z}\right\} = \frac{1}{2}|E_0|^2\frac{1}{\eta_0}(1 - |\Gamma|^2). \tag{1.111a}$$

and for $z > 0$, the time-average power flow through a 1 m^2 cross section is

$$P^+ = \frac{1}{2}\text{Re}\left\{\bar{S}^+ \cdot \hat{z}\right\} = \frac{1}{2}|E_0|^2\frac{1}{\eta_0}(1 - |\Gamma|^2) = P^-, \tag{1.111b}$$

so real power flow is conserved.

We now note a subtle point. When computing the complex Poynting vector for $z < 0$ in (1.110a), we used the total $\bar{E}$ and $\bar{H}$ fields. If we compute separately the Poynting vectors for the incident and reflected waves, we obtain

$$\bar{S}_i = \bar{E}_i \times \bar{H}_i^* = \hat{z}\frac{|E_0|^2}{\eta_0}, \tag{1.112a}$$

$$\bar{S}_r = \bar{E}_r \times \bar{H}_r^* = -\hat{z}\frac{|E_0|^2|\Gamma|^2}{\eta_0}, \tag{1.112b}$$

and we see that $\bar{S}_i + \bar{S}_r \neq \bar{S}^-$ of (1.110a). The missing cross-product terms account for stored reactive energy in the standing wave in the $z < 0$ region. Thus, the decomposition of a Poynting vector into incident and reflected components is not, in general, meaningful. It is possible to define a time-average Poynting vector as $(1/2)\text{Re}\{\bar{E} \times \bar{H}^*\}$, and in this case such a definition applied to the individual incident and reflected components will give the correct result since $P_i = (1/2)|\bar{E}_0|^2/\eta_0$ and $P_r = (-1/2)|E_0|^2|\Gamma|^2/\eta_0$, so $P_i + P_r = P^-$. However, this definition will fail to provide meaningful results when the medium for $z < 0$ is lossy.

Good Conductor

If the region for $z > 0$ is a good (but not perfect) conductor, the propagation constant can be written as discussed in Section 1.4:

$$\gamma = \alpha + j\beta = (1 + j)\sqrt{\frac{\omega\mu\sigma}{2}} = (1 + j)\frac{1}{\delta_s}. \tag{1.113}$$

Similarly, the intrinsic impedance of the conductor simplifies to

$$\eta = (1 + j)\sqrt{\frac{\omega\mu}{2\sigma}} = (1 + j)\frac{1}{\sigma\delta_s}. \tag{1.114}$$

Now the impedance is complex, with a phase angle of 45°, so $\bar{E}$ and $\bar{H}$ will be 45° out of phase, and Γ and T will be complex. In (1.113) and (1.114), $\delta_s = 1/\alpha$ is the skin depth, as defined in (1.60).

For $z < 0$ the complex Poynting vector can be evaluated at $z = 0$ to give

$$\bar{S}^-(z = 0) = \hat{z}|E_0|^2\frac{1}{\eta_0}(1 - |\Gamma|^2 + \Gamma - \Gamma^*). \tag{1.115a}$$

For $z > 0$ the complex Poynting vector is

$$\bar{S}^+ = \bar{E}_t \times \bar{H}_t^* = \hat{z}|E_0|^2|T|^2\frac{1}{\eta^*}e^{-2\alpha z},$$

and using (1.105) for T and Γ gives

$$\bar{S}^+ = \hat{z}|E_0|^2\frac{4\eta}{|\eta + \eta_0|^2}e^{-2\alpha z} = \hat{z}|E_0|^2\frac{1}{\eta_0}(1 - |\Gamma|^2 + \Gamma - \Gamma^*)e^{-2\alpha z}. \tag{1.115b}$$

So at the interface at $z = 0$, $\bar{S}^- = \bar{S}^+$, and complex power is conserved.

Observe that if we were to compute the separate incident and reflected Poynting vectors for $z < 0$ as

$$\bar{S}_i = \bar{E}_i \times \bar{H}_i^* = \hat{z}\frac{|E_0|^2}{\eta_0}, \tag{1.116a}$$

$$\bar{S}_r = \bar{E}_r \times \bar{H}_r^* = -\hat{z}\frac{|E_0|^2|\Gamma|^2}{\eta_0}, \tag{1.116b}$$

we would not obtain $\bar{S}_i + \bar{S}_r = \bar{S}^-$ of (1.115a), even for $z = 0$. It is possible, however, to consider real power flow in terms of the individual traveling wave components. Thus, the time-average power flows through a 1 m² cross section are

$$P^- = \frac{1}{2}\text{Re}(\bar{S}^- \cdot \hat{z}) = \frac{1}{2}|E_0|^2\frac{1}{\eta_0}(1 - |\Gamma|^2), \tag{1.117a}$$

$$P^+ = \frac{1}{2}\text{Re}(\bar{S}^- \cdot \hat{z}) = \frac{1}{2}|E_0|^2\frac{1}{\eta_0}(1 - |\Gamma|^2)e^{-2\alpha z}, \tag{1.117b}$$

which shows power balance at $z = 0$. In addition, $P_i = |E_0|^2/2\eta_0$ and $P_r = -|E_0|^2|\Gamma|^2/2\eta_0$, so that $P_i + P_r = P^-$, showing that the real power flow for $z < 0$ can be decomposed into incident and reflected wave components.

Notice that $\bar{S}^+$, the power density in the lossy conductor, decays exponentially according to the $e^{-2\alpha z}$ attenuation factor. This means that power is being dissipated in the lossy material as the wave propagates into the medium in the $+z$ direction. The power, and also the fields, decays to a negligibly small value within a few skin depths of the material, which for a reasonably good conductor is an extremely small distance at microwave frequencies.

The electric volume current density flowing in the conducting region is given as

$$\bar{J}_t = \sigma \bar{E}_t = \hat{x} \sigma E_0 T e^{-\gamma z} \, \text{A/m}^2,$$

(1.118)

and so the average power dissipated in (or transmitted into) a 1 m² cross-sectional volume of the conductor can be calculated from the conductor loss term of (1.92) (Joule's law) as

$$P^t = \frac{1}{2} \int_V \bar{E}_t \cdot \bar{J}_t^* \, dv = \frac{1}{2} \int_{x=0}^1 \int_{y=0}^1 \int_{z=0}^\infty (\hat{x} E_0 T e^{-\gamma z}) \cdot (\hat{x} \sigma E_0 T e^{-\gamma z})^* \, dz \, dy \, dx$$

$$= \frac{1}{2} \sigma |E_0|^2 |T|^2 \int_{z=0}^\infty e^{-2\alpha z} \, dz = \frac{\sigma |E_0|^2 |T|^2}{4\alpha}.$$

(1.119)

Since $1/\eta = \sigma \delta_s/(1+j) = (\sigma/2\alpha)(1-j)$, the real power entering the conductor through a 1 m² cross section [as given by $(1/2)\text{Re}\{\bar{S}^+ \cdot \hat{z}\}$ at $z = 0$] can be expressed using (1.115b) as $P^t = |E_0|^2 |T|^2 (\sigma/4\alpha)$, which is in agreement with (1.119).

Perfect Conductor

Now assume that the region $z > 0$ contains a perfect conductor. The above results can be specialized to this case by allowing $\sigma \to \infty$. Then, from (1.113), $\alpha \to \infty$; from (1.114), $\eta \to 0$; from (1.60), $\delta_s \to 0$; and from (1.105a, b), $T \to 0$ and $\Gamma \to -1$. The fields for $z > 0$ thus decay infinitely fast and are identically zero in the perfect conductor. The perfect conductor can be thought of as "shorting out" the incident electric field. For $z < 0$, from (1.99) and (1.100), the total $\bar{E}$ and $\bar{H}$ fields are, since $\Gamma = -1$,

$$\bar{E} = \bar{E}_i + \bar{E}_r = \hat{x} E_0 (e^{-jk_0 z} - e^{jk_0 z}) = -\hat{x} 2 j E_0 \sin k_0 z,$$

(1.120a)

$$\bar{H} = \bar{H}_i + \bar{H}_r = \hat{y} \frac{1}{\eta_0} E_0 (e^{-jk_0 z} + e^{jk_0 z}) = \hat{y} \frac{2}{\eta_0} E_0 \cos k_0 z.$$

(1.120b)

Observe that at $z = 0$, $\bar{E} = 0$ and $\bar{H} = \hat{y}(2/\eta_0)E_0$. The Poynting vector for $z < 0$ is

$$\bar{S}^- = \bar{E} \times \bar{H}^* = -\hat{z} j \frac{4}{\eta_0} |E_0|^2 \sin k_0 z \cos k_0 z,$$

(1.121)

which has a zero real part and thus indicates that no real power is delivered to the perfect conductor.

The volume current density of (1.118) for the lossy conductor reduces to an infinitely thin sheet of surface current in the limit of infinite conductivity:

$$\bar{J}_s = \hat{n} \times \bar{H} = -\hat{z} \times \left(\hat{y} \frac{2}{\eta_0} E_0 \cos k_0 z \right) \bigg|_{z=0} = \hat{x} \frac{2}{\eta_0} E_0 \, \text{A/m}.$$

(1.122)

The Surface Impedance Concept

In many problems, particularly those in which the effect of attenuation or conductor loss is needed, the presence of an imperfect conductor must be taken into account. The surface impedance concept allows us to do this in an approximate, but very convenient and accurate, manner. We will develop this method from the theory presented in the previous sections.

Consider a good conductor in the region $z > 0$. As we have seen, a plane wave normally incident on this conductor is mostly reflected, and the power that is transmitted into the conductor is dissipated as heat within a very short distance from the surface. There are three ways to compute this power.

First, we can use Joule's law, as in (1.119). For a 1 m² area of conductor surface, the power transmitted through this surface and dissipated as heat is given by (1.119). Using (1.105b) for T, (1.114) for η, and the fact that $\alpha = 1/\delta_s$ gives the following result:

$$\frac{\sigma |T|^2}{\alpha} = \frac{\sigma \delta_s 4 |\eta|^2}{|\eta + \eta_0|^2} \simeq \frac{8}{\sigma \delta_s \eta_0^2},$$

(1.123)

where we have assumed $\eta \ll \eta_0$, which is true for a good conductor. Then the power of (1.119) can be written as

$$P^t = \frac{\sigma |E_0|^2 |T|^2}{4\alpha} = \frac{2|E_0|^2}{\sigma \delta_s \eta_0^2} = \frac{2|E_0|^2 R_s}{\eta_0^2}, \tag{1.124}$$

where

$$R_s = \text{Re}\{\eta\} = \text{Re}\left\{\frac{1+j}{\sigma \delta_s}\right\} = \frac{1}{\sigma \delta_s} = \sqrt{\frac{\omega \mu}{2\sigma}} \tag{1.125}$$

is the surface resistance of the metal.

Another way to find the power loss is to compute the power flow into the conductor using the Poynting vector since all power entering the conductor at $z = 0$ is dissipated. As in (1.115b), we have

$$P^t = \frac{1}{2} \text{Re}\{\bar{S}^+ \cdot \hat{z}\}\Big|_{z=0} = \frac{2|E_0|^2 \text{Re}\{\eta\}}{|\eta + \eta_0|^2},$$

which for large conductivity becomes, since $\eta \ll \eta_0$,

$$P^t = \frac{2|E_0|^2 R_s}{\eta_0^2}, \tag{1.126}$$

which agrees with (1.124).

A third method uses an effective surface current density and the surface impedance, without the need for knowing the fields inside the conductor. From (1.118), the volume current density in the conductor is

$$\bar{J}_t = \hat{x}\sigma T E_0 e^{-\gamma z} \text{ A/m}^2, \tag{1.127}$$

so the total (surface) current flow per unit width in the x direction is

$$\bar{J}_s = \int_0^\infty \bar{J}_t \, dz = \hat{x}\sigma T E_0 \int_0^\infty e^{-\gamma z} \, dz = \frac{\hat{x}\sigma T E_0}{\gamma} \text{A/m}.$$

Approximating $\sigma T/\gamma$ for large σ and using (1.113), (1.105b), and (1.114) gives

$$\frac{\sigma T}{\gamma} = \frac{\sigma \delta_s}{(1+j)} \frac{2\eta}{(\eta + \eta_0)} \simeq \frac{\sigma \delta_s}{(1+j)} \frac{2(1+j)}{\sigma \delta_s \eta_0} = \frac{2}{\eta_0},$$

so

$$\bar{J}_s \simeq \hat{x}\frac{2E_0}{\eta_0} \text{ A/m}. \tag{1.128}$$

If the conductivity were infinite, then $\Gamma = -1$ and a true surface current density of

$$\bar{J}_s = \hat{n} \times \bar{H}|_{z=0} = -\hat{z} \times (\bar{H}_i + \bar{H}_r)|_{z=0} = \hat{x}E_0 \frac{1}{\eta_0}(1 - \Gamma) = \hat{x}\frac{2E_0}{\eta_0} \text{ A/m}$$

would flow, which is identical to the total current in (1.128).

Now replace the exponentially decaying volume current of (1.127) with a uniform volume current extending a distance of one skin depth. Thus, let

$$\bar{J}_t = \begin{cases} \bar{J}_s/\delta_s & \text{for } 0 < z < \delta_s \\ 0 & \text{for } z > \delta_s, \end{cases} \tag{1.129}$$

so that the total current flow is the same. Then Joule's law gives the power lost:

$$P^t = \frac{1}{2\sigma} \int_S \int_{z=0}^{\delta_s} \frac{|\bar{J}_s|^2}{\delta_s^2} \, dz \, ds = \frac{R_s}{2} \int_S |\bar{J}_s|^2 \, ds = \frac{2|E_0|^2 R_s}{\eta_0^2}, \tag{1.130}$$

where $\int_S$ denotes a surface integral over the conductor surface, in this case chosen as 1 m². The result of (1.130) agrees with our previous results for P^t in (1.126) and (1.124) and shows that the power loss in a good conductor can be accurately and simply calculated as

$$P^t = \frac{R_s}{2} \int_S |\bar{J}_s|^2 \, ds = \frac{R_s}{2} \int_S |\bar{H}_t|^2 \, ds, \tag{1.131}$$

in terms of the surface resistance R_s and the surface current $\bar{J}_s$, or tangential magnetic field $\bar{H}_t$. It is important to realize that the surface current can be found from $\bar{J}_s = \hat{n} \times \bar{H}$, as if the metal were a perfect conductor. This method is very general, applying to fields other than plane waves and to conductors of arbitrary shape, as long as bends or corners have radii on the order of a skin depth or larger. The method is also quite accurate, as the only approximation was that $\eta \ll \eta_0$, which is a good approximation. As an example, copper at 1 GHz has $|\eta| = 0.012 \ \Omega$, which is indeed much less than $\eta_0 = 377 \ \Omega$.

EXAMPLE 1.5   PLANE WAVE REFLECTION FROM A CONDUCTOR

Consider a plane wave normally incident on a half-space of copper. If $f = 4$ GHz, compute the propagation constant, intrinsic impedance, and skin depth for the conductor. Also compute the reflection and transmission coefficients.

Solution
For copper, $\sigma = 5.96 \times 10^7$ S/m at 20°C, so from (1.60), the skin depth is

$$\delta_s = \sqrt{\frac{2}{\omega\mu\sigma}} = 1.031 \times 10^{-6} \text{m},$$

and the propagation constant is, from (1.113),

$$\gamma = \frac{1+j}{\delta_s} = (9.701 + j9.701) \times 10^5 \text{ m}^{-1}.$$

The intrinsic impedance is, from (1.114),

$$\eta = \frac{1+j}{\sigma\delta_s} = (16.277 + j16.277) \times 10^{-3}\Omega,$$

which is quite small relative to the impedance of free-space ($\eta_0 = 377 \ \Omega$). The reflection coefficient is, from (1.105a),

$$\Gamma = \frac{\eta - \eta_0}{\eta + \eta_0} = 1.0\angle179.99°$$

(practically that of an ideal short circuit), and the transmission coefficient is

$$T = \frac{2\eta}{\eta + \eta_0} = 12.21 \times 10^{-5}\angle45°.$$

1.8   OBLIQUE INCIDENCE AT A DIELECTRIC INTERFACE

We continue our discussion of plane waves by considering the problem of a plane wave obliquely incident on a plane interface between two lossless dielectric regions, as shown in Figure 1.13. There are two canonical cases of this problem: the electric field is either in the xz plane (parallel polarization) or normal to the xz plane (perpendicular polarization). An arbitrary incident plane wave, of course, may have a polarization that is neither of these, but it can be expressed as a linear combination of these two individual cases.

The general method of solution is similar to the problem of normal incidence: we will write expressions for the incident, reflected, and transmitted fields in each region and match boundary conditions to find the unknown amplitude coefficients and angles.

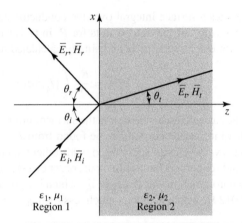

FIGURE 1.13 | Geometry for a plane wave obliquely incident at the interface between two dielectric regions.

Parallel Polarization

In this case the electric field vector lies in the xz plane, and the incident fields can be written as

$$\bar{E}_i = E_0(\hat{x}\cos\theta_i - \hat{z}\sin\theta_i)e^{-jk_1(x\sin\theta_i + z\cos\theta_i)}, \tag{1.132a}$$

$$\bar{H}_i = \frac{E_0}{\eta_1}\hat{y}e^{-jk_1(x\sin\theta_i + z\cos\theta_i)}, \tag{1.132b}$$

where $k_1 = \omega\sqrt{\mu_0\epsilon_1}$ and $\eta_1 = \sqrt{\mu_0/\epsilon_1}$ are the propagation constant and impedance of region 1. The reflected and transmitted fields can be written as

$$\bar{E}_r = E_0\Gamma(\hat{x}\cos\theta_r + \hat{z}\sin\theta_r)e^{-jk_1(x\sin\theta_r - z\cos\theta_r)}, \tag{1.133a}$$

$$\bar{H}_r = \frac{-E_0\Gamma}{\eta_1}\hat{y}e^{-jk_1(x\sin\theta_r - z\cos\theta_r)}, \tag{1.133b}$$

$$\bar{E}_t = E_0T(\hat{x}\cos\theta_t - \hat{z}\sin\theta_t)e^{-jk_2(x\sin\theta_t + z\cos\theta_t)}, \tag{1.134a}$$

$$\bar{H}_t = \frac{E_0T}{\eta_2}\hat{y}e^{-jk_2(x\sin\theta_t + z\cos\theta_t)}. \tag{1.134b}$$

Here, Γ and T are the reflection and transmission coefficients, and k_2 and η_2 are the propagation constant and impedance of region 2, defined as

$$k_2 = \omega\sqrt{\mu_0\epsilon_2}, \quad \eta_2 = \sqrt{\mu_0/\epsilon_2}.$$

At this point we have Γ, T, θ_r, and θ_t as unknowns.

We can obtain two complex equations for these unknowns by enforcing the continuity of E_x and H_y, the tangential field components, at the interface between the two regions at $z = 0$. We then obtain

$$\cos\theta_i e^{-jk_1 x\sin\theta_i} + \Gamma\cos\theta_r e^{-jk_1 x\sin\theta_r} = T\cos\theta_t e^{-jk_2 x\sin\theta_t}, \tag{1.135a}$$

$$\frac{1}{\eta_1}e^{-jk_1 x\sin\theta_i} - \frac{\Gamma}{\eta_1}e^{-jk_1 x\sin\theta_r} = \frac{T}{\eta_2}e^{-jk_2 x\sin\theta_t}. \tag{1.135b}$$

Both sides of (1.135a) and (1.135b) are functions of the coordinate x. If E_x and H_y are to be continuous at the interface $z = 0$ for all x, then this x variation must be the same on both sides of the equations, leading to the following condition:

$$k_1\sin\theta_i = k_1\sin\theta_r = k_2\sin\theta_t.$$

This results in the well-known *Snell's laws* of reflection and refraction:

$$\theta_i = \theta_r, \tag{1.136a}$$

$$k_1\sin\theta_i = k_2\sin\theta_t. \tag{1.136b}$$

The above argument ensures that the phase terms in (1.135) vary with x at the same rate on both sides of the interface, and so is often called the *phase matching condition.*

Using (1.136) in (1.135) allows us to solve for the reflection and transmission coefficients as

$$\Gamma = \frac{\eta_2 \cos \theta_t - \eta_1 \cos \theta_i}{\eta_2 \cos \theta_t + \eta_1 \cos \theta_i}, \tag{1.137a}$$

$$T = \frac{2\eta_2 \cos \theta_i}{\eta_2 \cos \theta_t + \eta_1 \cos \theta_i}. \tag{1.137b}$$

Observe that for normal incidence $\theta_i = 0$, we have $\theta_r = \theta_t = 0$, so then

$$\Gamma = \frac{\eta_2 - \eta_1}{\eta_2 + \eta_1} \quad \text{and} \quad T = \frac{2\eta_2}{\eta_2 + \eta_1},$$

which is in agreement with the results of Section 1.7.

For this polarization a special angle of incidence, θ_b, called the *Brewster angle*, exists where $\Gamma = 0$. This occurs when the numerator of (1.137a) goes to zero $(\theta_i = \theta_b)$: $\eta_2 \cos \theta_t = \eta_1 \cos \theta_b$, which can be rewritten using

$$\cos \theta_t = \sqrt{1 - \sin^2 \theta_t} = \sqrt{1 - \frac{k_1^2}{k_2^2} \sin^2 \theta_b},$$

to give

$$\sin \theta_b = \frac{1}{\sqrt{1 + \epsilon_1/\epsilon_2}}. \tag{1.138}$$

Perpendicular Polarization

In this case the electric field vector is perpendicular to the xz plane. The incident field can be written as

$$\bar{E}_i = E_0 \hat{y} e^{-jk_1(x \sin \theta_i + z \cos \theta_i)}, \tag{1.139a}$$

$$\bar{H}_i = \frac{E_0}{\eta_1}(-\hat{x} \cos \theta_i + \hat{z} \sin \theta_i) e^{-jk_1(x \sin \theta_i + z \cos \theta_i)}, \tag{1.139b}$$

where $k_1 = \omega\sqrt{\mu_0 \epsilon_1}$ and $\eta_1 = \sqrt{\mu_0/\epsilon_1}$ are the propagation constant and impedance for region 1, as before. The reflected and transmitted fields can be expressed as

$$\bar{E}_r = E_0 \Gamma \hat{y} e^{-jk_1(x \sin \theta_r - z \cos \theta_r)}, \tag{1.140a}$$

$$\bar{H}_r = \frac{E_0 \Gamma}{\eta_1}(\hat{x} \cos \theta_r + \hat{z} \sin \theta_r) e^{-jk_1(x \sin \theta_r - z \cos \theta_r)}, \tag{1.140b}$$

$$\bar{E}_t = E_0 T \hat{y} e^{-jk_2(x \sin \theta_t + z \cos \theta_t)}, \tag{1.141a}$$

$$\bar{H}_t = \frac{E_0 T}{\eta_2}(-\hat{x} \cos \theta_t + \hat{z} \sin \theta_t) e^{-jk_2(x \sin \theta_t + z \cos \theta_t)}, \tag{1.141b}$$

with $k_2 = \omega\sqrt{\mu_0 \epsilon_2}$ and $\eta_2 = \sqrt{\mu_0/\epsilon_2}$ being the propagation constant and impedance in region 2.

Equating the tangential field components E_y and H_x at $z = 0$ gives

$$e^{-jk_1 x \sin \theta_i} + \Gamma e^{-jk_1 x \sin \theta_r} = T e^{-jk_2 x \sin \theta_t}, \tag{1.142a}$$

$$\frac{-1}{\eta_1} \cos \theta_i e^{-jk_1 x \sin \theta_i} + \frac{\Gamma}{\eta_1} \cos \theta_r e^{-jk_2 x \sin \theta_r} = \frac{-T}{\eta_2} \cos \theta_t e^{-jk_2 x \sin \theta_t}. \tag{1.142b}$$

By the same phase matching argument that was used in the parallel case, we obtain Snell's laws

$$k_1 \sin \theta_i = k_1 \sin \theta_r = k_2 \sin \theta_t,$$

identical to (1.136).

Using (1.136) in (1.142) allows us to solve for the reflection and transmission coefficients as

$$\Gamma = \frac{\eta_2 \cos\theta_i - \eta_1 \cos\theta_t}{\eta_2 \cos\theta_i + \eta_1 \cos\theta_t}, \tag{1.143a}$$

$$T = \frac{2\eta_2 \cos\theta_i}{\eta_2 \cos\theta_i + \eta_1 \cos\theta_t}. \tag{1.143b}$$

Again, for the normally incident case, these results reduce to those of Section 1.7.

For this polarization no Brewster angle exists where $\Gamma = 0$, as we can see by examining the possibility that the numerator of (1.143a) could be zero:

$$\eta_2 \cos\theta_i = \eta_1 \cos\theta_t,$$

and using Snell's law to give

$$k_2^2\left(\eta_2^2 - \eta_1^2\right) = \left(k_2^2\eta_2^2 - k_1^2\eta_1^2\right)\sin^2\theta_i.$$

This leads to a contradiction since the term in parentheses on the right-hand side is identically zero for dielectric media. Thus, no Brewster angle exists for perpendicular polarization for dielectric media.

EXAMPLE 1.6   OBLIQUE REFLECTION FROM A DIELECTRIC INTERFACE

Plot the reflection coefficients versus incidence angle for parallel and perpendicular polarized plane waves incident from free-space onto a dielectric region with $\epsilon_r = 2.55$.

Solution
The impedances for the two regions are

$$\eta_1 = 377\ \Omega,$$

$$\eta_2 = \frac{\eta_0}{\sqrt{\epsilon_r}} = \frac{377}{\sqrt{2.55}} = 236\ \Omega.$$

We then evaluate (1.137a) and (1.143a) versus incidence angle; the results are shown in Figure 1.14.

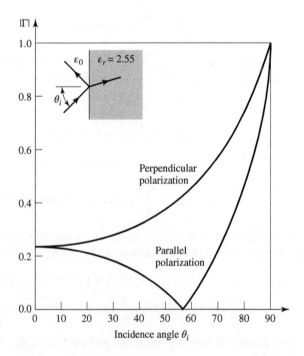

FIGURE 1.14 | Reflection coefficient magnitude for parallel and perpendicular polarizations of a plane wave obliquely incident on a dielectric half-space.

Total Reflection and Surface Waves

Snell's law of (1.136b) can be rewritten as

$$\sin \theta_t = \sqrt{\frac{\epsilon_1}{\epsilon_2}} \sin \theta_i. \tag{1.144}$$

Consider the case (for either parallel or perpendicular polarization) where $\epsilon_1 > \epsilon_2$. As θ_i increases, the refraction angle θ_t will increase, but at a faster rate than θ_i increases. The incidence angle θ_i for which $\theta_t = 90°$ is called the *critical angle*, θ_c, where

$$\sin \theta_c = \sqrt{\frac{\epsilon_2}{\epsilon_1}}. \tag{1.145}$$

At this angle and beyond, the incident wave will be totally reflected, as the transmitted wave will not propagate into region 2. Let us look at this situation more closely for the case of $\theta_i > \theta_c$ with parallel polarization.

When $\theta_i > \theta_c$ (1.144) shows that $\sin \theta_t > 1$, so that $\cos \theta_t = \sqrt{1 - \sin^2 \theta_t}$ must be imaginary, and the angle θ_t loses its physical significance. At this point, it is better to replace the expressions for the transmitted fields in region 2 with the following:

$$\bar{E}_t = E_0 T \left(\frac{-j\alpha}{k_2} \hat{x} - \frac{\beta}{k_2} \hat{z} \right) e^{-j\beta x} e^{-\alpha z}, \tag{1.146a}$$

$$\bar{H}_t = \frac{E_0 T}{\eta_2} \hat{y} e^{-j\beta x} e^{-\alpha z}. \tag{1.146b}$$

The form of these fields is derived from (1.134) after noting that $-jk_2 \sin \theta_t$ is still imaginary for $\sin \theta_t > 1$ but $-jk_2 \cos \theta_t$ is real, so we can replace $\sin \theta_t$ by β / k_2 and $\cos \theta_t$ by $-j\alpha / k_2$. Substituting (1.146b) into the Helmholtz wave equation for $\bar{H}$ gives

$$-\beta^2 + \alpha^2 + k_2^2 = 0. \tag{1.147}$$

Matching E_x and H_y of (1.146) with the $\hat{x}$ and $\hat{y}$ components of the incident and reflected fields of (1.132) and (1.133) at $z = 0$ gives

$$\cos \theta_i e^{-jk_1 x \sin \theta_i} + \Gamma \cos \theta_r e^{-jk_1 x \sin \theta_r} = \frac{-j\alpha}{k_2} T e^{-j\beta x}, \tag{1.148a}$$

$$\frac{1}{\eta_1} e^{-jk_1 x \sin \theta_i} - \frac{\Gamma}{\eta_1} e^{-jk_1 x \sin \theta_r} = \frac{T}{\eta_2} e^{-j\beta x}. \tag{1.148b}$$

To obtain phase matching at the $z = 0$ boundary, we must have

$$k_1 \sin \theta_i = k_1 \sin \theta_r = \beta,$$

which leads again to Snell's law for reflection, $\theta_i = \theta_r$, and to $\beta = k_1 \sin \theta_i$. Then α is determined from (1.147) as

$$\alpha = \sqrt{\beta^2 - k_2^2} = \sqrt{k_1^2 \sin^2 \theta_i - k_2^2}, \tag{1.149}$$

which is seen to be a positive real number since $\sin^2 \theta_i > \epsilon_2 / \epsilon_1$. The reflection and transmission coefficients can be obtained from (1.148) as

$$\Gamma = \frac{(-j\alpha/k_2)\eta_2 - \eta_1 \cos \theta_i}{(-j\alpha/k_2)\eta_2 + \eta_1 \cos \theta_i}, \tag{1.150a}$$

$$T = \frac{2\eta_2 \cos \theta_i}{(-j\alpha/k_2)\eta_2 + \eta_1 \cos \theta_i}. \tag{1.150b}$$

Since Γ is of the form $(ja - b)/(ja + b)$, its magnitude is unity, indicating that all incident power is reflected.

The transmitted fields of (1.146) show propagation in the x direction, along the interface, but exponential decay in the z direction. Such a field is known as a *surface wave*[3] since it is tightly bound to the interface. A surface wave is an example of a nonuniform plane wave, so called because it has an amplitude variation in the z direction, apart from the propagation factor in the x direction.

Finally, it is of interest to calculate the complex Poynting vector for the surface wave fields of (1.146):

$$\bar{S}_t = \bar{E}_t \times \bar{H}_t^* = \frac{|E_0|^2|T|^2}{\eta_2}\left(\hat{z}\frac{-j\alpha}{k_2} + \hat{x}\frac{\beta}{k_2}\right)e^{-2\alpha z}. \tag{1.151}$$

This shows that no real power flow occurs in the z direction. The real power flow in the x direction is that of the surface wave field, and it decays exponentially with distance into region 2. So even though no real power is transmitted into region 2, a nonzero field does exist there, in order to satisfy the boundary conditions at the interface.

1.9   SOME USEFUL THEOREMS

Finally, we discuss several theorems in electromagnetics that we will find useful for later discussions.

The Reciprocity Theorem

Reciprocity is a general concept that occurs in many areas of physics and engineering, and the reader may already be familiar with the reciprocity theorem of circuit theory. Here we will derive the Lorentz reciprocity theorem for electromagnetic fields in two different forms. This theorem will be used later in the book to obtain general properties of network matrices representing microwave circuits and to evaluate the coupling of waveguides from current probes and loops, as well as the coupling of waveguides through apertures. There are a number of other important uses of this powerful concept.

Consider the two separate sets of sources, $\bar{J}_1$, $\bar{M}_1$ and $\bar{J}_2$, $\bar{M}_2$, which generate the fields $\bar{E}_1$, $\bar{H}_1$, and $\bar{E}_2$, $\bar{H}_2$, respectively, in the volume V enclosed by the closed surface S, as shown in Figure 1.15. Maxwell's equations are satisfied individually for these two sets of sources and fields, so we can write

$$\nabla \times \bar{E}_1 = -j\omega\mu\bar{H}_1 - \bar{M}_1, \tag{1.152a}$$

$$\nabla \times \bar{H}_1 = j\omega\epsilon\bar{E}_1 + \bar{J}_1, \tag{1.152b}$$

$$\nabla \times \bar{E}_2 = -j\omega\mu\bar{H}_2 - \bar{M}_2, \tag{1.153a}$$

$$\nabla \times \bar{H}_2 = j\omega\epsilon\bar{E}_2 + \bar{J}_2. \tag{1.153b}$$

Now consider the quantity $\nabla \cdot (\bar{E}_1 \times \bar{H}_2 - \bar{E}_2 \times \bar{H}_1)$, which can be expanded using vector identity (B.8) to give

$$\nabla \cdot (\bar{E}_1 \times \bar{H}_2 - \bar{E}_2 \times \bar{H}_1) = \bar{J}_1 \cdot \bar{E}_2 - \bar{J}_2 \cdot \bar{E}_1 + \bar{M}_2 \cdot \bar{H}_1 - \bar{M}_1 \cdot \bar{H}_2. \tag{1.154}$$

Integrating over the volume V and applying the divergence theorem (B.15), gives

$$\int_V \nabla \cdot (\bar{E}_1 \times \bar{H}_2 - \bar{E}_2 \times \bar{H}_1)\,dv = \oint_S (\bar{E}_1 \times \bar{H}_2 - \bar{E}_2 \times \bar{H}_1) \cdot ds \tag{1.155}$$

$$= \int_V (\bar{E}_2 \cdot \bar{J}_1 - \bar{E}_1 \cdot \bar{J}_2 + \bar{H}_1 \cdot \bar{M}_2 - \bar{H}_2 \cdot \bar{M}_1)\,dv$$

Equation (1.155) represents a general form of the *reciprocity theorem*, but in practice a number of special situations often occur leading to some simplification. We will consider three cases.

---

[3] Some authors argue that the term "surface wave" should not be used for a field of this type since it exists only when plane wave fields exist in the $z < 0$ region, and so prefer the term "surface wave-like" field, or a "forced surface wave."

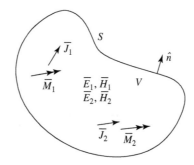

FIGURE 1.15 | Geometry for the Lorentz reciprocity theorem.

S encloses no sources: Then $\bar{J}_1 = \bar{J}_2 = \bar{M}_1 = \bar{M}_2 = 0$, and the fields $\bar{E}_1$, $\bar{H}_1$ and $\bar{E}_2$, $\bar{H}_2$ are source-free fields. In this case, the right-hand side of (1.155) vanishes, with the result that

$$\oint_S \bar{E}_1 \times \bar{H}_2 \cdot d\bar{s} = \oint_S \bar{E}_2 \times \bar{H}_1 \cdot d\bar{s}. \tag{1.156}$$

This result will be used in Chapter 4 when we demonstrate the symmetry of the impedance matrix for a reciprocal microwave network.

S bounds a perfect conductor: For example, S may be the inner surface of a perfectly conducting closed cavity. Then the surface integral of (1.155) vanishes since $\bar{E}_1 \times \bar{H}_2 \cdot \hat{n} = (\hat{n} \times \bar{E}_1) \cdot \bar{H}_2$ [by vector identity (B.3)], and $\hat{n} \times \bar{E}_1$ is zero on the surface of a perfect conductor (similarly for $\bar{E}_2$). The result is

$$\int_V (\bar{E}_1 \cdot \bar{J}_2 - \bar{H}_1 \cdot \bar{M}_2)\, dv = \int_V (\bar{E}_2 \cdot \bar{J}_1 - \bar{H}_2 \cdot \bar{M}_1)\, dv. \tag{1.157}$$

This result is analogous to the reciprocity theorem of circuit theory. In words, this result states that the system response $\bar{E}_1$ or $\bar{E}_2$ is not changed when the source and observation points are interchanged. That is, $\bar{E}_2$ (caused by $\bar{J}_2$) at $\bar{J}_1$ is the same as $\bar{E}_1$ (caused by $\bar{J}_1$) at $\bar{J}_2$.

S is a sphere at infinity: In this case the fields evaluated on S are very far from the sources and so can be considered locally as plane waves. Then the wave impedance relation $\bar{H} = \hat{n} \times \bar{E}/\eta$ applies to (1.155) to give

$$(\bar{E}_1 \times \bar{H}_2 - \bar{E}_2 \times \bar{H}_1) \cdot \hat{n} = (\hat{n} \times \bar{E}_1) \cdot \bar{H}_2 - (\hat{n} \times \bar{E}_2) \cdot \bar{H}_1$$

$$= \frac{1}{\eta} \bar{H}_1 \cdot \bar{H}_2 - \frac{1}{\eta} \bar{H}_2 \cdot \bar{H}_1 = 0,$$

so that the result of (1.157) is again obtained. This result can also be obtained for the case of a closed surface S where the surface impedance boundary condition applies.

Image Theory

In many problems a current source (electric or magnetic) is located in the vicinity of a conducting ground plane. Image theory permits the removal of the ground plane by placing a virtual *image source* on the other side of the ground plane. The reader should be familiar with this concept from electrostatics, so we will prove the result for an infinite current sheet next to an infinite ground plane and then summarize other possible cases.

Consider the surface current density $\bar{J}_s = J_{s0}\hat{x}$ parallel to a ground plane, as shown in Figure 1.16a. Because the current source is of infinite extent and is uniform in the x, y directions, it will excite plane waves traveling outward from it. The negatively traveling wave will reflect from the ground plane at $z = 0$ and then travel in the positive direction. Thus, there will be a standing wave field in the region

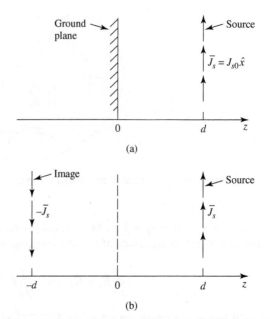

FIGURE 1.16 | Illustration of image theory as applied to an electric current source next to a ground plane. (a) An electric surface current density parallel to a ground plane. (b) The ground plane of (a) is replaced with image current at $z = -d$.

$0 < z < d$ and a positively traveling wave for $z > d$. The forms of the fields in these two regions can thus be written as

$$E_x^s = A(e^{jk_0z} - e^{-jk_0z}), \qquad \text{for } 0 < z < d, \tag{1.158a}$$

$$H_y^s = \frac{-A}{\eta_0}(e^{jk_0z} + e^{-jk_0z}), \qquad \text{for } 0 < z < d, \tag{1.158b}$$

$$E_x^+ = Be^{-jk_0z}, \qquad \text{for } z > d, \tag{1.159a}$$

$$H_y^+ = \frac{B}{\eta_0}e^{-jk_0z}, \qquad \text{for } z > d, \tag{1.159b}$$

where η_0 is the impedance of free-space. Note that the standing wave fields of (1.158) have been constructed to satisfy the boundary condition that $E_x = 0$ at $z = 0$. The remaining boundary conditions to satisfy are the continuity of $\bar{E}$ at $z = d$ and the discontinuity in the $\bar{H}$ field at $z = d$ due to the current sheet. From (1.36), since $\bar{M}_s = 0$,

$$E_x^s = E_x^+|_{z=d}, \tag{1.160a}$$

while from (1.37) we have

$$\bar{J}_s = \hat{z} \times \hat{y}(H_y^+ - H_y^s)|_{z=d}. \tag{1.160b}$$

Using (1.158) and (1.159) then gives

$$2jA \sin k_0 d = Be^{-jk_0d}$$

$$\text{and } J_{s0} = -\frac{B}{\eta_0}e^{-jk_0d} - \frac{2A}{\eta_0}\cos k_0 d,$$

which can be solved for A and B:

$$A = \frac{-J_{s0}\eta_0}{2}e^{-jk_0d},$$

$$B = -jJ_{s0}\eta_0 \sin k_0 d.$$

So the total fields are

$$E_x^s = -jJ_{s0}\eta_0 e^{-jk_0 d} \sin k_0 z, \qquad \text{for } 0 < z < d, \qquad (1.161a)$$

$$H_y^s = J_{s0} e^{-jk_0 d} \cos k_0 z, \qquad \text{for } 0 < z < d, \qquad (1.161b)$$

$$E_x^+ = -jJ_{s0}\eta_0 \sin k_0 d e^{-jk_0 z}, \qquad \text{for } z > d, \qquad (1.162a)$$

$$H_y^+ = -jJ_{s0} \sin k_0 d e^{-jk_0 z}, \qquad \text{for } z > d. \qquad (1.162b)$$

Now consider the application of image theory to this problem. As shown in Figure 1.16b, the ground plane is removed and an image source of $-\bar{J}_s$ is placed at $z = -d$. By superposition, the total fields for $z > 0$ can be found by combining the fields from the two sources individually. These fields can be derived by a procedure similar to that in the above, with the following results:

Fields due to source at $z = d$:

$$E_x = \begin{cases} \dfrac{-J_{s0}\eta_0}{2} e^{-jk_0(z-d)} & \text{for } z > d \\[2mm] \dfrac{-J_{s0}\eta_0}{2} e^{jk_0(z-d)} & \text{for } z < d, \end{cases} \qquad (1.163a)$$

$$H_y = \begin{cases} \dfrac{-J_{s0}}{2} e^{-jk_0(z-d)} & \text{for } z > d \\[2mm] \dfrac{J_{s0}}{2} e^{jk_0(z-d)} & \text{for } z < d. \end{cases} \qquad (1.163b)$$

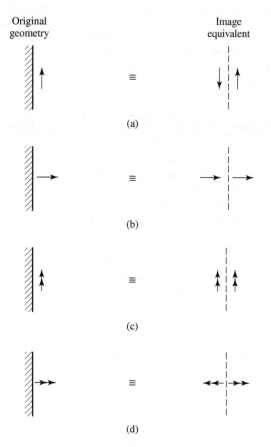

Original geometry

Image equivalent

(a)

(b)

(c)

(d)

FIGURE 1.17 | Electric and magnetic current images. (a) An electric current parallel to a ground plane. (b) An electric current normal to a ground plane. (c) A magnetic current parallel to a ground plane. (d) A magnetic current normal to a ground plane.

Fields due to source at $z = -d$:

$$E_x = \begin{cases} \dfrac{J_{s0}\eta_0}{2}e^{-jk_0(z+d)} & \text{for } z > -d \\[2mm] \dfrac{J_{s0}\eta_0}{2}e^{jk_0(z+d)} & \text{for } z < -d, \end{cases}$$ (1.164a)

$$H_y = \begin{cases} \dfrac{J_{s0}}{2}e^{-jk_0(z+d)} & \text{for } z > -d \\[2mm] \dfrac{-J_{s0}}{2}e^{jk_0(z+d)} & \text{for } z < -d. \end{cases}$$ (1.164b)

The reader can verify that this solution is identical to that of (1.161) for $0 < z < d$ and to that of (1.162) for $z > d$, thus verifying the validity of the image theory solution. Note that image theory only gives the correct fields to the right of the conducting plane. Figure 1.17 shows more general image theory results for electric and magnetic dipoles.

REFERENCES

[1] T. S. Sarkar, R. J. Mailloux, A. A. Oliner, M. Salazar-Palma, and D. Sengupta, *History of Wireless*, John Wiley & Sons, Hoboken, NJ, 2006.

[2] A. A. Oliner, "Historical Perspectives on Microwave Field Theory," *IEEE Transactions on Microwave Theory and Techniques*, vol. MTT-32, pp. 1022–1045, September 1984 [this special issue contains other articles on the history of microwave engineering].

[3] F. Ulaby, *Fundamentals of Applied Electromagnetics*, 6th edition, Prentice-Hall, Upper Saddle River, NJ, 2010.

[4] J. D. Kraus and D. A. Fleisch, *Electromagnetics*, 5th edition, McGraw-Hill, New York, 1999.

[5] S. Ramo, T. R. Whinnery, and T. van Duzer, *Fields and Waves in Communication Electronics*, 3rd edition, John Wiley & Sons, New York, 1994.

[6] R. E. Collin, *Foundations for Microwave Engineering*, 2nd edition, Wiley-IEEE Press, Hoboken, NJ, 2001.

[7] C. A. Balanis, *Advanced Engineering Electromagnetics*, John Wiley & Sons, New York, 1989.

[8] D. M. Pozar, *Microwave and RF Design of Wireless Systems*, John Wiley & Sons, Hoboken, NJ, 2001.

PROBLEMS

1.1 Who invented radio? Guglielmo Marconi often receives credit for the invention of modern radio, but there were several important developments by other workers before Marconi. Write a brief summary of the early work in wireless during the period of 1865–1900, particularly the work by Mahlon Loomis, Oliver Lodge, Nikola Tesla, JC Bose, and Marconi. Explain the difference between inductive communication schemes and wireless methods that involve wave propagation. Can the development of radio be attributed to a single individual? Reference [1] may be a good starting point.

1.2 A plane wave traveling along the x-axis in a polystyrene-filled region with $\epsilon_r = 2.56$ has an electric field given by $E_y = E_0 \cos(\omega t - kx)$. The frequency is 2.5 GHz, and $E_0 = 10$ V/m. Find the following: (a) the amplitude and direction of the magnetic field, (b) the phase velocity, (c) the wavelength, and (d) the phase shift between the positions $x_1 = 0.5$ m and $x_2 = 1.5$ m.

1.3 Show that a linearly polarized plane wave of the form $\bar{E} = E_0(a\hat{y} + b\hat{z})e^{-jk_0x}$, where a and b are real numbers, can be represented as the sum of an RHCP and an LHCP wave.

1.4 State and derive Poynting theorem (as in 1.92). Explain the significance of each term in the expression.

1.5 A plane wave is normally incident on a dielectric slab of permittivity ϵ_r and thickness d, where $d = \lambda_0/(4\sqrt{\epsilon_r})$ and λ_0 is the free-space wavelength of the incident wave, as shown in the accompanying figure. If free-space exists on both sides of the slab, find the reflection coefficient of the wave reflected from the front of the slab.

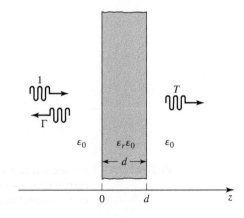

1.6 Consider an RHCP plane wave normally incident from free-space ($z < 0$) onto a half-space ($z > 0$) consisting of a good conductor. Let the incident electric field be of the form

$$\bar{E}_i = E_0(\hat{x} - j\hat{y})e^{-jk_0 z},$$

and find the electric and magnetic fields in the region $z > 0$. Compute the Poynting vectors for $z < 0$ and $z > 0$ and show that complex power is conserved. What is the polarization of the reflected wave?

1.7 Consider a plane wave propagating in a lossy dielectric medium for $y < 0$, with a perfectly conducting plate at $y = 0$. Assume that the lossy medium is characterized by $\epsilon = (5 - j2)\epsilon_0$, $\mu = \mu_0$, and that the frequency of the plane wave is 1.0 GHz, and let the amplitude of the incident electric field be 4 V/m at $y = 0$. Find the reflected electric field for $y < 0$ and plot the magnitude of the total electric field for $-0.5 \leq y \leq 0$.

1.8 A plane wave at 1.575 GHz is normally incident on a thin copper sheet of thickness t. (a) Compute the transmission losses, in dB, of the wave at the air–copper and the copper–air interfaces. (b) If the sheet is to be used as a shield to reduce the level of the transmitted wave by 250 dB, what is the minimum sheet thickness?

1.9 A uniform lossy medium with $\epsilon_r = 3.0$, $\tan \delta = 0.1$, and $\mu = \mu_0$ fills the region between $z = 0$ and $z = 20$ cm, with a ground plane at $z = 20$ cm, as shown in the accompanying figure. An incident plane wave with an electric field

$$\bar{E}_i = \hat{x}100 e^{-\gamma z} \text{ V/m}$$

is present at $z = 0$ and propagates in the $+z$ direction. The frequency is 3.0 GHz.

(a) Compute S_i, the power density of the incident wave, and S_r, the power density of the reflected wave, at $z = 0$.

(b) Compute the input power density, S_{in}, at $z = 0$ from the total fields at $z = 0$. Does $S_{in} = S_i - S_r$?

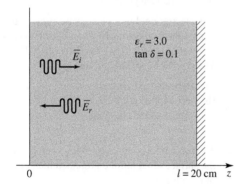

1.10 Assume that an infinite sheet of electric surface current density $\bar{J}_s = J_0 \hat{x}$ A/m is placed on the $z = 0$ plane between free-space for $z < 0$ and a dielectric with $\epsilon = \epsilon_r \epsilon_0$ for $z > 0$, as in the accompanying figure. Find the resulting $\bar{E}$ and $\bar{H}$ fields in the two regions. HINT: Assume plane wave solutions propagating away from the current sheet, and match boundary conditions to find the amplitudes, as in Example 1.3.

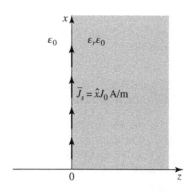

1.11 A perpendicularly polarized plane wave is obliquely incident from free-space onto a magnetic material with permittivity ϵ_0 and permeability $\mu_0 \mu_r$. Find the reflection and transmission coefficients. Does a Brewster angle exist for this case where the reflection coefficient vanishes for a particular angle of incidence?

1.12 An artificial anisotropic dielectric material has the tensor permittivity $[\epsilon]$ given as follows:

$$[\epsilon] = \epsilon_0 \begin{bmatrix} 1 & 4j & 0 \\ -4j & 2 & 0 \\ 0 & 0 & 3 \end{bmatrix}$$

At a certain point in the material the electric field is known to be $\bar{E} = 3\hat{x} + 4\hat{y} + 5\hat{z}$. What is $\bar{D}$ at this point?

1.13 The permittivity tensor for a gyrotropic dielectric material is

$$[\epsilon] = \epsilon_0 \begin{bmatrix} \epsilon_r & j\kappa & 0 \\ -j\kappa & \epsilon_r & 0 \\ 0 & 0 & 1 \end{bmatrix}.$$

Show that the transformations

$$E_+ = E_x - jE_y, \qquad D_+ = D_x - jD_y,$$
$$E_- = E_x + jE_y, \qquad D_- = D_x + jD_y,$$

allow the relation between $\bar{E}$ and $\bar{D}$ to be written as

$$\begin{bmatrix} D_+ \\ D_- \\ D_z \end{bmatrix} = [\epsilon'] \begin{bmatrix} E_+ \\ E_- \\ E_z \end{bmatrix},$$

where $[\epsilon']$ is now a diagonal matrix. What are the elements of $[\epsilon']$? Using this result, derive wave equations for E_+ and E_- and find the resulting propagation constants.

1.14 Show that the reciprocity theorem expressed in (1.157) also applies to a region enclosed by a closed surface S, where a surface impedance boundary condition applies.

1.15 Consider an electric surface current density of $\bar{J}_s = \hat{z}J_0 e^{-\beta y}$ A/m located on the $x = d$ plane. If a perfectly conducting ground plane is located at $x = 0$, use image theory to find the total fields for $x > 0$.

1.16 Let $\bar{E} = E_\rho \hat{\rho} + E_\phi \hat{\phi} + E_z \hat{z}$ be an electric field vector in cylindrical coordinates. Demonstrate that it is incorrect to interpret the expression $\nabla^2 \bar{E}$ in cylindrical coordinates as $\hat{\rho}\nabla^2 E_\rho + \hat{\phi}\nabla^2 E_\phi + \hat{z}\nabla^2 E_z$ by evaluating both sides of the vector identity $\nabla \times \nabla \times \bar{E} = \nabla(\nabla \cdot \bar{E}) - \nabla^2 \bar{E}$ for the given electric field.

1.17 A lossless dielectric for which $\eta = 30\pi$, $\mu_r = 1$ and $\bar{H} = 0.5$ $\cos(\omega t - z)\hat{x} - 0.1\sin(\omega t - z)\hat{y}$ A/m. Calculate ϵ_r, ω, and electric field.

1.18 A uniform plane wave propagating in a medium has $E = 4e^{-\alpha z}\sin(10^8 t - \beta z)\hat{y}$ V/m. If the medium is characterized by $\epsilon_r = 1$, $\mu_r = 20$, and $\sigma = 4$ mhos/m, find α, β, and H.

MULTIPLE CHOICE QUESTIONS

1.1 What is the electric flux ($\int \bar{E} \cdot d\hat{a}$) through a quarter-cylinder of height H (as shown in the figure) due to an infinitely long line charge along the axis of the cylinder with charge density of Q?

(a) $\dfrac{HQ}{\epsilon_0}$

(b) $\dfrac{HQ}{4\epsilon_0}$

(c) $\dfrac{H\epsilon_0}{4Q}$

(d) $\dfrac{4H}{Q\epsilon_0}$

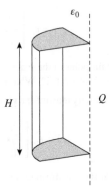

1.2 A uniform plane wave traveling in free space and having the electric field

$$\bar{E} = (\sqrt{2}a_x - a_z)\cos[6\sqrt{3}\pi \times 10^8 t - 2\pi(x + \sqrt{2}z)] \text{ V/m}$$

is incident on a dielectric medium (relative permittivity >1, relative permeability = 1) as shown in the figure and there is no reflected wave. The relative permittivity (correct to two decimal places) of the dielectric medium is _____.

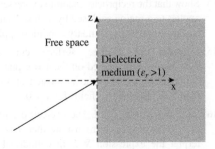

1.3 The expression for an electric field in free space is $\bar{E} = E_0(\hat{x} + \hat{y} + j2\hat{z})e^{-j(wt-kx+ky)}$, where x, y, z represent the

special coordinates, t represents time, and w, k are constants. This electric field

(a) does not represent a plane wave

(b) represents a circularly polarized plane wave propagating normal to the z-axis

(c) represents an elliptically polarized plane wave propagating along the x-y plane

(d) represents a linearly polarized plane wave

1.4 Faraday's law of electromagnetic induction is mathematically described by which one of the following equations?

(a) $\nabla \cdot \bar{B} = 0$

(b) $\nabla \cdot \bar{D} = \rho_v$

(c) $\nabla \times \bar{E} = -\dfrac{\partial \bar{B}}{\partial t}$

(d) $\nabla \times \bar{H} = \sigma\bar{E} + \dfrac{\partial \bar{D}}{\partial t}$

1.5 If a right-handed circularly polarized wave is incident normally on a plane perfect conductor, then the reflected wave will be

(a) right-handed circularly polarized

(b) left-handed circularly polarized

(c) elliptically polarized with a tilt angle of 45°

(d) horizontally polarized

1.6 The electric field of a uniform plane electromagnetic wave is $\bar{E} = (\hat{x} + j4\hat{y})\exp[j(2\pi \times 10^7 t - 0.2z)]$. The polarization of the wave is

(a) right-handed circular

(b) right-handed elliptical

(c) left-handed circular

(d) left-handed elliptical

1.7 The electric field of a plane wave propagating in a lossless nonmagnetic medium is given by the expression $E(z,t) = a_x 5\cos(2\pi \times 10^9 t + \beta z) + a_y 3\cos(2\pi \times 10^9 t + \beta z - \frac{\pi}{2})$. The type of the polarization is

(a) right-handed circular

(b) left-handed elliptical

(c) right-handed elliptical

(d) linear

1.8 The force on a point charge $+qq$ kept at a distance dd from the surface of an infinite grounded metal plate in a medium of permittivity ϵ is

(a) 0

(b) $\dfrac{q^2}{16\pi\epsilon d^2}$ away from the plate

(c) $\dfrac{q^2}{16\pi\epsilon d^2}$ towards the plate

(d) $\dfrac{q^2}{4\pi\epsilon d^2}$ towards the plate

1.9 If the electric field of a plane wave is

$$E(z,t) = a_x 3 \cos(\omega t - kz + 30°)$$
$$+ a_y 4 \sin(\omega t - kz + 45°)$$

The polarization state of the plane wave is

(a) left elliptical
(b) left circular
(c) right elliptical
(d) right circular

Common statement for Problems 1.10 and 1.11: A monochromatic plane wave of wavelength $\lambda = 600$ μm is propagating in the direction as shown in the figure below. E_i, E_r, and E_t denote incident, reflected, and transmitted electric field vectors associated with the wave

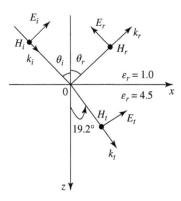

1.10 The angle of incidence θ and the expression for E_i are

(a) 60° and $\dfrac{E_0}{\sqrt{2}}(\hat{a}_x - \hat{a}_z)e^{-j\frac{\pi\times10^4(x+z)}{3\sqrt{2}}}$ V/m

(b) 45° and $\dfrac{E_0}{\sqrt{2}}(\hat{a}_x + \hat{a}_z)e^{-j\frac{\pi\times10^4 z}{3}}$ V/m

(c) 45° and $\dfrac{E_0}{\sqrt{2}}(\hat{a}_x - \hat{a}_z)e^{-j\frac{\pi\times10^4(x+z)}{3\sqrt{2}}}$ V/m

(d) 60° and $\dfrac{E_0}{\sqrt{2}}(\hat{a}_x - \hat{a}_z)e^{-j\frac{\pi\times10^4 z}{3}}$ V/m

1.11 The expression for E_r is

(a) $0.23\dfrac{E_0}{\sqrt{2}}(\hat{a}_x + \hat{a}_z)e^{-j\frac{\pi\times10^4(x-z)}{3\sqrt{2}}}$ V/m

(b) $-\dfrac{E_0}{\sqrt{2}}(\hat{a}_x + \hat{a}_z)e^{j\frac{\pi\times10^4 z}{3}}$ V/m

(c) $0.44\dfrac{E_0}{\sqrt{2}}(\hat{a}_x + \hat{a}_z)e^{-j\frac{\pi\times10^4(x-z)}{3\sqrt{2}}}$ V/m

(d) $\dfrac{E_0}{\sqrt{2}}(\hat{a}_x + \hat{a}_z)e^{-j\frac{\pi\times10^4(x+z)}{3}}$ V/m

1.12 In a specimen of ferromagnetic material with saturation magnetization as 8000 Gauss, as the flux density is increased from 0 to 2.5 T, μ_r will

(a) increase
(b) decrease
(c) first decrease then increase
(d) first increase then decrease

1.13 A parallel plate air-filled capacitor has plate area of 10^{-4} m^2 and plate separation of 10^{-3} m. It is connected to a 2 V, 1.8 GHz source. The magnitude of the displacement current is ($\varepsilon_0 = 1/36\pi \times 10^{-9}$ F/m)

(a) 200 mA   (c) 20 A
(b) 20 mA   (d) 2 mA

1.14 A uniform plane wave in air impinge at 45° angle on a lossless dielectric material with dielectric constant ϵ_r. The transmitted wave propagates in a 30° direction with respect to the normal. The value of ϵ_r is

(a) 1.5   (c) 2
(b) $\sqrt{1.5}$ (d) $\sqrt{2}$

1.15 Magnetic field intensity (H), within a magnetic material, where $M = 150$ A/m, $\mu = 1.5 \times 10^{-5}$ H/m, $\mu_r = 30$, is

(a) 14.921 A/m   (c) 5.17 A/m
(b) 14.138 A/m   (d) 13.715 A/m

1.16 A steel pipe is constructed of a material for which $\mu_r = 200$ and $\sigma = 5 \times 10^6/\Omega$/m. The outer and inner radii are 8 and 6 mm, respectively, and the length is 80 m. If the total current carried by the pipe is $2\cos(10^4\pi t)$ A, then the skin depth will be

(a) 0.225×10^{-3} m   (c) 0.352×10^{-3} m
(b) 0.300×10^{-3} m   (d) 0.125×10^{-3} m

1.17 Ground waves progress along the surface of the earth and must be polarized

(a) horizontally
(b) circularly
(c) elliptically
(d) vertically

1.18 In a certain material medium, a propagating electromagnetic wave attains 60% of the velocity of light. The distance at which the electromagnetic wave ($f = 10$ MHz) will have the same magnitude for the induction as well as the radiation fields is nearly

(a) 57.4 m   (c) 5.8 m
(b) 29.0 m   (d) 2.9 m

1.19 An electromagnetic wave is transmitted into a conducting medium of conductivity σ. The depth of penetration is

(a) directly proportional to frequency
(b) directly proportional to square root of frequency
(c) inversely proportional to frequency
(d) inversely proportional to square root of frequency

1.20 A plane $y = 2$ carries an infinite sheet of charge 4 nC/m^2. If the medium is free space, what is the force on a point charge of 5 mC located at origin?

(a) $0.54\pi\hat{y}$ N   (c) $-0.36\pi\hat{y}$ N
(b) $0.18\pi\hat{y}$ N   (d) $-0.18\pi\hat{y}$ N

ANSWER KEY				
1.1 (b)	**1.5** (b)	**1.9** (a)	**1.13** (b)	**1.17** (a)
1.2 (b)	**1.6** (d)	**1.10** (c)	**1.14** (c)	**1.18** (d)
1.3 (c)	**1.7** (b)	**1.11** (a)	**1.15** (c)	**1.19** (d)
1.4 (c)	**1.8** (c)	**1.12** (d)	**1.16** (a)	**1.20** (c)

Transmission Line Theory

■ **Learning Objectives**

After completing this chapter, you will be able to

- Learn the basic concepts of transmission lines
- Use the Smith chart in problem solving and other microwave measurements and applications
- Analyze and calculate the RF circuit parameters such as S-Parameter, VSWR, insertion loss, return loss, and

impedance transformation through derivations and the Smith chart
- Understand the importance of impedance matching, mismatching, and different losses in transmission lines

Transmission line theory bridges the gap between field analysis and basic circuit theory and therefore is of significant importance in the analysis of microwave circuits and devices. As we will see, the phenomenon of wave propagation on transmission lines can be approached from an extension of circuit theory or from a specialization of Maxwell's equations; we shall present both viewpoints and show how this wave propagation is described by equations very similar to those used in Chapter 1 for plane wave propagation.

2.1 THE LUMPED-ELEMENT CIRCUIT MODEL FOR A TRANSMISSION LINE

The key difference between circuit theory and transmission line theory is electrical size. Circuit analysis assumes that the physical dimensions of the network are much smaller than the electrical wavelength, while transmission lines may be a considerable fraction of a wavelength, or many wavelengths, in size. Thus a transmission line is a *distributed-parameter* network, where voltages and currents can vary in magnitude and phase over its length, while ordinary circuit analysis deals with *lumped elements*, where voltage and current do not vary appreciably over the physical dimension of the elements.

As shown in Figure 2.1a, a transmission line is often schematically represented as a two-wire line since transmission lines (for transverse electromagnetic [TEM] wave propagation) always have at least two conductors. The piece of line of infinitesimal length Δz of Figure 2.1a can be modeled as a lumped-element circuit, as shown in Figure 2.1b, where R, L, G, and C are per-unit-length quantities defined as follows:

R = series resistance per unit length, for both conductors, in Ω/m.

L = series inductance per unit length, for both conductors, in H/m.

G = shunt conductance per unit length, in S/m.

C = shunt capacitance per unit length, in F/m.

The series inductance L represents the total self-inductance of the two conductors, and the shunt capacitance C is due to the close proximity of the two conductors. The series resistance R represents the resistance due to the finite conductivity of the individual conductors, and the shunt conductance G is due to dielectric loss in the material between the conductors. R and G, therefore, represent loss. A finite length of transmission line can be viewed as a cascade of sections of the form shown in Figure 2.1b.

From the circuit of Figure 2.1b, Kirchhoff's voltage law can be applied to give

$$v(z, t) - R\Delta z i(z, t) - L\Delta z \frac{\partial i(z, t)}{\partial t} - v(z + \Delta z, t) = 0, \tag{2.1a}$$

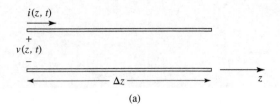

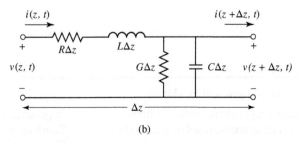

(b)

FIGURE 2.1 | Voltage and current definitions and equivalent circuit for an incremental length of transmission line. (a) Voltage and current definitions. (b) Lumped-element equivalent circuit.

and Kirchhoff's current law leads to

$$i(z,t) - G\Delta z v(z+\Delta z, t) - C\Delta z \frac{\partial v(z+\Delta z, t)}{\partial t} - i(z+\Delta z, t) = 0. \tag{2.1b}$$

Dividing (2.1a) and (2.1b) by Δz and taking the limit as $\Delta z \to 0$ gives the following differential equations:

$$\frac{\partial v(z,t)}{\partial z} = -Ri(z,t) - L\frac{\partial i(z,t)}{\partial t}, \tag{2.2a}$$

$$\frac{\partial i(z,t)}{\partial z} = -Gv(z,t) - C\frac{\partial v(z,t)}{\partial t}. \tag{2.2b}$$

These are the time domain form of the transmission line equations, also known as the *telegrapher equations*. For the sinusoidal steady-state condition, with cosine-based phasors, (2.2a) and (2.2b) simplify to

$$\frac{dV(z)}{dz} = -(R+j\omega L)I(z), \tag{2.3a}$$

$$\frac{dI(z)}{dz} = -(G+j\omega C)V(z). \tag{2.3b}$$

Note the similarity in the form of (2.3a) and (2.3b) and Maxwell's curl equations of (1.41a) and (1.41b).

Wave Propagation on a Transmission Line

The two equations (2.3a) and (2.3b) can be solved simultaneously to give wave equations for $V(z)$ and $I(z)$:

$$\frac{d^2 V(z)}{dz^2} - \gamma^2 V(z) = 0, \tag{2.4a}$$

$$\frac{d^2 I(z)}{dz^2} - \gamma^2 I(z) = 0, \tag{2.4b}$$

where

$$\gamma = \alpha + j\beta = \sqrt{(R+j\omega L)(G+j\omega C)} \tag{2.5}$$

is the complex propagation constant, which is a function of frequency. Traveling wave solutions to (2.4) can be found as

$$V(z) = V_o^+ e^{-\gamma z} + V_o^- e^{\gamma z}, \tag{2.6a}$$

$$I(z) = I_o^+ e^{-\gamma z} + I_o^- e^{\gamma z}, \tag{2.6b}$$

where the $e^{-\gamma z}$ term represents wave propagation in the $+z$ direction, and the $e^{\gamma z}$ term represents wave propagation in the $-z$ direction. Applying (2.3a) to the voltage of (2.6a) gives the current on the line:

$$I(z) = \frac{\gamma}{R + j\omega L} \left(V_o^+ e^{-\gamma z} - V_o^- e^{\gamma z} \right).$$

Comparison with (2.6b) shows that a *characteristic impedance*, Z_0, can be defined as

$$Z_0 = \frac{R + j\omega L}{\gamma} = \sqrt{\frac{R + j\omega L}{G + j\omega C}}, \tag{2.7}$$

to relate the voltage and current on the line as follows:

$$\frac{V_o^+}{I_o^+} = Z_0 = \frac{-V_o^-}{I_o^-}.$$

Then (2.6b) can be rewritten in the following form:

$$I(z) = \frac{V_o^+}{Z_0} e^{-\gamma z} - \frac{V_o^-}{Z_0} e^{\gamma z}. \tag{2.8}$$

Converting back to the time domain, we can express the voltage waveform as

$$v(z, t) = |V_o^+| \cos(\omega t - \beta z + \phi^+) e^{-\alpha z}$$
$$+ |V_o^-| \cos(\omega t + \beta z + \phi^-) e^{\alpha z}, \tag{2.9}$$

where $\phi^\pm$ is the phase angle of the complex voltage $V_o^\pm$. Using arguments similar to those in Section 1.4, we find that the wavelength on the line is

$$\lambda = \frac{2\pi}{\beta}, \tag{2.10}$$

and the phase velocity is

$$v_p = \frac{\omega}{\beta} = \lambda f. \tag{2.11}$$

The Lossless Line

The above solution is for a general transmission line, including loss effects, and it was seen that the propagation constant and characteristic impedance were complex. In many practical cases, however, the loss of the line is very small and so can be neglected, resulting in a simplification of the results. Setting $R = G = 0$ in (2.5) gives the propagation constant as

$$\gamma = \alpha + j\beta = j\omega\sqrt{LC},$$

or

$$\beta = \omega\sqrt{LC}, \tag{2.12a}$$
$$\alpha = 0. \tag{2.12b}$$

As expected for a lossless line, the attenuation constant α is zero. The characteristic impedance of (2.7) reduces to

$$Z_0 = \sqrt{\frac{L}{C}}, \tag{2.13}$$

which is now a real number. The general solutions for voltage and current on a lossless transmission line can then be written as

$$V(z) = V_o^+ e^{-j\beta z} + V_o^- e^{j\beta z}, \tag{2.14a}$$

$$I(z) = \frac{V_o^+}{Z_0} e^{-j\beta z} - \frac{V_o^-}{Z_0} e^{j\beta z}. \tag{2.14b}$$

The wavelength is

$$\lambda = \frac{2\pi}{\beta} = \frac{2\pi}{\omega\sqrt{LC}},$$

(2.15)

and the phase velocity is

$$v_p = \frac{\omega}{\beta} = \frac{1}{\sqrt{LC}}.$$

(2.16)

2.2   FIELD ANALYSIS OF TRANSMISSION LINES

In this section we will rederive the time-harmonic form of the telegrapher's equations starting from Maxwell's equations. We will begin by deriving the transmission line parameters (R, L, G, C) in terms of the electric and magnetic fields of the transmission line and then derive the telegrapher equations using these parameters for the specific case of a coaxial line.

Transmission Line Parameters

Consider a 1 m length of a uniform transmission line with fields $\bar{E}$ and $\bar{H}$, as shown in Figure 2.2, where S is the cross-sectional surface area of the line. Let the voltage between the conductors be $V_o e^{\pm j\beta z}$ and the current be $I_o e^{\pm j\beta z}$. The time-average stored magnetic energy for this 1 m length of line can be written, from (1.86), as

$$W_m = \frac{\mu}{4}\int_S \bar{H} \cdot \bar{H}^* ds,$$

while circuit theory gives $W_m = L|I_o|^2/4$ in terms of the current on the line. We can thus identify the self-inductance per unit length as

$$L = \frac{\mu}{|I_o|^2}\int_S \bar{H} \cdot \bar{H}^* ds \text{ H/m}.$$

(2.17)

Similarly, the time-average stored electric energy per unit length can be found from (1.84) as

$$W_e = \frac{\epsilon}{4}\int_S \bar{E} \cdot \bar{E}^* ds,$$

while circuit theory gives $W_e = C|V_o|^2/4$, resulting in the following expression for the capacitance per unit length:

$$C = \frac{\epsilon}{|V_o|^2}\int_S \bar{E} \cdot \bar{E}^* ds \text{ F/m}.$$

(2.18)

From (1.131), the power loss per unit length due to the finite conductivity of the metallic conductors is

$$P_c = \frac{R_s}{2}\int_{C_1+C_2} \bar{H} \cdot \bar{H}^* d\ell$$

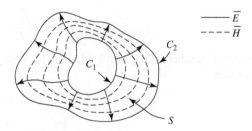

FIGURE 2.2 | Field lines on an arbitrary TEM transmission line.

(assuming $\bar{H}$ is tangential to S), while circuit theory gives $P_c = R|I_o|^2/2$, so the series resistance R per unit length of line is

$$R = \frac{R_s}{|I_o|^2} \int_{C_1+C_2} \bar{H} \cdot \bar{H}^* dl \ \Omega/\text{m}. \tag{2.19}$$

In (2.19), $R_s = 1/\sigma\delta_s$ is the surface resistance of the conductors, and $C_1 + C_2$ represent integration paths over the conductor boundaries. From (1.92), the time-average power dissipated per unit length in a lossy dielectric is

$$P_d = \frac{\omega\epsilon''}{2} \int_S \bar{E} \cdot \bar{E}^* ds,$$

where ϵ'' is the imaginary part of the complex permittivity $\epsilon = \epsilon' - j\epsilon'' = \epsilon'(1 - j\tan\delta)$. Circuit theory gives $P_d = G|V_o|^2/2$, so the shunt conductance per unit length can be written as

$$G = \frac{\omega\epsilon''}{|V_o|^2} \int_S \bar{E} \cdot \bar{E}^* ds \ \text{S/m}. \tag{2.20}$$

EXAMPLE 2.1 TRANSMISSION LINE PARAMETERS OF A COAXIAL LINE

The fields of a traveling TEM wave inside the coaxial line of Figure 2.3 can be expressed as

$$\bar{E} = \frac{V_o\hat{\rho}}{\rho \ln b/a} e^{-\gamma z},$$

$$\bar{H} = \frac{I_o\hat{\phi}}{2\pi\rho} e^{-\gamma z},$$

where γ is the propagation constant of the line. The conductors are assumed to have a surface resistivity R_s, and the material filling the space between the conductors is assumed to have a complex permittivity $\epsilon = \epsilon' - j\epsilon''$ and a permeability $\mu = \mu_0\mu_r$. Determine the transmission line parameters.

Solution
From (2.17)–(2.20) and the given fields the parameters of the coaxial line can be calculated as

$$L = \frac{\mu}{(2\pi)^2} \int_{\phi=0}^{2\pi} \int_{\rho=a}^{b} \frac{1}{\rho^2} \rho\,d\rho\,d\phi = \frac{\mu}{2\pi} \ln b/a \ \text{H/m},$$

$$C = \frac{\epsilon'}{(\ln b/a)^2} \int_{\phi=0}^{2\pi} \int_{\rho=a}^{b} \frac{1}{\rho^2} \rho\,d\rho\,d\phi = \frac{2\pi\epsilon'}{\ln b/a} \ \text{F/m},$$

$$R = \frac{R_s}{(2\pi)^2} \left\{ \int_{\phi=0}^{2\pi} \frac{1}{a^2} a\,d\phi + \int_{\phi=0}^{2\pi} \frac{1}{b^2} b\,d\phi \right\} = \frac{R_s}{2\pi} \left(\frac{1}{a} + \frac{1}{b} \right) \Omega/\text{m},$$

$$G = \frac{\omega\epsilon''}{(\ln b/a)^2} \int_{\phi=0}^{2\pi} \int_{\rho=a}^{b} \frac{1}{\rho^2} \rho\,d\rho\,d\phi = \frac{2\pi\omega\epsilon''}{\ln b/a} \ \text{S/m}.$$

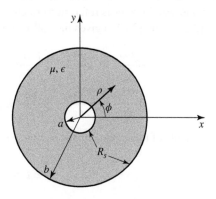

FIGURE 2.3 | Geometry of a coaxial line with surface resistance R_s on the inner and outer conductors.

TABLE 2.1 | **Transmission Line Parameters for Some Common Lines**

	Coax	Two-Wire	Parallel Plate
L	$\dfrac{\mu}{2\pi}\ln\dfrac{b}{a}$	$\dfrac{\mu}{\pi}\cosh^{-1}\left(\dfrac{D}{2a}\right)$	$\dfrac{\mu d}{w}$
C	$\dfrac{2\pi\epsilon'}{\ln b/a}$	$\dfrac{\pi\epsilon'}{\cosh^{-1}(D/2a)}$	$\dfrac{\epsilon' w}{d}$
R	$\dfrac{R_s}{2\pi}\left(\dfrac{1}{a}+\dfrac{1}{b}\right)$	$\dfrac{R_s}{\pi a}$	$\dfrac{2R_s}{w}$
G	$\dfrac{2\pi\omega\epsilon''}{\ln b/a}$	$\dfrac{\pi\omega\epsilon''}{\cosh^{-1}(D/2a)}$	$\dfrac{\omega\epsilon'' w}{d}$

Table 2.1 summarizes the parameters for coaxial, two-wire, and parallel plate lines. As we will see in the next chapter, the propagation constant, characteristic impedance, and attenuation of most transmission lines are usually derived directly from a field theory solution; the approach here of first finding the equivalent circuit parameters (L, C, R, G) is useful only for relatively simple lines. Nevertheless, it provides a helpful intuitive concept for understanding the properties of a transmission line and relates a transmission line to its equivalent circuit model.

The Telegrapher Equations Derived from Field Analysis of a Coaxial Line

We now show that the telegrapher equations of (2.3), derived using circuit theory, can also be obtained from Maxwell's equations. We will consider the specific geometry of the coaxial line of Figure 2.3. Although we will treat TEM wave propagation more generally in the next chapter, the present discussion should provide some insight into the relationship of circuit and field quantities.

A TEM wave on the coaxial line of Figure 2.3 will be characterized by $E_z = H_z = 0$; furthermore, due to azimuthal symmetry, the fields will have no ϕ variation, so $\partial/\partial\phi = 0$. The fields inside the coaxial line will satisfy Maxwell's curl equations,

$$\nabla \times \bar{E} = -j\omega\mu\bar{H}, \tag{2.21a}$$

$$\nabla \times \bar{H} = j\omega\epsilon\bar{E}, \tag{2.21b}$$

where $\epsilon = \epsilon' - j\epsilon''$ may be complex to allow for a lossy dielectric filling. Conductor loss will be ignored here. A rigorous field analysis of conductor loss can be carried out but at this point would tend to obscure our purpose; the interested reader is referred to references [1] and [2].

Expanding (2.21a) and (2.21b) gives the following two vector equations:

$$-\hat{\rho}\frac{\partial E_\phi}{\partial z} + \hat{\phi}\frac{\partial E_\rho}{\partial z} + \hat{z}\frac{1}{\rho}\frac{\partial}{\partial\rho}(\rho E_\phi) = -j\omega\mu(\hat{\rho}H_\rho + \hat{\phi}H_\phi), \tag{2.22a}$$

$$-\hat{\rho}\frac{\partial H_\phi}{\partial z} + \hat{\phi}\frac{\partial H_\rho}{\partial z} + \hat{z}\frac{1}{\rho}\frac{\partial}{\partial\rho}(\rho H_\phi) = j\omega\epsilon(\hat{\rho}E_\rho + \hat{\phi}E_\phi). \tag{2.22b}$$

Since the $\hat{z}$ components of these two equations must vanish, it is seen that E_ϕ and H_ϕ must have the forms

$$E_\phi = \frac{f(z)}{\rho}, \tag{2.23a}$$

$$H_\phi = \frac{g(z)}{\rho}. \tag{2.23b}$$

To satisfy the boundary condition that $E_\phi = 0$ at $\rho = a$, b, we must have $E_\phi = 0$ everywhere, due to the form of E_ϕ in (2.23a). Then from the $\hat{\rho}$ component of (2.22a), it is seen that $H_\rho = 0$. With these results, (2.22) can be reduced to

$$\frac{\partial E_\rho}{\partial z} = -j\omega\mu H_\phi, \tag{2.24a}$$

$$\frac{\partial H_\phi}{\partial z} = -j\omega\epsilon E_\rho. \tag{2.24b}$$

From the form of H_ϕ in (2.23b) and (2.24a), E_ρ must be of the form

$$E_\rho = \frac{h(z)}{\rho}. \tag{2.25}$$

Using (2.23b) and (2.25) in (2.24) gives

$$\frac{\partial h(z)}{\partial z} = -j\omega\mu g(z), \tag{2.26a}$$

$$\frac{\partial g(z)}{\partial z} = -j\omega\epsilon h(z). \tag{2.26b}$$

The voltage between the two conductors can be evaluated as

$$V(z) = \int_{\rho=a}^{b} E_\rho(\rho, z)d\rho = h(z) \int_{\rho=a}^{b} \frac{d\rho}{\rho} = h(z) \ln\frac{b}{a}, \tag{2.27a}$$

and the total current on the inner conductor at $\rho = a$ can be evaluated using (2.23b) as

$$I(z) = \int_{\phi=0}^{2\pi} H_\phi(a, z)a d\phi = 2\pi g(z). \tag{2.27b}$$

Then $h(z)$ and $g(z)$ can be eliminated from (2.26) by using (2.27) to give

$$\frac{\partial V(z)}{\partial z} = -j\frac{\omega\mu \ln b/a}{2\pi}I(z),$$

$$\frac{\partial I(z)}{\partial z} = -j\omega(\epsilon' - j\epsilon'')\frac{2\pi V(z)}{\ln b/a}.$$

Finally, using the results for L, G, and C for a coaxial line as derived earlier, we obtain the telegrapher equations as

$$\frac{\partial V(z)}{\partial z} = -j\omega LI(z), \tag{2.28a}$$

$$\frac{\partial I(z)}{\partial z} = -(G + j\omega C)V(z). \tag{2.28b}$$

This result excludes R, the series resistance, since the conductors were assumed to have perfect conductivity. A similar analysis can be carried out for other simple transmission lines.

Propagation Constant, Impedance, and Power Flow for the Lossless Coaxial Line

Equations (2.24a) and (2.24b) for E_ρ and H_ϕ can be simultaneously solved to yield a wave equation for E_ρ (or H_ϕ):

$$\frac{\partial^2 E_\rho}{\partial z^2} + \omega^2\mu\epsilon E_\rho = 0, \tag{2.29}$$

from which it is seen that the propagation constant is $\gamma^2 = -\omega^2\mu\epsilon$, which, for lossless media, reduces to

$$\beta = \omega\sqrt{\mu\epsilon} = \omega\sqrt{LC}, \tag{2.30}$$

where the last result is from (2.12). Observe that this propagation constant is of the same form as that for plane waves in a lossless dielectric medium. This is a general result for TEM transmission lines.

The wave impedance for the coaxial line is defined as $Z_w = E_\rho / H_\phi$, which can be calculated from (2.24a), assuming an $e^{-j\beta z}$ dependence, to give

$$Z_w = \frac{E_\rho}{H_\phi} = \frac{\omega\mu}{\beta} = \sqrt{\mu/\epsilon} = \eta. \tag{2.31}$$

This wave impedance is seen to be identical to the intrinsic impedance of the medium, η, and is a general result for TEM transmission lines.

The characteristic impedance of the coaxial line is defined as

$$Z_0 = \frac{V_o}{I_o} = \frac{E_\rho \ln b/a}{2\pi H_\phi} = \frac{\eta \ln b/a}{2\pi} = \sqrt{\frac{\mu}{\epsilon}} \frac{\ln b/a}{2\pi}, \tag{2.32}$$

where the forms for E_ρ and H_ϕ from Example 2.1 have been used. The characteristic impedance is geometry dependent and will be different for other transmission line configurations.

Finally, the power flow (in the z direction) on the coaxial line may be computed from the Poynting vector as

$$P = \frac{1}{2} \int_s \bar{E} \times \bar{H}^* \cdot d\bar{s} = \frac{1}{2} \int_{\phi=0}^{2\pi} \int_{\rho=a}^{b} \frac{V_o I_o^*}{2\pi\rho^2 \ln b/a} \rho\,d\rho\,d\phi = \frac{1}{2} V_o I_o^*, \tag{2.33}$$

a result that is in clear agreement with circuit theory. This shows that the flow of power in a transmission line takes place entirely via the electric and magnetic fields between the two conductors; power is not transmitted through the conductors themselves. As we will see later, for the case of finite conductivity, power may enter the conductors, but this power is then lost as heat and is not delivered to the load.

2.3   THE TERMINATED LOSSLESS TRANSMISSION LINE

Figure 2.4 shows a lossless transmission line terminated in an arbitrary load impedance Z_L. This problem will illustrate wave reflection on transmission lines, a fundamental property of distributed systems.

Assume that an incident wave of the form $V_o^+ e^{-j\beta z}$ is generated from a source at $z < 0$. We have seen that the ratio of voltage to current for such a traveling wave is Z_0, the characteristic impedance of the line. However, when the line is terminated in an arbitrary load $Z_L \neq Z_0$, the ratio of voltage to current at the load must be Z_L. Thus, a reflected wave must be excited with the appropriate amplitude to satisfy this condition. The total voltage on the line can then be written as in (2.14a), as a sum of incident and reflected waves:

$$V(z) = V_o^+ e^{-j\beta z} + V_o^- e^{j\beta z}. \tag{2.34a}$$

Similarly, the total current on the line is described by (2.14b):

$$I(z) = \frac{V_o^+}{Z_0} e^{-j\beta z} - \frac{V_o^-}{Z_0} e^{j\beta z}. \tag{2.34b}$$

The total voltage and current at the load are related by the load impedance, so at $z = 0$ we must have

$$Z_L = \frac{V(0)}{I(0)} = \frac{V_o^+ + V_o^-}{V_o^+ - V_o^-} Z_0.$$

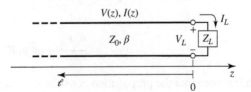

FIGURE 2.4 | A transmission line terminated in a load impedance Z_L.

Solving for V_o^- gives

$$V_o^- = \frac{Z_L - Z_0}{Z_L + Z_0} V_o^+.$$

The amplitude of the reflected voltage wave normalized to the amplitude of the incident voltage wave is defined as the *voltage reflection coefficient*, Γ:

$$\Gamma = \frac{V_o^-}{V_o^+} = \frac{Z_L - Z_0}{Z_L + Z_0}. \tag{2.35}$$

The total voltage and current waves on the line can then be written as

$$V(z) = V_o^+ \left(e^{-j\beta z} + \Gamma e^{j\beta z} \right), \tag{2.36a}$$

$$I(z) = \frac{V_o^+}{Z_0} \left(e^{-j\beta z} - \Gamma e^{j\beta z} \right). \tag{2.36b}$$

From these equations it is seen that the voltage and current on the line consist of a superposition of an incident and a reflected wave; such waves are called *standing waves*. Only when $\Gamma = 0$ is there no reflected wave. To obtain $\Gamma = 0$, the load impedance Z_L must be equal to the characteristic impedance Z_0 of the transmission line, as seen from (2.35). Such a load is said to be *matched* to the line since there is no reflection of the incident wave.

Now consider the time-average power flow along the line at the point z:

$$P_{\text{avg}} = \frac{1}{2} \text{Re} \left\{ V(z) I(z)^* \right\} = \frac{1}{2} \frac{|V_o^+|^2}{Z_0} \text{Re} \left\{ 1 - \Gamma^* e^{-2j\beta z} + \Gamma e^{2j\beta z} - |\Gamma|^2 \right\},$$

where (2.36) has been used. The middle two terms in the brackets are of the form $A - A^* = 2j \, \text{Im} \{A\}$ and so are purely imaginary. This simplifies the result to

$$P_{\text{avg}} = \frac{1}{2} \frac{|V_o^+|^2}{Z_0} \left(1 - |\Gamma|^2 \right), \tag{2.37}$$

which shows that the average power flow is constant at any point on the line and that the total power delivered to the load (P_{avg}) is equal to the incident power ($|V_o^+|^2 / 2Z_0$) minus the reflected power ($|V_o|^2 |\Gamma|^2 / 2Z_0$). If $\Gamma = 0$, maximum power is delivered to the load, while no power is delivered for $|\Gamma| = 1$. The above discussion assumes that the generator is matched, so that there is no re-reflection of the reflected wave from $z < 0$.

When the load is mismatched, not all of the available power from the generator is delivered to the load. This "loss" is called *return loss* (RL), and is defined (in dB) as

$$\text{RL} = -20 \log |\Gamma| \text{ dB}, \tag{2.38}$$

so that a matched load ($\Gamma = 0$) has a return loss of ∞ dB (no reflected power), while a total reflection ($|\Gamma| = 1$) has a return loss of 0 dB (all incident power is reflected). Note that return loss is a nonnegative number for reflection from a passive network.

If the load is matched to the line, $\Gamma = 0$ and the magnitude of the voltage on the line is $|V(z)| = |V_o^+|$, which is a constant. Such a line is sometimes said to be *flat*. When the load is mismatched, however, the presence of a reflected wave leads to standing waves, and the magnitude of the voltage on the line is not constant. Thus, from (2.36a),

$$|V(z)| = |V_o^+| |1 + \Gamma e^{2j\beta z}| = |V_o^+| |1 + \Gamma e^{-2j\beta \ell}|$$
$$= |V_o^+| |1 + |\Gamma| e^{j(\theta - 2\beta \ell)}|, \tag{2.39}$$

where $\ell = -z$ is the positive distance measured from the load at $z = 0$, and θ is the phase of the reflection coefficient ($\Gamma = |\Gamma| e^{j\theta}$). This result shows that the voltage magnitude oscillates with position z along the line. The maximum value occurs when the phase term $e^{j(\theta - 2\beta \ell)} = 1$ and is given by

$$V_{\text{max}} = |V_o^+| (1 + |\Gamma|). \tag{2.40a}$$

The minimum value occurs when the phase term $e^{j(\theta - 2\beta \ell)} = -1$ and is given by

$$V_{\text{min}} = |V_o^+| (1 - |\Gamma|). \tag{2.40b}$$

As $|\Gamma|$ increases, the ratio of $V_{\max}$ to $V_{\min}$ increases, so a measure of the mismatch of a line, called the *standing wave ratio* (SWR), can be defined as

$$\text{SWR} = \frac{V_{\max}}{V_{\min}} = \frac{1 + |\Gamma|}{1 - |\Gamma|}. \tag{2.41}$$

This quantity is also known as the *voltage standing wave ratio* and is sometimes identified as VSWR. From (2.41) it is seen that SWR is a real number such that $1 \leq \text{SWR} \leq \infty$, where SWR $= 1$ implies a matched load.

From (2.39), it is seen that the distance between two successive voltage maxima (or minima) is $\ell = 2\pi/2\beta = \pi\lambda/2\pi = \lambda/2$, while the distance between a maximum and a minimum is $\ell = \pi/2\beta = \lambda/4$, where λ is the wavelength on the transmission line.

The reflection coefficient of (2.35) was defined as the ratio of the reflected to the incident voltage wave amplitudes at the load ($\ell = 0$), but this quantity can be generalized to any point ℓ along the line as follows. From (2.34a), with $z = -\ell$, the ratio of the reflected component to the incident component is

$$\Gamma(\ell) = \frac{V_o^- e^{-j\beta\ell}}{V_o^+ e^{j\beta\ell}} = \Gamma(0)e^{-2j\beta\ell}, \tag{2.42}$$

where $\Gamma(0)$ is the reflection coefficient at $z = 0$, as given by (2.35). This result is useful when transforming the effect of a load mismatch down the line.

We have seen that the real power flow on the line is a constant (for a lossless line) but that the voltage amplitude, at least for a mismatched line, is oscillatory with position on the line. The perceptive reader may therefore have concluded that the impedance seen looking into the line must vary with position, and this is indeed the case. At a distance $\ell = -z$ from the load, the input impedance seen looking toward the load is

$$Z_{\text{in}} = \frac{V(-\ell)}{I(-\ell)} = \frac{V_o^+\left(e^{j\beta\ell} + \Gamma e^{-j\beta\ell}\right)}{V_o^+\left(e^{j\beta\ell} - \Gamma e^{-j\beta\ell}\right)} Z_0 = \frac{1 + \Gamma e^{-2j\beta\ell}}{1 - \Gamma e^{-2j\beta\ell}} Z_0, \tag{2.43}$$

where (2.36a,b) have been used for $V(z)$ and $I(z)$. A more usable form may be obtained by using (2.35) for Γ in (2.43):

$$\begin{aligned}
Z_{\text{in}} &= Z_0 \frac{(Z_L + Z_0)e^{j\beta\ell} + (Z_L - Z_0)e^{-j\beta\ell}}{(Z_L + Z_0)e^{j\beta\ell} - (Z_L - Z_0)e^{-j\beta\ell}} \\
&= Z_0 \frac{Z_L \cos\beta\ell + jZ_0 \sin\beta\ell}{Z_0 \cos\beta\ell + jZ_L \sin\beta\ell} \\
&= Z_0 \frac{Z_L + jZ_0 \tan\beta\ell}{Z_0 + jZ_L \tan\beta\ell}.
\end{aligned} \tag{2.44}$$

This is an important result giving the input impedance of a length of transmission line with an arbitrary load impedance. We will refer to this result as the *transmission line impedance equation;* some special cases will be considered next.

EXAMPLE 2.2   INPUT IMPEDANCE OF THE TRANSMISSION LINE

At an operating frequency of 300 MHz, a lossless 50 Ω air-spaced transmission line 2.5 m in length is terminated with an impedance $Z_L = (40 + j20)$ Ω. Find the input impedance.

Solution
Given a lossless transmission line, $Z_0 = 50$ Ω, $f = 300$ MHz, $\ell = 2.5$ m, and $Z_L = (40 + j20)$ Ω. Since the line is air filled, $v_p = c$. Therefore,

$$\beta = \frac{\omega}{v_p} = \frac{2\pi \times 300 \times 10^6}{3 \times 10^8} = 2\pi \text{ rad/m}$$

Since the line is lossless,

$$Z_{\text{in}} = Z_0 \left(\frac{Z_L + jZ_0 \tan\beta\ell}{Z_0 + jZ_L \tan\beta\ell}\right) = (40 + j20) \ \Omega$$

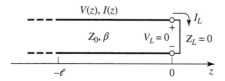

FIGURE 2.5 | A transmission line terminated in a short circuit.

Special Cases of Lossless Terminated Lines

A number of special cases of lossless terminated transmission lines will frequently appear in our work, so it is appropriate to consider the properties of such cases here.

Consider first the transmission line circuit shown in Figure 2.5, where a line is terminated in a short circuit, $Z_L = 0$. From (2.35) it is seen that the reflection coefficient for a short circuit load is $\Gamma = -1$; it then follows from (2.41) that the standing wave ratio is infinite. From (2.36) the voltage and current on the line are

$$V(z) = V_o^+ \left(e^{-j\beta z} - e^{j\beta z} \right) = -2jV_o^+ \sin \beta z, \tag{2.45a}$$

$$I(z) = \frac{V_o^+}{Z_0} \left(e^{-j\beta z} + e^{j\beta z} \right) = \frac{2V_o^+}{Z_0} \cos \beta z, \tag{2.45b}$$

which shows that $V = 0$ at the load (as expected, for a short circuit), while the current is a maximum there. From (2.44), or the ratio $V(-\ell)/I(-\ell)$, the input impedance is

$$Z_{\text{in}} = jZ_0 \tan \beta \ell, \tag{2.45c}$$

which is seen to be purely imaginary for any length ℓ and to take on all values between $+j\infty$ and $-j\infty$. For example, when $\ell = 0$ we have $Z_{\text{in}} = 0$, but for $\ell = \lambda/4$ we have $Z_{\text{in}} = \infty$ (open circuit). Equation (2.45c) also shows that the impedance is periodic in ℓ, repeating for multiples of $\lambda/2$. The voltage, current, and input reactance for the short-circuited line are plotted in Figure 2.6.

Next consider the open-circuited line shown in Figure 2.7, where $Z_L = \infty$. Dividing the numerator and denominator of (2.35) by Z_L and allowing $Z_L \rightarrow \infty$ shows that the reflection coefficient for this case is $\Gamma = 1$, and the standing wave ratio is again infinite. From (2.36) the voltage and current on the line are

$$V(z) = V_o^+ \left(e^{-j\beta z} + e^{j\beta z} \right) = 2V_o^+ \cos \beta z, \tag{2.46a}$$

$$I(z) = \frac{V_o^+}{Z_0} \left(e^{-j\beta z} - e^{j\beta z} \right) = \frac{-2jV_o^+}{Z_0} \sin \beta z, \tag{2.46b}$$

which shows that now $I = 0$ at the load, as expected for an open circuit, while the voltage is a maximum. The input impedance is

$$Z_{\text{in}} = -jZ_0 \cot \beta \ell, \tag{2.46c}$$

which is also purely imaginary for any length, ℓ. The voltage, current, and input reactance of the open-circuited line are plotted in Figure 2.8.

Now consider terminated transmission lines with some special lengths. If $\ell = \lambda/2$, (2.44) shows that

$$Z_{\text{in}} = Z_L, \tag{2.47}$$

meaning that a half-wavelength line (or any multiple of $\lambda/2$) does not alter or transform the load impedance, regardless of its characteristic impedance.

If the line is a quarter-wavelength long or, more generally, $\ell = \lambda/4 + n\lambda/2$, for $n = 1, 2, 3, \ldots$, (2.44) shows that the input impedance is given by

$$Z_{\text{in}} = \frac{Z_0^2}{Z_L}. \tag{2.48}$$

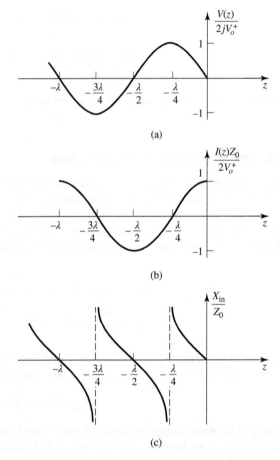

(a)

(b)

(c)

FIGURE 2.6 | (a) Voltage, (b) current, and (c) impedance ($R_{in} = 0$ or ∞) variation along a short-circuited transmission line.

Such a line is known as a *quarter-wave transformer* because it has the effect of transforming the load impedance in an inverse manner, depending on the characteristic impedance of the line. We will study this case more thoroughly in Section 5.4.

Next consider a transmission line of characteristic impedance Z_0 feeding a line of different characteristic impedance, Z_1, as shown in Figure 2.9. If the load line is infinitely long, or if it is terminated in its own characteristic impedance, so that there are no reflections from its far end, then the input impedance seen by the feed line is Z_1, so that the reflection coefficient Γ is

$$\Gamma = \frac{Z_1 - Z_0}{Z_1 + Z_0}. \tag{2.49}$$

Not all of the incident wave is reflected; some is transmitted onto the second line with a voltage amplitude given by a transmission coefficient.

From (2.36a) the voltage for $z < 0$ is

$$V(z) = V_o^+ \left(e^{-j\beta z} + \Gamma e^{j\beta z} \right), \quad z < 0, \tag{2.50a}$$

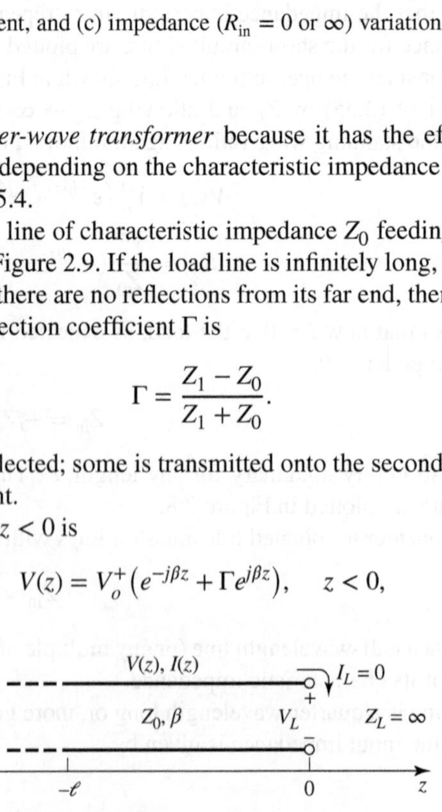

FIGURE 2.7 | A transmission line terminated in an open circuit.

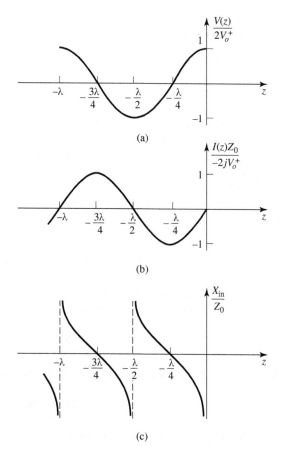

(a)

(b)

(c)

FIGURE 2.8 | (a) Voltage, (b) current, and (c) impedance ($R_{in} = 0$ or ∞) variation along an open-circuited transmission line.

where V_o^+ is the amplitude of the incident voltage wave on the feed line. The voltage wave for $z > 0$, in the absence of reflections, is outgoing only and can be written as

$$V(z) = V_o^+ T e^{-j\beta z} \quad \text{for } z > 0. \tag{2.50b}$$

Equating these voltages at $z = 0$ gives the *transmission coefficient*, T, as

$$T = 1 + \Gamma = 1 + \frac{Z_1 - Z_0}{Z_1 + Z_0} = \frac{2Z_1}{Z_1 + Z_0}. \tag{2.51}$$

The transmission coefficient between two points in a circuit is often expressed in dB as the *insertion loss*, IL,

$$\text{IL} = -20 \log |T| \, \text{dB}. \tag{2.52}$$

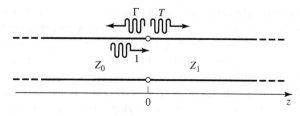

FIGURE 2.9 | Reflection and transmission at the junction of two transmission lines with different characteristic impedances.

POINT OF INTEREST: Decibels and Nepers

Often the ratio of two power levels P_1 and P_2 in a microwave system is expressed in decibels (dB) as

$$10 \log \frac{P_1}{P_2} \text{ dB.}$$

Thus, a power ratio of 2 is equivalent to 3 dB, while a power ratio of 0.1 is equivalent to -10 dB. Using power ratios in dB makes it easy to calculate power loss or gain through a series of components since multiplicative loss or gain factors can be accounted for by adding the loss or gain in dB for each stage. For example, a signal passing through a 6 dB attenuator followed by a 23 dB amplifier will have an overall gain of $23 - 6 = 17$ dB.

Decibels are used only to represent power ratios, but if $P_1 = V_1^2/R_1$ and $P_2 = V_2^2/R_2$, then the resulting power ratio in terms of voltage ratios is

$$10 \log \frac{V_1^2 R_2}{V_2^2 R_1} = 20 \log \frac{V_1}{V_2} \sqrt{\frac{R_2}{R_1}} \text{ dB,}$$

where R_1, R_2 are the load resistances and V_1, V_2 are the voltages appearing across these loads. If the load resistances are equal, then this formula simplifies to

$$20 \log \frac{V_1}{V_2} \text{ dB.}$$

The ratio of voltages across equal load resistances can also be expressed in terms of nepers (Np) as

$$\ln \frac{V_1}{V_2} \text{ Np.}$$

The corresponding expression in terms of powers is

$$\frac{1}{2} \ln \frac{P_1}{P_2} \text{ Np,}$$

since voltage is proportional to the square root of power. Transmission line attenuation is sometimes expressed in nepers. Since 1 Np corresponds to a power ratio of e^2, the conversion between nepers and decibels is

$$1 \text{ Np} = 10 \log e^2 = 8.686 \text{ dB.}$$

Absolute power can also be expressed in decibel notation if a reference power level is assumed. If we let $P_2 = 1$ mW, then the power P_1 can be expressed in dBm as

$$10 \log \frac{P_1}{1 \text{ mW}} \text{ dBm}$$

Thus a power of 1 mW is equivalent to 0 dBm, while a power of 1 W is equivalent to 30 dBm, and so on.

2.4   THE SMITH CHART

The Smith chart, shown in Figure 2.10, is a graphical aid that can be very useful for solving transmission line problems. Although there are a number of other impedance and reflection coefficient charts that can be used for such problems [3], the Smith chart is probably the best known and most widely used. It was developed in 1939 by P. Smith at the Bell Telephone Laboratories [4]. The reader might feel that, in this day of personal computers and computer-aided design (CAD) tools, graphical solutions have no place in modern engineering. The Smith chart, however, is more than just a graphical technique. Besides being an integral part of much of the current CAD software and test equipment for microwave design, the Smith chart provides a useful way of visualizing transmission line phenomenon without the need for detailed numerical calculations. A microwave engineer can develop a good intuition about transmission line and impedance-matching problems by learning to think in terms of the Smith chart.

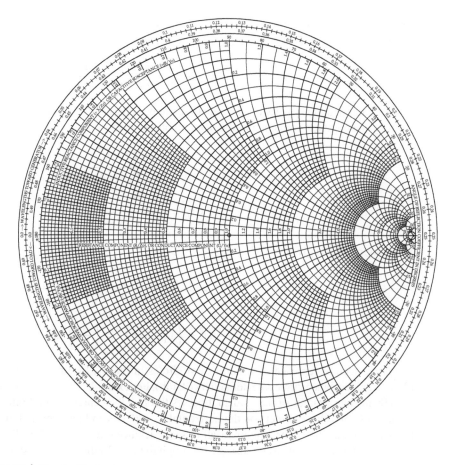

FIGURE 2.10 | The Smith chart.

At first glance the Smith chart may seem intimidating, but the key to its understanding is to realize that it is based on a polar plot of the voltage reflection coefficient, Γ. Let the reflection coefficient be expressed in magnitude and phase (polar) form as $\Gamma = |\Gamma|e^{j\theta}$. Then the magnitude $|\Gamma|$ is plotted as a radius ($|\Gamma| \leq 1$) from the center of the chart, and the angle θ ($-180° \leq \theta \leq 180°$) is measured counterclockwise from the right-hand side of the horizontal diameter. Any passively realizable ($|\Gamma| \leq 1$) reflection coefficient can then be plotted as a unique point on the Smith chart.

The real utility of the Smith chart, however, lies in the fact that it can be used to convert from reflection coefficients to normalized impedances (or admittances) and vice versa by using the impedance (or admittance) circles printed on the chart. When dealing with impedances on a Smith chart, normalized quantities are generally used, which we will denote by lowercase letters. The normalization constant is usually the characteristic impedance of the transmission line. Thus, $z = Z/Z_0$ represents the normalized version of the impedance Z.

If a lossless line of characteristic impedance Z_0 is terminated with a load impedance Z_L, the reflection coefficient at the load can be written from (2.35) as

$$\Gamma = \frac{z_L - 1}{z_L + 1} = |\Gamma|e^{j\theta}, \tag{2.53}$$

where $z_L = Z_L/Z_0$ is the normalized load impedance. This relation can be solved for z_L in terms of Γ to give [or, from (2.43) with $\ell = 0$]

$$z_L = \frac{1 + |\Gamma|e^{j\theta}}{1 - |\Gamma|e^{j\theta}}. \tag{2.54}$$

This complex equation can be reduced to two real equations by writing Γ and z_L in terms of their real and imaginary parts, $\Gamma = \Gamma_r + j\Gamma_i$, and $z_L = r_L + jx_L$, giving

$$r_L + jx_L = \frac{(1 + \Gamma_r) + j\Gamma_i}{(1 - \Gamma_r) - j\Gamma_i}.$$

The real and imaginary parts of this equation can be separated by multiplying the numerator and denominator by the complex conjugate of the denominator to give

$$r_L = \frac{1 - \Gamma_r^2 - \Gamma_i^2}{(1 - \Gamma_r)^2 + \Gamma_i^2}, \tag{2.55a}$$

$$x_L = \frac{2\Gamma_i}{(1 - \Gamma_r)^2 + \Gamma_i^2}. \tag{2.55b}$$

Rearranging (2.55) gives

$$\left(\Gamma_r - \frac{r_L}{1 + r_L} \right)^2 + \Gamma_i^2 = \left(\frac{1}{1 + r_L} \right)^2, \tag{2.56a}$$

$$(\Gamma_r - 1)^2 + \left(\Gamma_i - \frac{1}{x_L} \right)^2 = \left(\frac{1}{x_L} \right)^2, \tag{2.56b}$$

which are seen to represent two families of circles in the Γ_r, Γ_i plane. Resistance circles are defined by (2.56a) and reactance circles are defined by (2.56b). For example, the $r_L = 1$ circle has its center at $\Gamma_r = 0.5$, $\Gamma_i = 0$, and has a radius of 0.5, and so it passes through the center of the Smith chart. All of the resistance circles of (2.56a) have centers on the horizontal $\Gamma_i = 0$ axis and pass through the $\Gamma = 1$ point on the right-hand side of the chart. The centers of all of the reactance circles of (2.56b) lie on the vertical $\Gamma_r = 1$ line (off the chart), and these circles also pass through the $\Gamma = 1$ point. The resistance and reactance circles are orthogonal.

The Smith chart can also be used to graphically solve the transmission line impedance equation of (2.44) since this can be written in terms of the generalized reflection coefficient as

$$Z_{\text{in}} = Z_0 \frac{1 + \Gamma e^{-2j\beta\ell}}{1 - \Gamma e^{-2j\beta\ell}}, \tag{2.57}$$

where Γ is the reflection coefficient at the load and ℓ is the (positive) length of transmission line. We then see that (2.57) is of the same form as (2.54), differing only by the phase angles of the Γ terms. Thus, if we have plotted the reflection coefficient $|\Gamma|e^{j\theta}$ at the load, the normalized input impedance seen looking into a length ℓ of transmission line terminated with z_L can be found by rotating the point clockwise by an amount $2\beta\ell$ (subtracting $2\beta\ell$ from θ) around the center of the chart. The radius stays the same since the magnitude of Γ does not change with position along the line (assuming a lossless line).

To facilitate such rotations, the Smith chart has scales around its periphery calibrated in electrical wavelengths, toward and away from the "generator" (which simply means the direction away from the load). These scales are relative, so only the difference in wavelengths between two points on the Smith chart is meaningful. The scales cover a range of 0 to 0.5 wavelength, which reflects the fact that the Smith chart automatically includes the periodicity of transmission line phenomenon. Thus, a line of length $\lambda/2$ (or any multiple) requires a rotation of $2\beta\ell = 2\pi$ around the center of the chart, bringing the point back to its original position, showing that the input impedance of a load seen through a $\lambda/2$ line is unchanged.

We will now illustrate the use of the Smith chart for a variety of typical transmission line problems through examples.

EXAMPLE 2.3   BASIC SMITH CHART OPERATIONS

A load impedance of $50 + j25\ \Omega$ terminates a lossless $50\ \Omega$ transmission line. Find the following using the Smith chart: (a) the reflection coefficient at the load; (b) the standing wave ratio; (c) return loss; (d) the input impedance at 0.35λ from the load; and (e) the input admittance at 0.35λ from the load.

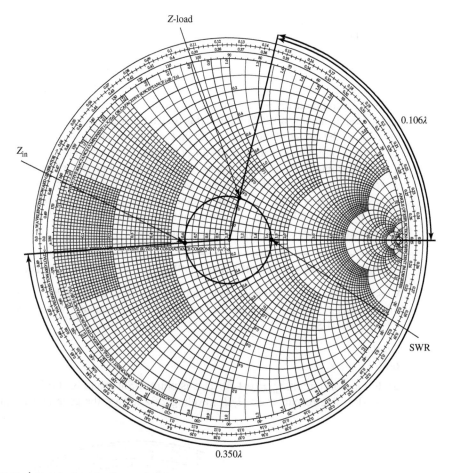

FIGURE 2.11 | Smith chart for Example 2.3.

Solution

The normalized load impedance is

$$z_L = \frac{Z_L}{Z_0} = 1 + j0.5$$

which can be plotted on the Smith chart as shown in Figure 2.11.

(a) By using a drawing compass and the voltage coefficient scale printed below the chart, one can read off the reflection coefficient magnitude at the load as $|\Gamma| = 0.24$.

(b) This same compass setting can then be applied to the SWR scale to read SWR = 1.64.

(c) The return loss (RL) (in dB) scale to read RL = 12.39 dB.

(d) Z_{in} is 0.35λ from the load, which is at 0.144λ on the wavelength to generator scale. Thus, point Z_{in} is at $0.144\lambda + 0.35\lambda = 0.494\lambda$ on the WTG scale. At point Z_{in}:

$$Z_{in} = z_{in}Z_0 = (0.61 - j0.022) \times 50\ \Omega = (30.5 - j1.09)\ \Omega$$

(e) At the point on the SWR circle opposite Z_{in}

$$Y_{in} = \frac{y_{in}}{Z_0} = \frac{1.64 + j0.06}{50} = 32.7 + j1.17\ \text{mS}$$

The Combined Impedance–Admittance Smith Chart

The Smith chart can be used for normalized admittance in the same way that it is used for normalized impedances, and it can be used to convert between impedance and admittance. The latter technique is based on the fact that, in normalized form, the input impedance of a load z_L connected to a $\lambda/4$ line is, from (2.44),

$$z_{in} = 1/z_L,$$

which has the effect of converting a normalized impedance to a normalized admittance.

Since a complete revolution around the Smith chart corresponds to a line length of $\lambda/2$, a $\lambda/4$ transformation is equivalent to a 180° rotation; this is also equivalent to imaging a given impedance (or admittance) point across the center of the chart to obtain the corresponding admittance (or impedance) point.

Thus, a Smith chart can be used for both impedance and admittance calculations during the solution of a given problem. At different stages of the solution, then, the chart may be either an *impedance Smith chart* or an *admittance Smith chart*. This procedure can be made less confusing by using a Smith chart that has a superposition of the scales for a regular Smith chart and the scales of a Smith chart that has been rotated by 180°. Such a chart is referred to as an *impedance and admittance Smith chart* and usually has different-colored scales for impedance and admittance.

EXAMPLE 2.4   BASIC SMITH CHART OPERATIONS: INPUT IMPEDANCE AND ADMITTANCE

A lossless 50 Ω transmission line is terminated in a short circuit. Use the Smith chart to find: (a) the input impedance at a distance 2.3λ from the load; and (b) the distance from the load at which the input admittance is $Y_{in} = -j0.04$ S.

Solution
Refer Figure 2.12.

(a) For a short circuit, $z_{in} = 0 + j0$. This is point *Z*-Short and is at $0.000 \, \lambda$ on the WTG scale. Since a lossless line repeats every $\lambda/2$, traveling 2.3λ toward the generator is equivalent to traveling 0.3λ

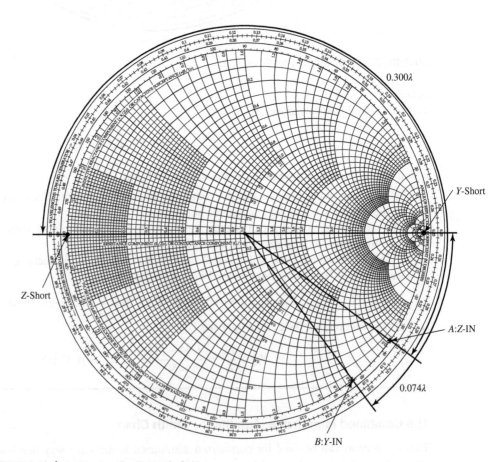

FIGURE 2.12 | Smith chart for Example 2.4.

toward the generator. This point is at A:Z-IN, and

$$Z_{in} = z_{in} Z_0 = (-j3.08) \times 50\ \Omega = -j154\ \Omega$$

(b) The admittance of a short is at point Y-Short and is at 0.25λ on the WTG scale:

$$y_{in} = Y_{in} Z_0 = (-j0.04\ S) \times 50\ \Omega = -j2$$

which is point B:Y-IN and is at 0.324λ on the WTG scale. Therefore, the line length is $0.324\lambda - 0.250\lambda = 0.074\lambda$. Any integer half wavelength farther is also valid.

The Slotted Line

A slotted line is a transmission line configuration (usually a waveguide or coaxial line) that allows the sampling of the electric field amplitude of a standing wave on a terminated line. With this device the SWR and the distance of the first voltage minimum from the load can be measured, and from these data the load impedance can be determined. Note that because the load impedance is, in general, a complex number (with two degrees of freedom), two distinct quantities must be measured with the slotted line to uniquely determine this impedance.

Although slotted lines used to be the principal way of measuring an unknown impedance at microwave frequencies, they have largely been superseded by the modern vector network analyzer in terms of accuracy, versatility, and convenience. The slotted line is still of some use, however, in certain applications such as high millimeter wave frequencies or where it is desired to avoid connector mismatches by connecting the unknown load directly to the slotted line, thus avoiding the use of imperfect transitions. Another reason for studying the slotted line is that it provides an unexcelled tool for learning the basic concepts of standing waves and mismatched transmission lines. We will derive expressions for finding the unknown load impedance from slotted line measurements and also show how the Smith chart can be used for the same purpose.

Assume that, for a certain terminated line, we have measured the SWR on the line and ℓ_{min}, the distance from the load to the first voltage minimum on the line. The load impedance Z_L can then be determined as follows. From (2.41) the magnitude of the reflection coefficient on the line is found from the standing wave ratio as

$$|\Gamma| = \frac{\text{SWR} - 1}{\text{SWR} + 1}. \tag{2.58}$$

From Section 2.3, we know that a voltage minimum occurs when $e^{j(\theta - 2\beta\ell)} = -1$, where θ is the phase angle of the reflection coefficient, $\Gamma = |\Gamma|e^{j\theta}$. The phase of the reflection coefficient is then

$$\theta = \pi + 2\beta\ell_{min}, \tag{2.59}$$

where ℓ_{min} is the distance from the load to the first voltage minimum. Actually, since the voltage minima repeat every $\lambda/2$, where λ is the wavelength on the line, any multiple of $\lambda/2$ can be added to ℓ_{min} without changing the result in (2.59) because this just amounts to adding $2\beta n\lambda/2 = 2\pi n$ to θ, which will not change Γ. Thus, the two quantities SWR and ℓ_{min} can be used to find the complex reflection coefficient Γ at the load. It is then straightforward to use (2.43) with $\ell = 0$ to find the load impedance from Γ:

$$Z_L = Z_0 \frac{1 + \Gamma}{1 - \Gamma}. \tag{2.60}$$

The use of the Smith chart in solving this problem is best illustrated by an example.

EXAMPLE 2.5 IMPEDANCE MEASUREMENT WITH A SLOTTED LINE

The following two-step procedure has been carried out with a 50 Ω coaxial slotted line to determine an unknown load impedance:

1. A short circuit is placed at the load plane, resulting in a standing wave on the line with infinite SWR and sharply defined voltage minima, as shown in Figure 2.13a. On the arbitrarily positioned scale on

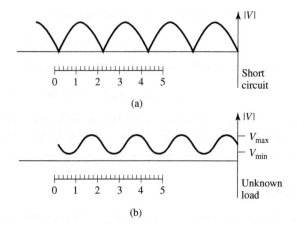

FIGURE 2.13 | Voltage standing wave patterns for Example 2.5. (a) Standing wave for short-circuit load. (b) Standing wave for unknown load.

the slotted line, voltage minima are recorded at

$$z = 0.2 \text{ cm}, \ 2.2 \text{ cm}, \ 4.2 \text{ cm}.$$

2. The short circuit is removed and replaced with the unknown load. The standing wave ratio is measured as SWR = 1.5, and voltage minima, which are not as sharply defined as those in step 1, are recorded at

$$z = 0.72 \text{ cm}, \ 2.72 \text{ cm}, \ 4.72 \text{ cm},$$

as shown in Figure 2.13b. Find the load impedance.

Solution
Knowing that voltage minima repeat every $\lambda/2$, we have from the data of step 1 that $\lambda = 4.0$ cm. In addition, because the reflection coefficient and input impedance also repeat every $\lambda/2$, we can consider the load terminals to be effectively located at any of the voltage minima locations listed in step 1. Thus, if we say the load is at 4.2 cm, then the data from step 2 show that the next voltage minimum away from the load occurs at 2.72 cm, giving $\ell_{min} = 4.2 - 2.72 = 1.48$ cm $= 0.37\lambda$.

Applying (2.58)–(2.60) to these data gives

$$|\Gamma| = \frac{1.5 - 1}{1.5 + 1} = 0.2,$$

$$\theta = \pi + \frac{4\pi}{4.0}(1.48) = 86.4°,$$

so

$$\Gamma = 0.2e^{j86.4°} = 0.0126 + j0.1996.$$

The load impedance is then

$$Z_L = 50\left(\frac{1 + \Gamma}{1 - \Gamma}\right) = 47.3 + j19.7 \ \Omega.$$

For the Smith chart version of the solution, we begin by drawing the SWR circle for SWR = 1.5, as shown in Figure 2.14; the unknown normalized load impedance must lie on this circle. The reference that we have is that the load is 0.37λ away from the first voltage minimum. On the Smith chart the position of a voltage minimum corresponds to the minimum impedance point (minimum voltage, maximum current), which is the horizontal axis (zero reactance) to the left of the origin. Thus, we begin at the voltage minimum point and move 0.37λ toward the load (counterclockwise), to the normalized load impedance point,

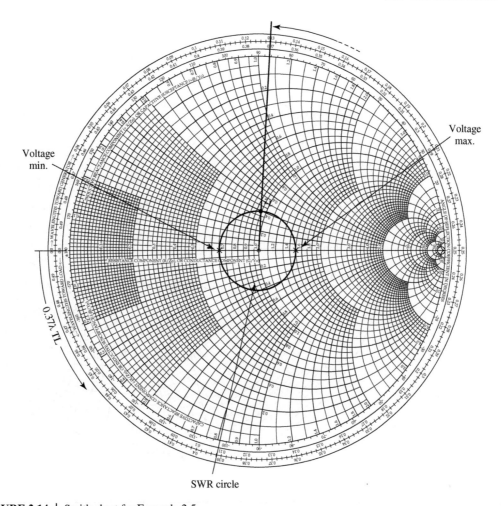

FIGURE 2.14 | Smith chart for Example 2.5.

$z_L = 0.95 + j0.4$, as shown in Figure 2.14. The actual load impedance is then $Z_L = 47.5 + j20\ \Omega$, in close agreement with the above result using equations.

Note that, in principle, voltage maxima locations could be used as well as voltage minima positions, but voltage minima are more sharply defined than voltage maxima and so usually result in greater accuracy.

EXAMPLE 2.6   IMPEDANCE MEASUREMENT USING SMITH CHART

Using a slotted line on a 50 Ω air-spaced lossless line, the following measurements were obtained: SWR = 1.6, V_{max} occurred only at 10 cm and 24 cm from the load. Find the unknown load impedance using the Smith chart.

Solution

Refer to Figure 2.15. The point SWR denotes $S = 1.6$. This point is also the location of a voltage maximum. From the locations of adjacent maxima, we can determine that $\lambda = 2 \times (24\ \text{cm} - 10\ \text{cm}) = 28\ \text{cm}$. Therefore, the load is $\frac{10\ \text{cm}}{28\ \text{cm}}\lambda = 0.357\lambda$ from the first voltage maximum, which is at 0.25 λ on the WTL scale. Traveling this far on the SWR circle, we find point Z-load at $0.25\lambda + 0.357\lambda - 0.5\lambda = 0.107\lambda$ on the WTL scale, and here

$$z_L = 0.82 - j0.39$$
$$Z_L = z_L Z_0 = (0.82 - j0.39) \times 50\ \Omega = (41 - j19.5)\ \Omega$$

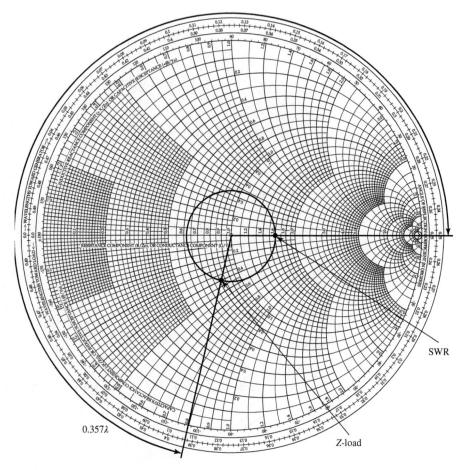

FIGURE 2.15 | Smith chart for Example 2.6.

Online Smith Chart

There are several online Smith chart tools (free and paid) available which can be used to

- Analyze transmission line properties (transmission line impedance, input impedance, VSWR, lossless and lossy lines with reactance)
- Design matching networks and obtain maximum power transfer between source and load (single stub, double stub, quarter-wave transformer, and *L* and *C* matching)
- Analyze broadband impedance matching, stub line analysis and practical transmission lines (two-wire transmission line, coaxial cable, stripline and microstrip line).

The manual Smith chart calculations can be verified and validated using these online tools. The main advantages of these online Smith chart tools are high accuracy, less calculation time, ease of handling and storage of results, and easy understanding with simulations.

Interested readers can access the website <http://www.amanogawa.com/archive/transmissionA.html> to solve problems and understand the concepts using the online Smith chart.

2.5 ___ GENERATOR AND LOAD MISMATCHES

In Section 2.3 we treated the terminated (mismatched) transmission line assuming that the generator was matched, so that no reflections occurred at the generator. In general, however, both generator and load may present mismatched impedances to the transmission line. We will study this case and also see that the condition for maximum power transfer from the generator to the load may, in some situations, involve a standing wave on the line.

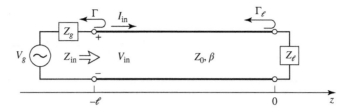

FIGURE 2.16 | Transmission line circuit for mismatched load and generator.

Figure 2.16 shows a transmission line circuit with arbitrary generator and load impedances Z_g and Z_ℓ, which may be complex. The transmission line is assumed to be lossless, with a length ℓ and characteristic impedance Z_0. This circuit is general enough to model most passive and active networks that occur in practice.

Because both the generator and load are mismatched, multiple reflections can occur on the line, as in the problem of the quarter-wave transformer. The present circuit could thus be analyzed using an infinite series to represent the multiple bounces, but we will use the easier and more useful method of impedance transformation. The input impedance looking into the terminated transmission line from the generator end is, from (2.43) and (2.44),

$$Z_{in} = Z_0 \frac{1 + \Gamma_\ell e^{-2j\beta\ell}}{1 - \Gamma_\ell e^{-2j\beta\ell}} = Z_0 \frac{Z_\ell + jZ_0 \tan\beta\ell}{Z_0 + jZ_\ell \tan\beta\ell}, \tag{2.61}$$

where Γ_ℓ is the reflection coefficient of the load:

$$\Gamma_\ell = \frac{Z_\ell - Z_0}{Z_\ell + Z_0}. \tag{2.62}$$

The voltage on the line can be written as

$$V(z) = V_o^+ \left(e^{-j\beta z} + \Gamma_\ell e^{j\beta z} \right), \tag{2.63}$$

and we can find V_o^+ from the voltage at the generator end of the line, where $z = -\ell$:

$$V(-\ell) = V_g \frac{Z_{in}}{Z_{in} + Z_g} = V_o^+ \left(e^{j\beta\ell} + \Gamma_\ell e^{-j\beta\ell} \right),$$

so that

$$V_o^+ = V_g \frac{Z_{in}}{Z_{in} + Z_g} \frac{1}{\left(e^{j\beta\ell} + \Gamma_\ell e^{-j\beta\ell} \right)}. \tag{2.64}$$

This can be rewritten, using (2.61), as

$$V_o^+ = V_g \frac{Z_0}{Z_0 + Z_g} \frac{e^{-j\beta\ell}}{\left(1 - \Gamma_\ell \Gamma_g e^{-2j\beta\ell} \right)}, \tag{2.65}$$

where Γ_g is the reflection coefficient seen looking into the generator:

$$\Gamma_g = \frac{Z_g - Z_0}{Z_g + Z_0}. \tag{2.66}$$

The standing wave ratio on the line is then

$$\text{SWR} = \frac{1 + |\Gamma_\ell|}{1 - |\Gamma_\ell|}. \tag{2.67}$$

The power delivered to the load is

$$P = \frac{1}{2}\text{Re}\{V_{in}I_{in}^*\} = \frac{1}{2}|V_{in}|^2\text{Re}\left\{\frac{1}{Z_{in}}\right\} = \frac{1}{2}|V_g|^2 \left|\frac{Z_{in}}{Z_{in} + Z_g}\right|^2 \text{Re}\left\{\frac{1}{Z_{in}}\right\}. \tag{2.68}$$

Now let $Z_{in} = R_{in} + jX_{in}$ and $Z_g = R_g + jX_g$; then (2.68) can be reduced to

$$P = \frac{1}{2}|V_g|^2 \frac{R_{in}}{(R_{in} + R_g)^2 + (X_{in} + X_g)^2}.$$
(2.69)

We now assume that the generator impedance, Z_g, is fixed, and consider three cases of load impedance.

Load Matched to Line

In this case we have $Z_l = Z_0$, so $\Gamma_\ell = 0$, and SWR = 1, from (2.62) and (2.67). Then the input impedance is $Z_{in} = Z_0$, and the power delivered to the load is, from (2.69),

$$P = \frac{1}{2}|V_g|^2 \frac{Z_0}{(Z_0 + R_g)^2 + X_g^2}.$$
(2.70)

Generator Matched to Loaded Line

In this case the load impedance Z_ℓ and/or the transmission line parameters $\beta\ell$, Z_0 are chosen to make the input impedance $Z_{in} = Z_g$, so that the generator is matched to the load presented by the terminated transmission line. Then the overall reflection coefficient, Γ, is zero:

$$\Gamma = \frac{Z_{in} - Z_g}{Z_{in} + Z_g} = 0.$$
(2.71)

There may, however, be a standing wave on the line since Γ_ℓ may not be zero. The power delivered to the load is

$$P = \frac{1}{2}|V_g|^2 \frac{R_g}{4(R_g^2 + X_g^2)}.$$
(2.72)

Observe that even though the loaded line is matched to the generator, the power delivered to the load may be less than that of (2.70), where the loaded line was not necessarily matched to the generator. Thus, we are led to the question of what is the optimum load impedance, or equivalently, what is the optimum input impedance, to achieve maximum power transfer to the load for a given generator impedance.

Conjugate Matching

Assuming that the generator series impedance Z_g is fixed, we may vary the input impedance Z_{in} until we achieve the maximum power delivered to the load. Knowing Z_{in}, it is then easy to find the corresponding load impedance Z_ℓ via an impedance transformation along the line. To maximize P, we differentiate with respect to the real and imaginary parts of Z_{in}. Using (2.69) gives

$$\frac{\partial P}{\partial R_{in}} = 0 \rightarrow \frac{1}{(R_{in} + R_g)^2 + (X_{in} + X_g)^2} + \frac{-2R_{in}(R_{in} + R_g)}{[(R_{in} + R_g)^2 + (X_{in} + X_g)^2]^2} = 0,$$

or

$$R_g^2 - R_{in}^2 + (X_{in} + X_g)^2 = 0,$$
(2.73a)

and

$$\frac{\partial P}{\partial X_{in}} = 0 \rightarrow \frac{-2R_{in}(X_{in} + X_g)}{[(R_{in} + R_g)^2 + (X_{in} + X_g)^2]^2} = 0,$$

or

$$X_{in}(X_{in} + X_g) = 0.$$
(2.73b)

Solving (2.73a) and (2.73b) simultaneously for R_{in} and X_{in} gives

$$R_{in} = R_g, \qquad X_{in} = -X_g,$$

or

$$Z_{\text{in}} = Z_g^*. \tag{2.74}$$

This condition is known as *conjugate matching*, and it results in maximum power transfer to the load for a fixed generator impedance. The power delivered is, from (2.69) and (2.74),

$$P = \frac{1}{2}|V_g|^2 \frac{1}{4R_g}, \tag{2.75}$$

which is seen to be greater than or equal to the powers of (2.70) or (2.72). This is also the maximum available power from the generator. Note that the reflection coefficients Γ_ℓ, Γ_g, and Γ may be nonzero. Physically, this means that in some cases the power in the multiple reflections on a mismatched line may add in phase to deliver more power to the load than would be delivered if the line were flat (no reflections). If the generator impedance is real ($X_g = 0$), then the last two cases reduce to the same result, which is that maximum power is delivered to the load when the loaded line is matched to the generator ($R_{\text{in}} = R_g$, with $X_{\text{in}} = X_g = 0$).

Finally, note that neither matching for zero reflection ($Z_\ell = Z_0$) nor conjugate matching ($Z_{\text{in}} = Z_g^*$) necessarily yields a system with the best efficiency. For example, if $Z_g = Z_\ell = Z_0$ then both load and generator are matched (no reflections), but only half the power produced by the generator is delivered to the load (the other half is lost in Z_g), for a transmission efficiency of 50%. This efficiency can only be improved by making Z_g as small as possible.

EXAMPLE 2.7   AVERAGE POWER DELIVERED TO THE LOAD

An antenna with load impedance $Z_L = (75 + j25)\ \Omega$ is connected to a transmitter through a 50 Ω lossless transmission line. Under matched conditions (50 Ω load), the transmitter can deliver 20 W to the load. How much power does it deliver to the antenna? Assume $Z_g = Z_0$.

Solution

$$V_0^+ = \left(\frac{V_g Z_{\text{in}}}{Z_g + Z_{\text{in}}} \right) \left(\frac{1}{e^{j\beta\ell} + \Gamma e^{-j\beta\ell}} \right)$$

After simplification,

$$V_0^+ = \frac{1}{2}V_g e^{-j\beta\ell}$$

The average power,

$$P_{\text{avg}} = \frac{|V_0^+|^2}{2Z_0}\left(1 - |\Gamma|^2\right)$$

After simplification,

$$P_{\text{avg}} = \frac{|V_g|^2}{8Z_0}\left(1 - |\Gamma|^2\right)$$

Under the matched condition, $|\Gamma| = 0$, $P_L = 20$ W, so $\dfrac{|V_g|^2}{8Z_0} = 20$ W. When $Z_L = (75 + j25)\ \Omega$,

$$\Gamma = \frac{Z_L - Z_0}{Z_L + Z_0} = 0.277\angle 33.6°$$

Thus,

$$P_{\text{avg}} = 20\text{ W}\left(1 - |\Gamma|^2\right) = 18.6\text{ W}$$

2.6   LOSSY TRANSMISSION LINES

In practice, transmission lines have losses due to finite conductivity and/or lossy dielectric, but these losses are usually small. In many practical problems loss may be neglected, but at other times the effect of loss may be very important, as when dealing with the attenuation of a transmission line, noise introduced by a lossy line, or the Q of a resonator, for example. In this section we will study the effects of loss on transmission line behavior and show how the attenuation constant can be calculated.

The Low-Loss Line

In most practical microwave and RF transmission lines the loss is small—if this were not the case, the line would be of little practical value. When the loss is small, some approximations can be made to simplify the expressions for the general transmission line parameters of $\gamma = \alpha + j\beta$ and Z_0.

The general expression for the complex propagation constant is, from (2.5),

$$\gamma = \sqrt{(R + j\omega L)(G + j\omega C)}, \tag{2.76}$$

which can be rearranged as

$$\gamma = \sqrt{(j\omega L)(j\omega C)\left(1 + \frac{R}{j\omega L}\right)\left(1 + \frac{G}{j\omega C}\right)}$$

$$= j\omega\sqrt{LC}\sqrt{1 - j\left(\frac{R}{\omega L} + \frac{G}{\omega C}\right) - \frac{RG}{\omega^2 LC}}. \tag{2.77}$$

For a low-loss line both conductor and dielectric loss will be small, and we can assume that $R \ll \omega L$ and $G \ll \omega C$. Then, $RG \ll \omega^2 LC$, and (2.77) reduces to

$$\gamma \simeq j\omega\sqrt{LC}\sqrt{1 - j\left(\frac{R}{\omega L} + \frac{G}{\omega C}\right)}. \tag{2.78}$$

If we were to ignore the $(R/\omega L + G/\omega C)$ term we would obtain the result that γ was purely imaginary (no loss), so we will instead use the first two terms of the Taylor series expansion for $\sqrt{1 + x} \simeq 1 + x/2 + \cdots$ to give the first higher order real term for γ:

$$\gamma \simeq j\omega\sqrt{LC}\left[1 - \frac{j}{2}\left(\frac{R}{\omega L} + \frac{G}{\omega C}\right)\right],$$

so that

$$\alpha \simeq \frac{1}{2}\left(R\sqrt{\frac{C}{L}} + G\sqrt{\frac{L}{C}}\right) = \frac{1}{2}\left(\frac{R}{Z_0} + GZ_0\right), \tag{2.79a}$$

$$\beta \simeq \omega\sqrt{LC}, \tag{2.79b}$$

where $Z_0 = \sqrt{L/C}$ is the characteristic impedance of the line in the absence of loss. Note from (2.79b) that the propagation constant β is identical to that of the lossless case of (2.12). By the same order of approximation, the characteristic impedance Z_0 can be approximated as a real quantity:

$$Z_0 = \sqrt{\frac{R + j\omega L}{G + j\omega C}} \simeq \sqrt{\frac{L}{C}}. \tag{2.80}$$

Equations (2.79)–(2.80) are known as the high-frequency, low-loss approximations for transmission lines, and they are important because they show that the propagation constant and characteristic impedance for a low-loss line can be closely approximated by considering the line as lossless.

EXAMPLE 2.8   ATTENUATION CONSTANT OF THE COAXIAL LINE

In Example 2.1 the L, C, R, and G parameters were derived for a lossy coaxial line. Assuming the loss is small, derive the attenuation constant from (2.79a) with the results from Example 2.1.

Solution
From (2.79a),

$$\alpha = \frac{1}{2}\left(R\sqrt{\frac{C}{L}} + G\sqrt{\frac{L}{C}}\right).$$

Using the results for R and G derived in Example 2.1 gives

$$\alpha = \frac{1}{2}\left[\frac{R_s}{\eta \ln b/a}\left(\frac{1}{a} + \frac{1}{b}\right) + \omega\epsilon''\eta\right],$$

where $\eta = \sqrt{\mu/\epsilon'}$ is the intrinsic impedance of the dielectric material filling the coaxial line. In addition, $\beta = \omega\sqrt{LC} = \omega\sqrt{\mu\epsilon'}$ and $Z_0 = \sqrt{L/C} = (\eta/2\pi)\ln b/a$.

This method for the calculation of attenuation requires that the line parameters L, C, R, and G be known. These can sometimes be derived using the formulas of (2.17)–(2.20), but a more direct and versatile procedure is to use the perturbation method, to be discussed shortly.

The Distortionless Line

As can be seen from the exact equations (2.76)–(2.77) for the propagation constant of a lossy line, the phase term β is generally a complicated function of frequency ω when loss is present. In particular, we note that β is generally not exactly a linear function of frequency, as in (2.79b), unless the line is lossless. If β is not a linear function of frequency (of the form $\beta = a\omega$), then the phase velocity $v_p = \omega/\beta$ will vary with frequency. The implication of this is that the various frequency components of a wideband signal will travel with different phase velocities and so arrive at the receiver end of the transmission line at slightly different times. This will lead to *dispersion*, a distortion of the signal, and is generally an undesirable effect. Granted, as we have argued, the departure of β from a linear function may be quite small, but the effect can be significant if the line is very long. This effect leads to the concept of group velocity, which we will address in detail in Section 3.10.

There is a special case, however, of a lossy line that has a linear phase factor as a function of frequency. Such a line is called a *distortionless* line, and it is characterized by line parameters that satisfy the relation

$$\frac{R}{L} = \frac{G}{C}. \tag{2.81}$$

From (2.77) the exact complex propagation constant, under the condition specified by (2.81), reduces to

$$\begin{aligned}
\gamma &= j\omega\sqrt{LC}\sqrt{1 - 2j\frac{R}{\omega L} - \frac{R^2}{\omega^2 L^2}} \\
&= j\omega\sqrt{LC}\left(1 - j\frac{R}{\omega L}\right) \\
&= R\sqrt{\frac{C}{L}} + j\omega\sqrt{LC} = \alpha + j\beta,
\end{aligned} \tag{2.82}$$

which shows that $\beta = \omega\sqrt{LC}$ is now a linear function of frequency. Equation (2.82) also shows that the attenuation constant, $\alpha = R\sqrt{C/L}$, does not depend on frequency, so that all frequency components of a signal will be attenuated by the same amount (actually, R is usually a weak function of frequency). Thus, the distortionless line is not loss free but is capable of passing a pulse or modulation envelope without distortion. To obtain a transmission line with parameters that satisfy (2.81) often requires that L be increased by adding series loading coils spaced periodically along the line.

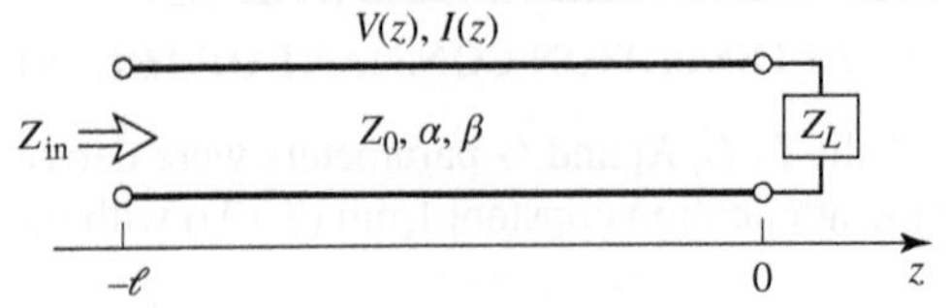

FIGURE 2.17 | A lossy transmission line terminated in the impedance Z_L.

The above theory for the distortionless line was first developed by Oliver Heaviside (1850–1925), who solved many problems in transmission line theory and reworked Maxwell's original theory of electromagnetism into the modern version that we are familiar with today [5].

The Terminated Lossy Line

Figure 2.17 shows a length ℓ of a lossy transmission line terminated in a load impedance Z_L. Thus, $\gamma = \alpha + j\beta$ is complex, but we assume the loss is small, so that Z_0 is approximately real, as in (2.80).

In (2.36), expressions for the voltage and current wave on a lossless line are given. The analogous expressions for the lossy case are

$$V(z) = V_o^+ \left(e^{-\gamma z} + \Gamma e^{\gamma z} \right), \tag{2.83a}$$

$$I(z) = \frac{V_o^+}{Z_0} \left(e^{-\gamma z} - \Gamma e^{\gamma z} \right), \tag{2.83b}$$

where Γ is the reflection coefficient of the load, as given in (2.35), and V_o^+ is the incident voltage amplitude referenced at $z = 0$. From (2.42) the reflection coefficient at a distance ℓ from the load is

$$\Gamma(\ell) = \Gamma e^{-2j\beta\ell} e^{-2\alpha\ell} = \Gamma e^{-2\gamma\ell}. \tag{2.84}$$

The input impedance Z_{in} at a distance ℓ from the load is then

$$Z_{\text{in}} = \frac{V(-\ell)}{I(-\ell)} = Z_0 \frac{Z_L + Z_0 \tanh \gamma\ell}{Z_0 + Z_L \tanh \gamma\ell}. \tag{2.85}$$

We can compute the power delivered to the input of the terminated line at $z = -\ell$ as

$$P_{\text{in}} = \frac{1}{2} \text{Re} \left\{ V(-\ell) I^*(-\ell) \right\} = \frac{|V_o^+|^2}{2Z_0} \left(e^{2\alpha\ell} - |\Gamma|^2 e^{-2\alpha\ell} \right)$$

$$= \frac{|V_o^+|^2}{2Z_0} \left(1 - |\Gamma(\ell)|^2 \right) e^{2\alpha\ell}, \tag{2.86}$$

where (2.83) has been used for $V(-\ell)$ and $I(-\ell)$. The power actually delivered to the load is

$$P_L = \frac{1}{2} \text{Re} \{ V(0) I^*(0) \} = \frac{|V_o^+|^2}{2Z_0} (1 - |\Gamma|^2). \tag{2.87}$$

The difference in these powers corresponds to the power lost in the line:

$$P_{\text{loss}} = P_{\text{in}} - P_L = \frac{|V_o^+|^2}{2Z_0} \left[\left(e^{2\alpha\ell} - 1 \right) + |\Gamma|^2 \left(1 - e^{-2\alpha\ell} \right) \right]. \tag{2.88}$$

The first term in (2.88) accounts for the power loss of the incident wave, while the second term accounts for the power loss of the reflected wave; note that both terms increase as α increases.

The Perturbation Method for Calculating Attenuation

Here we derive a useful and standard technique for finding the attenuation constant of a low-loss line. The method avoids the use of the transmission line parameters L, C, R, and G and instead relies on the fields of the lossless line, with the assumption that the fields of the lossy line are not greatly different from the fields of the lossless line—hence the term, *perturbation method*.

We have seen that the power flow along a lossy transmission line, in the absence of reflections, is of the form

$$P(z) = P_o e^{-2\alpha z}, \tag{2.89}$$

where P_o is the power at the $z = 0$ plane and α is the attenuation constant we wish to determine. Now define the power loss per unit length along the line as

$$P_\ell = -\frac{\partial P}{\partial z} = 2\alpha P_o e^{-2\alpha z} = 2\alpha P(z),$$

where the negative sign on the derivative was introduced so that P_ℓ would be a positive quantity. From this, the attenuation constant can be determined as

$$\alpha = \frac{P_\ell(z)}{2P(z)} = \frac{P_\ell(z=0)}{2P_o}. \tag{2.90}$$

This equation states that α can be determined from P_o, the power on the line, and P_ℓ, the power loss per unit length of line. It is important to realize that P_ℓ can be computed from the fields of the lossless line and can account for both conductor loss [using (1.131)] and dielectric loss [using (1.92)].

EXAMPLE 2.9 USING THE PERTURBATION METHOD TO FIND THE ATTENUATION CONSTANT

Use the perturbation method to find the attenuation constant of a coaxial line having a lossy dielectric and lossy conductors.

Solution
From Example 2.1 and (2.32), the fields of the lossless coaxial line are, for $a < \rho < b$,

$$\bar{E} = \frac{V_o \hat{\rho}}{\rho \ln b/a} e^{-j\beta z},$$

$$\bar{H} = \frac{V_o \hat{\phi}}{2\pi \rho Z_0} e^{-j\beta z},$$

where $Z_0 = (\eta/2\pi) \ln b/a$ is the characteristic impedance of the coaxial line and V_o is the voltage across the line at $z = 0$. The first step is to find P_o, the power flowing on the lossless line:

$$P_o = \frac{1}{2} \text{Re} \int_S \bar{E} \times \bar{H}^* \cdot d\bar{s} = \frac{|V_o|^2}{2Z_0} \int_{\rho=a}^{b} \int_{\phi=0}^{2\pi} \frac{\rho \, d\rho \, d\phi}{2\pi \rho^2 \ln b/a} = \frac{|V_o|^2}{2Z_0},$$

as expected from basic circuit theory.

The loss per unit length, P_ℓ, comes from conductor loss ($P_{\ell c}$) and dielectric loss ($P_{\ell d}$). From (1.131), the conductor loss in a 1 m length of line can be found as

$$P_{\ell c} = \frac{R_s}{2} \int_S |\bar{H}_t|^2 ds = \frac{R_s}{2} \int_{z=0}^{1} \left\{ \int_{\phi=0}^{2\pi} |H_\phi(\rho = a)|^2 a \, d\phi \right.$$

$$\left. + \int_{\phi=0}^{2\pi} |H_\phi(\rho = b)|^2 b \, d\phi \right\} dz$$

$$= \frac{R_s |V_o|^2}{4\pi Z_0^2} \left(\frac{1}{a} + \frac{1}{b} \right).$$

The dielectric loss in a 1 m length of line is, from (1.92),

$$P_{\ell d} = \frac{\omega \epsilon''}{2} \int_V |\bar{E}|^2 ds = \frac{\omega \epsilon''}{2} \int_{\rho=a}^{b} \int_{\phi=0}^{2\pi} \int_{z=0}^{1} |E_\rho|^2 \rho \, d\rho \, d\phi \, dz = \frac{\pi \omega \epsilon''}{\ln b/a} |V_o|^2,$$

where ϵ'' is the imaginary part of the complex permittivity, $\epsilon = \epsilon' - j\epsilon''$. Finally, applying (2.90) gives

$$\alpha = \frac{P_{\ell c} + P_{\ell d}}{2P_o} = \frac{R_s}{4\pi Z_0}\left(\frac{1}{a} + \frac{1}{b}\right) + \frac{\pi\omega\epsilon'' Z_0}{\ln b/a}$$

$$= \frac{R_s}{2\eta\ln b/a}\left(\frac{1}{a} + \frac{1}{b}\right) + \frac{\omega\epsilon''\eta}{2},$$

where $\eta = \sqrt{\mu/\epsilon'}$. This result is seen to agree with that of Example 2.8.

The Wheeler Incremental Inductance Rule

Another useful technique for the practical evaluation of attenuation due to conductor loss for TEM or quasi-TEM lines is the *Wheeler incremental inductance rule* [6]. This method is based on the similarity of the equations for the inductance per unit length and resistance per unit length of a transmission line, as given by (2.17) and (2.19), respectively. In other words, the conductor loss of a line is due to current flow inside the conductor, which, as was shown in Section 1.7, is directly related to the tangential magnetic field at the surface of the conductor and thus to the inductance of the line.

From (1.131), the power loss into a cross section S of a good (but not perfect) conductor is

$$P_\ell = \frac{R_s}{2}\int_S |\bar{J}_s|^2 ds = \frac{R_s}{2}\int_S |\bar{H}_t|^2 ds \text{ W/m}^2, \tag{2.91}$$

so the power loss per unit length of a uniform transmission line is

$$P_\ell = \frac{R_s}{2}\int_C |\bar{H}_t|^2 d\ell \text{ W/m}, \tag{2.92}$$

where the line integral of (2.92) is over the cross-sectional contours of both conductors. From (2.17), the inductance per unit length of the line is

$$L = \frac{\mu}{|I|^2}\int_S |\bar{H}|^2 ds, \tag{2.93}$$

which is computed assuming the conductors are lossless. When the conductors have a small loss, the $\bar{H}$ field in the conductor is no longer zero, and this field contributes a small additional "incremental" inductance, ΔL, to that of (2.93). As discussed in Chapter 1, the fields inside the conductor decay exponentially, so that the integration into the conductor dimension can be evaluated as

$$\Delta L = \frac{\mu_0\delta_s}{2|I|^2}\int_C |\bar{H}_t|^2 d\ell, \tag{2.94}$$

since $\int_0^\infty e^{-2z/\delta_s}dz = \delta_s/2$. (The skin depth is $\delta_s = \sqrt{2/\omega\mu\sigma}$.) Then P_ℓ from (2.92) can be written in terms of ΔL as

$$P_\ell = \frac{R_s|I|^2\Delta L}{\mu_0\delta_s} = \frac{|I|^2\Delta L}{\sigma\mu_0\delta_s^2} = \frac{|I|^2\omega\Delta L}{2}\text{W/m}, \tag{2.95}$$

since $R_s = \sqrt{\omega\mu_0/2\sigma} = 1/\sigma\delta_s$. Then from (2.90) the attenuation due to conductor loss can be evaluated as

$$\alpha_c = \frac{P_\ell}{2P_o} = \frac{\omega\Delta L}{2Z_0}, \tag{2.96}$$

since P_o, the total power flow down the line, is $P_o = |I|^2 Z_0/2$. In (2.96), ΔL is evaluated as the change in inductance when all conductor walls recede by an amount $\delta_s/2$.

Equation (2.96) can also be written in terms of the change in characteristic impedance since

$$Z_0 = \sqrt{\frac{L}{C}} = \frac{L}{\sqrt{LC}} = Lv_p, \tag{2.97}$$

so that

$$\alpha_c = \frac{\beta \Delta Z_0}{2Z_0},$$
(2.98)

where ΔZ_0 is the change in characteristic impedance when all conductor walls recede by an amount $\delta_s/2$. Yet another form of the incremental inductance rule can be obtained by using the first two terms of a Taylor series expansion for Z_0. Thus,

$$Z_0\left(\frac{\delta_s}{2}\right) \simeq Z_0 + \frac{\delta_s}{2}\frac{dZ_0}{d\ell},$$
(2.99)

so that

$$\Delta Z_0 = Z_0\left(\frac{\delta_s}{2}\right) - Z_0 = \frac{\delta_s}{2}\frac{dZ_0}{d\ell},$$

where $Z_0\left(\delta_s/2\right)$ refers to the characteristic impedance of the line when the walls recede by $\delta_s/2$, and ℓ refers to a distance into the conductors. Then (2.98) can be written as

$$\alpha_c = \frac{\beta\delta_s}{4Z_0}\frac{dZ_0}{d\ell} = \frac{R_s}{2Z_0\eta}\frac{dZ_0}{d\ell},$$
(2.100)

where $\eta = \sqrt{\mu_0/\epsilon}$ is the intrinsic impedance of the dielectric and R_s is the surface resistivity of the conductor. Equation (2.100) is one of the most practical forms of the incremental inductance rule because the characteristic impedance is known for a wide variety of transmission lines.

EXAMPLE 2.10   USING THE WHEELER INCREMENTAL INDUCTANCE RULE TO FIND THE ATTENUATION CONSTANT

Calculate the attenuation due to conductor loss of a coaxial line using the Wheeler incremental inductance rule.

Solution
From (2.32) the characteristic impedance of the coaxial line is

$$Z_0 = \frac{\eta}{2\pi}\ln\frac{b}{a}.$$

From the incremental inductance rule of the form given in (2.100), the attenuation due to conductor loss is

$$\alpha_c = \frac{R_s}{2Z_0\eta}\frac{dZ_0}{d\ell} = \frac{R_s}{4\pi Z_0}\left(\frac{d\ln b/a}{db} - \frac{d\ln b/a}{da}\right) = \frac{R_s}{4\pi Z_0}\left(\frac{1}{b} + \frac{1}{a}\right),$$

which is seen to be in agreement with the result of Example 2.9. The negative sign on the second differentiation in this equation is because the derivative for the inner conductor is in the $-\rho$ direction (receding wall).

Regardless of how attenuation is calculated, measured attenuation values for practical transmission lines are usually higher. One reason for this discrepancy is the fact that realistic transmission lines have metallic surfaces with a certain amount of roughness, which increases loss, while our theoretical calculations assume perfectly smooth conductors. A quasi-empirical formula that can be used to approximately account for surface roughness for a transmission line is [7]

$$\alpha_c' = \alpha_c\left[1 + \frac{2}{\pi}\tan^{-1}1.4\left(\frac{\Delta}{\delta_s}\right)^2\right],$$
(2.101)

where α_c is the attenuation due to perfectly smooth conductors, α_c' is the attenuation corrected for surface roughness, Δ is the rms surface roughness, and δ_s is the skin depth of the conductors.

2.7   TRANSIENTS ON TRANSMISSION LINES

So far we have concentrated on the behavior of transmission lines at a single frequency, and in many cases of practical interest this viewpoint is entirely satisfactory. In some situations, however, where short pulses or very wideband signals are propagating on a transmission line, it is useful to consider wave propagation from a transient, or time domain, point of view.

In this section we will discuss the reflection of transient pulses from terminated transmission lines, including the special cases of a matched line, a short-circuited line, and an open-circuited line. We will conclude with a description of bounce diagrams, which can be used to describe multiple reflections of pulses on transmission lines.

Reflection of Pulses from a Terminated Transmission Line

A transient transmission line circuit is shown in Figure 2.18a, where a DC source is switched on at $t = 0$. We first consider the case in which the line has a characteristic impedance of Z_0, the source impedance is Z_0, and the load impedance is Z_0. It is assumed that the voltage on the line is initially zero: $v(z, t) = 0$ for all z, for $t < 0$. We want to determine the voltage response on the transmission line as a function of time and position.

Because of the finite transit time of the line, its input impedance will appear to be equal to the characteristic impedance of the line for $t < 2\ell/v_p$, where v_p is the phase velocity of the line. In other words, the line looks infinitely long until the pulse has time to reach the load and (possibly) reflect back to the input. Therefore, when the switch closes at $t = 0$, the circuit appears as a voltage divider consisting of the source impedance and the input impedance, both being Z_0. The initial voltage on the line is thus $V_0/2$, and this voltage waveform propagates toward the load with a velocity v_p. The leading edge of the pulse will be at position z on the line at time $t = z/v_p$, as shown in Figure 2.18b.

The pulse reaches the load at time $t = \ell/v_p$. Since the load is matched to the line, there is no reflection of the pulse from the load. The circuit is now in a steady-state condition, and voltage on the line is constant: $v(z, t) = V_0/2$ for all $t > \ell/v_p$, as shown in Figure 2.18c. This is, of course, the DC value that we would expect for a voltage divider consisting of equal source and input impedances.

Next consider the transmission line circuit of Figure 2.19a, where the line is now terminated with a short circuit. Initially, the input impedance of the line again appears as Z_0, and the initial incident pulse again has an amplitude of $V_0/2$, as shown in Figure 2.19b. The short-circuit load has a reflection coefficient of $\Gamma = -1$, which has the effect of inverting the reflected pulse as it travels back toward the source. The superposition of the forward and reverse traveling pulses leads to cancellation, as shown in Figure 2.19c,

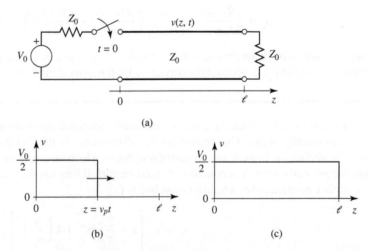

(a)

(b)   (c)

FIGURE 2.18 | Transient response of a transmission line terminated with a matched load. (a) Transmission line circuit with a step function voltage source. (b) Response for $0 < t < \ell/v_p$. (c) Response for $\ell/v_p < t < 2\ell/v_p$; there is no reflection from the load.

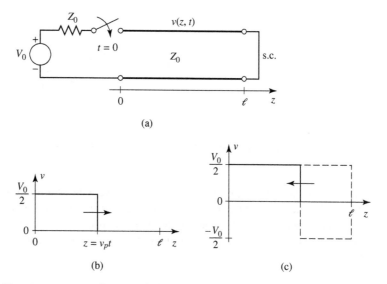

(a)

(b)   (c)

FIGURE 2.19 | Transient response of a transmission line terminated with a short circuit. (a) Transmission line circuit with a step function voltage source. (b) Response for $0 < t < \ell/v_p$. (c) Response for $\ell/v_p < t < 2\ell/v_p$; the incident pulse is reflected with $\Gamma = -1$.

for the period where $\ell/v_p < t < 2\ell/v_p$. When the return pulse reaches the source, at $t = 2\ell/v_p$, it will not be re-reflected because the source is matched to the line. The circuit is then in steady state, with zero voltage everywhere on the line. Again, this is consistent with DC circuit analysis, as the shorted line has zero electrical length at DC and thus appears as a short at its input, leading to a terminal voltage of zero. The voltage waveform at a fixed point z on the line will consist of a rectangular pulse of amplitude $V_0/2$ existing only over the time period $z/v_p < t < (2\ell - z)/v_p$. This effect can be used in practice to generate pulses of very short duration.

Finally, consider the effect of a transmission line with an open-circuit termination, as shown in Figure 2.20a. As in previous cases, the input impedance of the line initially appears as Z_0, and the initial incident pulse has an amplitude of $V_0/2$, as shown in Figure 2.20b. The open-circuit load has a reflection coefficient of $\Gamma = 1$, which reflects the incident waveform with the same polarity toward the source. The amplitudes

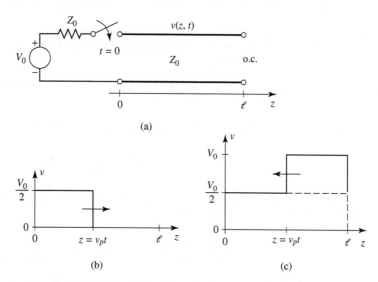

(a)

(b)   (c)

FIGURE 2.20 | Transient response of a transmission line terminated with an open circuit. (a) Transmission line circuit with a step function voltage source. (b) Response for $0 < t < \ell/v_p$. (c) Response for $\ell/v_p < t < 2\ell/v_p$; the incident pulse is reflected with $\Gamma = 1$.

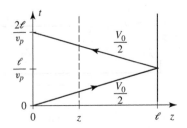

FIGURE 2.21 | Bounce diagram for the transient circuit of Figure 2.20a.

of the forward and reverse pulses add to create a wave with an amplitude of V_0, as shown in Figure 2.20c. At $t = 2\ell/v_p$ the return pulse reaches the source, but it is not re-reflected since the source is matched to the line. The circuit is then in steady state, with a constant voltage of V_0 on the line. By DC analysis, the open-circuited line presents an open circuit at its terminals, leading to a terminal voltage equal to the source voltage.

Bounce Diagrams for Transient Propagation

The plots in Figures 2.18–2.20 show the voltage of a propagating pulse versus position along the transmission line but do not directly show the time variable, nor do they show very clearly the contribution of reflections on the waveform (especially when multiple reflections are present). An alternative way of viewing the progress of a pulse propagating in time and position along a transmission line is with a *bounce diagram*.

As an example, Figure 2.21 shows the bounce diagram for the transient circuit of Figure 2.21a. The horizontal axis represents position on the line, while the vertical axis represents time. The ray representing the incident wave begins at $t = z = 0$ and travels to the right (increasing z) and up (for increasing t). This ray is labeled with the amplitude of the incident wave, $V_0/2$. At $t = \ell/v_p$ the incident wave reaches the open-circuit load and is reflected to produce a wave of amplitude $V_0/2$ traveling back to the source. The ray for this reflected wave thus moves to the left and up, until it reaches the source at $z = 0$ and $t = 2\ell/v_p$, at which point steady state is reached. The total voltage at any position z and time t can be easily found by drawing a vertical line through the point z and extending up from $t = 0$ to t. The total voltage is found by adding the voltages of each forward or reverse traveling wave component, as represented by the rays that intersect this vertical line.

The next example shows how a bounce diagram can be applied to circuits that have multiple reflections.

EXAMPLE 2.11 BOUNCE DIAGRAM FOR A TRANSIENT CIRCUIT WITH MULTIPLE REFLECTIONS

Draw the bounce diagram for the transient circuit of Figure 2.22, including the first three reflections.

Solution
The amplitude of the incident wave is given by a voltage divider as

$$v^+ = 12\frac{100}{50 + 100} = 8.0 \text{ V}$$

The incident ray can be plotted as a line from the origin to the point $z = \ell$ and $t = \ell/v_p$. The reflection coefficients at the generator and load are

$$\Gamma_g = \frac{50 - 100}{50 + 100} = -1/3 \quad \text{and} \quad \Gamma_L = \frac{200 - 100}{200 + 100} = 1/3,$$

so the amplitude of the wave reflected from the load is 8/3 V. When this wave reaches the source, it will be reflected to form a wave of amplitude −8/9 V. The next reflection from the load will have an amplitude of −8/27 V. These four waves are shown in the bounce diagram of Figure 2.23.

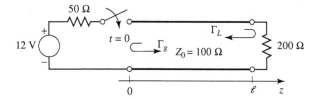

FIGURE 2.22 | Circuit for Example 2.11.

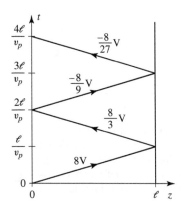

FIGURE 2.23 | Bounce diagram for Example 2.11.

REFERENCES

[1] S. Ramo, J. R. Winnery, and T. Van Duzer, *Fields and Waves in Communication Electronics*, 3rd edition, John Wiley & Sons, New York, 1994.

[2] J. A. Stratton, *Electromagnetic Theory*, McGraw-Hill, New York, 1941.

[3] H. A. Wheeler, "Reflection Charts Relating to Impedance Matching," *IEEE Transactions on Microwave Theory and Techniques*, vol. MTT-32, pp. 1008–1021, September 1984.

[4] P. H. Smith, "Transmission Line Calculator," *Electronics*, vol. 12, No. 1, pp. 29–31, January 1939.

[5] P. J. Nahin, *Oliver Heaviside: Sage in Solitude*, IEEE Press, New York, 1988.

[6] H. A. Wheeler, "Formulas for the Skin Effect," *Proceedings of the IRE*, vol. 30, pp. 412–424, September 1942.

[7] T. C. Edwards, *Foundations for Microstrip Circuit Design*, John Wiley & Sons, New York, 1987.

PROBLEMS

2.1 A 75 Ω coaxial line has a current $i(t, z) = 1.5 \cos(1.85 \times 10^{10} t - 80 z)$ mA. Determine (a) the frequency, (b) the phase velocity, (c) the wavelength, (d) the relative permittivity of the line, (e) the phasor form of the current, and (f) the time domain voltage on the line.

2.2 A transmission line has the following per-unit-length parameters: $L = 1.5 \mu$H/m, $C = 150$ pF/m, $R = 5.0$ Ω/m, and $G = 0.05$ S/m. Calculate the propagation constant and characteristic impedance of this line at 900 MHz. If the line is 30 cm long, what is the attenuation in dB? Recalculate these quantities in the absence of loss ($R = G = 0$).

2.3 RG-402U semirigid coaxial cable has an inner conductor diameter of 0.91 mm and a dielectric diameter (equal to the inner diameter of the outer conductor) of 3.02 mm. Both conductors are copper, and the dielectric material is Teflon. Compute the $R, L, G,$ and C parameters of this line at 1 GHz, and use these results to find the characteristic impedance and attenuation of the line at 2.4 GHz. Compare your results

to the manufacturer's specifications of 50 Ω and 0.43 dB/m, and discuss reasons for the difference.

2.4 Compute and plot the attenuation of the coaxial line of Problem 2.3, in dB/m, over a frequency range of 1 MHz to 100 GHz. Use log-log graph paper.

2.5 For the parallel plate line shown in the accompanying figure, derive the $R, L, G,$ and C parameters. Assume $W \gg d$.

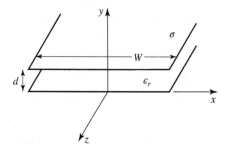

2.6 For the parallel plate line of Problem 2.5, derive the telegrapher equations using the field theory approach.

2.7 Show that the T-model of a transmission line shown in the accompanying figure also yields the telegrapher equations derived in Section 2.1.

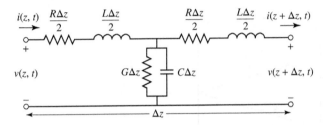

2.8 A lossless transmission line of electrical length $\ell = 0.4\lambda$ is terminated with a complex load impedance as shown in the accompanying figure. Find the reflection coefficient at the load, the SWR on the line, the reflection coefficient at the input of the line, and the input impedance to the line.

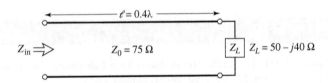

2.9 A 75 Ω coaxial transmission line has a length of 2.5 cm and is terminated with a load impedance of $37.5 + j75$ Ω. If the relative permittivity of the line is 2.2 and the frequency is 6.0 GHz, find the input impedance to the line, the reflection coefficient at the load, the reflection coefficient at the input, and the SWR on the line.

2.10 A terminated transmission line with $Z_0 = 50$ Ω has a reflection coefficient at the load of $\Gamma = 0.5\angle 75°$. (a) What is the load impedance? (b) What is the reflection coefficient 0.325λ away from the load? (c) What is the input impedance at this point?

2.11 A 100 Ω transmission line has an effective dielectric constant of 1.65. Find the shortest open-circuited length of this line that appears at its input as a capacitor of 5 pF at 2.5 GHz. Repeat for an inductance of 5 nH.

2.12 A lossless transmission line is terminated with a 200 Ω load. If the SWR on the line is 3, find the two possible values for the characteristic impedance of the line.

2.13 Let Z_{sc} be the input impedance of a length of coaxial line when one end is short-circuited, and let Z_{oc} be the input impedance of the line when one end is open-circuited. Derive an expression for the characteristic impedance of the cable in terms of Z_{sc} and Z_{oc}.

2.14 A radio transmitter is connected to an antenna having an impedance $100 + j50$ Ω with a 75 Ω coaxial cable. If the 75 Ω transmitter can deliver 50 W when connected to a 75 Ω load, how much power is delivered to the antenna?

2.15 Calculate standing wave ratio, reflection coefficient magnitude, and return loss values to complete the entries in the following table:

SWR	$|\Gamma|$	RL (dB)
1.00	0.00	∞
1.01	—	—
—	0.01	—
1.05	—	—
—	—	30.0
1.10	—	—
1.20	—	—
—	0.10	—
1.50	—	—
—	—	10.0
2.00	—	—
2.50	—	—

2.16 The transmission line circuit in the accompanying figure has $V_g = 15$ V rms, $Z_g = 75$ Ω, $Z_0 = 75$ Ω, $Z_L = 60 - j40$ Ω, and $\ell = 0.7\lambda$. Compute the power delivered to the load using three different techniques:

(a) Find Γ and compute

$$P_L = \left(\frac{V_g}{2}\right)^2 \frac{1}{Z_0}(1 - |\Gamma|^2);$$

(b) find Z_{in} and compute

$$P_L = \left|\frac{V_g}{Z_g + Z_{in}}\right|^2 \text{Re}\{Z_{in}\};$$

(c) find V_L and compute

$$P_L = \left|\frac{V_L}{Z_L}\right|^2 \text{Re}\{Z_L\}.$$

Discuss the rationale for each of these methods. Which of these methods can be used if the line is not lossless?

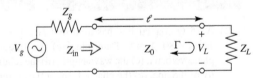

2.17 For a purely reactive load impedance of the form $Z_L = jX$, show that the reflection coefficient magnitude $|\Gamma|$ is always unity. Assume that the characteristic impedance Z_0 is real.

2.18 Consider the transmission line circuit shown in the accompanying figure. Compute the incident power, the reflected power, and the power transmitted into the infinite 75 Ω line. Show that power conservation is satisfied.

2.19 A generator is connected to a transmission line as shown in the accompanying figure. Find the voltage as a function of z along the transmission line. Plot the magnitude of this voltage for $-\ell \leq z \leq 0$.

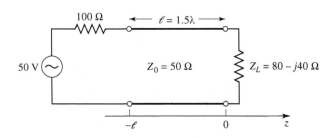

2.20 Use the Smith chart to find the following quantities for the transmission line circuit shown in the accompanying figure:

(a) The SWR on the line.
(b) The reflection coefficient at the load.
(c) The load admittance.
(d) The input impedance of the line.
(e) The distance from the load to the first voltage minimum.
(f) The distance from the load to the first voltage maximum.

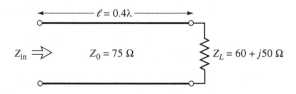

2.21 Use the Smith chart to find the shortest lengths of an open-circuited 50 Ω line to give the following input impedance:

(a) $Z_{in} = 0$.
(b) $Z_{in} = \infty$.
(c) $Z_{in} = j50 \ \Omega$.
(d) $Z_{in} = -j25 \ \Omega$.
(e) $Z_{in} = j10 \ \Omega$.

2.22 A slotted-line experiment is performed with the following results: distance between successive minima = 2.5 cm; distance of first voltage minimum from load = 1.2 cm; SWR of load = 2.0. If $Z_0 = 50 \ \Omega$, find the load impedance.

2.23 Design a quarter-wave matching transformer to match a 35 Ω load to a 75 Ω line. Plot the SWR for $0.5 \leq f/f_o \leq 2.0$, where f_o is the frequency at which the line is $\lambda/4$ long.

2.24 Consider the quarter-wave matching transformer circuit shown in the accompanying figure. Derive expressions for V^+ and V^-, the respective amplitudes of the forward and reverse traveling waves on the quarter-wave line section, in terms of V^i, the incident voltage amplitude.

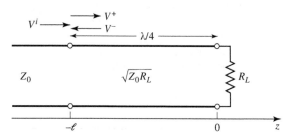

2.25 Derive equation (2.65) from (2.64).

2.26 In Example 2.9, the attenuation of a coaxial line due to finite conductivity is

$$\alpha_c = \frac{R_s}{2\eta \ln b/a}\left(\frac{1}{a} + \frac{1}{b}\right).$$

Show that α_c is minimized for conductor radii such that $x \ln x = 1 + x$, where $x = b/a$. Solve this equation for x, and show that the corresponding characteristic impedance for $\epsilon_r = 1$ is 77 Ω.

2.27 Compute and plot the factor by which attenuation is increased due to surface roughness, for rms roughness ranging from 0 to 0.01 mm. Assume copper conductors at 12 GHz.

2.28 A 50 Ω transmission line is matched to a 20 V source and feeds a load $Z_L = 150 \ \Omega$. If the line is 2.8λ long and has an attenuation constant $\alpha = 0.5$ dB/λ, find the powers that are delivered by the source, lost in the line, and delivered to the load.

2.29 Consider a nonreciprocal transmission line having different propagation constants, β^+ and β^-, for propagation in the forward and reverse directions, with corresponding characteristic impedances Z_0^+ and Z_0^-. (An example of such a line could be a microstrip transmission line on a magnetized ferrite substrate.) If the line is terminated as shown in the accompanying figure, derive expressions for the reflection coefficient and impedance seen at the input of the line.

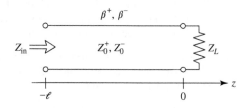

2.30 Plot the bounce diagram for the transient circuit shown in the accompanying figure. Include at least three reflections. What is the total voltage at the midpoint of the line ($z = \ell/2$), at time $t = 3\ell/v_p$?

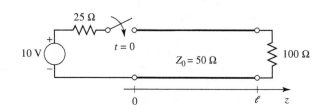

2.1 Two identical wires $W1$ and $W2$ placed in parallel, as shown in the figure, carry currents I and $2I$, respectively, in opposite directions. If the two wires are separated by a distance of $4r$, then the magnitude of the magnetic field $\bar{B}$ between the wires at a distance r from $W1$ is

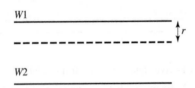

(a) $\dfrac{\mu_0 I}{6\pi r}$

(c) $\dfrac{5\mu_0 I}{6\pi r}$

(b) $\dfrac{6\mu_0 I}{5\pi r}$

(d) $\dfrac{\mu_0^2 I^2}{2\pi r^2}$

2.2 Points P, Q, and R shown on the Smith chart (normalized impedance chart) in the following figure represent

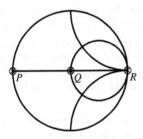

(a) P: open circuit, Q: short circuit, R: matched load
(b) P: open circuit, Q: matched load, R: short circuit
(c) P: short circuit, Q: matched load, R: open circuit
(d) P: short circuit, Q: open circuit, R: matched load

2.3 A lossy transmission line has resistance per unit length $R = 0.05$ Ω/m. The line is distortionless and has characteristic impedance of 50 Ω. The attenuation constant (in Np/m, correct to three decimal places) of the line is _____.

2.4 A two-wire transmission line terminates in a television set. The VSWR measured of the line is 5.8. The percentage of the power that is reflected from the television set is _____.

2.5 The voltage of an electromagnetic wave propagating in a coaxial cable with uniform characteristic impedance is $V(l) = e^{-\gamma l + j\omega t}$ V, where l is the distance along the length of the cable in meters, $\gamma = (0.1 + j40)$ m^{-1} is the complex propagation constant, and $\omega = 2\pi \times 10^9$ rad/sec is the angular frequency. The absolute value of the attenuation in the cable in dB/meter is _____.

2.6 A microwave circuit consisting of lossless transmission lines T_1 and T_2 is shown in the figure. The plot shows the magnitude of the input reflection coefficient Γ as a function of frequency f. The phase velocity of the signal in the transmission lines is 2×10^8 m/sec.

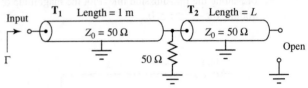

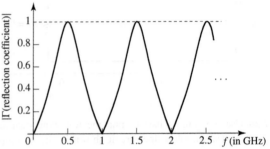

The length L (in meters) of T_2 is _____.

2.7 The propagation constant of a lossy transmission line is $(2 + j5)$ m^{-1} and its characteristic impedance is $(50 + j0)$ Ω at $\omega = 10^6$ rad/sec. The values of the line constants L, C, R, G are, respectively,

(a) $L = 200$ μH/m, $C = 0.1$ μF/m, $R = 50$ Ω/m, $G = 0.02$ S/m
(b) $L = 250$ μH/m, $C = 0.1$ μF/m, $R = 100$ Ω/m, $G = 0.04$ S/m
(c) $L = 200$ μH/m, $C = 0.2$ μF/m, $R = 100$ Ω/m, $G = 0.02$ S/m
(d) $L = 250$ μH/m, $C = 0.2$ μF/m, $R = 50$ Ω/m, $G = 0.04$ S/m

2.8 A coaxial cable is made of two brass conductors. The spacing between the conductors is filled with Teflon ($\epsilon_r = 2.1, \tan \delta = 0$). Which one of the following circuits can represent the lumped element model of a small piece of this cable having length Δz?

(a)

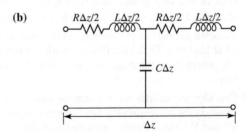

(b)

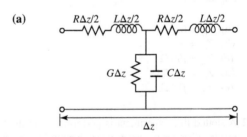

(c)

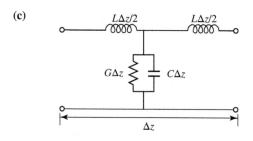

(d)

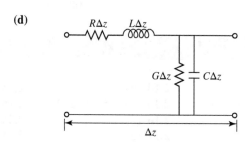

2.9 A 200 m long transmission line having parameters shown in the figure is terminated into a load R_L. The line is connected to a 400 V source having source resistance R_S through a switch which is closed at $t = 0$. The transient response of the circuit at the input of the line ($z = 0$) is also drawn in the figure. The value of R_L (in Ω) is _____.

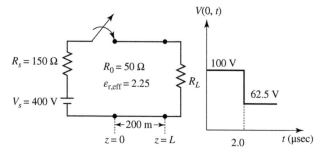

2.10 Consider the 3 m long lossless air-filled transmission line shown in the figure. It has a characteristic impedance of $120\pi\ \Omega$, is terminated by a short circuit, and is excited with a frequency of 37.5 MHz. What is the nature of the input impedance (Z_{in})?

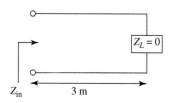

(a) open   **(c)** inductive
(b) short   **(d)** capacitive

2.11 Design a single section, quarter-wave impedance transformer at 5 GHz from 3.75 cm × 2 cm guide to

3.75 cm × 1 cm guide. Assume air-filled waveguide with transformer section having same width as that of the input and output sections.

(a) height = 1.414 cm, length = 3 cm
(b) height = 1.5 cm, length = 2.5 cm
(c) height = 1.414 cm, length = 2.5 cm
(d) height = 1.5 cm, length = 3 cm

2.12 An electromagnetic wave propagates through a lossless insulator with a velocity 1.5×10^{10} cm/sec. Calculate the electric and magnetic properties of the insulator if its intrinsic impedance is 90π ohms.

(a) $\epsilon_r = 2.66,\ \mu_r = 1.5$
(b) $\epsilon_r = 1.5,\ \mu_r = 2.66$
(c) $\epsilon_r = 1.2,\ \mu_r = 2$
(d) $\epsilon_r = 2,\ \mu_r = 1.2$

2.13 A transmission line with a characteristic impedance of $100\ \Omega$ is used to match a $50\ \Omega$ section to a $200\ \Omega$ section. If the matching is to be done both at 500 MHz and 1.2 GHz, the length of the transmission line can be approximately

(a) 1.75 m   **(c)** 1.35 m
(b) 1.0 m   **(d)** 1.5 m

2.14 If the reflected wave at the load of a transmission line is 20 dB below the incident wave, the SWR at the load is

(a) 1.5   **(c)** 3.0
(b) 1.22   **(d)** 4.0

2.15 A lossless transmission line with characteristic impedance $Z_0 = 50\ \Omega$ is 30 m long and operates at 2 MHz. The line is shorted at the load. If the phase velocity = 0.6 times the velocity of light, the input impedance of the line is

(a) $75\angle 90°\ \Omega$
(b) $\dfrac{50}{\sqrt{3}}\angle 180°\ \Omega$
(c) $\dfrac{100}{\sqrt{2}}\angle 180°\ \Omega$
(d) $\dfrac{75}{\sqrt{2}}\angle 270°\ \Omega$

2.16 In a two-wire transmission line, two consecutive voltage minima are found at 20.6 cm and 25.6 cm. The operating frequency is

(a) 3.5 GHz   **(c)** 1 GHz
(b) 3 GHz   **(d)** 2.5 GHz

2.17 TEM mode transmission line has distributed circuit parameters as $R = 1\ \Omega/m$, $L = 200$ nH/m, $G = 300\ \mu$S/m, $C = 60$ pF. The line is

(a) lossless   **(c)** distortionless
(b) lossy   **(d)** none of the above

2.18 A lossless transmission line has the distributed circuit parameters of inductance and capacitance per meter as 625 nH/m and 64 pF/m, respectively. The phase constant of the line at 100 MHz is

(a) 3.97 rad/m   **(c)** 1.56 rad/m
(b) 18.42 rad/m   **(d)** 9.21 rad/m

2.19 For a lossless line terminated in a short circuit, the stationary voltage minima and maxima are separated by

(a) $\dfrac{\lambda}{8}$

(c) $\dfrac{\lambda}{3}$

(b) $\dfrac{\lambda}{2}$

(d) $\dfrac{\lambda}{4}$

2.20 The characteristic impedance of an 80 cm long lossless transmission line having $L = 0.25\ \mu$H/m and $C = 100$ pF/m will be

(a) 25 Ω

(c) 50 Ω

(b) 40 Ω

(d) 80 Ω

2.21 It is required to match a 200 Ω load to a 300 Ω transmission line to reduce the SWR along the line to 1. If it is connected directly to the load, the characteristic impedance of the quarter-wave transformer used for this purpose will be

(a) 275 Ω

(c) 245 Ω

(b) 260 Ω

(d) 230 Ω

ANSWER KEY

2.1 (c)	**2.6** 0.09 to 0.11	**2.10** (d)	**2.14** (b)	**2.18** (a)
2.2 (c)	**2.7** (b)	**2.11** none; length = 1.5 cm	**2.15** none; $5\sqrt{3}\angle 270°$	**2.19** (d)
2.3 0.001	**2.8** (b)	**2.12** (a)	**2.16** (b)	**2.20** (c)
2.4 48.0 to 51.0	**2.9** 30	**2.13** (c)	**2.17** (c)	**2.21** (c)
2.5 0.85 to 0.88				

Learning Objectives

After completing this chapter, you will be able to

- Learn the concepts of TEM, TE, and TM waves and waveguides
- Design and develop mathematical models of parallel plate, rectangular, circular, and coaxial waveguides and their various design parameters

- Understand the concepts of stripline and microstrip line
- Learn the concepts of different types of wave velocities and dispersion

One of the early milestones in microwave engineering was the development of waveguide and other transmission lines for the low-loss transmission of power at high frequencies. Although Heaviside considered the possibility of propagation of electromagnetic waves inside a closed hollow tube in 1893, he rejected the idea because he believed that two conductors were necessary for the transfer of electromagnetic energy [1]. In 1897, Lord Rayleigh (John William Strutt) mathematically proved that wave propagation in waveguides was possible for both circular and rectangular cross sections [2]. Rayleigh also noted the infinite set of waveguide modes of the TE and TM type that were possible and the existence of a cutoff frequency, but no experimental verification was made at the time. The waveguide was then essentially forgotten until it was rediscovered independently in 1936 by two researchers [3]. After preliminary experiments in 1932, George C. Southworth of the AT&T Company in New York presented a paper on the waveguide in 1936. At the same meeting, W. L. Barrow of MIT presented a paper on the circular waveguide, with experimental confirmation of propagation.

Early RF and microwave systems relied on waveguides, two-wire lines, and coaxial lines for transmission. Waveguides have the advantage of high power-handling capability and low loss but are bulky and expensive, especially at low frequencies. Two-wire lines are inexpensive but lack shielding. Coaxial lines are shielded but are a difficult medium in which to fabricate complex microwave components. Planar transmission lines provide an alternative, in the form of stripline, microstrip lines, slotlines, coplanar waveguides, and several other types of related geometries. Such transmission lines are compact, low in cost, and capable of being easily integrated with active circuit devices, such as diodes and transistors, to form microwave integrated circuits. The first planar transmission line may have been a flat-strip coaxial line, similar to a stripline, used in a production power divider network in World War II [4], but planar lines did not see intensive development until the 1950s. Microstrip lines were developed at ITT laboratories [5] and were competitors of stripline. The first microstrip lines used a relatively thick dielectric substrate, which accentuated the non-TEM mode behavior and frequency dispersion of the line. This characteristic made it less desirable than stripline until the 1960s, when much thinner substrates began to be used. This reduced the frequency dependence of the line, and now microstrip lines are often the preferred medium for microwave integrated circuits.

In this chapter we will study the properties of several types of transmission lines and waveguides that are in common use. As we know from Chapter 2, a transmission line is characterized by a propagation constant, an attenuation constant, and a characteristic impedance. These quantities will be derived by field theory analysis for the various lines and waveguides treated here.

We begin with a discussion of the different types of wave propagation and modes that can exist on general transmission lines and waveguides. Transmission lines that consist of two or more conductors may support *transverse electromagnetic* (TEM) waves, characterized by the lack of longitudinal field components. Such lines have a uniquely defined voltage, current, and characteristic impedance. Waveguides, often consisting of a single conductor, support *transverse electric* (TE) and/or *transverse magnetic* (TM) waves, characterized by the presence of longitudinal magnetic or electric field components. As we will see in Chapter 4, a unique definition of characteristic impedance is not possible for such waves, although definitions can be chosen so that the characteristic impedance concept can be extended to waveguides with meaningful results.

3.1 GENERAL SOLUTIONS FOR TEM, TE, AND TM WAVES

In this section we will find general solutions to Maxwell's equations for the specific cases of TEM, TE, and TM wave propagation in cylindrical transmission lines or waveguides. The geometry of an arbitrary transmission line or waveguide is shown in Figure 3.1 and is characterized by conductor boundaries that are parallel to the z-axis. These structures are assumed to be uniform in shape and dimension in the z direction and infinitely long. The conductors will initially be assumed to be perfectly conducting, but attenuation can be found by the perturbation method discussed in Chapter 2.

We assume time-harmonic fields with an $e^{j\omega t}$ dependence and wave propagation along the z-axis. The electric and magnetic fields can then be written as

$$\bar{E}(x, y, z) = [\bar{e}(x, y) + \hat{z}e_z(x, y)]e^{-j\beta z}, \tag{3.1a}$$

$$\bar{H}(x, y, z) = [\bar{h}(x, y) + \hat{z}h_z(x, y)]e^{-j\beta z}, \tag{3.1b}$$

where $\bar{e}(x, y)$ and $\bar{h}(x, y)$ represent the transverse ($\hat{x}$, $\hat{y}$) electric and magnetic field components, and e_z and h_z are the longitudinal electric and magnetic field components. In (3.1) the wave is propagating in the $+z$ direction; $-z$ propagation can be obtained by replacing β with $-\beta$. In addition, if conductor or dielectric loss is present, the propagation constant will be complex; $j\beta$ should then be replaced with $\gamma = \alpha + j\beta$.

Assuming that the transmission line or waveguide region is source free, we can write Maxwell's equations as

$$\nabla \times \bar{E} = -j\omega\mu\bar{H}, \tag{3.2a}$$

$$\nabla \times \bar{H} = j\omega\epsilon\bar{E}. \tag{3.2b}$$

With an $e^{-j\beta z}$ z dependence, the three components of each of these vector equations can be reduced to the following:

$$\frac{\partial E_z}{\partial y} + j\beta E_y = -j\omega\mu H_x, \tag{3.3a}$$

$$-j\beta E_x - \frac{\partial E_z}{\partial x} = -j\omega\mu H_y, \tag{3.3b}$$

$$\frac{\partial E_y}{\partial x} - \frac{\partial E_x}{\partial y} = -j\omega\mu H_z, \tag{3.3c}$$

$$\frac{\partial H_z}{\partial y} + j\beta H_y = j\omega\epsilon E_x, \tag{3.4a}$$

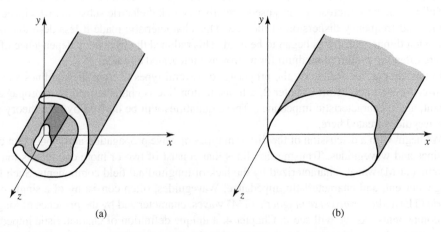

(a)   (b)

FIGURE 3.1 | (a) General two-conductor transmission line and (b) closed waveguide.

$$-j\beta H_x - \frac{\partial H_z}{\partial x} = j\omega\epsilon E_y, \tag{3.4b}$$

$$\frac{\partial H_y}{\partial x} - \frac{\partial H_x}{\partial y} = j\omega\epsilon E_z. \tag{3.4c}$$

These six equations can be solved for the four transverse field components in terms of E_z and H_z [e.g., H_x can be derived by eliminating E_y from (3.3a) and (3.4b)] as follows:

$$H_x = \frac{j}{k_c^2}\left(\omega\epsilon\frac{\partial E_z}{\partial y} - \beta\frac{\partial H_z}{\partial x}\right), \tag{3.5a}$$

$$H_y = \frac{-j}{k_c^2}\left(\omega\epsilon\frac{\partial E_z}{\partial x} + \beta\frac{\partial H_z}{\partial y}\right), \tag{3.5b}$$

$$E_x = \frac{-j}{k_c^2}\left(\beta\frac{\partial E_z}{\partial x} + \omega\mu\frac{\partial H_z}{\partial y}\right), \tag{3.5c}$$

$$E_y = \frac{j}{k_c^2}\left(-\beta\frac{\partial E_z}{\partial y} + \omega\mu\frac{\partial H_z}{\partial x}\right), \tag{3.5d}$$

where

$$k_c^2 = k^2 - \beta^2 \tag{3.6}$$

is defined as the *cutoff wave number*; the reason for this terminology will become clear later. As in previous chapters,

$$k = \omega\sqrt{\mu\epsilon} = 2\pi/\lambda \tag{3.7}$$

is the wave number of the material filling the transmission line or waveguide region. If dielectric loss is present, ϵ can be made complex by using $\epsilon = \epsilon_o\epsilon_r(1 - j\tan\delta)$, where $\tan\delta$ is the loss tangent of the material.

Equations (3.5a)–(3.5d) are general results that can be applied to a variety of waveguiding systems. We will now specialize these results to specific wave types.

TEM Waves

Transverse electromagnetic (TEM) waves are characterized by $E_z = H_z = 0$. Observe from (3.5) that if $E_z = H_z = 0$, then the transverse fields are also all zero, unless $k_c^2 = 0(k^2 = \beta^2)$, in which case we have an indeterminate result. However, we can return to (3.3)–(3.4) and apply the condition that $E_z = H_z = 0$. Then from (3.3a) and (3.4b), we can eliminate H_x to obtain

$$\beta^2 E_y = \omega^2\mu\epsilon E_y,$$

or

$$\beta = \omega\sqrt{\mu\epsilon} = k, \tag{3.8}$$

as noted earlier. [This result can also be obtained from (3.3b) and (3.4a).] The cutoff wave number, $k_c = \sqrt{k^2 - \beta^2}$, is thus zero for TEM waves.

The Helmholtz wave equation for E_x is, from (1.42),

$$\left(\frac{\partial^2}{\partial x^2} + \frac{\partial^2}{\partial y^2} + \frac{\partial^2}{\partial z^2} + k^2\right)E_x = 0, \tag{3.9}$$

but for $e^{-j\beta z}$ dependence, $(\partial^2/\partial z^2)E_x = -\beta^2 E_x = -k^2 E_x$, so (3.9) reduces to

$$\left(\frac{\partial^2}{\partial x^2} + \frac{\partial^2}{\partial y^2}\right)E_x = 0. \tag{3.10}$$

A similar result also applies to E_y, so using the form of $\bar{E}$ assumed in (3.1a), we can write

$$\nabla_t^2 \bar{e}(x, y) = 0, \tag{3.11}$$

where $\nabla_t^2 = \partial^2/\partial x^2 + \partial^2/\partial y^2$ is the Laplacian operator in the two transverse dimensions.

The result of (3.11) shows that the transverse electric fields, $\bar{e}(x, y)$, of a TEM wave satisfy Laplace's equation. It is easy to show in the same way that the transverse magnetic fields also satisfy Laplace's equation:

$$\nabla_t^2 \bar{h}(x, y) = 0. \tag{3.12}$$

The transverse fields of a TEM wave are thus the same as the static fields that can exist between the conductors. In the electrostatic case, we know that the electric field can be expressed as the gradient of a scalar potential, $\Phi(x, y)$:

$$\bar{e}(x, y) = -\nabla_t \Phi(x, y), \tag{3.13}$$

where $\nabla_t = \hat{x}(\partial/\partial x) + \hat{y}(\partial/\partial y)$ is the transverse gradient operator in two dimensions. For the relation in (3.13) to be valid, the curl of $\bar{e}$ must vanish, and this is indeed the case here since

$$\nabla_t \times \bar{e} = -j\omega\mu h_z \hat{z} = 0.$$

Using the fact that $\nabla \cdot \bar{D} = \epsilon \nabla_t \cdot \bar{e} = 0$ with (3.13) shows that $\Phi(x, y)$ also satisfies Laplace's equation,

$$\nabla_t^2 \Phi(x, y) = 0, \tag{3.14}$$

as expected from electrostatics. The voltage between two conductors can be found as

$$V_{12} = \Phi_1 - \Phi_2 = \int_1^2 \bar{E} \cdot d\bar{\ell}, \tag{3.15}$$

where Φ_1 and Φ_2 represent the potential at conductors 1 and 2, respectively. The current flow on a given conductor can be found from Ampere's law as

$$I = \oint_C \bar{H} \cdot d\bar{\ell}, \tag{3.16}$$

where C is the cross-sectional contour of the conductor.

The wave impedance of a TEM mode can be found as the ratio of the transverse electric and magnetic fields:

$$Z_{\text{TEM}} = \frac{E_x}{H_y} = \frac{\omega\mu}{\beta} = \sqrt{\frac{\mu}{\epsilon}} = \eta, \tag{3.17a}$$

where (3.4a) was used. The other pair of transverse field components, from (3.3a), gives

$$Z_{\text{TEM}} = \frac{-E_y}{H_x} = \sqrt{\frac{\mu}{\epsilon}} = \eta. \tag{3.17b}$$

Combining the results of (3.17a) and (3.17b) gives a general expression for the transverse fields as

$$\bar{h}(x, y) = \frac{1}{Z_{\text{TEM}}} \hat{z} \times \bar{e}(x, y). \tag{3.18}$$

Note that the wave impedance is the same as that for a plane wave in a lossless medium, as derived in Chapter 1; the reader should not confuse this impedance with the characteristic impedance, Z_0, of a transmission line. The latter relates traveling voltage and current and is a function of the line geometry as well as the material filling the line, while the wave impedance relates transverse field components and is dependent only on the material constants. From (2.32), the characteristic impedance of the TEM line is $Z_0 = V/I$, where V and I are the amplitudes of the incident voltage and current waves.

The procedure for analyzing a TEM line can be summarized as follows:

1. Solve Laplace's equation, (3.14), for $\Phi(x, y)$. The solution will contain several unknown constants.
2. Find these constants by applying the boundary conditions for the known voltages on the conductors.
3. Compute $\bar{e}$ and $\bar{E}$ from (3.13) and (3.1a). Compute $\bar{h}$ and $\bar{H}$ from (3.18) and (3.1b).
4. Compute V from (3.15) and I from (3.16).
5. The propagation constant is given by (3.8), and the characteristic impedance is given by $Z_0 = V/I$.

Impossibility of TEM Mode

TEM waves are characterized by $E_z = H_z = 0$ resulting from (3.5a)–(3.5d); the field components are $H_x = H_y = E_x = E_y = 0$. TEM waves cannot sustain in such waveguide environments.

TEM waves can exist when two or more conductors are present. Plane waves are also examples of TEM waves since there are no field components in the direction of propagation; in this case the transmission line conductors may be considered to be two infinitely large plates separated to infinity. The above results show that a closed conductor (such as a rectangular waveguide) cannot support TEM waves since the corresponding static potential in such a region would be zero (or possibly a constant), leading to $\bar{e} = 0$.

TEM mode is described by electric and magnetic fields which are perpendicular to each other and also perpendicular to the direction of propagation. The electric field E and the magnetic field H are supported by induced voltages and induced currents in the walls, respectively. For wave propagation, both the fields must reinforce each other. In transmission lines, two separate conductors are linked capacitively and inductively in order to have the voltage and current in the conductors to reinforce each other. The only way to propagate a TEM mode through a specific structure is by flowing a physical current through that particular structure. To produce a current, a physical voltage is required. Therefore, two conductors are essential for TEM mode propagation. Since waveguide is a single conductor, the fields cancel each other, resulting in impossibility of TEM mode propagation in a waveguide.

TE Waves

Transverse electric (TE) waves, (also referred to as H-waves) are characterized by $E_z = 0$ and $H_z \neq 0$. Equations (3.5) then reduce to

$$H_x = \frac{-j\beta}{k_c^2} \frac{\partial H_z}{\partial x}, \tag{3.19a}$$

$$H_y = \frac{-j\beta}{k_c^2} \frac{\partial H_z}{\partial y}, \tag{3.19b}$$

$$E_x = \frac{-j\omega\mu}{k_c^2} \frac{\partial H_z}{\partial y}, \tag{3.19c}$$

$$E_y = \frac{j\omega\mu}{k_c^2} \frac{\partial H_z}{\partial x}. \tag{3.19d}$$

In this case $k_c \neq 0$, and the propagation constant $\beta = \sqrt{k^2 - k_c^2}$ is generally a function of frequency and the geometry of the line or guide. To apply (3.19), one must first find H_z from the Helmholtz wave equation,

$$\left(\frac{\partial^2}{\partial x^2} + \frac{\partial^2}{\partial y^2} + \frac{\partial^2}{\partial z^2} + k^2 \right) H_z = 0, \tag{3.20}$$

which, since $H_z(x, y, z) = h_z(x, y)e^{-j\beta z}$, can be reduced to a two-dimensional wave equation for h_z:

$$\left(\frac{\partial^2}{\partial x^2} + \frac{\partial^2}{\partial y^2} + k_c^2 \right) h_z = 0, \tag{3.21}$$

since $k_c^2 = k^2 - \beta^2$. This equation must be solved subject to the boundary conditions of the specific guide geometry.

The TE wave impedance can be found as

$$Z_{TE} = \frac{E_x}{H_y} = \frac{-E_y}{H_x} = \frac{\omega\mu}{\beta} = \frac{k\eta}{\beta}, \tag{3.22}$$

which is seen to be frequency dependent. TE waves can be supported inside closed conductors, as well as between two or more conductors.

TM Waves

Transverse magnetic (TM) waves (also referred to as E-waves) are characterized by $E_z \neq 0$ and $H_z = 0$. Equations (3.5) then reduce to

$$H_x = \frac{j\omega\epsilon}{k_c^2} \frac{\partial E_z}{\partial y}, \tag{3.23a}$$

$$H_y = \frac{-j\omega\epsilon}{k_c^2} \frac{\partial E_z}{\partial x}, \tag{3.23b}$$

$$E_x = \frac{-j\beta}{k_c^2} \frac{\partial E_z}{\partial x}, \tag{3.23c}$$

$$E_y = \frac{-j\beta}{k_c^2} \frac{\partial E_z}{\partial y}. \tag{3.23d}$$

As in the TE case, $k_c \neq 0$, and the propagation constant $\beta = \sqrt{k^2 - k_c^2}$ is a function of frequency and the geometry of the line or guide. E_z is found from the Helmholtz wave equation,

$$\left(\frac{\partial^2}{\partial x^2} + \frac{\partial^2}{\partial y^2} + \frac{\partial^2}{\partial z^2} + k^2 \right) E_z = 0, \tag{3.24}$$

which, since $E_z(x, y, z) = e_z(x, y)e^{-j\beta z}$, can be reduced to a two-dimensional wave equation for e_z:

$$\left(\frac{\partial^2}{\partial x^2} + \frac{\partial^2}{\partial y^2} + k_c^2 \right) e_z = 0, \tag{3.25}$$

since $k_c^2 = k^2 - \beta^2$. This equation must be solved subject to the boundary conditions of the specific guide geometry.

The TM wave impedance can be found as

$$Z_{TM} = \frac{E_x}{H_y} = \frac{-E_y}{H_x} = \frac{\beta}{\omega\epsilon} = \frac{\beta\eta}{k}, \tag{3.26}$$

which is frequency dependent. As for TE waves, TM waves can be supported inside closed conductors, as well as between two or more conductors.

The procedure for analyzing TE and TM waveguides can be summarized as follows:

1. Solve the reduced Helmholtz equation, (3.21) or (3.25), for h_z or e_z. The solution will contain several unknown constants and the unknown cutoff wave number, k_c.
2. Use (3.19) or (3.23) to find the transverse fields from h_z or e_z.
3. Apply the boundary conditions to the appropriate field components to find the unknown constants and k_c.
4. The propagation constant is given by (3.6) and the wave impedance by (3.22) or (3.26).

Attenuation Due to Dielectric Loss

Attenuation in a transmission line or waveguide can be caused by either dielectric loss or conductor loss. If α_d is the attenuation constant due to dielectric loss and α_c is the attenuation constant due to conductor loss, then the total attenuation constant is $\alpha = \alpha_d + \alpha_c$.

Attenuation caused by conductor loss can be calculated using the perturbation method of Section 2.6; this loss depends on the field distribution in the guide and so must be evaluated separately for each type of transmission line or waveguide. However, if the line or guide is completely filled with a homogeneous dielectric, the attenuation due to a lossy dielectric material can be calculated from the propagation constant, and this result will apply to any guide or line with a homogeneous dielectric filling.

Thus, use of the complex permittivity allows the complex propagation constant to be written as

$$\gamma = \alpha_d + j\beta = \sqrt{k_c^2 - k^2}$$

$$= \sqrt{k_c^2 - \omega^2 \mu_0 \epsilon_0 \epsilon_r (1 - j \tan \delta)}. \tag{3.27}$$

In practice, most dielectric materials have small losses ($\tan \delta \ll 1$), and so this expression can be simplified by using the first two terms of the Taylor expansion,

$$\sqrt{a^2 + x^2} \simeq a + \frac{1}{2}\left(\frac{x^2}{a}\right), \quad \text{for } x \ll a.$$

Then (3.27) reduces to

$$\gamma = \sqrt{k_c^2 - k^2 + jk^2 \tan \delta}$$

$$\simeq \sqrt{k_c^2 - k^2} + \frac{jk^2 \tan \delta}{2\sqrt{k_c^2 - k^2}}$$

$$= \frac{k^2 \tan \delta}{2\beta} + j\beta, \tag{3.28}$$

since $\sqrt{k_c^2 - k^2} = j\beta$. In these results, $k = \omega\sqrt{\mu_0 \epsilon_0 \epsilon_r}$ is the (real) wave number in the absence of loss. Equation (3.28) shows that when the loss is small the phase constant β is unchanged, while the attenuation constant due to dielectric loss is given by

$$\alpha_d = \frac{k^2 \tan \delta}{2\beta} \text{ Np/m (TE or TM waves).} \tag{3.29}$$

This result applies to any TE or TM wave, as long as the guide is completely filled with the dielectric material. It can also be used for TEM lines, where $k_c = 0$, by letting $\beta = k$:

$$\alpha_d = \frac{k \tan \delta}{2}\text{Np/m (TEM waves).} \tag{3.30}$$

3.2   PARALLEL PLATE WAVEGUIDE

The parallel plate waveguide is the simplest type of guide that can support TM and TE modes; it can also support a TEM mode since it is formed from two flat conducting plates, or strips, as shown in Figure 3.2. Although it is an idealization, understanding the parallel plate guide can be useful because its operation is similar to that of many other waveguides. The parallel plate guide can also be useful for modeling the propagation of higher order modes in stripline.

In the geometry of the parallel plate waveguide of Figure 3.2, the strip width, W, is assumed to be much greater than the separation, d, so that fringing fields and any x variation can be ignored. A material with permittivity ϵ and permeability μ is assumed to fill the region between the two plates. We will derive solutions for TEM, TM, and TE waves.

TEM Modes

As discussed in Section 3.1, the TEM mode solution can be obtained by solving Laplace's equation, (3.14), for the electrostatic potential $\Phi(x, y)$ between the two plates. Thus,

$$\nabla_t^2 \Phi(x, y) = 0, \quad \text{for } 0 \leq x \leq W, \ 0 \leq y \leq d. \tag{3.31}$$

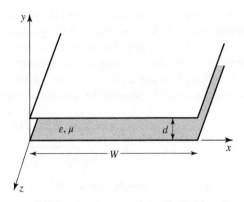

FIGURE 3.2 | Geometry of a parallel plate waveguide.

If we assume that the bottom plate is at ground (zero) potential and the top plate at a potential of V_o, then the boundary conditions for $\Phi(x, y)$ are

$$\Phi(x, 0) = 0, \tag{3.32a}$$

$$\Phi(x, d) = V_o. \tag{3.32b}$$

Because there is no variation in x, the general solution to (3.31) for $\Phi(x, y)$ is

$$\Phi(x, y) = A + By,$$

and the constants A, B can be evaluated from the boundary conditions of (3.32) to give the final solution as

$$\Phi(x, y) = V_o y / d. \tag{3.33}$$

The transverse electric field is, from (3.13),

$$\bar{e}(x, y) = -\nabla_t \Phi(x, y) = -\hat{y}\frac{V_o}{d}, \tag{3.34}$$

so that the total electric field is

$$\bar{E}(x, y, z) = \bar{e}(x, y)e^{-jkz} = -\hat{y}\frac{V_o}{d}e^{-jkz}, \tag{3.35}$$

where $k = \omega\sqrt{\mu\epsilon}$ is the propagation constant of the TEM wave, as in (3.8). The magnetic field, from (3.18), is

$$\bar{H}(x, y, z) = \bar{h}(x, y)\, e^{-jkz} = \frac{1}{\eta}\hat{z} \times \bar{E}(x, y, z) = \hat{x}\frac{V_o}{\eta d}e^{-jkz}, \tag{3.36}$$

where $\eta = \sqrt{\mu/\epsilon}$ is the intrinsic impedance of the medium between the parallel plates. Note that $E_z = H_z = 0$ and that the fields are similar in form to a plane wave in a homogeneous region.

The voltage of the top plate with respect to the bottom plate can be calculated from (3.15) and (3.35) as

$$V = -\int_{y=0}^{d} E_y \, dy = V_o e^{-jkz}, \tag{3.37}$$

as expected. The total current on the top plate can be found from Ampere's law or the surface current density:

$$I = \int_{x=0}^{W} \bar{J}_s \cdot \hat{z}\, dx = \int_{x=0}^{W} (-\hat{y} \times \bar{H}) \cdot \hat{z}\, dx = \int_{x=0}^{W} H_x dx = \frac{WV_o}{\eta d}e^{-jkz}. \tag{3.38}$$

Then the characteristic impedance is

$$Z_0 = \frac{V}{I} = \frac{\eta d}{W}, \tag{3.39}$$

which is seen to be a constant dependent only on the geometry and material parameters of the guide. The phase velocity is also a constant:

$$v_p = \frac{\omega}{\beta} = \frac{1}{\sqrt{\mu\epsilon}},$$ (3.40)

which is the speed of light in the material medium.

Attenuation due to dielectric loss is given by (3.30). The formula for conductor attenuation will be derived in the next subsection as a special case of TM mode attenuation.

TM Modes

As discussed in Section 3.1, TM waves are characterized by $H_z = 0$ and a nonzero E_z field that satisfies the reduced wave equation of (3.25), with $\partial/\partial x = 0$:

$$\left(\frac{\partial^2}{\partial y^2} + k_c^2 \right) e_z(x, y) = 0,$$ (3.41)

where $k_c = \sqrt{k^2 - \beta^2}$ is the cutoff wave number, and $E_z(x, y, z) = e_z(x, y)e^{-j\beta z}$. The general solution to (3.41) is of the form

$$e_z(x, y) = A \sin k_c y + B \cos k_c y,$$ (3.42)

subject to the boundary conditions that

$$e_z(x, y) = 0, \quad \text{at } y = 0, d.$$ (3.43)

This implies that $B = 0$ and $k_c d = n\pi$ for $n = 0, 1, 2, 3 \dots$, or

$$k_c = \frac{n\pi}{d}, \quad n = 0, 1, 2, 3, \dots.$$ (3.44)

Thus the cutoff wave number, k_c, is constrained to discrete values as given by (3.44); this implies that the propagation constant, β, is given by

$$\beta = \sqrt{k^2 - k_c^2} = \sqrt{k^2 - (n\pi/d)^2}.$$ (3.45)

The solution for $e_z(x, y)$ is then

$$e_z(x, y) = A_n \sin \frac{n\pi y}{d},$$ (3.46)

and thus,

$$E_z(x, y, z) = A_n \sin \frac{n\pi y}{d} e^{-j\beta z}.$$ (3.47)

The transverse field components can be found, using (3.23), to be

$$H_x = \frac{j\omega\epsilon}{k_c} A_n \cos \frac{n\pi y}{d} e^{-j\beta z},$$ (3.48a)

$$E_y = \frac{-j\beta}{k_c} A_n \cos \frac{n\pi y}{d} e^{-j\beta z},$$ (3.48b)

$$E_x = H_y = 0.$$ (3.48c)

Observe that for $n = 0$, $\beta = k = \omega\sqrt{\mu\epsilon}$, and that $E_z = 0$. The E_y and H_x fields are then constant in y, so that the TM_0 mode is actually identical to the TEM mode. For $n > 0$, however, the situation is different. Each value of n corresponds to a different TM mode, denoted as the TM_n mode, and each mode has its own propagation constant given by (3.45) and field expressions given by (3.48).

From (3.45) it can be seen that β is real only when $k > k_c$. Because $k = \omega\sqrt{\mu\epsilon}$ is proportional to frequency, the TM_n modes (for $n > 0$) exhibit a cutoff phenomenon, whereby no propagation will occur until the frequency is such that $k > k_c$. The *cutoff frequency* of the TM$_n$ mode can be found as

$$f_c = \frac{k_c}{2\pi\sqrt{\mu\epsilon}} = \frac{n}{2d\sqrt{\mu\epsilon}}. \tag{3.49}$$

Thus, the TM mode (for $n > 0$) that propagates at the lowest frequency is the TM$_1$ mode, with a cutoff frequency of $f_c = 1/2d\sqrt{\mu\epsilon}$; the TM$_2$ mode has a cutoff frequency equal to twice this value, and so on. At frequencies below the cutoff frequency of a given mode, the propagation constant is purely imaginary, corresponding to a rapid exponential decay of the fields. Such modes are referred to as *cutoff modes*, or *evanescent modes*. Because of the cutoff frequency, below which propagation cannot occur, waveguide mode propagation is analogous to a high-pass filter response.

The wave impedance of a TM mode, from (3.26), is a function of frequency:

$$Z_{\text{TM}} = \frac{-E_y}{H_x} = \frac{\beta}{\omega\epsilon} = \frac{\beta\eta}{k}, \tag{3.50}$$

which we see is pure real when $f > f_c$ but pure imaginary when $f < f_c$. The phase velocity is also a function of frequency:

$$v_p = \frac{\omega}{\beta}, \tag{3.51}$$

and is seen to be greater than $1/\sqrt{\mu\epsilon} = \omega/k$, the speed of light in the medium, since $\beta < k$. The guide wavelength is defined as

$$\lambda_g = \frac{2\pi}{\beta}, \tag{3.52}$$

and is the distance between equiphase planes along the z-axis. Note that $\lambda_g > \lambda = 2\pi/k$, the wavelength of a plane wave in the material. The phase velocity and guide wavelength are defined only for a propagating mode, for which β is real. One may also define a *cutoff wavelength* for the TM$_n$ mode as

$$\lambda_c = \frac{2d}{n}. \tag{3.53}$$

It is instructive to compute the Poynting vector to see how power propagates in the TM$_n$ mode. From (1.91), the time-average power passing a transverse cross section of the parallel plate guide is

$$P_o = \frac{1}{2}\text{Re}\int_{x=0}^{W}\int_{y=0}^{d}\bar{E}\times\bar{H}^*\cdot\hat{z}\,dy\,dx = -\frac{1}{2}\text{Re}\int_{x=0}^{W}\int_{y=0}^{d}E_y H_x^*\,dy\,dx$$

$$= \frac{W\text{Re}(\beta)\omega\epsilon}{2k_c^2}|A_n|^2\int_{y=0}^{d}\cos^2\frac{n\pi y}{d}\,dy = \begin{cases} \dfrac{W\text{Re}(\beta)\omega\epsilon d}{4k_c^2}|A_n|^2 & \text{for } n > 0 \\[12pt] \dfrac{W\text{Re}(\beta)\omega\epsilon d}{2k_c^2}|A_n|^2 & \text{for } n = 0 \end{cases}$$

$$\tag{3.54}$$

where (3.48a, b) were used for E_y, H_x. Thus, P_o is positive and nonzero when β is real, which occurs when $f > f_c$. When the mode is below cutoff, β is imaginary, and then $P_o = 0$.

TM (or TE) waveguide mode propagation has an interesting interpretation when viewed as a pair of bouncing plane waves. For example, consider the dominant TM$_1$ mode, which has propagation constant

$$\beta_1 = \sqrt{k^2 - (\pi/d)^2}, \tag{3.55}$$

and E_z field

$$E_z = A_1\sin\frac{\pi y}{d}e^{-j\beta_1 z},$$

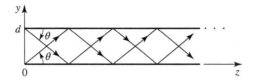

FIGURE 3.3 | Bouncing plane wave interpretation of the TM_1 parallel plate waveguide mode.

which can be rewritten as

$$E_z = \frac{A_1}{2j} \left[e^{j(\pi y/d - \beta_1 z)} - e^{-j(\pi y/d + \beta_1 z)} \right].$$ (3.56)

This result is in the form of two plane waves traveling obliquely in the $-y, +z$ and $+y, +z$ directions, respectively, as shown in Figure 3.3. By comparison with the phase factor of (1.132), the angle θ that each plane wave makes with the z-axis satisfies the relations

$$k \sin \theta = \frac{\pi}{d},$$ (3.57a)

$$k \cos \theta = \beta_1,$$ (3.57b)

so that $(\pi/d)^2 + \beta_1^2 = k^2$, as in (3.55). For $f > f_c$, β is real and less than k_1, so θ is some angle between $0°$ and $90°$, and the mode can be thought of as two plane waves alternately bouncing off of the top and bottom plates.

The phase velocity of each plane wave along its direction of propagation (θ direction) is $\omega/k = 1/\sqrt{\mu\epsilon}$, which is the speed of light in the material filling the guide. However, the phase velocity of the plane waves in the z direction is $\omega/\beta_1 = 1/\sqrt{\mu\epsilon} \cos \theta$, which is greater than the speed of light in the material. (This situation is analogous to ocean waves hitting a shoreline: the intersection point of the shore and an obliquely incident wave crest moves faster than the wave crest itself.) The superposition of the two plane wave fields is such that complete cancellation occurs at $y = 0$ and $y = d$, to satisfy the boundary condition that $E_z = 0$ at these planes. As f decreases to f_c, β_1 approaches zero, so that, by (3.57b), θ approaches $90°$. The two plane waves are then bouncing up and down, with no motion in the $+z$ direction, and no real power flow occurs in the z direction.

Attenuation due to dielectric loss can be found from (3.29). Conductor loss can be treated using the perturbation method. Thus,

$$\alpha_c = \frac{P_\ell}{2P_o},$$ (3.58)

where P_o is the power flow down the guide in the absence of conductor loss, as given by (3.54). P_ℓ is the power dissipated per unit length in the two lossy conductors and can be found from (2.91) as

$$P_\ell = 2\left(\frac{R_s}{2}\right) \int_{x=0}^{W} |\bar{J}_s|^2 \, dx = \frac{\omega^2 \epsilon^2 R_s W}{k_c^2} |A_n|^2,$$ (3.59)

where R_s is the surface resistivity of the conductors. Using (3.54) and (3.59) in (3.58) gives the attenuation due to conductor loss as

$$\alpha_c = \frac{2\omega\epsilon R_s}{\beta d} = \frac{2kR_s}{\beta\eta d} \text{ Np/m}, \quad \text{for } n > 0.$$ (3.60)

As discussed previously, the TEM mode is identical to the TM_0 mode for the parallel plate waveguide, so the above attenuation results for the TM_n mode can be used to obtain the TEM mode attenuation by letting $n = 0$. For this case, the $n = 0$ result of (3.54) must be used in (3.58), to obtain

$$\alpha_c = \frac{R_s}{\eta d} \text{ Np/m}.$$ (3.61)

TE Modes

TE modes, characterized by $E_z = 0$, can also propagate in a parallel plate waveguide. From (3.21), with $\partial/\partial x = 0$, H_z must satisfy the reduced wave equation,

$$\left(\frac{\partial^2}{\partial y^2} + k_c^2 \right) h_z(x, y) = 0, \tag{3.62}$$

where $k_c = \sqrt{k^2 - \beta^2}$ is the cutoff wave number and $H_z(x, y, z) = h_z(x, y)e^{-j\beta z}$. The general solution to (3.62) is

$$h_z(x, y) = A \sin k_c y + B \cos k_c y. \tag{3.63}$$

The boundary conditions are that $E_x = 0$ at $y = 0, d$; E_z is identically zero for TE modes. From (3.19c) we have

$$E_x = \frac{-j\omega\mu}{k_c} \left(A \cos k_c y - B \sin k_c y \right) e^{-j\beta z}, \tag{3.64}$$

and applying the boundary conditions shows that $A = 0$ and

$$k_c = \frac{n\pi}{d}, \quad n = 1, 2, 3 \ldots, \tag{3.65}$$

as for the TM case. The final solution for H_z is then

$$H_z(x, y) = B_n \cos \frac{n\pi y}{d} e^{-j\beta z}. \tag{3.66}$$

The transverse fields can be computed from (3.19) as

$$E_x = \frac{j\omega\mu}{k_c} B_n \sin \frac{n\pi y}{d} e^{-j\beta z}, \tag{3.67a}$$

$$H_y = \frac{j\beta}{k_c} B_n \sin \frac{n\pi y}{d} e^{-j\beta z}, \tag{3.67b}$$

$$E_y = H_x = 0. \tag{3.67c}$$

The propagation constant of the TE_n mode is given as

$$\beta = \sqrt{k^2 - \left(\frac{n\pi}{d} \right)^2}, \tag{3.68}$$

which is the same as the propagation constant of the TM_n mode. The cutoff frequency of the TE_n mode is

$$f_c = \frac{n}{2d\sqrt{\mu\epsilon}}, \tag{3.69}$$

which is also identical to that of the TM_n mode. The wave impedance of the TE_n mode is, from (3.22),

$$Z_{\text{TE}} = \frac{E_x}{H_y} = \frac{\omega\mu}{\beta} = \frac{k\eta}{\beta}, \tag{3.70}$$

which is seen to be real for propagating modes and imaginary for nonpropagating, or cutoff, modes. The phase velocity, guide wavelength, and cutoff wavelength are similar to the results obtained for the TM modes.

The power flow down the guide for a TE_n mode can be calculated as

$$P_o = \frac{1}{2}\text{Re} \int_{x=0}^{W} \int_{y=0}^{d} \bar{E} \times \bar{H}^* \cdot \hat{z}\, dy\, dx = \frac{1}{2}\text{Re} \int_{x=0}^{W} \int_{y=0}^{d} E_x H_y^*\, dy\, dx$$

$$= \frac{\omega\mu dW}{4k_c^2} |B_n|^2 \text{Re}(\beta), \qquad \text{for } n > 0, \tag{3.71}$$

which is zero if the operating frequency is below the cutoff frequency (β imaginary).

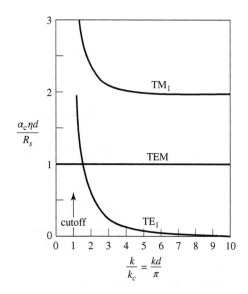

FIGURE 3.4 | Attenuation due to conductor loss for the TEM, TM_1, and TE_1 modes of a parallel plate waveguide.

Note that if $n = 0$, then $E_x = H_y = 0$ from (3.67), and thus $P_o = 0$, implying that there is no TE_0 mode.

Attenuation can be calculated in the same way as for the TM modes. The attenuation due to dielectric loss is given by (3.29). It is left as a problem to show that the attenuation due to conductor loss for TE modes is given by

$$\alpha_c = \frac{2k_c^2 R_s}{\omega \mu \beta d} = \frac{2k_c^2 R_s}{k \beta \eta d} \text{ Np/m.} \tag{3.72}$$

Figure 3.4 shows attenuation versus frequency due to conductor loss for the TEM, TM_1, and TE_1 modes. Observe that $\alpha_c \to \infty$ as cutoff is approached for the TM and TE modes.

Table 3.1 summarizes a number of useful results for parallel plate waveguide modes. Field lines for the TEM, TM_1, and TE_1 modes are shown in Figure 3.5.

TABLE 3.1 | Summary of Results for Parallel Plate Waveguide

Quantity	TEM Mode	TM_n Mode	TE_n Mode
k	$\omega\sqrt{\mu\epsilon}$	$\omega\sqrt{\mu\epsilon}$	$\omega\sqrt{\mu\epsilon}$
k_c	0	$n\pi/d$	$n\pi/d$
β	$k = \omega\sqrt{\mu\epsilon}$	$\sqrt{k^2 - k_c^2}$	$\sqrt{k^2 - k_c^2}$
λ_c	∞	$2\pi/k_c = 2d/n$	$2\pi/k_c = 2d/n$
λ_g	$2\pi/k$	$2\pi/\beta$	$2\pi/\beta$
v_p	$\omega/k = 1/\sqrt{\mu\epsilon}$	ω/β	ω/β
α_d	$(k\tan\delta)/2$	$(k^2\tan\delta)/2\beta$	$(k^2\tan\delta)/2\beta$
α_c	$R_s/\eta d$	$2kR_s/\beta\eta d$	$2k_c^2 R_s/k\beta\eta d$
E_z	0	$A\sin(n\pi y/d)e^{-j\beta z}$	0
H_z	0	0	$B\cos(n\pi y/d)e^{-j\beta z}$
E_x	0	0	$(j\omega\mu/k_c)B\sin(n\pi y/d)e^{-j\beta z}$
E_y	$(-V_o/d)e^{-j\beta z}$	$(-j\beta/k_c)A\cos(n\pi y/d)e^{-j\beta z}$	0
H_x	$(V_o/\eta d)e^{-j\beta z}$	$(j\omega\epsilon/k_c)A\cos(n\pi y/d)e^{-j\beta z}$	0
H_y	0	0	$(j\beta/k_c)B_n\sin(n\pi y/d)e^{-j\beta z}$
Z	$Z_{\text{TEM}} = \eta d/W$	$Z_{\text{TM}} = \beta\eta/k$	$Z_{\text{TE}} = k\eta/\beta$

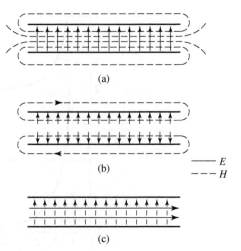

———— E
----- H

FIGURE 3.5 | Field lines for the (a) TEM, (b) TM_1, and (c) TE_1 modes of a parallel plate waveguide. There is no variation across the width of the waveguide.

EXAMPLE 3.1   CHARACTERISTICS OF A PARALLEL PLATE WAVEGUIDE

Consider a dielectric-filled parallel plate waveguide with $a = 2$ cm. The permeability of the dielectric filling is μ_0 and its refractive index is $n = 1.5$. (a) Which TE_m and TM_m modes can propagate a 12 GHz signal in the waveguide? (b) What would be the associated cutoff wavelengths in each case? (c) What would be the associated group velocities in each case? Assume a modulated 12 GHz carrier with a narrow modulation bandwidth.

Solution
Unguided propagation velocity for the dielectric filling the waveguide is,

$$v = \frac{c}{n} = 2 \times 10^8 \text{ m/sec}$$

Using v in place of c in the cutoff frequency formula for TE_m and TM_m modes, we find

$$f_c = \frac{mv}{2a} = 5m \text{ GHz}$$

$f = 12$ GHz exceeds the cutoff frequencies of TE_m and TM_m for $m = 1$ and 2, but not 3. Therefore, the propagating modes at $f = 12$ GHz are TM_0, TE_1, TM_1, TE_2, and TM_2.
   Cutoff wavelength,

$$\lambda_c = \frac{2a}{m}$$

and do not depend on the dielectric filling. They are (with $a = 2$ cm) $\lambda_c = 4$ cm for $TE_1 = TM_1$ and $\lambda_c = 2$ cm for $TE_2 = TM_2$. The cutoff wavelength is ∞ for TEM mode (which does not have a cutoff condition).
   Group velocities,

$$v_g = v\sqrt{1 - \frac{f_c^2}{f^2}}$$

where $v = c/n$. For the nondispersive TEM mode with $f_c = 0$, the group velocity is $v_g = 2 \times 10^8$ m/sec. For TE_1 and TM_1 modes,

$$v_g = 1.82 \times 10^8 \text{ m/sec}$$

For TE_2 and TM_2 modes,

$$v_g = 1.11 \times 10^8 \text{ m/sec}$$

3.3   RECTANGULAR WAVEGUIDE

Rectangular waveguides were one of the earliest types of transmission lines used to transport microwave signals, and they are still used for many applications. A large variety of components such as couplers, detectors, isolators, attenuators, and slotted lines are commercially available for various standard waveguide bands from 1 to 220 GHz. Figure 3.6 shows some of the standard rectangular waveguide components that are available. Because of the trend toward miniaturization and integration, most modern microwave circuitry is fabricated using planar transmission lines such as microstrips and stripline rather than waveguides. There is, however, still a need for waveguides in many cases, including high-power systems, millimeter wave applications, satellite systems, and some precision test applications.

The hollow rectangular waveguide can propagate TM and TE modes but not TEM waves since only one conductor is present. We will see that the TM and TE modes of a rectangular waveguide have cutoff frequencies below which propagation is not possible, similar to the TM and TE modes of the parallel plate guide.

TE Modes

The geometry of a rectangular waveguide is shown in Figure 3.7, where it is assumed that the guide is filled with a material of permittivity ϵ and permeability μ. It is standard convention to have the longest side of the waveguide along the x-axis, so that $a > b$.

TE waveguide modes are characterized by fields with $E_z = 0$, while H_z must satisfy the reduced wave equation of (3.21):

$$\left(\frac{\partial^2}{\partial x^2} + \frac{\partial^2}{\partial y^2} + k_c^2 \right) h_z(x, y) = 0, \tag{3.73}$$

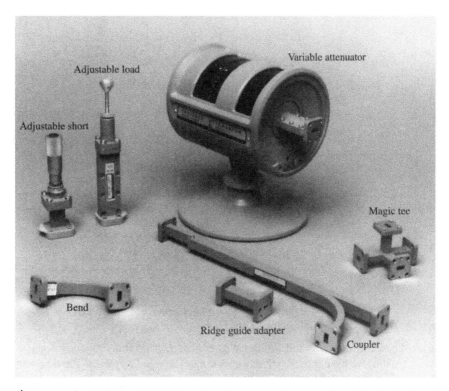

FIGURE 3.6 | Photograph of Ka-band (WR-28) rectangular waveguide components. Clockwise from top: a variable attenuator, an E-H (magic) tee junction, a directional coupler, an adaptor to ridge waveguide, an E-plane swept bend, an adjustable short, and a sliding matched load.

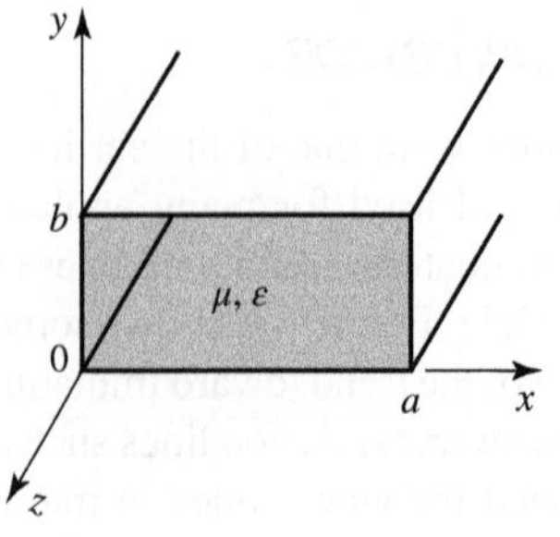

FIGURE 3.7 | Geometry of a rectangular waveguide.

with $H_z(x, y, z) = h_z(x, y)e^{-j\beta z}$; here $k_c = \sqrt{k^2 - \beta^2}$ is the cutoff wave number. The partial differential equation (3.73) can be solved by the method of separation of variables by letting

$$h_z(x, y) = X(x)Y(y) \tag{3.74}$$

and substituting into (3.73) to obtain

$$\frac{1}{X}\frac{d^2X}{dx^2} + \frac{1}{Y}\frac{d^2Y}{dy^2} + k_c^2 = 0. \tag{3.75}$$

Then, by the usual separation-of-variables argument (see Section 1.5), each of the terms in (3.75) must be equal to a constant, so we define separation constants k_x and k_y such that

$$\frac{d^2X}{dx^2} + k_x^2 X = 0, \tag{3.76a}$$

$$\frac{d^2Y}{dy^2} + k_y^2 Y = 0, \tag{3.76b}$$

and

$$k_x^2 + k_y^2 = k_c^2. \tag{3.77}$$

The general solution for h_z can then be written as

$$h_z(x, y) = (A\cos k_x x + B\sin k_x x)(C\cos k_y y + D\sin k_y y). \tag{3.78}$$

To evaluate the constants in (3.78) we must apply the boundary conditions on the electric field components tangential to the waveguide walls. That is,

$$e_x(x, y) = 0, \qquad \text{at } y = 0, b, \tag{3.79a}$$

$$e_y(x, y) = 0, \qquad \text{at } x = 0, a. \tag{3.79b}$$

We therefore cannot use h_z of (3.78) directly but must first use (3.19c) and (3.19d) to find e_x and e_y from h_z:

$$e_x = \frac{-j\omega\mu}{k_c^2}k_y(A\cos k_x x + B\sin k_x x)(-C\sin k_y y + D\cos k_y y), \tag{3.80a}$$

$$e_y = \frac{j\omega\mu}{k_c^2}k_x(-A\sin k_x x + B\cos k_x x)(C\cos k_y y + D\sin k_y y). \tag{3.80b}$$

Then from (3.79a) and (3.80a) we see that $D = 0$, and $k_y = n\pi/b$ for $n = 0, 1, 2\ldots$. From (3.79b) and (3.80b) we have that $B = 0$ and $k_x = m\pi/a$ for $m = 0, 1, 2\ldots$. The final solution for H_z is then

$$H_z(x, y, z) = A_{mn}\cos\frac{m\pi x}{a}\cos\frac{n\pi y}{b}e^{-j\beta z}, \tag{3.81}$$

where A_{mn} is an arbitrary amplitude constant composed of the remaining constants A and C of (3.78).

The transverse field components of the TE_{mn} mode can be found using (3.19) and (3.81):

$$E_x = \frac{j\omega\mu n\pi}{k_c^2 b} A_{mn} \cos\frac{m\pi x}{a} \sin\frac{n\pi y}{b} e^{-j\beta z}, \tag{3.82a}$$

$$E_y = \frac{-j\omega\mu m\pi}{k_c^2 a} A_{mn} \sin\frac{m\pi x}{a} \cos\frac{n\pi y}{b} e^{-j\beta z}, \tag{3.82b}$$

$$H_x = \frac{j\beta m\pi}{k_c^2 a} A_{mn} \sin\frac{m\pi x}{a} \cos\frac{n\pi y}{b} e^{-j\beta z}, \tag{3.82c}$$

$$H_y = \frac{j\beta n\pi}{k_c^2 b} A_{mn} \cos\frac{m\pi x}{a} \sin\frac{n\pi y}{b} e^{-j\beta z}. \tag{3.82d}$$

The propagation constant is

$$\beta = \sqrt{k^2 - k_c^2} = \sqrt{k^2 - \left(\frac{m\pi}{a}\right)^2 - \left(\frac{n\pi}{b}\right)^2}, \tag{3.83}$$

which is seen to be real, corresponding to a propagating mode, when

$$k > k_c = \sqrt{\left(\frac{m\pi}{a}\right)^2 + \left(\frac{n\pi}{b}\right)^2}.$$

Each mode (each combination of m and n) has a cutoff frequency $f_{c_{mn}}$ given by

$$f_{c_{mn}} = \frac{k_c}{2\pi\sqrt{\mu\epsilon}} = \frac{1}{2\pi\sqrt{\mu\epsilon}} \sqrt{\left(\frac{m\pi}{a}\right)^2 + \left(\frac{n\pi}{b}\right)^2}. \tag{3.84}$$

The mode with the lowest cutoff frequency is called the dominant mode; because we have assumed $a > b$, the lowest cutoff frequency occurs for the $\text{TE}_{10}(m = 1, n = 0)$ mode:

$$f_{c_{10}} = \frac{1}{2a\sqrt{\mu\epsilon}}. \tag{3.85}$$

Thus the TE_{10} mode is the dominant TE mode and, as we will see, the overall dominant mode of the rectangular waveguide. Observe that the field expressions for $\bar{E}$ and $\bar{H}$ in (3.82) are all zero if both $m = n = 0$; there is no TE_{00} mode.

At a given operating frequency f only those modes having $f > f_c$ will propagate; modes with $f < f_c$ will lead to an imaginary β (or real α), meaning that all field components will decay exponentially away from the source of excitation. Such modes are referred to as *cutoff modes*, or *evanescent modes*. If more than one mode is propagating, the waveguide is said to be *overmoded*.

From (3.22) the wave impedance that relates the transverse electric and magnetic fields is

$$Z_{\text{TE}} = \frac{E_x}{H_y} = \frac{-E_y}{H_x} = \frac{k\eta}{\beta}, \tag{3.86}$$

where $\eta = \sqrt{\mu/\epsilon}$ is the intrinsic impedance of the material filling the waveguide. Note that Z_{TE} is real when β is real (a propagating mode) but is imaginary when β is imaginary (a cutoff mode).

The guide wavelength is defined as the distance between two equal-phase planes along the waveguide and is equal to

$$\lambda_g = \frac{2\pi}{\beta} > \frac{2\pi}{k} = \lambda, \tag{3.87}$$

which is thus greater than λ, the wavelength of a plane wave in the medium filling the guide. The phase velocity is

$$v_p = \frac{\omega}{\beta} > \frac{\omega}{k} = 1/\sqrt{\mu\epsilon}, \tag{3.88}$$

which is greater than $1/\sqrt{\mu\epsilon}$, the speed of light (plane wave) in the medium.

In the vast majority of waveguide applications the operating frequency and guide dimensions are chosen so that only the dominant TE_{10} mode will propagate. Because of the practical importance of the TE_{10} mode, we will list the field components and derive the attenuation due to conductor loss for this case.

Specializing (3.81) and (3.82) to the $m = 1, \ n = 0$ case gives the following results for the TE_{10} mode fields:

$$H_z = A_{10} \cos \frac{\pi x}{a} e^{-j\beta z}, \tag{3.89a}$$

$$E_y = \frac{-j\omega\mu a}{\pi} A_{10} \sin \frac{\pi x}{a} e^{-j\beta z}, \tag{3.89b}$$

$$H_x = \frac{j\beta a}{\pi} A_{10} \sin \frac{\pi x}{a} e^{-j\beta z}, \tag{3.89c}$$

$$E_x = E_z = H_y = 0. \tag{3.89d}$$

The cutoff wave number and propagation constant for the TE_{10} mode are, respectively,

$$k_c = \pi/a, \tag{3.90}$$

$$\beta = \sqrt{k^2 - (\pi/a)^2}. \tag{3.91}$$

The power flow down the guide for the TE_{10} mode can be calculated as

$$
\begin{aligned}
P_{10} &= \frac{1}{2} \text{Re} \int_{x=0}^{a} \int_{y=0}^{b} \bar{E} \times \bar{H}^* \cdot \hat{z} \, dy \, dx \\
&= \frac{1}{2} \text{Re} \int_{x=0}^{a} \int_{y=0}^{b} E_y H_x^* \, dy \, dx \\
&= \frac{\omega\mu a^2}{2\pi^2} \text{Re}(\beta)|A_{10}|^2 \int_{x=0}^{a} \int_{y=0}^{b} \sin^2 \frac{\pi x}{a} \, dy \, dx \\
&= \frac{\omega\mu a^3 |A_{10}|^2 b}{4\pi^2} \text{Re}(\beta).
\end{aligned}
\tag{3.92}
$$

Note that this result gives nonzero real power only when β is real, corresponding to a propagating mode.

Attenuation in a rectangular waveguide may occur due to dielectric loss or conductor loss. Dielectric loss can be treated by making ϵ complex and using the general result given in (3.29). Conductor loss is best treated using the perturbation method. The power lost per unit length due to finite wall conductivity is, from (1.131),

$$P_\ell = \frac{R_s}{2} \int_C |\bar{J}_s|^2 d\ell, \tag{3.93}$$

where R_s is the wall surface resistance, and the integration contour C encloses the inside perimeter of the guide walls. There are surface currents on all four walls, but from symmetry the currents on the top and bottom walls are identical, as are the currents on the left and right side walls. So we can compute the power lost in the walls at $x = 0$ and $y = 0$ and double their sum to obtain the total power loss. The surface current on the $x = 0$ (left) wall is

$$\bar{J}_s = \hat{n} \times \bar{H}|_{x=0} = \hat{x} \times \hat{z} H_z|_{x=0} = -\hat{y} H_z|_{x=0} = -\hat{y} A_{10} e^{-j\beta z}, \tag{3.94a}$$

and the surface current on the $y = 0$ (bottom) wall is

$$
\begin{aligned}
\bar{J}_s &= \hat{n} \times \bar{H}|_{y=0} = \hat{y} \times (\hat{x} H_x|_{y=0} + \hat{z} H_z|_{y=0}) \\
&= -\hat{z} \frac{j\beta a}{\pi} A_{10} \sin \frac{\pi x}{a} e^{-j\beta z} + \hat{x} A_{10} \cos \frac{\pi x}{a} e^{-j\beta z}.
\end{aligned}
\tag{3.94b}
$$

Substituting (3.94) into (3.93) gives

$$P_\ell = R_s \int_{y=0}^{b} |J_{sy}|^2 dy + R_s \int_{x=0}^{a} \left[|J_{sx}|^2 + |J_{sz}|^2 \right] dx$$

$$= R_s |A_{10}|^2 \left(b + \frac{a}{2} + \frac{\beta^2 a^3}{2\pi^2} \right). \tag{3.95}$$

The attenuation due to conductor loss for the TE_{10} mode is then

$$\alpha_c = \frac{P_\ell}{2P_{10}} = \frac{2\pi^2 R_s (b + a/2 + \beta^2 a^3/2\pi^2)}{\omega \mu a^3 b \beta}$$

$$= \frac{R_s}{a^3 b \beta k \eta} (2b\pi^2 + a^3 k^2) \text{ Np/m}. \tag{3.96}$$

TM Modes

TM modes are characterized by fields with $H_z = 0$, while E_z must satisfy the reduced wave equation (3.25):

$$\left(\frac{\partial^2}{\partial x^2} + \frac{\partial^2}{\partial y^2} + k_c^2 \right) e_z(x, y) = 0, \tag{3.97}$$

with $E_z(x, y, z) = e_z(x, y)e^{-j\beta z}$ and $k_c^2 = k^2 - \beta^2$. Equation (3.97) can be solved by the separation-of-variables procedure that was used for TE modes. The general solution is

$$e_z(x, y) = (A \cos k_x x + B \sin k_x x)(C \cos k_y y + D \sin k_y y). \tag{3.98}$$

The boundary conditions can be applied directly to e_z:

$$e_z(x, y) = 0, \qquad \text{at } x = 0, a, \tag{3.99a}$$

$$e_z(x, y) = 0, \qquad \text{at } y = 0, b. \tag{3.99b}$$

We will see that satisfaction of these conditions on e_z will lead to satisfaction of the boundary conditions by e_x and e_y.

Applying (3.99a) to (3.98) shows that $A = 0$ and $k_x = m\pi/a$ for $m = 1, 2, 3\ldots$. Similarly, applying (3.99b) to (3.98) shows that $C = 0$ and $k_y = n\pi/b$ for $n = 1, 2, 3\ldots$. The solution for E_z then reduces to

$$E_z(x, y, z) = B_{mn} \sin \frac{m\pi x}{a} \sin \frac{n\pi y}{b} e^{-j\beta z}, \tag{3.100}$$

where B_{mn} is an arbitrary amplitude constant.

The transverse field components for the TM_{mn} mode can be computed from (3.23) and (3.100) as

$$E_x = \frac{-j\beta m\pi}{a k_c^2} B_{mn} \cos \frac{m\pi x}{a} \sin \frac{n\pi y}{b} e^{-j\beta z}, \tag{3.101a}$$

$$E_y = \frac{-j\beta n\pi}{b k_c^2} B_{mn} \sin \frac{m\pi x}{a} \cos \frac{n\pi y}{b} e^{-j\beta z}, \tag{3.101b}$$

$$H_x = \frac{j\omega\epsilon n\pi}{b k_c^2} B_{mn} \sin \frac{m\pi x}{a} \cos \frac{n\pi y}{b} e^{-j\beta z}, \tag{3.101c}$$

$$H_y = \frac{-j\omega\epsilon m\pi}{a k_c^2} B_{mn} \cos \frac{m\pi x}{a} \sin \frac{n\pi y}{b} e^{-j\beta z}. \tag{3.101d}$$

As for the TE modes, the propagation constant is

$$\beta = \sqrt{k^2 - k_c^2} = \sqrt{k^2 - \left(\frac{m\pi}{a} \right)^2 - \left(\frac{n\pi}{b} \right)^2} \tag{3.102}$$

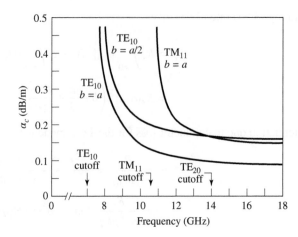

FIGURE 3.8 | Attenuation of various modes in a rectangular brass waveguide with $a = 2.0$ cm.

and is real for propagating modes and imaginary for cutoff modes. The cutoff frequencies for the TM_{mn} modes are also the same as those of the TE_{mn} modes, as given in (3.84). The guide wavelength and phase velocity for TM modes are also the same as those for TE modes.

Observe that the field expressions for $\bar{E}$ and $\bar{H}$ in (3.101) are identically zero if either m or n is zero. Thus there is no TM_{00}, TM_{01}, or TM_{10} mode, and the lowest order TM mode to propagate (lowest f_c) is the TM_{11} mode, having a cutoff frequency of

$$f_{c_{11}} = \frac{1}{2\pi\sqrt{\mu\epsilon}}\sqrt{\left(\frac{\pi}{a}\right)^2 + \left(\frac{\pi}{b}\right)^2}, \tag{3.103}$$

which is seen to be larger than $f_{c_{10}}$, the cutoff frequency of the TE_{10} mode.

The wave impedance relating the transverse electric and magnetic fields for TM modes is, from (3.26),

$$Z_{TM} = \frac{E_x}{H_y} = \frac{-E_y}{H_x} = \frac{\beta\eta}{k}. \tag{3.104}$$

Attenuation due to dielectric loss is computed in the same way as for TE modes, with the same result. The calculation of attenuation due to conductor loss is left as a problem. Figure 3.8 shows attenuation versus frequency for some TE and TM modes in a rectangular waveguide. Table 3.2 summarizes results for TE and TM wave propagation in rectangular waveguides, and Figure 3.9 shows the field lines for several of the lowest order TE and TM modes.

EXAMPLE 3.2 CHARACTERISTICS OF A RECTANGULAR WAVEGUIDE

Consider a length of Teflon-filled, copper K-band rectangular waveguide having dimensions $a = 1.07$ cm and $b = 0.43$ cm. Find the cutoff frequencies of the first five propagating modes. If the operating frequency is 15 GHz, find the attenuation due to dielectric and conductor losses.

Solution
From Appendix H, for Teflon, $\epsilon_r = 2.08$ and $\tan\delta = 0.0004$. From (3.84) the cutoff frequencies are given by

$$f_{c_{mn}} = \frac{c}{2\pi\sqrt{\epsilon_r}}\sqrt{\left(\frac{m\pi}{a}\right)^2 + \left(\frac{n\pi}{b}\right)^2}.$$

TABLE 3.2 | **Summary of Results for Rectangular Waveguide**

Quantity	TE_{mn} Mode	TM_{mn} Mode
k	$\omega\sqrt{\mu\epsilon}$	$\omega\sqrt{\mu\epsilon}$
k_c	$\sqrt{(m\pi/a)^2 + (n\pi/b)^2}$	$\sqrt{(m\pi/a)^2 + (n\pi/b)^2}$
β	$\sqrt{k^2 - k_c^2}$	$\sqrt{k^2 - k_c^2}$
λ_c	$\dfrac{2\pi}{k_c}$	$\dfrac{2\pi}{k_c}$
λ_g	$\dfrac{2\pi}{\beta}$	$\dfrac{2\pi}{\beta}$
v_p	$\dfrac{\omega}{\beta}$	$\dfrac{\omega}{\beta}$
α_d	$\dfrac{k^2 \tan\delta}{2\beta}$	$\dfrac{k^2 \tan\delta}{2\beta}$
E_z	0	$B\sin\dfrac{m\pi x}{a}\sin\dfrac{n\pi y}{b}e^{-j\beta z}$
H_z	$A\cos\dfrac{m\pi x}{a}\cos\dfrac{n\pi y}{b}e^{-j\beta z}$	0
E_x	$\dfrac{j\omega\mu n\pi}{k_c^2 b}A\cos\dfrac{m\pi x}{a}\sin\dfrac{n\pi y}{b}e^{-j\beta z}$	$\dfrac{-j\beta m\pi}{k_c^2 a}B\cos\dfrac{m\pi x}{a}\sin\dfrac{n\pi y}{b}e^{-j\beta z}$
E_y	$\dfrac{-j\omega\mu m\pi}{k_c^2 a}A\sin\dfrac{m\pi x}{a}\cos\dfrac{n\pi y}{b}e^{-j\beta z}$	$\dfrac{-j\beta n\pi}{k_c^2 b}B\sin\dfrac{m\pi x}{a}\cos\dfrac{n\pi y}{b}e^{-j\beta z}$
H_x	$\dfrac{j\beta m\pi}{k_c^2 a}A\sin\dfrac{m\pi x}{a}\cos\dfrac{n\pi y}{b}e^{-j\beta z}$	$\dfrac{j\omega\epsilon n\pi}{k_c^2 b}B\sin\dfrac{m\pi x}{a}\cos\dfrac{n\pi y}{b}e^{-j\beta z}$
H_y	$\dfrac{j\beta n\pi}{k_c^2 b}A\cos\dfrac{m\pi x}{a}\sin\dfrac{n\pi y}{b}e^{-j\beta z}$	$\dfrac{-j\omega\epsilon m\pi}{k_c^2 a}B\cos\dfrac{m\pi x}{a}\sin\dfrac{n\pi y}{b}e^{-j\beta z}$
Z	$Z_{TE} = \dfrac{k\eta}{\beta}$	$Z_{TM} = \dfrac{\beta\eta}{k}$

Computing f_c for the first few values of m and n gives the following results:

Mode	m	n	f_c(GHz)
TE	1	0	9.72
TE	2	0	19.44
TE	0	1	24.19
TE, TM	1	1	26.07
TE, TM	2	1	31.03

Thus the TE_{10}, TE_{20}, TE_{01}, TE_{11}, and TM_{11} modes will be the first five modes to propagate. At 15 GHz, $k = 453.1$ m^{-1}, and the propagation constant for the TE_{10} mode is

$$\beta = \sqrt{\left(\frac{2\pi f\sqrt{\epsilon_r}}{c}\right)^2 - \left(\frac{\pi}{a}\right)^2} = \sqrt{k^2 - \left(\frac{\pi}{a}\right)^2} = 345.1 \text{ m}^{-1}.$$

From (3.29), the attenuation due to dielectric loss is

$$\alpha_d = \frac{k^2 \tan\delta}{2\beta} = 0.119 \text{ Np/m} = 1.03 \text{ dB/m}.$$

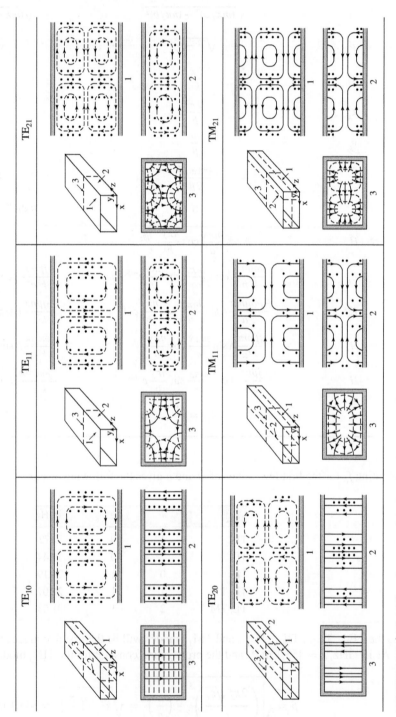

FIGURE 3.9 | Field lines for some of the lower order modes of a rectangular waveguide. Reprinted with permission from S. Ramo, J. R. Whinnery, and T. Van Duzer, *Fields and Waves in Communication Electronics.* Copyright © 1965 by John Wiley & Sons, Inc. Table 8.02.

The surface resistivity of the copper walls is ($\sigma = 5.8 \times 10^7$ S/m)

$$R_s = \sqrt{\frac{\omega \mu_0}{2\sigma}} = 0.032 \; \Omega,$$

and the attenuation due to conductor loss, from (3.96), is

$$\alpha_c = \frac{R_s}{a^3 b \beta k \eta}(2b\pi^2 + a^3 k^2) = 0.050 \; \text{Np/m} = 0.434 \; \text{dB/m}.$$

TE$_{m0}$ Modes of a Partially Loaded Waveguide

The above results apply to an empty waveguide as well as one filled with a homogeneous dielectric or magnetic material, but in some cases of practical interest (such as impedance matching or phase-shifting sections) a waveguide is used with a partial dielectric filling. In this case an additional set of boundary conditions are introduced at the material interface, necessitating a new analysis. To illustrate the technique we will consider the TE$_{m0}$ modes of a rectangular waveguide that is partially filled with a dielectric slab, as shown in Figure 3.10. The analysis still follows the basic procedure outlined at the end of Section 3.1.

Since the geometry is uniform in the y direction and $n = 0$, the TE$_{m0}$ modes have no y dependence. Then the wave equation of (3.21) for h_z can be written separately for the dielectric and air regions as

$$\left(\frac{\partial^2}{\partial x^2} + k_d^2\right)h_z = 0, \qquad \text{for } 0 \leq x \leq t, \tag{3.105a}$$

$$\left(\frac{\partial^2}{\partial x^2} + k_a^2\right)h_z = 0, \qquad \text{for } t \leq x \leq a, \tag{3.105b}$$

where k_d and k_a are the cutoff wave numbers for the dielectric and air regions, defined as follows:

$$\beta = \sqrt{\epsilon_r k_0^2 - k_d^2}, \tag{3.106a}$$

$$\beta = \sqrt{k_0^2 - k_a^2}. \tag{3.106b}$$

These relations incorporate the fact that the propagation constant, β, must be the same in both regions to ensure phase matching (see Section 1.8) of the fields along the interface at $x = t$. The solutions to (3.105) can be written as

$$h_z = \begin{cases} A\cos k_d x + B\sin k_d x & \text{for } 0 \leq x \leq t \\ C\cos k_a(a-x) + D\sin k_a(a-x) & \text{for } t \leq x \leq a, \end{cases} \tag{3.107}$$

where the form of the solution for $t < x < a$ was chosen to simplify the evaluation of boundary conditions at $x = a$.

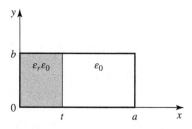

FIGURE 3.10 | Geometry of a partially loaded rectangular waveguide.

We need $\hat{y}$ and $\hat{z}$ electric and magnetic field components to apply the boundary conditions at $x = 0$, t, and a. $E_z = 0$ for TE modes, and $H_y = 0$ since $\partial/\partial y = 0$. E_y is found from (3.19d) as

$$e_y = \begin{cases} \dfrac{j\omega\mu_0}{k_d}(-A \sin k_d x + B \cos k_d x) & \text{for } 0 \leq x \leq t \\[4mm] \dfrac{j\omega\mu_0}{k_a}[C \sin k_a(a - x) - D \cos k_a(a - x)] & \text{for } t \leq x \leq a. \end{cases} \tag{3.108}$$

To satisfy the boundary conditions that $E_y = 0$ at $x = 0$ and $x = a$ requires that $B = D = 0$. We next enforce continuity of tangential fields (E_y, H_z) at $x = t$. Equations (3.107) and (3.108) then give the following:

$$\frac{-A}{k_d} \sin k_d t = \frac{C}{k_a} \sin k_a(a - t),$$

$$A \cos k_d t = C \cos k_a(a - t).$$

Because this is a homogeneous set of equations, the determinant must vanish in order to have a nontrivial solution. Thus,

$$k_a \tan k_d t + k_d \tan k_a(a - t) = 0. \tag{3.109}$$

Using (3.106) allows k_a and k_d to be expressed in terms of β, so (3.109) can be solved numerically for β. There is an infinite number of solutions to (3.109), corresponding to the propagation constants of the TE_{m0} modes.

This technique can be applied to many other waveguide geometries involving dielectric or magnetic material inhomogeneities, such as the surface waveguide of Section 3.6 or the ferrite-loaded waveguide of Section 9.3. In some cases, however, it will be impossible to satisfy all the necessary boundary conditions with only TE- or TM-type modes, and a hybrid combination of both types of modes may be required.

POINT OF INTEREST: Waveguide Flanges

There are two commonly used waveguide flanges: the cover flange and the choke flange. As shown in the accompanying figure, two waveguides with cover-type flanges can be bolted together to form a contacting joint. To avoid reflections and resistive loss at this joint it is necessary that the contacting surfaces be smooth, clean, and square because RF currents must flow across this discontinuity. In high-power applications voltage breakdown may occur at an imperfect junction. Otherwise, the simplicity of the cover-to-cover connection makes it preferable for general use. The SWR from such a joint is typically less than 1.03.

An alternative waveguide connection uses a cover flange against a choke flange, as shown in the figure. The choke flange is machined to form an effective radial transmission line in the narrow gap between the two flanges; this line is approximately $\lambda_g/4$ in length between the guide and the point of contact for the two flanges. Another $\lambda_g/4$ line is formed by a circular axial groove in the choke flange. Then the short circuit at the right-hand end of this groove is transformed into an open circuit at the contact point of the flanges. Any resistance in this contact is in series with an infinite (or very high) impedance and thus has little effect. This high impedance is transformed back into a short circuit (or very low impedance) at the edges of the waveguides to provide an effective low-resistance path for current flow across the joint. Because there is a negligible voltage drop across the ohmic contact between the flanges, voltage breakdown is avoided. Thus, the cover-to-choke connection can be useful for high-power applications. The SWR for this joint is typically less than 1.05 but is more frequency dependent than that of the cover-to-cover joint.

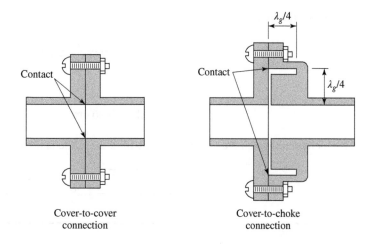

Reference: C. G. Montgomery, R. H. Dicke, and E. M. Purcell, *Principles of Microwave Circuits*, McGraw-Hill, New York, 1948.

EXAMPLE 3.3   RECTANGULAR WAVEGUIDE DESIGN

A hollow rectangular waveguide is to be used to transmit signals at a carrier frequency of 6 GHz. Choose its dimensions so that the cutoff frequency of the dominant TE mode is lower than the carrier by 25% and that of the next mode is at least 25% higher than the carrier.

Solution
For $m = 1$ and $n = 0$ (TE$_{10}$ mode),

$$f_{10} = \frac{c}{2a}$$

Denoting the carrier frequency as $f_0 = 6$ GHz. Setting

$$f_{10} = 0.75 f_0 = 4.5 \text{ GHz}$$
$$a = \frac{c}{2f_{10}} = 3.3 \text{ cm}$$

If b is chosen such that $a > b > a/2$, the second mode will be TE$_{10}$, followed by TE$_{20}$ at $f_{20} = 9$ GHz. For TE$_{01}$,

$$f_{01} = \frac{c}{2b}$$

Setting $f_{01} = 1.25, f_0 = 7.5$ GHz, we get

$$b = \frac{c}{2f_{01}} = 2 \text{ cm}$$

EXAMPLE 3.4   CUTOFF WAVELENGTH OF A RECTANGULAR WAVEGUIDE

A rectangular waveguide carries an electromagnetic wave having a frequency of 4000 MHz. A standing wave indicator shows that the wavelength of the wave in the guide is 11.4 cm. What is the cutoff wavelength of the waveguide and the velocity at which energy is propagated along the guide?

Solution
Here, $f = 4000$ MHz, $\lambda_g = 11.4$ cm
Wavelength of the impressed signal is

$$\lambda = \frac{c}{f} = 7.5 \text{ cm}$$

We have the relation,

$$\lambda_g = \frac{\lambda}{\sqrt{1 - \left(\frac{\lambda}{\lambda_c}\right)^2}}$$

$$\lambda_c = 9.96 \text{ cm}$$

The velocity with which the energy travels within the guide is the group velocity,

$$v_g = v\sqrt{1 - \frac{f_c^2}{f^2}} = 1.298 \times 10^8 \text{ m/sec}$$

3.4   CIRCULAR WAVEGUIDE

A hollow, round metal pipe also supports TE and TM waveguide modes. Figure 3.11 shows the geometry of such a circular waveguide, with inner radius a. Because cylindrical geometry is involved, it is appropriate to employ cylindrical coordinates. As in the rectangular coordinate case, the transverse fields in cylindrical coordinates can be derived from E_z or H_z field components for TM and TE modes, respectively. Paralleling the development of Section 3.1, we can derive the cylindrical components of the transverse fields from the longitudinal components as

$$E_\rho = \frac{-j}{k_c^2}\left(\beta\frac{\partial E_z}{\partial\rho} + \frac{\omega\mu}{\rho}\frac{\partial H_z}{\partial\phi}\right), \tag{3.110a}$$

$$E_\phi = \frac{-j}{k_c^2}\left(\frac{\beta}{\rho}\frac{\partial E_z}{\partial\phi} - \omega\mu\frac{\partial H_z}{\partial\rho}\right), \tag{3.110b}$$

$$H_\rho = \frac{j}{k_c^2}\left(\frac{\omega\epsilon}{\rho}\frac{\partial E_z}{\partial\phi} - \beta\frac{\partial H_z}{\partial\rho}\right), \tag{3.110c}$$

$$H_\phi = \frac{-j}{k_c^2}\left(\omega\epsilon\frac{\partial E_z}{\partial\rho} + \frac{\beta}{\rho}\frac{\partial H_z}{\partial\phi}\right), \tag{3.110d}$$

where $k_c^2 = k^2 - \beta^2$, and $e^{-j\beta z}$ propagation has been assumed. For $e^{+j\beta z}$ propagation, replace β with $-\beta$ in all expressions.

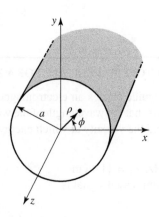

FIGURE 3.11 | Geometry of a circular waveguide.

TE Modes

For TE modes, $E_z = 0$, and H_z is a solution to the wave equation,

$$\nabla^2 H_z + k^2 H_z = 0. \tag{3.111}$$

If $H_z(\rho, \phi, z) = h_z(\rho, \phi)e^{-j\beta z}$, (3.111) can be expressed in cylindrical coordinates as

$$\left(\frac{\partial^2}{\partial \rho^2} + \frac{1}{\rho} \frac{\partial}{\partial \rho} + \frac{1}{\rho^2} \frac{\partial^2}{\partial \phi^2} + k_c^2 \right) h_z(\rho, \phi) = 0. \tag{3.112}$$

As before, we apply the method of separation of variables. Thus, let

$$h_z(\rho, \phi) = R(\rho)P(\phi), \tag{3.113}$$

and substitute into (3.112) to obtain

$$\frac{1}{R} \frac{d^2 R}{d\rho^2} + \frac{1}{\rho R} \frac{dR}{d\rho} + \frac{1}{\rho^2 P} \frac{d^2 P}{d\phi^2} + k_c^2 = 0,$$

or

$$\frac{\rho^2}{R} \frac{d^2 R}{d\rho^2} + \frac{\rho}{R} \frac{dR}{d\rho} + \rho^2 k_c^2 = \frac{-1}{P} \frac{d^2 P}{d\phi^2}. \tag{3.114}$$

The left side of this equation depends only on ρ (not ϕ), while the right side depends only on ϕ. Thus, each side must be equal to a constant, which we will call k_ϕ^2. Then,

$$\frac{-1}{P} \frac{d^2 P}{d\phi^2} = k_\phi^2,$$

or

$$\frac{d^2 P}{d\phi^2} + k_\phi^2 P = 0. \tag{3.115}$$

In addition,

$$\rho^2 \frac{d^2 R}{d\rho^2} + \rho \frac{dR}{d\rho} + \left(\rho^2 k_c^2 - k_\phi^2 \right) R = 0. \tag{3.116}$$

The general solution to (3.115) is

$$P(\phi) = A \sin k_\phi \phi + B \cos k_\phi \phi. \tag{3.117}$$

Because the solution to h_z must be periodic in ϕ [i.e., $h_z(\rho, \phi) = h_z(\rho, \phi \pm 2m\pi)$], k_ϕ must be an integer, n. Thus (3.117) becomes

$$P(\phi) = A \sin n\phi + B \cos n\phi, \tag{3.118}$$

and (3.116) becomes

$$\rho^2 \frac{d^2 R}{d\rho^2} + \rho \frac{dR}{d\rho} + \left(\rho^2 k_c^2 - n^2 \right) R = 0, \tag{3.119}$$

which is recognized as Bessel's differential equation. The solution is

$$R(\rho) = C J_n(k_c \rho) + D Y_n(k_c \rho), \tag{3.120}$$

where $J_n(x)$ and $Y_n(x)$ are the Bessel functions of first and second kinds, respectively. Because $Y_n(k_c \rho)$ becomes infinite at $\rho = 0$, this term is physically unacceptable for a circular waveguide, so $D = 0$. The solution for h_z can then be simplified to

$$h_z(\rho, \phi) = (A \sin n\phi + B \cos n\phi) J_n(k_c \rho), \tag{3.121}$$

where the constant C of (3.120) has been absorbed into the constants A and B of (3.121). We must still determine the cutoff wave number k_c, which we can do by enforcing the boundary condition that $E_{\text{tan}} = 0$ on the waveguide wall. Because $E_z = 0$, we must have that

$$E_\phi(\rho, \phi) = 0 \quad \text{at } \rho = a. \tag{3.122}$$

From (3.110b), we find E_ϕ from H_z as

$$E_\phi(\rho, \phi, z) = \frac{j\omega\mu}{k_c}(A \sin n\phi + B \cos n\phi)J_n'(k_c\rho)e^{-j\beta z}, \tag{3.123}$$

where the notation $J_n'(k_c\rho)$ refers to the derivative of J_n with respect to its argument. For E_ϕ to vanish at $\rho = a$, we must have

$$J_n'(k_c a) = 0. \tag{3.124}$$

If the roots of $J_n'(x)$ are defined as p_{nm}', so that $J_n'(p_{nm}') = 0$, where p_{nm}' is the mth root of J_n', then k_c must have the value

$$k_{c_{nm}} = \frac{p_{nm}'}{a}. \tag{3.125}$$

Values of p_{nm}' are given in mathematical tables; the first few values are listed in Table 3.3.

The TE$_{nm}$ modes are thus defined by the cutoff wave number $k_{c_{nm}} = p_{nm}'/a$, where n refers to the number of circumferential (ϕ) variations and m refers to the number of radial (ρ) variations. The propagation constant of the TE$_{nm}$ mode is

$$\beta_{nm} = \sqrt{k^2 - k_c^2} = \sqrt{k^2 - \left(\frac{p_{nm}'}{a}\right)^2}, \tag{3.126}$$

with a cutoff frequency of

$$f_{c_{nm}} = \frac{k_c}{2\pi\sqrt{\mu\epsilon}} = \frac{p_{nm}'}{2\pi a\sqrt{\mu\epsilon}}. \tag{3.127}$$

The first TE mode to propagate is the mode with the smallest p_{nm}', which from Table 3.3 is seen to be the TE$_{11}$ mode. This mode is therefore the dominant circular waveguide mode and the one most frequently used. Because $m \geq 1$, there is no TE$_{10}$ mode, but there is a TE$_{01}$ mode.

The transverse field components are, from (3.110) and (3.121),

$$E_\rho = \frac{-j\omega\mu n}{k_c^2\rho}(A \cos n\phi - B \sin n\phi)J_n(k_c\rho)e^{-j\beta z}, \tag{3.128a}$$

$$E_\phi = \frac{j\omega\mu}{k_c}(A \sin n\phi + B \cos n\phi)J_n'(k_c\rho)e^{-j\beta z}, \tag{3.128b}$$

$$H_\rho = \frac{-j\beta}{k_c}(A \sin n\phi + B \cos n\phi)J_n'(k_c\rho)e^{-j\beta z}, \tag{3.128c}$$

$$H_\phi = \frac{-j\beta n}{k_c^2\rho}(A \cos n\phi - B \sin n\phi)J_n(k_c\rho)e^{-j\beta z}. \tag{3.128d}$$

TABLE 3.3 | Values of p_{nm}' for TE Modes of a Circular Waveguide

n	p_{n1}'	p_{n2}'	p_{n3}'
0	3.832	7.016	10.174
1	1.841	5.331	8.536
2	3.054	6.706	9.970

The wave impedance is

$$Z_{\text{TE}} = \frac{E_\rho}{H_\phi} = \frac{-E_\phi}{H_\rho} = \frac{\eta k}{\beta}.\tag{3.129}$$

In the above solutions there are two remaining arbitrary amplitude constants, A and B. These constants control the amplitude of the $\sin n\phi$ and $\cos n\phi$ terms, which are independent. That is, because of the azimuthal symmetry of the circular waveguide, both the $\sin n\phi$ and $\cos n\phi$ terms represent valid solutions, and both may be present in a specific problem. The actual amplitudes of these terms will depend on the excitation of the waveguide. From a different viewpoint, the coordinate system can be rotated about the z-axis to obtain an h_z with either $A = 0$ or $B = 0$.

Now consider the dominant TE_{11} mode with an excitation such that $B = 0$. The fields can be written as

$$H_z = A \sin \phi J_1(k_c\rho)e^{-j\beta z},\tag{3.130a}$$

$$E_\rho = \frac{-j\omega\mu}{k_c^2\rho}A \cos \phi J_1(k_c\rho)e^{-j\beta z},\tag{3.130b}$$

$$E_\phi = \frac{j\omega\mu}{k_c}A \sin \phi J_1'(k_c\rho)e^{-j\beta z},\tag{3.130c}$$

$$H_\rho = \frac{-j\beta}{k_c}A \sin \phi J_1'(k_c\rho)e^{-j\beta z},\tag{3.130d}$$

$$H_\phi = \frac{-j\beta}{k_c^2\rho}A \cos \phi J_1(k_c\rho)e^{-j\beta z},\tag{3.130e}$$

$$E_z = 0.\tag{3.130f}$$

The power flow down the guide can be computed as

$$\begin{aligned}
P_o &= \frac{1}{2}\text{Re}\int_{\rho=0}^{a}\int_{\phi=0}^{2\pi} \bar{E} \times \bar{H}^* \cdot \hat{z}\rho \, d\phi \, d\rho\\
&= \frac{1}{2}\text{Re}\int_{\rho=0}^{a}\int_{\phi=0}^{2\pi} \left(E_\rho H_\phi^* - E_\phi H_\rho^*\right)\rho \, d\phi \, d\rho\\
&= \frac{\omega\mu|A|^2\text{Re}(\beta)}{2k_c^4}\int_{\rho=0}^{a}\int_{\phi=0}^{2\pi}\left[\frac{1}{\rho^2}\cos^2\phi J_1^2(k_c\rho) + k_c^2 \sin^2\phi J_1'^2(k_c\rho)\right]\rho \, d\phi \, d\rho\\
&= \frac{\pi\omega\mu|A|^2\text{Re}(\beta)}{2k_c^4}\int_{\rho=0}^{a}\left[\frac{1}{\rho}J_1^2(k_c\rho) + \rho k_c^2 J_1'^2(k_c\rho)\right] d\rho\\
&= \frac{\pi\omega\mu|A|^2\text{Re}(\beta)}{4k_c^4}\left(p_{11}'^2 - 1\right) J_1^2(k_c a),
\end{aligned}\tag{3.131}$$

which is seen to be nonzero only when β is real, corresponding to a propagating mode. (The required integral for this result is given in Appendix C.)

Attenuation due to dielectric loss is given by (3.29). The attenuation due to a lossy waveguide conductor can be found by computing the power loss per unit length of guide:

$$\begin{aligned}
P_\ell &= \frac{R_s}{2}\int_{\phi=0}^{2\pi}|\bar{J}_s|^2 a \, d\phi\\
&= \frac{R_s}{2}\int_{\phi=0}^{2\pi}\left(|H_\phi|^2 + |H_z|^2\right)a \, d\phi\\
&= \frac{|A|^2 R_s}{2}\int_{\phi=0}^{2\pi}\left(\frac{\beta^2}{k_c^4 a^2}\cos^2\phi + \sin^2\phi\right)J_1^2(k_c a)a \, d\phi\\
&= \frac{\pi|A|^2 R_s a}{2}\left(1 + \frac{\beta^2}{k_c^4 a^2}\right)J_1^2(k_c a).
\end{aligned}\tag{3.132}$$

The attenuation constant is then

$$\alpha_c = \frac{P_\ell}{2P_o} = \frac{R_s \left(k_c^4 a^2 + \beta^2\right)}{\eta k \beta a (p_{11}'^2 - 1)}$$

$$= \frac{R_s}{ak\eta\beta} \left(k_c^2 + \frac{k^2}{p_{11}'^2 - 1}\right) \text{Np/m}. \tag{3.133}$$

TM Modes

For the TM modes of the circular waveguide, we must solve for E_z from the wave equation in cylindrical coordinates:

$$\left(\frac{\partial^2}{\partial \rho^2} + \frac{1}{\rho}\frac{\partial}{\partial \rho} + \frac{1}{\rho^2}\frac{\partial^2}{\partial \phi^2} + k_c^2\right) e_z = 0, \tag{3.134}$$

where $E_z(\rho, \phi, z) = e_z(\rho, \phi)e^{-j\beta z}$, and $k_c^2 = k^2 - \beta^2$. Because this equation is identical to (3.107), the general solutions are the same. Thus, from (3.121),

$$e_z(\rho, \phi) = (A \sin n\phi + B \cos n\phi)J_n(k_c\rho). \tag{3.135}$$

The difference between the TE solution and the present solution is that the boundary conditions can now be applied directly to e_z of (3.135) since

$$E_z(\rho, \phi) = 0 \quad \text{at } \rho = a. \tag{3.136}$$

Thus, we must have

$$J_n(k_c a) = 0, \tag{3.137}$$

or

$$k_c = p_{nm}/a, \tag{3.138}$$

where p_{nm} is the mth root of $J_n(x)$, that is, $J_n(p_{nm}) = 0$. Values of p_{nm} are given in mathematical tables; the first few values are listed in Table 3.4.

The propagation constant of the TM_{nm} mode is

$$\beta_{nm} = \sqrt{k^2 - k_c^2} = \sqrt{k^2 - (p_{nm}/a)^2}, \tag{3.139}$$

and the cutoff frequency is

$$f_{c_{nm}} = \frac{k_c}{2\pi\sqrt{\mu\epsilon}} = \frac{p_{nm}}{2\pi a\sqrt{\mu\epsilon}}. \tag{3.140}$$

Thus, the first TM mode to propagate is the TM_{01} mode, with $p_{01} = 2.405$. Because this is greater than $p_{11}' = 1.841$ for the lowest order TE_{11} mode, the TE_{11} mode is the dominant mode of the circular waveguide. As with the TE modes, $m \geq 1$, so there is no TM_{10} mode.

TABLE 3.4 | Values of p_{nm} for TM Modes of a Circular Waveguide

n	p_{n1}	p_{n2}	p_{n3}
0	2.405	5.520	8.654
1	3.832	7.016	10.174
2	5.135	8.417	11.620

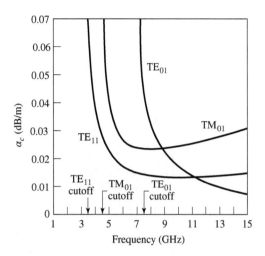

FIGURE 3.12 | Attenuation of various modes in a circular copper waveguide with $a = 2.54$ cm.

From (3.110), the transverse fields can be derived as

$$E_\rho = \frac{-j\beta}{k_c}(A \sin n\phi + B \cos n\phi)J'_n(k_c\rho)e^{-j\beta z}, \tag{3.141a}$$

$$E_\phi = \frac{-j\beta n}{k_c^2\rho}(A \cos n\phi - B \sin n\phi)J_n(k_c\rho)e^{-j\beta z}, \tag{3.141b}$$

$$H_\rho = \frac{j\omega\epsilon n}{k_c^2\rho}(A \cos n\phi - B \sin n\phi)J_n(k_c\rho)e^{-j\beta z}, \tag{3.141c}$$

$$H_\phi = \frac{-j\omega\epsilon}{k_c}(A \sin n\phi + B \cos n\phi)J'_n(k_c\rho)e^{-j\beta z}. \tag{3.141d}$$

The wave impedance is

$$Z_{TM} = \frac{E_\rho}{H_\phi} = \frac{-E_\phi}{H_\rho} = \frac{\eta\beta}{k}. \tag{3.142}$$

Calculation of the attenuation for TM modes is left as a problem. Figure 3.12 shows the attenuation due to conductor loss versus frequency for various modes of a circular waveguide. Observe that the attenuation of the TE_{01} mode decreases to a very small value with increasing frequency. This property makes the TE_{01} mode of interest for low-loss transmission over long distances. Unfortunately, this mode is not the dominant mode of the circular waveguide, so in practice power can be lost from the TE_{01} mode to lower order propagating modes.

Figure 3.13 shows the relative cutoff frequencies of the TE and TM modes, and Table 3.5 summarizes results for wave propagation in circular waveguide. Field lines for some of the lowest order TE and TM modes are shown in Figure 3.14.

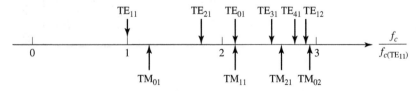

FIGURE 3.13 | Cutoff frequencies of the first few TE and TM modes of a circular waveguide relative to the cutoff frequency of the dominant TE_{11} mode.

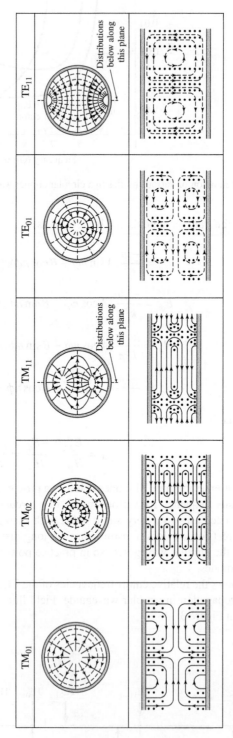

FIGURE 3.14 | Field lines for some of the lower order modes of a circular waveguide.
Reprinted with permission from S. Ramo, J. R. Whinnery, and T. Van Duzer, *Fields and Waves in Communication Electronics*, Copyright © 1965 by John Wiley & Sons, Inc. Table 8.04.

TABLE 3.5 | **Summary of Results for Circular Waveguide**

Quantity	TE_{nm} Mode	TM_{nm} Mode
k	$\omega\sqrt{\mu\epsilon}$	$\omega\sqrt{\mu\epsilon}$
k_c	$\dfrac{p'_{nm}}{a}$	$\dfrac{p_{nm}}{a}$
β	$\sqrt{k^2 - k_c^2}$	$\sqrt{k^2 - k_c^2}$
λ_c	$\dfrac{2\pi}{k_c}$	$\dfrac{2\pi}{k_c}$
λ_g	$\dfrac{2\pi}{\beta}$	$\dfrac{2\pi}{\beta}$
v_p	$\dfrac{\omega}{\beta}$	$\dfrac{\omega}{\beta}$
α_d	$\dfrac{k^2 \tan\delta}{2\beta}$	$\dfrac{k^2 \tan\delta}{2\beta}$
E_z	0	$(A\sin n\phi + B\cos n\phi)J_n(k_c\rho)e^{-j\beta z}$
H_z	$(A\sin n\phi + B\cos n\phi)J_n(k_c\rho)e^{-j\beta z}$	0
E_ρ	$\dfrac{-j\omega\mu n}{k_c^2\rho}(A\cos n\phi - B\sin n\phi)J_n(k_c\rho)e^{-j\beta z}$	$\dfrac{-j\beta}{k_c}(A\sin n\phi + B\cos n\phi)J'_n(k_c\rho)e^{-j\beta z}$
E_ϕ	$\dfrac{j\omega\mu}{k_c}(A\sin n\phi + B\cos n\phi)J'_n(k_c\rho)e^{-j\beta z}$	$\dfrac{-j\beta n}{k_c^2\rho}(A\cos n\phi - B\sin n\phi)J_n(k_c\rho)e^{-j\beta z}$
H_ρ	$\dfrac{-j\beta}{k_c}(A\sin n\phi + B\cos n\phi)J'_n(k_c\rho)e^{-j\beta z}$	$\dfrac{j\omega\epsilon n}{k_c^2\rho}(A\cos n\phi - B\sin n\phi)J_n(k_c\rho)e^{-j\beta z}$
H_ϕ	$\dfrac{-j\beta n}{k_c^2\rho}(A\cos n\phi - B\sin n\phi)J_n(k_c\rho)e^{-j\beta z}$	$\dfrac{-j\omega\epsilon}{k_c}(A\sin n\phi + B\cos n\phi)J'_n(k_c\rho)e^{-j\beta z}$
Z	$Z_{\text{TE}} = \dfrac{k\eta}{\beta}$	$Z_{\text{TM}} = \dfrac{\beta\eta}{k}$

EXAMPLE 3.5   CHARACTERISTICS OF A CIRCULAR WAVEGUIDE

Find the cutoff frequencies of the first two propagating modes of a Teflon-filled circular waveguide with $a = 0.5$ cm. If the interior of the guide is gold plated, calculate the overall loss in dB for a 30 cm length operating at 14 GHz.

Solution
From Figure 3.13, the first two propagating modes of a circular waveguide are the TE_{11} and TM_{01} modes. The cutoff frequencies can be found using (3.127) and (3.140):

$$\text{TE}_{11}: \qquad f_c = \frac{p'_{11}c}{2\pi a\sqrt{\epsilon_r}} = \frac{1.841(3\times 10^8)}{2\pi(0.005)\sqrt{2.08}} = 12.19 \text{ GHz},$$

$$\text{TM}_{01}: \qquad f_c = \frac{p_{01}c}{2\pi a\sqrt{\epsilon_r}} = \frac{2.405(3\times 10^8)}{2\pi(0.005)\sqrt{2.08}} = 15.92 \text{ GHz}.$$

So only the TE_{11} mode is propagating at 14 GHz. The wave number is

$$k = \frac{2\pi f\sqrt{\epsilon_r}}{c} = \frac{2\pi(14\times 10^9)\sqrt{2.08}}{3\times 10^8} = 422.9 \text{ m}^{-1},$$

and the propagation constant of the TE_{11} mode is

$$\beta = \sqrt{k^2 - \left(\frac{p'_{11}}{a}\right)^2} = \sqrt{(422.9)^2 - \left(\frac{1.841}{0.005}\right)^2} = 208.0 \text{ m}^{-1}.$$

The attenuation due to dielectric loss is calculated from (3.29) as

$$\alpha_d = \frac{k^2 \tan \delta}{2\beta} = \frac{(422.9)^2(0.0004)}{2(208.0)} = 0.172 \text{ Np/m} = 1.49 \text{ dB/m}.$$

The conductivity of gold is $\sigma = 4.1 \times 10^7$ S/m, so the surface resistance is

$$R_s = \sqrt{\frac{\omega \mu_0}{2\sigma}} = 0.0367 \ \Omega.$$

Then from (3.133) the attenuation due to conductor loss is

$$\alpha_c = \frac{R_s}{ak\eta\beta}\left(k_c^2 + \frac{k^2}{p'^2_{11} - 1}\right) = 0.0672 \text{ Np/m} = 0.583 \text{ dB/m}.$$

The total attenuation is $\alpha = \alpha_d + \alpha_c = 2.07 \text{dB/m}$, and the loss in the 30 cm length of guide is

$$\text{attenuation (dB)} = \alpha(\text{dB/m}) \times L \text{ (m)} = (2.07)(0.3) = 0.62 \text{ dB}.$$

EXAMPLE 3.6   CIRCULAR WAVEGUIDE

Evaluate the ratio of the area of a circular waveguide to that of a rectangular waveguide if both have the same cutoff frequency for dominant mode.

Solution
Let r be the radius of the circular waveguide and a be the larger dimension of the rectangular waveguide. The dominant mode in a circular waveguide is TE_{11} mode. For this mode, the cutoff frequency is given as,

$$f_c = \frac{v_0}{2\pi}h_{mn} = \frac{v_0}{2\pi r} \times 1.84$$

For TE_{10} mode rectangular waveguide, the cutoff frequency is given as,

$$f_c = \frac{v_0}{2}\sqrt{\left(\frac{m}{a}\right)^2 + \left(\frac{n}{b}\right)^2} = \frac{v_0}{2a}$$

Since both will have the same cutoff frequency, we get

$$\frac{v_0}{2\pi r} \times 1.84 = \frac{v_0}{2a}$$

$$\frac{r}{a} = 0.586$$

The area of the circular waveguide is

$$A_c = \pi r^2$$

For standard rectangular waveguide $a : b = 2 : 1$, so that the area of the rectangular waveguide is,

$$A_r = \frac{a^2}{2}$$

$$\frac{A_c}{A_r} = 2.155$$

3.5 COAXIAL LINE

TEM Modes

Although we have already discussed TEM mode propagation on a coaxial line in Chapter 2, we will briefly reconsider it here in the context of the general framework that is being used in this chapter.

The coaxial transmission line geometry is shown in Figure 3.15, where the inner conductor is at a potential of V_o volts and the outer conductor is at zero volts. From Section 3.1 we know that the fields can be derived from a scalar potential function, $\Phi(\rho, \phi)$, which is a solution to Laplace's equation (3.14). In cylindrical coordinates Laplace's equation takes the form

$$\frac{1}{\rho}\frac{\partial}{\partial \rho}\left(\rho\frac{\partial \Phi(\rho,\phi)}{\partial \rho}\right) + \frac{1}{\rho^2}\frac{\partial^2 \Phi(\rho,\phi)}{\partial \phi^2} = 0. \tag{3.143}$$

This equation must be solved for $\Phi(\rho, \phi)$ subject to the boundary conditions

$$\Phi(a, \phi) = V_o, \tag{3.144a}$$

$$\Phi(b, \phi) = 0. \tag{3.144b}$$

By the method of separation of variables, let $\Phi(\rho, \phi)$ be expressed in product form as

$$\Phi(\rho, \phi) = R(\rho)P(\phi). \tag{3.145}$$

Substituting (3.145) into (3.143) and dividing by RP gives

$$\frac{\rho}{R}\frac{\partial}{\partial \rho}\left(\rho\frac{dR}{d\rho}\right) + \frac{1}{P}\frac{d^2P}{d\phi^2} = 0. \tag{3.146}$$

By the usual separation-of-variables argument, the two terms in (3.146) must be equal to constants, so that

$$\frac{\rho}{R}\frac{\partial}{\partial \rho}\left(\rho\frac{dR}{d\rho}\right) = -k_\rho^2, \tag{3.147}$$

$$\frac{1}{P}\frac{d^2P}{d\phi^2} = -k_\phi^2, \tag{3.148}$$

$$k_\rho^2 + k_\phi^2 = 0. \tag{3.149}$$

The general solution to (3.148) is

$$P(\phi) = A\cos n\phi + B\sin n\phi, \tag{3.150}$$

where $k_\phi = n$ must be an integer since increasing ϕ by a multiple of 2π should not change the result. Now, because the boundary conditions of (3.144) do not vary with ϕ, the potential $\Phi(\rho, \phi)$ should not vary with ϕ.

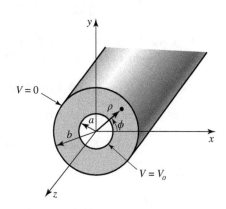

FIGURE 3.15 | Coaxial line geometry.

Thus, n must be zero. By (3.149), this implies that k_ρ must also be zero, so that the equation for $R(\rho)$ in (3.147) reduces to

$$\frac{\partial}{\partial \rho}\left(\rho \frac{dR}{d\rho}\right) = 0.$$

The solution for $R(\rho)$ is then

$$R(\rho) = C \ln \rho + D,$$

and so

$$\Phi(\rho, \phi) = C \ln \rho + D. \tag{3.151}$$

Applying the boundary conditions of (3.144) gives two equations for the constants C and D:

$$\Phi(a, \phi) = V_o = C \ln a + D, \tag{3.152a}$$

$$\Phi(b, \phi) = 0 = C \ln b + D. \tag{3.152b}$$

After solving for C and D, we can write the final solution for $\Phi(\rho, \phi)$ as

$$\Phi(\rho, \phi) = \frac{V_o \ln b/\rho}{\ln b/a}. \tag{3.153}$$

The $\bar{E}$ and $\bar{H}$ fields can now be found using (3.13) and (3.18), and the voltage, current, and characteristic impedance can be determined as in Chapter 2. Attenuation due to dielectric or conductor loss has already been treated in Chapter 2.

Higher Order Modes

The coaxial line, like the parallel plate waveguide, can also support TE and TM waveguide modes in addition to the TEM mode. In practice, these modes are usually cut off (evanescent), and so have only a reactive effect near discontinuities or sources, where they may be excited. It is important in practice, however, to be aware of the cutoff frequency of the lowest order waveguide-type modes to avoid the propagation of these modes. Undesirable effects can occur if two or more modes with different propagation constants are propagating at the same time. Avoiding propagation of higher order modes sets an upper limit on the size of a coaxial cable or, equivalently, an upper limit on the frequency of operation for a given cable. This also affects the power handling capacity of a coaxial line (see the Point of Interest on power capacity of transmission lines).

We will derive the solution for the TE modes of the coaxial line; the TE_{11} mode is the dominant waveguide mode of the coaxial line and so is of primary importance.

For TE modes, $E_z = 0$, and H_z satisfies the wave equation of (3.112):

$$\left(\frac{\partial^2}{\partial \rho^2} + \frac{1}{\rho}\frac{\partial}{\partial \rho} + \frac{1}{\rho^2}\frac{\partial^2}{\partial \phi^2} + k_c^2\right) h_z(\rho, \phi) = 0, \tag{3.154}$$

where $H_z(\rho, \phi, z) = h_z(\rho, \phi)e^{-j\beta z}$, and $k_c^2 = k^2 - \beta^2$. The general solution to this equation, as derived in Section 3.4, is given by the product of (3.118) and (3.120):

$$h_z(\rho, \phi) = (A \sin n\phi + B \cos n\phi)(CJ_n(k_c\rho) + DY_n(k_c\rho)). \tag{3.155}$$

In this case, $a \le \rho \le b$, so we have no reason to discard the Y_n term. The boundary conditions are

$$E_\phi(\rho, \phi, z) = 0 \text{ for } \rho = a, b. \tag{3.156}$$

Using (3.110b) to find E_ϕ from H_z gives

$$E_\phi = \frac{j\omega\mu}{k_c}(A \sin n\phi + B \cos n\phi)[CJ_n'(k_c\rho) + DY_n'(k_c\rho)]e^{-j\beta z}. \tag{3.157}$$

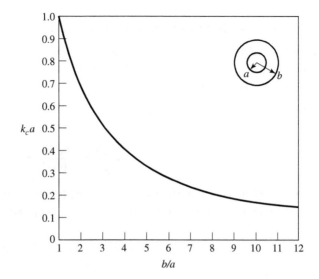

FIGURE 3.16 | Normalized cutoff frequency of the dominant TE_{11} waveguide mode for a coaxial line.

Applying (3.156) to (3.157) gives two equations:

$$CJ'_n(k_c a) + DY'_n(k_c a) = 0, \tag{3.158a}$$

$$CJ'_n(k_c b) + DY'_n(k_c b) = 0. \tag{3.158b}$$

Because this is a homogeneous set of equations, the only nontrivial ($C \neq 0, D \neq 0$) solution occurs when the determinant is zero. Thus we must have

$$J'_n(k_c a)Y'_n(k_c b) = J'_n(k_c b)Y'_n(k_c a). \tag{3.159}$$

This is a characteristic (or eigenvalue) equation for k_c. The values of k_c that satisfy (3.159) then define the TE_{nm} modes of the coaxial line.

Equation (3.159) is a transcendental equation, which must be solved numerically for k_c. Figure 3.16 shows the result of such a solution for $n = 1$ for various b/a ratios. An approximate solution that is often used in practice is

$$k_c = \frac{2}{a+b}.$$

Once k_c is known, the propagation constant or cutoff frequency can be determined. Solutions for the TM modes can be found in a similar manner; the required determinantal equation is the same as (3.159), except for the derivatives. Field lines for the TEM and TE_{11} modes of the coaxial line are shown in Figure 3.17.

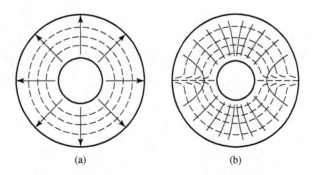

FIGURE 3.17 | Field lines for the (a) TEM and (b) TE_{11} modes of a coaxial line.

EXAMPLE 3.7 HIGHER ORDER MODE OF A COAXIAL LINE

Consider a RG-401U semirigid coaxial cable, with inner and outer conductor diameters of 0.0645 in. and 0.215 in., and a dielectric with $\epsilon_r = 2.2$. What is the highest usable frequency before the TE_{11} waveguide mode starts to propagate?

Solution
We have

$$\frac{b}{a} = \frac{2b}{2a} = \frac{0.215}{0.0645} = 3.33.$$

From Figure 3.16 this value of b/a gives $k_c a = 0.45$ [the approximate result is $k_c a = 2/(1 + b/a) = 0.462$]. Thus, $k_c = 549.4$ m^{-1}, and the cutoff frequency of the TE_{11} mode is

$$f_c = \frac{ck_c}{2\pi\sqrt{\epsilon_r}} = 17.7 \text{ GHz.}$$

In practice, a 5% safety margin is usually recommended, so

$$f_{max} = (0.95)(17.7 \text{ GHz}) = 16.8 \text{ GHz.}$$

POINT OF INTEREST: Coaxial Connectors

Most coaxial cables and connectors in common use have a 50 Ω characteristic impedance, with an exception being the 75 Ω cable used in television systems. The reasoning behind these choices is that an air-filled coaxial line has minimum attenuation for a characteristic impedance of about 77 Ω (Problem 2.26), while maximum power capacity occurs for a characteristic impedance of about 30 Ω (Problem 3.28). A 50 Ω characteristic impedance thus represents a compromise between minimum attenuation and maximum power capacity. Other requirements for coaxial connectors include low SWR, higher-order-mode–free operation at a high frequency, high repeatability after a connect–disconnect cycle, and mechanical

strength. Connectors are used in pairs, with a male end and a female end (or plug and jack). The accompanying photo shows several types of commonly used coaxial connectors and adapters. From top left: Type-N, TNC, SMA, APC-7, and 2.4 mm.

Type-N: This connector was developed in 1942 and is named after its inventor, P. Neil, of Bell Labs. The outer diameter of the female end is about 0.625 in. The recommended upper frequency limit ranges from 11 to 18 GHz, depending on cable size. This rugged but large connector is often found on older equipment.

TNC: This is a threaded version of the very common BNC connector. Its use is limited to frequencies below 1 GHz.

SMA: The need for smaller and lighter connectors led to the development of this connector in the 1960s. The outer diameter of the female end is about 0.25 in. It can be used up to frequencies in the range of 18–25 GHz and is probably the most commonly used microwave connector today.

APC-7:   This is a precision connector (Amphenol Precision Connector) that can repeatedly achieve SWR less than 1.04 at frequencies up to 18 GHz. The connectors are "sexless," with butt contact between both inner conductors and outer conductors. This connector is most commonly used for measurement and instrumentation applications.

2.4 mm: The need for connectors at millimeter wave frequencies led to the development of several variations of the SMA connector. One of the most common is the 2.4 mm connector, which is useful to about 50 GHz. The size of this connector is similar to that of the SMA connector.

3.6   SURFACE WAVES ON A GROUNDED DIELECTRIC SHEET

We briefly discussed surface waves in Chapter 1 in connection with the field of a plane wave totally reflected from a dielectric interface, but surface waves can exist in a variety of geometries involving dielectric interfaces. Here we consider the TM and TE surface waves that can be excited along a grounded dielectric sheet. Other geometries that can be used as surface waveguides include an ungrounded dielectric sheet, a dielectric rod, a corrugated conductor, and a dielectric-coated conducting rod.

Surface waves are typified by a field that decays exponentially away from the dielectric surface, with most of the field contained in or near the dielectric. At higher frequencies the field generally becomes more tightly bound to the dielectric, making such waveguides practical. Because of the presence of the dielectric, the phase velocity of a surface wave is less than the velocity of light in a vacuum. Another reason for studying surface waves is that they may be excited on some types of planar transmission lines, such as microstrip line and slotline.

TM Modes

Figure 3.18 shows the geometry of a grounded dielectric slab waveguide. The dielectric sheet, of thickness d and relative permittivity ϵ_r, is assumed to be of infinite extent in the y and z directions. We will assume propagation in the $+z$ direction with an $e^{-j\beta z}$ propagation factor and no variation in the y direction ($\partial/\partial y = 0$).

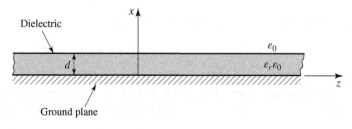

FIGURE 3.18 | Geometry of a grounded dielectric sheet.

Because there are two distinct regions, with and without a dielectric, we must separately consider the field in these regions and then match tangential fields across the interface. E_z must satisfy the wave equation of (3.25) in each region:

$$\left(\frac{\partial^2}{\partial x^2} + \epsilon_r k_0^2 - \beta^2 \right) e_z(x, y) = 0, \qquad \text{for } 0 \le x \le d, \tag{3.160a}$$

$$\left(\frac{\partial^2}{\partial x^2} + k_0^2 - \beta^2 \right) e_z(x, y) = 0, \qquad \text{for } d \le x < \infty, \tag{3.160b}$$

where $E_z(x, y, z) = e_z(x, y)e^{-j\beta z}$.

We define the cutoff wave numbers for the two regions as

$$k_c^2 = \epsilon_r k_0^2 - \beta^2, \tag{3.161a}$$

$$h^2 = \beta^2 - k_0^2, \tag{3.161b}$$

where the sign on h^2 has been selected in anticipation of an exponentially decaying result for $x > d$. Observe that the same propagation constant, β, has been used for both regions. This must be the case to achieve phase matching of the tangential fields at the $x = d$ interface for all values of z.

The general solutions to (3.160) are

$$e_z(x, y) = A \sin k_c x + B \cos k_c x, \qquad \text{for } 0 \le x \le d \tag{3.162a}$$

$$e_z(x, y) = Ce^{hx} + De^{-hx}, \qquad \text{for } d \le x < \infty \tag{3.162b}$$

Note that these solutions are valid for k_c and h either real or imaginary; it will turn out that both k_c and h are real because of the choice of definitions in (3.161).

The boundary conditions that must be satisfied are

$$E_z(x, y, z) = 0, \qquad \text{at } x = 0, \tag{3.163a}$$

$$E_z(x, y, z) < \infty, \qquad \text{as } x \to \infty, \tag{3.163b}$$

$$E_z(x, y, z) \text{ continuous at } x = d, \tag{3.163c}$$

$$H_y(x, y, z) \text{ continuous at } x = d. \tag{3.163d}$$

From (3.23), $H_x = E_y = H_z = 0$. Condition (3.163a) implies that $B = 0$ in (3.162a). Condition (3.163b) is a result of a requirement for finite fields (and energy) infinitely far away from a source and implies that $C = 0$. The continuity of E_z leads to

$$A \sin k_c d = De^{-hd}, \tag{3.164a}$$

while (3.23b) must be used to apply continuity to H_y, to obtain

$$\frac{\epsilon_r A}{k_c} \cos k_c d = \frac{D}{h} e^{-hd}. \tag{3.164b}$$

For a nontrivial solution, the determinant of the two equations of (3.164) must vanish, leading to

$$k_c \tan k_c d = \epsilon_r h. \tag{3.165}$$

Eliminating β from (3.161a) and (3.161b) gives

$$k_c^2 + h^2 = (\epsilon_r - 1)k_0^2. \tag{3.166}$$

Equations (3.165) and (3.166) constitute a set of simultaneous transcendental equations that must be solved for the propagation constants k_c and h, given k_o and ϵ_r. These equations are easily solved numerically, but Figure 3.19 shows a graphical representation of the solutions. Multiplying both sides of (3.166) by d^2 gives

$$(k_c d)^2 + (hd)^2 = (\epsilon_r - 1)(k_0 d)^2,$$

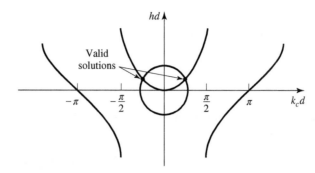

FIGURE 3.19 | Graphical solution of the transcendental equation for the cutoff frequency of a TM surface wave mode of the grounded dielectric sheet.

which is the equation of a circle in the $k_c d$, hd plane, as shown in Figure 3.19. The radius of the circle is $\sqrt{\epsilon_r - 1}k_0 d$, which is proportional to the electrical thickness of the dielectric sheet. Multiplying (3.165) by d gives

$$k_c d \tan k_c d = \epsilon_r hd,$$

which is also plotted in Figure 3.19. The intersection of these curves implies a solution to both (3.165) and (3.166). Observe that k_c may be positive or negative; from (3.162a) this is seen to merely change the sign of the constant A. As $\sqrt{\epsilon_r - 1}k_0 d$ becomes larger, the circle may intersect more than one branch of the tangent function, implying that more than one TM mode can propagate. Solutions for negative h, however, must be excluded since we assumed h was positive real when applying boundary condition (3.163b).

For any nonzero-thickness sheet with a relative permittivity greater than unity, there is at least one propagating TM mode, which we will call the TM_0 mode. This is the dominant mode of the dielectric slab waveguide, and it has a zero cutoff frequency. (Although for $k_0 = 0$, $k_c = h = 0$, and all fields vanish.) From Figure 3.19 it can be seen that the next TM mode, the TM_1 mode, will not begin to propagate until the radius of the circle becomes greater than π. The cutoff frequency of the TM_n mode can then be derived as

$$f_c = \frac{nc}{2d\sqrt{\epsilon_r - 1}}, \quad n = 0, 1, 2, \dots. \tag{3.167}$$

Once k_c and h have been found for a particular surface wave mode, the field expressions can be found as

$$E_z(x, y, z) = \begin{cases} A \sin k_c x e^{-j\beta z} & \text{for } 0 \le x \le d \\ A \sin k_c d e^{-h(x-d)} e^{-j\beta z} & \text{for } d \le x < \infty, \end{cases} \tag{3.168a}$$

$$E_x(x, y, z) = \begin{cases} \dfrac{-j\beta}{k_c} A \cos k_c x e^{-j\beta z} & \text{for } 0 \le x \le d \\ \dfrac{-j\beta}{h} A \sin k_c d e^{-h(x-d)} e^{-j\beta z} & \text{for } d \le x < \infty, \end{cases} \tag{3.168b}$$

$$H_y(x, y, z) = \begin{cases} \dfrac{-j\omega\epsilon_0\epsilon_r}{k_c} A \cos k_c x e^{-j\beta z} & \text{for } 0 \le x \le d \\ \dfrac{-j\omega\epsilon_0}{h} A \sin k_c d e^{-h(x-d)} e^{-j\beta z} & \text{for } d \le x < \infty. \end{cases} \tag{3.168c}$$

TE Modes

TE modes can also be supported by the grounded dielectric sheet. The H_z field satisfies the wave equations

$$\left(\frac{\partial^2}{\partial x^2} + k_c^2\right) h_z(x, y) = 0, \quad \text{for } 0 \le x \le d, \tag{3.169a}$$

$$\left(\frac{\partial^2}{\partial x^2} - h^2\right) h_z(x, y) = 0, \quad \text{for } d \le x < \infty, \tag{3.169b}$$

with $H_z(x, y, z) = h_z(x, y)e^{-j\beta z}$ and k_c^2 and h^2 defined in (3.161a) and (3.161b). As for the TM modes, the general solutions to (3.169) are

$$h_z(x, y) = A \sin k_c x + B \cos k_c x, \qquad (3.170a)$$

$$h_z(x, y) = Ce^{hx} + De^{-hx}. \qquad (3.170b)$$

To satisfy the radiation condition, $C = 0$. Using (3.19d) to find E_y from H_z leads to $A = 0$ for $E_y = 0$ at $x = 0$ and to the equation

$$\frac{-B}{k_c} \sin k_c d = \frac{D}{h} e^{-hd} \qquad (3.171a)$$

for continuity of E_y at $x = d$. Continuity of H_z at $x = d$ gives

$$B \cos k_c d = De^{-hd}. \qquad (3.171b)$$

Simultaneously solving (3.171a) and (3.171b) leads to the determinantal equation

$$-k_c \cot k_c d = h. \qquad (3.172)$$

From (3.161a) and (3.161b) we also have that

$$k_c^2 + h^2 = (\epsilon_r - 1)k_0^2. \qquad (3.173)$$

Equations (3.172) and (3.173) must be solved simultaneously for the variables k_c and h. Equation (3.173) again represents circles in the $k_c d$, hd plane, while (3.172) can be rewritten as

$$-k_c d \cot k_c d = hd,$$

and plotted as a family of curves in the $k_c d$, hd plane, as shown in Figure 3.20. Because negative values of h must be excluded, we see from Figure 3.20 that the first TE mode does not start to propagate until the radius of the circle, $\sqrt{\epsilon_r - 1}k_0 d$, becomes greater than $\pi/2$. The cutoff frequency of the TE_n modes can then be found as

$$f_c = \frac{(2n - 1)c}{4d\sqrt{\epsilon_r - 1}} \qquad \text{for } n = 1, 2, 3, \ldots. \qquad (3.174)$$

Comparing with (3.167) shows that the order of propagation for the TM_n and TE_n modes is TM_0, TE_1, TM_1, TE_2, TM_2, $\ldots$.

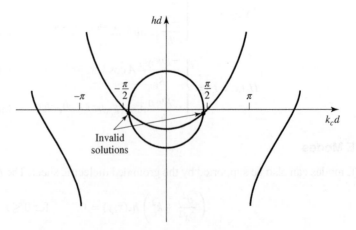

FIGURE 3.20 | Graphical solution of the transcendental equation for the cutoff frequency of a TE surface wave mode. The figure depicts a mode below cutoff.

After finding the constants k_c and h, the field expressions can be derived as

$$H_z(x,y,z) = \begin{cases} B \cos k_c x e^{-j\beta z} & \text{for } 0 \le x \le d \\ B \cos k_c d e^{-h(x-d)} e^{-j\beta z} & \text{for } d \le x < \infty, \end{cases} \tag{3.175a}$$

$$H_x(x,y,z) = \begin{cases} \dfrac{j\beta}{k_c} B \sin k_c x e^{-j\beta z} & \text{for } 0 \le x \le d \\[2mm] \dfrac{-j\beta}{h} B \cos k_c d e^{-h(x-d)} e^{-j\beta z} & \text{for } d \le x < \infty, \end{cases} \tag{3.175b}$$

$$E_y(x,y,z) = \begin{cases} \dfrac{-j\omega\mu_0}{k_c} B \sin k_c x e^{-j\beta z} & \text{for } 0 \le x \le d \\[2mm] \dfrac{j\omega\mu_0}{h} B \cos k_c d e^{-h(x-d)} e^{-j\beta z} & \text{for } d \le x < \infty. \end{cases} \tag{3.175c}$$

EXAMPLE 3.8   SURFACE WAVE PROPAGATION CONSTANTS

Calculate and plot the propagation constants of the first three propagating surface wave modes of a grounded dielectric sheet with $\epsilon_r = 2.55$, for $d/\lambda_0 = 0$ to 1.2.

Solution
The first three propagating surface wave modes are the TM_0, TE_1, and TM_1 modes. The cutoff frequencies for these modes can be found from (3.167) and (3.174) as

$$TM_0 : f_c = 0 \implies \frac{d}{\lambda_0} = 0,$$

$$TE_1 : f_c = \frac{c}{4d\sqrt{\epsilon_r - 1}} \implies \frac{d}{\lambda_0} = \frac{1}{(4\sqrt{\epsilon_r - 1})},$$

$$TM_1 : f_c = \frac{c}{2d\sqrt{\epsilon_r - 1}} \implies \frac{d}{\lambda_0} = \frac{1}{(2\sqrt{\epsilon_r - 1})}.$$

The propagation constants can be found from the numerical solution of (3.165) and (3.166) for the TM modes and (3.172) and (3.173) for the TE modes. This can be done with a relatively simple root-finding algorithm (see the Point of Interest on root-finding algorithms); the results are shown in Figure 3.21.

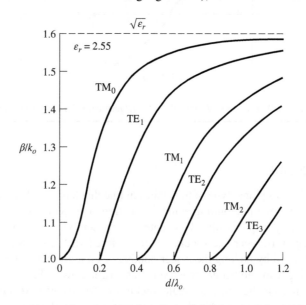

FIGURE 3.21 | Surface wave propagation constants for a grounded dielectric slab with $\epsilon_r = 2.55$.

POINT OF INTEREST: Root-Finding Algorithms

In several examples throughout this book we will need to numerically find the root of a transcendental equation, so it may be useful to review two relatively simple but effective algorithms for doing this. Both methods can be easily programmed.

In the *interval-halving method* the root of $f(x) = 0$ is first bracketed between the values x_1 and x_2. These values can often be estimated from the problem under consideration. If a single root lies between x_1 and x_2, then $f(x_1)f(x_2) < 0$. An estimate, x_3, of the root is made by halving the interval between x_1 and x_2. Thus,

$$x_3 = \frac{x_1 + x_2}{2}.$$

If $f(x_1)f(x_3) < 0$, then the root must lie in the interval $x_1 < x < x_3$; if $f(x_3)f(x_2) < 0$, the root must be in the interval $x_3 < x < x_2$. A new estimate, x_4, can be made by halving the appropriate interval, and this process is repeated until the location of the root has been determined with the desired accuracy. The accompanying figure illustrates this algorithm for several iterations.

The *Newton–Raphson* method begins with an estimate, x_1, of the root of $f(x) = 0$. Then a new estimate, x_2, is obtained from the formula

$$x_2 = x_1 - \frac{f(x_1)}{f'(x_1)},$$

where $f'(x_1)$ is the derivative of $f(x)$ at x_1. This result is easily derived from a two-term Taylor series expansion of $f(x)$ near $x = x_1$: $f(x) = f(x_1) + (x - x_1)f'(x_1)$. It can also be interpreted geometrically as fitting a straight line at $x = x_1$ with the same slope as $f(x)$ at this point; this line then intercepts the x-axis at $x = x_2$, as shown in the figure. Reapplying the above formula gives improved estimates of the root. Convergence is generally much faster than with the interval-halving method, but a disadvantage is that the derivative of $f(x)$ is required; this can often be computed numerically. The Newton–Raphson technique can easily be applied to the case where the root is complex (a situation that occurs, for example, when finding the propagation constant of a line or guide with loss).

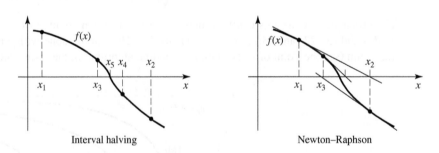

Interval halving   Newton–Raphson

Reference: R. W. Hornbeck, *Numerical Methods*, Quantum Publishers, New York, 1975.

3.7 STRIPLINE

Stripline is a planar type of transmission line that lends itself well to microwave integrated circuitry, miniaturization, and photolithographic fabrication. The geometry of stripline is shown in Figure 3.22a. A thin conducting strip of width W is centered between two wide conducting ground planes of separation b, and the region between the ground planes is filled with a dielectric material. In practice stripline is usually constructed by etching the center conductor on a grounded dielectric substrate of thickness $b/2$ and then covering with another grounded substrate. Variations of the basic geometry of Figure 3.22a include stripline with differing dielectric substrate thicknesses (*asymmetric stripline*) or different dielectric constants (*inhomogeneous stripline*). Air dielectric is sometimes used when it is necessary to minimize loss. An example of a stripline circuit is shown in Figure 3.23.

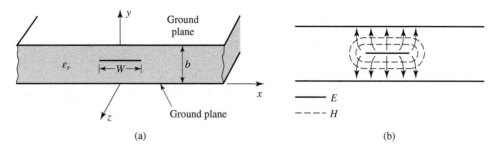

FIGURE 3.22 | Stripline transmission line. (a) Geometry. (b) Electric and magnetic field lines.

Because stripline has two conductors and a homogeneous dielectric, it supports a TEM wave, and this is the usual mode of operation. Like parallel plate guide and coaxial line, however, stripline can also support higher order waveguide modes. These can usually be avoided in practice by restricting both the ground plane spacing and the sidewall width to less than $\lambda_d/2$. Shorting vias between the ground planes are often used to enforce this condition relative to the sidewall width. Shorting vias should also be used to eliminate higher order modes that can be generated when an asymmetry is introduced between the ground planes (e.g., when a surface-mounted coaxial transition is used).

Intuitively, one can think of stripline as a sort of "flattened-out" coax—both have a center conductor completely enclosed by an outer conductor and are uniformly filled with a dielectric medium. A sketch of the field lines for stripline is shown in Figure 3.22b.

The geometry of stripline does not lend itself to the simple analyses that were used for previously treated transmission lines and waveguides. Because we will be concerned primarily with the TEM mode of stripline, an electrostatic analysis is sufficient to give the propagation constant and characteristic impedance. An exact

FIGURE 3.23 | Photograph of a stripline circuit assembly (cover removed), showing four quadrature hybrids, open-circuit tuning stubs, and coaxial transitions.

solution of Laplace's equation is possible by a conformal mapping approach [6], but the procedure and results are cumbersome. Instead, we will present closed-form expressions that give good approximations to the exact results and then discuss an approximate numerical technique for solving Laplace's equation for a geometry similar to stripline.

Formulas for Propagation Constant, Characteristic Impedance, and Attenuation

From Section 3.1 we know that the phase velocity of a TEM mode is given by

$$v_p = 1/\sqrt{\mu_0 \epsilon_0 \epsilon_r} = c/\sqrt{\epsilon_r}, \tag{3.176}$$

and thus the propagation constant of stripline is

$$\beta = \frac{\omega}{v_p} = \omega \sqrt{\mu_0 \epsilon_0 \epsilon_r} = \sqrt{\epsilon_r} k_0. \tag{3.177}$$

In (3.176), $c = 3 \times 10^8$ m/sec is the speed of light in free-space. Using (2.13) and (2.16) allows us to write the characteristic impedance of a transmission line as

$$Z_0 = \sqrt{\frac{L}{C}} = \frac{\sqrt{LC}}{C} = \frac{1}{v_p C}, \tag{3.178}$$

where L and C are the inductance and capacitance per unit length of the line. Thus, we can find Z_0 if we know C. As mentioned previously, Laplace's equation can be solved by conformal mapping to find the capacitance per unit length of stripline, but the resulting solution involves complicated special functions [6], so for practical computations simple formulas have been developed by curve fitting to the exact solution [6, 7]. The resulting formula for characteristic impedance is

$$Z_0 = \frac{30\pi}{\sqrt{\epsilon_r}} \frac{b}{W_e + 0.441b}, \tag{3.179a}$$

where W_e is the *effective width* of the center conductor given by

$$\frac{W_e}{b} = \frac{W}{b} - \begin{cases} 0 & \text{for } \dfrac{W}{b} > 0.35 \\[2mm] (0.35 - W/b)^2 & \text{for } \dfrac{W}{b} < 0.35. \end{cases} \tag{3.179b}$$

These formulas assume a strip with zero thickness and are quoted as being accurate to about 1% of the exact results. It is seen from (3.179) that the characteristic impedance decreases as the strip width W increases.

When designing stripline circuits one usually needs to find the strip width, given the characteristic impedance (and height b and relative permittivity ϵ_r), which requires the inverse of the formulas in (3.179). Such formulas have been derived as

$$\frac{W}{b} = \begin{cases} x & \text{for } \sqrt{\epsilon_r} Z_0 < 120 \; \Omega \\[2mm] 0.85 - \sqrt{0.6 - x} & \text{for } \sqrt{\epsilon_r} Z_0 > 120 \; \Omega, \end{cases} \tag{3.180a}$$

where

$$x = \frac{30\pi}{\sqrt{\epsilon_r} Z_0} - 0.441. \tag{3.180b}$$

Since stripline is a TEM line, the attenuation due to dielectric loss is of the same form as that for other TEM lines and is given in (3.30). The attenuation due to conductor loss can be found by the perturbation method or Wheeler's incremental inductance rule. An approximate result is

$$\alpha_c = \begin{cases} \dfrac{2.7 \times 10^{-3} R_s \epsilon_r Z_0}{30\pi(b-t)} A & \text{for } \sqrt{\epsilon_r} Z_0 < 120 \; \Omega \\[4mm] \dfrac{0.16 R_s}{Z_0 b} B & \text{for } \sqrt{\epsilon_r} Z_0 > 120 \; \Omega \end{cases} \quad \text{Np/m}, \tag{3.181}$$

with

$$A = 1 + \frac{2W}{b-t} + \frac{1}{\pi}\frac{b+t}{b-t}\ln\left(\frac{2b-t}{t}\right),$$

$$B = 1 + \frac{b}{(0.5W + 0.7t)}\left(0.5 + \frac{0.414t}{W} + \frac{1}{2\pi}\ln\frac{4\pi W}{t}\right),$$

where t is the thickness of the strip.

EXAMPLE 3.9 STRIPLINE DESIGN

Find the width of a 50 Ω copper stripline conductor with $b = 0.5$ cm and $\epsilon_r = 3.0$. The loss tangent is 0.002 and the operating frequency is 8 GHz. Calculate the attenuation in dB/λ. Assume a conductor thickness of $t = 0.003$ cm, and the surface resistance of 0.03 Ω.

Solution

Because $\sqrt{\epsilon_r}Z_0 = \sqrt{3} \times 50 = 86.603 < 120$ and $x = \frac{30\pi}{\sqrt{\epsilon_r}Z_0} - 0.441 = \frac{30\pi}{\sqrt{3} \times 50} - 0.441 = 0.6474$

The strip width is

$$\frac{W}{b} = x \Rightarrow W = bx = (0.5 \times 10^{-2})(0.6474) = 0.003237 \text{ m}$$

The wave number at $f = 8$ GHz

$$k = \frac{2\pi f \sqrt{\epsilon_r}}{c} = \frac{2\pi(8 \times 10^9)\sqrt{3}}{3 \times 10^8} = 290.246 \text{ m}^{-1}$$

The dielectric attenuation is

$$\alpha_d = \frac{k\tan\delta}{2} = \frac{(290.246)(0.002)}{2} = 0.29 \text{ Np/m}$$

The conductor attenuation is

$$A = 1 + \frac{2W}{b-t} + \frac{1}{\pi}\frac{b+t}{b-t}\ln\left(\frac{2b-t}{t}\right)$$

$$A = 1 + \frac{2(0.003237)}{(0.005 - 0.003 \times 10^{-2})} + \frac{1}{\pi}\frac{(0.005 + 0.003 \times 10^{-2})}{(0.005 - 0.003 \times 10^{-2})}\ln\left(\frac{2(0.005) - 0.003 \times 10^{-2}}{0.003 \times 10^{-2}}\right)$$

$$A = 4.88(4.173)$$

$$\alpha_c = \frac{2.7 \times 10^{-3}R_s\epsilon_r Z_0 A}{30\pi(b-t)}$$

$$\alpha_c = \frac{2.7 \times 10^{-3}(0.03)(3)(50)(4.88)}{30\pi(0.005 - 0.003 \times 10^{-2})} = 0.1259 \text{ Np/m}$$

Total loss:

$$\alpha = \alpha_d + \alpha_c = 0.29 + 0.1259 = 0.4159 \text{ Np/m}$$

In dB:

$$\alpha(\text{dB/m}) = 20\log e^\alpha = 20\log e^{0.4159} = 3.6123 \text{ dB/m}$$

The guided wavelength in m

$$\lambda_g = \frac{c}{\sqrt{\epsilon_r}f} = \frac{3 \times 10^8}{\sqrt{3} \times 8 \times 10^9} = 0.02165 \text{ m}$$

The attenuation in dB/λ:

$$\alpha(\text{dB}/\lambda) = (3.6123)(0.02165) = 0.0782 \text{ dB}/\lambda$$

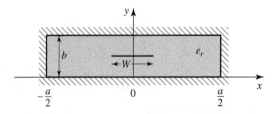

FIGURE 3.24 | Geometry of enclosed stripline.

An Approximate Electrostatic Solution

Many practical problems in microwave engineering are very complicated and do not lend themselves to straightforward analytical solutions but require some sort of numerical approach. Thus it is useful for the student to become aware of such techniques; we will introduce such methods when appropriate throughout this book, beginning with a numerical solution for the characteristic impedance of stripline.

We know that the fields of the TEM mode on stripline must satisfy Laplace's equation, (3.11), in the region between the two parallel plates. The idealized stripline geometry of Figure 3.22a extends to $\pm\infty$, which makes the analysis more difficult. Because we suspect, from the field line drawing of Figure 3.22b, that the field lines do not extend very far away from the center conductor, we can simplify the geometry by truncating the plates beyond some distance, say $|x| > a/2$, and placing metal walls on the sides. Thus, the geometry we will analyze is shown in Figure 3.24, where $a \gg b$, so that the fields around the center conductor are not perturbed by the sidewalls. We then have a closed finite region in which the potential $\Phi(x, y)$ satisfies Laplace's equation,

$$\nabla_t^2 \Phi(x, y) = 0 \quad \text{for } |x| \leq a/2, \, 0 \leq y \leq b, \tag{3.182}$$

with the boundary conditions

$$\Phi(x, y) = 0, \qquad \text{at } x = \pm a/2, \tag{3.183a}$$

$$\Phi(x, y) = 0, \qquad \text{at } y = 0, b. \tag{3.183b}$$

Laplace's equation can be solved by the method of separation of variables. Because the center conductor at $y = b/2$ will contain a surface charge density, the potential $\Phi(x, y)$ will have a slope discontinuity there because $\bar{D} = -\epsilon_0 \epsilon_r \nabla_t \Phi$ is discontinuous at $y = b/2$. Therefore, separate solutions for $\Phi(x, y)$ must be found for $0 < y < b/2$ and $b/2 < y < b$. The general solutions for $\Phi(x, y)$ in these two regions can be written as

$$\Phi(x, y) = \begin{cases} \displaystyle\sum_{\substack{n=1 \\ \text{odd}}}^{\infty} A_n \cos\frac{n\pi x}{a} \sinh\frac{n\pi y}{a} & \text{for } 0 \leq y \leq b/2 \\[4mm] \displaystyle\sum_{\substack{n=1 \\ \text{odd}}}^{\infty} B_n \cos\frac{n\pi x}{a} \sinh\frac{n\pi}{a}(b - y) & \text{for } b/2 \leq y \leq b. \end{cases} \tag{3.184}$$

Only the odd-n terms are needed in (3.184) because the solution is an even function of x. The reader can verify by substitution that (3.184) satisfies Laplace's equation in the two regions and satisfies the boundary conditions of (3.183).

The potential must be continuous at $y = b/2$, which from (3.184) leads to

$$A_n = B_n. \tag{3.185}$$

The remaining set of unknown coefficients, A_n, can be found by solving for the charge density on the center strip. Because $E_y = -\partial\Phi/\partial y$, we have

$$E_y = \begin{cases} -\sum_{\substack{n=1 \\ \text{odd}}}^{\infty} A_n\left(\frac{n\pi}{a}\right)\cos\frac{n\pi x}{a}\cosh\frac{n\pi y}{a} & \text{for } 0 \le y \le b/2 \\ \sum_{\substack{n=1 \\ \text{odd}}}^{\infty} A_n\left(\frac{n\pi}{a}\right)\cos\frac{n\pi x}{a}\cosh\frac{n\pi}{a}(b-y) & \text{for } b/2 \le y \le b. \end{cases}$$

(3.186)

The surface charge density on the strip at $y = b/2$ is

$$\rho_s = D_y(x, y = b/2^+) - D_y(x, y = b/2^-)$$

$$= \epsilon_0\epsilon_r[E_y(x, y = b/2^+) - E_y(x, y = b/2^-)]$$

$$= 2\epsilon_0\epsilon_r \sum_{\substack{n=1 \\ \text{odd}}}^{\infty} A_n\left(\frac{n\pi}{a}\right)\cos\frac{n\pi x}{a}\cosh\frac{n\pi b}{2a},$$

(3.187)

which is seen to be a Fourier series in x for the surface charge density, ρ_s, on the strip at $y = b/2$. If we know the surface charge density we could easily find the unknown constants, A_n, and then the capacitance. We do not know the exact surface charge density, but we can make a good guess by approximating it as a constant over the width of the strip,

$$\rho_s(x) = \begin{cases} 1 & \text{for } |x| < W/2 \\ 0 & \text{for } |x| > W/2. \end{cases}$$

(3.188)

Equating this to (3.187) and using the orthogonality properties of the $\cos(n\pi x/a)$ functions gives the constants A_n as

$$A_n = \frac{2a\sin(n\pi W/2a)}{(n\pi)^2\epsilon_0\epsilon_r\cosh(n\pi b/2a)}.$$

(3.189)

The voltage of the strip conductor relative to the bottom conductor can be found by integrating the vertical electric field from $y = 0$ to $b/2$. Because the solution is approximate, this voltage is not constant over the width of the strip but varies with position, x. Rather than choosing the voltage at an arbitrary position, we can obtain an improved result by averaging the voltage over the width of the strip:

$$V_{\text{avg}} = \frac{1}{W}\int_{-W/2}^{W/2}\int_0^{b/2} E_y(x, y)\, dy\, dx = \sum_{\substack{n=1 \\ \text{odd}}}^{\infty} A_n\left(\frac{2a}{n\pi W}\right)\sin\frac{n\pi W}{2a}\sinh\frac{n\pi b}{2a}.$$

(3.190)

The total charge per unit length on the center conductor is

$$Q = \int_{-W/2}^{W/2} \rho_s(x)\, dx = W \text{ Coul/m},$$

(3.191)

so the capacitance per unit length of the stripline is

$$C = \frac{Q}{V_{\text{avg}}} = \frac{W}{\displaystyle\sum_{\substack{n=1 \\ \text{odd}}}^{\infty} A_n\left(\frac{2a}{n\pi W}\right)\sin\frac{n\pi W}{2a}\sinh\frac{n\pi b}{2a}} \text{ F/m}.$$

(3.192)

Finally, the characteristic impedance is given by

$$Z_0 = \sqrt{\frac{L}{C}} = \frac{\sqrt{LC}}{C} = \frac{1}{v_p C} = \frac{\sqrt{\epsilon_r}}{cC},$$

where $c = 3 \times 10^8$ m/sec.

EXAMPLE 3.10 NUMERICAL CALCULATION OF STRIPLINE IMPEDANCE

Evaluate the above expressions for a stripline having $\epsilon_r = 2.55$ and $a = 100b$ to find the characteristic impedance for $W/b = 0.25$ to 5.0. Compare with the results from (3.179).

Solution

A computer program was written to evaluate (3.192). The series was truncated after 500 terms, and the results for Z_0 are as follows.

W/b	Z_0, Ω Numerical, Eq. (3.192)	Z_0, Ω Formula, Eq. (3.179)	Z_0, Ω Commercial CAD
0.25	90.9	86.6	85.3
0.50	66.4	62.7	61.7
1.0	43.6	41.0	40.2
2.0	25.5	24.2	24.4
5.0	11.1	10.8	11.9

We see that the results are in reasonable agreement with the closed-form equations of (3.179) and the results from a commercial CAD package, particularly for wider strips where the charge density is closer to uniform. Better results could be obtained if more sophisticated estimates were used for the charge density.

3.8 MICROSTRIP LINE

Microstrip line is one of the most popular types of planar transmission lines primarily because it can be fabricated by photolithographic processes and is easily miniaturized and integrated with both passive and active microwave devices. The geometry of a microstrip line is shown in Figure 3.25a. A conductor of width W is printed on a thin, grounded dielectric substrate of thickness d and relative permittivity ϵ_r; a sketch of the field lines is shown in Figure 3.25b.

If the dielectric substrate were not present ($\epsilon_r = 1$), we would have a two-wire line consisting of a flat strip conductor over a ground plane, embedded in a homogeneous medium (air). This would constitute a simple TEM transmission line with phase velocity $v_p = c$ and propagation constant $\beta = k_0$.

The presence of the dielectric, particularly the fact that the dielectric does not fill the region above the strip ($y > d$), complicates the behavior and analysis of microstrip line. Unlike stripline, where all the fields are contained within a homogeneous dielectric region, microstrip has some (usually most) of its field lines in the dielectric region between the strip conductor and the ground plane and some fraction in the air region above the substrate. For this reason microstrip line cannot support a pure TEM wave since the phase velocity of TEM fields in the dielectric region would be $c/\sqrt{\epsilon_r}$, while the phase velocity of TEM fields in the air region would be c, so a phase-matching condition at the dielectric–air interface would be impossible to enforce.

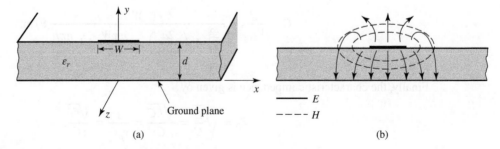

FIGURE 3.25 | Microstrip transmission line. (a) Geometry. (b) Electric and magnetic field lines.

In actuality, the exact fields of a microstrip line constitute a hybrid TM-TE wave and require more advanced analysis techniques than we are prepared to deal with here. In most practical applications, however, the dielectric substrate is electrically very thin ($d \ll \lambda$), and so the fields are quasi-TEM. In other words, the fields are essentially the same as those of the static (DC) case. Thus, good approximations for the phase velocity, propagation constant, and characteristic impedance can be obtained from static, or *quasi-static*, solutions. Then the phase velocity and propagation constant can be expressed as

$$v_p = \frac{c}{\sqrt{\epsilon_e}}, \tag{3.193}$$

$$\beta = k_0 \sqrt{\epsilon_e}, \tag{3.194}$$

where ϵ_e is the *effective dielectric constant* of the microstrip line. Because some of the field lines are in the dielectric region and some are in air, the effective dielectric constant satisfies the relation

$$1 < \epsilon_e < \epsilon_r$$

and depends on the substrate dielectric constant, the substrate thickness, the conductor width, and the frequency.

We will present approximate design formulas for the effective dielectric constant, characteristic impedance, and attenuation of microstrip line; these results are curve-fit approximations to rigorous quasi-static solutions [8, 9]. Then we will discuss additional aspects of microstrip lines, including frequency-dependent effects, higher order modes, and parasitic effects.

Formulas for Effective Dielectric Constant, Characteristic Impedance, and Attenuation

The effective dielectric constant of a microstrip line is given approximately by

$$\epsilon_e = \frac{\epsilon_r + 1}{2} + \frac{\epsilon_r - 1}{2} \frac{1}{\sqrt{1 + 12d/W}}. \tag{3.195}$$

The effective dielectric constant can be interpreted as the dielectric constant of a homogeneous medium that equivalently replaces the air and dielectric regions of the microstrip line, as shown in Figure 3.26. The phase velocity and propagation constant are then given by (3.193) and (3.194).

Given the dimensions of the microstrip line, the characteristic impedance can be calculated as

$$Z_0 = \begin{cases} \dfrac{60}{\sqrt{\epsilon_e}} \ln\left(\dfrac{8d}{W} + \dfrac{W}{4d}\right) & \text{for } W/d \leq 1 \\[3ex] \dfrac{120\pi}{\sqrt{\epsilon_e}\left[W/d + 1.393 + 0.667\ln\left(W/d + 1.444\right)\right]} & \text{for } W/d \geq 1. \end{cases} \tag{3.196}$$

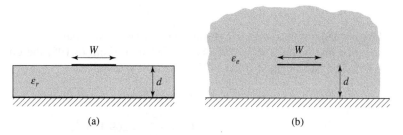

(a)   (b)

FIGURE 3.26 | Equivalent geometry of a quasi-TEM microstrip line. (a) Original geometry. (b) Equivalent geometry, where the dielectric substrate of relative permittivity ϵ_r is replaced with a homogeneous medium of effective relative permittivity ϵ_e.

For a given characteristic impedance Z_0 and dielectric constant ϵ_r, the W/d ratio can be found as

$$\frac{W}{d} = \begin{cases} \dfrac{8e^A}{e^{2A} - 2} & \text{for } W/d < 2 \\[2ex] \dfrac{2}{\pi}\left[B - 1 - \ln(2B - 1) + \dfrac{\epsilon_r - 1}{2\epsilon_r}\left\{ \ln(B - 1) + 0.39 - \dfrac{0.61}{\epsilon_r} \right\} \right] & \text{for } W/d > 2, \end{cases}$$ (3.197)

where

$$A = \frac{Z_0}{60}\sqrt{\frac{\epsilon_r + 1}{2}} + \frac{\epsilon_r - 1}{\epsilon_r + 1}\left(0.23 + \frac{0.11}{\epsilon_r} \right)$$

$$B = \frac{377\pi}{2Z_0\sqrt{\epsilon_r}}.$$

Considering a microstrip line as a quasi-TEM line, we can determine the attenuation due to dielectric loss as

$$\alpha_d = \frac{k_0\epsilon_r(\epsilon_e - 1)\tan\delta}{2\sqrt{\epsilon_e}(\epsilon_r - 1)} \text{ Np/m,}$$ (3.198)

where $\tan\delta$ is the loss tangent of the dielectric. This result is derived from (3.30) by multiplying by a "filling factor,"

$$\frac{\epsilon_r(\epsilon_e - 1)}{\epsilon_e(\epsilon_r - 1)},$$

which accounts for the fact that the fields around the microstrip line are partly in air (lossless) and partly in the dielectric (lossy). The attenuation due to conductor loss is given approximately by [8]

$$\alpha_c = \frac{R_s}{Z_0 W} \text{ Np/m,}$$ (3.199)

where $R_s = \sqrt{\omega\mu_0/2\sigma}$ is the surface resistivity of the conductor. For most microstrip substrates, conductor loss is more significant than dielectric loss; exceptions may occur, however, with some semiconductor substrates.

EXAMPLE 3.11   MICROSTRIP LINE DESIGN

Design a microstrip line on a 0.5 mm alumina substrate ($\epsilon_r = 9.9, \tan\delta = 0.001$) for a 50 Ω characteristic impedance. Find the length of this line required to produce a phase delay of 270° at 10 GHz, and compute the total loss on this line, assuming copper conductors. Compare the results obtained from the approximate formulas of (3.195)–(3.199) with those from a microwave CAD package.

Solution
First find W/d for $Z_0 = 50 \Omega$, and initially guess that $W/d < 2$. From (3.197),

$$A = 2.142, \quad W/d = 0.9654.$$

So the condition that $W/d < 2$ is satisfied; otherwise we would use the expression for $W/d > 2$. Then the required line width is $W = 0.9654d = 0.483$ mm. From (3.195) the effective dielectric constant is $\epsilon_e = 6.665$. The line length, ℓ, for a 270° phase shift is found as

$$\phi = 270° = \beta\ell = \sqrt{\epsilon_e}k_0\ell,$$

$$k_0 = \frac{2\pi f}{c} = 209.4 \text{ m}^{-1},$$

$$\ell = \frac{270°(\pi/180°)}{\sqrt{\epsilon_e}k_0} = 8.72 \text{ mm.}$$

Attenuation due to dielectric loss is found from (3.198) as $\alpha_d = 0.255$ Np/m $= 0.022$ dB/cm. The surface resistivity for copper at 10 GHz is 0.026 Ω, and the attenuation due to conductor loss is, from (3.199), $\alpha_c = 0.0108$ Np/cm $= 0.094$dB/cm. The total loss on the line is then 0.101 dB.

A commercial microwave CAD package gives the following results: $W = 0.478$ mm, $\epsilon_e = 6.83$, $\ell = 8.61$ mm, $\alpha_d = 0.022$ dB/cm, and $\alpha_c = 0.054$ dB/cm. The approximate formulas give results that are within a few percent of the CAD data for linewidth, effective dielectric constant, line length, and dielectric attenuation. The greatest discrepancy occurs for the attenuation constant for conductor loss.

Frequency-Dependent Effects and Higher Order Modes

The results for the parameters of microstrip line presented in the previous section were based on the quasi-static approximation and are strictly valid only at DC (or very low frequencies). At higher frequencies a number of effects can occur that lead to variations from the quasi-static results for effective dielectric constant, characteristic impedance, and attenuation of microstrip line. In addition, new effects can arise, such as higher order modes and parasitic reactances.

Because microstrip line is not a true TEM line, its propagation constant is not a linear function of frequency, meaning that the effective dielectric constant varies with frequency. The electromagnetic field that exists on microstrip line involves a hybrid coupling of TM and TE modes, complicated by the boundary condition imposed by the air and dielectric substrate interface. In addition, the current on the strip conductor is not uniform across the width of the strip, and this distribution varies with frequency. The thickness of the strip conductor also has an effect on the current distribution and hence affects the line parameters (especially the conductor loss).

The variation with frequency of the parameters of a transmission line is important for several reasons. First, if the variation is significant it becomes important to know and use the parameters at the particular frequency of interest to avoid errors in design or analysis. Typically, for microstrip line, the frequency variation of the effective dielectric constant is more significant than the variation of characteristic impedance, both in terms of relative change and the relative effect on performance. A change in the effective dielectric constant may have a substantial effect on the phase delay through a long section of line, while a small change in characteristic impedance has the primary effect of introducing a small impedance mismatch. Second, a variation in line parameters with frequency means that different frequency components of a broadband signal will propagate differently. A variation in phase velocity, for example, means that different frequency components will arrive at the output of the line at different times, leading to *signal dispersion* and distortion of the input signal. Third, because of the complexity of modeling these effects, approximate formulas are generally useful only for a limited range of frequency and line parameters, and numerical computer models are usually more accurate and useful.

There are a number of approximate formulas, developed from numerical computer solutions and/or experimental data, that have been suggested for predicting the frequency variation of microstrip line parameters [8, 9]. A popular frequency-dependent model for the effective dielectric constant has a form similar to the following formula [8]:

$$\epsilon_e(f) = \epsilon_r - \frac{\epsilon_r - \epsilon_e(0)}{1 + G(f)}, \tag{3.200}$$

where $\epsilon_e(f)$ represents the frequency-dependent effective dielectric constant, ϵ_r is the relative permittivity of the substrate, and $\epsilon_e(0)$ is the effective dielectric constant of the line at DC, as given by (3.195). The function $G(f)$ can take various forms, but one suggested in reference [8] is that $G(f) = g\left(f/f_p\right)^2$, with $g = 0.6 + 0.009 Z_0$ and $f_p = Z_0/8\pi d$ (Z_0 is in ohms, f is in GHz, and d is in cm). It can be seen from the form of (3.200) that $\epsilon_e(f)$ reduces to the DC value $\epsilon_e(0)$ when $f = 0$ and increases toward ϵ_r as frequency increases.

Approximate formulas like the above were primarily developed in the years before computer-aided design tools for RF and microwave engineering became commonly available (see the Point of Interest on computer-aided design in Chapter 4). Such tools usually give accurate results for a wide range of line parameters and today are usually preferred over closed-form approximations.

Another potential difficulty with microstrip line is that it may support several types of higher order modes, particularly at higher frequencies. Some of these are directly related to the TM and TE surface waves modes that were discussed in Section 3.6, while others are related to waveguide-type modes in the cross section of the line.

The TM_0 surface wave mode for a grounded dielectric substrate has a zero cutoff frequency, as we know from (3.167). Because some of the field lines of this mode are aligned with the field lines of the quasi-TEM mode of a microstrip line, it is possible for coupling to occur from the desired microstrip mode to a surface wave, leading to excess power loss and possibly undesired coupling to adjacent microstrip elements. Because the fields of the TM_0 surface wave are zero at DC, there is little coupling to the quasi-TEM microstrip mode until a critical frequency is reached. Studies have shown that this threshold frequency is greater than zero and less than the cutoff frequency of the TM_1 surface wave mode. A commonly used approximation is [8]

$$f_{T1} \simeq \frac{c}{2\pi d}\sqrt{\frac{2}{\epsilon_r - 1}}\tan^{-1}\epsilon_r. \tag{3.201}$$

For ϵ_r ranging from 1 to 10, (3.201) gives a frequency that is 35% to 66% of f_{c1}, the cutoff frequency of the TM_1 surface wave mode.

When a microstrip circuit has transverse discontinuities (such as bends, junctions, or even step changes in width), the transverse currents on the conductors that are generated may allow coupling to TE surface wave modes. Most practical microstrip circuits involve such discontinuities, so this type of coupling is often important. The minimum threshold frequency where such coupling becomes important is given by the cutoff of the TE_1 surface wave, from (3.174):

$$f_{T2} \simeq \frac{c}{4d\sqrt{\epsilon_r - 1}}. \tag{3.202}$$

For wide microstrip lines, it is possible to excite a transverse resonance along the x axis of the microstrip line below the strip in the dielectric region because the sides below the strip conductor appear approximately as magnetic walls. This condition occurs when the width is about $\lambda/2$ in the dielectric, but because of field fringing the effective width of the strip is somewhat larger than the physical width. A rough approximation for the effective width is $W + d/2$, so the approximate threshold frequency for transverse resonance is

$$f_{T3} \simeq \frac{c}{\sqrt{\epsilon_r}(2W + d)}. \tag{3.203}$$

It is rare that a microstrip line is wide enough to approach this limit in practice.

Finally, a parallel plate–type waveguide mode may propagate when the vertical spacing between the strip conductor and ground plane approaches $\lambda/2$ in the dielectric. Thus, an approximation for the threshold frequency for this mode (valid for wide microstrip lines) can be given as

$$f_{T4} \simeq \frac{c}{2d\sqrt{\epsilon_r}}. \tag{3.204}$$

Thinner microstrip lines will have more fringing field that effectively lengthens the path between the strip and ground plane, thus reducing the threshold frequency by as much as 50%.

The net effect of the threshold frequencies given in (3.201)–(3.204) is to impose an upper frequency limit of operation for a given microstrip geometry. This limit is a function of the substrate thickness, dielectric constant, and strip width.

EXAMPLE 3.12 FREQUENCY DEPENDENCE OF EFFECTIVE DIELECTRIC CONSTANT

Use the approximate formula of (3.200) to plot the change in effective dielectric constant over frequency for a 25 Ω microstrip line on a substrate having a relative permittivity of 10.0 and a thickness of 0.65 mm. Compare the approximate data with results from a CAD model for frequencies up to 20 GHz. Compare the calculated phase delay at 10 GHz through a 1.093 cm length of line when using $\epsilon_e(0)$ versus $\epsilon_e(10 \text{ GHz})$.

Solution

The required linewidth for a 25 Ω impedance is $w = 2.00$ mm. The effective dielectric constant for this line at low frequencies can be found from (3.195) to be $\epsilon_e(0) = 7.53$. A short computer program was used to calculate the effective dielectric constant as a function of frequency using (3.200), and the result is shown in Figure 3.27. Comparison with a commercial microwave CAD package shows that the approximate model is reasonably accurate up to about 10 GHz but gives an overestimate at higher frequencies.

Using an effective dielectric constant of $\epsilon_e(0) = 7.53$, we find the phase delay through a 1.093 cm length of line to be $\phi_0 = \sqrt{\epsilon_e(0)}k_0\ell = 360°$. The effective dielectric constant at 10 GHz is 8.120 (CAD), with a corresponding phase delay of $\phi_{10} = \sqrt{\epsilon_e(10\ \text{GHz})}k_0\ell = 374°$—an error of about 14°.

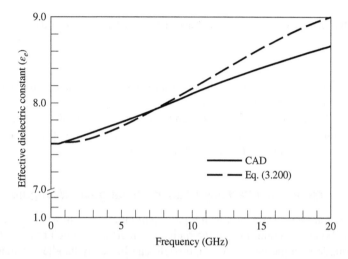

FIGURE 3.27 | Effective dielectric constant versus frequency for the microstrip line of Example 3.12, comparing the approximate model of (3.200) with data from a computer-aided design package.

3.9   THE TRANSVERSE RESONANCE TECHNIQUE

According to the general solutions of Maxwell's equations for TE or TM waves given in Section 3.1, a uniform waveguide structure always has a propagation constant of the form

$$\beta = \sqrt{k^2 - k_c^2} = \sqrt{k^2 - k_x^2 - k_y^2}, \tag{3.205}$$

where $k_c = \sqrt{k_x^2 + k_y^2}$ is the cutoff wave number of the guide and, for a given mode, is a fixed function of the cross-sectional geometry of the guide. Thus, if we know k_c we can determine the propagation constant of the guide. In previous sections we determined k_c by solving the wave equation in the guide, subject to the appropriate boundary conditions. Although this technique is very powerful and general, it can be complicated for complex waveguides, especially if dielectric layers are present. In addition, the wave equation solution gives a complete field description inside the waveguide, which is often more information than we really need if we are only interested in the propagation constant of the guide. The *transverse resonance technique* employs a transmission line model of the transverse cross section of the waveguide and gives a much simpler and more direct solution for the cutoff frequency. This is another example where circuit and transmission line theory offers a simplified alternative to a field theory solution.

The transverse resonance procedure is based on the fact that in a waveguide at cutoff, the fields form standing waves in the transverse plane of the guide, as can be inferred from the "bouncing plane wave" interpretation of waveguide modes discussed in Section 3.2. This situation can be modeled with an equivalent transmission line circuit operating at resonance. One of the conditions of such a resonant line is the fact that, at any point on the line, the sum of the input impedances seen looking to either side must be zero. That is,

$$Z_{\text{in}}^r(x) + Z_{\text{in}}^\ell(x) = 0 \quad \text{for all } x, \tag{3.206}$$

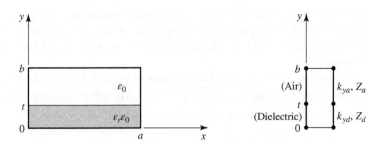

FIGURE 3.28 | A rectangular waveguide partially filled with dielectric and the transverse resonance equivalent circuit.

where $Z_{in}^{r}(x)$ and $Z_{in}^{\ell}(x)$ are the input impedances seen looking to the right and left, respectively, at any point x on the resonant line.

The transverse resonance technique only gives results for the cutoff frequency of the guide. If fields or attenuation due to conductor loss are needed, the complete field theory solution will be required. The procedure will now be illustrated with an example.

TE$_{0n}$ Modes of a Partially Loaded Rectangular Waveguide

The transverse resonance technique is particularly useful when the guide contains dielectric layers because the boundary conditions at the dielectric interfaces, which require the solution of simultaneous algebraic equations in the field theory approach, can be easily handled as junctions of different transmission lines. As an example, consider a rectangular waveguide partially filled with dielectric, as shown in Figure 3.28. To find the cutoff frequencies for the TE$_{0n}$ modes, the equivalent transverse resonance circuit shown in the figure can be used. The line for $0 < y < t$ represents the dielectric-filled part of the guide and has a transverse propagation constant k_{yd} and a characteristic impedance for TE modes given by

$$Z_d = \frac{k\eta}{k_{yd}} = \frac{k_0\eta_0}{k_{yd}}, \tag{3.207a}$$

where $k_0 = \omega\sqrt{\mu_0\epsilon_0}$ and $\eta_0 = \sqrt{\mu_0/\epsilon_0}$. For $t < y < b$, the guide is air filled and has a transverse propagation constant k_{ya} and an equivalent characteristic impedance given by

$$Z_a = \frac{k_0\eta_0}{k_{ya}}. \tag{3.207b}$$

Applying condition (3.206) yields

$$k_{ya} \tan k_{yd}t + k_{yd} \tan k_{ya}(b-t) = 0. \tag{3.208}$$

This equation contains two unknowns, k_{ya} and k_{yd}. An additional equation is obtained from the fact that the longitudinal propagation constant, β, must be the same in both regions for phase matching of the tangential fields at the dielectric interface. Thus, with $k_x = 0$,

$$\beta = \sqrt{\epsilon_r k_0^2 - k_{yd}^2} = \sqrt{k_0^2 - k_{ya}^2},$$

or

$$\epsilon_r k_0^2 - k_{yd}^2 = k_0^2 - k_{ya}^2. \tag{3.209}$$

Equations (3.208) and (3.209) can be solved (numerically or graphically) to obtain k_{yd} and k_{ya}. There will be an infinite number of solutions, corresponding to the n dependence (number of variations in y) of the TE$_{0n}$ mode.

3.10 WAVE VELOCITIES AND DISPERSION

We have so far encountered two types of velocities related to the propagation of electromagnetic waves:

- The speed of light in a medium ($1/\sqrt{\mu\epsilon}$)
- The phase velocity ($v_p = \omega/\beta$)

The speed of light in a medium is the velocity at which a plane wave would propagate in that medium, while the phase velocity is the speed at which a constant phase point travels. For a TEM plane wave, these two velocities are identical, but for other types of guided wave propagation the phase velocity may be greater or less than the speed of light.

If the phase velocity and attenuation of a line or guide are constants that do not change with frequency, then the phase of a signal that contains more than one frequency component will not be distorted. If the phase velocity is different for different frequencies, then the individual frequency components will not maintain their original phase relationships as they propagate down the transmission line or waveguide, and signal distortion will occur. Such an effect is called *dispersion* since different phase velocities allow the "faster" waves to lead in phase relative to the "slower" waves, and the original phase relationships will gradually be dispersed as the signal propagates down the line. In such a case, there is no single phase velocity that can be attributed to the signal as a whole. However, if the bandwidth of the signal is relatively small or if the dispersion is not too severe, a *group velocity* can be defined in a meaningful way. This velocity can be used to describe the speed at which the signal propagates.

Group Velocity

As discussed earlier, the physical interpretation of group velocity is the velocity at which a narrowband signal propagates. We will derive the relation of group velocity to the propagation constant by considering a signal $f(t)$ in the time domain. The Fourier transform of this signal is defined as

$$F(\omega) = \int_{-\infty}^{\infty} f(t)e^{-j\omega t}dt, \tag{3.210a}$$

and the inverse transform is

$$f(t) = \frac{1}{2\pi}\int_{-\infty}^{\infty} F(\omega)e^{j\omega t}d\omega. \tag{3.210b}$$

Now consider the transmission line or waveguide on which the signal $f(t)$ is propagating as a linear system, with a transfer function $Z(\omega)$ that relates the output, $F_o(\omega)$, of the line to the input, $F(\omega)$, of the line, as shown in Figure 3.29. Thus,

$$F_o(\omega) = Z(\omega)F(\omega). \tag{3.211}$$

For a lossless matched transmission line or waveguide, the transfer function $Z(\omega)$ can be expressed as

$$Z(\omega) = Ae^{-j\beta z} = |Z(\omega)|e^{-j\psi}, \tag{3.212}$$

where A is a constant and β is the propagation constant of the line or guide.

The time domain representation of the output signal, $f_o(t)$, can then be written as

$$f_o(t) = \frac{1}{2\pi}\int_{-\infty}^{\infty} F(\omega)|Z(\omega)|e^{j(\omega t - \psi)}d\omega. \tag{3.213}$$

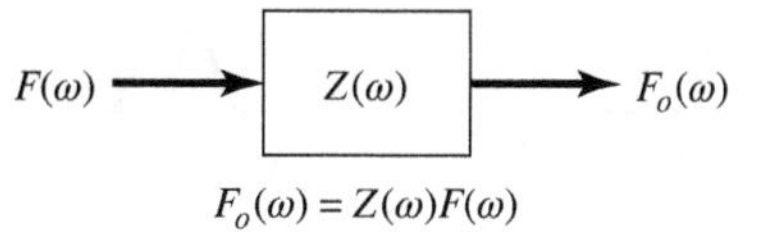

$$F(\omega) \longrightarrow \boxed{Z(\omega)} \longrightarrow F_o(\omega)$$

$$F_o(\omega) = Z(\omega)F(\omega)$$

FIGURE 3.29 | A transmission line or waveguide represented as a linear system with transfer function $Z(\omega)$.

If $|Z(\omega)| = A$ is a constant and the phase ψ of $Z(\omega)$ is a linear function of ω, say $\psi = a\omega$, the output can be expressed as

$$f_o(t) = \frac{1}{2\pi} \int_{-\infty}^{\infty} AF(\omega)e^{j\omega(t-a)}d\omega = Af(t - a), \tag{3.214}$$

which is seen to be a replica of $f(t)$, except for an amplitude factor A and time shift a. Thus, a transfer function of the form $Z(\omega) = Ae^{-j\omega a}$ does not distort the input signal. A lossless TEM wave has a propagation constant $\beta = \omega/c$, which is of this form, so a TEM line is *dispersionless* and does not lead to signal distortion. If the TEM line is lossy, however, the attenuation may be a function of frequency, which could lead to signal distortion.

Now consider a narrowband input signal of the form

$$s(t) = f(t) \cos \omega_0 t = \text{Re} \left\{ f(t)e^{j\omega_0 t} \right\}, \tag{3.215}$$

which represents an amplitude-modulated carrier wave of frequency ω_0. Assume that the highest frequency component of $f(t)$ is ω_m, where $\omega_m \ll \omega_0$. The Fourier transform, $S(\omega)$, of $s(t)$, is

$$S(\omega) = \int_{-\infty}^{\infty} f(t)e^{-j\omega_0 t}e^{j\omega t}dt = F(\omega - \omega_0), \tag{3.216}$$

where we have used the complex form of the input signal as expressed in (3.215). We will need to take the real part of the output inverse transform to obtain the time domain output signal. The spectra of $F(\omega)$ and $S(\omega)$ are depicted in Figure 3.30.

The output signal spectrum is

$$S_o(\omega) = AF(\omega - \omega_0)e^{-j\beta z}, \tag{3.217}$$

and in the time domain,

$$\begin{aligned} s_o(t) &= \frac{1}{2\pi}\text{Re} \int_{-\infty}^{\infty} S_o(\omega)e^{j\omega t}d\omega \\ &= \frac{1}{2\pi}\text{Re} \int_{\omega_0-\omega_m}^{\omega_0+\omega_m} AF(\omega - \omega_0)e^{j(\omega t - \beta z)}d\omega. \end{aligned} \tag{3.218}$$

In general, the propagation constant β may be a complicated function of ω. However, if $F(\omega)$ is narrowband ($\omega_m \ll \omega_0$), then β can often be linearized by using a Taylor series expansion about ω_0:

$$\beta(\omega) = \beta(\omega_0) + \frac{d\beta}{d\omega}\bigg|_{\omega=\omega_0}(\omega - \omega_0) + \frac{1}{2}\frac{d^2\beta}{d\omega^2}\bigg|_{\omega=\omega_0}(\omega - \omega_0)^2 + \cdots. \tag{3.219}$$

Retaining the first two terms gives

$$\beta(\omega) \simeq \beta_0 + \beta_o'(\omega - \omega_0), \tag{3.220}$$

where

$$\beta_o = \beta(\omega_0),$$

$$\beta_o' = \frac{d\beta}{d\omega}\bigg|_{\omega=\omega_0}.$$

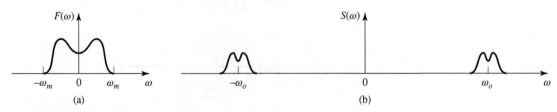

FIGURE 3.30 | Fourier spectra of the signals (a) $f(t)$ and (b) $s(t)$.

After a change of variables to $y = \omega - \omega_o$, the expression for $s_0(t)$ becomes

$$
\begin{aligned}
s_o(t) &= \frac{A}{2\pi} \operatorname{Re} \left\{ e^{j(\omega_o t - \beta_o z)} \int_{-\omega_m}^{\omega_m} F(y) e^{j(t - \beta_o' z) y} \, dy \right\} \\
&= A \operatorname{Re} \left\{ f(t - \beta_o' z) e^{j(\omega_o t - \beta_o z)} \right\} \\
&= A f(t - \beta_o' z) \cos(\omega_o t - \beta_o z),
\end{aligned}
\tag{3.221}
$$

which is a time-shifted replica of the original modulation envelope, $f(t)$, of (3.215). The velocity of this envelope is the group velocity, v_g:

$$
v_g = \frac{1}{\beta_o'} = \left(\frac{d\beta}{d\omega} \right)^{-1} \Bigg|_{\omega = \omega_o}.
\tag{3.222}
$$

EXAMPLE 3.13   WAVEGUIDE WAVE VELOCITIES

Calculate the group velocity for a waveguide mode propagating in an air-filled guide. Compare this velocity to the phase velocity and speed of light.

Solution
The propagation constant for a mode in an air-filled waveguide is

$$
\beta = \sqrt{k_0^2 - k_c^2} = \sqrt{(\omega/c)^2 - k_c^2}.
$$

Taking the derivative with respect to frequency gives

$$
\frac{d\beta}{d\omega} = \frac{\omega/c^2}{\sqrt{(\omega/c)^2 - k_c^2}} = \frac{k_o}{c\beta},
$$

so from (3.222) the group velocity is

$$
v_g = \left(\frac{d\beta}{d\omega} \right)^{-1} = \frac{c\beta}{k_0}.
$$

The phase velocity is $v_p = \omega/\beta = ck_0/\beta$. Since $\beta < k_0$, we have that $v_g < c < v_p$, which indicates that the phase velocity of a waveguide mode may be greater than the speed of light, but the group velocity (the velocity of a narrowband signal) will be less than the speed of light.

3.11 SUMMARY OF TRANSMISSION LINES AND WAVEGUIDES

We have discussed a variety of transmission lines and waveguides in this chapter, and here we will summarize some of the basic properties of these transmission media and their relative advantages in a broader context.

We made a distinction between TEM, TM, and TE waves and saw that transmission lines and waveguides can be categorized according to which type of waves they can support. We saw that TEM waves are nondispersive, with no cutoff frequency, while TM and TE waves exhibit dispersion and generally have nonzero cutoff frequencies. Other electrical considerations include bandwidth, attenuation, and power-handling capacity. Mechanical factors are also very important, however, and include such considerations as physical size (volume and weight), ease of fabrication (cost), and the ability to be integrated with other devices (active or passive). Table 3.6 compares several types of transmission media with regard to these considerations; this table only gives general guidelines, as specific cases may give better or worse results than those indicated.

TABLE 3.6 | Comparison of Common Transmission Lines and Waveguides

Characteristic	Coax	Waveguide	Stripline	Microstrip
Modes: Preferred	TEM	TE_{10}	TEM	Quasi-TEM
Other	TM,TE	TM,TE	TM,TE	Hybrid TM,TE
Dispersion	None	Medium	None	Low
Bandwidth	High	Low	High	High
Loss	Medium	Low	High	High
Power capacity	Medium	High	Low	Low
Physical size	Large	Large	Medium	Small
Ease of fabrication	Medium	Medium	Easy	Easy
Integration with	Hard	Hard	Fair	Easy

Other Types of Lines and Guides

Although we have discussed the most common types of waveguides and transmission lines, there are many other guides and lines (and many variations) that we are not able to present in detail. A few of the more popular types are briefly mentioned here.

Ridge waveguide: The practical bandwidth of rectangular waveguide is slightly less than an octave (a 2:1 frequency range). This is because the TE_{20} mode begins to propagate at a frequency equal to twice the cutoff frequency of the TE_{10} mode. The ridge waveguide, shown in Figure 3.31, consists of a rectangular waveguide loaded with conducting ridges on the top and/or bottom walls. This loading tends to lower the cutoff frequency of the dominant mode, leading to increased bandwidth and better (more constant) impedance characteristics. Ridge waveguides are often used for impedance matching purposes, where the ridge may be tapered along the length of the guide. The presence of the ridge, however, reduces the power-handling capacity of the waveguide.

Dielectric waveguide: As we have seen from our study of surface waves, metallic conductors are not necessary to confine and support a propagating electromagnetic field. The dielectric waveguide shown in Figure 3.32 is another example of such a guide, where ϵ_{r2}, the dielectric constant of the ridge, is usually greater than ϵ_{r1}, the dielectric constant of the substrate. The fields are thus mostly confined to the ridge and the surrounding area. This type of guide supports TM and TE modes, and is convenient for miniaturization and integration with active devices. Its small size makes it useful for millimeter wave to optical frequencies,

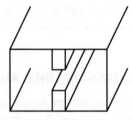

FIGURE 3.31 | Cross section of a ridge waveguide.

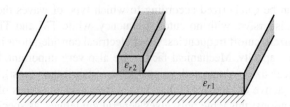

FIGURE 3.32 | Dielectric waveguide geometry.

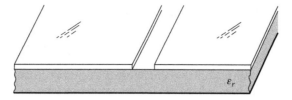

FIGURE 3.33 | Geometry of a printed slotline.

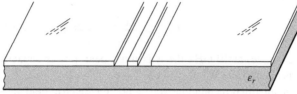

FIGURE 3.34 | Coplanar waveguide geometry.

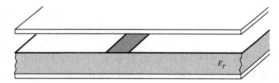

FIGURE 3.35 | Covered microstrip line.

although it can be very lossy at bends or junctions in the ridge line. Many variations in this basic geometry are possible.

Slotline: Slotline is another one of the many possible types of planar transmission lines. The geometry of a slotline is shown in Figure 3.33. It consists of a thin slot in the ground plane on one side of a dielectric substrate. Thus, like microstrip line, the two conductors of slotline lead to a quasi-TEM type of mode. The width of the slot controls the characteristic impedance of the line.

Coplanar waveguide: The coplanar waveguide, shown in Figure 3.34, is similar to the slotline, and can be viewed as a slotline with a third conductor centered in the slot region. Because of the presence of this additional conductor, this type of line can support even or odd quasi-TEM modes, depending on whether the electric fields in the two slots are in the opposite direction or the same direction. Coplanar waveguides are particularly useful for fabricating active circuitry due to the presence of the center conductor and the close proximity of the ground planes.

Covered microstrip: Many variations of the basic microstrip line geometry are possible, but one of the more common is the covered microstrip, shown in Figure 3.35. The metallic cover plate is often used for electrical shielding and physical protection of the microstrip circuitry and is usually situated several substrate thicknesses away from the circuit. Its presence, however, can perturb the operation of the circuit enough so that its effect must be taken into account during design.

POINT OF INTEREST: Power Capacity of Transmission Lines

The power-handling capacity of an air-filled transmission line or waveguide is usually limited by voltage breakdown, which occurs at a field strength of about $E_d = 3 \times 10^6$ V/m for room temperature air at sea level pressure. Thermal effects may also serve to limit the power capacity of some types of lines.

In an air-filled coaxial line the electric field varies as $E_\rho = V_o/(\rho \ln b/a)$, which has a maximum at $\rho = a$ (at the inner conductor). Thus the maximum voltage before breakdown is

$$V_{\max} = E_d a \ln \frac{b}{a} \quad \text{(peak-to-peak)},$$

and the maximum power capacity is then

$$P_{max} = \frac{V_{max}^2}{2Z_0} = \frac{\pi a^2 E_d^2}{\eta_0} \ln \frac{b}{a}.$$

As might be expected, this result shows that power capacity can be increased by using a larger coaxial cable (larger a, b with fixed b/a for the same characteristic impedance). However, propagation of higher order modes limits the maximum operating frequency for a given cable size. Thus, there is an upper limit on the power capacity of a coaxial line for a given maximum operating frequency, f_{max}, which can be shown to be given by

$$P_{max} = \frac{0.025}{\eta_0} \left(\frac{cE_d}{f_{max}} \right)^2 = 5.8 \times 10^{12} \left(\frac{E_d}{f_{max}} \right)^2.$$

As an example, at 10 GHz the maximum peak power capacity of any coaxial line with no higher order modes is about 520 kW.

In an air-filled rectangular waveguide the electric field varies as $E_y = E_o \sin(\pi x/a)$, which has a maximum value of E_o at $x = a/2$ (the middle of the guide). Thus the maximum power capacity before breakdown is

$$P_{max} = \frac{abE_o^2}{4Z_w} = \frac{abE_d^2}{4Z_w},$$

which shows that power capacity increases with guide size. For most standard waveguides, $b \simeq 2a$. To avoid propagation of the TE_{20} mode we must have $a < c/f_{max}$, where f_{max} is the maximum operating frequency. Then the maximum power capacity of the guide can be shown to be

$$P_{max} = \frac{0.11}{\eta_0} \left(\frac{cE_d}{f_{max}} \right)^2 = 2.6 \times 10^{13} \left(\frac{E_d}{f_{max}} \right)^2.$$

As an example, at 10 GHz the maximum peak power capacity of a rectangular waveguide operating in the TE_{10} mode is about 2300 kW, which is considerably higher than the power capacity of a coaxial cable at the same frequency.

Because arcing and voltage breakdown are high-speed transient effects, these voltage and power limits are peak values; average power capacity is lower. In addition, it is good engineering practice to provide a safety factor of at least two, so the maximum powers that can be safely transmitted should be limited to about half of the above values. If there are reflections on the line or guide, the power capacity is further reduced. In the worst case, a reflection coefficient magnitude of unity will double the maximum voltage on the line, so the power capacity will be reduced by a factor of four.

The power capacity of a line can be increased by pressurizing the line with air or an inert gas or by using a dielectric. The dielectric strength (E_d) of most dielectric materials is greater than that of air, but the power capacity may be further limited by the heating of the dielectric due to ohmic loss.

Reference: P. A. Rizzi, *Microwave Engineering—Passive Circuits*, Prentice-Hall, Englewood Cliffs, NJ, 1988.

REFERENCES

[1] O. Heaviside, *Electromagnetic Theory*, Vol. 1, 1893. Reprinted by Dover, New York, 1950.

[2] Lord Rayleigh, "On the Passage of Electric Waves through Tubes," *Philosophical Magazine*, vol. 43, pp. 125–132, 1897. Reprinted in *Collected Papers*, Cambridge University Press, Cambridge, 1903.

[3] K. S. Packard, "The Origin of Waveguides: A Case of Multiple Rediscovery," *IEEE Transactions on Microwave Theory and Techniques*, vol. MTT-32, pp. 961–969, September 1984.

[4] R. M. Barrett, "Microwave Printed Circuits—An Historical Perspective," *IEEE Transactions on Microwave Theory and Techniques*, vol. MTT-32, pp. 983–990, September 1984.

[5] D. D. Grieg and H. F. Englemann, "Microstrip—A New Transmission Technique for the Kilomegacycle Range," *Proceedings of the IRE*, vol. 40, pp. 1644–1650, December 1952.

[6] H. Howe, Jr., *Stripline Circuit Design*, Artech House, Dedham, MA, 1974.

[7] I. J. Bahl and R. Garg, "A Designer's Guide to Stripline Circuits," *Microwaves*, January 1978, pp. 90–96.

[8] I. J. Bahl and D. K. Trivedi, "A Designer's Guide to Microstrip Line," *Microwaves*, May 1977, pp. 174–182.

[9] K. C. Gupta, R. Garg, and I. J. Bahl, *Microstrip Lines and Slotlines*, Artech House, Dedham, MA, 1979.

PROBLEMS

3.1 Devise at least two variations of the basic coaxial transmission line geometry of Section 3.5, and discuss the advantages and disadvantages of your proposed lines in terms of size, loss, cost, higher order modes, dispersion, or other considerations. Repeat this exercise for the microstrip line geometry of Section 3.8.

3.2 Derive equations (3.5a)–(3.5d) from equations (3.3) and (3.4).

3.3 Calculate the attenuation due to conductor loss for the TE_n mode of a parallel plate waveguide.

3.4 Consider a section of air-filled K-band waveguide. From the dimensions given in Appendix J, determine the cutoff frequencies of the first two propagating modes. From the recommended operating range given in Appendix J for this guide, determine the percentage reduction in bandwidth that this operating range represents, relative to the theoretical bandwidth for a single propagating mode.

3.5 A 10 cm length of an X-band copper waveguide is filled with a dielectric material with $\epsilon_r = 2.2$ and $\tan\delta = 0.0009$. If the operating frequency is 10 GHz, find the total loss through the guide and the phase delay from the input to the output of the guide.

3.6 An attenuator can be made using a section of waveguide operating below cutoff, as shown in the accompanying figure. If $a = 2.286$ cm and the operating frequency is 10 GHz, determine the required length of the below-cutoff section of waveguide to achieve an attenuation of 90 dB between the input and output guides. Ignore the effect of reflections at the step discontinuities.

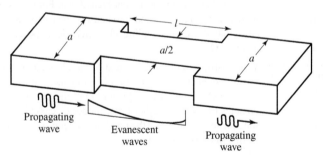

3.7 Find expressions for the electric surface current density on the walls of a rectangular waveguide for a TE_{10} mode. Why can a narrow slot be cut along the centerline of the broad wall of a rectangular waveguide without perturbing the operation of the guide? (Such a slot is often used in a slotted line for a probe to sample the standing wave field inside the guide.)

3.8 Derive the expression for the attenuation of the TM_{mn} mode of a rectangular waveguide due to imperfectly conducting walls.

3.9 For the partially loaded rectangular waveguide shown in the accompanying figure, solve (3.109) with $\beta = 0$ to find the cutoff frequency of the TE_{10} mode. Assume $a = 2.286$ cm, $t = a/2$, and $\epsilon_r = 2.55$.

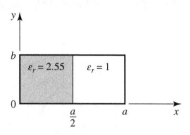

3.10 Consider the partially filled parallel plate waveguide shown in the accompanying figure. Derive the solution (fields and cutoff frequency) for the lowest order TE mode of this structure. Assume the metal plates are infinitely wide. Can a TEM wave propagate on this structure?

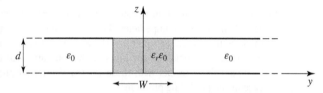

3.11 Derive equations (3.110a)–(3.110d) for the transverse field components in terms of longitudinal fields, in cylindrical coordinates.

3.12 Derive the expression for the attenuation of the TM_{nm} mode in a circular waveguide with finite conductivity.

3.13 A circular copper waveguide has a radius of 0.5 cm and is filled with a dielectric material having $\epsilon_r = 2.2$ and $\tan\delta = 0.0009$. Identify the first four propagating modes and their cutoff frequencies. For the dominant mode, calculate the total attenuation at 22 GHz.

3.14 Derive the $\bar{E}$ and $\bar{H}$ fields of a coaxial line from the expression for the potential given in (3.153). Also find expressions for the voltage and current on the line and the characteristic impedance.

3.15 Derive a transcendental equation for the cutoff frequency of the TM modes of a coaxial waveguide. Using tables, obtain an approximate value of $k_c a$ for the TM_{01} mode if $b/a = 2.5$.

3.16 Derive an expression for the attenuation of a TE surface wave on a grounded dielectric substrate when the ground plane has finite conductivity.

3.17 Consider the grounded magnetic substrate shown in the accompanying figure. Derive a solution for the TM surface waves that can propagate on this structure.

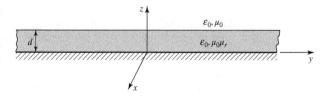

3.18 Consider the partially filled coaxial line shown in the accompanying figure. Can a TEM wave propagate on this line? Derive the solution for the TM_{0m} (no azimuthal variation) modes of this geometry.

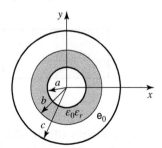

3.19 A copper stripline transmission line is to be designed for a 150 Ω characteristic impedance. The ground plane separation is 1.2 mm and the dielectric constant is 2.25, with $\tan \delta = 0.002$. At 2.4 GHz, find the guide wavelength on the line and the total attenuation.

3.20 A copper microstrip transmission line is to be designed for a 150 Ω characteristic impedance. The substrate is 1.575 mm thick, with $\epsilon_r = 2.20$ and $\tan \delta = 0.0009$. At 2.4 GHz, find the guide wavelength on the line and the total attenuation. Compare these results with those for the similar stripline case of the preceding problem.

3.21 A 125 Ω microstrip line is printed on a substrate of thickness 0.0752 cm with a dielectric constant of 2.2. Ignoring losses and fringing fields, find the shortest length of this line that appears at its input as a capacitor of 5 pF at 2.4 GHz. Repeat for an inductance of 5 nH. Using a microwave CAD package with a physical model for the microstrip line, compute the actual input impedance seen when losses are included (assume copper conductors and $\tan \delta = 0.0009$).

3.22 A microwave antenna feed network operating at 5 GHz requires a 50 Ω printed transmission line that is 16λ long. Possible choices are (1) copper microstrip, with $d = 0.16$ cm, $\epsilon_r = 2.20$, and $\tan \delta = 0.0009$, or (2) copper stripline, with $b = 0.32$ cm, $\epsilon_r = 2.20$, $t = 0.01$ mm, and $\tan \delta = 0.0009$. Which line should be used if attenuation is to be minimized?

3.23 Consider the TE modes of an arbitrary uniform waveguiding structure in which the transverse fields are related to H_z as in (3.19). If H_z is of the form $H_z(x, y, z) = h_z(x, y)e^{-j\beta z}$, where

$h_z(x, y)$ is a real function, compute the Poynting vector and show that real power flow occurs only in the z direction. Assume that β is real, corresponding to a propagating mode.

3.24 A piece of rectangular waveguide is air filled for $y < 0$ and dielectric filled for $y > 0$. Assume that both regions can support only the dominant TE_{10} mode and that a TE_{10} mode is incident on the interface from $y < 0$. Using a field analysis, write general expressions for the transverse field components of the incident, reflected, and transmitted waves in the two regions and enforce the boundary conditions at the dielectric interface to find the reflection and transmission coefficients. Compare these results to those obtained with an impedance approach, using Y_{TE} for each region.

3.25 Use the transverse resonance technique to derive a transcendental equation for the propagation constant of the TM modes of a rectangular waveguide that is air filled for $0 < z < d$ and dielectric filled for $d < z < a$.

3.26 Apply the transverse resonance technique to find the propagation constants for the TE surface waves that can be supported by the structure of Problem 3.17.

3.27 An X-band waveguide filled with Rexolite is operating at 10.0 GHz. Calculate the speed of light in this material and the phase and group velocities in the waveguide.

3.28 As discussed in the Point of Interest on the power-handling capacity of transmission lines, the maximum power capacity of a coaxial line is limited by voltage breakdown and is given by

$$P_{max} = \frac{\pi a^2 E_d^2}{\eta_0} \ln \frac{b}{a},$$

where E_d is the field strength at breakdown. Find the value of b/a that maximizes the maximum power capacity and show that the corresponding characteristic impedance is about 30 Ω.

3.29 A microstrip circuit is fabricated on an alumina substrate having a dielectric constant of 9.8, a thickness of 1.8 mm, and a 50 Ω linewidth of 1.95 mm. Find the threshold frequencies of the four higher order modes discussed in Section 3.8, and recommend the maximum operating frequency for this microstrip circuit.

MULTIPLE CHOICE QUESTIONS

3.1 The dispersion equation of a waveguide, which relates the wave number k to the frequency ω, is

$$k(\omega) = (1/c)\sqrt{\omega^2 - \omega_o^2}$$

where the speed of light $c = 3 \times 10^8$ m/sec, and ω_o is a constant. If the group velocity is 2×10^8 m/sec, then the phase velocity is

(a) 1.5×10^8 m/sec   **(c)** 3×10^8 m/sec
(b) 2×10^8 m/sec   **(d)** 4.5×10^8 m/sec

3.2 A rectangular waveguide of width w and height h has cutoff frequencies for TE_{10} and TE_{11} modes in the ratio 1:2. The aspect ratio w/h, rounded off to two decimal places, is _____.

3.3 The cutoff frequency of TE_{01} mode of an air-filled rectangular waveguide having inner dimensions a cm $\times b$ cm $(a > b)$ is twice that of the dominant TE_{10} mode. When the waveguide is operated at a frequency which is 25% higher than the cutoff frequency of the dominant mode, the guide wavelength is found to be 4 cm. The value of b (in cm, correct to two decimal places) is _____.

3.4 Standard air-filled rectangular waveguides of dimensions $a = 2.29$ cm and $b = 1.02$ cm are designed for radar applications. It is desired that these waveguides operate only in the dominant TE_{10} mode with the operating frequency at least 25% above the cutoff frequency of the TE_{10} mode but not higher than 95% of the next higher cutoff frequency. The range of the allowable operating frequency f is

(a) 8.19 GHz $\le f \le$ 13.1 GHz
(b) 8.19 GHz $\le f \le$ 12.45 GHz
(c) 6.55 GHz $\le f \le$ 13.1 GHz
(d) 1.64 GHz $\le f \le$ 10.24 GHz

3.5 Consider an air-filled rectangular waveguide with dimensions $a = 2.286$ cm and $b = 1.016$ cm. At 10 GHz operating frequency, the value of the propagation constant (per meter) of the corresponding propagating mode is _____.

3.6 Consider an air-filled rectangular waveguide with dimensions $a = 2.286$ cm and $b = 1.016$ cm. The increasing order of the cutoff frequencies for different modes is

(a) $TE_{01} < TE_{10} < TE_{11} < TE_{20}$
(b) $TE_{20} < TE_{11} < TE_{10} < TE_{01}$
(c) $TE_{10} < TE_{20} < TE_{01} < TE_{11}$
(d) $TE_{10} < TE_{11} < TE_{20} < TE_{01}$

3.7 A coaxial capacitor of inner radius 1 mm and outer radius 5 mm has a capacitance per unit length of 172 pF/m. If the ratio of outer radius to inner is doubled, the capacitance per unit length (in pF/m) is _____.

3.8 An air-filled rectangular waveguide of internal dimensions a cm $\times b$ cm $(a > b)$ has a cutoff frequency of 6 GHz for the dominant TE_{10} mode. For the same waveguide, if the cutoff frequency of the TM_{11} mode is 15 GHz, the cutoff frequency of the TE_{01} mode in GHz is _____.

3.9 An air-filled rectangular waveguide with dimensions $a = 75$ mm, $b = 37.5$ mm has same guide wavelength at frequencies f_1 and f_2 when operated at TE_{10} and TE_{20} modes, respectively. If the frequency f_1 is 3 GHz, what is frequency f_2 in GHz?

(a) 10
(b) 5
(c) $\sqrt{13}/2$
(d) $2\sqrt{13}$

3.10 A waveguide of dimensions $a = 15$ mm and $b = 7.5$ mm is used as a high-pass filter. If the stop band attenuation required at 8 GHz is ~109.2 dB, what is the length of the filter? (Assume conductor losses to be zero, approximate $\pi = 3.14$ and 1 Np ~ 8.69 dB.) ($\log_{10} e = 0.4343$)

(a) 100 mm
(b) 869 mm
(c) 86.9 mm
(d) 54.6 mm

3.11 An air-filled rectangular waveguide R_1 is operating at the frequency 2 GHz and another air-filled rectangular waveguide R_2 is operating at 4 GHz. The guide wavelengths of these waveguides at their respective frequencies are equal. If the cutoff frequency of waveguide R_1 is 1 GHz, what is the cutoff frequency of the waveguide R_2 in GHz?

(a) $\sqrt{10}$
(b) $\sqrt{11}$
(c) $\sqrt{12}$
(d) $\sqrt{13}$

3.12 A square waveguide carries TE_{11} mode whose axial magnetic field is given by

$$H_z = H_0 \times \cos\left(\frac{\pi x}{\sqrt{8}}\right) \cos\left(\frac{\pi y}{\sqrt{8}}\right) A/m$$

where waveguide dimensions are in cm. What is the cutoff frequency of the mode?

(a) 5.5 GHz
(b) 6.5 GHz
(c) 7.5 GHz
(d) 8.5 GHz

3.13 Two rectangular waveguides have dimensions of 1 cm $\times$ 0.5 cm and 1 cm $\times$ 0.25 cm, respectively. Their respective cutoff frequencies will be

(a) 15 GHz and 30 GHz
(b) 30 GHz and 60 GHz
(c) 15 GHz and 15 GHz
(d) 30 GHz and 30 GHz

3.14 The highest frequency for which a circular coaxial transmission line having outer diameter = 3.1 mm and inner diameter = 1.3 mm can be operated in pure TEM mode (assuming free space medium between the two conductors) should be less than

(a) 12.2 GHz
(b) 18.6 GHz
(c) 26.5 GHz
(d) 43.4 GHz

3.15 A rectangular waveguide with air medium has dimensions $a = 22.86$ mm and $b = 10.6$ mm is fed by a 3 GHz carrier from a coaxial cable. Which of the following is a false statement for TE_{01} mode?

(a) propagating mode
(b) nonpropagating mode
(c) propagating mode in case filled fully with dielectric material of proper dielectric constant
(d) none of the above

3.16 TM_{01} mode in rectangular perfect metallic waveguide is

(a) propagating mode
(b) evanescent mode
(c) nonexistent mode
(d) none of the above

3.17 For a standard rectangular waveguide having an aspect ratio of 2:1, cutoff wavelength for TM_{11} mode will be nearly

(a) $0.9a$
(b) $0.7a$
(c) $0.5a$
(d) $0.3a$

3.18 The irises in the rectangular metallic waveguide may be

1. inductive
2. resistive
3. capacitive

Select the correct answer using the code given below:

(a) 1, 2, and 3
(b) 1 and 2 only
(c) 1 and 3 only
(d) 2 and 3 only

3.19 A 10 GHz signal is propagated in a waveguide whose wall separation is 6 cm. The greatest number of halfwaves of electric intensity will be possible to establish between the two walls. The guide wavelength for this mode of propagation will be

(a) 6.48 cm
(b) 4.54 cm
(c) 2.48 cm
(d) 1.54 cm

3.20 In $TE_{m,n}$ mode, m and n are integers denoting the number of

(a) $\dfrac{1}{2}$ the wavelengths of intensity between each pair of walls

(b) $\dfrac{1}{3}$ the wavelengths of intensity between each pair of walls

(c) $\dfrac{1}{4}$ the wavelengths of intensity between each pair of walls

(d) $\dfrac{1}{8}$ the wavelengths of intensity between each pair of walls

ANSWER KEY				
3.1 (d)	**3.5** 157.0 to 159.0	**3.9** (b)	**3.13** (c)	**3.17** (a)
3.2 1.73	**3.6** (c)	**3.10** (a)	**3.14** (d)	**3.18** (a)
3.3 0.75	**3.7** 120.22	**3.11** (d)	**3.15** (b)	**3.19** (b)
3.4 (b)	**3.8** 13.7	**3.12** (c)	**3.16** (c)	**3.20** (a)

CHAPTER

4   Microwave Network Analysis

Learning Objectives

After completing this chapter, you will be able to

- Understand the concepts of impedance, admittance, transmission, and S parameters
- Design and analyze multiport microwave networks using signal flow graph method

- Understand the different concepts of excitation of waveguides
- Interpret and apply the concepts for solving multiport problems

Circuits operating at low frequencies, for which the circuit dimensions are small relative to the wavelength, can be treated as an interconnection of lumped passive or active components with unique voltages and currents defined at any point in the circuit. In this situation the circuit dimensions are small enough such that there is negligible phase delay from one point in the circuit to another. In addition, the fields can be considered as TEM fields supported by two or more conductors. This leads to a quasi-static type of solution to Maxwell's equations and to the well-known Kirchhoff voltage and current laws and impedance concepts of circuit theory [1]. As the reader is aware, there is a powerful and useful set of techniques for analyzing low-frequency circuits. In general, these techniques cannot be directly applied to microwave circuits, but it is the purpose of the present chapter to show how basic circuit and network concepts can be extended to handle many microwave analysis and design problems of practical interest.

The main reason for doing this is that it is usually much easier to apply the simple and intuitive ideas of circuit analysis to a microwave problem than it is to solve Maxwell's equations for the same problem. In a way, field analysis gives us much more information about the particular problem under consideration than we really want or need. That is, because the solution to Maxwell's equations for a given problem is complete, it gives the electric and magnetic fields at all points in space. However, usually we are only interested in the voltage or current at a set of terminals, the power flow through a device, or some other type of "terminal" quantity, as opposed to a minute description of the fields at all points in space. Another reason for using circuit or network analysis is that it is then very easy to modify the original problem, or combine several elements together and find the response, without having to reanalyze in detail the behavior of each element in combination with its neighbors. A field analysis using Maxwell's equations for such problems would be hopelessly difficult. There are situations, however, in which such circuit analysis techniques are an oversimplification and may lead to erroneous results. In such cases one must resort to a field analysis approach, using Maxwell's equations. Fortunately, there are a number of commercially available computer-aided design packages that can model RF and microwave problems using both field theory analysis and network analysis. It is part of the education of a microwave engineer to be able to determine when network analysis concepts apply and when they should be cast aside in favor of more rigorous analysis.

The basic procedure for microwave network analysis is as follows. We first treat a set of basic, canonical problems rigorously, using field analysis and Maxwell's equations (as we have done in Chapters 2 and 3, for a variety of transmission line and waveguide problems). When so doing, we try to obtain quantities that can be directly related to a circuit or transmission line parameter. For example, when we treated various transmission lines and waveguides in Chapter 3 we derived the propagation constant and characteristic impedance of the line. This allowed the transmission line or waveguide to be treated as an idealized distributed component characterized by its length, propagation constant, and characteristic impedance. At this point, we can interconnect various components and use network and/or transmission line theory to analyze the behavior of the entire system of components, including effects such as multiple reflections, loss, impedance transformations, and transitions from one type of transmission medium to another (e.g., coax to microstrip). As we will see, a transition between different transmission lines, or a discontinuity on a transmission line, generally cannot be treated as a simple junction between two transmission lines, but typically includes some type of equivalent circuit to account for reactances associated with the transition or discontinuity.

Microwave network theory was originally developed in the service of radar system and component development at the MIT Radiation Lab in the 1940s. This work was continued at the Polytechnic Institute of Brooklyn and other locations by researchers such as E. Weber, N. Marcuvitz, A. A. Oliner, L. B. Felsen, A. Hessel, and many others [2].

4.1   IMPEDANCE AND EQUIVALENT VOLTAGES AND CURRENTS

Equivalent Voltages and Currents

At microwave frequencies the measurement of voltage or current is difficult (or impossible), unless a clearly defined terminal pair is available. Such a terminal pair may be present in the case of TEM-type lines (such as coaxial cable, microstrip line, or stripline), but does not strictly exist for non-TEM lines (such as rectangular, circular, or surface waveguides).

Figure 4.1 shows the electric and magnetic field lines for an arbitrary two-conductor TEM transmission line. As in Chapter 3, the voltage, V, of the $+$ conductor relative to the $-$ conductor can be found as

$$V = \int_{+}^{-} \bar{E} \cdot d\bar{\ell}, \tag{4.1}$$

where the integration path begins on the $+$ conductor and ends on the $-$ conductor. It is important to realize that, because of the electrostatic nature of the transverse fields between the two conductors, the voltage defined in (4.1) is unique and does not depend on the shape of the integration path. The total current flowing on the $+$ conductor can be determined from an application of Ampere's law as

$$I = \oint_{C^+} \bar{H} \cdot d\bar{\ell}, \tag{4.2}$$

where the integration contour is any closed path enclosing the $+$ conductor (but not the $-$ conductor). A characteristic impedance Z_0 can then be defined for traveling waves as

$$Z_0 = \frac{V}{I}. \tag{4.3}$$

At this point, after having defined and determined a voltage, current, and characteristic impedance (and assuming we know the propagation constant for the line), we can proceed to apply the circuit theory for transmission lines developed in Chapter 2 to characterize this line as a circuit element.

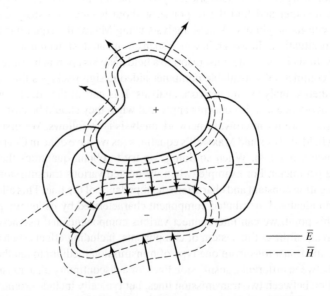

FIGURE 4.1 | Electric and magnetic field lines for an arbitrary two-conductor TEM line.

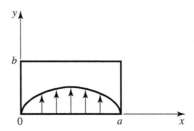

FIGURE 4.2 | Electric field lines for the TE_{10} mode of a rectangular waveguide.

The situation is more difficult for waveguides. To see why, we will look at the case of a rectangular waveguide, as shown in Figure 4.2. For the dominant TE_{10} mode, the transverse fields can be written, from Table 3.2, as

$$E_y(x,y,z) = \frac{j\omega\mu a}{\pi}A\sin\frac{\pi x}{a}e^{-j\beta z} = Ae_y(x,y)e^{-j\beta z}, \tag{4.4a}$$

$$H_x(x,y,z) = \frac{j\beta a}{\pi}A\sin\frac{\pi x}{a}e^{-j\beta z} = Ah_x(x,y)e^{-j\beta z}. \tag{4.4b}$$

Applying (4.1) to the electric field of (4.4a) gives

$$V = \frac{-j\omega\mu a}{\pi}A\sin\frac{\pi x}{a}e^{-j\beta z}\int_y dy. \tag{4.5}$$

Thus it is seen that this voltage depends on the position, x, as well as the length of the integration contour along the y direction. For example, integrating from $y = 0$ to b for $x = a/2$ gives a voltage that is quite different from that obtained by integrating from $y = 0$ to b for $x = 0$. What, then, is the correct voltage? The answer is that there is no "correct" voltage in the sense of being unique or pertinent for all applications. A similar problem arises with current, and also impedance. We will now show how we can define *equivalent* voltages, currents, and impedances that can be useful for non-TEM lines.

There are many ways to define equivalent voltage, current, and impedance for waveguides since these quantities are not unique for non-TEM lines, but the following considerations usually lead to the most useful results [1, 3, 4]:

- Voltage and current are defined only for a particular waveguide mode, and are defined so that the voltage is proportional to the transverse electric field and the current is proportional to the transverse magnetic field.
- In order to be useful in a manner similar to voltages and currents of circuit theory, the equivalent voltages and currents should be defined so that their product gives the power flow of the waveguide mode.
- The ratio of the voltage to the current for a single traveling wave should be equal to the characteristic impedance of the line. This impedance may be chosen arbitrarily, but is usually selected as equal to the wave impedance of the line, or else normalized to unity.

For an arbitrary waveguide mode with both positively and negatively traveling waves, the transverse fields can be written as

$$\bar{E}_t(x,y,z) = \bar{e}(x,y)\left(A^+e^{-j\beta z} + A^-e^{j\beta z}\right) = \frac{\bar{e}(x,y)}{C_1}\left(V^+e^{-j\beta z} + V^-e^{j\beta z}\right), \tag{4.6a}$$

$$\bar{H}_t(x,y,z) = \bar{h}(x,y)\left(A^+e^{-j\beta z} - A^-e^{j\beta z}\right) = \frac{\bar{h}(x,y)}{C_2}\left(I^+e^{-j\beta z} - I^-e^{j\beta z}\right), \tag{4.6b}$$

where $\bar{e}$ and $\bar{h}$ are the transverse field variations of the mode, and A^+, A^- are the field amplitudes of the traveling waves. Because $\bar{E}_t$ and $\bar{H}_t$ are related by the wave impedance, Z_w, according to (3.22) or (3.26), we also have that

$$\bar{h}(x, y) = \frac{\hat{z} \times \bar{e}(x, y)}{Z_w}. \tag{4.7}$$

Equation (4.6) also defines equivalent voltage and current waves as

$$V(z) = V^+ e^{-j\beta z} + V^- e^{j\beta z}, \tag{4.8a}$$

$$I(z) = I^+ e^{-j\beta z} - I^- e^{j\beta z}, \tag{4.8b}$$

with $V^+/I^+ = V^-/I^- = Z_0$. This definition embodies the idea of making the equivalent voltage and current proportional to the transverse electric and magnetic fields, respectively. The proportionality constants for this relationship are $C_1 = V^+/A^+ = V^-/A^-$ and $C_2 = I^+/A^+ = I^-/A^-$, and can be determined from the remaining two conditions for power and impedance.

The complex power flow for the incident wave is given by

$$P^+ = \frac{1}{2}|A^+|^2 \int_S \bar{e} \times \bar{h}^* \cdot \hat{z} ds = \frac{V^+ I^{+*}}{2 C_1 C_2^*} \int_S \bar{e} \times \bar{h}^* \cdot \hat{z} ds. \tag{4.9}$$

Because we want this power to be equal to $(1/2)V^+ I^{+*}$, we have the result that

$$C_1 C_2^* = \int_S \bar{e} \times \bar{h}^* \cdot \hat{z} ds, \tag{4.10}$$

where the surface integration is over the cross section of the waveguide. The characteristic impedance is

$$Z_0 = \frac{V^+}{I^+} = \frac{V^-}{I^-} = \frac{C_1}{C_2}, \tag{4.11}$$

since $V^+ = C_1 A$ and $I^+ = C_2 A$, from (4.6a) and (4.6b). If it is desired to have $Z_0 = Z_w$, the wave impedance (Z_{TE} or Z_{TM}) of the mode, then

$$\frac{C_1}{C_2} = Z_w \ (Z_{\text{TE}} \text{ or } Z_{\text{TM}}). \tag{4.12a}$$

Alternatively, it may be desirable to normalize the characteristic impedance to unity ($Z_0 = 1$), in which case we have

$$\frac{C_1}{C_2} = 1. \tag{4.12b}$$

For a given waveguide mode, (4.10) and (4.12) can be solved for the constants C_1 and C_2, and equivalent voltages and currents defined. Higher order modes can be treated in the same way, so that a general field in a waveguide can be expressed in the following form:

$$\bar{E}_t(x, y, z) = \sum_{n=1}^{N} \left(\frac{V_n^+}{C_{1n}} e^{-j\beta_n z} + \frac{V_n^-}{C_{1n}} e^{j\beta_n z} \right) \bar{e}_n(x, y), \tag{4.13a}$$

$$\bar{H}_t(x, y, z) = \sum_{n=1}^{N} \left(\frac{I_n^+}{C_{2n}} e^{-j\beta_n z} - \frac{I_n^-}{C_{2n}} e^{j\beta_n z} \right) \bar{h}_n(x, y), \tag{4.13b}$$

where $V_n^\pm$ and $I_n^\pm$ are the equivalent voltages and currents for the nth mode, and C_{1n} and C_{2n} are the proportionality constants for each mode.

EXAMPLE 4.1   EQUIVALENT VOLTAGE AND CURRENT FOR A RECTANGULAR WAVEGUIDE

Find the equivalent voltages and currents for a TE_{10} mode in a rectangular waveguide.

Solution

The transverse field components and power flow of the TE_{10} rectangular waveguide mode and the equivalent transmission line model of this mode can be written as follows:

Waveguide Fields	Transmission Line Model
$E_y = (A^+ e^{-j\beta z} + A^- e^{j\beta z}) \sin \dfrac{\pi x}{a}$	$V(z) = V^+ e^{-j\beta z} + V^- e^{j\beta z}$
$H_x = \dfrac{-1}{Z_{TE}}(A^+ e^{-j\beta z} - A^- e^{j\beta z}) \sin \dfrac{\pi x}{a}$	$\begin{aligned} I(z) &= I^+ e^{-j\beta z} - I^- e^{j\beta z} \\ &= \dfrac{1}{Z_0}(V^+ e^{-j\beta z} - V^- e^{j\beta z}) \end{aligned}$
$P^+ = \dfrac{-1}{2} \displaystyle\int_S E_y H_x^* \, dx\, dy = \dfrac{ab}{4Z_{TE}}|A^+|^2$	$P^+ = \dfrac{1}{2} V^+ I^{+*}$

We now find the constants $C_1 = V^+/A^+ = V^-/A^-$ and $C_2 = I^+/A^+ = I^-/A^-$ that relate the equivalent voltages $V^\pm$ and currents $I^\pm$ to the field amplitudes, $A^\pm$. Equating incident powers gives

$$\frac{ab\,|A^+|^2}{4Z_{TE}} = \frac{1}{2}V^+ I^{+*} = \frac{1}{2}|A^+|^2 C_1 C_2^*.$$

If we choose $Z_0 = Z_{TE}$, then we also have that

$$\frac{V^+}{I^+} = \frac{C_1}{C_2} = Z_{TE}.$$

Solving for C_1, C_2 gives

$$C_1 = \sqrt{\frac{ab}{2}},$$

$$C_2 = \frac{1}{Z_{TE}}\sqrt{\frac{ab}{2}},$$

which completes the transmission line equivalence for the TE_{10} mode.

The Concept of Impedance

We have used the idea of impedance in several different ways, so it may be useful at this point to summarize this important concept. The term *impedance* was first used by Oliver Heaviside in the nineteenth century to describe the complex ratio V/I in AC circuits consisting of resistors, inductors, and capacitors; the impedance concept quickly became indispensable in the analysis of AC circuits. It was then applied to transmission lines, in terms of lumped-element equivalent circuits and the distributed series impedance and shunt admittance of the line. In the 1930s, S. A. Schelkunoff recognized that the impedance concept could be extended to electromagnetic fields in a systematic way, and noted that impedance should be regarded as characteristic of the type of field, as well as of the medium [2]. In addition, in relation to the analogy between transmission lines and plane wave propagation, impedance may even be dependent on direction. The concept of impedance, then, forms an important link between field theory and transmission line or circuit theory.

We summarize the various types of impedance we have used so far, and their notation:

- $\eta = \sqrt{\mu/\epsilon}$ = intrinsic impedance of the medium. This impedance is dependent only on the material parameters of the medium, and is equal to the wave impedance for plane waves.

- $Z_w = E_t/H_t = 1/Y_w =$ wave impedance. This impedance is a characteristic of the particular type of wave. TEM, TM, and TE waves each have different wave impedances (Z_{TEM}, Z_{TM}, Z_{TE}), which may depend on the type of line or guide, the material, and the operating frequency.
- $Z_0 = 1/Y_0 = V^+/I^+ =$ characteristic impedance. Characteristic impedance is the ratio of voltage to current for a traveling wave on a transmission line. Because voltage and current are uniquely defined for TEM waves, the characteristic impedance of a TEM wave is unique. TE and TM waves, however, do not have a uniquely defined voltage and current, so the characteristic impedance for such waves may be defined in different ways.

EXAMPLE 4.2   APPLICATION OF WAVEGUIDE IMPEDANCE

Consider a rectangular waveguide with $a = 2.286$ cm and $b = 1.016$ cm (X-band guide), air filled for $x < 0$ and Polyimide filled ($\epsilon_r = 3.4$) for $x > 0$, as shown in Figure 4.3. If the operating frequency is 10 GHz, use an equivalent transmission line model to compute the reflection coefficient of a TE_{10} wave incident on the interface from $x < 0$.

Solution

The waveguide propagation constants in the air ($x < 0$) and the dielectric ($x > 0$) regions are

$$\beta_a = \sqrt{k_0^2 - \left(\frac{\pi}{a}\right)^2} = 158.0 \text{ m}^{-1},$$

$$\beta_d = \sqrt{\epsilon_r k_0^2 - \left(\frac{\pi}{a}\right)^2} = 360.8 \text{ m}^{-1},$$

where $k_0 = 209.4$ m^{-1}.

The reader may verify that the TE_{10} mode is the only propagating mode in either waveguide region. We can set up an equivalent transmission line for the TE_{10} mode in each waveguide, and treat the problem as the reflection of an incident voltage wave at the junction of two infinite transmission lines.

By Example 4.1 and Table 3.2, the equivalent characteristic impedances for the two lines are

$$Z_{0_a} = \frac{k_0 \eta_0}{\beta_a} = \frac{(209.4)(377)}{158.0} = 500.0 \ \Omega,$$

$$Z_{0_d} = \frac{k\eta}{\beta_d} = \frac{k_0 \eta_0}{\beta_d} = \frac{(209.4)(377)}{360.8} = 218.8 \ \Omega.$$

The reflection coefficient seen looking into the dielectric-filled region is then

$$\Gamma = \frac{Z_{0_d} - Z_{0_a}}{Z_{0_d} + Z_{0_a}} = -0.391.$$

With this result, expressions for the incident, reflected, and transmitted waves can be written in terms of fields, or in terms of equivalent voltages and currents.

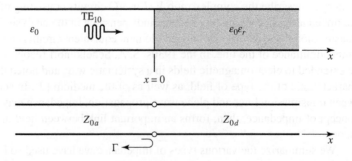

FIGURE 4.3 | Geometry of a partially filled waveguide and its transmission line equivalent for Example 4.2.

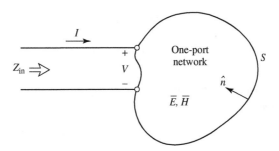

FIGURE 4.4 | An arbitrary one-port network.

We now consider the arbitrary one-port network shown in Figure 4.4 and derive a general relation between its impedance properties and electromagnetic energy stored in, and the power dissipated by, the network. The complex power delivered to this network is given by (1.91):

$$P = \frac{1}{2} \oint_S \bar{E} \times \bar{H}^* \cdot d\bar{s} = P_\ell + 2j\omega(W_m - W_e), \tag{4.14}$$

where P_ℓ is real and represents the average power dissipated by the network, and W_m and W_e represent the stored magnetic and electric energy, respectively. Note that the unit normal vector in Figure 4.4 is pointing into the volume.

If we define real transverse modal fields $\bar{e}$ and $\bar{h}$ over the terminal plane of the network such that

$$\bar{E}_t(x, y, z) = V(z)\bar{e}(x, y)e^{-j\beta z}, \tag{4.15a}$$

$$\bar{H}_t(x, y, z) = I(z)\bar{h}(x, y)e^{-j\beta z}, \tag{4.15b}$$

with a normalization such that

$$\int_S \bar{e} \times \bar{h} \cdot d\bar{s} = 1,$$

then we can express (4.14) in terms of the terminal voltage and current:

$$P = \frac{1}{2} \int_S VI^* \bar{e} \times \bar{h} \cdot d\bar{s} = \frac{1}{2} VI^*. \tag{4.16}$$

Then the input impedance is

$$Z_{\text{in}} = R + jX = \frac{V}{I} = \frac{VI^*}{|I|^2} = \frac{P}{\frac{1}{2}|I|^2} = \frac{P_\ell + 2j\omega(W_m - W_e)}{\frac{1}{2}|I|^2}. \tag{4.17}$$

Thus we see that the real part, R, of the input impedance is related to the dissipated power, while the imaginary part, X, is related to the net energy stored in the network. If the network is lossless, then $P_\ell = 0$ and $R = 0$. Then Z_{in} is purely imaginary, with a reactance

$$X = \frac{4\omega(W_m - W_e)}{|I|^2}, \tag{4.18}$$

which is positive for an inductive load ($W_m > W_e$), and negative for a capacitive load ($W_m < W_e$).

Even and Odd Properties of $Z(\omega)$ and $\Gamma(\omega)$

Consider the driving point impedance, $Z(\omega)$, at the input port of an electrical network. The voltage and current at this port are related as $V(\omega) = Z(\omega)I(\omega)$. For an arbitrary frequency dependence, we can find the time-domain voltage by taking the inverse Fourier transform of $V(\omega)$:

$$v(t) = \frac{1}{2\pi} \int_{-\infty}^{\infty} V(\omega)e^{j\omega t} d\omega. \tag{4.19}$$

Because $v(t)$ must be real, we have that $v(t) = v^*(t)$, or

$$\int_{-\infty}^{\infty} V(\omega)e^{j\omega t}\, d\omega = \int_{-\infty}^{\infty} V^*(\omega)e^{-j\omega t}\, d\omega = \int_{-\infty}^{\infty} V^*(-\omega)e^{j\omega t}\, d\omega,$$

where the last term was obtained by a change of variable from ω to $-\omega$. This shows that $V(\omega)$ must satisfy the relation

$$V(-\omega) = V^*(\omega), \tag{4.20}$$

which means that $\text{Re}\{V(\omega)\}$ is even in ω, while $\text{Im}\{V(\omega)\}$ is odd in ω. Similar results hold for $I(\omega)$, and for $Z(\omega)$ since

$$V^*(-\omega) = Z^*(-\omega)I^*(-\omega) = Z^*(-\omega)I(\omega) = V(\omega) = Z(\omega)I(\omega).$$

Thus, if $Z(\omega) = R(\omega) + jX(\omega)$, then $R(\omega)$ is even in ω and $X(\omega)$ is odd in ω. These results can also be inferred from (4.17).

Now consider the reflection coefficient at the input port:

$$\Gamma(\omega) = \frac{Z(\omega) - Z_0}{Z(\omega) + Z_0} = \frac{R(\omega) - Z_0 + jX(\omega)}{R(\omega) + Z_0 + jX(\omega)}. \tag{4.21}$$

Then

$$\Gamma(-\omega) = \frac{R(\omega) - Z_0 - jX(\omega)}{R(\omega) + Z_0 - jX(\omega)} = \Gamma^*(\omega), \tag{4.22}$$

which shows that the real and imaginary parts of $\Gamma(\omega)$ are even and odd, respectively, in ω. Finally, the magnitude of the reflection coefficient is

$$|\Gamma(\omega)|^2 = \Gamma(\omega)\Gamma^*(\omega) = \Gamma(\omega)\Gamma(-\omega) = |\Gamma(-\omega)|^2, \tag{4.23}$$

which shows that $|\Gamma(\omega)|^2$ and $|\Gamma(\omega)|$ are even functions of ω. This result implies that only even series of the form $a + b\omega^2 + c\omega^4 + \cdots$ can be used to represent $|\Gamma(\omega)|$ or $|\Gamma(\omega)|^2$.

4.2   IMPEDANCE AND ADMITTANCE MATRICES

In the previous section we have seen how equivalent voltages and currents can be defined for TEM and non-TEM waves. Once such voltages and currents have been defined at various points in a microwave network, we can use the impedance and/or admittance matrices of circuit theory to relate these terminal or *port* quantities to each other, and thus to essentially arrive at a matrix description of the network. This type of representation lends itself to the development of equivalent circuits of arbitrary networks, which will be quite useful when we discuss the design of passive components such as couplers and filters. (The term *port* was introduced by H. A. Wheeler in the 1950s to replace the less descriptive and more cumbersome phrase "two-terminal pair" [2, 3].)

We begin by considering an arbitrary N-port microwave network, as depicted in Figure 4.5. The ports in Figure 4.5 may be any type of transmission line or transmission line equivalent of a single propagating waveguide mode. If one of the physical ports of the network is a waveguide supporting more than one propagating mode, additional electrical ports can be added to account for these modes. At a specific point on the nth port, a terminal plane, t_n, is defined along with equivalent voltages and currents for the incident (V_n^+, I_n^+) and reflected (V_n^-, I_n^-) waves. The terminal planes are important in providing a phase reference for the voltage and current phasors. Now, at the nth terminal plane, the total voltage and current are given by

$$V_n = V_n^+ + V_n^-, \tag{4.24a}$$

$$I_n = I_n^+ - I_n^-, \tag{4.24b}$$

as seen from (4.8) when $z = 0$.

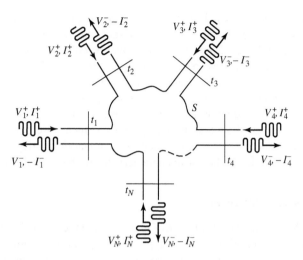

FIGURE 4.5 | An arbitrary N-port microwave network.

The impedance matrix $[Z]$ of the microwave network then relates these voltages and currents:

$$
\begin{bmatrix} V_1 \\ V_2 \\ \vdots \\ V_N \end{bmatrix} = \begin{bmatrix} Z_{11} & Z_{12} & \cdots & Z_{1N} \\ Z_{21} & & & \vdots \\ \vdots & & & \vdots \\ Z_{N1} & \cdots & \cdots & Z_{NN} \end{bmatrix} \begin{bmatrix} I_1 \\ I_2 \\ \vdots \\ I_N \end{bmatrix},
$$

or in matrix form as

$$[V] = [Z][I]. \tag{4.25}$$

Similarly, we can define an admittance matrix $[Y]$ as

$$
\begin{bmatrix} I_1 \\ I_2 \\ \vdots \\ I_N \end{bmatrix} = \begin{bmatrix} Y_{11} & Y_{12} & \cdots & Y_{1N} \\ Y_{21} & & & \vdots \\ \vdots & & & \vdots \\ Y_{N1} & \cdots & \cdots & Y_{NN} \end{bmatrix} \begin{bmatrix} V_1 \\ V_2 \\ \vdots \\ V_N \end{bmatrix},
$$

or in matrix form as

$$[I] = [Y][V]. \tag{4.26}$$

Of course, the $[Z]$ and $[Y]$ matrices are the inverses of each other:

$$[Y] = [Z]^{-1}. \tag{4.27}$$

Note that both the $[Z]$ and $[Y]$ matrices relate the total port voltages and currents.

From (4.25), we see that Z_{ij} can be found as

$$Z_{ij} = \left.\frac{V_i}{I_j}\right|_{I_k=0 \text{ for } k \neq j}. \tag{4.28}$$

In words, (4.28) states that Z_{ij} can be found by driving port j with the current I_j, open-circuiting all other ports (so $I_k = 0$ for $k \neq j$), and measuring the open-circuit voltage at port i. Thus, Z_{ii} is the input impedance seen looking into port i when all other ports are open-circuited, and Z_{ij} is the transfer impedance between ports i and j when all other ports are open-circuited.

Similarly, from (4.26), Y_{ij} can be found as

$$Y_{ij} = \left.\frac{I_i}{V_j}\right|_{V_k=0 \text{ for } k \neq j}, \tag{4.29}$$

which states that Y_{ij} can be determined by driving port j with the voltage V_j, short-circuiting all other ports (so $V_k = 0$ for $k \neq j$), and measuring the short-circuit current at port i.

In general, each Z_{ij} or Y_{ij} element may be complex. For an arbitrary N-port network, the impedance and admittance matrices are $N \times N$ in size, so there are $2N^2$ independent quantities or degrees of freedom. In practice, however, many networks are either reciprocal or lossless, or both. If the network is reciprocal (not containing any active devices or nonreciprocal media, such as ferrites or plasmas), we will show that the impedance and admittance matrices are symmetric, so that $Z_{ij} = Z_{ji}$, and $Y_{ij} = Y_{ji}$. If the network is lossless, we can show that all the Z_{ij} or Y_{ij} elements are purely imaginary. Either of these special cases serves to reduce the number of independent quantities or degrees of freedom that an N-port network may have. We now derive the above characteristics for reciprocal and lossless networks.

Reciprocal Networks

Consider the arbitrary network of Figure 4.5 to be reciprocal (no active devices, ferrites, or plasmas), with short circuits placed at all terminal planes except those of ports 1 and 2. Let $\bar{E}_a$, $\bar{H}_a$ and $\bar{E}_b$, $\bar{H}_b$ be the fields anywhere in the network due to two independent sources, a and b, located somewhere in the network. Then the reciprocity theorem of (1.156) states that

$$\oint_S \bar{E}_a \times \bar{H}_b \cdot d\bar{s} = \oint_S \bar{E}_b \times \bar{H}_a \cdot d\bar{s}, \tag{4.30}$$

where S is the closed surface along the boundaries of the network and through the terminal planes of the ports. If the boundary walls of the network and transmission lines are metal, then $\bar{E}_{\tan} = 0$ on these walls (assuming perfect conductors). If the network or the transmission lines are open structures, like microstrip line or slotline, the boundaries of the network can be taken arbitrarily far from the lines so that $\bar{E}_{\tan}$ is negligible. Then the only nonzero contribution to the integrals of (4.30) come from the cross-sectional areas of ports 1 and 2.

From Section 4.1, the fields due to sources a and b can be evaluated at the terminal planes t_1 and t_2 as

$$\bar{E}_{1a} = V_{1a}\bar{e}_1, \quad \bar{H}_{1a} = I_{1a}\bar{h}_1, \tag{4.31a}$$

$$\bar{E}_{1b} = V_{1b}\bar{e}_1, \quad \bar{H}_{1b} = I_{1b}\bar{h}_1, \tag{4.31b}$$

$$\bar{E}_{2a} = V_{2a}\bar{e}_2, \quad \bar{H}_{2a} = I_{2a}\bar{h}_2, \tag{4.31c}$$

$$\bar{E}_{2b} = V_{2b}\bar{e}_2, \quad \bar{H}_{2b} = I_{2b}\bar{h}_2, \tag{4.31d}$$

where $\bar{e}_1$, $\bar{h}_1$ and $\bar{e}_2$, $\bar{h}_2$ are the transverse modal fields of ports 1 and 2, respectively, and the Vs and Is are the equivalent total voltages and currents. (For instance, $\bar{E}_{1b}$ is the transverse electric field at terminal plane t_1 of port 1 due to source b.) Substituting the fields of (4.31) into (4.30) gives

$$(V_{1a}I_{1b} - V_{1b}I_{1a})\int_{S_1} \bar{e}_1 \times \bar{h}_1 \cdot d\bar{s} + (V_{2a}I_{2b} - V_{2b}I_{2a})\int_{S_2} \bar{e}_2 \times \bar{h}_2 \cdot d\bar{s} = 0, \tag{4.32}$$

where S_1 and S_2 are the cross-sectional areas at the terminal planes of ports 1 and 2.

As in Section 4.1, the equivalent voltages and currents have been defined so that the power through a given port can be expressed as $VI^*/2$; then, comparing (4.31) to (4.6) implies that $C_1 = C_2 = 1$ for each port, so that

$$\int_{S_1} \bar{e}_1 \times \bar{h}_1 \cdot d\bar{s} = \int_{S_2} \bar{e}_2 \times \bar{h}_2 \cdot d\bar{s} = 1. \tag{4.33}$$

This reduces (4.32) to

$$V_{1a}I_{1b} - V_{1b}I_{1a} + V_{2a}I_{2b} - V_{2b}I_{2a} = 0. \tag{4.34}$$

Now use the 2×2 admittance matrix of the (effectively) two-port network to eliminate the Is:

$$I_1 = Y_{11}V_1 + Y_{12}V_2,$$
$$I_2 = Y_{21}V_1 + Y_{22}V_2.$$

Substitution into (4.34) gives

$$(V_{1a}V_{2b} - V_{1b}V_{2a})(Y_{12} - Y_{21}) = 0. \tag{4.35}$$

Because the sources a and b are independent, the voltages V_{1a}, V_{1b}, V_{2a}, and V_{2b} can take on arbitrary values. So in order for (4.35) to be satisfied for any choice of sources, we must have $Y_{12} = Y_{21}$, and since the choice of which ports are labeled as 1 and 2 is arbitrary, we have the general result that

$$Y_{ij} = Y_{ji}. \tag{4.36}$$

Then if $[Y]$ is a symmetric matrix, its inverse, $[Z]$, is also symmetric.

Lossless Networks

Now consider a reciprocal lossless N-port junction; we will show that the elements of the impedance and admittance matrices must be pure imaginary. If the network is lossless, then the net real power delivered to the network must be zero. Thus, $\text{Re}\{P_{\text{avg}}\} = 0$, where

$$\begin{aligned}
P_{\text{avg}} &= \frac{1}{2}[V]^t[I]^* = \frac{1}{2}([Z][I])^t[I]^* = \frac{1}{2}[I]^t[Z][I]^* \\
&= \frac{1}{2}(I_1 Z_{11} I_1^* + I_1 Z_{12} I_2^* + I_2 Z_{21} I_1^* + \cdots) \\
&= \frac{1}{2}\sum_{n=1}^{N}\sum_{m=1}^{N} I_m Z_{mn} I_n^*.
\end{aligned} \tag{4.37}$$

We have used the result from matrix algebra that $([A][B])^t = [B]^t[A]^t$. Because the I_n are independent, we must have the real part of each self term $(I_n Z_{nn} I_n^*)$ equal to zero, since we could set all port currents equal to zero except for the nth current. So,

$$\text{Re}\left\{I_n Z_{nn} I_n^*\right\} = |I_n|^2 \text{Re}\left\{Z_{nn}\right\} = 0,$$

or

$$\text{Re}\left\{Z_{nn}\right\} = 0. \tag{4.38}$$

Now let all port currents be zero except for I_m and I_n. Then (4.37) reduces to

$$\text{Re}\left\{(I_n I_m^* + I_m I_n^*)Z_{mn}\right\} = 0,$$

since $Z_{mn} = Z_{nm}$. However, $(I_n I_m^* + I_m I_n^*)$ is a purely real quantity that is, in general, nonzero. Thus we must have that

$$\text{Re}\left\{Z_{mn}\right\} = 0. \tag{4.39}$$

Then (4.38) and (4.39) imply that $\text{Re}\left\{Z_{mn}\right\} = 0$ for any m, n. The reader can verify that this also leads to an imaginary $[Y]$ matrix.

EXAMPLE 4.3   EVALUATION OF IMPEDANCE PARAMETERS

Find the Z parameters of the two-port T-network shown in Figure 4.6.

Solution
From (4.28), Z_{11} can be found as the input impedance of port 1 when port 2 is open-circuited:

$$Z_{11} = \left.\frac{V_1}{I_1}\right|_{I_2=0} = Z_A + Z_C.$$

The transfer impedance Z_{12} can be found measuring the open-circuit voltage at port 1 when a current I_2 is applied at port 2. By voltage division,

$$Z_{12} = \left.\frac{V_1}{I_2}\right|_{I_1=0} = \frac{V_2}{I_2}\frac{Z_C}{Z_B + Z_C} = Z_C.$$

The reader can verify that $Z_{21} = Z_{12}$, indicating that the circuit is reciprocal. Finally, Z_{22} is found as

$$Z_{22} = \left.\frac{V_2}{I_2}\right|_{I_1=0} = Z_B + Z_C.$$

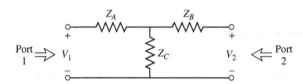

FIGURE 4.6 | A two-port T-network.

EXAMPLE 4.4   EVALUATION OF ADMITTANCE PARAMETERS

Find the Y parameters of the two-port π-network shown in Figure 4.7.

Solution
From (4.28), Y_{11} can be found as the input admittance of port 1 when port 2 is short-circuited:

$$Y_{11} = \left.\frac{I_1}{V_1}\right|_{V_2=0} = \left.\frac{V_1(Y_A) + (V_1 - V_2)Y_B}{V_1}\right|_{V_2=0} = Y_A + Y_B$$

The transfer admittance Y_{12} can be found measuring the short-circuit current at port 1 when a voltage V_2 is applied at port 2.

$$Y_{12} = \left.\frac{I_1}{V_2}\right|_{V_1=0} = \left.\frac{V_1(Y_A) + (V_1 - V_2)Y_B}{V_2}\right|_{V_1=0} = -Y_B$$

The reader can verify that $Y_{21} = Y_{12}$, indicating that the circuit is reciprocal. Finally, Y_{22} is found as

$$Y_{22} = \left.\frac{I_2}{V_2}\right|_{V_1=0} = \left.\frac{V_2(Y_C) + (V_2 - V_1)Y_B}{V_2}\right|_{V_1=0} = Y_B + Y_C$$

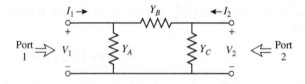

FIGURE 4.7 | A two-port π-network.

4.3   THE SCATTERING MATRIX

We have already discussed the difficulty in defining voltages and currents for non-TEM lines. In addition, a practical problem exists when trying to measure voltages and currents at microwave frequencies because direct measurements usually involve the magnitude (inferred from power) and phase of a wave traveling in a given direction or of a standing wave. Thus, equivalent voltages and currents, and the related impedance and admittance matrices, become somewhat an abstraction when dealing with high-frequency networks.

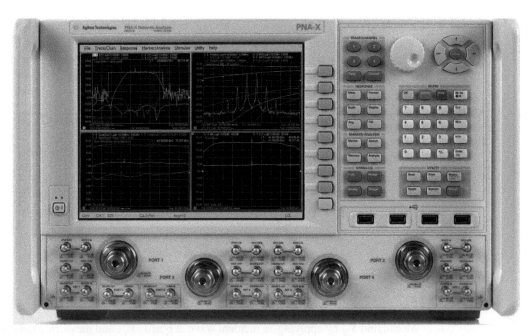

FIGURE 4.8 | Photograph of the Agilent N5247A Programmable Network Analyzer. This instrument is used to measure the scattering parameters of RF and microwave networks from 10 MHz to 67 GHz. The instrument is programmable, performs error correction, and has a wide variety of display formats and data conversions.

Courtesy of Agilent Technologies.

A representation more in accord with direct measurements, and with the ideas of incident, reflected, and transmitted waves, is given by the scattering matrix.

Like the impedance or admittance matrix for an N-port network, the scattering matrix provides a complete description of the network as seen at its N ports. While the impedance and admittance matrices relate the total voltages and currents at the ports, the scattering matrix relates the voltage waves incident on the ports to those reflected from the ports. For some components and circuits, the scattering parameters can be calculated using network analysis techniques. Otherwise, the scattering parameters can be measured directly with a vector network analyzer; a photograph of a modern network analyzer is shown in Figure 4.8. Once the scattering parameters of the network are known, conversion to other matrix parameters can be performed, if needed.

Consider the N-port network shown in Figure 4.5, where V_n^+ is the amplitude of the voltage wave incident on port n and V_n^- is the amplitude of the voltage wave reflected from port n. The scattering matrix, or $[S]$ matrix, is defined in relation to these incident and reflected voltage waves as

$$
\begin{bmatrix} V_1^- \\ V_2^- \\ \vdots \\ V_N^- \end{bmatrix} = \begin{bmatrix} S_{11} & S_{12} & \cdots & S_{1N} \\ S_{21} & & & \vdots \\ S_{N1} & \cdots & & S_{NN} \\ \vdots & & & \end{bmatrix} \begin{bmatrix} V_1^+ \\ V_2^+ \\ \vdots \\ V_N^+ \end{bmatrix},
$$

or

$$[V^-] = [S][V^+]. \qquad (4.40)$$

A specific element of the scattering matrix can be determined as

$$S_{ij} = \frac{V_i^-}{V_j^+} \bigg|_{V_k^+=0 \text{ for } k \neq j}. \qquad (4.41)$$

In words, (4.41) says that S_{ij} is found by driving port j with an incident wave of voltage V_j^+ and measuring the reflected wave amplitude V_i^- coming out of port i. The incident waves on all ports except the jth port are set to zero, which means that all ports should be terminated in matched loads to avoid reflections. Thus, S_{ii} is the reflection coefficient seen looking into port i when all other ports are terminated in matched loads, and S_{ij} is the transmission coefficient from port j to port i when all other ports are terminated in matched loads.

EXAMPLE 4.5 EVALUATION OF SCATTERING PARAMETERS

Find the scattering parameters of the 3 dB attenuator circuit shown in Figure 4.9.

Solution
From (4.41), S_{11} can be found as the reflection coefficient seen at port 1 when port 2 is terminated in a matched load ($Z_0 = 50\ \Omega$):

$$S_{11} = \left.\frac{V_1^-}{V_1^+}\right|_{V_2^+=0} = \Gamma^{(1)}|_{V_2^+=0} = \left.\frac{Z_{\text{in}}^{(1)} - Z_0}{Z_{\text{in}}^{(1)} + Z_0}\right|_{Z_0 \text{ on port 2}},$$

but $Z_{\text{in}}^{(1)} = 8.6 + [142(8.6 + 50)]/(142 + 8.6 + 50) = 50\ \Omega$, so $S_{11} = 0$. Because of the symmetry of the circuit, $S_{22} = 0$.

We can find S_{21} by applying an incident wave at port 1, V_1^+, and measuring the outcoming wave at port 2, V_2^-. This is equivalent to the transmission coefficient from port 1 to port 2:

$$S_{21} = \left.\frac{V_2^-}{V_1^+}\right|_{V_2^+=0}.$$

From the fact that $S_{11} = S_{22} = 0$, we know that $V_1^- = 0$ when port 2 is terminated in $Z_0 = 50\ \Omega$, and that $V_2^+ = 0$. In this case we have that $V_1^+ = V_1$ and $V_2^- = V_2$. By applying a voltage V_1 at port 1 and using voltage division twice we find $V_2^- = V_2$ as the voltage across the 50 Ω load resistor at port 2:

$$V_2^- = V_2 = V_1 \left(\frac{41.44}{41.44 + 8.6}\right)\left(\frac{50}{50 + 8.6}\right) = 0.707V_1,$$

where $41.48 = 142(58.6)/(142 + 58.6)$ is the resistance of the parallel combination of the 50 Ω load and the 8.6 Ω resistor with the 142 Ω resistor. Thus, $S_{12} = S_{21} = 0.707$.

If the input power is $|V_1^+|^2/2Z_0$, then the output power is $|V_2^-|^2/2Z_0 = |S_{21}V_1^+|^2/2Z_0 = |S_{21}|^2/2Z_0|V_1^+|^2 = |V_1^+|^2/4Z_0$, which is one-half ($-3$ dB) of the input power.

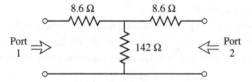

FIGURE 4.9 | A matched 3 dB attenuator with a 50 Ω characteristic impedance (Example 4.5).

We now show how the scattering matrix can be determined from the $[Z]$ (or $[Y]$) matrix and vice versa. First, we must assume that the characteristic impedances, Z_{0n}, of all the ports are identical. (This restriction will be removed when we discuss generalized scattering parameters.) Then, for convenience, we can set $Z_{0n} = 1$. From (4.24) the total voltage and current at the nth port can be written as

$$V_n = V_n^+ + V_n^-, \tag{4.42a}$$

$$I_n = I_n^+ - I_n^- = V_n^+ - V_n^-. \tag{4.42b}$$

Using the definition of $[Z]$ from (4.25) with (4.42) gives

$$[Z][I] = [Z][V^+] - [Z][V^-] = [V] = [V^+] + [V^-],$$

which can be rewritten as

$$([Z] + [U])[V^-] = ([Z] - [U])[V^+], \tag{4.43}$$

where $[U]$ is the unit, or identity, matrix defined as

$$[U] = \begin{bmatrix} 1 & 0 & \cdots & 0 \\ 0 & 1 & & \vdots \\ \vdots & & \ddots & \\ 0 & & \cdots & 1 \end{bmatrix}.$$

Comparing (4.43) to (4.40) suggests that

$$[S] = ([Z] + [U])^{-1} ([Z] - [U]), \tag{4.44}$$

giving the scattering matrix in terms of the impedance matrix. Note that for a one-port network (4.44) reduces to

$$S_{11} = \frac{z_{11} - 1}{z_{11} + 1},$$

in agreement with the result for the reflection coefficient seen looking into a load with a normalized input impedance of z_{11}.

To find $[Z]$ in terms of $[S]$, rewrite (4.44) as $[Z][S] + [U][S] = [Z] - [U]$, and solve for $[Z]$ to give

$$[Z] = ([U] + [S])([U] - [S])^{-1}. \tag{4.45}$$

Reciprocal Networks and Lossless Networks

As we discussed in Section 4.2, the impedance and admittance matrices are symmetric for reciprocal networks, and are purely imaginary for lossless networks. The scattering matrices for these particular types of networks also have special properties. We will show that the scattering matrix for a reciprocal network is symmetric, and that the scattering matrix for a lossless network is unitary.

By adding (4.42a) and (4.42b) we obtain

$$V_n^+ = \frac{1}{2}(V_n + I_n),$$

or

$$[V^+] = \frac{1}{2}([Z] + [U])[I]. \tag{4.46a}$$

By subtracting (4.42a) and (4.42b) we obtain

$$V_n^- = \frac{1}{2}(V_n - I_n),$$

or

$$[V^-] = \frac{1}{2}([Z] - [U])[I]. \tag{4.46b}$$

Eliminating $[I]$ from (4.46a) and (4.46b) gives

$$[V^-] = ([Z] - [U])([Z] + [U])^{-1}[V^+],$$

so that

$$[S] = ([Z] - [U])([Z] + [U])^{-1}. \tag{4.47}$$

Taking the transpose of (4.47) gives

$$[S]^t = \{([Z] + [U])^{-1}\}^t([Z] - [U])^t.$$

Now $[U]$ is diagonal, so $[U]^t = [U]$, and if the network is reciprocal, $[Z]$ is symmetric. So that $[Z]^t = [Z]$. The above equation then reduces to

$$[S]^t = ([Z] + [U])^{-1}([Z] - [U]),$$

which is equivalent to (4.44). We have thus shown that

$$[S] = [S]^t, \tag{4.48}$$

so the scattering matrix is symmetric for reciprocal networks.

If the network is lossless, no real power can be delivered to the network. Thus, if the characteristic impedances of all the ports are identical and assumed to be unity, the average power delivered to the network is

$$
\begin{aligned}
P_{\text{avg}} &= \frac{1}{2}\text{Re}\{[V]^t[I]^*\} = \frac{1}{2}\text{Re}\{([V^+]^t + [V^-]^t)([V^+]^* - [V^-]^*)\} \\
&= \frac{1}{2}\text{Re}\{[V^+]^t[V^+]^* - [V^+]^t[V^-]^* + [V^-]^t[V^+]^* - [V^-]^t[V^-]^*\} \\
&= \frac{1}{2}[V^+]^t[V^+]^* - \frac{1}{2}[V^-]^t[V^-]^* = 0,
\end{aligned}
\tag{4.49}
$$

since the terms $-[V^+]^t[V^-]^* + [V^-]^t[V^+]^*$ are of the form $A - A^*$, and so are purely imaginary. Of the remaining terms in (4.49), $(1/2)[V^+]^t[V^+]^*$ represents the total incident power, while $(1/2)[V^-]^t[V^-]^*$ represents the total reflected power. So, for a lossless junction, we have the intuitive result that the incident and reflected powers are equal:

$$[V^+]^t[V^+]^* = [V^-]^t[V^-]^*. \tag{4.50}$$

Using $[V^-] = [S][V^+]$ in (4.50) gives

$$[V^+]^t[V^+]^* = [V^+]^t[S]^t[S]^*[V^+]^*,$$

so that, for nonzero $[V^+]$,

$$[S]^t[S]^* = [U], \tag{4.51}$$

or

$$[S]^* = \{[S]^t\}^{-1}.$$

A matrix that satisfies the condition of (4.51) is called a *unitary matrix*.

The matrix equation of (4.51) can be written in summation form as

$$\sum_{k=1}^{N} S_{ki}S_{kj}^* = \delta_{ij}, \text{ for all } i, j, \tag{4.52}$$

where $\delta_{ij} = 1$ if $i = j$, and $\delta_{ij} = 0$ if $i \neq j$, is the Kronecker delta symbol. Thus, if $i = j$, (4.52) reduces to

$$\sum_{k=1}^{N} S_{ki}S_{ki}^* = 1, \tag{4.53a}$$

while if $i \neq j$, (4.52) reduces to

$$\sum_{k=1}^{N} S_{ki}S_{kj}^* = 0, \text{ for } i \neq j. \tag{4.53b}$$

In words, (4.53a) states that the dot product of any column of $[S]$ with the conjugate of that same column gives unity, while (4.53b) states that the dot product of any column with the conjugate of a different column gives zero (the columns are orthonormal). From (4.51) we also have that

$$[S][S]^{*t} = [U],$$

so the same statements can be made about the rows of the scattering matrix.

EXAMPLE 4.6   APPLICATION OF SCATTERING PARAMETERS

A two-port network is known to have the following scattering matrix:

$$[S] = \begin{bmatrix} 0.25\angle 0° & 0.75\angle -45° \\ 0.75\angle -45° & 0.25\angle 0° \end{bmatrix}$$

Determine if the network is reciprocal and lossless. If port 2 is terminated with a matched load, what is the return loss seen at port 1? If port 2 is terminated with a short circuit, what is the return loss seen at port 1?

Solution

Because [S] is not symmetric, the network is not reciprocal. To be lossless, the scattering parameters must satisfy (4.53). Taking the first column [$i = 1$ in (4.53a)] gives

$$|S_{11}|^2 + |S_{21}|^2 = (0.25)^2 + (0.75)^2 = 0.625 \neq 1,$$

so the network is not lossless.

When port 2 is terminated with a matched load, the reflection coefficient seen at port 1 is $\Gamma = S_{11} = 0.25$. So the return loss is

$$RL = -20 \log |\Gamma| = -20 \log(0.25) = 12.04 \text{ dB}.$$

When port 2 is terminated with a short circuit, the reflection coefficient seen at port 1 can be found as follows. From the definition of the scattering matrix and the fact that $V_2^+ = -V_2^-$ (for a short circuit at port 2), we can write

$$V_1^- = S_{11} V_1^+ + S_{12} V_2^+ = S_{11} V_1^+ - S_{12} V_2^-,$$
$$V_2^- = S_{21} V_1^+ + S_{22} V_2^+ = S_{21} V_1^+ - S_{22} V_2^-.$$

The second equation gives

$$V_2^- = \frac{S_{21}}{1 + S_{22}} V_1^+.$$

Dividing the first equation by V_1^+ and using the above result gives the reflection coefficient seen at port 1 as

$$\Gamma = \frac{V_1^-}{V_1^+} = S_{11} - S_{12} \frac{V_2^-}{V_1^+} = S_{11} - \frac{S_{12} S_{21}}{1 + S_{22}}$$
$$= 0.25 - \frac{(0.75\angle -45°)(0.75\angle -45°)}{1 + 0.25} = 0.512 \angle 60.95°.$$

So the return loss is $RL = -20 \log |\Gamma| = -20 \log(0.512) = 5.8$ dB.

An important point to understand about scattering parameters is that the reflection coefficient looking into port n is not equal to S_{nn} unless all other ports are matched (this is illustrated in the above example). Similarly, the transmission coefficient from port m to port n is not equal to S_{nm} unless all other ports are matched. The scattering parameters of a network are properties only of the network itself (assuming the network is linear), and are defined under the condition that all ports are matched. Changing the terminations or excitations of a network does not change its scattering parameters, but may change the reflection coefficient seen at a given port, or the transmission coefficient between two ports.

A Shift in Reference Planes

Because scattering parameters relate amplitudes (magnitude and phase) of traveling waves incident on and reflected from a microwave network, phase reference planes must be specified for each port of the network. We now show how scattering parameters are transformed when the reference planes are moved from their original locations.

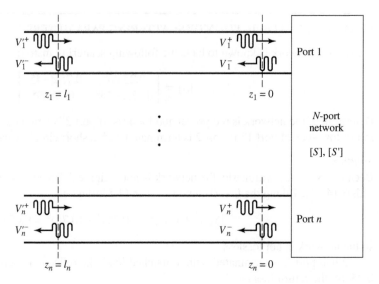

FIGURE 4.10 | Shifting reference planes for an N-port network.

Consider the N-port microwave network shown in Figure 4.10, where the original terminal planes are assumed to be located at $z_n = 0$ for the nth port, where z_n is an arbitrary coordinate measured along the transmission line feeding the nth port. The scattering matrix for the network with this set of terminal planes is denoted by $[S]$. Now consider a new set of reference planes defined at $z_n = \ell_n$ for the nth port, and let the new scattering matrix be denoted as $[S']$. Then in terms of the incident and reflected port voltages we have that

$$[V^-] = [S][V^+], \tag{4.54a}$$

$$[V'^-] = [S'][V'^+], \tag{4.54b}$$

where the unprimed quantities are referenced to the original terminal planes at $z_n = 0$, and the primed quantities are referenced to the new terminal planes at $z_n = \ell_n$.

From the theory of traveling waves on lossless transmission lines we can relate the new wave amplitudes to the original ones as

$$V'^+_n = V^+_n e^{j\theta_n}, \tag{4.55a}$$

$$V'^-_n = V^-_n e^{-j\theta_n}, \tag{4.55b}$$

where $\theta_n = \beta_n \ell_n$ is the electrical length of the outward shift of the reference plane of port n. Writing (4.55) in matrix form and substituting into (4.54a) gives

$$
\begin{bmatrix}
e^{j\theta_1} & & & 0 \\
& e^{j\theta_2} & & \\
& & \ddots & \\
0 & & & e^{j\theta_N}
\end{bmatrix}
[V'^-] = [S]
\begin{bmatrix}
e^{-j\theta_1} & & & 0 \\
& e^{-j\theta_2} & & \\
& & \ddots & \\
0 & & & e^{-j\theta_N}
\end{bmatrix}
[V'^+].
$$

Multiplying by the inverse of the first matrix on the left gives

$$
[V'^-] =
\begin{bmatrix}
e^{-j\theta_1} & & & 0 \\
& e^{-j\theta_2} & & \\
& & \ddots & \\
0 & & & e^{-j\theta_N}
\end{bmatrix}
[S]
\begin{bmatrix}
e^{-j\theta_1} & & & 0 \\
& e^{-j\theta_2} & & \\
& & \ddots & \\
0 & & & e^{-j\theta_N}
\end{bmatrix}
[V'^+].
$$

Comparing with (4.54b) shows that

$$
[S'] =
\begin{bmatrix}
e^{-j\theta_1} & & & 0 \\
& e^{-j\theta_2} & & \\
& & \ddots & \\
0 & & & e^{-j\theta_N}
\end{bmatrix}
[S]
\begin{bmatrix}
e^{-j\theta_1} & & & 0 \\
& e^{-j\theta_2} & & \\
& & \ddots & \\
0 & & & e^{-j\theta_N}
\end{bmatrix},
\tag{4.56}
$$

which is the desired result. Note that $S'_{nn} = e^{-2j\theta_n}S_{nn}$, meaning that the phase of S_{nn} is shifted by twice the electrical length of the shift in terminal plane n because the wave travels twice over this length upon incidence and reflection. This result is consistent with (2.42), which gives the change in the reflection coefficient on a transmission line due to a shift in the reference plane.

Power Waves and Generalized Scattering Parameters

We previously expressed the total voltage and current on a transmission line in terms of the incident and reflected voltage wave amplitudes, as in (2.34) or (4.42):

$$V = V_0^+ + V_0^-, \tag{4.57a}$$

$$I = \frac{1}{Z_0}(V_0^+ - V_0^-), \tag{4.57b}$$

with Z_0 being the characteristic impedance of the line. Inverting (4.57) gives the incident and reflected voltage wave amplitudes in terms of the total voltage and current:

$$V_0^+ = \frac{V + Z_0 I}{2}, \tag{4.58a}$$

$$V_0^- = \frac{V - Z_0 I}{2}. \tag{4.58b}$$

The average power delivered to a load can be expressed as

$$P_L = \frac{1}{2}\text{Re}\{VI^*\} = \frac{1}{2Z_0}\text{Re}\{|V_0^+|^2 - V_0^+V_0^{-*} + V_0^{+*}V_0^- - |V_0^-|^2\}$$

$$= \frac{1}{2Z_0}(|V_0^+|^2 - |V_0^-|^2), \tag{4.59}$$

where the last step follows because the quantity $V_0^{+*}V_0^- - V_0^+V_0^{-*}$ is pure imaginary. This is a physically satisfying result since it expresses the net power delivered to the load as the difference between the incident and reflected powers. Unfortunately, this result is only valid when the characteristic impedance is real; it does not apply when Z_0 is complex, as in the case of a lossy line. In addition, these results are not useful when no transmission line is present between the generator and load, as in the circuit shown in Figure 4.11.

In the circuit of Figure 4.11 there is no defined characteristic impedance, nor is there a voltage reflection coefficient, or incident and reflected voltage or current waves. It is possible, however, to define a new set of waves, called *power waves*, which have useful properties when dealing with power transfer between a generator and a load, and can be applied to circuits like that of Figure 4.11, as well as to problems with lossless or lossy transmission lines. We will also see how power waves lead to a generalization of scattering parameters.

The incident and reflected power wave amplitudes a and b are defined as the following linear transformations of the total voltage and current:

$$a = \frac{V + Z_R I}{2\sqrt{R_R}}, \tag{4.60a}$$

$$b = \frac{V - Z_R^* I}{2\sqrt{R_R}}, \tag{4.60b}$$

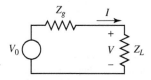

FIGURE 4.11 | A generator with impedance Z_g connected to a load impedance Z_L.

where $Z_R = R_R + jX_R$ is known as the *reference impedance*, and may be complex. Note that the power wave amplitudes of (4.60) are similar in form to the voltage waves of (4.58), but do not have units of power, voltage, or current.

Inverting (4.60) gives the total voltage and current in terms of the power wave amplitudes:

$$V = \frac{Z_R^* a + Z_R b}{\sqrt{R_R}}, \tag{4.61a}$$

$$I = \frac{a - b}{\sqrt{R_R}}. \tag{4.61b}$$

Then the power delivered to the load can be expressed as

$$P_L = \frac{1}{2}\text{Re}\left\{VI^*\right\} = \frac{1}{2R_R}\text{Re}\left\{Z_R^*|a|^2 - Z_R^* ab^* + Z_R a^* b - Z_R|b|^2\right\}$$

$$= \frac{1}{2}|a|^2 - \frac{1}{2}|b|^2, \tag{4.62}$$

since the quantity $Z_R a^* b - Z_R^* ab^*$ is pure imaginary. Once again we have the satisfying result that the load power is the difference between the powers of the incident and reflected power waves. It is important to note that this result is valid for any reference impedance Z_R.

The reflection coefficient, Γ_p, for the reflected power wave can be found by using (4.60) and the fact that $V = Z_L I$ at the load:

$$\Gamma_p = \frac{b}{a} = \frac{V - Z_R^* I}{V + Z_R I} = \frac{Z_L - Z_R^*}{Z_L + Z_R}. \tag{4.63}$$

Observe that this reflection coefficient reduces to our usual voltage reflection coefficient of (2.35) when $Z_R = Z_0$ is a real characteristic impedance. Equation (4.63) suggests that choosing the reference impedance as the conjugate of the load impedance [5],

$$Z_R = Z_L^*, \tag{4.64}$$

will have the useful effect of making the reflected power wave amplitude go to zero.[1]

From basic circuit theory, the voltage, current, and load power for the circuit of Figure 4.11 are

$$V = V_0\frac{Z_L}{Z_L + Z_g}, \quad I = \frac{V_0}{Z_L + Z_g}, \quad P_L = \frac{V_0^2}{2}\frac{R_L}{|Z_L + Z_g|^2}, \tag{4.65a, b, c}$$

where $Z_L = R_L + jX_L$. Then the power wave amplitudes can be found from (4.60), with $Z_R = Z_L^*$, as

$$a = \frac{V + Z_R I}{2\sqrt{R_R}} = V_0\frac{\dfrac{Z_L}{Z_L + Z_g} + \dfrac{Z_L^*}{Z_L + Z_g}}{2\sqrt{R_R}} = V_0\frac{\sqrt{R_L}}{Z_L + Z_g}, \tag{4.66a}$$

$$b = \frac{V - Z_R^* I}{2\sqrt{R_R}} = V_0\frac{\dfrac{Z_L}{Z_L + Z_g} - \dfrac{Z_L}{Z_L + Z_g}}{2\sqrt{R_R}} = 0. \tag{4.66b}$$

---

[1] Some authors choose the reference impedance equal to the generator impedance. This has the same effect as (4.64) when the generator and load are conjugately matched, but the choice of (4.64) leads to a zero reflected wave even when the conjugate matching condition is not satisfied, and so can be more useful in general.

From (4.62) the power delivered to the load is

$$P_L = \frac{1}{2} |a|^2 = \frac{V_0^2}{2} \frac{R_L}{|Z_L + Z_g|^2},$$

in agreement with (4.65c).

When the load is conjugately matched to the generator, so that $Z_g = Z_L^*$, we have $P_L = V_0^2/8R_L$. Note that selecting the reference impedance as $Z_R = Z_L^*$ results in the condition that $b = 0$ (and $\Gamma_p = 0$), but this does not necessarily mean that the load is conjugately matched to the generator, nor that maximum power is delivered to the load. The incident power wave amplitude of (4.66a) depends on Z_L and Z_g, and is maximum only when $Z_g = Z_L^*$.

To define the scattering matrix for power waves for an N-port network, we assume the reference impedance for port i is Z_{Ri}. Then, analogous to (4.60), we define the power wave amplitude vectors in terms of the total voltage and current vectors:

$$[a] = [F] \left([V] + [Z_R][I]\right), \tag{4.67a}$$

$$[b] = [F] \left([V] - [Z_R]^*[I]\right), \tag{4.67b}$$

where $[F]$ is a diagonal matrix with elements $1/2\sqrt{\text{Re}\left\{Z_{Ri}\right\}}$ and $[Z_R]$ is a diagonal matrix with elements Z_{Ri}. By the impedance matrix relation that $[V] = [Z][I]$, (4.67) can be written as

$$[b] = [F] \left([Z] - [Z_R]^*\right) \left([Z] + [Z_R]\right)^{-1} [F]^{-1} [a].$$

Because the scattering matrix for power waves, $[S_p]$, should relate $[b]$ to $[a]$, we have

$$[S_p] = [F] \left([Z] - [Z_R]^*\right) \left([Z] + [Z_R]\right)^{-1} [F]^{-1}. \tag{4.68}$$

The ordinary scattering matrix for a network can first be converted to an impedance matrix, using a relation similar to (4.45), then converted to the generalized power wave scattering matrix using (4.68). The generalized scattering matrix has the useful property that the diagonal elements can be made to be zero by proper selection of the reference impedances.

POINT OF INTEREST: The Vector Network Analyzer

The scattering parameters of passive and active networks can be measured with a *vector network analyzer*, which is a two-channel (or four-channel) microwave receiver designed to process the magnitude and phase of the transmitted and reflected waves from the network. A simplified block diagram of a network analyzer is shown in the accompanying figure. In operation, the RF source is usually set to sweep over a specified bandwidth. A four-port reflectometer samples the incident, reflected, and transmitted RF waves; a switch allows the network to be driven from either port 1 or port 2. Four dual-conversion channels convert these signals to 100-kHz IF frequencies, which are then detected and converted to digital form. An internal computer is used to calculate and display the magnitude and phase of the scattering parameters or other quantities that can be derived from these data, such as SWR, return loss, group delay, impedance, etc. An important feature of the network analyzer is the substantial improvement in accuracy made possible with error-correcting software. Errors caused by directional coupler mismatch, imperfect directivity, loss, and variations in the frequency response of the analyzer system are accounted for by using a 12-term error model and a calibration procedure. Another useful feature is the ability to determine the time-domain response of the network by calculating the inverse Fourier transform of the frequency-domain data.

4.4 THE TRANSMISSION (*ABCD*) MATRIX

The Z, Y, and S parameter representations can be used to characterize a microwave network with an arbitrary number of ports, but in practice many microwave networks consist of a cascade connection of two or more two-port networks. In this case it is convenient to define a 2×2 *transmission*, or *ABCD*, *matrix*, for each two-port network. We will see that the *ABCD* matrix of the cascade connection of two or more two-port networks can be easily found by multiplying the *ABCD* matrices of the individual two-ports.

The *ABCD* matrix is defined for a two-port network in terms of the total voltages and currents as shown in Figure 4.12a and the following:

$$V_1 = AV_2 + BI_2,$$

$$I_1 = CV_2 + DI_2,$$

or in matrix form as

$$\begin{bmatrix} V_1 \\ I_1 \end{bmatrix} = \begin{bmatrix} A & B \\ C & D \end{bmatrix} \begin{bmatrix} V_2 \\ I_2 \end{bmatrix}. \tag{4.69}$$

It is important to note from Figure 4.12a that a change in the sign convention of I_2 has been made from our previous definitions, which had I_2 as the current flowing *into* port 2. The convention that I_2 flows *out* of port 2 will be used when dealing with *ABCD* matrices so that in a cascade network I_2 will be the same current that flows into the adjacent network, as shown in Figure 4.12b. Then the left-hand side of (4.69) represents the voltage and current at port 1 of the network, while the column on the right-hand side of (4.69) represents the voltage and current at port 2.

In the cascade connection of two two-port networks shown in Figure 4.12b we have that

$$\begin{bmatrix} V_1 \\ I_1 \end{bmatrix} = \begin{bmatrix} A_1 & B_1 \\ C_1 & D_1 \end{bmatrix} \begin{bmatrix} V_2 \\ I_2 \end{bmatrix},$$

$$\tag{4.70a, b}$$

$$\begin{bmatrix} V_2 \\ I_2 \end{bmatrix} = \begin{bmatrix} A_2 & B_2 \\ C_2 & D_2 \end{bmatrix} \begin{bmatrix} V_3 \\ I_3 \end{bmatrix}.$$

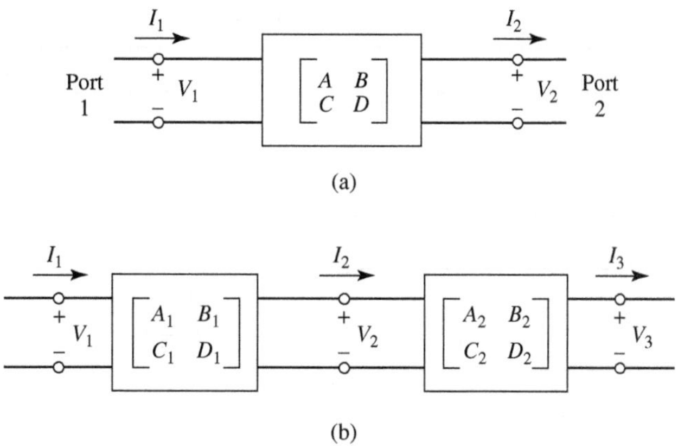

(a)

(b)

FIGURE 4.12 | (a) A two-port network. (b) A cascade connection of two-port networks.

Substituting (4.70b) into (4.70a) gives

$$\begin{bmatrix} V_1 \\ I_1 \end{bmatrix} = \begin{bmatrix} A_1 & B_1 \\ C_1 & C_1 \end{bmatrix} \begin{bmatrix} A_2 & B_2 \\ C_2 & D_2 \end{bmatrix} \begin{bmatrix} V_3 \\ I_3 \end{bmatrix}, \tag{4.71}$$

which shows that the *ABCD* matrix of the cascade connection of the two networks is equal to the product of the *ABCD* matrices representing the individual two-ports. Note that the order of multiplication of the matrix must be the same as the order in which the networks are arranged since matrix multiplication is not, in general, commutative.

The usefulness of the *ABCD* matrix representation lies in the fact that a library of *ABCD* matrices for elementary two-port networks can be built up, and applied in building-block fashion to more complicated microwave networks that consist of cascades of these simpler two-ports. Table 4.1 lists a number of useful two-port networks and their *ABCD* matrices.

EXAMPLE 4.7   EVALUATION OF *ABCD* PARAMETERS

Find the *ABCD* parameters of a two-port network consisting of a series impedance Z between ports 1 and 2 (the first entry in Table 4.1).

Solution
From the defining relations of (4.69), we have that

$$A = \left. \frac{V_1}{V_2} \right|_{I_2=0},$$

which indicates that A is found by applying a voltage V_1 at port 1, and measuring the open-circuit voltage V_2 at port 2. Thus, $A = 1$. Similarly,

$$B = \left. \frac{V_1}{I_2} \right|_{V_2=0} = \frac{V_1}{V_1/Z} = Z,$$

$$C = \left. \frac{I_1}{V_2} \right|_{I_2=0} = 0,$$

$$D = \left. \frac{I_1}{I_2} \right|_{V_2=0} = \frac{I_1}{I_1} = 1.$$

Relation to Impedance Matrix

The impedance parameters of a network can be easily converted to *ABCD* parameters. Thus, from the definition of the *ABCD* parameters in (4.69), and from the defining relations for the *Z* parameters of (4.25)

TABLE 4.1 | *ABCD* Parameters of Some Useful Two-Port Circuits

Circuit	*ABCD* Parameters
	$A = 1 \qquad B = Z$ $C = 0 \qquad D = 1$
	$A = 1 \qquad B = 0$ $C = Y \qquad D = 1$
	$A = \cos \beta \ell \qquad B = j Z_0 \sin \beta \ell$ $C = j Y_0 \sin \beta \ell \qquad D = \cos \beta \ell$
	$A = N \qquad B = 0$ $C = 0 \qquad D = \dfrac{1}{N}$
	$A = 1 + \dfrac{Y_2}{Y_3} \qquad B = \dfrac{1}{Y_3}$ $C = Y_1 + Y_2 + \dfrac{Y_1 Y_2}{Y_3} \qquad D = 1 + \dfrac{Y_1}{Y_3}$
	$A = 1 + \dfrac{Z_1}{Z_3} \qquad B = Z_1 + Z_2 + \dfrac{Z_1 Z_2}{Z_3}$ $C = \dfrac{1}{Z_3} \qquad D = 1 + \dfrac{Z_2}{Z_3}$

for a two-port network with I_2 to be consistent with the sign convention used with *ABCD* parameters,

$$V_1 = I_1 Z_{11} - I_2 Z_{12}, \tag{4.72a}$$

$$V_2 = I_1 Z_{21} - I_2 Z_{22}, \tag{4.72b}$$

we have that

$$A = \frac{V_1}{V_2} \bigg|_{I_2=0} = \frac{I_1 Z_{11}}{I_1 Z_{21}} = Z_{11}/Z_{21}, \tag{4.73a}$$

$$B = \frac{V_1}{I_2} \bigg|_{V_2=0} = \frac{I_1 Z_{11} - I_2 Z_{12}}{I_2} \bigg|_{V_2=0} = Z_{11} \frac{I_1}{I_2} \bigg|_{V_2=0} - Z_{12}$$

$$= Z_{11} \frac{I_1 Z_{22}}{I_1 Z_{21}} - Z_{12} = \frac{Z_{11} Z_{22} - Z_{12} Z_{21}}{Z_{21}}, \tag{4.73b}$$

$$C = \frac{I_1}{V_2} \bigg|_{I_2=0} = \frac{I_1}{I_1 Z_{21}} = 1/Z_{21}, \tag{4.73c}$$

$$D = \frac{I_1}{I_2} \bigg|_{V_2=0} = \frac{I_2 Z_{22}/Z_{21}}{I_2} = Z_{22}/Z_{21}. \tag{4.73d}$$

If the network is reciprocal, then $Z_{12} = Z_{21}$ and (4.73) can be used to show that $AD - BC = 1$.

Equivalent Circuits for Two-Port Networks

The special case of a two-port microwave network occurs so frequently in practice that it deserves further attention. Here we will discuss the use of equivalent circuits to represent an arbitrary two-port network. Useful conversions between two-port network parameters are given in Table 4.2.

Figure 4.13a shows a transition between a coaxial line and a microstrip line, and is an example of a two-port network. Terminal planes can be defined at arbitrary points on the two transmission lines; a convenient choice might be as shown in the figure. However, because of the physical discontinuity in the transition from a coaxial line to a microstrip line, electric and/or magnetic energy can be stored in the vicinity of the junction, leading to reactive effects. Characterization of such effects can be obtained by measurement or by numerical analysis (such analysis may be quite complicated), and represented by the two-port "black box" shown in Figure 4.13b. The properties of the transition can then be expressed in terms of the network parameters (Z, Y, S, or $ABCD$) of the two-port network. This type of treatment can be applied to a variety of two-port junctions, such as transitions from one type of transmission line to another, transmission line discontinuities such as step changes in width or bends, etc. When modeling a microwave junction in this way, it is often useful to replace the two-port "black box" with an equivalent circuit containing a few idealized components, as shown in Figure 4.13c. This is particularly useful if the component values can be related to some physical features of the actual junction. There is an unlimited number of ways in which such equivalent circuits can be defined; we will discuss some of the most common and useful types below.

As we have seen, an arbitrary two-port network can be described in terms of impedance parameters as

$$V_1 = Z_{11}I_1 + Z_{12}I_2,$$
$$V_2 = Z_{21}I_1 + Z_{22}I_2, \qquad (4.74a)$$

or in terms of admittance parameters as

$$I_1 = Y_{11}V_1 + Y_{12}V_2,$$
$$I_2 = Y_{21}V_1 + Y_{22}V_2. \qquad (4.74b)$$

If the network is reciprocal, then $Z_{12} = Z_{21}$ and $Y_{12} = Y_{21}$. These representations lead naturally to the T and π equivalent circuits shown in Figures 4.14a and 4.14b. The relations in Table 4.2 can be used to relate the component values to other network parameters.

Other equivalent circuits can also be used to represent a two-port network. If the network is reciprocal, there are six degrees of freedom (the real and imaginary parts of three matrix elements), so the equivalent circuit should have six independent parameters. A nonreciprocal network cannot be represented by a passive equivalent circuit using reciprocal elements.

If the network is lossless, which is a good approximation for many practical two-port junctions, some simplifications can be made in the equivalent circuit. As was shown in Section 4.2, the impedance or admittance matrix elements are purely imaginary for a lossless network. This reduces the degrees of freedom for such a network to three, and implies that the T and π equivalent circuits of Figure 4.14 can be constructed from purely reactive elements.

4.5   SIGNAL FLOW GRAPHS

We have seen how transmitted and reflected waves can be represented by scattering parameters, and how the interconnection of sources, networks, and loads can be treated with various matrix representations. In this section we discuss the *signal flow graph*, which is an additional technique that is very useful for the analysis of microwave networks in terms of transmitted and reflected waves. We first discuss the features and the construction of the flow graph itself, and then present a technique for the reduction, or solution, of the flow graph.

The primary components of a signal flow graph are nodes and branches:

- Nodes: Each port i of a microwave network has two nodes, a_i and b_i. Node a_i is identified with a wave entering port i, while node b_i is identified with a wave reflected from port i. The voltage at a node is equal to the sum of all signals entering that node.
- Branches: A branch is a directed path between two nodes representing signal flow from one node to another. Every branch has an associated scattering parameter or reflection coefficient.

TABLE 4.2 | Conversions Between Two-Port Network Parameters

	S	Z	Y	ABCD
S_{11}	S_{11}	$\dfrac{(Z_{11}-Z_0)(Z_{22}+Z_0)-Z_{12}Z_{21}}{\Delta Z}$	$\dfrac{(Y_0-Y_{11})(Y_0+Y_{22})+Y_{12}Y_{21}}{\Delta Y}$	$\dfrac{A+B/Z_0-CZ_0-D}{A+B/Z_0+CZ_0+D}$
S_{12}	S_{12}	$\dfrac{2Z_{12}Z_0}{\Delta Z}$	$\dfrac{-2Y_{12}Y_0}{\Delta Y}$	$\dfrac{2(AD-BC)}{A+B/Z_0+CZ_0+D}$
S_{21}	S_{21}	$\dfrac{2Z_{21}Z_0}{\Delta Z}$	$\dfrac{-2Y_{21}Y_0}{\Delta Y}$	$\dfrac{2}{A+B/Z_0+CZ_0+D}$
S_{22}	S_{22}	$\dfrac{(Z_{11}+Z_0)(Z_{22}-Z_0)-Z_{12}Z_{21}}{\Delta Z}$	$\dfrac{(Y_0+Y_{11})(Y_0-Y_{22})+Y_{12}Y_{21}}{\Delta Y}$	$\dfrac{-A+B/Z_0-CZ_0+D}{A+B/Z_0+CZ_0+D}$
Z_{11}	$Z_0\dfrac{(1+S_{11})(1-S_{22})+S_{12}S_{21}}{(1-S_{11})(1-S_{22})-S_{12}S_{21}}$	Z_{11}	$\dfrac{Y_{22}}{|Y|}$	$\dfrac{A}{C}$
Z_{12}	$Z_0\dfrac{2S_{12}}{(1-S_{11})(1-S_{22})-S_{12}S_{21}}$	Z_{12}	$\dfrac{-Y_{12}}{|Y|}$	$\dfrac{AD-BC}{C}$
Z_{21}	$Z_0\dfrac{2S_{21}}{(1-S_{11})(1-S_{22})-S_{12}S_{21}}$	Z_{21}	$\dfrac{-Y_{21}}{|Y|}$	$\dfrac{1}{C}$
Z_{22}	$Z_0\dfrac{(1-S_{11})(1+S_{22})+S_{12}S_{21}}{(1-S_{11})(1-S_{22})-S_{12}S_{21}}$	Z_{22}	$\dfrac{Y_{11}}{|Y|}$	$\dfrac{D}{C}$
Y_{11}	$Y_0\dfrac{(1-S_{11})(1+S_{22})+S_{12}S_{21}}{(1+S_{11})(1+S_{22})-S_{12}S_{21}}$	$\dfrac{Z_{22}}{|Z|}$	Y_{11}	$\dfrac{D}{B}$
Y_{12}	$Y_0\dfrac{-2S_{12}}{(1+S_{11})(1+S_{22})-S_{12}S_{21}}$	$\dfrac{-Z_{12}}{|Z|}$	Y_{12}	$\dfrac{BC-AD}{B}$
Y_{21}	$Y_0\dfrac{-2S_{21}}{(1+S_{11})(1+S_{22})-S_{12}S_{21}}$	$\dfrac{-Z_{21}}{|Z|}$	Y_{21}	$\dfrac{-1}{B}$
Y_{22}	$Y_0\dfrac{(1+S_{11})(1-S_{22})+S_{12}S_{21}}{(1+S_{11})(1+S_{22})-S_{12}S_{21}}$	$\dfrac{Z_{11}}{|Z|}$	Y_{22}	$\dfrac{A}{B}$
A	$\dfrac{(1+S_{11})(1-S_{22})+S_{12}S_{21}}{2S_{21}}$	$\dfrac{Z_{11}}{Z_{21}}$	$\dfrac{-Y_{22}}{Y_{21}}$	A
B	$Z_0\dfrac{(1+S_{11})(1+S_{22})-S_{12}S_{21}}{2S_{21}}$	$\dfrac{|Z|}{Z_{21}}$	$\dfrac{-1}{Y_{21}}$	B
C	$\dfrac{1}{Z_0}\dfrac{(1-S_{11})(1-S_{22})-S_{12}S_{21}}{2S_{21}}$	$\dfrac{1}{Z_{21}}$	$\dfrac{-|Y|}{Y_{21}}$	C
D	$\dfrac{(1-S_{11})(1+S_{22})+S_{12}S_{21}}{2S_{21}}$	$\dfrac{Z_{22}}{Z_{21}}$	$\dfrac{-Y_{11}}{Y_{21}}$	D

$|Z|=Z_{11}Z_{22}-Z_{12}Z_{21}$; $|Y|=Y_{11}Y_{22}-Y_{12}Y_{21}$; $\Delta Z=(Z_{11}+Z_0)(Z_{22}+Z_0)-Z_{12}Z_{21}$; $\Delta Y=(Y_{11}+Y_0)(Y_{22}+Y_0)-Y_{12}Y_{21}$; $Y_0=1/Z_0$.

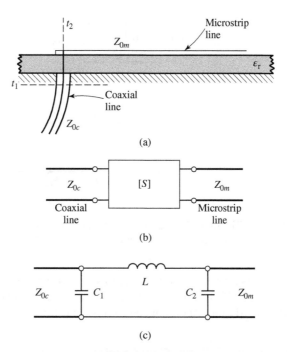

FIGURE 4.13 | A coax-to-microstrip transition and equivalent circuit representations. (a) Geometry of the transition. (b) Representation of the transition by a "black box." (c) A possible equivalent circuit for the transition [6].

At this point it is useful to consider the flow graph of an arbitrary two-port network, as shown in Figure 4.15. Figure 4.15a shows a two-port network with incident and reflected waves at each port, and Figure 4.15b shows the corresponding signal flow graph representation. The flow graph gives an intuitive graphical illustration of the network behavior.

For example, a wave of amplitude a_1 incident at port 1 is split, with part going through S_{11} and out port 1 as a reflected wave, and part transmitted through S_{21} to node b_2. At node b_2, the wave goes out port 2; if a load with nonzero reflection coefficient is connected at port 2, this wave will be at least partly reflected and reenter the two-port network at node a_2. Part of this wave can be reflected back out port 2 via S_{22}, and part can be transmitted out port 1 through S_{12}.

Two other special networks—a one-port network and a voltage source—are shown in Figure 4.16, along with their signal flow graph representations. Once a microwave network has been represented in signal flow

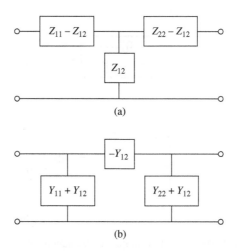

FIGURE 4.14 | Equivalent circuits for a reciprocal two-port network. (a) T equivalent. (b) π equivalent.

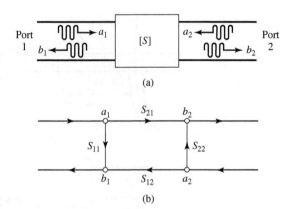

FIGURE 4.15 | The signal flow graph representation of a two-port network. (a) Definition of incident and reflected waves. (b) Signal flow graph.

graph form, it is a relatively easy matter to solve for the ratio of any combination of wave amplitudes. We will discuss how this can be done using four basic decomposition rules, but the same results can also be obtained using Mason's rule from control system theory.

Decomposition of Signal Flow Graphs

A signal flow graph can be reduced to a single branch between two nodes using the following four basic decomposition rules to obtain any desired wave amplitude ratio.

- **Rule 1** (Series Rule). Two branches, whose common node has only one incoming and one outgoing wave (branches in series), may be combined to form a single branch whose coefficient is the product of the coefficients of the original branches. Figure 4.17a shows the flow graphs for this rule. Its derivation follows from the basic relation

$$V_3 = S_{32}V_2 = S_{32}S_{21}V_1. \tag{4.75}$$

- **Rule 2** (Parallel Rule). Two branches from one common node to another common node (branches in parallel) may be combined into a single branch whose coefficient is the sum of the coefficients of the original branches. Figure 4.17b shows the flow graphs for this rule. The derivation follows from the obvious relation

$$V_2 = S_aV_1 + S_bV_1 = (S_a + S_b)V_1. \tag{4.76}$$

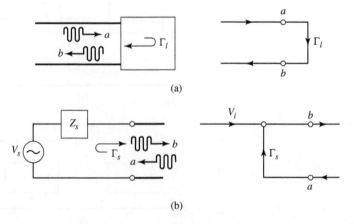

FIGURE 4.16 | The signal flow graph representations of a one-port network and a source. (a) A one-port network and its flow graph. (b) A source and its flow graph.

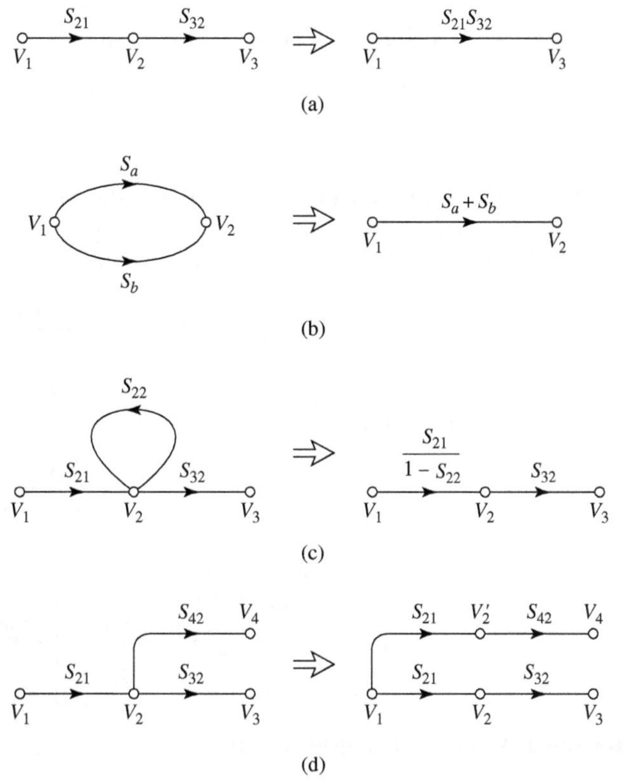

FIGURE 4.17 | Decomposition rules. (a) Series rule. (b) Parallel rule. (c) Self-loop rule. (d) Splitting rule.

- **Rule 3** (Self-Loop Rule). When a node has a self-loop (a branch that begins and ends on the same node) of coefficient S, the self-loop can be eliminated by multiplying coefficients of the branches feeding that node by $1/(1 - S)$. Figure 4.17c shows the flow graphs for this rule, which can be derived as follows. From the original network we have

$$V_2 = S_{21} V_1 + S_{22} V_2, \tag{4.77a}$$
$$V_3 = S_{32} V_2. \tag{4.77b}$$

Eliminating V_2 gives

$$V_3 = \frac{S_{32} S_{21}}{1 - S_{22}} V_1, \tag{4.78}$$

which is seen to be the transfer function for the reduced graph of Figure 4.17c.

- **Rule 4** (Splitting Rule). A node may be split into two separate nodes as long as the resulting flow graph contains, once and only once, each combination of separate (not self-loops) input and output branches that connect to the original node. This rule is illustrated in Figure 4.17d and follows from the observation that

$$V_4 = S_{42} V_2 = S_{21} S_{42} V_1 \tag{4.79}$$

in both the original flow graph and the flow graph with the split node.

We now illustrate the use of each of these rules with an example.

EXAMPLE 4.8 APPLICATION OF SIGNAL FLOW GRAPH

Use signal flow graphs to derive expressions for Γ_{in} and Γ_{out} for the microwave network shown in Figure 4.18.

Solution
The signal flow graph for the circuit of Figure 4.18 is shown in Figure 4.19. In terms of node voltages, Γ_{in} is given by the ratio b_1/a_1. The first two steps of the required decomposition of the flow graph are shown in Figures 4.20a and 4.20b, from which the desired result follows by inspection:

$$\Gamma_{in} = \frac{b_1}{a_1} = S_{11} + \frac{S_{12}S_{21}\Gamma_\ell}{1 - S_{22}\Gamma_\ell}.$$

Next, Γ_{out} is given by the ratio b_2/a_2. The first two steps for this decomposition are shown in Figures 4.20c and 4.20d. The desired result is

$$\Gamma_{out} = \frac{b_2}{a_2} = S_{22} + \frac{S_{12}S_{21}\Gamma_s}{1 - S_{11}\Gamma_s}$$

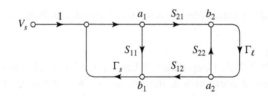

FIGURE 4.18 | A terminated two-port network.

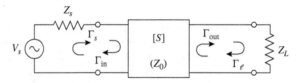

FIGURE 4.19 | Signal flow graph for the two-port network with general source and load impedances of Figure 4.18.

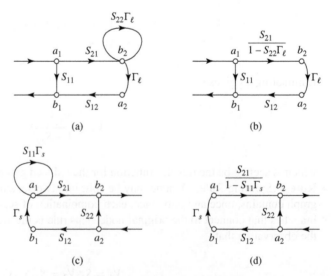

FIGURE 4.20 | Decompositions of the flow graph of Figure 4.19 to find $\Gamma_{in} = b_1/a_1$ and $\Gamma_{out} = b_2/a_2$. (a) Using Rule 4 on node a_2. (b) Using Rule 3 for the self-loop at node b_2. (c) Using Rule 4 on node b_1. (d) Using Rule 3 for the self-loop at node a_1.

Application to Thru-Reflect-Line Network Analyzer Calibration

As a further application of signal flow graphs we consider the calibration of a network analyzer using the *Thru-Reflect-Line* (TRL) technique [7]. The general problem is shown in Figure 4.21, where it is intended to measure the scattering parameters of a two-port device at the indicated reference planes. As discussed in the previous Point of Interest, a network analyzer measures scattering parameters as ratios of complex voltage amplitudes. The primary reference plane for such measurements is generally at some point within the analyzer itself, so the measurement will include losses and phase delays caused by the effects of the connectors, cables, and transitions that must be used to connect the device under test (DUT) to the analyzer. In the block diagram of Figure 4.21 these effects are lumped together in a two-port *error box* placed at each port between the actual measurement reference plane and the desired reference plane for the two-port DUT. A calibration procedure is used to characterize the error boxes before measurement of the DUT; then the actual error-corrected scattering parameters of the DUT can be calculated from the measured data. Measurement of a one-port network can be considered as a reduced version of the two-port case.

The simplest way to calibrate a network analyzer is to use three or more known loads, such as shorts, opens, and matched loads. The problem with this approach is that such standards are always imperfect to some degree, and therefore introduce errors into the measurement. These errors become increasingly significant at higher frequencies and as the quality of the measurement system improves. The TRL calibration scheme does not rely on known standard loads, but uses three simple connections to allow the error boxes to be characterized completely. These three connections are shown in Figure 4.22. The *Thru* connection is made by directly connecting port 1 to port 2 at the desired reference planes. The *Reflect* connection uses a load having a large reflection coefficient, Γ_L, such as a nominal open or short. It is not necessary to know the exact value of Γ_L, as this will be determined by the TRL calibration procedure. The *Line* connection involves connecting ports 1 and 3 together through a length of matched transmission line. It is not necessary to know the length of the line, and it is not required that the line be lossless; these parameters will be determined by the TRL procedure.

We can use signal flow graphs to derive the set of equations necessary to find the scattering parameters for the error boxes in the TRL calibration procedure. With reference to Figure 4.21, we will apply the *Thru*, *Reflect*, and *Line* connections at the reference plane for the DUT, and measure the scattering parameters for these three cases at the measurement planes. For simplicity, we assume the same characteristic impedance for ports 1 and 2, and that the error boxes are reciprocal and identical for both ports. The error boxes are characterized by a scattering matrix $[S]$ and, alternatively, by an *ABCD* matrix. Thus $S_{21} = S_{12}$ for both error boxes. Also note that ports 1 and 2 of the error boxes are in opposite positions since they are symmetrically connected, as shown in the figure. To avoid confusion in notation we will denote the measured scattering parameters for the *Thru*, *Reflect*, and *Line* connections as the $[T]$, $[R]$, and $[L]$ matrices, respectively.

Figure 4.22a shows the arrangement for the *Thru* connection and the corresponding signal flow graph. Observe that we have made use of the fact that $S_{21} = S_{12}$ and that the error boxes are identical and symmetrically arranged. The signal flow graph can be easily reduced using the decomposition rules to give the

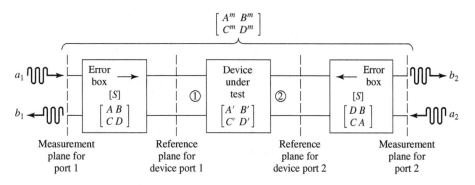

FIGURE 4.21 | Block diagram of a network analyzer measurement of a two-port device.

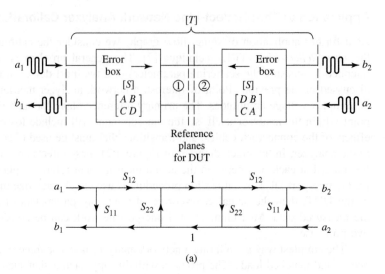

FIGURE 4.22a | Block diagram and signal flow graph for the *Thru* connection.

measured scattering parameters at the measurement planes in terms of the scattering parameters of the error boxes as

$$T_{11} = \frac{b_1}{a_1}\bigg|_{a_2=0} = S_{11} + \frac{S_{22}S_{12}^2}{1 - S_{22}^2} \tag{4.80a}$$

$$T_{12} = \frac{b_1}{a_2}\bigg|_{a_1=0} = \frac{S_{12}^2}{1 - S_{22}^2} \tag{4.80b}$$

By symmetry we have $T_{22} = T_{11}$, and by reciprocity we have $T_{21} = T_{12}$.

The *Reflect* connection is shown in Figure 4.22b, with the corresponding signal flow graph. Note that this arrangement effectively decouples the two measurement ports, so $R_{12} = R_{21} = 0$. The signal flow graph can be easily reduced to show that

$$R_{11} = \frac{b_1}{a_1}\bigg|_{a_2=0} = S_{11} + \frac{S_{12}^2 \Gamma_L}{1 - S_{22}\Gamma_L}. \tag{4.81}$$

By symmetry we have $R_{22} = R_{11}$.

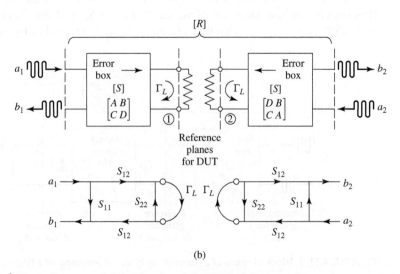

FIGURE 4.22b | Block diagram and signal flow graph for the *Reflect* connection.

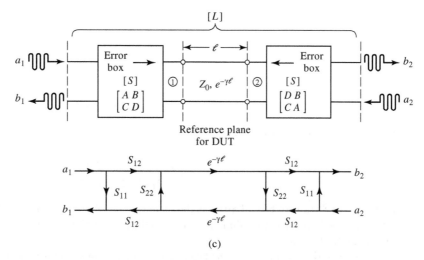

FIGURE 4.22c | Block diagram and signal flow graph for the *Line* connection.

The *Line* connection is shown in Figure 4.22c, with its corresponding signal flow graph. A reduction similar to that used for the *Thru* case gives

$$L_{11} = \left.\frac{b_1}{a_1}\right|_{a_2=0} = S_{11} + \frac{S_{22}S_{12}^2 e^{-2\gamma\ell}}{1 - S_{22}^2 e^{-2\gamma\ell}}, \tag{4.82a}$$

$$L_{12} = \left.\frac{b_1}{a_2}\right|_{a_1=0} = \frac{S_{12}^2 e^{-\gamma\ell}}{1 - S_{22}^2 e^{-2\gamma\ell}}. \tag{4.82b}$$

By symmetry and reciprocity we have $L_{22} = L_{11}$ and $L_{21} = L_{12}$.

We now have five equations (4.80)–(4.82) for the five unknowns $S_{11}, S_{12}, S_{22}, \Gamma_L,$ and $e^{\gamma\ell}$; the solution is straightforward but lengthy. Because (4.81) is the only equation that contains Γ_L, we can first solve the four equations in (4.80) and (4.82) for the other four unknowns. Equation (4.80b) can be used to eliminate S_{12} from (4.80a) and (4.82), and then S_{11} can be eliminated from (4.80a) and (4.82a). This leaves two equations for S_{22} and $e^{\gamma\ell}$:

$$L_{12}e^{2\gamma\ell} - L_{12}S_{22}^2 = T_{12}e^{\gamma\ell} - T_{12}S_{22}^2 e^{\gamma\ell}, \tag{4.83a}$$

$$e^{2\gamma\ell}(T_{11} - S_{22}T_{12}) - T_{11}S_{22}^2 = L_{11}\left(e^{2\gamma\ell} - S_{22}^2\right) - S_{22}T_{12}. \tag{4.83b}$$

Equation (4.83a) can be solved for S_{22} and substituted into (4.83b) to give a quadratic equation for $e^{\gamma\ell}$. Application of the quadratic formula then gives the solution for $e^{\gamma\ell}$ in terms of the measured TRL scattering parameters as

$$e^{\gamma\ell} = \frac{L_{12}^2 + T_{12}^2 - (T_{11} - L_{11})^2 \pm \sqrt{\left[L_{12}^2 + T_{12}^2 - (T_{11} - L_{11})^2\right]^2 - 4L_{12}^2 T_{12}^2}}{2L_{12}T_{12}}. \tag{4.84}$$

The choice of sign can be determined by the requirement that the real and imaginary parts of γ be positive, or by knowing the phase of Γ_L [as determined from (4.83)] to within 180°.

Now multiply (4.80b) by S_{22} and subtract from (4.80a) to get

$$T_{11} = S_{11} + S_{22}T_{12}, \tag{4.85a}$$

and similarly multiply (4.82b) by $S_{22}e^{-\gamma\ell}$ and subtract from (4.82a) to get

$$L_{11} = S_{11} + S_{22}L_{12}e^{-\gamma\ell}. \tag{4.85b}$$

Eliminating S_{11} from these two equations gives S_{22} in terms of $e^{-\gamma\ell}$ as

$$S_{22} = \frac{T_{11} - L_{11}}{T_{12} - L_{12}e^{-\gamma\ell}}. \tag{4.86}$$

Solving (4.85a) for S_{11} gives

$$S_{11} = T_{11} - S_{22}T_{12}, \tag{4.87}$$

and solving (4.80b) for S_{12} gives

$$S_{12}^2 = T_{12}\left(1 - S_{22}^2\right). \tag{4.88}$$

Finally, (4.81) can be solved for Γ_L to give

$$\Gamma_L = \frac{R_{11} - S_{11}}{S_{12}^2 + S_{22}(R_{11} - S_{11})}. \tag{4.89}$$

Equations (4.84) and (4.86)–(4.89) give the scattering parameters for the error boxes, as well as the unknown reflection coefficient Γ_L (to within the sign), and the propagation factor $e^{-\gamma\ell}$. This completes the calibration procedure for the TRL method.

The scattering parameters of the DUT can now be measured at the measurement reference planes shown in Figure 4.21, and corrected using the above TRL error box parameters to give the scattering parameters at the reference planes of the DUT. Because we are working with a cascade of three two-port networks, it is convenient to use $ABCD$ parameters. Thus, we convert the error box scattering parameters to the corresponding $ABCD$ parameters, and convert the measured scattering parameters of the cascade to the corresponding $A^m B^m C^m D^m$ parameters. If we use $A'B'C'D'$ to denote the parameters for the DUT, then we have

$$\begin{bmatrix} A^m & B^m \\ C^m & D^m \end{bmatrix} = \begin{bmatrix} A & B \\ C & D \end{bmatrix} \begin{bmatrix} A' & B' \\ C' & D' \end{bmatrix} \begin{bmatrix} D & B \\ C & A \end{bmatrix},$$

where the change in the elements of the last matrix account for the reversal of ports for the error box at port 2 of the DUT (see Problem 4.25). Then the $ABCD$ parameters for the DUT can be determined as

$$\begin{bmatrix} A' & B' \\ C' & D' \end{bmatrix} = \begin{bmatrix} A & B \\ C & D \end{bmatrix}^{-1} \begin{bmatrix} A^m & B^m \\ C^m & D^m \end{bmatrix} \begin{bmatrix} D & B \\ C & A \end{bmatrix}^{-1}. \tag{4.90}$$

POINT OF INTEREST: Computer-Aided Design for Microwave Circuits

Computer-aided design (CAD) software packages have become essential tools for the analysis, design, and optimization of RF and microwave circuits and systems. Several microwave CAD products are commercially available, including Microwave Office (Applied Wave Research), ADS (Agilent Technologies), Microwave Studio (CST), Designer (Ansoft), and many others. RF and microwave CAD packages can be divided into two types: those that use "physics-based" solutions, where Maxwell's equations are numerically solved for physical geometries such as printed circuit geometries or waveguides, and "circuit-based" solutions, which use equivalent circuits for various elements, including distributed elements, discontinuities, coupled lines, and active devices. Some packages combine these two approaches. Both linear and nonlinear modeling, as well as circuit optimization, are generally possible. Although such computer programs can be fast, powerful, and accurate, they cannot serve as a substitute for engineering experience and a good understanding of microwave principles.

A typical design process usually begins with specifications or design goals for the circuit or system. Based on previous designs and his or her experience, an engineer can develop an initial design, including specific components and a circuit layout. CAD can then be used to model and analyze the design, using data for each of the components and including effects such as loss and discontinuities. The software can be used to optimize the design by adjusting some of the circuit parameters to achieve the best performance. If the specifications are not met, the design may have to be revised. CAD tools can also be used to study the effects of component tolerances and errors to improve circuit reliability and robustness. When the design meets the specifications, an engineering prototype can be built and tested. If the measured results satisfy the specifications, the design process is completed. Otherwise the design will need to be revised and the procedure repeated.

Without CAD tools the design process would require the construction and measurement of laboratory prototypes at each iteration, which is expensive and time consuming. Thus, CAD can greatly decrease the time and cost of a design while enhancing its quality. The simulation and optimization process is especially important for monolithic microwave integrated circuits because these circuits cannot easily be tuned or trimmed after fabrication.

CAD techniques are not without limitations, however. Of primary importance is the fact that any computer model is only an approximation to a "real-world" physical circuit and cannot completely account for the inevitable differences due to component and fabrication tolerances, surface roughness, spurious coupling, higher order modes, junction discontinuities, thermal effects, and a number of other practical issues that can occur with a physical circuit or device.

4.6 DISCONTINUITIES AND MODAL ANALYSIS

By either necessity or design, microwave circuits and networks often consist of transmission lines with various types of discontinuities. In some cases discontinuities are an unavoidable result of mechanical or electrical transitions from one medium to another (e.g., a junction between two waveguides, or a coax-to-microstrip transition), and the discontinuity effect is unwanted but may be significant enough to warrant characterization. In other cases discontinuities may be deliberately introduced into the circuit to perform a certain electrical function (e.g., reactive diaphragms in waveguide, or stubs on a microstrip line for matching or filter circuits). In any event, a transmission line discontinuity can be represented as an equivalent circuit at some point on the transmission line. Depending on the type of discontinuity, the equivalent circuit may be a simple shunt or series element across the line or, in the more general case, a T- or π-equivalent circuit may be required. The component values of an equivalent circuit depend on the parameters of the line and the discontinuity, as well as on the frequency of operation. In some cases the equivalent circuit involves a shift in the phase reference planes on the transmission lines. Once the equivalent circuit of a given discontinuity is known, its effect can be incorporated into the analysis or design of the network using the theory developed previously in this chapter.

The purpose of the present section is to discuss how equivalent circuits are obtained for transmission line discontinuities; we will see that one approach is to start with a field theory solution to a canonical discontinuity problem and develop a circuit model with component values. This is thus another example of our objective of replacing complicated field analyses with circuit concepts. In other cases, it may be easier to measure the network parameters of an isolated discontinuity.

Figures 4.23 and 4.24 show some common transmission line discontinuities and their equivalent circuits. As shown in Figures 4.23a–4.23c, thin metallic diaphragms (or "irises") can be placed in the cross section of a waveguide to yield equivalent shunt inductance, capacitance, or a resonant combination. Similar effects occur with step changes in the height or width of the waveguide, as shown in Figures 4.23d and 4.23e. Similar discontinuities can also be made in circular waveguide. The classic reference for waveguide discontinuities and their equivalent circuits is the *Waveguide Handbook* [8].

Some typical microstrip discontinuities and transitions are shown in Figure 4.24; similar geometries exist for stripline and other printed transmission lines such as slotline, covered microstrip, and coplanar waveguide. Although approximate equivalent circuits have been developed for some printed transmission line discontinuities [9], many do not lend themselves to easy or accurate modeling, and must be treated by numerical analysis. Modern CAD tools are usually capable of accurately modeling such problems.

Modal Analysis of an *H*-Plane Step in Rectangular Waveguide

The field analysis of most transmission line discontinuity problems is difficult, and beyond the scope of this book. The technique of waveguide *modal analysis*, however, is relatively straightforward and similar in principle to the reflection/transmission problems that were discussed in Chapters 1 and 2. In addition, modal analysis is a rigorous and versatile technique that can be applied to a number of waveguide and coax discontinuity problems, and lends itself well to computer implementation. We will illustrate the technique by applying it to the problem of finding the equivalent circuit of an *H*-plane step (change in width) in a rectangular waveguide.

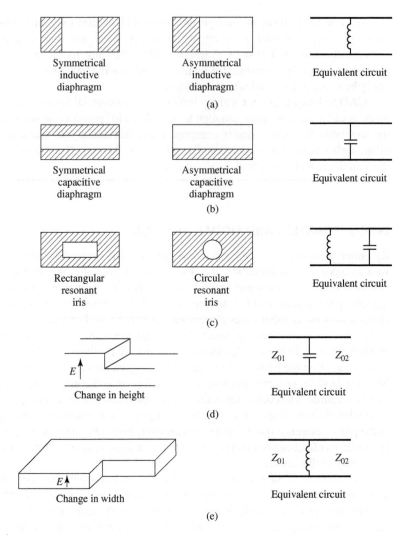

FIGURE 4.23 | Rectangular waveguide discontinuities.

The geometry of the *H*-plane waveguide step is shown in Figure 4.25. It is assumed that only the dominant TE_{10} mode is propagating in guide 1 ($z < 0$) and is incident on the junction from $z < 0$. It is also assumed that no modes are propagating in guide 2, although the analysis to follow is still valid if propagation can occur in guide 2. From Section 3.3, the transverse components of the incident TE_{10} mode can be written, for $z < 0$, as

$$E_y^i = \sin \frac{\pi x}{a} e^{-j\beta_1^a z}, \tag{4.91a}$$

$$H_x^i = \frac{-1}{Z_1^a} \sin \frac{\pi x}{a} e^{-j\beta_1^a z}, \tag{4.91b}$$

where

$$\beta_n^a = \sqrt{k_0^2 - \left(\frac{n\pi}{a}\right)^2} \tag{4.92}$$

is the propagation constant of the TE_{n0} mode in guide 1 (of width *a*), and

$$Z_n^a = \frac{k_0 \eta_0}{\beta_n^a} \tag{4.93}$$

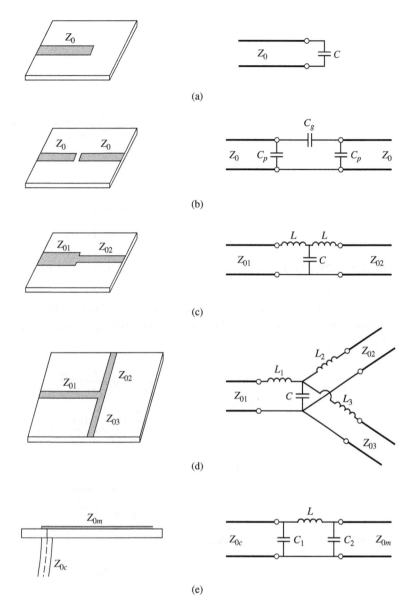

FIGURE 4.24 | Some common microstrip discontinuities. (a) Open-ended microstrip. (b) Gap in microstrip. (c) Change in width. (d) T-junction. (e) Coax-to-microstrip junction.

is the wave impedance of the TE_{n0} mode in guide 1. Because of the discontinuity at $z = 0$ there will be reflected and transmitted waves in both guides, consisting of infinite sets of TE_{n0} modes in guides 1 and 2. Only the TE_{10} mode will propagate in guide 1, but higher order modes are also important in this problem because they account for stored energy, localized near $z = 0$. Because there is no y variation introduced by this discontinuity, TE_{nm} modes for $m \neq 0$ are not excited, nor are any TM modes. A more general discontinuity, however, may excite such modes.

The reflected modes in guide 1 may be written, for $z < 0$, as

$$E_y^r = \sum_{n=1}^{\infty} A_n \sin \frac{n\pi x}{a} e^{j\beta_n^a z}, \tag{4.94a}$$

$$H_x^r = \sum_{n=1}^{\infty} \frac{A_n}{Z_n^a} \sin \frac{n\pi x}{a} e^{j\beta_n^a z}, \tag{4.94b}$$

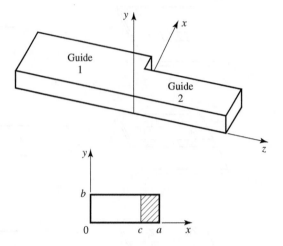

FIGURE 4.25 | Geometry of an *H*-plane step (change in width) in a rectangular waveguide.

where A_n is the unknown amplitude coefficient of the reflected TE_{n0} mode in guide 1. The reflection coefficient of the incident TE_{10} mode is then A_1. Similarly, the transmitted modes into guide 2 can be written, for $z > 0$, as

$$E_y^t = \sum_{n=1}^{\infty} B_n \sin \frac{n\pi x}{c} e^{-j\beta_n^c z}, \tag{4.95a}$$

$$H_x^t = -\sum_{n=1}^{\infty} \frac{B_n}{Z_n^c} \sin \frac{n\pi x}{c} e^{-j\beta_n^c z}, \tag{4.95b}$$

where the propagation constant in guide 2 is

$$\beta_n^c = \sqrt{k_0^2 - \left(\frac{n\pi}{c}\right)^2}, \tag{4.96}$$

and the wave impedance in guide 2 is

$$Z_n^c = \frac{k_0 \eta_0}{\beta_n^c}. \tag{4.97}$$

At $z = 0$, the transverse fields (E_y, H_x) must be continuous for $0 < x < c$; in addition, E_y must be zero for $c < x < a$ because of the step. Enforcing these boundary conditions leads to the following equations:

$$E_y = \sin \frac{\pi x}{a} + \sum_{n=1}^{\infty} A_n \sin \frac{n\pi x}{a} = \begin{cases} \sum_{n=1}^{\infty} B_n \sin \frac{n\pi x}{c} & \text{for } 0 < x < c, \\ 0 & \text{for } c < x < a, \end{cases} \tag{4.98a}$$

$$H_x = \frac{-1}{Z_1^a} \sin \frac{\pi x}{a} + \sum_{n=1}^{\infty} \frac{A_n}{Z_n^a} \sin \frac{n\pi x}{a} = -\sum_{n=1}^{\infty} \frac{B_n}{Z_n^c} \sin \frac{n\pi x}{c} \quad \text{for } 0 < x < c. \tag{4.98b}$$

Equations (4.98a) and (4.98b) constitute a doubly infinite set of linear equations for the modal coefficients A_n and B_n. We will first eliminate the B_n and then truncate the resulting equation to a finite number of terms and solve for the A_n.

Multiplying (4.98a) by $\sin(m\pi x/a)$, integrating from $x = 0$ to a, and using the orthogonality relations from Appendix E yields

$$\frac{a}{2}\delta_{m1} + \frac{a}{2}A_m = \sum_{n=1}^{\infty} B_n I_{mn} = \sum_{k=1}^{\infty} B_k I_{mk}, \tag{4.99}$$

where

$$I_{mn} = \int_{x=0}^{c} \sin \frac{m\pi x}{a} \sin \frac{n\pi x}{c} dx \tag{4.100}$$

is an integral that can be easily evaluated, and

$$\delta_{mn} = \begin{cases} 1 & \text{if } m = n \\ 0 & \text{if } m \neq n \end{cases} \tag{4.101}$$

is the Kronecker delta symbol. Now solve (4.98b) for B_k by multiplying (4.98b) by $\sin(k\pi x/c)$ and integrating from $x = 0$ to c. After using orthogonality relations, we obtain

$$\frac{-1}{Z_1^a} I_{k1} + \sum_{n=1}^{\infty} \frac{A_n}{Z_n^a} I_{kn} = \frac{-cB_k}{2Z_k^c}. \tag{4.102}$$

Substituting B_k from (4.102) into (4.99) gives an infinite set of linear equations for the A_n, where $m = 1, 2, \ldots,$

$$\frac{a}{2} A_m + \sum_{n=1}^{\infty} \sum_{k=1}^{\infty} \frac{2Z_k^c I_{mk} I_{kn} A_n}{c Z_n^a} = \sum_{k=1}^{\infty} \frac{2Z_k^c I_{mk} I_{k1}}{c Z_1^a} - \frac{a}{2} \delta_{m1}. \tag{4.103}$$

For numerical calculation we can truncate these summations to N terms, which will result in N linear equations for the first N coefficients, A_n. For example, let $N = 1$. Then (4.103) reduces to

$$\frac{a}{2} A_1 + \frac{2Z_1^c I_{11}^2}{c Z_1^a} A_1 = \frac{2Z_1^c I_{11}^2}{c Z_1^a} - \frac{a}{2}. \tag{4.104}$$

Solving for A_1 (the reflection coefficient of the incident TE_{10} mode) gives

$$A_1 = \frac{Z_\ell - Z_1^a}{Z_\ell + Z_1^a} \quad \text{for } N = 1, \tag{4.105}$$

where $Z_\ell = 4Z_1^c I_{11}^2 / ac$, which looks like an effective load impedance to guide 1. Accuracy is improved by using larger values of N and leads to a set of equations that can be written in matrix form as

$$[Q][A] = [P], \tag{4.106}$$

where $[Q]$ is a square $N \times N$ matrix of coefficients,

$$Q_{mn} = \frac{a}{2} \delta_{mn} + \sum_{k=1}^{N} \frac{2Z_k^c I_{mk} I_{kn}}{c Z_n^a}, \tag{4.107}$$

$[P]$ is an $N \times 1$ column vector of coefficients given by

$$P_m = \sum_{k=1}^{N} \frac{2Z_k^c I_{mk} I_{k1}}{c Z_1^a} - \frac{a}{2} \delta_{m1}, \tag{4.108}$$

and $[A]$ is an $N \times 1$ column vector of the coefficients A_n. After the A_n are found, the B_n can be calculated from (4.102), if desired. Equations (4.106)–(4.108) lend themselves well to computer implementation, and Figure 4.26 shows the results of such a calculation for various matrix sizes.

    If the width c of guide 2 is such that all modes are cut off (evanescent), then no real power can be transmitted into guide 2, and all the incident power is reflected back into guide 1. The evanescent fields on both sides of the discontinuity store reactive power, however, which implies that the step discontinuity and guide 2 beyond the discontinuity look like a reactance (in this case an inductive reactance) to an incident TE_{10} mode in guide 1. Thus the equivalent circuit of the H-plane step looks like a shunt inductor at the $z = 0$ plane of guide 1, as shown in Figure 4.23e. The equivalent reactance can be found from the reflection coefficient A_1 [after solving (4.106)] as

$$X = -jZ_1^a \frac{1 + A_1}{1 - A_1}. \tag{4.109}$$

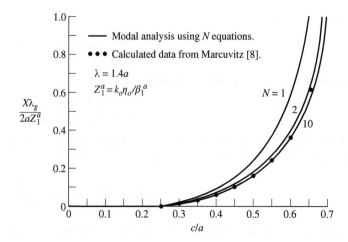

FIGURE 4.26 | Equivalent inductance of an H-plane asymmetric step.

Figure 4.26 shows the normalized equivalent inductance versus the ratio of the guide widths c/a for a free-space wavelength $\lambda = 1.4a$ and for $N = 1, 2$, and 10 equations. The modal analysis results are compared to data from reference [8]. Note that the solution converges very quickly (because of the fast exponential decay of the higher order evanescent modes), and that the result using just two modes is very close to the data of reference [8].

The fact that the H-plane step appears inductive is a result of the actual value of the reflection coefficient, A_1, but we can verify the inductive nature of the discontinuity by computing the complex power flow into the evanescent modes on either side of the discontinuity. For example, the complex power flow into guide 2 can be found as

$$
P = \int_{x=0}^{c} \int_{y=0}^{b} \bar{E} \times \bar{H}^* \Big|_{z=0^+} \cdot \hat{z}\,dx\,dy
$$

$$
= -b \int_{x=0}^{c} E_y H_x^* \, dx
$$

$$
= -b \int_{x=0}^{c} \left[\sum_{n=1}^{\infty} B_n \sin \frac{n\pi x}{c} \right] \left[-\sum_{m=1}^{\infty} \frac{B_m^*}{Z_m^{c*}} \sin \frac{m\pi x}{c} \right] dx
$$

$$
= \frac{bc}{2} \sum_{n=1}^{\infty} \frac{|B_n|^2}{Z_n^{c*}}
$$

$$
= \frac{jbc}{2k_0 \eta_0} \sum_{n=1}^{\infty} |B_n|^2 |\beta_n^c|, \tag{4.110}
$$

where the orthogonality property of the sine functions was used, as well as (4.95)–(4.97). Equation (4.110) shows that the complex power flow into guide 2 is positive imaginary, implying stored magnetic energy and an inductive reactance. A similar result can be derived for the evanescent modes in guide 1; this is left as a problem.

POINT OF INTEREST: Microstrip Discontinuity Compensation

Because a microstrip circuit is easy to fabricate and allows the convenient integration of passive and active components, many types of microwave circuits and subsystems are made in microstrip form. One problem with microstrip circuits (and other planar circuits) is that the inevitable discontinuities at bends, step changes in widths, and junctions can cause degradation in circuit performance. This is because such discontinuities introduce parasitic reactances that can lead to phase and amplitude errors, input and output mismatch, and possibly spurious coupling or radiation. One approach for eliminating such effects is to construct an equivalent circuit for the discontinuity (perhaps by measurement), including it in the design of the circuit, and compensating for its effect by adjusting other circuit parameters (such as line lengths and characteristic impedances, or tuning stubs). Another approach is to minimize the effect of a discontinuity by compensating the discontinuity directly, often by chamfering or mitering the conductor.

Consider the case of a bend in a microstrip line. The straightforward right-angle bend shown below has a parasitic discontinuity capacitance caused by the increased conductor area at the corner of the bend. This effect could be eliminated by making a smooth, "swept" bend with a radius $r \geq 3W$, but this takes up more space. Alternatively, the right-angle bend can be compensated by mitering the corner, which has the effect of reducing the excess capacitance at the bend. As shown later, this technique can be applied to bends of arbitrary angle. The optimum value of the miter length, a, depends on the characteristic impedance and the bend angle, but a value of $a = 1.8W$ is often used in practice. The technique of mitering can also be used to compensate step and T-junction discontinuities, as shown in the figure.

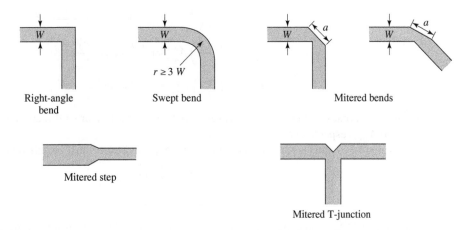

Right-angle bend   Swept bend   Mitered bends

Mitered step   Mitered T-junction

Reference: T. C. Edwards, *Foundations for Microwave Circuit Design*, John Wiley & Sons, New York, 1981.

4.7 EXCITATION OF WAVEGUIDES—ELECTRIC AND MAGNETIC CURRENTS

So far we have considered the propagation, reflection, and transmission of guided waves in the absence of sources, but obviously the waveguide or transmission line must be coupled to a generator or some other source of power. For TEM or quasi-TEM lines, there is usually only one propagating mode that can be excited by a given source, although there may be reactance (stored energy) associated with a given feed. In the waveguide case, it may be possible for several propagating modes to be excited, along with evanescent modes that store energy. In this section we will develop a formalism for determining the excitation of a given waveguide mode due to an arbitrary electric or magnetic current source. This theory can then be used to find the excitation and input impedance of probe and loop feeds.

Current Sheets That Excite Only One Waveguide Mode

Consider an infinitely long rectangular waveguide with a transverse sheet of electric surface current density at $z = 0$, as shown in Figure 4.27. First assume that this current has $\hat{x}$ and $\hat{y}$ components given as

$$\bar{J}_s^{\text{TE}}(x, y) = -\hat{x}\frac{2A_{mn}^+ n\pi}{b} \cos\frac{m\pi x}{a} \sin\frac{n\pi y}{b} + \hat{y}\frac{2A_{mn}^+ m\pi}{a} \sin\frac{m\pi x}{a} \cos\frac{n\pi y}{b}. \tag{4.111}$$

We will show that such a current excites a single TE_{mn} waveguide mode traveling away from the current source in both the $+z$ and $-z$ directions.

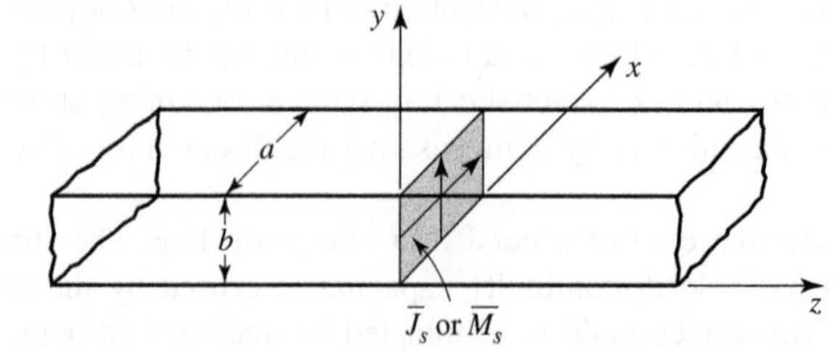

FIGURE 4.27 | An infinitely long rectangular waveguide with surface current densities at $z = 0$.

From Table 3.2, the transverse fields for positive and negative traveling TE_{mn} waveguide modes can be written as

$$E_x^{\pm} = Z_{TE}\left(\frac{n\pi}{b}\right) A_{mn}^{\pm} \cos\frac{m\pi x}{a} \sin\frac{n\pi y}{b} e^{\mp j\beta z}, \tag{4.112a}$$

$$E_y^{\pm} = -Z_{TE}\left(\frac{m\pi}{a}\right) A_{mn}^{\pm} \sin\frac{m\pi x}{a} \cos\frac{n\pi y}{b} e^{\mp j\beta z}, \tag{4.112b}$$

$$H_x^{\pm} = \pm\left(\frac{m\pi}{a}\right) A_{mn}^{\pm} \sin\frac{m\pi x}{a} \cos\frac{n\pi y}{b} e^{\mp j\beta z}, \tag{4.112c}$$

$$H_y^{\pm} = \pm\left(\frac{n\pi}{b}\right) A_{mn}^{\pm} \cos\frac{m\pi x}{a} \sin\frac{n\pi y}{b} e^{\mp j\beta z}, \tag{4.112d}$$

where the $\pm$ notation refers to waves traveling in the $+z$ direction or $-z$ direction with amplitude coefficients A_{mn}^+ and A_{mn}^-, respectively.

From (1.36) and (1.37), the following boundary conditions must be satisfied at $z = 0$:

$$(\bar{E}^+ - \bar{E}^-) \times \hat{z} = 0, \tag{4.113a}$$

$$\hat{z} \times (\bar{H}^+ - \bar{H}^-) = \bar{J}_s. \tag{4.113b}$$

Equation (4.112a) states that the transverse components of the electric field must be continuous at $z = 0$, which when applied to (4.112a) and (4.112b), gives

$$A_{mn}^+ = A_{mn}^-. \tag{4.114}$$

Equation (4.113b) states that the discontinuity in the transverse magnetic field is equal to the electric surface current density. Thus, the surface current density at $z = 0$ must be

$$\begin{aligned}
\bar{J}_s &= \hat{y}(H_x^+ - H_x^-) - \hat{x}(H_y^+ - H_y^-) \\
&= -\hat{x}\frac{2A_{mn}^+ n\pi}{b} \cos\frac{m\pi x}{a} \sin\frac{n\pi y}{b} + \hat{y}\frac{2A_{mn}^+ m\pi}{a} \sin\frac{m\pi x}{a} \cos\frac{n\pi y}{b},
\end{aligned} \tag{4.115}$$

where (4.114) was used. This current is seen to be the same as the current of (4.111), which shows, by the uniqueness theorem, that such a current will excite only the TE_{mn} mode propagating in each direction, since Maxwell's equations and all boundary conditions are satisfied.

The analogous electric current that excites only the TM_{mn} mode can be shown to be

$$\bar{J}_s^{TM}(x, y) = \hat{x}\frac{2B_{mn}^+ m\pi}{a} \cos\frac{m\pi x}{a} \sin\frac{n\pi y}{b} + \hat{y}\frac{2B_{mn}^+ n\pi}{b} \sin\frac{m\pi x}{a} \cos\frac{n\pi y}{b}. \tag{4.116}$$

It is left as a problem to verify that this current excites TM_{mn} modes that satisfy the appropriate boundary conditions.

Similar results can be derived for magnetic surface current sheets. From (1.36) and (1.37) the appropriate boundary conditions are

$$(\bar{E}^+ - \bar{E}^-) \times \hat{z} = \bar{M}_s, \tag{4.117a}$$

$$\hat{z} \times (\bar{H}^+ - \bar{H}^-) = 0. \tag{4.117b}$$

For a magnetic current sheet at $z = 0$, the TE_{mn} waveguide mode fields of (4.112) must now have continuous H_x and H_y field components, due to (4.117b). This results in the condition that

$$A_{mn}^+ = -A_{mn}^-. \tag{4.118}$$

Then applying (4.117a) gives the source current as

$$\bar{M}_s^{TE} = \frac{-\hat{x}2Z_{TE}A_{mn}^+ m\pi}{a} \sin\frac{m\pi x}{a} \cos\frac{n\pi y}{b} - \hat{y}\frac{2Z_{TE}A_{mn}^+ n\pi}{b} \cos\frac{m\pi x}{a} \sin\frac{n\pi y}{b}. \tag{4.119}$$

The corresponding magnetic surface current that excites only the TM_{mn} mode can be shown to be

$$\bar{M}_s^{TM} = \frac{-\hat{x}2B_{mn}^+ n\pi}{b} \sin\frac{m\pi x}{a} \cos\frac{n\pi y}{b} + \frac{\hat{y}2B_{mn}^+ m\pi}{a} \cos\frac{m\pi x}{a} \sin\frac{n\pi y}{b}. \tag{4.120}$$

These results show that a single waveguide mode can be selectively excited, to the exclusion of all other modes, by either an electric or magnetic current sheet of the appropriate form. In practice, however, such currents are difficult to generate and are usually only approximated with one or two probes or loops. In this case many modes may be excited, but usually most of these modes are evanescent.

Mode Excitation from an Arbitrary Electric or Magnetic Current Source

We now consider the excitation of waveguide modes by an arbitrary electric or magnetic current source [4]. With reference to Figure 4.28, first consider an electric current source $\bar{J}$ located between two transverse planes at z_1 and z_2, which generates the fields $\bar{E}^+, \bar{H}^+$ traveling in the $+z$ direction, and the fields $\bar{E}^-, \bar{H}^-$ traveling in the $-z$ direction. These fields can be expressed in terms of the waveguide modes as follows:

$$\bar{E}^+ = \sum_n A_n^+ \bar{E}_n^+ = \sum_n A_n^+(\bar{e}_n + \hat{z}e_{zn})e^{-j\beta_n z}, \; z > z_2, \tag{4.121a}$$

$$\bar{H}^+ = \sum_n A_n^+ \bar{H}_n^+ = \sum_n A_n^+(\bar{h}_n + \hat{z}h_{zn})e^{-j\beta_n z}, \; z > z_2, \tag{4.121b}$$

$$\bar{E}^- = \sum_n A_n^- \bar{E}_n^- = \sum_n A_n^-(\bar{e}_n - \hat{z}e_{zn})e^{j\beta_n z}, \; z < z_1, \tag{4.121c}$$

$$\bar{H}^- = \sum_n A_n^- \bar{H}_n^- = \sum_n A_n^-(-\bar{h}_n + \hat{z}h_{zn})e^{j\beta_n z}, \; z < z_1, \tag{4.121d}$$

where the single index n is used to represent any possible TE or TM mode. For a given current $\bar{J}$, we can determine the unknown amplitude A_n^+ by using the Lorentz reciprocity theorem of (1.155) with $\bar{M}_1 = \bar{M}_2 = 0$ (since here we are only considering an electric current source),

$$\oint_S (\bar{E}_1 \times \bar{H}_2 - \bar{E}_2 \times \bar{H}_1) \cdot d\bar{s} = \int_V (\bar{E}_2 \cdot \bar{J}_1 - \bar{E}_1 \cdot \bar{J}_2)dv,$$

where S is a closed surface enclosing the volume V, and $\bar{E}_i, \bar{H}_i$ are the fields due to the current source $\bar{J}_i$ (for $i = 1$ or 2).

To apply the reciprocity theorem to the present problem we let the volume V be the region between the waveguide walls and the transverse cross-section planes at z_1 and z_2. Then let $\bar{E}_1 = \bar{E}^\pm$ and $\bar{H}_1 = \bar{H}^\pm$,

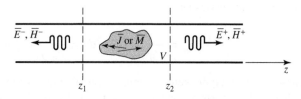

FIGURE 4.28 | An arbitrary electric or magnetic current source in an infinitely long waveguide.

depending on whether $z \geq z_2$ or $z \leq z_1$, and let $\bar{E}_2, \bar{H}_2$ be the nth waveguide mode traveling in the negative z direction:

$$\bar{E}_2 = \bar{E}_n^- = (\bar{e}_n - \hat{z}e_{zn})e^{j\beta_n z},$$

$$\bar{H}_2 = \bar{H}_n^- = (-\bar{h}_n + \hat{z}h_{zn})e^{j\beta_n z}.$$

Substitution into the above form of the reciprocity theorem gives, with $\bar{J}_1 = \bar{J}$ and $\bar{J}_2 = 0$,

$$\oint_S (\bar{E}^\pm \times \bar{H}_n^- - \bar{E}_n^- \times \bar{H}^\pm) \cdot d\bar{s} = \int_V \bar{E}_n^- \cdot \bar{J}dv. \tag{4.122}$$

The portion of the surface integral over the waveguide walls vanishes because the tangential electric field is zero there; that is, $\bar{E} \times \bar{H} \cdot \hat{z} = \bar{H} \cdot (\hat{z} \times \bar{E}) = 0$ on the waveguide walls. This reduces the integration to the guide cross section, S_0, at the planes z_1 and z_2. In addition, the waveguide modes are orthogonal over the guide cross section:

$$\int_{S_0} \bar{E}_m^\pm \times \bar{H}_n^\pm \cdot d\bar{s} = \int_{S_0} (\bar{e}_m \pm \hat{z}e_{zn}) \times (\pm\bar{h}_n + \hat{z}h_{zn}) \cdot \hat{z}ds$$

$$= \pm \int_{S_0} \bar{e}_m \times \bar{h}_n \cdot \hat{z}ds = 0, \text{ for } m \neq n. \tag{4.123}$$

Using (4.121) and (4.123) then reduces (4.122) to

$$A_n^+ \int_{z_2} (\bar{E}_n^+ \times \bar{H}_n^- - \bar{E}_n^- \times \bar{H}_n^+) \cdot d\bar{s} + A_n^- \int_{z_1} (\bar{E}_n^- \times \bar{H}_n^- - \bar{E}_n^- \times \bar{H}_n^-) \cdot d\bar{s}$$

$$= \int_V \bar{E}_n^- \cdot \bar{J}dv.$$

Because the second integral vanishes, this further reduces to

$$A_n^+ \int_{z_2} [(\bar{e}_n + \hat{z}e_{zn}) \times (-\bar{h}_n + \hat{z}h_{zn}) - (\bar{e}_n - \hat{z}e_{zn}) \times (\bar{h}_n + \hat{z}h_{zn})] \cdot \hat{z}ds$$

$$= -2A_n^+ \int_{z_2} \bar{e}_n \times \bar{h}_n \cdot \hat{z}ds = \int_V \bar{E}_n^- \cdot \bar{J}dv,$$

or

$$A_n^+ = \frac{-1}{P_n} \int_V \bar{E}_n^- \cdot \bar{J}dv = \frac{-1}{P_n} \int_V (\bar{e}_n - \hat{z}e_{zn}) \cdot \bar{J}e^{j\beta_n z}dv, \tag{4.124}$$

where

$$P_n = 2 \int_{S_0} \bar{e}_n \times \bar{h}_n \cdot \hat{z}ds \tag{4.125}$$

is a normalization constant proportional to the power flow of the nth mode.

By repeating the above procedure with $\bar{E}_2 = \bar{E}_n^+$ and $\bar{H}_2 = \bar{H}_n^+$, we can derive the amplitude of the negatively traveling waves as

$$A_n^- = \frac{-1}{P_n} \int_V \bar{E}_n^+ \cdot \bar{J}dv = \frac{-1}{P_n} \int_V (\bar{e}_n + \hat{z}e_{zn}) \cdot \bar{J}e^{-j\beta_n z}dv. \tag{4.126}$$

These results are quite general, being applicable to any type of waveguide (including planar lines such as stripline and microstrip), where modal fields can be defined. Example 4.9 applies this theory to the problem of a probe-fed rectangular waveguide.

EXAMPLE 4.9   PROBE-FED RECTANGULAR WAVEGUIDE

For the probe-fed rectangular waveguide shown in Figure 4.29, determine the amplitudes of the forward and backward traveling TE_{10} modes, and the input resistance seen by the probe. Assume that the TE_{10} mode is the only propagating mode.

Solution

If the current probe is assumed to have an infinitesimal diameter, the source volume current density $\bar{J}$ can be written as

$$\bar{J}(x, y, z) = I_0 \delta \left(x - \frac{a}{2} \right) \delta(z)\hat{y} \text{ for } 0 \le y \le b.$$

From Chapter 3 the TE_{10} modal fields can be written as

$$\bar{e}_1 = \hat{y} \sin \frac{\pi x}{a},$$

$$\bar{h}_1 = \frac{-\hat{x}}{Z_1} \sin \frac{\pi x}{a},$$

where $Z_1 = k_0 \eta_0 / \beta_1$ is the TE_{10} wave impedance. From (4.125) the normalization constant P_1 is

$$P_1 = \frac{2}{Z_1} \int_{x=0}^{a} \int_{y=0}^{b} \sin^2 \frac{\pi x}{a} dx dy = \frac{ab}{Z_1}.$$

Then from (4.124) the amplitude A_1^+ is

$$A_1^+ = \frac{-1}{P_1} \int_V \sin \frac{\pi x}{a} e^{j\beta_1 z} I_0 \delta \left(x - \frac{a}{2} \right) \delta(z) dx dy dz = \frac{-I_0 b}{P_1} = \frac{-Z_1 I_0}{a}.$$

Similarly,

$$A_1^- = \frac{-Z_1 I_0}{a}.$$

If the TE_{10} mode is the only propagating mode in the waveguide, then this mode carries all of the average power, which can be calculated for real Z_1 as

$$P = \frac{1}{2} \int_{S_0} \bar{E}^+ \times \bar{H}^{+*} \cdot d\bar{s} + \frac{1}{2} \int_{S_0} \bar{E}^- \times \bar{H}^{-*} \cdot d\bar{s}$$

$$= \int_{S_0} \bar{E}^+ \times \bar{H}^{+*} \cdot d\bar{s}$$

$$= \int_{x=0}^{a} \int_{y=0}^{b} \frac{|A_1^+|^2}{Z_1} \sin^2 \frac{\pi x}{a} dx dy$$

$$= \frac{ab|A_1^+|^2}{2Z_1}.$$

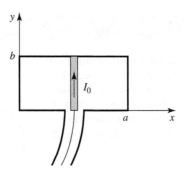

FIGURE 4.29 | A uniform current probe in a rectangular waveguide.

If the input resistance seen looking into the probe is R_{in}, and the terminal current is I_0, then $P = I_0^2 R_{\text{in}}/2$, so that the input resistance is

$$R_{\text{in}} = \frac{2P}{I_0^2} = \frac{ab|A_1^+|^2}{I_0^2 Z_1} = \frac{bZ_1}{a},$$

which is real for real Z_1 (corresponding to a propagating TE_{10} mode).

A similar derivation can be carried out for a magnetic current source $\bar{M}$ (e.g., a small loop). This source will also generate positively and negatively traveling waves, which can be expressed as a superposition of waveguide modes, as in (4.121). For $\bar{J}_1 = \bar{J}_2 = 0$, the reciprocity theorem of (1.155) reduces to

$$\oint_S (\bar{E}_1 \times \bar{H}_2 - \bar{E}_2 \times \bar{H}_1) \cdot d\bar{s} = \int_V (\bar{H}_1 \cdot \bar{M}_2 - \bar{H}_2 \cdot \bar{M}_1)dv. \tag{4.127}$$

By following the same procedure as for the electric current case, we can derive the excitation coefficients of the nth waveguide mode as

$$A_n^+ = \frac{1}{P_n} \int_V \bar{H}_n^- \cdot \bar{M}dv = \frac{1}{P_n} \int_V (-\bar{h}_n + \hat{z}h_{zn}) \cdot \bar{M}e^{j\beta_n z}dv, \tag{4.128}$$

$$A_n^- = \frac{1}{P_n} \int_V \bar{H}_n^+ \cdot \bar{M}dv = \frac{1}{P_n} \int_V (\bar{h}_n + \hat{z}h_{zn}) \cdot \bar{M}e^{-j\beta_n z}dv, \tag{4.129}$$

where P_n is defined in (4.125).

REFERENCES

[1] S. Ramo, T. R. Whinnery, and T. van Duzer, *Fields and Waves in Communication Electronics*, John Wiley & Sons, New York, 1965.

[2] A. A. Oliner, "Historical Perspectives on Microwave Field Theory," *IEEE Transactions on Microwave Theory and Techniques*, vol. MTT-32, pp. 1022–1045, September 1984.

[3] C. G. Montgomery, R. H. Dicke, and E. M. Purcell, eds., *Principles of Microwave Circuits*, MIT Radiation Laboratory Series, Vol. 8, McGraw-Hill, New York, 1948.

[4] R. E. Collin, *Foundations for Microwave Engineering*, 2nd edition, McGraw-Hill, New York, 1992.

[5] J. Rahola, "Power Waves and Conjugate Matching," *IEEE Transactions on Circuits and Systems*, vol. 55, pp. 92–96, January 2008.

[6] J. S. Wright, O. P. Jain, W. J. Chudobiak, and V. Makios, "Equivalent Circuits of Microstrip Impedance Discontinuities and Launchers," *IEEE Transactions on Microwave Theory and Techniques*, vol. MTT-22, pp. 48–52, January 1974.

[7] G. F. Engen and C. A. Hoer, "Thru-Reflect-Line: An Improved Technique for Calibrating the Dual Six-Port Automatic Network Analyzer," *IEEE Transactions on Microwave Theory and Techniques*, vol. MTT-27, pp. 987–998, December 1979.

[8] N. Marcuvitz, ed., *Waveguide Handbook*, MIT Radiation Laboratory Series, Vol. 10, McGraw-Hill, New York, 1948.

[9] K. C. Gupta, R. Garg, and I. J. Bahl, *Microstrip Lines and Slotlines*, Artech House, Dedham, MA, 1979.

PROBLEMS

4.1 Consider the reflection of a TE_{10} mode, incident from $x < 0$, at a step change in the height of a rectangular waveguide, as shown in the figure. Show that if the method of Example 4.2 is used, the result $\Gamma = 0$ is obtained. Do you think this is the correct solution? Why? (This problem shows that the one-mode impedance viewpoint does not always provide a correct analysis.)

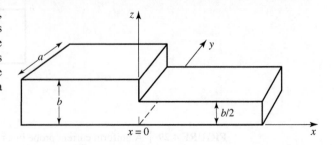

4.2 Consider a series *RLC* circuit with a current *I*. Calculate the power lost and the stored electric and magnetic energies, and show that the input impedance can be expressed as in (4.17).

4.3 Show that the input impedance Z of a parallel *RLC* circuit satisfies the condition that $Z(-\omega) = Z^*(\omega)$.

4.4 A two-port network is driven at both ports such that the port voltages and currents have the following values ($Z_0 = 50\ \Omega$):

$$V_1 = 20\angle 90°, \quad I_1 = 0.4\angle 90°,$$
$$V_2 = 16\angle 0°, \quad I_2 = 0.32\angle -90°.$$

Determine the input impedance seen at each port, and find the incident and reflected voltages at each port.

4.5 Show that the admittance matrix of a lossless *N*-port network has purely imaginary elements.

4.6 Does a nonreciprocal lossless network always have a purely imaginary impedance matrix?

4.7 Derive the [Z] and [Y] matrices for the two-port networks shown in the figure below.

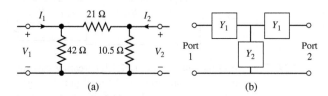

(a)   (b)

4.8 Consider a two-port network, and let $Z_{\text{SC}}^{(1)}, Z_{\text{SC}}^{(2)}, Z_{\text{OC}}^{(1)},$ and $Z_{\text{OC}}^{(2)}$ be the input impedance seen when port 2 is short-circuited, when port 1 is short-circuited, when port 2 is open-circuited, and when port 1 is open-circuited, respectively. Show that the impedance matrix elements are given by

$$Z_{11} = Z_{\text{OC}}^{(1)}, \quad Z_{22} = Z_{\text{OC}}^{(2)}, \quad Z_{12}^2 = Z_{21}^2 = \left(Z_{\text{OC}}^{(1)} - Z_{\text{SC}}^{(1)}\right) Z_{\text{OC}}^{(2)}.$$

4.9 Find the impedance parameters of a section of transmission line with length ℓ, characteristic impedance Z_0, and propagation constant β.

4.10 Show that the admittance matrix of the two parallel-connected two-port π networks shown below can be found by adding the admittance matrices of the individual two-ports. Apply this result to find the admittance matrix of the bridged-T circuit shown. What is the corresponding result for the impedance matrix of two series-connected T-networks?

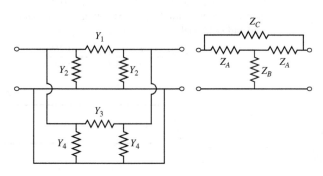

4.11 Find the scattering parameters for the shunt and series loads shown below. Show that $S_{12} = 1 + S_{11}$ for the shunt case, and that $S_{12} = 1 - S_{11}$ for the series case. Assume a characteristic impedance $Z_0 = 50\ \Omega$.

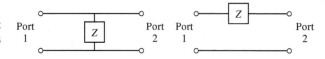

4.12 Consider two two-port networks with individual scattering matrices $[S^A]$ and $[S^B]$. Show that the overall S_{21} parameter of the cascade of these networks is given by

$$S_{21} = \frac{S_{21}^A S_{21}^B}{1 - S_{22}^A S_{11}^B}.$$

4.13 Consider a lossless two-port network. (a) If the network is reciprocal, show that $|S_{21}|^2 = 1 - |S_{11}|^2$. (b) If the network is nonreciprocal, show that it is impossible to have unidirectional transmission, where $S_{12} = 0$ and $S_{21} \neq 0$.

4.14 A four-port network has the scattering matrix shown as follows. (a) Is this network lossless? (b) Is this network reciprocal? (c) What is the return loss at port 1 when all other ports are terminated with matched loads? (d) What is the insertion loss and phase delay between ports 2 and 4 when all other ports are terminated with matched loads? (e) What is the reflection coefficient seen at port 1 if a short circuit is placed at the terminal plane of port 3 and all other ports are terminated with matched loads?

$$[S] = \begin{bmatrix} 0.8\angle 90° & 0.1\angle -45° & 0.3\angle -45° & 0 \\ 0.1\angle -45° & 0 & 0 & 0.6\angle 45° \\ 0.3\angle -45° & 0 & 0 & 0.4\angle -45° \\ 0 & 0.6\angle 45° & 0.4\angle -45° & 0 \end{bmatrix}.$$

4.15 Show that it is impossible to construct a three-port network that is lossless, reciprocal, and matched at all ports. Is it possible to construct a nonreciprocal three-port network that is lossless and matched at all ports?

4.16 Prove the following *decoupling theorem*: For any lossless reciprocal three-port network, one port (say port 3) can be terminated in a reactance so that the other two ports (say ports 1 and 2) are decoupled (no power flow from port 1 to port 2, or from port 2 to port 1).

4.17 A certain three-port network is lossless and reciprocal, and has $S_{13} = S_{23}$ and $S_{11} = S_{22}$. Show that if port 2 is terminated with a matched load, then port 1 can be matched by placing an appropriate reactance at port 3.

4.18 A four-port network has the scattering matrix shown as follows. If ports 3 and 4 are connected with a lossless matched transmission line with an electrical length of 45°, find the resulting insertion loss and phase delay between ports 1 and 2.

$$[S] = \begin{bmatrix} 0.5\angle 45° & 0 & 0 & 0.4\angle -45° \\ 0 & 0.7\angle -45° & 0.6\angle 45° & 0 \\ 0 & 0.6\angle 45° & 0.7\angle -45° & 0 \\ 0.4\angle -45° & 0 & 0 & 0.2\angle 45° \end{bmatrix}.$$

4.19 When normalized to a single characteristic impedance Z_0, a certain two-port network has scattering parameters S_{ij}. Find the generalized scattering parameters, S_{ij}^p, in terms of the real reference impedances, R_{01} and R_{02}, at ports 1 and 2, respectively.

4.20 At reference plane A, for the circuit shown below, choose an appropriate reference impedance, find the power wave amplitudes, and compute the power delivered to the load. Repeat this procedure for reference plane B. Assume the transmission line is lossless.

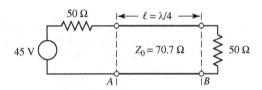

4.21 The $ABCD$ parameters of the first entry in Table 4.1 were derived in Example 4.7. Verify the $ABCD$ parameters for the second, third, and fourth entries.

4.22 Derive expressions that give the impedance parameters in terms of the $ABCD$ parameters.

4.23 Find the $ABCD$ matrix for the circuit shown below by direct calculation using the definition of the $ABCD$ matrix, and compare with the $ABCD$ matrix of the appropriate cascade of canonical circuits from Table 4.1.

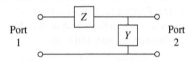

4.24 Use $ABCD$ matrices to find the voltage V_L across the load resistor in the circuit shown below.

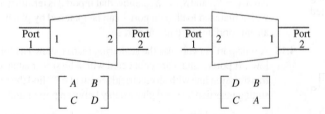

4.25 A reciprocal two-port network with its $ABCD$ matrix is shown below at left. Prove that the network with ports 1 and 2 in reversed positions has the $ABCD$ matrix shown below at right. Choose a simple asymmetrical network to demonstrate this result.

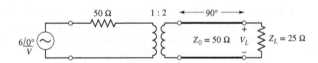

4.26 Derive the expressions for S parameters in terms of the $ABCD$ parameters, as given in Table 4.2.

4.27 As shown in the accompanying figure, a variable attenuator can be implemented using a four-port 90° hybrid coupler by terminating ports 2 and 3 with equal but adjustable loads.

(a) Using the given scattering matrix for the coupler, show that the transmission coefficient between the input (port 1) and the output (port 4) is given as $T = j\Gamma$, where Γ is the reflection coefficient of the mismatch at ports 2 and 3. Also show that the input port is matched for all values of Γ.
(b) Plot the attenuation, in dB, from the input to the output as a function of Z_L/Z_0, for $0 \le Z_L/Z_0 \le 20$ (let Z_L be real).

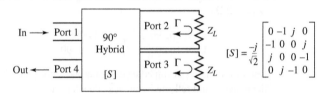

4.28 Use signal flow graphs to find the power ratios P_2/P_1 and P_3/P_1 for the mismatched three-port network shown in the accompanying figure.

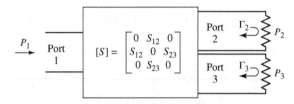

4.29 The $ABCD$ parameters are useful for treating cascades of two-port networks in terms of the total port voltages and currents, but it is also possible to use incident and reflected voltages to treat cascades. One way of doing this is with the *transfer*, or *T-*, *parameters*, defined as follows:

$$\begin{bmatrix} a_1 \\ b_1 \end{bmatrix} = \begin{bmatrix} T_{11} & T_{12} \\ T_{21} & T_{22} \end{bmatrix} \begin{bmatrix} b_2 \\ a_2 \end{bmatrix},$$

where a_1, b_1 and a_2, b_2 are the incident and reflected voltages at ports 1 and 2, respectively. Derive the T-parameters in terms of the scattering parameters of a two-port network. Show how the T-parameters can be used for a cascade of two two-port networks.

4.30 The end of an open-circuited microstrip line has fringing fields that can be modeled as a shunt capacitor, C_f, at the end of the line, as shown below. This capacitance can be replaced with an additional length, Δ, of microstrip line. Derive an expression for the length extension in terms of the fringing capacitance. Evaluate the length extension for a 50 Ω open-circuited microstrip line on a substrate with $d = 1.6$ mm and $\epsilon_r = 2.2$ ($w = 4.87$ mm, $\epsilon_e = 1.894$), if the fringing capacitance is known to be $C_f = 0.075$ pF. Compare your result with the approximation given by Hammerstad and Bekkadal:

$$\Delta = 0.412d\left(\frac{\epsilon_e + 0.3}{\epsilon_e - 0.258}\right)\left(\frac{w + 0.262d}{w + 0.813d}\right).$$

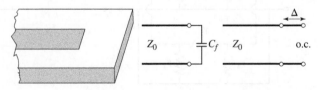

4.31 For the *H*-plane step analysis of Section 4.6, compute the complex power flow in the reflected modes in guide 1, and show that the reactive power is inductive.

4.32 Derive the modal analysis equations for the symmetric *H*-plane step shown below. (HINT: Because of symmetry, only the TE_{n0} modes for *n* odd will be excited.)

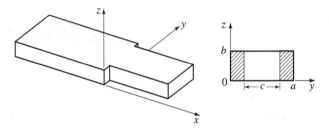

4.33 Find the transverse $\bar{E}$ and $\bar{H}$ fields excited by the current of (4.116) by postulating traveling TM_{mn} modes on either side of the source at $z = 0$ and applying the appropriate boundary conditions.

4.34 An infinitely long rectangular waveguide is fed with a probe of length *d* as shown below. The current on this probe can be approximated as $I(z) = I_0 \sin k(d - z) / \sin kd$. If the TE_{10} mode is the only propagating mode in the waveguide, compute the input resistance seen at the probe terminals.

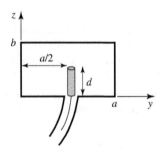

4.35 Consider the infinitely long waveguide fed with two probes driven 180° out of phase, as shown below. What are the resulting excitation coefficients for the TE_{10} and TE_{20} modes? What other modes can be excited by this feeding arrangement?

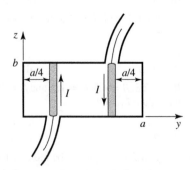

4.36 Consider a small current loop on the sidewall of a rectangular waveguide, as shown in the accompanying figure. Find the TE_{10} fields excited by this loop if the loop is of radius r_0.

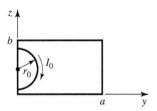

4.37 A rectangular waveguide is shorted at $x = 0$ and has an electric current sheet, J_{sz}, located at $x = d$, where

$$J_{sz} = \frac{2\pi A}{a} \sin \frac{\pi y}{a}$$

(see the accompanying figure). Find expressions for the fields generated by this current by assuming standing wave fields for $0 < x < d$, and traveling wave fields for $x > d$, and applying boundary conditions at $x = 0$ and $x = d$. Now solve the problem using image theory, by placing a current sheet $-J_{sz}$ at $x = -d$, and removing the shorting wall at $x = 0$. Use the results of Section 4.7 and superposition to find the fields radiated by these two currents, which should be the same as the first results for $x > 0$.

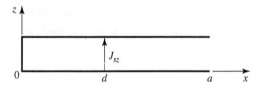

4.38 A four-port network has the scattering matrix as shown below. If ports 3 and 4 are connected with a lossless matched transmission line with an electrical length of 60°, find the resulting S_{21} magnitude and phase delay.

$$[S] = \begin{bmatrix} 0.3\angle-30° & 0 & 0 & 0.8\angle-45° \\ 0 & 0.7\angle-30° & 0.6\angle-45° & 0 \\ 0 & 0.6\angle-45° & 0.7\angle-30° & 0 \\ 0.8\angle-45° & 0 & 0 & 0.3\angle-30° \end{bmatrix}$$

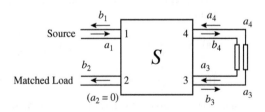

4.39 An engineer purchased a device that comes with the [*S*] matrix as shown. However, the [*S*] parameters are given in terms of dB and phase angle

$$[S] = \begin{bmatrix} -43\angle66° & -3\angle53° & -3\angle-143° & -43\angle80° \\ -3\angle53° & -43\angle66° & -43\angle80° & -3\angle-143° \\ -3\angle-143° & -43\angle80° & -43\angle66° & -3\angle53° \\ -43\angle80° & -3\angle-143° & -3\angle53° & -43\angle66° \end{bmatrix}$$

(a) Is this a lossless network?
(b) Is this a reciprocal network?

(c) Is this a symmetric network?

(d) What is the input return loss if ports 3 and 4 are properly matched and port 2 is opened?

(e) What is the function of this device?

4.40 For a four-port X-network, write its four-port S parameters and determine the reflection coefficient at port 1 when:

(a) ports 2 and 3 are open-circuited and port 4 is matched

(b) ports 2 and 3 are loaded with 50 Ω each and port 4 is open-circuited

(c) ports 2 and 3 are connected to each other and port 4 is loaded with 50 Ω

Do not solve the problem using nodal analysis, and solve it using the [S] matrix and boundary conditions.

MULTIPLE CHOICE QUESTIONS

4.1 In a two-port network, if voltage at port 1 is 5 V and current at port 2 is 2 A, then the impedance Z_{12} is

(a) 0.4 Ω
(b) 2.5 Ω
(c) 10 Ω
(d) insufficient data

4.2 In a two-port network, if current at port 2 is 3 A and voltage at port 1 is 6 V, then the admittance Y_{12} is

(a) 0.5 ℧
(b) 1 ℧
(c) 2 ℧
(d) 4 ℧

4.3 Z matrix for a reciprocal network is a

(a) null matrix
(b) unit matrix
(c) symmetric matrix
(d) skew symmetric matrix

4.4 For a lossless network, the admittance and impedance matrices are

(a) real
(b) purely imaginary
(c) rational
(d) complex

4.5 The matrix with impedance parameters $Z_{11} = 2 + j2$, $Z_{12} = 2 + j1, Z_{22} = 1, Z_{21} = 2 + j1$ is said to be a

(a) lossy network
(b) lossless network
(c) reciprocal network
(d) none of the above

4.6 In a reciprocal network the scattering matrix is _____ in nature.

(a) unitary
(b) symmetric
(c) skew symmetric
(d) identity matrix

4.7 The specific element of the scattering matrix S_{ij} can be determined as

(a) $S_{ij} = \dfrac{V_{i-}}{V_{j+}}$
(b) $S_{ij} = \dfrac{V_{i+}}{V_{j-}}$
(c) $S_{ij} = \dfrac{V_{j+}}{V_{i-}}$
(d) none of the above

4.8 The reflection coefficient of a two-port network is given as 0.75. The return loss in the network in dB is

(a) 6.02 dB
(b) 1.25 dB
(c) 2.5 dB
(d) 12 dB

4.9 For a lossless matrix the scattering matrix is given to be

(a) unitary
(b) identity matrix
(c) symmetric
(d) null matrix

4.10 An $ABCD$ matrix is used

(a) to represent a two-port network
(b) to represent the impedance of a microwave network

(c) when there are two or more port networks in the cascade

(d) none of the above

4.11 When two-port networks are in cascade connection, then the $ABCD$ matrix is equal to

(a) sum of the $ABCD$ matrices representing the individual two ports

(b) difference of the $ABCD$ matrices representing the individual two ports

(c) product of the $ABCD$ matrices representing the individual two ports

(d) sum of the transpose of $ABCD$ matrices representing the individual two ports

4.12 Simple impedance or an equivalent impedance of a network is represented as an $ABCD$ matrix, then $ABCD$ parameters are

(a) $A = 1, B = Z, C = 0, D = 1$
(b) $A = 1, B = 0, C = 1, D = Z$
(c) $A = Z, B = 1, C = 1, D = 0$
(d) $A = 0, B = 1, C = Z, D = 1$

4.13 Given a transmission line of characteristic impedance Z_0, the C parameter given for the phase constant β and length ℓ is

(a) $jY_0 \tan \beta\ell$
(b) $jZ_0 \sin \beta\ell$
(c) $jZ_0 \tan \beta\ell$
(d) $jY_0 \sin \beta\ell$

4.14 For a two-port network, if the admittance parameter $Y_{12} = 0.5$, then the parameter B of $ABCD$ parameters for this network is

(a) 4.5
(b) 4
(c) 2
(d) 2.5

4.15 If $D = 1.8$ and $B = 3.6$ for a two-port network, then $Y_{11} = ?$

(a) 0.5
(b) 0.75
(c) 1.5
(d) 2.0

4.16 If $A = 2.8, B = 4.5, C = 1.4$, and $D = 2.8$ for a two-port network then $Z_{11} = ?$

(a) 1.6
(b) 2
(c) 0.5
(d) 1.0

4.17 A signal flow graph can be reduced to a single branch between two nodes using the _____ number of basic decomposition rules.

(a) 2
(b) 5
(c) 3
(d) 4

4.18 An L network is required to match a load impedance of 35 Ω to a transmission line of characteristic impedance of 50 Ω. The components of the L network are

(a) $22.91 + j0\ \Omega$

(b) $22.91 + j22.91\ \Omega$

(c) $50\ \Omega$

(d) $45.82\ \Omega$

4.19 The imaginary part of a matching network is given by the relation

(a) $\pm\left(\sqrt{\dfrac{Z_0 - R_L}{R_L}}\right)Z_0$

(b) $\sqrt{\dfrac{Z_0 - R_L}{R_L}}$

(c) $\pm\left(\sqrt{\dfrac{Z_0 - R_L}{Z_0}}\right)$

(d) none of the above

4.20 The most important factor to be considered in the selection of a matching network is

(a) amplification factor

(b) bandwidth

(c) noise component

(d) none of the above

ANSWER KEY				
4.1 (b)	**4.5** (c)	**4.9** (a)	**4.13** (d)	**4.17** (d)
4.2 (a)	**4.6** (b)	**4.10** (c)	**4.14** (c)	**4.18** (a)
4.3 (c)	**4.7** (a)	**4.11** (c)	**4.15** (a)	**4.19** (a)
4.4 (b)	**4.8** (c)	**4.12** (a)	**4.16** (b)	**4.20** (b)

Impedance Matching and Tuning

Learning Objectives

After completing this chapter, you will be able to

- Learn about the concepts of impedance matching with lumped elements (*L* networks)
- Develop and design impedance matching circuits using single-stub tuning and double-stub tuning

- Understand the concept of the quarter-wave transformer and the theory of small reflections
- Learn about the concept of binomial multisection matching transformers and Chebyshev multisection matching transformers

This chapter marks a turning point, in that we now begin to apply the theory and techniques of previous chapters to practical problems in microwave engineering. We start with the topic of *impedance matching*, which is often an important part of a larger design process for a microwave component or system. The basic idea of impedance matching is illustrated in Figure 5.1, which shows an impedance matching network placed between a load impedance and a transmission line. The matching network is ideally lossless, to avoid unnecessary loss of power, and is usually designed so that the impedance seen looking into the matching network is Z_0. Then reflections will be eliminated on the transmission line to the left of the matching network, although there will usually be multiple reflections between the matching network and the load. This procedure is sometimes referred to as *tuning*. Impedance matching or tuning is important for the following reasons:

- Maximum power is delivered when the load is matched to the line (assuming the generator is matched), and power loss in the feed line is minimized.
- Impedance matching sensitive receiver components (antenna, low-noise amplifier, etc.) may improve the signal-to-noise ratio of the system.
- Impedance matching in a power distribution network (such as an antenna array feed network) may reduce amplitude and phase errors.

As long as the load impedance, Z_L, has a positive real part, a matching network can always be found. Many choices are available, however, and we will discuss the design and performance of several types of practical matching networks. Factors that may be important in the selection of a particular matching network include the following:

- *Complexity*—As with most engineering solutions, the simplest design that satisfies the required specifications is generally preferable. A simpler matching network is usually cheaper, smaller, more reliable, and less lossy than a more complex design.
- *Bandwidth*—Any type of matching network can ideally give a perfect match (zero reflection) at a single frequency. In many applications, however, it is desirable to match a load over a band of frequencies. There are several ways of doing this, with, of course, a corresponding increase in complexity.
- *Implementation*—Depending on the type of transmission line or waveguide being used, one type of matching network may be preferable to another. For example, tuning stubs are much easier to implement in waveguide than are multisection quarter-wave transformers.
- *Adjustability*—In some applications the matching network may require adjustment to match a variable load impedance. Some types of matching networks are more amenable than others in this regard.

FIGURE 5.1 | A lossless network matching an arbitrary load impedance to a transmission line.

5.1 ____ MATCHING WITH LUMPED ELEMENTS (*L* NETWORKS)

Probably the simplest type of matching network is the *L-section*, which uses two reactive elements to match an arbitrary load impedance to a transmission line. There are two possible configurations for this network, as shown in Figure 5.2. If the normalized load impedance, $z_L = Z_L/Z_0$, is inside the $1 + jx$ circle on the Smith chart, then the circuit of Figure 5.2a should be used. If the normalized load impedance is outside the $1 + jx$ circle on the Smith chart, the circuit of Figure 5.2b should be used. The $1 + jx$ circle is the resistance circle on the impedance Smith chart for which $r = 1$.

In either of the configurations of Figure 5.2, the reactive elements may be either inductors or capacitors, depending on the load impedance. Thus, there are eight distinct possibilities for the matching circuit for various load impedances. If the frequency is low enough and/or the circuit size is small enough, actual lumped-element capacitors and inductors can be used. This may be feasible for frequencies up to about 1 GHz or so, although modern microwave integrated circuits may be small enough such that lumped elements can be used at higher frequencies as well. There is, however, a large range of frequencies and circuit sizes where lumped elements may not be realizable. This is a limitation of the *L*-section matching technique. We will first derive analytical expressions for the matching network elements of the two cases in Figure 5.2, and then illustrate an alternative design procedure using the Smith chart.

Analytical Solutions

Although we will discuss a simple graphical solution using the Smith chart, it is also useful to have simple expressions for the *L*-section matching network components. These expressions can be used in a computer-aided design program for *L*-section matching, or when it is necessary to have more accuracy than the Smith chart can provide.

Consider first the circuit of Figure 5.2a, and let $Z_L = R_L + jX_L$. We stated that this circuit would be used when $z_L = Z_L/Z_0$ is inside the $1 + jx$ circle on the Smith chart, which implies that $R_L > Z_0$ for this case. The impedance seen looking into the matching network, followed by the load impedance, must be equal to Z_0 for an impedance-matched condition:

$$Z_0 = jX + \frac{1}{jB + 1/(R_L + jX_L)}. \tag{5.1}$$

Rearranging and separating into real and imaginary parts gives two equations for the two unknowns, X and B:

$$B(XR_L - X_L Z_0) = R_L - Z_0, \tag{5.2a}$$

$$X(1 - BX_L) = BZ_0 R_L - X_L. \tag{5.2b}$$

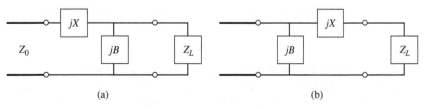

(a)   (b)

FIGURE 5.2 | *L*-section matching networks. (a) Network for z_L inside the $1 + jx$ circle. (b) Network for z_L outside the $1 + jx$ circle.

Solving (5.2a) for X and substituting into (5.2b) gives a quadratic equation for B. The solution is

$$B = \frac{X_L \pm \sqrt{R_L/Z_0}\sqrt{R_L^2 + X_L^2 - Z_0 R_L}}{R_L^2 + X_L^2}. \tag{5.3a}$$

Note that since $R_L > Z_0$, the argument of the second square root is always positive. Then the series reactance can be found as

$$X = \frac{1}{B} + \frac{X_L Z_0}{R_L} - \frac{Z_0}{B R_L}. \tag{5.3b}$$

Equation (5.3a) indicates that two solutions are possible for B and X. Both of these solutions are physically realizable since both positive and negative values of B and X are possible (positive X implies an inductor and negative X implies a capacitor, while positive B implies a capacitor and negative B implies an inductor). One solution, however, may result in significantly smaller values for the reactive components, or may be the preferred solution if the bandwidth of the match is better, or if the SWR on the line between the matching network and the load is smaller.

Next consider the circuit of Figure 5.2b. This circuit is used when z_L is outside the $1 + jx$ circle on the Smith chart, which implies that $R_L < Z_0$. The admittance seen looking into the matching network, followed by the load impedance, must be equal to $1/Z_0$ for an impedance-matched condition:

$$\frac{1}{Z_0} = jB + \frac{1}{R_L + j(X + X_L)}. \tag{5.4}$$

Rearranging and separating into real and imaginary parts gives two equations for the two unknowns, X and B:

$$BZ_0(X + X_L) = Z_0 - R_L, \tag{5.5a}$$

$$(X + X_L) = BZ_0 R_L. \tag{5.5b}$$

Solving for X and B gives

$$X = \pm\sqrt{R_L(Z_0 - R_L)} - X_L, \tag{5.6a}$$

$$B = \pm\frac{\sqrt{(Z_0 - R_L)/R_L}}{Z_0}. \tag{5.6b}$$

Because $R_L < Z_0$, the arguments of the square roots are always positive. Again, note that two solutions are possible.

In order to match an arbitrary complex load to a line of characteristic impedance Z_0, the real part of the input impedance to the matching network must be Z_0, while the imaginary part must be zero. This implies that a general matching network must have at least two degrees of freedom; in the L-section matching circuit these two degrees of freedom are provided by the values of the two reactive components.

Smith Chart Solutions

Instead of the above formulas, the Smith chart can be used to quickly and accurately design L-section matching networks. The procedure is best illustrated by an example.

EXAMPLE 5.1   *L*-SECTION IMPEDANCE MATCHING

Design an L-section matching network to match a series RC load with an impedance $Z_L = 100 + j150\ \Omega$ to a 75 Ω line at a frequency of 3 GHz.

Solution
The normalized load impedance is $z_L = 1.33 + j2$, which is plotted on the Smith chart of Figure 5.3a. This point is inside the $1 + jx$ circle, so we use the matching circuit of Figure 5.2a. Because the first element from the load is a shunt susceptance, it makes sense to convert to admittance by drawing the SWR circle through the load, and a straight line from the load through the center of the chart, as shown in Figure 5.3a. After we add the shunt susceptance and convert back to impedance, we want to be on the $1 + jx$ circle so

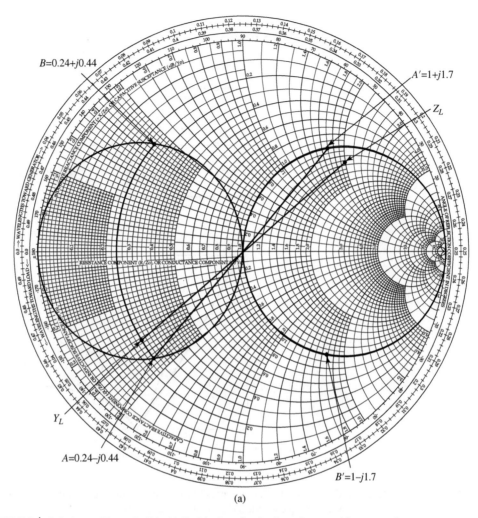

FIGURE 5.3 | Solution to Example 5.1. (a) Smith chart for the *L*-section matching networks.

that we can add a series reactance to cancel jx and match the load. This means that the shunt susceptance must move us from y_L to the $1 + jx$ circle on the *admittance* Smith chart. Thus, we construct the rotated $1 + jx$ circle as shown in Figure 5.3a (center at $r = 0.333$). (A combined *ZY* chart may be convenient to use here, if it is not too confusing.) Then we see that adding a susceptance of $jb = -j0.09$ will move us along a constant-conductance circle to $y = 0.24 - j0.44$ (this choice is the shortest distance from y_L to the shifted $1 + jx$ circle). Converting back to impedance leaves us at $z = 1 + j1.7$, indicating that a series reactance of $x = -j1.7$ will bring us to the center of the chart. For comparison, the formulas (5.3a) and (5.3b) give the solution as $b = 0.79$, $x = 1.7$.

This matching circuit consists of a shunt capacitor and a series inductor, as shown in Figure 5.3b. For a matching frequency of 3 GHz, the capacitor has a value of

$$C = \frac{b}{2\pi f Z_0} = 0.558 \text{ pF},$$

and the inductor has a value of

$$L = \frac{xZ_0}{2\pi f} = 6.76 \text{ nH}.$$

It is also interesting to look at the second solution to this matching problem. If instead of adding a shunt susceptance of $b = 0.3$, we use a shunt susceptance of $jb = j0.79$, we will move to a point on the lower half of the shifted $1 + jx$ circle, to $y = 0.24 + j0.44$. Then converting to impedance and adding a series reactance of $x = j1.7$ leads to a match as well. Formulas (5.3a) and (5.3b) give this solution as $b = -0.09$, $x = -1.7$. This matching circuit is also shown in Figure 5.3b, and is seen to have the positions of the inductor and

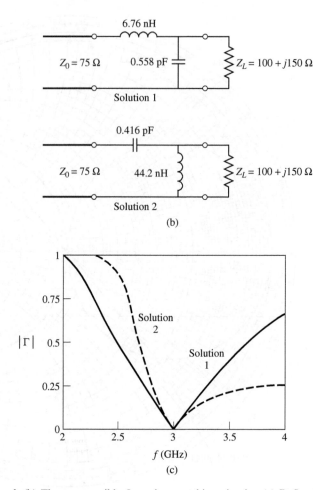

FIGURE 5.3 | Continued. (b) The two possible *L*-section matching circuits. (c) Reflection coefficient magnitudes versus frequency for the matching circuits of (b).

capacitor reversed from the first matching network. At a frequency of $f = 3$ GHz, the capacitor has a value of

$$C = \frac{-1}{2\pi f x Z_0} = 0.416 \text{ pF},$$

while the inductor has a value of

$$L = \frac{-Z_0}{2\pi f b} = 44.20 \text{ nH}.$$

Figure 5.3c shows the reflection coefficient magnitude versus frequency for these two matching networks, assuming that the load impedance of $Z_L = 100 + j150\ \Omega$ at 3 GHz consists of a 100 Ω resistor and a 7.95 nH capacitor in series. There is not a substantial difference in bandwidth for these two solutions.

POINT OF INTEREST: Lumped Elements for Microwave Integrated Circuits

Lumped R, L, and C elements can be practically realized at microwave frequencies if the length, ℓ, of the component is very small relative to the operating wavelength. Over a limited range of values, such components can be used in hybrid and monolithic microwave integrated circuits at frequencies up to 60 GHz, or higher, if the condition that $\ell < \lambda/10$ is satisfied. Usually, however, the characteristics of such an element are far from ideal, requiring that undesirable effects such as parasitic capacitance and/or inductance, spurious resonances, fringing fields, loss, and perturbations caused by a ground plane be incorporated in the design via a CAD model (see the Point of Interest concerning CAD).

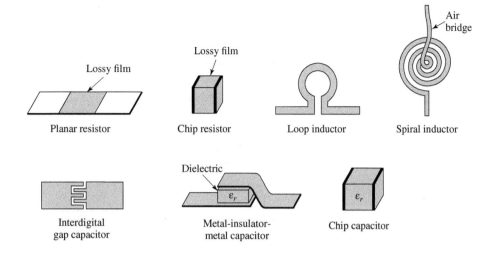

Resistors are fabricated with thin films of lossy material such as nichrome, tantalum nitride, or doped semiconductor material. In monolithic circuits such films can be deposited or grown, whereas chip resistors made from a lossy film deposited on a ceramic chip can be bonded or soldered in a hybrid circuit. Low resistances are hard to obtain.

Small values of inductance can be realized with a short length or loop of transmission line, and larger values (up to about 10 nH) can be obtained with a spiral inductor, as shown in the figures. Larger inductance values generally incur more loss and more shunt capacitance; this leads to a resonance that limits the maximum operating frequency.

Capacitors can be fabricated in several ways. A short transmission line stub can provide a shunt capacitance in the range of 0–0.1 pF. A single gap, or an interdigital set of gaps, in a transmission line can provide a series capacitance up to about 0.5 pF. Greater values (up to about 25 pF) can be obtained using a metal-insulator-metal sandwich in either monolithic or chip (hybrid) form.

5.2 SINGLE-STUB TUNING

Another popular matching technique uses a single open-circuited or short-circuited length of transmission line (a *stub*) connected either in parallel or in series with the transmission feed line at a certain distance from the load, as shown in Figure 5.4. Such a *single-stub tuning* circuit is often very convenient because the stub can be fabricated as part of the transmission line media of the circuit, and lumped elements are avoided. Shunt stubs are preferred for microstrip line or stripline, while series stubs are preferred for slotline or coplanar waveguide.

In single-stub tuning the two adjustable parameters are the distance, d, from the load to the stub position, and the value of susceptance or reactance provided by the stub. For the shunt-stub case, the basic idea is to select d so that the admittance, Y, seen looking into the line at distance d from the load is of the form $Y_0 + jB$. Then the stub susceptance is chosen as $-jB$, resulting in a matched condition. For the series-stub case, the distance d is selected so that the impedance, Z, seen looking into the line at a distance d from the load is of the form $Z_0 + jX$. Then the stub reactance is chosen as $-jX$, resulting in a matched condition.

As discussed in Chapter 2, the proper length of an open or shorted transmission line section can provide any desired value of reactance or susceptance. For a given susceptance or reactance, the difference in lengths of an open- or short-circuited stub is $\lambda/4$. For transmission line media such as microstrip or stripline, open-circuited stubs are easier to fabricate since a via hole through the substrate to the ground plane is not needed. For lines like coax or waveguide, however, short-circuited stubs are usually preferred because the cross-sectional area of such an open-circuited line may be large enough (electrically) to radiate, in which case the stub is no longer purely reactive.

We will discuss both Smith chart and analytical solutions for shunt- and series-stub tuning. The Smith chart solutions are fast, intuitive, and usually accurate enough in practice. The analytical expressions are more precise, and are useful for computer analysis.

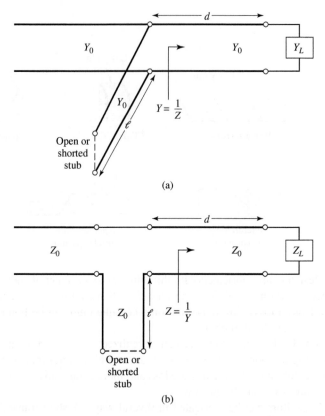

FIGURE 5.4 | Single-stub tuning circuits. (a) Shunt stub. (b) Series stub.

Shunt Stubs

The single-stub shunt tuning circuit is shown in Figure 5.4a. We will first discuss an example illustrating the Smith chart solution and then derive formulas for d and ℓ.

EXAMPLE 5.2   SINGLE-STUB SHUNT TUNING

For a load impedance $Z_L = 40 + j30\ \Omega$, design two single-stub (short circuit) shunt tuning networks to match this load to a $100\ \Omega$ line. Assuming that the load is matched at 2 GHz and that the load consists of a resistor and a capacitor in series, plot the reflection coefficient magnitude from 1 to 3 GHz for each solution.

Solution
The first step is to plot the normalized load impedance $z_L = 0.4 + j0.3$, construct the appropriate SWR circle, and convert to the load admittance, y_L, as shown on the Smith chart in Figure 5.5a. For the remaining steps we consider the Smith chart as an admittance chart. Notice that the SWR circle intersects the $1 + jb$ circle at two points, denoted as A and B in Figure 5.5a. Thus the distance d from the load to the stub is given by either of these two intersections. Reading the WTG scale, we obtain

$$d_1 = 0.336 - 0.302 = 0.034\lambda,$$

$$d_2 = (0.5 - 0.302) + 0.163 = 0.361\lambda.$$

Actually, there is an infinite number of distances d around the SWR circle that intersect the $1 + jb$ circle. Usually it is desired to keep the matching stub as close as possible to the load to improve the bandwidth of the match and to reduce losses caused by a possibly large standing wave ratio on the line between the stub and the load.

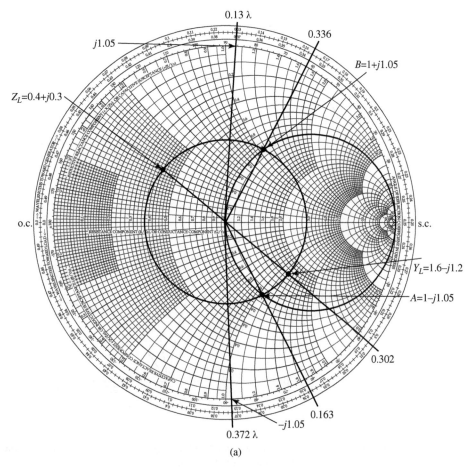

FIGURE 5.5 | Solution to Example 5.2. (a) Smith chart for the shunt-stub tuners.

At the two intersection points, the normalized admittances are

$$A = 1.00 - j1.05,$$
$$B = 1.00 + j1.05.$$

Thus, the first tuning solution requires a stub with a susceptance of $j1.05$. The length of a short-circuited stub that gives this susceptance can be found on the Smith chart by starting at $y = \infty$ (the short circuit) and moving along the outer edge of the chart ($g = 0$) toward the generator to the $j1.05$ point. The stub length is then

$$\ell_1 = 0.372\lambda.$$

Similarly, the required short-circuit stub length for the second solution is

$$\ell_2 = 0.13\lambda.$$

This completes the two tuner designs.

To analyze the frequency dependence of these two designs, we need to know the load impedance as a function of frequency. The series-RC load impedance is $Z_L = 40 + j30\ \Omega$ at 2 GHz, so $R = 40\ \Omega$ and $L = 2.39$ nH. The two tuning circuits are shown in Figure 5.5b. Figure 5.5c shows the calculated reflection coefficient magnitudes for these two solutions. Observe that solution 1 has a significantly better bandwidth than solution 2; this is because both d and ℓ are shorter for solution 1, which reduces the frequency variation of the match.

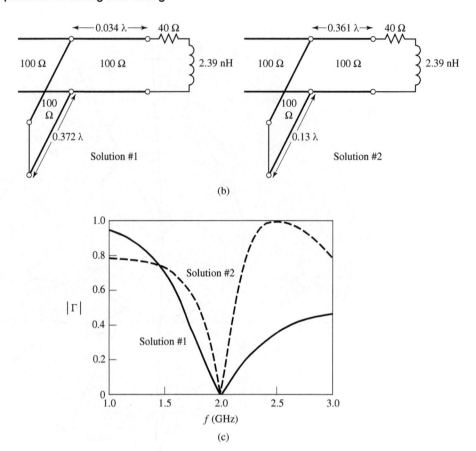

FIGURE 5.5 | Continued. (b) The two shunt-stub tuning solutions. (c) Reflection coefficient magnitudes versus frequency for the tuning circuits of (b).

To derive formulas for d and ℓ, let the load impedance be written as $Z_L = 1/Y_L = R_L + jX_L$. Then the impedance Z down a length d of line from the load is

$$Z = Z_0 \frac{(R_L + jX_L) + jZ_0 t}{Z_0 + j(R_L + jX_L)t}, \tag{5.7}$$

where $t = \tan \beta d$. The admittance at this point is

$$Y = G + jB = \frac{1}{Z},$$

where

$$G = \frac{R_L(1 + t^2)}{R_L^2 + (X_L + Z_0 t)^2}, \tag{5.8a}$$

$$B = \frac{R_L^2 t - (Z_0 - X_L t)(X_L + Z_0 t)}{Z_0 \left[R_L^2 + (X_L + Z_0 t)^2 \right]}. \tag{5.8b}$$

Now d (which implies t) is chosen so that $G = Y_0 = 1/Z_0$. From (5.8a), this results in a quadratic equation for t:

$$Z_0(R_L - Z_0)t^2 - 2X_L Z_0 t + \left(R_L Z_0 - R_L^2 - X_L^2 \right) = 0.$$

Solving for t gives

$$t = \frac{X_L \pm \sqrt{R_L \left[(Z_0 - R_L)^2 + X_L^2 \right] / Z_0}}{R_L - Z_0} \qquad \text{for } R_L \neq Z_0. \tag{5.9}$$

If $R_L = Z_0$, then $t = -X_L/2Z_0$. Thus, the two principal solutions for d are

$$\frac{d}{\lambda} = \begin{cases} \dfrac{1}{2\pi} \tan^{-1} t & \text{for } t \geq 0 \\[2mm] \dfrac{1}{2\pi} (\pi + \tan^{-1} t) & \text{for } t < 0. \end{cases} \tag{5.10}$$

To find the required stub lengths, first use t in (5.8b) to find the stub susceptance, $B_s = -B$. Then, for an open-circuited stub,

$$\frac{\ell_o}{\lambda} = \frac{1}{2\pi} \tan^{-1} \left(\frac{B_s}{Y_0} \right) = \frac{-1}{2\pi} \tan^{-1} \left(\frac{B}{Y_0} \right), \tag{5.11a}$$

and for a short-circuited stub,

$$\frac{\ell_s}{\lambda} = \frac{-1}{2\pi} \tan^{-1} \left(\frac{Y_0}{B_s} \right) = \frac{1}{2\pi} \tan^{-1} \left(\frac{Y_0}{B} \right). \tag{5.11b}$$

If the length given by (5.11a) or (5.11b) is negative, $\lambda/2$ can be added to give a positive result.

Series Stubs

The series-stub tuning circuit is shown in Figure 5.4b. We will illustrate the Smith chart solution by an example, and then derive expressions for d and ℓ.

EXAMPLE 5.3   SINGLE-STUB SERIES TUNING

Match a load impedance of $Z_L = 90 + j60$ to a 75 Ω line using a single series open-circuit stub. Assuming that the load is matched at 2 GHz and that the load consists of a resistor and an inductor in series, plot the reflection coefficient magnitude from 1 to 3 GHz.

Solution
First plot the normalized load impedance, $z_L = 1.2 + j0.8$, and draw the SWR circle. For the series-stub design the chart is an impedance chart. Note that the SWR circle intersects the $1 + jx$ circle at two points, denoted as A and B in Figure 5.6a. The shortest distance, d_1, from the load to the stub is, from the WTG scale,

$$d_1 = 0.347 - 0.174 = 0.173\lambda,$$

and the second distance is

$$d_2 = (0.5 - 0.174) + 0.156 = 0.482\lambda.$$

As in the shunt-stub case, additional rotations around the SWR circle lead to additional solutions, but these are usually not of practical interest.

The normalized impedances at the two intersection points are

$$A = 1 - j0.725,$$

$$B = 1 + j0.76.$$

Thus, the first solution requires a stub with a reactance of $j0.725$. The length of an open-circuited stub that gives this reactance can be found on the Smith chart by starting at $z = \infty$ (open circuit), and moving along the outer edge of the chart ($r = 0$) toward the generator to the $j0.725$ point. This gives a stub length of

$$\ell_1 = 0.398\lambda.$$

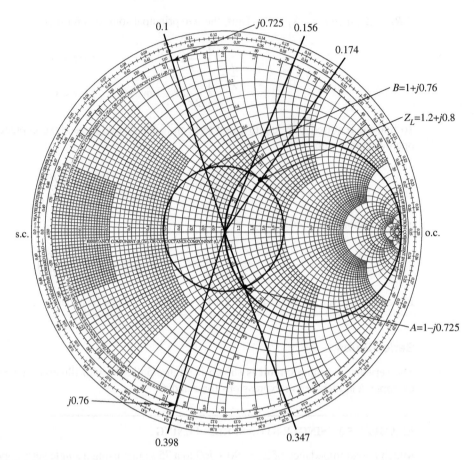

FIGURE 5.6 | Solution to Example 5.3. (a) Smith chart for the series-stub tuners.

Similarly, the required open-circuited stub length for the second solution is

$$\ell_2 = 0.1\lambda.$$

This completes the tuner designs.

If the load is a series resistor and inductor with $Z_L = 90 + j60$ Ω at 2 GHz, then $R = 90$ Ω and $L = 4.77$ nH. The two matching circuits are shown in Figure 5.6b. Figure 5.6c shows the calculated reflection coefficient magnitudes versus frequency for the two solutions.

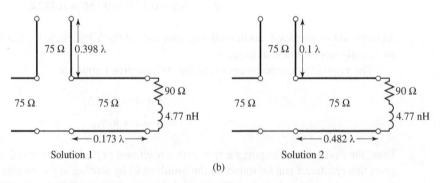

FIGURE 5.6 | Continued. (b) The two series-stub tuning solutions.

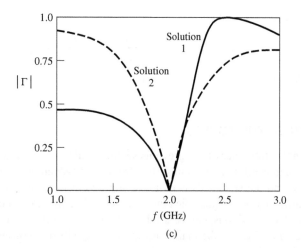

f (GHz)

(c)

FIGURE 5.6 | Continued. (c) Reflection coefficient magnitudes versus frequency for the tuning circuits of (b).

To derive formulas for d and ℓ for the series-stub tuner, let the load admittance be written as $Y_L = 1/Z_L = G_L + jB_L$. Then the admittance Y down a length d of line from the load is

$$Y = Y_0 \frac{(G_L + jB_L) + jtY_0}{Y_0 + jt(G_L + jB_L)}, \tag{5.12}$$

where $t = \tan \beta d$ and $Y_0 = 1/Z_0$. The impedance at this point is

$$Z = R + jX = \frac{1}{Y},$$

where

$$R = \frac{G_L(1 + t^2)}{G_L^2 + (B_L + Y_0 t)^2}, \tag{5.13a}$$

$$X = \frac{G_L^2 t - (Y_0 - tB_L)(B_L + tY_0)}{Y_0 \left[G_L^2 + (B_L + Y_0 t)^2 \right]}. \tag{5.13b}$$

Now d (which implies t) is chosen so that $R = Z_0 = 1/Y_0$. From (5.13a), this results in a quadratic equation for t:

$$Y_0(G_L - Y_0)t^2 - 2B_L Y_0 t + \left(G_L Y_0 - G_L^2 - B_L^2 \right) = 0.$$

Solving for t gives

$$t = \frac{B_L \pm \sqrt{G_L \left[(Y_0 - G_L)^2 + B_L^2 \right] / Y_0}}{G_L - Y_0} \quad \text{for } G_L \neq Y_0. \tag{5.14}$$

If $G_L = Y_0$, then $t = -B_L/2Y_0$. Then the two principal solutions for d are

$$d/\lambda = \begin{cases} \dfrac{1}{2\pi} \tan^{-1} t & \text{for } t \geq 0 \\[2mm] \dfrac{1}{2\pi} (\pi + \tan^{-1} t) & \text{for } t < 0. \end{cases} \tag{5.15}$$

The required stub lengths are determined by first using t in (5.13b) to find the reactance X. This reactance is the negative of the necessary stub reactance, X_s. Thus, for a short-circuited stub,

$$\frac{\ell_s}{\lambda} = \frac{1}{2\pi} \tan^{-1} \left(\frac{X_s}{Z_0} \right) = \frac{-1}{2\pi} \tan^{-1} \left(\frac{X}{Z_0} \right), \tag{5.16a}$$

and for an open-circuited stub,

$$\frac{\ell_o}{\lambda} = \frac{-1}{2\pi} \tan^{-1}\left(\frac{Z_0}{X_s}\right) = \frac{1}{2\pi} \tan^{-1}\left(\frac{Z_0}{X}\right). \tag{5.16b}$$

If the length given by (5.16a) or (5.16b) is negative, $\lambda/2$ can be added to give a positive result.

5.3   DOUBLE-STUB TUNING

The single-stub tuner of the previous section is able to match any load impedance (having a positive real part) to a transmission line, but suffers from the disadvantage of requiring a variable length of line between the load and the stub. This may not be a problem for a fixed matching circuit, but would probably pose some difficulty if an adjustable tuner was desired. In this case, the *double-stub tuner*, which uses two tuning stubs in fixed positions, can be used. Such tuners are often fabricated in coaxial line with adjustable stubs connected in shunt to the main coaxial line. We will see, however, that a double-stub tuner cannot match all load impedances.

The double-stub tuner circuit is shown in Figure 5.7a, where the load may be an arbitrary distance from the first stub. Although this is more representative of a practical situation, the circuit of Figure 5.7b, where the load Y'_L has been transformed back to the position of the first stub, is easier to deal with and does not lose any generality. The shunt stubs shown in Figure 5.7 can be conveniently implemented for some types of transmission lines, while series stubs are more appropriate for other types of lines. In either case, the stubs can be open-circuited or short-circuited.

Smith Chart Solution

The Smith chart of Figure 5.8 illustrates the basic operation of the double-stub tuner. As in the case of the single-stub tuner, two solutions are possible. The susceptance of the first stub, b_1 (or b'_1, for the second solution), moves the load admittance to y_1 (or y'_1). These points lie on the rotated $1 + jb$ circle; the amount of rotation is d wavelengths toward the load, where d is the electrical distance between the two stubs. Then transforming y_1 (or y'_1) toward the generator through a length d of line leaves us at the point y_2 (or y'_2),

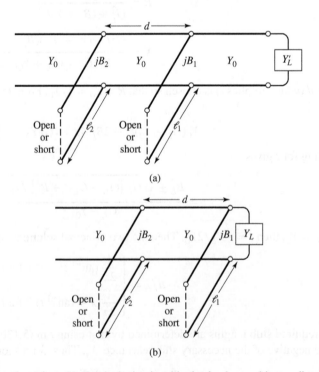

FIGURE 5.7 | Double-stub tuning. (a) Original circuit with the load an arbitrary distance from the first stub. (b) Equivalent circuit with the load transformed to the first stub.

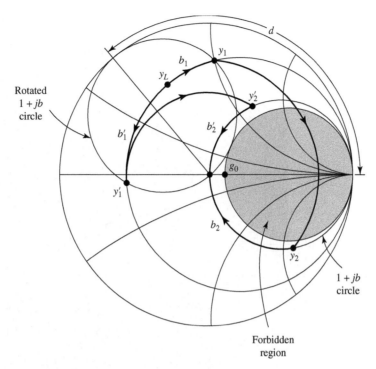

FIGURE 5.8 | Smith chart diagram for the operation of a double-stub tuner.

which must be on the $1 + jb$ circle. The second stub then adds a susceptance b_2 (or b'_2), which brings us to the center of the chart and completes the match.

Notice from Figure 5.8 that if the load admittance, y_L, were inside the shaded region of the $g_0 + jb$ circle, no value of stub susceptance b_1 could ever bring the load point to intersect the rotated $1 + jb$ circle. This shaded region thus forms a forbidden range of load admittances that cannot be matched with this particular double-stub tuner. A simple way of reducing the forbidden range is to reduce the distance d between the stubs. This has the effect of swinging the rotated $1 + jb$ circle back toward the $y = \infty$ point, but d must be kept large enough for the practical purpose of fabricating the two separate stubs. In addition, stub spacings near 0 or $\lambda/2$ lead to matching networks that are very frequency sensitive. In practice, stub spacings are usually chosen as $\lambda/8$ or $3\lambda/8$. If the length of line between the load and the first stub can be adjusted, then the load admittance y_L can always be moved out of the forbidden region.

EXAMPLE 5.4   DOUBLE-STUB TUNING

Design a double-stub shunt tuner to match a load impedance $Z_L = 60 - j80 \ \Omega$ to a 50 Ω line. The stubs are to be open-circuited stubs and are spaced $\lambda/8$ apart. Assuming that this load consists of a series resistor and capacitor and that the match frequency is 2 GHz, plot the reflection coefficient magnitude versus frequency from 1 to 3 GHz.

Solution
The normalized load admittance is $y_L = 0.3 + j0.4$, which is plotted on the Smith chart of Figure 5.9a. Next we construct the rotated $1 + jb$ conductance circle by moving every point on the $g = 1$ circle $\lambda/8$ toward the load. We then find the susceptance of the first stub, which can be one of two possible values:

$$b_1 = 1.314 \quad \text{or} \quad b'_1 = -0.114.$$

We now transform through the $\lambda/8$ section of line by rotating along a constant-radius (SWR) circle $\lambda/8$ toward the generator. This brings the two solutions to the following points:

$$y_2 = 1 - j3.38 \quad \text{or} \quad y'_2 = 1 + j1.38.$$

Then the susceptance of the second stub should be

$$b_2 = 3.38 \quad \text{or} \quad b'_2 = -1.38.$$

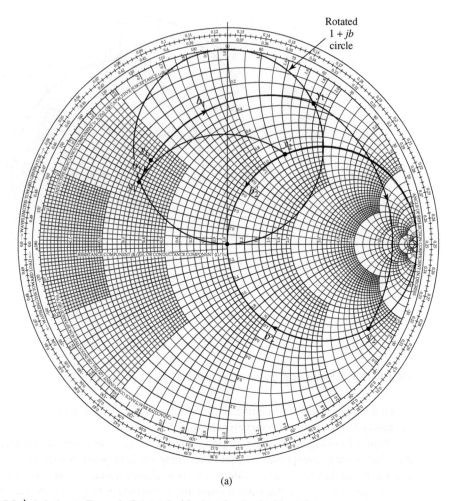

(a)

FIGURE 5.9 | Solution to Example 5.4. (a) Smith chart for the double-stub tuners.

The lengths of the open-circuited stubs are then found as

$$\ell_1 = 0.146\lambda, \ell_2 = 0.204\lambda \quad \text{or} \quad \ell_1' = 0.482\lambda, \ell_2' = 0.350\lambda.$$

This completes both solutions for the double-stub tuner design.

At $f = 2$ GHz the resistor-capacitor load of $Z_L = 60 - j80$ Ω implies that $R = 60$ Ω and $C = 0.995$ pF. The two tuning circuits are then as shown in Figure 5.9b, and the reflection coefficient magnitudes are plotted versus frequency in Figure 5.9c. Note that the first solution has a much narrower bandwidth than the second (primed) solution due to the fact that both stubs for the first solution are somewhat longer (and closer to $\lambda/2$) than the stubs of the second solution.

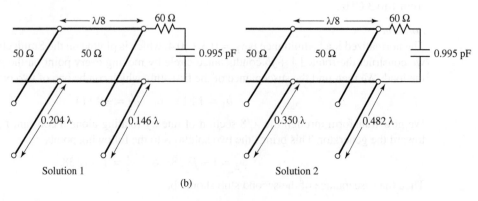

(b)

FIGURE 5.9 | Continued. (b) The two double-stub tuning solutions.

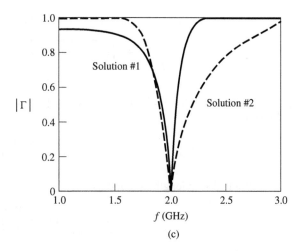

(c)

FIGURE 5.9 | Continued. (c) Reflection coefficient magnitudes versus frequency for the tuning circuits of (b).

Analytical Solution

The admittance just to the left of the first stub in Figure 5.7b is

$$Y_1 = G_L + j(B_L + B_1), \tag{5.17}$$

where $Y_L = G_L + jB_L$ is the load admittance, and B_1 is the susceptance of the first stub. After transforming through a length d of transmission line, we find that the admittance just to the right of the second stub is

$$Y_2 = Y_0 \frac{G_L + j(B_L + B_1 + Y_0 t)}{Y_0 + jt(G_L + jB_L + jB_1)}, \tag{5.18}$$

where $t = \tan \beta d$ and $Y_0 = 1/Z_0$. At this point the real part of Y_2 must equal Y_0, which leads to the equation

$$G_L^2 - G_L Y_0 \frac{1 + t^2}{t^2} + \frac{(Y_0 - B_L t - B_1 t)^2}{t^2} = 0. \tag{5.19}$$

Solving for G_L gives

$$G_L = Y_0 \frac{1 + t^2}{2t^2} \left[1 \pm \sqrt{1 - \frac{4t^2(Y_0 - B_L t - B_1 t)^2}{Y_0^2 (1 + t^2)^2}} \right]. \tag{5.20}$$

Because G_L is real, the quantity within the square root must be nonnegative, and so

$$0 \le \frac{4t^2(Y_0 - B_L t - B_1 t)^2}{Y_0^2 (1 + t^2)^2} \le 1.$$

This implies that

$$0 \le G_L \le Y_0 \frac{1 + t^2}{t^2} = \frac{Y_0}{\sin^2 \beta d}, \tag{5.21}$$

which gives the range on G_L that can be matched for a given stub spacing d. After d has been set, the first stub susceptance can be determined from (5.19) as

$$B_1 = -B_L + \frac{Y_0 \pm \sqrt{(1 + t^2)G_L Y_0 - G_L^2 t^2}}{t}. \tag{5.22}$$

Then the second stub susceptance can be found from the negative of the imaginary part of (5.18) to be

$$B_2 = \frac{\pm Y_0 \sqrt{Y_0 G_L(1 + t^2) - G_L^2 t^2} + G_L Y_0}{G_L t}.$$ (5.23)

The upper and lower signs in (5.22) and (5.23) correspond to the same solutions. The open-circuited stub length is found as

$$\frac{\ell_o}{\lambda} = \frac{1}{2\pi} \tan^{-1}\left(\frac{B}{Y_0}\right),$$ (5.24a)

and the short-circuited stub length is found as

$$\frac{\ell_s}{\lambda} = \frac{-1}{2\pi} \tan^{-1}\left(\frac{Y_0}{B}\right),$$ (5.24b)

where $B = B_1$ or B_2.

5.4   THE QUARTER-WAVE TRANSFORMER

The quarter-wave transformer is a useful and practical circuit for impedance matching and also provides a simple transmission line circuit that further illustrates the properties of standing waves on a mismatched line. We will first approach the problem from the impedance viewpoint and then show how this result can also be interpreted in terms of an infinite set of multiple reflections on the matching section.

The Impedance Viewpoint

Figure 5.10 shows a circuit employing a quarter-wave transformer. The load resistance R_L and the feedline characteristic impedance Z_0 are both real and assumed to be known. These two components are connected with a lossless piece of transmission line of (unknown) characteristic impedance Z_1 and length $\lambda/4$. It is desired to match the load to the Z_0 line by using the $\lambda/4$ section of line and so make $\Gamma = 0$ looking into the $\lambda/4$ matching section. From (2.44) the input impedance Z_{in} can be found as

$$Z_{in} = Z_1 \frac{R_L + jZ_1 \tan \beta\ell}{Z_1 + jR_L \tan \beta\ell}.$$ (5.25)

To evaluate this for $\beta\ell = (2\pi/\lambda)(\lambda/4) = \pi/2$, we can divide the numerator and denominator by $\tan \beta\ell$ and take the limit as $\beta\ell \to \pi/2$ to get

$$Z_{in} = \frac{Z_1^2}{R_L}.$$ (5.26)

In order for $\Gamma = 0$, we must have $Z_{in} = Z_0$, which yields the characteristic impedance Z_1 as

$$Z_1 = \sqrt{Z_0 R_L},$$ (5.27)

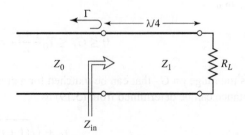

FIGURE 5.10 | The quarter-wave matching transformer.

the geometric mean of the load and source impedances. Then there will be no standing waves on the feedline (SWR = 1), although there will be standing waves on the $\lambda/4$ matching section. Also, the above condition applies only when the length of the matching section is $\lambda/4$, or an odd multiple $(2n + 1)$ of $\lambda/4$ long, so that a perfect match may be achieved at one frequency, but mismatch will occur at other frequencies.

EXAMPLE 5.5   FREQUENCY RESPONSE OF A QUARTER-WAVE TRANSFORMER

Consider a load resistance $R_L = 100 \, \Omega$ to be matched to a 50 Ω line with a quarter-wave transformer. Find the characteristic impedance of the matching section and plot the magnitude of the reflection coefficient versus normalized frequency, f/f_o, where f_o is the frequency at which the line is $\lambda/4$ long.

Solution
From (5.27), the necessary characteristic impedance is

$$Z_1 = \sqrt{(50)(100)} = 70.71 \, \Omega.$$

The reflection coefficient magnitude is given as

$$|\Gamma| = \left| \frac{Z_{in} - Z_0}{Z_{in} + Z_0} \right|,$$

where the input impedance Z_{in} is a function of frequency as given by (2.44). The frequency dependence in (2.44) comes from the $\beta\ell$ term, which can be written in terms of f/f_o as

$$\beta\ell = \left(\frac{2\pi}{\lambda} \right) \left(\frac{\lambda_0}{4} \right) = \left(\frac{2\pi f}{v_p} \right) \left(\frac{v_p}{4f_o} \right) = \frac{\pi f}{2f_o},$$

where it is seen that $\beta\ell = \pi/2$ for $f = f_o$, as expected. For higher frequencies the matching section looks electrically longer, and for lower frequencies it looks shorter. The magnitude of the reflection coefficient is plotted versus f/f_o in Figure 5.11.

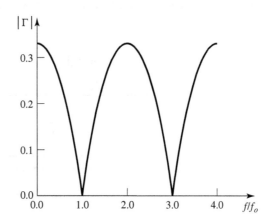

FIGURE 5.11 | Reflection coefficient versus normalized frequency for the quarter-wave transformer of Example 5.5.

This method of impedance matching is limited to real load impedances, although a complex load impedance can easily be made real, at a single frequency, by transformation through an appropriate length of line.

The above analysis shows how useful the impedance concept can be when solving transmission line problems, and this method is probably the preferred method in practice. It may aid our understanding of the quarter-wave transformer (and other transmission line circuits), however, if we now look at it from the viewpoint of multiple reflections.

The Multiple-Reflection Viewpoint

Figure 5.12 shows the quarter-wave transformer circuit with reflection and transmission coefficients defined as follows:

Γ = overall, or total, reflection coefficient of a wave incident on the $\lambda/4$ transformer (same as Γ in Example 5.5).
Γ_1 = partial reflection coefficient of a wave incident on a load Z_1, from the Z_0 line.
Γ_2 = partial reflection coefficient of a wave incident on a load Z_0, from the Z_1 line.
Γ_3 = partial reflection coefficient of a wave incident on a load R_L, from the Z_1 line.
T_1 = partial transmission coefficient of a wave from the Z_0 line into the Z_1 line.
T_2 = partial transmission coefficient of a wave from the Z_1 line into the Z_0 line.

These coefficients can be expressed as

$$\Gamma_1 = \frac{Z_1 - Z_0}{Z_1 + Z_0}, \tag{5.28a}$$

$$\Gamma_2 = \frac{Z_0 - Z_1}{Z_0 + Z_1} = -\Gamma_1, \tag{5.28b}$$

$$\Gamma_3 = \frac{R_L - Z_1}{R_L + Z_1}, \tag{5.28c}$$

$$T_1 = \frac{2Z_1}{Z_1 + Z_0}, \tag{5.28d}$$

$$T_2 = \frac{2Z_0}{Z_1 + Z_0}. \tag{5.28e}$$

Now think of the quarter-wave transformer of Figure 5.12 in the time domain, and imagine a wave traveling down the Z_0 feedline toward the transformer. When the wave first hits the junction with the Z_1 line, it sees only an impedance Z_1 since it has not yet traveled to the load R_L and cannot see that effect. Part of the wave is reflected with a coefficient Γ_1, and part is transmitted onto the Z_1 line with a coefficient T_1. The transmitted wave then travels $\lambda/4$ to the load, is reflected with a coefficient Γ_3, and travels another $\lambda/4$ back

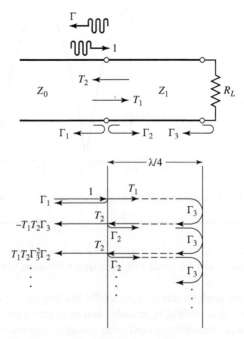

FIGURE 5.12 | Multiple reflection analysis of the quarter-wave transformer.

to the junction with the Z_0 line. Part of this wave is transmitted through (to the left) to the Z_0 line, with coefficient T_2, and part is reflected back toward the load with coefficient Γ_2. Clearly, this process continues with an infinite number of bouncing waves, and the total reflection coefficient, Γ, is the sum of all of these partial reflections. Since each round trip path up and down the $\lambda/4$ transformer section results in a 180° phase shift, the total reflection coefficient can be expressed as

$$\Gamma = \Gamma_1 - T_1 T_2 \Gamma_3 + T_1 T_2 \Gamma_2 \Gamma_3^2 - T_1 T_2 \Gamma_2^2 \Gamma_3^3 + \cdots$$

$$= \Gamma_1 - T_1 T_2 \Gamma_3 \sum_{n=0}^{\infty} \left(-\Gamma_2 \Gamma_3 \right)^n. \tag{5.29}$$

Since $|\Gamma_3| < 1$ and $|\Gamma_2| < 1$, the infinite series in (5.29) can be summed using the geometric series result that

$$\sum_{n=0}^{\infty} x^n = \frac{1}{1-x}, \qquad \text{for } |x| < 1,$$

to give

$$\Gamma = \Gamma_1 - \frac{T_1 T_2 \Gamma_3}{1 + \Gamma_2 \Gamma_3} = \frac{\Gamma_1 + \Gamma_1 \Gamma_2 \Gamma_3 - T_1 T_2 \Gamma_3}{1 + \Gamma_2 \Gamma_3}. \tag{5.30}$$

The numerator of this expression can be simplified using (5.28) to give

$$\Gamma_1 - \Gamma_3 \left(\Gamma_1^2 + T_1 T_2 \right) = \Gamma_1 - \Gamma_3 \left[\frac{(Z_1 - Z_0)^2}{(Z_1 + Z_0)^2} + \frac{4 Z_1 Z_0}{(Z_1 + Z_0)^2} \right]$$

$$= \Gamma_1 - \Gamma_3 = \frac{(Z_1 - Z_0)(R_L + Z_1) - (R_L - Z_1)(Z_1 + Z_0)}{(Z_1 + Z_0)(R_L + Z_1)}$$

$$= \frac{2 \left(Z_1^2 - Z_0 R_L \right)}{(Z_1 + Z_0)(R_L + Z_1)},$$

which is seen to vanish if we choose $Z_1 = \sqrt{Z_0 R_L}$, as in (5.27). Then Γ of (5.30) is zero, and the line is matched. This analysis shows that the matching property of the quarter-wave transformer comes about by properly selecting the characteristic impedance and length of the matching section so that the superposition of all of the partial reflections adds to zero. Under steady-state conditions, an infinite sum of waves traveling in the same direction with the same phase velocity can be combined into a single traveling wave. Thus, the infinite set of waves traveling in the forward and reverse directions on the matching section can be reduced to two waves traveling in opposite directions.

Impedance Matching of the Quarter-Wave Transformer

An additional feature of the quarter-wave transformer is that it can be extended to multisection designs in a methodical manner to provide broader bandwidth. If only a narrowband impedance match is required, a single-section transformer may suffice. However, as we will see in the next few sections, multisection quarter-wave transformer designs can be synthesized to yield optimum matching characteristics over a desired frequency band. We will see in Chapter 8 that such networks are closely related to bandpass filters.

One drawback of the quarter-wave transformer is that it can only match a real load impedance. A complex load impedance can always be transformed into a real impedance, however, by using an appropriate length of transmission line between the load and the transformer, or an appropriate series or shunt reactive element. These techniques will usually alter the frequency dependence of the load, and this often has the effect of reducing the bandwidth of the match.

The single-section quarter-wave matching transformer circuit is shown in Figure 5.13, with the characteristic impedance of the matching section given as

$$Z_1 = \sqrt{Z_0 Z_L}. \tag{5.31}$$

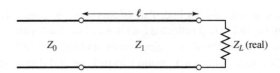

FIGURE 5.13 | A single-section quarter-wave matching transformer. $\ell = \lambda_0/4$ at the design frequency f_0.

At the design frequency, f_0, the electrical length of the matching section is $\lambda_0/4$, but at other frequencies the length is different, so a perfect match is no longer obtained. We will derive an approximate expression for the resulting impedance mismatch versus frequency.

The input impedance seen looking into the matching section is

$$Z_{\text{in}} = Z_1 \frac{Z_L + jZ_1 t}{Z_1 + jZ_L t}, \tag{5.32}$$

where $t = \tan \beta\ell = \tan\theta$, and $\beta\ell = \theta = \pi/2$ at the design frequency f_0. The resulting reflection coefficient is

$$\Gamma = \frac{Z_{\text{in}} - Z_0}{Z_{\text{in}} + Z_0} = \frac{Z_1(Z_L - Z_0) + jt\left(Z_1^2 - Z_0 Z_L\right)}{Z_1(Z_L + Z_0) + jt\left(Z_1^2 + Z_0 Z_L\right)}. \tag{5.33}$$

Because $Z_1^2 = Z_0 Z_L$, this reduces to

$$\Gamma = \frac{Z_L - Z_0}{Z_L + Z_0 + j2t\sqrt{Z_0 Z_L}}. \tag{5.34}$$

The reflection coefficient magnitude is

$$\begin{aligned}
|\Gamma| &= \frac{|Z_L - Z_0|}{\left[(Z_L + Z_0)^2 + 4t^2 Z_0 Z_L\right]^{1/2}} \\
&= \frac{1}{\left\{(Z_L + Z_0)^2/(Z_L - Z_0)^2 + [4t^2 Z_0 Z_L/(Z_L - Z_0)^2]\right\}^{1/2}} \\
&= \frac{1}{\left\{1 + [4Z_0 Z_L/(Z_L - Z_0)^2] + [4Z_0 Z_L t^2/(Z_L - Z_0)^2]\right\}^{1/2}} \\
&= \frac{1}{\left\{1 + [4Z_0 Z_L/(Z_L - Z_0)^2]\ \sec^2\theta\right\}^{1/2}},
\end{aligned} \tag{5.35}$$

since $1 + t^2 = 1 + \tan^2\theta = \sec^2\theta$.

If we assume that the operating frequency is near the design frequency f_0, then $\ell \simeq \lambda_0/4$ and $\theta \simeq \pi/2$. Then $\sec^2\theta \gg 1$, and (5.35) simplifies to

$$|\Gamma| \simeq \frac{|Z_L - Z_0|}{2\sqrt{Z_0 Z_L}} |\cos\theta| \quad \text{for } \theta \text{ near } \pi/2. \tag{5.36}$$

This result gives the approximate mismatch of the quarter-wave transformer near the design frequency, as sketched in Figure 5.14.

If we set a maximum value, Γ_m, for an acceptable reflection coefficient magnitude, then the bandwidth of the matching transformer can be defined as

$$\Delta\theta = 2\left(\frac{\pi}{2} - \theta_m\right), \tag{5.37}$$

since the response of (5.35) is symmetric about $\theta = \pi/2$, and $\Gamma = \Gamma_m$ at $\theta = \theta_m$ and at $\theta = \pi - \theta_m$. Equating Γ_m to the exact expression for the reflection coefficient magnitude in (5.35) allows us to solve for θ_m:

$$\frac{1}{\Gamma_m^2} = 1 + \left(\frac{2\sqrt{Z_0 Z_L}}{Z_L - Z_0} \sec\theta_m\right)^2,$$

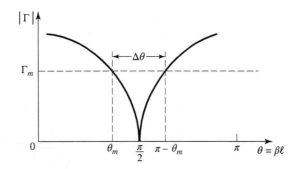

FIGURE 5.14 | Approximate behavior of the reflection coefficient magnitude for a single-section quarter-wave transformer operating near its design frequency.

or

$$\cos \theta_m = \frac{\Gamma_m}{\sqrt{1 - \Gamma_m^2}} \frac{2\sqrt{Z_0 Z_L}}{|Z_L - Z_0|}. \tag{5.38}$$

If we assume TEM lines, then

$$\theta = \beta \ell = \frac{2\pi f}{v_p} \frac{v_p}{4f_0} = \frac{\pi f}{2f_0},$$

and so the frequency of the lower band edge at $\theta = \theta_m$ is

$$f_m = \frac{2\theta_m f_0}{\pi},$$

and the fractional bandwidth is, using (5.38),

$$\frac{\Delta f}{f_0} = \frac{2(f_0 - f_m)}{f_0} = 2 - \frac{2f_m}{f_0} = 2 - \frac{4\theta_m}{\pi}$$

$$= 2 - \frac{4}{\pi} \cos^{-1} \left[\frac{\Gamma_m}{\sqrt{1 - \Gamma_m^2}} \frac{2\sqrt{Z_0 Z_L}}{|Z_L - Z_0|} \right]. \tag{5.39}$$

Fractional bandwidth is usually expressed as a percentage, $100\Delta f/f_0\%$. Note that the bandwidth of the transformer increases as Z_L becomes closer to Z_0 (a less mismatched load).

The above results are strictly valid only for TEM lines. When non-TEM lines (such as waveguides) are used, the propagation constant is no longer a linear function of frequency, and the wave impedance will be frequency dependent. These factors serve to complicate the general behavior of quarter-wave transformers for non-TEM lines, but in practice the bandwidth of the transformer is often small enough that these complications do not substantially affect the result. Another factor ignored in the above analysis is the effect of reactances associated with discontinuities when there is a step change in the dimensions of a transmission line. This can often be compensated by making a small adjustment in the length of the matching section.

Figure 5.15 shows a plot of the reflection coefficient magnitude versus normalized frequency for various mismatched loads. Note the trend of increased bandwidth for smaller load mismatches.

EXAMPLE 5.6   QUARTER-WAVE TRANSFORMER BANDWIDTH

Design a single-section quarter-wave matching transformer to match a 20 Ω load to a 50 Ω transmission line at $f_0 = 2.4$ GHz. Determine the percent bandwidth for which the SWR ≤ 1.3.

Solution
From (5.31), the characteristic impedance of the matching section is

$$Z_1 = \sqrt{Z_0 Z_L} = \sqrt{(50)(20)} = 31.62 \ \Omega,$$

and the length of the matching section is $\lambda/4$ at 2 GHz (the physical length depends on the dielectric constant of the line). An SWR of 1.3 corresponds to a reflection coefficient magnitude of

$$\Gamma_m = \frac{\text{SWR} - 1}{\text{SWR} + 1} = \frac{1.3 - 1}{1.3 + 1} = 0.13.$$

The fractional bandwidth is computed from (5.39) as

$$\frac{\Delta f}{f_0} = 2 - \frac{4}{\pi} \cos^{-1}\left[\frac{\Gamma_m}{\sqrt{1 - \Gamma_m^2}} \frac{2\sqrt{Z_0 Z_L}}{|Z_L - Z_0|} \right]$$

$$= 2 - \frac{4}{\pi} \cos^{-1}\left[\frac{0.13}{\sqrt{1 - (0.13)^2}} \frac{2\sqrt{(50)(20)}}{|50 - 20|} \right]$$

$$= 0.36 \text{ or } 36\%.$$

5.5 THE THEORY OF SMALL REFLECTIONS

The quarter-wave transformer provides a simple means of matching any real load impedance to any transmission line impedance. For applications requiring more bandwidth than a single quarter-wave section can provide, *multisection transformers* can be used. The design of such transformers is the subject of the next two sections, but prior to that material we need to derive some approximate results for the total reflection coefficient caused by the partial reflections from several small discontinuities. This topic is generally referred to as the *theory of small reflections* [1].

Single-Section Transformer

We will derive an approximate expression for the overall reflection coefficient, Γ, for the single-section matching transformer shown in Figure 5.16. The partial reflection and transmission coefficients are

$$\Gamma_1 = \frac{Z_2 - Z_1}{Z_2 + Z_1}, \tag{5.40}$$

$$\Gamma_2 = -\Gamma_1, \tag{5.41}$$

$$\Gamma_3 = \frac{Z_L - Z_2}{Z_L + Z_2}, \tag{5.42}$$

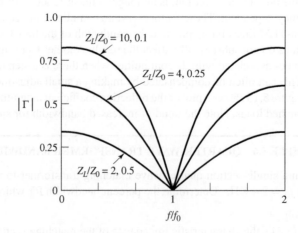

FIGURE 5.15 | Reflection coefficient magnitude versus frequency for a single-section quarter-wave matching transformer with various load mismatches.

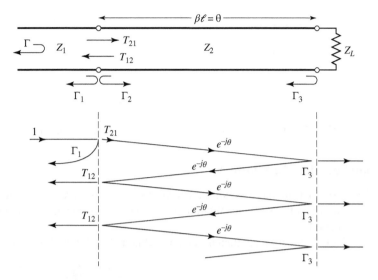

FIGURE 5.16 | Partial reflections and transmissions on a single-section matching transformer.

$$T_{21} = 1 + \Gamma_1 = \frac{2Z_2}{Z_1 + Z_2}, \tag{5.43}$$

$$T_{12} = 1 + \Gamma_2 = \frac{2Z_1}{Z_1 + Z_2}. \tag{5.44}$$

We can compute the total reflection, Γ, seen by the feed line using either the impedance method, or the multiple reflection method, as discussed in Section 5.4. For our present purpose the latter technique is preferred, so we express the total reflection as an infinite sum of partial reflections and transmissions as follows:

$$\Gamma = \Gamma_1 + T_{12}T_{21}\Gamma_3 e^{-2j\theta} + T_{12}T_{21}\Gamma_3^2\Gamma_2 e^{-4j\theta} + \cdots$$

$$= \Gamma_1 + T_{12}T_{21}\Gamma_3 e^{-2j\theta} \sum_{n=0}^{\infty} \Gamma_2^n \Gamma_3^n e^{-2jn\theta}. \tag{5.45}$$

The summation of the geometric series

$$\sum_{n=0}^{\infty} x^n = \frac{1}{1-x} \quad \text{for } |x| < 1$$

allows us to express (5.45) in closed form as

$$\Gamma = \Gamma_1 + \frac{T_{12}T_{21}\Gamma_3 e^{-2j\theta}}{1 - \Gamma_2\Gamma_3 e^{-2j\theta}}. \tag{5.46}$$

From (5.41), (5.43), and (5.44), we use $\Gamma_2 = -\Gamma_1, T_{21} = 1 + \Gamma_1$, and $T_{12} = 1 - \Gamma_1$ in (5.46) to give

$$\Gamma = \frac{\Gamma_1 + \Gamma_3 e^{-2j\theta}}{1 + \Gamma_1\Gamma_3 e^{-2j\theta}}. \tag{5.47}$$

If the discontinuities between the impedances Z_1, Z_2 and Z_2, Z_L are small, then $|\Gamma_1\Gamma_3| = 1$, so we can approximate (5.47) as

$$\Gamma \simeq \Gamma_1 + \Gamma_3 e^{-2j\theta}. \tag{5.48}$$

This result expresses the intuitive idea that the total reflection is dominated by the reflection from the initial discontinuity between Z_1 and Z_2 (Γ_1), and the first reflection from the discontinuity between Z_2 and Z_L ($\Gamma_3 e^{-2j\theta}$). The $e^{-2j\theta}$ term accounts for the phase delay when the incident wave travels up and down the line. The accuracy of this approximation is illustrated in Problem 5.11.

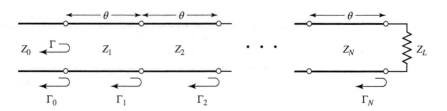

FIGURE 5.17 | Partial reflection coefficients for a multisection matching transformer.

Multisection Transformer

Now consider the multisection transformer shown in Figure 5.17, which consists of N equal-length (*commensurate*) sections of transmission lines. We will derive an approximate expression for the total reflection coefficient Γ.

Partial reflection coefficients can be defined at each junction, as follows:

$$\Gamma_0 = \frac{Z_1 - Z_0}{Z_1 + Z_0}, \tag{5.49a}$$

$$\Gamma_n = \frac{Z_{n+1} - Z_n}{Z_{n+1} + Z_n}, \tag{5.49b}$$

$$\Gamma_N = \frac{Z_L - Z_N}{Z_L + Z_N}. \tag{5.49c}$$

We also assume that all Z_n increase or decrease monotonically across the transformer and that Z_L is real. This implies that all Γ_n will be real and of the same sign ($\Gamma_n > 0$ if $Z_L > Z_0$; $\Gamma_n < 0$ if $Z_L < Z_0$). Using the results of the previous section allows us to approximate the overall reflection coefficient as

$$\Gamma(\theta) = \Gamma_0 + \Gamma_1 e^{-2j\theta} + \Gamma_2 e^{-4j\theta} + \cdots + \Gamma_N e^{-2jN\theta}. \tag{5.50}$$

Further assume that the transformer can be made symmetrical, so that $\Gamma_0 = \Gamma_N$, $\Gamma_1 = \Gamma_{N-1}$, $\Gamma_2 = \Gamma_{N-2}$, and so on. (Note that this does *not* imply that the Z_n are symmetrical.) Then (5.50) can be written as

$$\Gamma(\theta) = e^{-jN\theta} \left\{ \Gamma_0 [e^{jN\theta} + e^{-jN\theta}] + \Gamma_1 [e^{j(N-2)\theta} + e^{-j(N-2)\theta}] + \cdots \right\}. \tag{5.51}$$

If N is odd, the last term is $\Gamma_{(N-1)/2}(e^{j\theta} + e^{-j\theta})$, while if N is even, the last term is $\Gamma_{N/2}$. Equation (5.51) is seen to be of the form of a finite Fourier cosine series in θ, which can be written as

$$\Gamma(\theta) = 2e^{-jN\theta} \left[\Gamma_0 \cos N\theta + \Gamma_1 \cos(N-2)\theta + \cdots + \Gamma_n \cos(N-2n)\theta \right.$$

$$\left. + \cdots + \frac{1}{2}\Gamma_{N/2} \right] \quad \text{for } N \text{ even,} \tag{5.52a}$$

$$\Gamma(\theta) = 2e^{-jN\theta} [\Gamma_0 \cos N\theta + \Gamma_1 \cos(N-2)\theta + \cdots + \Gamma_n \cos(N-2n)\theta$$

$$+ \cdots + \Gamma_{(N-1)/2} \cos\theta] \quad \text{for } N \text{ odd.} \tag{5.52b}$$

The importance of these results lies in the fact that we can synthesize any desired reflection coefficient response as a function of frequency (θ) by properly choosing the Γ_n and using enough sections (N). This should be clear from the realization that a Fourier series can approximate an arbitrary smooth function if enough terms are used. In the next two sections we will show how to use this theory to design multisection transformers for two of the most commonly used passband responses: the *binomial* (maximally flat) response, and the *Chebyshev* (equal-ripple) response.

5.6   BINOMIAL MULTISECTION MATCHING TRANSFORMERS

The passband response (the frequency band where a good impedance match is achieved) of a binomial matching transformer is optimum in the sense that, for a given number of sections, the response is as flat as possible near the design frequency. This type of response, which is also known as *maximally flat*, is determined for an N-section transformer by setting the first $N-1$ derivatives of $|\Gamma(\theta)|$ to zero at the center frequency, f_0. Such a response can be obtained with a reflection coefficient of the following form:

$$\Gamma(\theta) = A(1 + e^{-2j\theta})^N. \tag{5.53}$$

Then the reflection coefficient magnitude is

$$|\Gamma(\theta)| = |A||e^{-j\theta}|^N|e^{j\theta} + e^{-j\theta}|^N$$
$$= 2^N|A||\cos\theta|^N \tag{5.54}$$

Note that $|\Gamma(\theta)| = 0$ for $\theta = \pi/2$, and that $d^n|\Gamma(\theta)|/d\theta^n = 0$ at $\theta = \pi/2$ for $n = 1, 2, \ldots, N-1$. ($\theta = \pi/2$ corresponds to the center frequency, f_0, for which $\ell = \lambda/4$ and $\theta = \beta\ell = \pi/2$.)

We can determine the constant A by letting $f \to 0$. Then $\theta = \beta\ell = 0$, and (5.53) reduces to

$$\Gamma(0) = 2^N A = \frac{Z_L - Z_0}{Z_L + Z_0},$$

since for $f = 0$ all sections are of zero electrical length. The constant A can then be written as

$$A = 2^{-N}\frac{Z_L - Z_0}{Z_L + Z_0}. \tag{5.55}$$

Next we expand $\Gamma(\theta)$ in (5.53) according to the binomial expansion:

$$\Gamma(\theta) = A(1 + e^{-2j\theta})^N = A\sum_{n=0}^{N} C_n^N e^{-2jn\theta}, \tag{5.56}$$

where

$$C_n^N = \frac{N!}{(N-n)!n!} \tag{5.57}$$

are the binomial coefficients. Note that $C_n^N = C_{N-n}^N$, $C_0^N = 1$, and $C_1^N = N = C_{N-1}^N$. The key step is now to equate the desired passband response, given by (5.56), to the actual response as given (approximately) by (5.50):

$$\Gamma(\theta) = A\sum_{n=0}^{N} C_n^N e^{-2jn\theta} = \Gamma_0 + \Gamma_1 e^{-2j\theta} + \Gamma_2 e^{-4j\theta} + \cdots + \Gamma_N e^{-2jN\theta}.$$

This shows that the Γ_n must be chosen as

$$\Gamma_n = AC_n^N. \tag{5.58}$$

where A is given by (5.55) and C_n^N is a binomial coefficient.

At this point, the characteristic impedances, Z_n, can be found via (5.49), but a simpler solution can be obtained using the following approximation [1]. Because we assumed that the Γ_n are small, we can write

$$\Gamma_n = \frac{Z_{n+1} - Z_n}{Z_{n+1} + Z_n} \simeq \frac{1}{2}\ln\frac{Z_{n+1}}{Z_n},$$

since $\ln x \simeq 2(x-1)/(x+1)$ for x close to unity. Then, using (5.58) and (5.55) gives

$$\ln\frac{Z_{n+1}}{Z_n} \simeq 2\Gamma_n = 2AC_n^N = 2(2^{-N})\frac{Z_L - Z_0}{Z_L + Z_0}C_n^N \simeq 2^{-N}C_n^N\ln\frac{Z_L}{Z_0}, \tag{5.59}$$

which can be used to find Z_{n+1}, starting with $n = 0$. This technique has the advantage of ensuring self-consistency, in that Z_{N+1} computed from (5.59) will be equal to Z_L, as it should.

Exact design results, including the effect of multiple reflections in each section, can be found by using the transmission line equations for each section and numerically solving for the characteristic impedances [2]. The results of such calculations are listed in Table 5.1, which gives the exact line impedances for $N = 2$-, 3-, 4-, 5-, and 6-section binomial matching transformers for various ratios of load impedance, Z_L, to feed

TABLE 5.1 | Binomial Transformer Design

	N = 2		N = 3			N = 4			
Z_L/Z_0	Z_1/Z_0	Z_2/Z_0	Z_1/Z_0	Z_2/Z_0	Z_3/Z_0	Z_1/Z_0	Z_2/Z_0	Z_3/Z_0	Z_4/Z_0
1.0	1.0000	1.0000	1.0000	1.0000	1.0000	1.0000	1.0000	1.0000	1.0000
1.5	1.1067	1.3554	1.0520	1.2247	1.4259	1.0257	1.1351	1.3215	1.4624
2.0	1.1892	1.6818	1.0907	1.4142	1.8337	1.0444	1.2421	1.6102	1.9150
3.0	1.3161	2.2795	1.1479	1.7321	2.6135	1.0718	1.4105	2.1269	2.7990
4.0	1.4142	2.8285	1.1907	2.0000	3.3594	1.0919	1.5442	2.5903	3.6633
6.0	1.5651	3.8336	1.2544	2.4495	4.7832	1.1215	1.7553	3.4182	5.3500
8.0	1.6818	4.7568	1.3022	2.8284	6.1434	1.1436	1.9232	4.1597	6.9955
10.0	1.7783	5.6233	1.3409	3.1623	7.4577	1.1613	2.0651	4.8424	8.6110

	N = 5					N = 6					
Z_L/Z_0	Z_1/Z_0	Z_2/Z_0	Z_3/Z_0	Z_4/Z_0	Z_5/Z_0	Z_1/Z_0	Z_2/Z_0	Z_3/Z_0	Z_4/Z_0	Z_5/Z_0	Z_6/Z_0
1.0	1.0000	1.0000	1.0000	1.0000	1.0000	1.0000	1.0000	1.0000	1.0000	1.0000	1.0000
1.5	1.0128	1.0790	1.2247	1.3902	1.4810	1.0064	1.0454	1.1496	1.3048	1.4349	1.4905
2.0	1.0220	1.1391	1.4142	1.7558	1.9569	1.0110	1.0790	1.2693	1.5757	1.8536	1.9782
3.0	1.0354	1.2300	1.7321	2.4390	2.8974	1.0176	1.1288	1.4599	2.0549	2.6577	2.9481
4.0	1.0452	1.2995	2.0000	3.0781	3.8270	1.0225	1.1661	1.6129	2.4800	3.4302	3.9120
6.0	1.0596	1.4055	2.4495	4.2689	5.6625	1.0296	1.2219	1.8573	3.2305	4.9104	5.8275
8.0	1.0703	1.4870	2.8284	5.3800	7.4745	1.0349	1.2640	2.0539	3.8950	6.3291	7.7302
10.0	1.0789	1.5541	3.1623	6.4346	9.2687	1.0392	1.2982	2.2215	4.5015	7.7030	9.6228

line impedance, Z_0. The table gives results only for $Z_L/Z_0 > 1$; if $Z_L/Z_0 < 1$, the results for Z_0/Z_L should be used but with Z_1 starting at the load end. This is because the solution is symmetric about $Z_L/Z_0 = 1$; the same transformer that matches Z_L to Z_0 can be reversed and used to match Z_0 to Z_L. More extensive tables can be found in reference [2].

The bandwidth of the binomial transformer can be evaluated as follows. As in Section 5.4, let Γ_m be the maximum value of reflection coefficient that can be tolerated over the passband. Then from (5.54),

$$\Gamma_m = 2^N |A| \cos^N \theta_m,$$

where $\theta_m < \pi/2$ is the lower edge of the passband, as shown in Figure 5.14. Thus,

$$\theta_m = \cos^{-1}\left[\frac{1}{2}\left(\frac{\Gamma_m}{|A|}\right)^{1/N}\right], \tag{5.60}$$

and using (5.39) gives the fractional bandwidth as

$$\frac{\Delta f}{f_0} = \frac{2(f_0 - f_m)}{f_0} = 2 - \frac{4\theta_m}{\pi}$$

$$= 2 - \frac{4}{\pi}\cos^{-1}\left[\frac{1}{2}\left(\frac{\Gamma_m}{|A|}\right)^{1/N}\right]. \tag{5.61}$$

EXAMPLE 5.7 BINOMIAL TRANSFORMER DESIGN

Design a three-section binomial transformer to match a 50 Ω load to a 100 Ω line and calculate the bandwidth for $\Gamma_m = 0.05$. Plot the reflection coefficient magnitude versus normalized frequency for the exact designs using 1, 2, 3, 4, and 5 sections.

Solution

For $N = 3, Z_L = 50 \ \Omega$, and $Z_0 = 100 \ \Omega$ we have, from (5.55) and (5.59),

$$A = 2^{-N} \frac{Z_L - Z_0}{Z_L + Z_0} \simeq \frac{1}{2^{N+1}} \ln \frac{Z_L}{Z_0} = -0.0433.$$

From (5.61) the bandwidth is

$$\frac{\Delta f}{f_0} = 2 - \frac{4}{\pi} \cos^{-1} \left[\frac{1}{2} \left(\frac{\Gamma_m}{|A|} \right)^{1/N} \right]$$

$$= 2 - \frac{4}{\pi} \cos^{-1} \left[\frac{1}{2} \left(\frac{0.05}{0.0433} \right)^{1/3} \right] = 0.70, \text{ or } 70\%.$$

The necessary binomial coefficients are

$$C_0^3 = \frac{3!}{3!0!} = 1 \quad C_1^3 = \frac{3!}{2!1!} = 3 \quad C_2^3 = \frac{3!}{1!2!} = 3.$$

Using (5.59) gives the required characteristic impedances as

$$n = 0: \quad \ln Z_1 = \ln Z_0 + 2^{-N} C_0^3 \ln \frac{Z_L}{Z_0} = \ln 100 + 2^{-3}(1) \ln \frac{50}{100} = 4.518,$$

$$Z_1 = 91.7 \ \Omega;$$

$$n = 1: \quad \ln Z_2 = \ln Z_1 + 2^{-N} C_1^3 \ln \frac{Z_L}{Z_0} = \ln 91.7 + 2^{-3}(3) \ln \frac{50}{100} = 4.26,$$

$$Z_2 = 70.7 \ \Omega;$$

$$n = 2: \quad \ln Z_3 = \ln Z_2 + 2^{-N} C_2^3 \ln \frac{Z_L}{Z_0} = \ln 70.7 + 2^{-3}(3) \ln \frac{50}{100} = 4.00,$$

$$Z_3 = 54.5 \ \Omega.$$

To use the data in Table 5.1 we reverse the source and load impedances and consider the problem of matching a 100 Ω load to a 50 Ω line. Then $Z_L/Z_0 = 2.0$, and we obtain the exact characteristic impedances as $Z_1 = 91.7 \ \Omega, Z_2 = 70.7 \ \Omega$, and $Z_3 = 54.5 \ \Omega$, which agree with the approximate results to three significant digits. Figure 5.18 shows the reflection coefficient magnitude versus frequency for exact designs using $N = 1, 2, 3, 4$, and 5 sections. Observe that greater bandwidth is obtained for transformers using more sections.

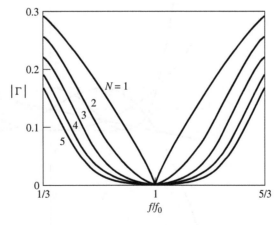

FIGURE 5.18 | Reflection coefficient magnitude versus frequency for multisection binomial matching transformers of Example 5.7. $Z_L = 50 \ \Omega$ and $Z_0 = 100 \ \Omega$.

5.7 CHEBYSHEV MULTISECTION MATCHING TRANSFORMERS

In contrast with the binomial transformer, the multisection *Chebyshev matching transformer* optimizes bandwidth at the expense of passband ripple. Compromising on the flatness of the passband response leads to a bandwidth that is substantially better than that of the binomial transformer for a given number of sections. The Chebyshev transformer is designed by equating $\Gamma(\theta)$ to a Chebyshev polynomial, which has the optimum characteristics needed for this type of transformer. We will first discuss the properties of Chebyshev polynomials and then derive a design procedure for Chebyshev matching transformers using the small-reflection theory of Section 5.5.

Chebyshev Polynomials

The nth-order Chebyshev polynomial is a polynomial of degree n, denoted by $T_n(x)$. The first four Chebyshev polynomials are

$$T_1(x) = x, \tag{5.62a}$$

$$T_2(x) = 2x^2 - 1, \tag{5.62b}$$

$$T_3(x) = 4x^3 - 3x, \tag{5.62c}$$

$$T_4(x) = 8x^4 - 8x^2 + 1. \tag{5.62d}$$

Higher order polynomials can be found using the following recurrence formula:

$$T_n(x) = 2xT_{n-1}(x) - T_{n-2}(x). \tag{5.63}$$

The first four Chebyshev polynomials are plotted in Figure 5.19, from which the following very useful properties of Chebyshev polynomials can be noted:

- For $-1 \leq x \leq 1$, $|T_n(x)| \leq 1$. In this range the Chebyshev polynomials oscillate between ± 1. This is the *equal-ripple* property, and this region will be mapped to the passband of the matching transformer.
- For $|x| > 1$, $|T_n(x)| > 1$. This region will map to the frequency range outside the passband.
- For $|x| > 1$, the $|T_n(x)|$ increases faster with x as n increases.

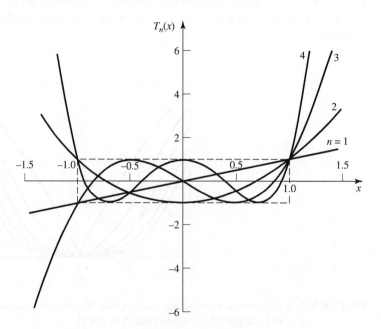

FIGURE 5.19 | The first four Chebyshev polynomials, $T_n(x)$.

Now let $x = \cos\theta$ for $|x| < 1$. Then it can be shown that the Chebyshev polynomials can be expressed as

$$T_n(\cos\theta) = \cos n\theta,$$

or more generally as

$$T_n(x) = \cos(n\cos^{-1}x) \qquad \text{for } |x| < 1, \tag{5.64a}$$

$$T_n(x) = \cosh(n\cosh^{-1}x) \quad \text{for } x > 1. \tag{5.64b}$$

We desire equal ripple for the passband response of the transformer, so it is necessary to map θ_m to $x = 1$ and $\pi - \theta_m$ to $x = -1$, where θ_m and $\pi - \theta_m$ are the lower and upper edges of the passband, respectively, as shown in Figure 5.14. This can be accomplished by replacing $\cos\theta$ in (5.64a) with $\cos\theta/\cos\theta_m$:

$$T_n\left(\frac{\cos\theta}{\cos\theta_m}\right) = T_n(\sec\theta_m\cos\theta) = \cos n\left[\cos^{-1}\left(\frac{\cos\theta}{\cos\theta_m}\right)\right]. \tag{5.65}$$

Then $|\sec\theta_m\cos\theta| \le 1$ for $\theta_m < \theta < \pi - \theta_m$, so $|T_n(\sec\theta_m\cos\theta)| \le 1$ over this same range.

Because $\cos^n\theta$ can be expanded into a sum of terms of the form $\cos(n - 2m)\theta$, the Chebyshev polynomials of (5.62) can be rewritten in the following useful form:

$$T_1(\sec\theta_m\cos\theta) = \sec\theta_m\cos\theta, \tag{5.66a}$$

$$T_2(\sec\theta_m\cos\theta) = \sec^2\theta_m(1 + \cos 2\theta) - 1, \tag{5.66b}$$

$$T_3(\sec\theta_m\cos\theta) = \sec^3\theta_m(\cos 3\theta + 3\cos\theta) - 3\sec\theta_m\cos\theta, \tag{5.66c}$$

$$T_4(\sec\theta_m\cos\theta) = \sec^4\theta_m(\cos 4\theta + 4\cos 2\theta + 3)$$
$$-4\sec^2\theta_m(\cos 2\theta + 1) + 1. \tag{5.66d}$$

These results can be used to design matching transformers with up to four sections, and will also be used in later chapters for the design of directional couplers and filters.

Design of Chebyshev Transformers

We can now synthesize a Chebyshev equal-ripple passband by making $\Gamma(\theta)$ proportional to $T_N(\sec\theta_m\cos\theta)$, where N is the number of sections in the transformer. Thus, using (5.52), we have

$$\Gamma(\theta) = 2e^{-jN\theta}[\Gamma_0\cos N\theta + \Gamma_1\cos(N - 2)\theta + \cdots + \Gamma_n\cos(N - 2n)\theta + \cdots]$$
$$= Ae^{-jN\theta}T_N(\sec\theta_m\cos\theta), \tag{5.67}$$

where the last term in the series of (5.67) is $(1/2)\Gamma_{N/2}$ for N even and $\Gamma_{(N-1)/2}\cos\theta$ for N odd. As in the binomial transformer case, we can find the constant A by letting $\theta = 0$, corresponding to zero frequency. Thus,

$$\Gamma(0) = \frac{Z_L - Z_0}{Z_L + Z_0} = AT_N(\sec\theta_m),$$

so we have

$$A = \frac{Z_L - Z_0}{Z_L + Z_0}\frac{1}{T_N(\sec\theta_m)}. \tag{5.68}$$

If the maximum allowable reflection coefficient magnitude in the passband is Γ_m, then from (5.67) $\Gamma_m = |A|$ since the maximum value of $T_n(\sec\theta_m\cos\theta)$ in the passband is unity. Then (5.68) gives

$$T_N(\sec\theta_m) = \frac{1}{\Gamma_m}\left|\frac{Z_L - Z_0}{Z_L + Z_0}\right|,$$

which, after using (5.66b) and the approximations introduced in Section 5.6, allows us to determine θ_m as

$$\sec \theta_m = \cosh \left[\frac{1}{N} \cosh^{-1} \left(\frac{1}{\Gamma_m} \left| \frac{Z_L - Z_0}{Z_L + Z_0} \right| \right) \right]$$
$$\simeq \cosh \left[\frac{1}{N} \cosh^{-1} \left(\left| \frac{\ln Z_L/Z_0}{2\Gamma_m} \right| \right) \right]. \tag{5.69}$$

Once θ_m is known, the fractional bandwidth can be calculated from (5.39) as

$$\frac{\Delta f}{f_0} = 2 - \frac{4\theta_m}{\pi}. \tag{5.70}$$

From (5.67), the Γ_n can be determined using the results of (5.66) to expand $T_N(\sec \theta_m \cos \theta)$ and equating similar terms of the form $\cos(N - 2n)\theta$. The characteristic impedances Z_n can be found from (5.49), although, as in the case of the binomial transformer, accuracy can be improved and self-consistency can be achieved by using the approximation that

$$\Gamma_n \simeq \frac{1}{2} \ln \frac{Z_{n+1}}{Z_n}.$$

This procedure will be illustrated in Example 5.8.

The above results are approximate because of the reliance on small-reflection theory but are general enough to design transformers with an arbitrary ripple level, Γ_m. Table 5.2 gives exact results [2] for a few specific values of Γ_m for $N = 2$, 3, and 4 sections; more extensive tables can be found in reference [2].

TABLE 5.2 | Chebyshev Transformer Design

	$N = 2$				$N = 3$					
	$\Gamma_m = 0.05$		$\Gamma_m = 0.20$		$\Gamma_m = 0.05$			$\Gamma_m = 0.20$		
Z_L/Z_0	Z_1/Z_0	Z_2/Z_0	Z_1/Z_0	Z_2/Z_0	Z_1/Z_0	Z_2/Z_0	Z_3/Z_0	Z_1/Z_0	Z_2/Z_0	Z_3/Z_0
1.0	1.0000	1.0000	1.0000	1.0000	1.0000	1.0000	1.0000	1.0000	1.0000	1.0000
1.5	1.1347	1.3219	1.2247	1.2247	1.1029	1.2247	1.3601	1.2247	1.2247	1.2247
2.0	1.2193	1.6402	1.3161	1.5197	1.1475	1.4142	1.7429	1.2855	1.4142	1.5558
3.0	1.3494	2.2232	1.4565	2.0598	1.2171	1.7321	2.4649	1.3743	1.7321	2.1829
4.0	1.4500	2.7585	1.5651	2.5558	1.2662	2.0000	3.1591	1.4333	2.0000	2.7908
6.0	1.6047	3.7389	1.7321	3.4641	1.3383	2.4495	4.4833	1.5193	2.4495	3.9492
8.0	1.7244	4.6393	1.8612	4.2983	1.3944	2.8284	5.7372	1.5766	2.8284	5.0742
10.0	1.8233	5.4845	1.9680	5.0813	1.4385	3.1623	6.9517	1.6415	3.1623	6.0920

	$N = 4$							
	$\Gamma_m = 0.05$				$\Gamma_m = 0.20$			
Z_L/Z_0	Z_1/Z_0	Z_2/Z_0	Z_3/Z_0	Z_4/Z_0	Z_1/Z_0	Z_2/Z_0	Z_3/Z_0	Z_4/Z_0
1.0	1.0000	1.0000	1.0000	1.0000	1.0000	1.0000	1.0000	1.0000
1.5	1.0892	1.1742	1.2775	1.3772	1.2247	1.2247	1.2247	1.2247
2.0	1.1201	1.2979	1.5409	1.7855	1.2727	1.3634	1.4669	1.5715
3.0	1.1586	1.4876	2.0167	2.5893	1.4879	1.5819	1.8965	2.0163
4.0	1.1906	1.6414	2.4369	3.3597	1.3692	1.7490	2.2870	2.9214
6.0	1.2290	1.8773	3.1961	4.8820	1.4415	2.0231	2.9657	4.1623
8.0	1.2583	2.0657	3.8728	6.3578	1.4914	2.2428	3.5670	5.3641
10.0	1.2832	2.2268	4.4907	7.7930	1.5163	2.4210	4.1305	6.5950

EXAMPLE 5.8   CHEBYSHEV TRANSFORMER DESIGN

Design a three-section Chebyshev transformer to match a 100 Ω load to a 50 Ω line with $\Gamma_m = 0.05$, using the above theory. Plot the reflection coefficient magnitude versus normalized frequency for exact designs using 1, 2, 3, and 4 sections.

Solution
From (5.67) with $N = 3$,

$$\Gamma(\theta) = 2e^{-j3\theta} \left(\Gamma_0 \cos 3\theta + \Gamma_1 \cos \theta \right) = Ae^{-j3\theta} T_3(\sec \theta_m \cos \theta).$$

Then $A = \Gamma_m = 0.05$, and from (5.69),

$$\sec \theta_m = \cosh \left[\frac{1}{N} \cosh^{-1} \left(\frac{\ln Z_L/Z_0}{2\Gamma_m} \right) \right]$$

$$= \cosh \left[\frac{1}{3} \cosh^{-1} \left(\frac{\ln(100/50)}{2(0.05)} \right) \right]$$

$$= 1.408,$$

so $\theta_m = 44.7°$.

Using (5.66c) for T_3 gives

$$2 \left(\Gamma_0 \cos 3\theta + \Gamma_1 \cos \theta \right) = A \sec^3 \theta_m (\cos 3\theta + 3 \cos \theta) - 3A \sec \theta_m \cos \theta.$$

Equating similar terms in $\cos n\theta$ gives the following results:

$$\cos 3\theta: \quad 2\Gamma_0 = A \sec^3 \theta_m,$$
$$\Gamma_0 = 0.0698.$$
$$\cos \theta: \quad 2\Gamma_1 = 3A(\sec^3 \theta_m - \sec \theta_m),$$
$$\Gamma_1 = 0.1037.$$

From symmetry we also have that

$$\Gamma_3 = \Gamma_0 = 0.0698,$$
$$\Gamma_2 = \Gamma_1 = 0.1037.$$

Then the characteristic impedances are:

$$n = 0: \quad \ln Z_1 = \ln Z_0 + 2\Gamma_0$$
$$= \ln 50 + 2(0.0698) = 4.051$$
$$Z_1 = 57.5 \ \Omega;$$
$$n = 1: \quad \ln Z_2 = \ln Z_1 + 2\Gamma_1$$
$$= \ln 57.5 + 2(0.1037) = 4.259$$
$$Z_2 = 70.7 \ \Omega;$$
$$n = 2: \quad \ln Z_3 = \ln Z_2 + 2\Gamma_2$$
$$= \ln 70.7 + 2(0.1037) = 4.466$$
$$Z_3 = 87.0 \ \Omega.$$

These values can be compared to the exact values from Table 5.2 of $Z_1 = 57.37 \ \Omega$, $Z_2 = 70.71 \ \Omega$, and $Z_3 = 87.15 \ \Omega$. The bandwidth, from (5.70), is

$$\frac{\Delta f}{f_0} = 2 - \frac{4\theta_m}{\pi} = 2 - 4 \left(\frac{44.7°}{180°} \right) = 1.01,$$

or 101%. This is significantly greater than the bandwidth of the binomial transformer of Example 5.7 (70%), which involved the same impedance mismatch. The trade-off, of course, is a nonzero ripple in the passband of the Chebyshev transformer.

Figure 5.20 shows reflection coefficient magnitudes versus frequency for the exact designs from Table 5.2 for $N = 1, 2, 3,$ and 4 sections.

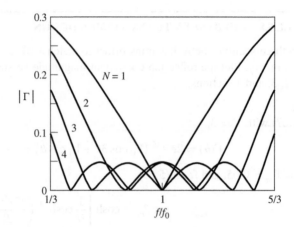

FIGURE 5.20 | Reflection coefficient magnitude versus frequency for the multisection matching transformers of Example 5.8.

5.8   TAPERED LINES

In the preceding sections we discussed how an arbitrary real load impedance could be matched to a line over a desired bandwidth by using multisection matching transformers. As the number N of discrete transformer sections increases, the step changes in characteristic impedance between the sections become smaller, and the transformer geometry approaches a continuously tapered line. In practice, of course, a matching transformer must be of finite length—often no more than a few sections long. This suggests that, instead of discrete sections, the transformer can be continuously tapered, as shown in Figure 5.21a. Different passband characteristics can be obtained by using different types of taper.

In this section we will derive an approximate theory, again based on the theory of small reflections, to predict the reflection coefficient response as a function of the impedance taper versus position, $Z(z)$. We will apply these results to a few common types of impedance tapers.

Consider the continuously tapered line of Figure 5.21a as being made up of a number of incremental sections of length Δz, with an impedance change $\Delta Z(z)$ from one section to the next, as shown in Figure 5.21b. The incremental reflection coefficient from the impedance step at z is given by

$$\Delta\Gamma = \frac{(Z + \Delta Z) - Z}{(Z + \Delta Z) + Z} \simeq \frac{\Delta Z}{2Z}. \tag{5.71}$$

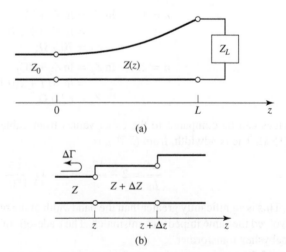

FIGURE 5.21 | A tapered transmission line matching section and the model for an incremental length of tapered line. (a) The tapered transmission line matching section. (b) Model for an incremental step change in impedance of the tapered line.

In the limit as $\Delta z \to 0$ we have an exact differential:

$$d\Gamma = \frac{dZ}{2Z} = \frac{1}{2}\frac{d(\ln Z/Z_0)}{dz}dz,\tag{5.72}$$

since

$$\frac{d(\ln f(z))}{dz} = \frac{1}{f}\frac{df(z)}{dz}.$$

By using the theory of small reflections, we can find the total reflection coefficient at $z = 0$ by summing all the partial reflections with their appropriate phase shifts:

$$\Gamma(\theta) = \frac{1}{2}\int_{z=0}^{L}e^{-2j\beta z}\frac{d}{dz}\ln\left(\frac{Z}{Z_0}\right)dz,\tag{5.73}$$

where $\theta = 2\beta\ell$. If $Z(z)$ is known, $\Gamma(\theta)$ can be found as a function of frequency. Alternatively, if $\Gamma(\theta)$ is specified, then in principle $Z(z)$ can be found by inversion. This latter procedure is difficult, and is generally avoided in practice; the reader is referred to references [1] and [4] for further discussion of this topic. Here we will consider three special cases of $Z(z)$ impedance tapers, and evaluate the resulting responses.

Exponential Taper

Consider first an *exponential taper*, where

$$Z(z) = Z_0 e^{az}\quad\text{for } 0 < z < L,\tag{5.74}$$

as indicated in Figure 5.22a. At $z = 0$, $Z(0) = Z_0$, as desired. At $z = L$ we wish to have $Z(L) = Z_L = Z_0 e^{aL}$, which determines the constant a as

$$a = \frac{1}{L}\ln\left(\frac{Z_L}{Z_0}\right).\tag{5.75}$$

We find $\Gamma(\theta)$ by using (5.74) and (5.75) in (5.73):

$$\begin{aligned}\Gamma &= \frac{1}{2}\int_0^L e^{-2j\beta z}\frac{d}{dz}(\ln e^{az})dz\\ &= \frac{\ln Z_L/Z_0}{2L}\int_0^L e^{-2j\beta z}dz\\ &= \frac{\ln Z_L/Z_0}{2}e^{-j\beta L}\frac{\sin\beta L}{\beta L}.\end{aligned}\tag{5.76}$$

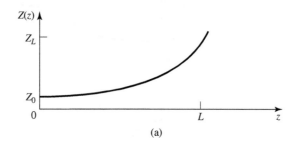

(a)

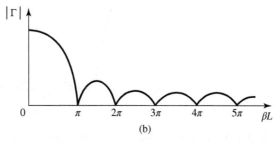

(b)

FIGURE 5.22 | A matching section with an exponential impedance taper. (a) Variation of impedance. (b) Resulting reflection coefficient magnitude response.

Observe that this derivation assumes that β, the propagation constant of the tapered line, is not a function of z—an assumption generally valid only for TEM lines.

The magnitude of the reflection coefficient in (5.76) is sketched in Figure 5.22b; note that the peaks in $|\Gamma|$ decrease with increasing length, as one might expect, and that the length should be greater than $\lambda/2$ $(\beta L > \pi)$ to minimize the mismatch at low frequencies.

Triangular Taper

Next consider a *triangular taper* for $d \ln \left(Z/Z_0 \right) /dz$, that is,

$$Z(z) = \begin{cases} Z_0 e^{2(z/L)^2 \ln Z_L/Z_0} & \text{for } 0 \leq z \leq L/2 \\ Z_0 e^{(4z/L - 2z^2/L^2 - 1) \ln Z_L/Z_0} & \text{for } L/2 \leq z \leq L, \end{cases} \tag{5.77}$$

so that the derivative is triangular in form:

$$\frac{d(\ln Z/Z_0)}{dz} = \begin{cases} 4z/L^2 \ln Z_L/Z_0 & \text{for } 0 \leq z \leq L/2 \\ (4/L - 4z/L^2) \ln Z_L/Z_0 & \text{for } L/2 \leq z \leq L. \end{cases} \tag{5.78}$$

$Z(z)$ is plotted in Figure 5.23a. Evaluating Γ from (5.73) gives

$$\Gamma(\theta) = \frac{1}{2} e^{-j\beta L} \ln \left(\frac{Z_L}{Z_0} \right) \left[\frac{\sin(\beta L/2)}{\beta L/2} \right]^2 . \tag{5.79}$$

The magnitude of this result is sketched in Figure 5.23b. Note that, for $\beta L > 2\pi$, the peaks of the triangular taper are lower than the corresponding peaks of the exponential case. However, the first null for the triangular taper occurs at $\beta L = 2\pi$, whereas for the exponential taper it occurs at $\beta L = \pi$.

Klopfenstein Taper

Considering the fact that there is an infinite number of possibilities for choosing an impedance matching taper, it is logical to ask if there is a design that is "best." For a given taper length (greater than some critical

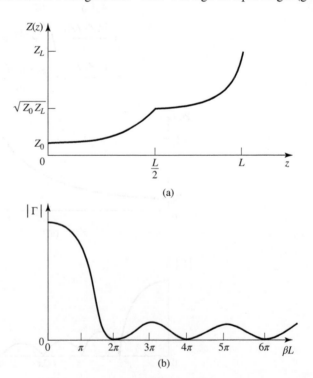

(a)

(b)

FIGURE 5.23 | A matching section with a triangular taper for $d(\ln Z/Z_0)/dz$. (a) Variation of impedance. (b) Resulting reflection coefficient magnitude response.

value), the *Klopfenstein impedance taper* [4, 5] has been shown to be optimum in the sense that the reflection coefficient is minimum over the passband. Alternatively, for a maximum reflection coefficient specification in the passband, the Klopfenstein taper yields the shortest matching section.

The Klopfenstein taper is derived from a stepped Chebyshev transformer as the number of sections increases to infinity, and is analogous to the Taylor distribution of antenna array theory. We will not present the details of this derivation, which can be found in references [1] and [4]; only the necessary results for the design of Klopfenstein tapers are given in what follows.

The logarithm of the characteristic impedance variation for the Klopfenstein taper is given by

$$\ln Z(z) = \frac{1}{2} \ln Z_0 Z_L + \frac{\Gamma_0}{\cosh A} A^2 \phi(2z/L - 1, A) \quad \text{for } 0 \le z \le L, \tag{5.80}$$

where the function $\phi(x, A)$ is defined as

$$\phi(x, A) = -\phi(-x, A) = \int_0^x \frac{I_1(A\sqrt{1 - y^2})}{A\sqrt{1 - y^2}} dy \quad \text{for } |x| \le 1, \tag{5.81}$$

where $I_1(x)$ is the modified Bessel function. The function of (5.81) has the following special values:

$$\phi(0, A) = 0$$
$$\phi(x, 0) = \frac{x}{2}$$
$$\phi(1, A) = \frac{\cosh A - 1}{A^2},$$

but otherwise (5.81) must be calculated numerically. A simple and efficient method for doing this is available [6].

The resulting reflection coefficient is given by

$$\Gamma(\theta) = \Gamma_0 e^{-j\beta L} \frac{\cos \sqrt{(\beta L)^2 - A^2}}{\cosh A} \quad \text{for } \beta L > A. \tag{5.82}$$

If $\beta L < A$, the $\cos \sqrt{(\beta L)^2 - A^2}$ term becomes $\cosh \sqrt{A^2 - (\beta L)^2}$.

In (5.80) and (5.82), Γ_0 is the reflection coefficient at zero frequency, given as

$$\Gamma_0 = \frac{Z_L - Z_0}{Z_L + Z_0} \simeq \frac{1}{2} \ln \left(\frac{Z_L}{Z_0} \right). \tag{5.83}$$

The passband is defined as $\beta L \ge A$, and so the maximum ripple in the passband is

$$\Gamma_m = \frac{\Gamma_0}{\cosh A} \tag{5.84}$$

because $\Gamma(\theta)$ oscillates between $\pm \Gamma_0 / \cosh A$ for $\beta L > A$.

It is interesting to note that the impedance taper of (5.80) has steps at $z = 0$ and L (the ends of the tapered section) and so does not smoothly join the source and load impedances. A typical Klopfenstein impedance taper and its response are given in the following example.

EXAMPLE 5.9   DESIGN OF TAPERED MATCHING SECTIONS

Design a triangular taper, an exponential taper, and a Klopfenstein taper (with $\Gamma_m = 0.02$) to match a 50 Ω load to a 100 Ω line. Plot the impedance variations and resulting reflection coefficient magnitudes versus βL.

Solution

Triangular taper: From (5.77) the impedance variation is

$$Z(z) = Z_0 \begin{cases} e^{2(z/L)^2 \ln Z_L/Z_0} & \text{for } 0 \le z \le L/2 \\ e^{(4z/L - 2z^2/L^2 - 1) \ln Z_L/Z_0} & \text{for } L/2 \le z \le L, \end{cases}$$

with $Z_0 = 100\ \Omega$ and $Z_L = 50\ \Omega$. The resulting reflection coefficient response is given by (5.79):

$$|\Gamma(\theta)| = \frac{1}{2} \ln\left(\frac{Z_L}{Z_0}\right) \left[\frac{\sin(\beta L/2)}{\beta L/2}\right]^2.$$

Exponential taper: From (5.74) the impedance variation is

$$Z(z) = Z_0 e^{az} \quad \text{for } 0 < z < L,$$

with $a = (1/L)\ln Z_L/Z_0 = 0.693/L$. The reflection coefficient response is, from (5.76),

$$|\Gamma(\theta)| = \frac{1}{2} \ln\left(\frac{Z_L}{Z_0}\right) \frac{\sin \beta L}{\beta L}.$$

Klopfenstein taper: Using (5.83) gives Γ_0 as

$$\Gamma_0 = \frac{1}{2}\ln\left(\frac{Z_L}{Z_0}\right) = 0.346,$$

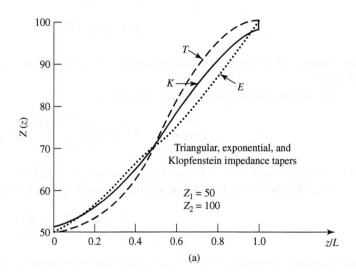

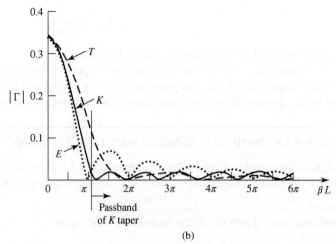

FIGURE 5.24 | Solution to Example 5.9. (a) Impedance variations for the triangular, exponential, and Klopfenstein tapers. (b) Resulting reflection coefficient magnitude versus frequency for the tapers of (a).

and (5.84) gives A as

$$A = \cosh^{-1}\left(\frac{\Gamma_0}{\Gamma_m}\right) = \cosh^{-1}\left(\frac{0.346}{0.02}\right) = 3.543.$$

The impedance taper must be numerically evaluated from (5.80). The reflection coefficient magnitude is given by (5.82):

$$|\Gamma(\theta)| = \Gamma_0 \frac{\cos\sqrt{(\beta L)^2 - A^2}}{\cosh A}.$$

The passband for the Klopfenstein taper is defined as $\beta L > A = 3.543 = 1.13\pi$.

Figure 5.24 shows the impedance variations (vs. z/L), and the resulting reflection coefficient magnitude (vs. βL) for the three types of tapers. The Klopfenstein taper gives the desired response of $|\Gamma| \leq \Gamma_m = 0.02$ for $\beta L \geq 1.13\pi$, which is smaller than the corresponding lengths of either the triangular or the exponential taper transformer. Also note that, like the stepped-Chebyshev matching transformer, the response of the Klopfenstein taper has equal-ripple lobes versus frequency in its passband.

REFERENCES

[1] R. E. Collin, *Foundations for Microwave Engineering*, 2nd edition, McGraw-Hill, New York, 1992.

[2] G. L. Matthaei, L. Young, and E. M. T. Jones, *Microwave Filters, Impedance-Matching Networks, and Coupling Structures,* Artech House Books, Dedham, MA, 1980.

[3] P. Bhartia and I. J. Bahl, *Millimeter Wave Engineering and Applications*, Wiley Interscience, New York, 1984.

[4] R. E. Collin, "The Optimum Tapered Transmission Line Matching Section," *Proceedings of the IRE,* vol. 44, pp. 539–548, April 1956.

[5] R. W. Klopfenstein, "A Transmission Line Taper of Improved Design," *Proceedings of the IRE,* vol. 44, pp. 31–15, January 1956.

[6] M. A. Grossberg, "Extremely Rapid Computation of the Klopfenstein Impedance Taper," *Proceedings of the IEEE,* vol. 56, pp. 1629–1630, September 1968.

PROBLEMS

5.1 Design two lossless *L*-section matching circuits to match each of the following loads to a 100 Ω generator at 2.4 GHz. (a) $Z_L = 20 - j90$ Ω and (b) $Z_L = 150 - j200$ Ω.

5.2 We have seen that the matching of an arbitrary load impedance requires a network with at least two degrees of freedom. Determine the types of load impedances/admittances that can be matched with the two single-element networks shown below.

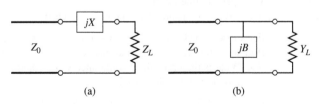

(a)   (b)

5.3 A load impedance $Z_L = 100 + j60$ Ω is to be matched to a 50 Ω line using a single shunt-stub tuner. Find two designs using short-circuited stubs.

5.4 A load impedance $Z_L = 95 + j75$ Ω is to be matched to a 75 Ω line using a single series-stub tuner. Find two designs using open-circuited stubs.

5.5 In the circuit shown below a load $Z_L = 180 + j100$ Ω is to be matched to a 45 Ω line, using a length ℓ of lossless transmission line of characteristic impedance Z_1. Find ℓ and Z_1. Determine, in general, what type of load impedances can be matched using such a circuit.

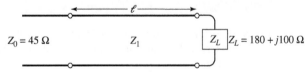

5.6 An open-circuit tuning stub is to be made from a lossy transmission line with an attenuation constant α. What is the maximum value of normalized reactance that can be obtained with this stub? What is the maximum value of normalized reactance that can be obtained with a shorted stub of the same type of transmission line? Assume $\alpha\ell$ is small.

5.7 Design a double-stub tuner using open-circuited stubs with a $\lambda/8$ spacing to match a load admittance $Z_L = (90 + j60)$ Ω to a 75 Ω line.

5.8 Derive the design equations for a double-stub tuner using two series stubs spaced a distance d apart. Assume the load impedance is $Z_L = R_L + jX_L$.

5.9 Consider matching a load $Z_L = 200\ \Omega$ to a $100\ \Omega$ line, using single shunt-stub, single series stub, and double shunt-stub tuners, with short-circuited stubs. Which tuner will give the best bandwidth? Justify your answer by calculating the reflection coefficient for all six solutions at $1.1f_0$, where f_0 is the match frequency, or use CAD to plot the reflection coefficient versus frequency.

5.10 Design a single-section quarter-wave matching transformer to match a $300\ \Omega$ load to a $100\ \Omega$ line. What is the percent bandwidth of this transformer, for SWR ≤ 2? If the design frequency is 5 GHz, sketch the layout of a microstrip circuit, including dimensions, to implement this matching transformer. Assume the substrate is 1.6 mm thick, with a relative permittivity of 2.2.

5.11 Consider the quarter-wave transformer of Figure 5.16 with $Z_1 = 100\ \Omega$, $Z_2 = 150\ \Omega$, and $Z_L = 225\ \Omega$. Evaluate the worst-case percent error in computing $|\Gamma|$ from the approximate expression (5.48), compared to the exact result.

5.12 A waveguide load with an equivalent TE_{10} wave impedance of $377\ \Omega$ must be matched to an air-filled X-band rectangular guide at 10 GHz. A quarter-wave matching transformer is to be used, and is to consist of a section of guide filled with dielectric. Find the required dielectric constant and physical length of the matching section. What restrictions on the load impedance apply to this technique?

5.13 A four-section binomial matching transformer is to be used to match a $15\ \Omega$ load to a $50\ \Omega$ line at a center frequency of 1.575 GHz. (a) Design the matching transformer, and compute the bandwidth for $\Gamma_m = 0.05$. Use CAD to plot the input reflection coefficient versus frequency. (b) Lay out the microstrip implementation of this circuit on an FR4 substrate having $\epsilon_r = 4.4$, $d = 1.5$ mm, and $\tan \delta = 0.02$, with copper conductors 2 mil thick. Use CAD to plot the insertion loss versus frequency.

5.14 Derive the exact characteristic impedance for a two-section binomial matching transformer for a normalized load impedance $Z_L/Z_0 = 1.5$. Check your results with Table 5.1.

5.15 Calculate and plot the percent bandwidth for an $N = 1$-, 2-, and 4-section binomial matching transformer versus $Z_L/Z_0 = 1.5$ to 6 for $\Gamma_m = 0.2$.

5.16 Design a four-section Chebyshev matching transformer to match a $50\ \Omega$ line to a $25\ \Omega$ load. The maximum permissible SWR over the passband is 1.5. What is the resulting bandwidth? Use the approximate theory developed in the text, as opposed to the tables. Use CAD to plot the input SWR versus frequency.

5.17 Derive the exact characteristic impedances for a two-section Chebyshev matching transformer for a normalized load impedance $Z_L/Z_0 = 1.5$. Check your results with Table 5.2 for $\Gamma_m = 0.05$.

5.18 A load of $Z_L/Z_0 = 1.5$ is to be matched to a feed line using a multisection transformer, and it is desired to have a passband response with $|\Gamma(\theta)| = A(0.1 + \cos^2 \theta)$ for $0 \leq \theta \leq \pi$. Use the approximate theory for multisection transformers to design a two-section transformer.

5.19 A tapered matching section has $d \ln (Z/Z_0)/dz = A \sin \pi z/L$. Find the constant A so that $Z(0) = Z_0$ and $Z(L) = Z_L$. Compute Γ, and plot $|\Gamma|$ versus βL.

5.20 Design an exponentially tapered matching transformer to match a $100\ \Omega$ load to a $50\ \Omega$ line. Plot $|\Gamma|$ versus βL, and find the length of the matching section (at the center frequency) required to obtain $|\Gamma| \leq 0.05$ over a 100% bandwidth. How many sections would be required if a Chebyshev matching transformer were used to achieve the same specifications?

5.21 An ultra wideband (UWB) radio transmitter, operating from 3.75 to 11.25 GHz, drives a parallel RC load with $R = 80\ \Omega$ and $C = 0.79$ pF. What is the best return loss that can be obtained with an optimum matching network?

5.22 Design an L-section matching circuit to match $Z_L = 20 + j50\ \Omega$ to $50\ \Omega$. Calculate X_L and X_C at 3 GHz by clearly showing your work on the Smith chart.

5.23 A $75\ \Omega$ transmission line is connected to load impedance $Z_L = 90 + j60\ \Omega$. Find the position and length of a short-circuited series stub required to match the line.

MULTIPLE CHOICE QUESTIONS

5.1 If SWR $= 1.4$ with a wavelength of 5 cm and the distance between load and first minima is 1.5 cm, then the reflection coefficient is

(a) $0.135 + j0.098$ (c) $0.167 + j0.098$
(b) 0.167 (d) none of the above

5.2 If the characteristic impedance of a transmission line is $50\ \Omega$ and the reflection coefficient is $0.015 + j0.25$, then the load impedance is

(a) $4.538 + j2.4\ \Omega$ (c) $2.4 + j4.54\ \Omega$
(b) $45.38 + j24.21\ \Omega$ (d) none of the above

5.3 For waveguides and coaxial lines, _____ is more preferred.

(a) slotted section
(b) short-circuited stub

(c) open-circuited stub
(d) waveguides cannot be impedance matched

5.4 For a load impedance of $Z_L = 85 - j60$, design two single-stub shunt tuning networks and match this load to a $50\ \Omega$ line. What is the normalized admittance value obtained so as to plot it on the Smith chart?

(a) $1.5 + j1.5$ (c) $0.39 + j0.28$
(b) $0.28 + j0.39$ (d) $0.39 - j0.28$

5.5 If the normalized admittance at a point on a transmission line to be matched is $1.2 + j1.57$. Then the normalized susceptance of the stub used for shunt stub matching is

(a) $1.2\ \Omega$ (c) $-1.57\ \Omega$
(b) $1.57\ \Omega$ (d) $-1.2\ \Omega$

5.6 Single-stub tuning has a major disadvantage as

(a) it requires a variable length of line between the load and the stub

(b) it involves two variable parameters

(c) it involves complex calculation

(d) none of the above

5.7 The simplest method for reducing the forbidden range of impedances is by

(a) increasing the distance between the stubs

(b) reducing the distance between the stubs

(c) increasing the length of the stubs

(d) reducing the length of the stubs

5.8 The standard spacing used between the stubs are

(a) $\lambda/8, 3\lambda/8$ (c) $0, \lambda/2$

(b) $\lambda/4, \lambda/8$ (d) none of the above

5.9 For impedance matching if a single-section quarter-wave transformer is used at some frequency, then the length of the matching line

(a) is a constant

(b) is $\lambda/2$ for other frequencies

(c) is different at different frequencies

(d) none of the above

5.10 If a load of 35 Ω has to be matched to a transmission line of characteristic impedance of 75 Ω, then the characteristic impedance of the matching section of the transmission line is

(a) 50 Ω (c) 51.23 Ω

(b) 60 Ω (d) 102.47 Ω

5.11 The term $e^{-2j\theta}$ in the expression for total reflection in a single-section quarter-wave transformer impedance matching network $\Gamma = \Gamma_1 + \Gamma_3 e^{-2j\theta}$ signifies

(a) narrowing bandwidth   (c) phase delay

(b) frequency change   (d) none of the above

5.12 The third and the first reflection coefficients of a matched line are 0.01 and 0.2, respectively. If the quarter-wave transformer is used for impedance matching, then the total reflection coefficient is

(a) 0.15   (c) 0.91

(b) 0.10   (d) 0.19

5.13 For a binomial multisection transformer, the condition $|\Gamma(\theta)| = 0$ for $\theta = \pi/2$ corresponds to

(a) lower cutoff frequency

(b) upper cutoff frequency

(c) center frequency

(d) none of the above

5.14 For a binomial multisection transformer, the reflection coefficient magnitude is

(a) $2^N|A||\cos(\theta)|^N$ (c) $2^N|\cos(\theta)|^N$

(b) $2^N|A|$ (d) none of the above

5.15 In a binomial matching transformer, the reflection coefficient Γ_N in terms of successive impedances Z_n and Z_{n+1} when multisection transformers are used is given by

(a) Z_n/Z_{n+1} (c) $0.5\ln(Z_n/Z_{n+1})$

(b) $\ln(Z_{n+1}/Z_n)$ (d) $0.5\ln(Z_{n+1}/Z_n)$

5.16 A three-section binomial transformer is used to match a 75 Ω transmission line to a 50 Ω transmission line. Then the value of the constant A for this design is

(a) 0.0253   (c) 0.051

(b) −0.0253   (d) −0.051

5.17 The advantage of Chebyshev matching transformers over binomial transformers is

(a) low power losses

(b) higher gain

(c) higher roll-off in the characteristic curve

(d) higher bandwidth

5.18 In a three-section Chebyshev matching network, if $Z_3 = 75\ \Omega$ and $Z_2 = 60\ \Omega$, then the reflection coefficient Γ_2 is

(a) 0.112   (c) 0.335

(b) 0.223   (d) 0.669

5.19 The value of constant A for an exponentially tapered line of length 6 cm with load impedance being 75 Ω and characteristic impedance of the line being 50 Ω is

(a) 0.0675   (c) 0.75

(b) 0.675   (d) 0.5

5.20 The incremental reflection coefficient for a continually tapered line is

(a) $\dfrac{2Z}{\Delta Z}$ (c) $\dfrac{\Delta Z_0}{2Z_0}$

(b) $\dfrac{\Delta Z}{2Z}$ (d) none of the above

ANSWER KEY

5.1 (a)	**5.5** (c)	**5.9** (c)	**5.13** (c)	**5.17** (d)
5.2 (b)	**5.6** (a)	**5.10** (c)	**5.14** (a)	**5.18** (a)
5.3 (b)	**5.7** (b)	**5.11** (c)	**5.15** (d)	**5.19** (a)
5.4 (c)	**5.8** (a)	**5.12** (d)	**5.16** (b)	**5.20** (b)

Microwave Resonators

After completing this chapter, you will be able to

- Learn the concepts of series and parallel resonant circuits
- Develop and design transmission line resonators

- Describe rectangular and circular waveguide cavity resonators
- Understand the concept of excitation of different types of resonators

$\mathbf{M}$icrowave resonators are used in a variety of applications, including filters, oscillators, frequency meters, and tuned amplifiers. Because the operation of microwave resonators is very similar to that of lumped-element resonators of circuit theory, we will begin by reviewing the basic characteristics of series and parallel RLC resonant circuits. We will then discuss various implementations of resonators at microwave frequencies using distributed elements such as transmission lines, rectangular and circular waveguides, and dielectric cavities. We will also discuss the excitation of resonators using apertures and current sheets.

6.1 SERIES AND PARALLEL RESONANT CIRCUITS

At frequencies near resonance, a microwave resonator can usually be modeled by either a series or parallel RLC lumped-element equivalent circuit, and so we will now review some of the basic properties of these circuits.

Series Resonant Circuit

A series RLC resonant circuit is shown in Figure 6.1a. The input impedance is

$$Z_{in} = R + j\omega L - j\frac{1}{\omega C}, \tag{6.1}$$

and the complex power delivered to the resonator is

$$P_{in} = \frac{1}{2}VI^* = \frac{1}{2}Z_{in}|I|^2 = \frac{1}{2}Z_{in}\left|\frac{V}{Z_{in}}\right|^2$$

$$= \frac{1}{2}|I|^2\left(R + j\omega L - j\frac{1}{\omega C}\right). \tag{6.2}$$

The power dissipated by the resistor R is

$$P_{loss} = \frac{1}{2}|I|^2 R, \tag{6.3a}$$

the average magnetic energy stored in the inductor L is

$$W_m = \frac{1}{4}|I|^2 L, \tag{6.3b}$$

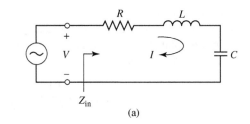

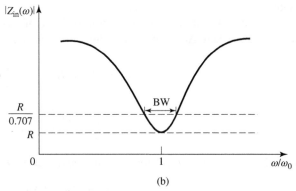

(a)

(b)

FIGURE 6.1 | A series *RLC* resonator and its response. (a) A series *RLC* resonator circuit. (b) Input impedance magnitude versus frequency.

and the average electric energy stored in the capacitor C is

$$W_e = \frac{1}{4}|V_c|^2 C = \frac{1}{4}|I|^2 \frac{1}{\omega^2 C}, \tag{6.3c}$$

where V_c is the voltage across the capacitor. Then the complex power of (6.2) can be rewritten as

$$P_{in} = P_{loss} + 2j\omega(W_m - W_e), \tag{6.4}$$

and the input impedance of (6.1) can be rewritten as

$$Z_{in} = \frac{2P_{in}}{|I|^2} = \frac{P_{loss} + 2j\omega(W_m - W_e)}{\frac{1}{2}|I|^2}. \tag{6.5}$$

Resonance occurs when the average stored magnetic and electric energies are equal, or $W_m = W_e$. Then from (6.5) and (6.3a), the input impedance at resonance is

$$Z_{in} = \frac{P_{loss}}{\frac{1}{2}|I|^2} = R,$$

which is purely real. From (6.3b,c), $W_m = W_e$ implies that the resonant frequency, ω_0, can be defined as

$$\omega_0 = \frac{1}{\sqrt{LC}}. \tag{6.6}$$

Another important parameter of a resonant circuit is its Q, or *quality factor*, which is defined as

$$Q = \omega \frac{\text{average energy stored}}{\text{energy loss/second}}$$

$$= \omega \frac{W_m + W_e}{P_{loss}}. \tag{6.7}$$

Thus Q is a measure of the loss of a resonant circuit—lower loss implies a higher Q. Resonator losses may be due to conductor loss, dielectric loss, or radiation loss, and are represented by the resistance, R, of the equivalent circuit. An external connecting network may introduce additional loss. Each of these loss mechanisms will have the effect of lowering the Q. The Q of the resonator itself, disregarding external loading effects, is called the *unloaded Q*, denoted as Q_0.

For the series resonant circuit of Figure 6.1a, the unloaded Q can be evaluated from (6.7), using (6.3) and the fact that $W_m = W_e$ at resonance, to give

$$Q_0 = \omega_0 \frac{2W_m}{P_{\text{loss}}} = \frac{\omega_0 L}{R} = \frac{1}{\omega_0 RC}, \tag{6.8}$$

which shows that Q increases as R decreases.

Next, consider the behavior of the input impedance of this resonator near its resonant frequency [1]. Let $\omega = \omega_0 + \Delta\omega$, where $\Delta\omega$ is small. The input impedance can then be rewritten from (6.1) as

$$Z_{\text{in}} = R + j\omega L \left(1 - \frac{1}{\omega^2 LC} \right)$$

$$= R + j\omega L \left(\frac{\omega^2 - \omega_0^2}{\omega^2} \right),$$

since $\omega_0^2 = 1/LC$. Now $\omega^2 - \omega_0^2 = (\omega - \omega_0)(\omega + \omega_0) = \Delta\omega(2\omega - \Delta\omega) \simeq 2\omega\Delta\omega$ for small $\Delta\omega$. Thus,

$$Z_{\text{in}} \simeq R + j2L\Delta\omega$$

$$\simeq R + j\frac{2RQ_0\Delta\omega}{\omega_0}. \tag{6.9}$$

This form will be useful for identifying equivalent circuits with distributed element resonators.

Alternatively, a resonator with loss can be modeled as a lossless resonator whose resonant frequency, ω_0, has been replaced with a complex effective resonant frequency:

$$\omega_0 \leftarrow \omega_0 \left(1 + \frac{j}{2Q_0} \right). \tag{6.10}$$

This can be seen by considering the input impedance of a series resonator with no loss, as given by (6.9) with $R = 0$:

$$Z_{\text{in}} = j2L(\omega - \omega_0).$$

Then substituting the complex frequency of (6.10) for ω_0 gives

$$Z_{\text{in}} = j2L \left(\omega - \omega_0 - j\frac{\omega_0}{2Q_0} \right)$$

$$= \frac{\omega_0 L}{Q_0} + j2L(\omega - \omega_0) = R + j2L\Delta\omega,$$

which is identical to (6.9). This is a useful procedure because for most practical resonators the loss is very small, so the Q can be found using the perturbation method, beginning with the solution for the lossless case. Then the effect of loss can be added to the input impedance by replacing ω_0 with the complex resonant frequency given in (6.10).

Finally, consider the *half-power fractional bandwidth* of the resonator. Figure 6.1b shows the variation of the magnitude of the input impedance versus frequency. When the frequency is such that $|Z_{\text{in}}|^2 = 2R^2$, then by (6.2) the average (real) power delivered to the circuit is one-half that delivered at resonance. If BW is the fractional bandwidth, then $\Delta\omega/\omega_0 = \text{BW}/2$ at the upper band edge. Using (6.9) gives

$$|R + jRQ_0(\text{BW})|^2 = 2R^2,$$

or

$$\text{BW} = \frac{1}{Q_0}. \tag{6.11}$$

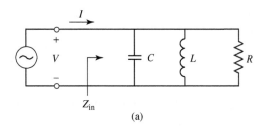

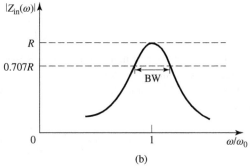

FIGURE 6.2 | A parallel *RLC* resonator and its response. (a) A parallel *RLC* circuit. (b) Input impedance magnitude versus frequency.

Parallel Resonant Circuit

The parallel *RLC* resonant circuit, shown in Figure 6.2a, is the dual of the series *RLC* circuit. The input impedance is

$$Z_{in} = \left(\frac{1}{R} + \frac{1}{j\omega L} + j\omega C \right)^{-1}, \tag{6.12}$$

and the complex power delivered to the resonator is

$$P_{in} = \frac{1}{2} V I^* = \frac{1}{2} Z_{in} |I|^2 = \frac{1}{2} |V|^2 \frac{1}{Z_{in}^*}$$

$$= \frac{1}{2} |V|^2 \left(\frac{1}{R} + \frac{j}{\omega L} - j\omega C \right). \tag{6.13}$$

The power dissipated by the resistor, *R*, is

$$P_{loss} = \frac{1}{2} \frac{|V|^2}{R}, \tag{6.14a}$$

the average electric energy stored in the capacitor, *C*, is

$$W_e = \frac{1}{4} |V|^2 C, \tag{6.14b}$$

and the average magnetic energy stored in the inductor, *L*, is

$$W_m = \frac{1}{4} |I_L|^2 L = \frac{1}{4} |V|^2 \frac{1}{\omega^2 L}, \tag{6.14c}$$

where I_L is the current through the inductor. Then the complex power of (6.13) can be rewritten as

$$P_{in} = P_{loss} + 2j\omega (W_m - W_e), \tag{6.15}$$

which is identical to (6.4). Similarly, the input impedance can be expressed as

$$Z_{in} = \frac{2P_{in}}{|I|^2} = \frac{P_{loss} + 2j\omega (W_m - W_e)}{\frac{1}{2} |I|^2}, \tag{6.16}$$

which is identical to (6.5).

As in the series case, resonance occurs when $W_m = W_e$. Then from (6.16) and (6.14a) the input impedance at resonance is

$$Z_{in} = \frac{P_{loss}}{\frac{1}{2}|I|^2} = R,$$

which is a purely real impedance. From (6.14b) and (6.14c), $W_m = W_e$ implies that the resonant frequency, ω_0, can be defined as

$$\omega_0 = \frac{1}{\sqrt{LC}}, \tag{6.17}$$

which is identical to the series resonant circuit case. Resonance in the case of a parallel RLC circuit is sometimes referred to as an *antiresonance*.

From the definition of (6.7), and the results in (6.14), the unloaded Q of the parallel resonant circuit can be expressed as

$$Q_0 = \omega_0 \frac{2W_m}{P_{loss}} = \frac{R}{\omega_0 L} = \omega_0 RC, \tag{6.18}$$

since $W_m = W_e$ at resonance. This result shows that the Q of the parallel resonant circuit increases as R increases.

Near resonance, the input impedance of (6.12) can be simplified using the series expansion result that

$$\frac{1}{1+x} \simeq 1 - x + \cdots .$$

Again letting $\omega = \omega_0 + \Delta\omega$, where $\Delta\omega$ is small, allows (6.12) to be rewritten as [1]

$$Z_{in} \simeq \left(\frac{1}{R} + \frac{1 - \Delta\omega/\omega_0}{j\omega_0 L} + j\omega_0 C + j\Delta\omega C \right)^{-1}$$

$$\simeq \left(\frac{1}{R} + j\frac{\Delta\omega}{\omega_0^2 L} + j\Delta\omega C \right)^{-1}$$

$$\simeq \left(\frac{1}{R} + 2j\Delta\omega C \right)^{-1}$$

$$\simeq \frac{R}{1 + 2j\Delta\omega RC} = \frac{R}{1 + 2jQ_0\Delta\omega/\omega_0}, \tag{6.19}$$

since $\omega_0^2 = 1/LC$. When $R = \infty$ (6.19) reduces to

$$Z_{in} = \frac{1}{j2C(\omega - \omega_0)}.$$

As in the series resonator case, the effect of loss can be accounted for by replacing ω_0 in this expression with a complex effective resonant frequency:

$$\omega_0 \leftarrow \omega_0 \left(1 + \frac{j}{2Q_0} \right). \tag{6.20}$$

Figure 6.2b shows the behavior of the magnitude of the input impedance versus frequency. The half-power bandwidth edges occur at frequencies ($\Delta\omega/\omega_0 = BW/2$) such that

$$|Z_{in}|^2 = \frac{R^2}{2},$$

which, from (6.19), implies that

$$BW = \frac{1}{Q_0}, \tag{6.21}$$

as in the series resonance case.

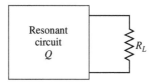

FIGURE 6.3 | A resonant circuit connected to an external load, R_L.

Loaded and Unloaded Q

The unloaded Q, Q_0, defined in the preceding sections is a characteristic of the resonator itself, in the absence of any loading effects caused by external circuitry. In practice, however, a resonator is invariably coupled to other circuitry, which will have the effect of lowering the overall, or *loaded Q*, Q_L, of the circuit. Figure 6.3 depicts a resonator coupled to an external load resistor, R_L. If the resonator is a series *RLC* circuit, the load resistor R_L adds in series with R, so the effective resistance in (6.8) is $R + R_L$. If the resonator is a parallel *RLC* circuit, the load resistor R_L combines in parallel with R, so the effective resistance in (6.18) is $R R_L/(R + R_L)$. If we define an *external Q*, Q_e, as

$$Q_e = \begin{cases} \dfrac{\omega_0 L}{R_L} & \text{for series circuits} \\[2ex] \dfrac{R_L}{\omega_0 L} & \text{for parallel circuits,} \end{cases} \tag{6.22}$$

then the loaded Q can be expressed as

$$\frac{1}{Q_L} = \frac{1}{Q_e} + \frac{1}{Q_0}. \tag{6.23}$$

Table 6.1 summarizes the above results for series and parallel resonant circuits.

6.2 TRANSMISSION LINE RESONATORS

As we have seen, ideal lumped circuit elements are often unattainable at microwave frequencies, so distributed elements are frequently used. In this section we will study the use of transmission line sections with various lengths and terminations (usually open- or short-circuited) to form resonators. Because we are interested in the Q of these resonators, we must consider transmission lines with losses.

TABLE 6.1 | **Summary of Results for Series and Parallel Resonators**

Quantity	Series Resonator	Parallel Resonator
Input impedance/admittance	$Z_{\text{in}} = R + j\omega L - j\dfrac{1}{\omega C}$	$Y_{\text{in}} = \dfrac{1}{R} + j\omega C - j\dfrac{1}{\omega L}$
	$\simeq R + j\dfrac{2 R Q_0 \Delta\omega}{\omega_0}$	$\simeq \dfrac{1}{R} + j\dfrac{2 Q_0 \Delta\omega}{R\omega_0}$
Power loss	$P_{\text{loss}} = \dfrac{1}{2}|I|^2 R$	$P_{\text{loss}} = \dfrac{1}{2}\dfrac{|V|^2}{R}$
Stored magnetic energy	$W_m = \dfrac{1}{4}|I|^2 L$	$W_m = \dfrac{1}{4}|V|^2 \dfrac{1}{\omega^2 L}$
Stored electric energy	$W_e = \dfrac{1}{4}|I|^2 \dfrac{1}{\omega^2 C}$	$W_e = \dfrac{1}{4}|V|^2 C$
Resonant frequency	$\omega_0 = \dfrac{1}{\sqrt{LC}}$	$\omega_0 = \dfrac{1}{\sqrt{LC}}$
Unloaded Q	$Q_0 = \dfrac{\omega_0 L}{R} = \dfrac{1}{\omega_0 R C}$	$Q_0 = \omega_0 R C = \dfrac{R}{\omega_0 L}$
External Q	$Q_e = \dfrac{\omega_0 L}{R_L}$	$Q_e = \dfrac{R_L}{\omega_0 L}$

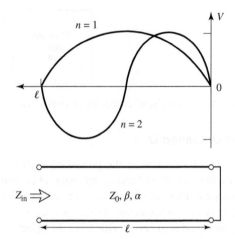

FIGURE 6.4 | A short-circuited length of lossy transmission line, and the voltage distributions for $n = 1$ ($\ell = \lambda/2$) and $n = 2$ ($\ell = \lambda$) resonators.

Short-Circuited $\lambda/2$ Line

A length of lossy transmission line, short circuited at one end, is shown in Figure 6.4. The line has a characteristic impedance, Z_0, propagation constant, β, and attenuation constant, α. At the resonant frequency $\omega = \omega_0$, the length of the line is $\ell = \lambda/2$. From (2.85), the input impedance is

$$Z_{\text{in}} = Z_0 \tanh(\alpha + j\beta)\ell.$$

Using an identity for the hyperbolic tangent gives

$$Z_{\text{in}} = Z_0 \frac{\tanh \alpha\ell + j\tan \beta\ell}{1 + j\tan \beta\ell \tanh \alpha\ell}. \tag{6.24}$$

Observe that $Z_{\text{in}} = jZ_0 \tan \beta\ell$ if $\alpha = 0$ (a lossless line).

In practice it is usually desirable to use a low-loss transmission line, so we assume that $\alpha\ell \ll 1$, and then $\tanh \alpha\ell \simeq \alpha\ell$. Again let $\omega = \omega_0 + \Delta\omega$, where $\Delta\omega$ is small. Then, assuming a TEM line, we have

$$\beta\ell = \frac{\omega\ell}{v_p} = \frac{\omega_0\ell}{v_p} + \frac{\Delta\omega\ell}{v_p},$$

where v_p is the phase velocity of the transmission line. Because $\ell = \lambda/2 = \pi v_p/\omega_0$ for $\omega = \omega_0$, we have

$$\beta\ell = \pi + \frac{\Delta\omega\pi}{\omega_0},$$

and then

$$\tan \beta\ell = \tan\left(\pi + \frac{\Delta\omega\pi}{\omega_0}\right) = \tan\frac{\Delta\omega\pi}{\omega_0} \simeq \frac{\Delta\omega\pi}{\omega_0}.$$

Using these results in (6.24) gives

$$Z_{\text{in}} \simeq Z_0 \frac{\alpha\ell + j(\Delta\omega\pi/\omega_0)}{1 + j(\Delta\omega\pi/\omega_0)\alpha\ell} \simeq Z_0\left(\alpha\ell + j\frac{\Delta\omega\pi}{\omega_0}\right), \tag{6.25}$$

since $\Delta\omega\alpha\ell/\omega_0 \ll 1$.

Equation (6.25) is of the form

$$Z_{\text{in}} = R + 2jL\Delta\omega,$$

which is the input impedance of a series *RLC* resonant circuit, as given by (6.9). We can identify the resistance of the equivalent circuit as

$$R = Z_0\alpha\ell, \tag{6.26a}$$

and the inductance of the equivalent circuit as

$$L = \frac{Z_0 \pi}{2\omega_0}. \tag{6.26b}$$

The capacitance of the equivalent circuit can be found from (6.6) as

$$C = \frac{1}{\omega_0^2 L}. \tag{6.26c}$$

The resonator of Figure 6.4 thus resonates for $\Delta\omega = 0$ ($\ell = \lambda/2$), and its input impedance at resonance is $Z_{\text{in}} = R = Z_0 \alpha\ell$. Resonance also occurs for $\ell = n\lambda/2$, $n = 1, 2, 3, \ldots$. The voltage distributions for the $n = 1$ and $n = 2$ resonant modes are shown in Figure 6.4. The unloaded Q of this resonator can be found from (6.8) and (6.26) as

$$Q_0 = \frac{\omega_0 L}{R} = \frac{\pi}{2\alpha\ell} = \frac{\beta}{2\alpha}, \tag{6.27}$$

since $\beta\ell = \pi$ at the first resonance. This result shows that the Q decreases as the attenuation of the line increases, as expected.

EXAMPLE 6.1 Q OF HALF-WAVE COAXIAL LINE RESONATORS

A $\lambda/2$ resonator is made from a piece of copper coaxial line having an inner conductor radius of 2 mm and an outer conductor radius of 4 mm. If the resonant frequency is 6 GHz, compare the unloaded Q of an air-filled coaxial line resonator to that of a Teflon-filled coaxial line resonator.

Solution
We first compute the attenuation of the coaxial line, using the results of Examples 2.8 or 2.9. From Appendix G, the conductivity of copper is $\sigma = 5.813 \times 10^7$ S/m. The surface resistivity at 6 GHz is

$$R_s = \sqrt{\frac{\omega\mu_0}{2\sigma}} = 2.02 \times 10^{-2} \ \Omega.$$

The attenuation due to conductor loss for the air-filled line is

$$\alpha_c = \frac{R_s}{2\eta \ln b/a} \left(\frac{1}{a} + \frac{1}{b} \right)$$

$$= \frac{2.02 \times 10^{-2}}{2(377) \ln (0.004/0.002)} \left(\frac{1}{0.002} + \frac{1}{0.004} \right) = 0.029 \ \text{Np/m}.$$

For Teflon, $\epsilon_r = 2.1$ and $\tan\delta = 0.0004$, so the attenuation due to conductor loss for the Teflon-filled line is

$$\alpha_c = \frac{R_s \sqrt{\epsilon_r}}{2\eta \ln b/a} \left(\frac{1}{a} + \frac{1}{b} \right) = \frac{2.02 \times 10^{-2} \sqrt{2.1}}{2(377) \ln (0.004/0.002)} \left(\frac{1}{0.002} + \frac{1}{0.004} \right) = 0.042 \ \text{Np/m}.$$

The dielectric loss of the air-filled line is zero, but the dielectric loss of the Teflon-filled line is

$$\alpha_d = k_0 \frac{\sqrt{\epsilon_r}}{2} \tan\delta$$

$$= \frac{(104.7)\sqrt{2.1}(0.0004)}{2} = 0.0303 \ \text{Np/m}.$$

Finally, from (6.27), the unloaded Q_s can be computed as

$$Q_{\text{air}} = \frac{\beta}{2\alpha} = \frac{104.7}{2(0.029)} = 1805,$$

$$Q_{\text{Teflon}} = \frac{\beta}{2\alpha} = \frac{104.7\sqrt{2.1}}{2(0.042 + 0.0303)} = 1049.$$

Thus it is seen that the Q of the air-filled line is almost twice that of the Teflon-filled line. The Q can be further increased by using silver-plated conductors.

Short-Circuited $\lambda/4$ Line

A parallel type of resonance (antiresonance) can be achieved using a short-circuited transmission line of length $\lambda/4$. The input impedance of a shorted line of length ℓ is

$$
\begin{aligned}
Z_{\text{in}} &= Z_0 \tanh(\alpha + j\beta)\ell \\
&= Z_0 \frac{\tanh \alpha\ell + j\tan \beta\ell}{1 + j\tan \beta\ell \tanh \alpha\ell} \\
&= Z_0 \frac{1 - j\tanh \alpha\ell \cot \beta\ell}{\tanh \alpha\ell - j\cot \beta\ell},
\end{aligned}
\tag{6.28}
$$

where the last result was obtained by multiplying both numerator and denominator by $-j\cot \beta\ell$. Now assume that $\ell = \lambda/4$ at $\omega = \omega_0$, and let $\omega = \omega_0 + \Delta\omega$. Then, for a TEM line,

$$
\beta\ell = \frac{\omega_0\ell}{v_p} + \frac{\Delta\omega\ell}{v_p} = \frac{\pi}{2} + \frac{\pi\Delta\omega}{2\omega_0},
$$

and so

$$
\cot \beta\ell = \cot\left(\frac{\pi}{2} + \frac{\pi\Delta\omega}{2\omega_0}\right) = -\tan \frac{\pi\Delta\omega}{2\omega_0} \simeq \frac{-\pi\Delta\omega}{2\omega_0}.
$$

Also, as before, $\tanh \alpha\ell \simeq \alpha\ell$ for small loss. Using these results in (6.28) gives

$$
Z_{\text{in}} = Z_0 \frac{1 + j\alpha\ell\pi\Delta\omega/2\omega_0}{\alpha\ell + j\pi\Delta\omega/2\omega_0} \simeq \frac{Z_0}{\alpha\ell + j\pi\Delta\omega/2\omega_0},
\tag{6.29}
$$

since $\alpha\ell\pi\Delta\omega/2\omega_0 \ll 1$. This result is of the same form as the impedance of a parallel RLC circuit, as given in (6.19):

$$
Z_{\text{in}} = \frac{1}{(1/R) + 2j\Delta\omega C}.
$$

We can identify the resistance of the equivalent circuit as

$$
R = \frac{Z_0}{\alpha\ell}
\tag{6.30a}
$$

and the capacitance of the equivalent circuit as

$$
C = \frac{\pi}{4\omega_0 Z_0}.
\tag{6.30b}
$$

The inductance of the equivalent circuit can be found as

$$
L = \frac{1}{\omega_0^2 C}.
\tag{6.30c}
$$

The resonator of Figure 6.4 therefore has a parallel-type resonance for $\ell = \lambda/4$, with an input impedance at resonance of $Z_{\text{in}} = R = Z_0/\alpha\ell$. From (6.18) and (6.30) the unloaded Q of this resonator is

$$
Q_0 = \omega_0 RC = \frac{\pi}{4\alpha\ell} = \frac{\beta}{2\alpha},
\tag{6.31}
$$

since $\ell = \pi/2\beta$ at resonance.

Open-Circuited $\lambda/2$ Line

A practical resonator that is often used in microstrip circuits consists of an open-circuited length of transmission line, as shown in Figure 6.5. This resonator will behave as a parallel resonant circuit when the length is $\lambda/2$, or multiples of $\lambda/2$.

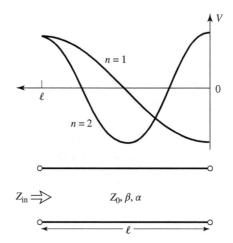

FIGURE 6.5 | An open-circuited length of lossy transmission line, and the voltage distributions for $n = 1$ ($\ell = \lambda/2$) and $n = 2$ ($\ell = \lambda$) resonators.

The input impedance of an open-circuited lossy transmission line of length ℓ is

$$Z_{in} = Z_0 \coth(\alpha + j\beta)\ell = Z_0 \frac{1 + j \tan \beta\ell \tanh \alpha\ell}{\tanh \alpha\ell + j \tan \beta\ell}. \tag{6.32}$$

As before, assume that $\ell = \lambda/2$ at $\omega = \omega_0$, and let $\omega = \omega_0 + \Delta\omega$. Then,

$$\beta\ell = \pi + \frac{\pi\Delta\omega}{\omega_0},$$

and so

$$\tan \beta\ell = \tan \frac{\Delta\omega\pi}{\omega} \simeq \frac{\Delta\omega\pi}{\omega_0},$$

and $\tanh \alpha\ell \simeq \alpha\ell$. Using these results in (6.32) gives

$$Z_{in} = \frac{Z_0}{\alpha\ell + j(\Delta\omega\pi/\omega_0)}. \tag{6.33}$$

Comparison with the input impedance of a parallel resonant circuit, as given by (6.19), suggests that the resistance of the equivalent RLC circuit is

$$R = \frac{Z_0}{\alpha\ell}, \tag{6.34a}$$

and the capacitance of the equivalent circuit is

$$C = \frac{\pi}{2\omega_0 Z_0}. \tag{6.34b}$$

The inductance of the equivalent circuit is

$$L = \frac{1}{\omega_0^2 C}. \tag{6.34c}$$

From (6.18) and (6.34) the unloaded Q is

$$Q_0 = \omega_0 RC = \frac{\pi}{2\alpha\ell} = \frac{\beta}{2\alpha}, \tag{6.35}$$

since $\ell = \pi/\beta$ at resonance.

EXAMPLE 6.2   A HALF-WAVE MICROSTRIP RESONATOR

Consider a microstrip resonator constructed from a $\lambda/2$ length of 50 Ω open-circuited microstrip line. The substrate is Teflon ($\epsilon_r = 2.1, \tan\delta = 0.0004$), with a thickness of 1.6 mm, and the conductors are copper. Compute the required length of the line for resonance at 6 GHz, and the unloaded Q of the resonator. Ignore fringing fields at the end of the line.

Solution
From (3.197), the width of a 50 Ω microstrip line on this substrate is found to be $W = 0.508$ cm, and the effective permittivity is $\epsilon_e = 1.802$. The resonant length can then be calculated as

$$\ell = \frac{\lambda}{2} = \frac{v_p}{2f} = \frac{c}{2f\sqrt{\epsilon_e}} = \frac{3\times10^8}{2(6\times10^9)\sqrt{1.802}} = 1.86 \text{ cm.}$$

The propagation constant is

$$\beta = \frac{2\pi f}{v_p} = \frac{2\pi f\sqrt{\epsilon_e}}{c} = \frac{2\pi(6\times10^9)\sqrt{1.802}}{3\times10^8} = 168.69 \text{ rad/m.}$$

From (3.199), the attenuation due to conductor loss is

$$\alpha_c = \frac{R_s}{Z_0 W} = \frac{2.02\times10^{-2}}{50(0.00508)} = 0.0795 \text{ Np/m,}$$

where we used R_s from Example 6.1. From (3.198), the attenuation due to dielectric loss is

$$\alpha_d = \frac{k_0\epsilon_r(\epsilon_e - 1)\tan\delta}{2\sqrt{\epsilon_e}(\epsilon_r - 1)} = \frac{(104.7)(2.1)(0.802)(0.0004)}{2\sqrt{1.802}(1.1)} = 0.0238 \text{ Np/m.}$$

Then from (6.35) the unloaded Q is

$$Q_0 = \frac{\beta}{2\alpha} = \frac{168.69}{2(0.0795 + 0.0238)} = 816.5.$$

6.3   RECTANGULAR WAVEGUIDE CAVITY RESONATORS

Microwave resonators can also be constructed from closed sections of waveguide. Because radiation loss from an open-ended waveguide can be significant, waveguide resonators are usually short circuited at both ends, thus forming a closed box, or *cavity*. Electric and magnetic energy is stored within the cavity enclosure, and power is dissipated in the metallic walls of the cavity as well as in the dielectric material that may fill the cavity. Coupling to a cavity resonator may be by a small aperture, or a small probe or loop. We will see that there are many possible resonant modes for a cavity resonator, corresponding to field variations along the three dimensions of the structure.

We will first derive the resonant frequencies for a general TE or TM resonant mode of a rectangular cavity, and then derive an expression for the unloaded Q of the $TE_{10\ell}$ mode. A complete treatment of the unloaded Q for arbitrary TE and TM modes can be made using the same procedure, but is not included here because of its length and complexity.

Resonant Frequencies

The geometry of a rectangular cavity is shown in Figure 6.6. It consists of a length, d, of rectangular waveguide shorted at both ends ($z = 0, d$). We will find the resonant frequencies of this cavity under the assumption that the cavity is lossless, then determine the unloaded Q using the perturbation method outlined in Section 2.6. Although we could begin with the Helmholtz wave equation and the method of separation of variables to solve for the electric and magnetic fields that satisfy the boundary conditions of the cavity, it is easier to start with the fields of the TE or TM waveguide modes since these already satisfy the necessary boundary conditions on the side walls ($x = 0, a$ and $y = 0, b$) of the cavity. Then it is only necessary to enforce the boundary conditions that $E_x = E_y = 0$ on the end walls at $z = 0, d$.

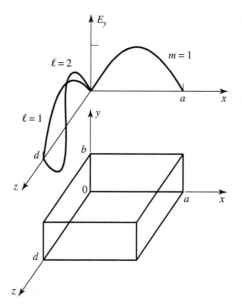

FIGURE 6.6 | A rectangular cavity resonator, and the electric field variations for the TE_{101} and TE_{102} resonant modes.

From Table 3.2 the transverse electric fields (E_x, E_y) of the TE_{mn} or TM_{mn} rectangular waveguide mode can be written as

$$\bar{E}_t(x, y, z) = \bar{e}(x, y) \left(A^+ e^{-j\beta_{mn}z} + A^- e^{j\beta_{mn}z} \right), \tag{6.36}$$

where $\bar{e}(x, y)$ is the transverse variation of the mode, and A^+, A^- are arbitrary amplitudes of the forward and backward traveling waves. The propagation constant of the m, nth TE or TM mode is

$$\beta_{mn} = \sqrt{k^2 - \left(\frac{m\pi}{a}\right)^2 - \left(\frac{n\pi}{b}\right)^2}, \tag{6.37}$$

where $k = \omega\sqrt{\mu\epsilon}$, and μ and ϵ are the permeability and permittivity of the material filling the cavity.

Applying the condition that $\bar{E}_t = 0$ at $z = 0$ to (6.36) implies that $A^+ = -A^-$ (as we should expect for reflection from a perfectly conducting wall). Then the condition that $\bar{E}_t = 0$ at $z = d$ leads to the equation

$$\bar{E}_t(x, y, d) = -\bar{e}(x, y)A^+ 2j \sin \beta_{mn}d = 0.$$

The only nontrivial $(A^+ \neq 0)$ solution occurs for

$$\beta_{mn}d = \ell\pi, \quad \ell = 1, 2, 3, \ldots, \tag{6.38}$$

which implies that the cavity must be an integer multiple of a half-guide wavelength long at the resonant frequency. No nontrivial solutions are possible for other lengths, or for frequencies other than the resonant frequencies.

A resonance wave number for the rectangular cavity can be defined as

$$k_{mn\ell} = \sqrt{\left(\frac{m\pi}{a}\right)^2 + \left(\frac{n\pi}{b}\right)^2 + \left(\frac{\ell\pi}{d}\right)^2}. \tag{6.39}$$

Then we can refer to the $TE_{mn\ell}$ or $TM_{mn\ell}$ resonant mode of the cavity, where the indices m, n, ℓ indicate the number of variations in the standing wave pattern in the x, y, z directions, respectively. The resonant frequency of the $TE_{mn\ell}$ or $TM_{mn\ell}$ mode is given by

$$f_{mn\ell} = \frac{ck_{mn\ell}}{2\pi\sqrt{\mu_r\epsilon_r}} = \frac{c}{2\pi\sqrt{\mu_r\epsilon_r}} \sqrt{\left(\frac{m\pi}{a}\right)^2 + \left(\frac{n\pi}{b}\right)^2 + \left(\frac{\ell\pi}{d}\right)^2}. \tag{6.40}$$

If $b < a < d$, the dominant resonant mode (lowest resonant frequency) will be the TE_{101} mode, corresponding to the TE_{10} dominant waveguide mode in a shorted guide of length $\lambda_g/2$, and is similar to the short-circuited $\lambda/2$ transmission line resonator. The dominant TM resonant mode is the TM_{110} mode.

Unloaded Q of the $TE_{10\ell}$ Mode

From Table 3.2, (6.36), and the fact that $A^- = -A^+$, the total fields for the $TE_{10\ell}$ resonant mode can be written as

$$E_y = A^+ \sin \frac{\pi x}{a} \left(e^{-j\beta z} - e^{j\beta z}\right), \tag{6.41a}$$

$$H_x = \frac{-A^+}{Z_{TE}} \sin \frac{\pi x}{a} \left(e^{-j\beta z} + e^{j\beta z}\right), \tag{6.41b}$$

$$H_z = \frac{j\pi A^+}{k\eta a} \cos \frac{\pi x}{a} \left(e^{-j\beta z} - e^{j\beta z}\right). \tag{6.41c}$$

Letting $E_0 = -2jA^+$ and using (6.38) allows these expressions to be simplified to

$$E_y = E_0 \sin \frac{\pi x}{a} \sin \frac{\ell \pi z}{d}, \tag{6.42a}$$

$$H_x = \frac{-jE_0}{Z_{TE}} \sin \frac{\pi x}{a} \cos \frac{\ell \pi z}{d}, \tag{6.42b}$$

$$H_z = \frac{j\pi E_0}{k\eta a} \cos \frac{\pi x}{a} \sin \frac{\ell \pi z}{d}, \tag{6.42c}$$

which clearly show that the fields form standing waves inside the cavity. We can now compute the unloaded Q of this mode by finding the stored electric and magnetic energies, and the power lost in the conducting walls and the dielectric filling.

The stored electric energy is, from (1.84),

$$W_e = \frac{\epsilon}{4} \int_V E_y E_y^* dv = \frac{\epsilon abd}{16} E_0^2, \tag{6.43a}$$

while the stored magnetic energy is, from (1.86),

$$W_m = \frac{\mu}{4} \int_V (H_x H_x^* + H_z H_z^*) dv$$

$$= \frac{\mu abd}{16} E_0^2 \left(\frac{1}{Z_{TE}^2} + \frac{\pi^2}{k^2 \eta^2 a^2}\right). \tag{6.43b}$$

Because $Z_{TE} = k\eta/\beta$, with $\beta = \beta_{10} = \sqrt{k^2 - (\pi/a)^2}$, the quantity in parentheses in (6.43b) can be reduced to

$$\left(\frac{1}{Z_{TE}^2} + \frac{\pi^2}{k^2 \eta^2 a^2}\right) = \frac{\beta^2 + (\pi/a)^2}{k^2 \eta^2} = \frac{1}{\eta^2} = \frac{\epsilon}{\mu},$$

showing that $W_e = W_m$ at resonance. The condition of equal stored electric and magnetic energies at resonance also applied to the *RLC* resonant circuits of Section 6.1.

For small losses we can find the power dissipated in the cavity walls using the perturbation method of Section 2.6. Thus, the power lost in the conducting walls is given by (1.131) as

$$P_c = \frac{R_s}{2} \int_{walls} |H_t|^2 ds, \tag{6.44}$$

where $R_s = \sqrt{\omega\mu_0/2\sigma}$ is the surface resistivity of the metallic walls, and H_t is the tangential magnetic field at the surface of the walls. Using (6.42b), (6.42c) in (6.44) gives

$$P_c = \frac{R_s}{2}\left\{ 2\int_{y=0}^{b}\int_{x=0}^{a}|H_x(z=0)|^2\,dxdy + 2\int_{z=0}^{d}\int_{y=0}^{b}|H_z(x=0)|^2\,dydz\right.$$

$$\left. + 2\int_{z=0}^{d}\int_{x=0}^{a}\left[|H_x(y=0)|^2 + |H_z(y=0)|^2\right]dxdz\right\}$$

$$= \frac{R_s E_0^2 \lambda^2}{8\eta^2}\left(\frac{\ell^2 ab}{d^2} + \frac{bd}{a^2} + \frac{\ell^2 a}{2d} + \frac{d}{2a}\right), \tag{6.45}$$

where use has been made of the symmetry of the cavity in doubling the contributions from the walls at $x=0, y=0$, and $z=0$ to account for the contributions from the walls at $x=a, y=b$, and $z=d$, respectively. The relations $k = 2\pi/\lambda$ and $Z_{TE} = k\eta/\beta = 2d\eta/\ell\lambda$ were also used in simplifying (6.45). Then, from (6.7), the unloaded Q of the cavity with lossy conducting walls but lossless dielectric can be found as

$$Q_c = \frac{2\omega_0 W_e}{P_c}$$

$$= \frac{k^3 abd\eta}{4\pi^2 R_s}\frac{1}{[(\ell^2 ab/d^2) + (bd/a^2) + (\ell^2 a/2d) + (d/2a)]}$$

$$= \frac{(kad)^3 bn}{2\pi^2 R_s}\frac{1}{(2\ell^2 a^3 b + 2bd^3 + \ell^2 a^3 d + ad^3)}. \tag{6.46}$$

Next we compute the power lost in the dielectric material that may fill the cavity. As discussed in Chapter 1, a lossy dielectric has an effective conductivity $\sigma = \omega\epsilon'' = \omega\epsilon_r\epsilon_0\tan\delta$, where $\epsilon = \epsilon' - j\epsilon'' = \epsilon_r\epsilon_0(1 - j\tan\delta)$, and $\tan\delta$ is the loss tangent of the material. The power dissipated in the dielectric is, from (1.92),

$$P_d = \frac{1}{2}\int_V \bar{J}\cdot\bar{E}^*\,dv = \frac{\omega\epsilon''}{2}\int_V |\bar{E}|^2\,dv = \frac{abd\omega\epsilon''|E_0|^2}{8}, \tag{6.47}$$

where $\bar{E}$ is given by (6.42a). Then from (6.7) the unloaded Q of the cavity with a lossy dielectric filling, but with perfectly conducting walls, is

$$Q_d = \frac{2\omega W_e}{P_d} = \frac{\epsilon'}{\epsilon''} = \frac{1}{\tan\delta}. \tag{6.48}$$

The simplicity of this result is due to the fact that the integral in (6.43a) for W_e cancels with the identical integral in (6.47) for P_d. This result therefore applies to Q_d for an arbitrary resonant cavity mode. When both wall losses and dielectric losses are present, the total power loss is $P_c + P_d$, so (6.7) gives the total unloaded Q as

$$Q_0 = \left(\frac{1}{Q_c} + \frac{1}{Q_d}\right)^{-1}. \tag{6.49}$$

EXAMPLE 6.3   DESIGN OF A RECTANGULAR CAVITY RESONATOR

A rectangular waveguide cavity is made from a piece of copper WR-187 H-band waveguide, with $a = 4.755$ cm and $b = 2.215$ cm. The cavity is filled with polyethylene ($\epsilon_r = 2.25$, $\tan\delta = 0.0004$). If resonance is to occur at $f = 6$ GHz, find the required length, d, and the resulting unloaded Q for the $\ell = 1$ and $\ell = 2$ resonant modes.

Solution
The wave number k is

$$k = \frac{2\pi f\sqrt{\epsilon_r}}{c} = \frac{2\pi(6\times10^9)\sqrt{2.25}}{3\times10^8} = 188.5 \text{ m}^{-1}.$$

From (6.40) the required length for resonance can be found as ($m = 1, n = 0$)

$$d = \frac{\ell \pi}{\sqrt{k^2 - (\pi/a)^2}},$$

$$\text{for } \ell = 1, \quad d = \frac{\pi}{\sqrt{(188.5)^2 - (\pi/0.04755)^2}} = 1.78 \text{ cm},$$

$$\text{for } \ell = 2, \quad d = \frac{2\pi}{\sqrt{(188.5)^2 - (\pi/0.04755)^2}} = 3.56 \text{ cm}.$$

From Example 6.1, the surface resistivity of copper at 6 GHz is $R_s = 2.02 \times 10^{-2} \; \Omega$. The intrinsic impedance is

$$\eta = \frac{377}{\sqrt{\epsilon_r}} = 251.3 \; \Omega.$$

Then from (6.46) the Q due to conductor loss only is

$$\text{for } \ell = 1, \quad Q_c = \frac{(kad)^3 b\eta}{2\pi^2 R_s} \frac{1}{(2\ell^2 a^3 b + 2bd^3 + \ell^2 a^3 d + ad^3)} = 7838,$$

$$\text{for } \ell = 2, \quad Q_c = \frac{(kad)^3 b\eta}{2\pi^2 R_s} \frac{1}{(2\ell^2 a^3 b + 2bd^3 + \ell^2 a^3 d + ad^3)} = 11{,}684.$$

From (6.48) the Q due to dielectric loss only is, for both $\ell = 1$ and $\ell = 2$,

$$Q_d = \frac{1}{\tan \delta} = \frac{1}{0.0004} = 2500.$$

Then total unloaded Qs are, from (6.49)

$$\text{for } \ell = 1, \quad Q_0 = \left(\frac{1}{7838} + \frac{1}{2500} \right)^{-1} = 1896,$$

$$\text{for } \ell = 2, \quad Q_0 = \left(\frac{1}{11{,}684} + \frac{1}{2500} \right)^{-1} = 2060.$$

Note that the dielectric loss has the dominant effect on the Q; higher Q could be obtained using an air-filled cavity. These results can be compared to those of Examples 6.1 and 6.2, which used similar types of materials at the same frequency.

6.4   CIRCULAR WAVEGUIDE CAVITY RESONATORS

A cylindrical cavity resonator can be constructed from a section of circular waveguide shorted at both ends, similar to rectangular cavities. Because the dominant circular waveguide mode is the TE_{11} mode, the dominant cylindrical cavity mode is the TE_{111} mode. We will derive the resonant frequencies for the $TE_{nm\ell}$ and $TM_{nm\ell}$ circular cavity modes, and an expression for the unloaded Q of the $TE_{nm\ell}$ mode.

Circular cavities are often used for microwave frequency meters. The cavity is constructed with a movable top wall to allow mechanical tuning of the resonant frequency, and the cavity is loosely coupled to a waveguide through a small aperture. In operation, power will be absorbed by the cavity as it is tuned to the operating frequency of the system; this absorption can be monitored with a power meter elsewhere in the system. The mechanical tuning dial is usually directly calibrated in frequency, as in the model shown in Figure 6.7. Because frequency resolution is determined by the Q of the resonator, the TE_{011} mode is often used for frequency meters because its Q is much higher than the Q of the dominant circular cavity mode. This is also the reason for a loose coupling to the cavity.

Resonant Frequencies

The geometry of a cylindrical cavity is shown in Figure 6.8. As in the case of the rectangular cavity, the solution is simplified by beginning with the circular waveguide modes, which already satisfy the necessary

FIGURE 6.7 | Photograph of a W-band waveguide frequency meter. The knob rotates to change the length of the circular cavity resonator; the scale gives a readout of the frequency.

Photograph courtesy of Millitech Inc., Northampton, Mass.

boundary conditions on the wall of the circular waveguide. From Table 3.5, the transverse electric fields (E_ρ, E_ϕ) of the TE_{nm} or TM_{nm} circular waveguide mode can be written as

$$\bar{E}_t(\rho, \phi, z) = \bar{e}(\rho, \phi)\left(A^+ e^{-j\beta_{nm}z} + A^- e^{j\beta_{nm}z}\right), \tag{6.50}$$

where $\bar{e}(\rho, \phi)$ represents the transverse variation of the mode, and A^+ and A^- are arbitrary amplitudes of the forward and backward traveling waves. The propagation constant of the TE_{nm} mode is, from (3.126),

$$\beta_{nm} = \sqrt{k^2 - \left(\frac{p'_{nm}}{a}\right)^2}, \tag{6.51a}$$

while the propagation constant of the TM_{nm} mode is, from (3.139),

$$\beta_{nm} = \sqrt{k^2 - \left(\frac{p_{nm}}{a}\right)^2}, \tag{6.51b}$$

where $k = \omega\sqrt{\mu\epsilon}$.

In order to have $\bar{E}_t = 0$ at $z = 0, d$, we must choose $A^+ = -A^-$, and $A^+ \sin\beta_{nm}d = 0$,

or
$$\beta_{nm}d = \ell\pi, \quad \text{for } \ell = 0, 1, 2, 3, \dots, \tag{6.52}$$

which implies that the waveguide must be an integer number of half-guide wavelengths long. Thus, the resonant frequency of the $TE_{nm\ell}$ mode is

$$f_{nm\ell} = \frac{c}{2\pi\sqrt{\mu_r\epsilon_r}}\sqrt{\left(\frac{p'_{nm}}{a}\right)^2 + \left(\frac{\ell\pi}{d}\right)^2}, \tag{6.53a}$$

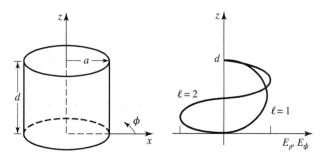

FIGURE 6.8 | A cylindrical resonant cavity, and the electric field distribution for resonant modes with $\ell = 1$ or $\ell = 2$.

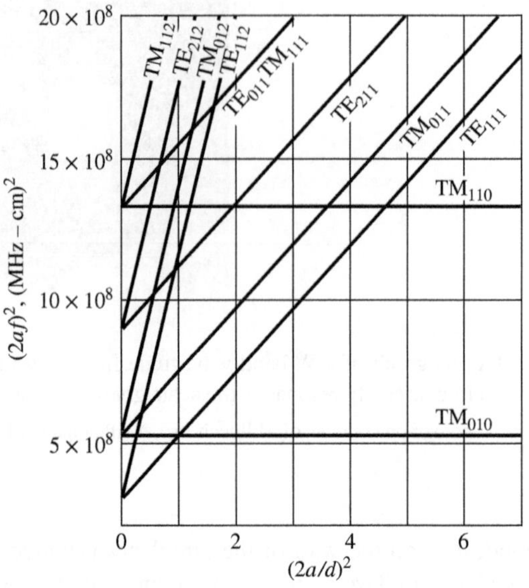

FIGURE 6.9 | Resonant mode chart for a cylindrical cavity.

Adapted from data from R. E. Collin, *Foundations for Microwave Engineering,* 2nd edition, Wiley–IEEE Press, Hoboken, NJ, 2001. Used with permission.

and the resonant frequency of the $\text{TM}_{nm\ell}$ mode is

$$f_{nm\ell} = \frac{c}{2\pi \sqrt{\mu_r \epsilon_r}} \sqrt{\left(\frac{p_{nm}}{a}\right)^2 + \left(\frac{\ell \pi}{d}\right)^2}. \tag{6.53b}$$

Thus the dominant TE mode is the TE_{111} mode, while the dominant TM mode is the TM_{010} mode. Figure 6.9 shows a *mode chart* for the lower order resonant modes of a cylindrical cavity. Such a chart is useful for the design of circular cavity resonators, as it shows what modes can be excited at a given frequency for a given cavity size.

Unloaded Q of the TE$_{nm\ell}$ Mode

From Table 3.5, (6.50), and the fact that $A^+ = -A^-$, the fields of the $\text{TE}_{nm\ell}$ mode can be written as

$$H_z = H_0 J_n \left(\frac{p'_{nm}\rho}{a}\right) \cos n\phi \sin \frac{\ell \pi z}{d}, \tag{6.54a}$$

$$H_\rho = \frac{\beta a H_0}{p'_{nm}} J'_n \left(\frac{p'_{nm}\rho}{a}\right) \cos n\phi \cos \frac{\ell \pi z}{d}, \tag{6.54b}$$

$$H_\phi = \frac{-\beta a^2 n H_0}{(p'_{nm})^2 \rho} J_n \left(\frac{p'_{nm}\rho}{a}\right) \sin n\phi \cos \frac{\ell \pi z}{d}, \tag{6.54c}$$

$$E_\rho = \frac{jk\eta a^2 n H_0}{(p'_{nm})^2 \rho} J_n \left(\frac{p'_{nm}\rho}{a}\right) \sin n\phi \sin \frac{\ell \pi z}{d}, \tag{6.54d}$$

$$E_\phi = \frac{jk\eta a H_0}{p'_{nm}} J'_n \left(\frac{p'_{nm}\rho}{a}\right) \cos n\phi \sin \frac{\ell \pi z}{d}, \tag{6.54e}$$

$$E_z = 0, \tag{6.54f}$$

where $\eta = \sqrt{\mu/\epsilon}$ and $H_0 = -2jA^+$.

Because the time-average stored electric and magnetic energies are equal, the total stored energy is

$$
\begin{aligned}
W = 2W_e &= \frac{\epsilon}{2} \int_{z=0}^{d} \int_{\phi=0}^{2\pi} \int_{\rho=0}^{a} \left(|E_\rho|^2 + |E_\phi|^2 \right) \rho \, d\rho \, d\phi \, dz \\
&= \frac{\epsilon k^2 \eta^2 a^2 \pi d H_0^2}{4(p'_{nm})^2} \int_{\rho=0}^{a} \left[J_n'^2 \left(\frac{p'_{nm}\rho}{a} \right) + \left(\frac{na}{p'_{nm}\rho} \right)^2 J_n^2 \left(\frac{p'_{nm}\rho}{a} \right) \right] \rho \, d\rho \\
&= \frac{\epsilon k^2 \eta^2 a^4 H_0^2 \pi d}{8(p'_{nm})^2} \left[1 - \left(\frac{n}{p'_{nm}} \right)^2 \right] J_n^2(p'_{nm}),
\end{aligned}
\tag{6.55}
$$

where the integral identity of Appendix C.17 has been used. The power loss in the conducting walls is

$$
\begin{aligned}
P_c &= \frac{R_s}{2} \int_S |\bar{H}_{\tan}|^2 ds \\
&= \frac{R_s}{2} \left\{ \int_{z=0}^{d} \int_{\phi=0}^{2\pi} \left[|H_\phi(\rho = a)|^2 + |H_z(\rho = a)|^2 \right] a \, d\phi \, dz \right. \\
&\quad \left. + 2 \int_{\phi=0}^{2\pi} \int_{\rho=0}^{a} \left[|H_\rho(z = 0)|^2 + |H_\phi(z = 0)|^2 \right] \rho \, d\rho \, d\phi \right\} \\
&= \frac{R_s}{2} \pi H_0^2 J_n^2(p'_{nm}) \left\{ \frac{da}{2} \left[1 + \left(\frac{\beta a n}{(p'_{nm})^2} \right)^2 \right] + \left(\frac{\beta a^2}{p'_{nm}} \right)^2 \left(1 - \frac{n^2}{(p'_{nm})^2} \right) \right\}.
\end{aligned}
\tag{6.56}
$$

Then, from (6.8), the unloaded Q of the cavity with imperfectly conducting walls but lossless dielectric is

$$
Q_c = \frac{\omega_0 W}{P_c} = \frac{(ka)^3 \eta ad}{4(p'_{nm})^2 R_s} \frac{1 - \left(\dfrac{n}{p'_{nm}} \right)^2}{\left\{ \dfrac{ad}{2} \left[1 + \left(\dfrac{\beta a n}{(p'_{nm})^2} \right)^2 \right] + \left(\dfrac{\beta a^2}{p'_{nm}} \right)^2 \left(1 - \dfrac{n^2}{(p'_{nm})^2} \right) \right\}}.
\tag{6.57}
$$

From (6.52) and (6.51) we see that $\beta = \ell \pi/d$ and $(ka)^2$ are constants that do not vary with frequency, for a cavity with fixed dimensions. Thus, the frequency dependence of Q_c is given by k/R_s, which varies as $1/\sqrt{f}$; this gives the variation in Q_c for a given resonant mode and cavity shape (fixed n, m, ℓ, and a/d).

Figure 6.10 shows the normalized unloaded Q due to conductor loss for various resonant modes of a cylindrical cavity. Observe that the TE_{011} mode has an unloaded Q significantly higher than that of the lower order TE_{111}, TM_{010}, or TM_{111} mode.

To compute the unloaded Q due to dielectric loss, we must compute the power dissipated in the dielectric. Thus,

$$
\begin{aligned}
P_d &= \frac{1}{2} \int_V \bar{J} \cdot \bar{E}^* dv = \frac{\omega \epsilon''}{2} \int_V \left[|E_\rho|^2 + |E_\phi|^2 \right] dv \\
&= \frac{\omega \epsilon'' k^2 \eta^2 a^2 H_0^2 \pi d}{4(p'_{nm})^2} \int_{\rho=0}^{a} \left[\left(\frac{na}{p'_{nm}\rho} \right)^2 J_n^2 \left(\frac{p'_{nm}\rho}{a} \right) + J_n'^2 \left(\frac{p'_{nm}\rho}{a} \right) \right] \rho \, d\rho \\
&= \frac{\omega \epsilon'' k^2 \eta^2 a^4 H_0^2}{8(p'_{nm})^2} \left[1 - \left(\frac{n}{p'_{nm}} \right)^2 \right] J_n^2(p'_{nm}).
\end{aligned}
\tag{6.58}
$$

Then (6.8) gives the unloaded Q due to dielectric loss as

$$
Q_d = \frac{\omega W}{P_d} = \frac{\epsilon}{\epsilon''} = \frac{1}{\tan \delta},
\tag{6.59}
$$

where $\tan \delta$ is the loss tangent of the dielectric. This is the same as the result for Q_d of (6.48) for the rectangular cavity. When both conductor and dielectric losses are present, the total unloaded cavity Q can be found from (6.49).

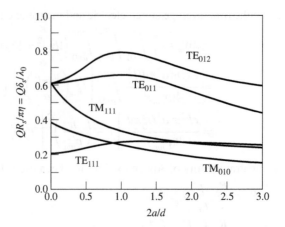

FIGURE 6.10 | Normalized unloaded Q for various cylindrical cavity modes (air filled).

Adapted from data from R. E. Collin, *Foundations for Microwave Engineering*, 2nd edition, Wiley–IEEE Press, Hoboken, NJ, 2001. Used with permission.

EXAMPLE 6.4   DESIGN OF A CIRCULAR CAVITY RESONATOR

A circular cavity resonator with $d = 2a$ is to be designed to resonate at 5.0 GHz in the TE_{011} mode. If the cavity is made from copper and is Teflon filled ($\epsilon_r = 2.1$, $\tan \delta = 0.0004$), find its dimensions and unloaded Q.

Solution

$$k = \frac{2\pi f_{011} \sqrt{\epsilon_r}}{c} = \frac{2\pi(5 \times 10^9)\sqrt{2.1}}{3 \times 10^8} = 151.8 \ \text{m}^{-1}$$

From (6.53a) the resonant frequency of the TE_{011} mode is

$$f_{011} = \frac{c}{2\pi\sqrt{\epsilon_r}} \sqrt{\left(\frac{p'_{01}}{a}\right)^2 + \left(\frac{\pi}{d}\right)^2},$$

with $p'_{01} = 3.832$. Then, since $d = 2a$

$$\frac{2\pi f_{011}\sqrt{\epsilon_r}}{c} = k = \sqrt{\left(\frac{p'_{01}}{a}\right)^2 + \left(\frac{\pi}{d}\right)^2}.$$

Solving for a gives

$$a = \frac{\sqrt{(p'_{01})^2 + (\pi/2)^2}}{k} = \frac{\sqrt{(3.832)^2 + (\pi/2)^2}}{151.8} = 2.73 \ \text{cm},$$

so we have $d = 5.48$ cm.

The surface resistivity of copper at 5 GHz is $R_s = 0.0184 \ \Omega$. Then from (6.57), with $n = 0$, $m = \ell = 1$, and $d = 2a$, the unloaded Q due to conductor losses is

$$Q_c = \frac{(ka)^3 \eta ad}{4(p'_{01})^2 R_s} \frac{1}{[ad/2 + (\beta a^2/p'_{01})^2]} = \frac{ka\eta}{2R_s} = 29,302,$$

where (6.51a) was used to simplify the expression. From (6.59) the unloaded Q due to dielectric loss is

$$Q_d = \frac{1}{\tan \delta} = \frac{1}{0.0004} = 2500,$$

and the total unloaded Q of the cavity is

$$Q_0 = \left(\frac{1}{Q_c} + \frac{1}{Q_d}\right)^{-1} = 2303.$$

This result can be compared with the rectangular cavity case of Example 6.3, which had $Q_0 = 1896$ for the TE_{101} mode and $Q_0 = 2060$ for the TE_{102} mode. If this cavity were air filled, the Q would increase to 42,400.

6.5   DIELECTRIC RESONATORS

A small disc or cube (or other shape) of dielectric material can also be used as a microwave resonator. The operation of such a *dielectric resonator* is similar in principle to the rectangular or cylindrical cavity resonators previously discussed. Dielectric resonators typically use materials with low loss and a high dielectric constant, ensuring that most of the fields will be contained within the dielectric. Unlike metallic cavities, however, there is some field fringing or leakage from the sides and ends of a dielectric resonator (which are not metalized), leading to a small radiation loss and consequent lowering of Q. A dielectric resonator is generally smaller in size, cost, and weight than an equivalent metallic cavity, and it can easily be incorporated into microwave integrated circuits and coupled to planar transmission lines. Materials with dielectric constants in the range of 10–100 are generally used, with barium tetratitanate and titanium dioxide being typical examples. Conductor losses are absent, but dielectric loss usually increases with dielectric constant; Qs of up to several thousand can sometimes be achieved, however. By using an adjustable metal plate above the resonator, the resonant frequency can be mechanically tuned. Because of these desirable features, dielectric resonators have become key components for integrated microwave filters and oscillators.

Below we present an approximate analysis for the resonant frequencies of the $TE_{01\delta}$ mode of a cylindrical dielectric resonator; this mode is the one most commonly used in practice, and is analogous to the TE_{011} mode of a circular metallic cavity.

Resonant Frequencies of $TE_{01\delta}$ Mode

The geometry of a cylindrical dielectric resonator is shown in Figure 6.11. The basic operation of the $TE_{01\delta}$ mode can be explained as follows. The dielectric resonator is considered as a short length, L, of dielectric waveguide open at both ends. The lowest order TE mode of this guide is the TE_{01} mode, and is the dual of the TM_{01} mode of a circular metallic waveguide. Because of the high permittivity of the resonator, propagation along the z-axis can occur inside the dielectric at the resonant frequency, but the fields will be cut off (evanescent) in the air regions around the dielectric. Thus the H_z field will look like that sketched in Figure 6.12; higher order resonant modes will have more variations in the z direction inside the resonator. Because the resonant length for the $TE_{01\delta}$ mode is less than $\lambda_g/2$ (where λ_g is the guide wavelength of the TE_{01} dielectric waveguide mode), the symbol $\delta = 2L/\lambda_g < 1$ is used to denote the z variation of the

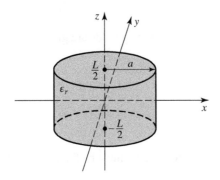

FIGURE 6.11 | Geometry of a cylindrical dielectric resonator.

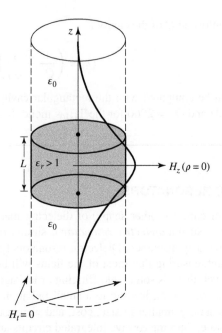

FIGURE 6.12 | Magnetic wall boundary condition approximation and distribution of H_z versus z for $\rho = 0$ of the first mode of a cylindrical dielectric resonator.

resonant mode. The equivalent circuit of the resonator looks like a length of transmission line terminated in purely reactive loads at both ends.

Our analysis follows that of Cohn [2], and involves the assumption that a magnetic wall boundary condition can be imposed at $\rho = a$. This approximation is based on the fact that the reflection coefficient of a wave in a high dielectric constant region incident on an air-filled region approaches $+1$:

$$\Gamma = \frac{\eta_0 - \eta}{\eta_0 + \eta} = \frac{\sqrt{\epsilon_r} - 1}{\sqrt{\epsilon_r} + 1} \to 1 \quad \text{as } \epsilon_r \to \infty.$$

This reflection coefficient is the same as that obtained at an ideal magnetic wall boundary condition, or a perfect open circuit.

We begin by finding the fields of the TE_{01} dielectric waveguide mode with a magnetic wall boundary condition at $\rho = a$. For TE modes, $E_z = 0$, and H_z must satisfy the wave equation

$$(\nabla^2 + k^2)H_z = 0, \tag{6.60}$$

where

$$k = \begin{cases} \sqrt{\epsilon_r}k_0 & \text{for } |z| < L/2 \\ k_0 & \text{for } |z| > L/2. \end{cases} \tag{6.61}$$

Because $\partial/\partial\phi = 0$, the transverse fields are given by (3.110) as follows:

$$E_\phi = \frac{j\omega\mu_0}{k_c^2} \frac{\partial H_z}{\partial \rho}, \tag{6.62a}$$

$$H_\rho = \frac{-j\beta}{k_c^2} \frac{\partial H_z}{\partial \rho}, \tag{6.62b}$$

where $k_c^2 = k^2 - \beta^2$. Because H_z must be finite at $\rho = 0$ and zero at $\rho = a$ (the magnetic wall), we have

$$H_z = H_0 J_0(k_c\rho)e^{\pm j\beta z}, \tag{6.63}$$

where $k_c = p_{01}/a$, and $J_0(p_{01}) = 0$ ($p_{01} = 2.405$). Then from (6.62) the transverse fields are

$$E_\phi = \frac{j\omega\mu_0 H_0}{k_c} J_0'(k_c\rho)e^{\pm j\beta z}, \tag{6.64a}$$

$$H_\rho = \frac{\mp j\beta H_0}{k_c} J_0'(k_c\rho)e^{\pm j\beta z}. \tag{6.64b}$$

In the dielectric region, for $|z| < L/2$, the propagation constant is real:

$$\beta = \sqrt{\epsilon_r k_0^2 - k_c^2} = \sqrt{\epsilon_r k_0^2 - \left(\frac{p_{01}}{a}\right)^2}, \tag{6.65a}$$

and a wave impedance can be defined as

$$Z_d = \frac{E_\phi}{H_\rho} = \frac{\omega\mu_0}{\beta}. \tag{6.65b}$$

In the air region, for $|z| > L/2$, the propagation constant will be imaginary, so it is convenient to write

$$\alpha = \sqrt{k_c^2 - k_0^2} = \sqrt{\left(\frac{p_{01}}{a}\right)^2 - k_0^2}, \tag{6.66a}$$

and to define a wave impedance in the air region as

$$Z_a = \frac{j\omega\mu_0}{\alpha}, \tag{6.66b}$$

which is seen to be imaginary.

From symmetry, the H_z and E_ϕ field distributions for the lowest order mode will be even functions about $z = 0$. Then the transverse fields for the $TE_{01\delta}$ mode can be written for $|z| < L/2$ as

$$E_\phi = AJ_0'(k_c\rho)\cos\beta z, \tag{6.67a}$$

$$H_\rho = \frac{-jA}{Z_d} J_0'(k_c\rho)\sin\beta z, \tag{6.67b}$$

and for $|z| > L/2$ as

$$E_\phi = BJ_0'(k_c\rho)e^{-\alpha|z|}, \tag{6.68a}$$

$$H_\rho = \frac{\pm B}{Z_a} J_0'(k_c\rho)e^{-\alpha|z|}, \tag{6.68b}$$

where A and B are unknown amplitude coefficients. In (6.68b), the $\pm$ sign is used for $z > L/2$ or $z < -L/2$, respectively.

Matching tangential fields at $z = L/2$ (or $z = -L/2$) leads to the following two equations:

$$A\cos\frac{\beta L}{2} = Be^{-\alpha L/2}, \tag{6.69a}$$

$$\frac{-jA}{Z_d}\sin\frac{\beta L}{2} = \frac{B}{Z_a}e^{-\alpha L/2}, \tag{6.69b}$$

which can be reduced to a single transcendental equation:

$$-jZ_a\sin\frac{\beta L}{2} = Z_d\cos\frac{\beta L}{2}.$$

Using (6.65b) and (6.66b) allows this to be simplified as

$$\tan\frac{\beta L}{2} = \frac{\alpha}{\beta}, \tag{6.70}$$

where β is given by (6.65a) and α is given by (6.66a). This equation can be solved numerically for k_0, which determines the resonant frequency.

This solution is approximate since it ignores fringing fields at the sides of the resonator, and it yields accuracies only on the order of 10% (usually not accurate enough for practical purposes), but it serves to illustrate the basic behavior of dielectric resonators. More accurate solutions are available in the literature [3].

The unloaded Q of the resonator can be calculated by determining the stored energy (inside and outside the dielectric cylinder), and the power dissipated in the dielectric and possibly lost to radiation. If the latter is small, the unloaded Q can be approximated as $1/\tan \delta$, as in the case of the metallic cavity resonators.

EXAMPLE 6.5   RESONANT FREQUENCY AND Q OF A DIELECTRIC RESONATOR

Find the resonant frequency and approximate unloaded Q for the $TE_{01\delta}$ mode of a dielectric resonator made from titania, with $\epsilon_r = 95$ and $\tan \delta = 0.001$. The resonator dimensions are $a = 0.413$ cm and $L = 0.8255$ cm.

Solution
The transcendental equation of (6.70) must be solved for k_0, with β and α given by (6.65a) and (6.66a). Thus,

$$\tan \frac{\beta L}{2} = \frac{\alpha}{\beta},$$

where

$$\alpha = \sqrt{(2.405/a)^2 - k_0^2},$$

$$\beta = \sqrt{\epsilon_r k_0^2 - (2.405/a)^2},$$

and

$$k_0 = \frac{2\pi f}{c}.$$

Because α and β must both be real, the possible frequency range is from f_1 to f_2, where

$$f_1 = \frac{c k_0}{2\pi} = \frac{c(2.405)}{2\pi \sqrt{\epsilon_r} a} = 2.853 \text{ GHz},$$

$$f_2 = \frac{c k_0}{2\pi} = \frac{c(2.405)}{2\pi a} = 27.804 \text{ GHz}.$$

Using the interval-halving method (see the Point of Interest on root-finding algorithms in Chapter 3) to find the root of the above equation gives a resonant frequency of about 3.152 GHz. This compares with a measured value of about 3.4 GHz from reference [2], indicating a 10% error. The approximate unloaded Q, due to dielectric loss, is

$$Q_d = \frac{1}{\tan \delta} = 1000.$$

6.6   EXCITATION OF RESONATORS

Resonators are not useful unless they are coupled to external circuitry, so we now discuss how resonators can be coupled to transmission lines and waveguides. In practice, the way in which this is done depends on the type of resonator under consideration; some examples of resonator coupling techniques are shown in Figure 6.13. We will discuss the operation of a common coupling technique, that is, gap coupling. We begin by discussing the coupling coefficient for a resonator connected to a feed line, and the subject of critical coupling. A related topic of practical interest is how the unloaded Q of a resonator can be determined from the two-port response of a resonator coupled to a transmission line.

The Coupling Coefficient and Critical Coupling

The level of coupling required between a resonator and its attached circuitry depends on the application. A waveguide cavity used as a frequency meter, for example, is usually loosely coupled to its feed guide in

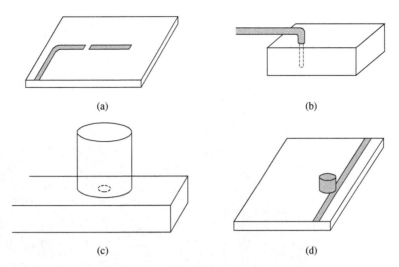

FIGURE 6.13 | Coupling to microwave resonators. (a) A microstrip transmission line resonator gap coupled to a microstrip feedline. (b) A rectangular cavity resonator fed by a coaxial probe. (c) A circular cavity resonator aperture coupled to a rectangular waveguide. (d) A dielectric resonator coupled to a microstrip line.

order to maintain high Q and good accuracy. A resonator used in an oscillator or tuned amplifier, however, may be tightly coupled in order to achieve maximum power transfer. A measure of the level of coupling between a resonator and a feed is given by the *coupling coefficient*. To obtain maximum power transfer between a resonator and a feed line, the resonator should be matched to the line at the resonant frequency; the resonator is then said to be *critically coupled* to the feed. We will illustrate these concepts by considering the series resonant circuit shown in Figure 6.14.

From (6.9), the input impedance near resonance of the series resonant circuit of Figure 6.14 is given by

$$Z_{in} = R + j2L\Delta\omega = R + j\frac{2RQ_0\Delta\omega}{\omega_0}, \tag{6.71}$$

and the unloaded Q is, from (6.8),

$$Q_0 = \frac{\omega_0 L}{R}. \tag{6.72}$$

At resonance, $\Delta\omega = 0$, so from (6.71) the input impedance is $Z_{in} = R$. In order to match the resonator to the line we must have

$$R = Z_0. \tag{6.73}$$

In this case the unloaded Q is

$$Q_0 = \frac{\omega_0 L}{Z_0}. \tag{6.74}$$

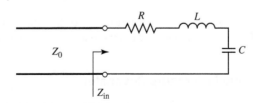

FIGURE 6.14 | A series resonant circuit coupled to a feedline.

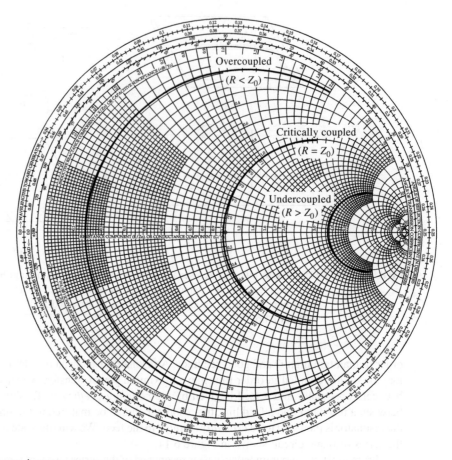

FIGURE 6.15 | Smith chart illustrating coupling to a series *RLC* circuit.

From (6.22), the external Q is

$$Q_e = \frac{\omega_0 L}{Z_0} = Q_0, \tag{6.75}$$

which shows that the external and unloaded Qs are equal under the condition of critical coupling. The loaded Q is half this value.

We can define the coupling coefficient, g, as

$$g = \frac{Q_0}{Q_e}, \tag{6.76}$$

which can be applied to both series ($g = Z_0/R$) and parallel ($g = R/Z_0$) resonant circuits, when connected to a transmission line of characteristic impedance Z_0. Three cases can be distinguished:

1. $g < 1$: The resonator is said to be *undercoupled* to the feedline.
2. $g = 1$: The resonator is *critically coupled* to the feedline.
3. $g > 1$: The resonator is said to be *overcoupled* to the feedline.

Figure 6.15 shows a Smith chart sketch of the impedance loci for the series resonant circuit, as given by (6.71), for various values of R corresponding to the above cases.

A Gap-Coupled Microstrip Resonator

Consider a $\lambda/2$ open-circuited microstrip resonator proximity coupled to the open end of a microstrip transmission line, as shown in Figure 6.13a. The gap between the resonator and the microstrip line can be

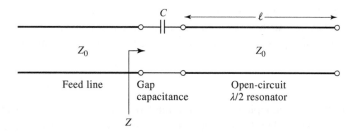

FIGURE 6.16 | Equivalent circuit of the gap-coupled microstrip resonator of Figure 6.13a.

modeled as a series capacitor, so the equivalent circuit can be constructed as shown in Figure 6.16. The normalized input impedance seen by the feedline is

$$z = \frac{Z}{Z_0} = -j\frac{(1/\omega C + Z_0 \cot \beta\ell)}{Z_0} = -j\left(\frac{\tan \beta\ell + b_c}{b_c \tan \beta\ell}\right), \tag{6.77}$$

where $b_c = Z_0\omega C$ is the normalized susceptance of the coupling capacitor, C. Resonance occurs with $z = 0$, or when

$$\tan \beta\ell + b_c = 0. \tag{6.78}$$

The solutions to this transcendental equation are shown in the graph of Figure 6.17. In practice, $b_c \ll 1$, so the first resonant frequency, ω_1, will be close to the frequency for which $\beta\ell = \pi$ (the first resonant frequency of the unloaded resonator). The coupling of the resonator to the feedline has the effect of lowering its resonant frequency.

We now wish to simplify the driving point impedance of (6.77) to relate this resonator to a series *RLC* equivalent circuit. This can be accomplished by expanding $z(\omega)$ in a Taylor series about the resonant frequency, ω_1, and assuming that b_c is small. Thus,

$$z(\omega) = z\left(\omega_1\right) + (\omega - \omega_1)\frac{dz(\omega)}{d\omega}\bigg|_{\omega_1} + \cdots = (\omega - \omega_1)\frac{dz(\omega)}{d(\beta\ell)}\frac{d(\beta\ell)}{d\omega}\bigg|_{\omega_1} + \cdots, \tag{6.79}$$

since, from (6.77) and (6.78), $z(\omega_1) = 0$. Then,

$$\frac{dz}{d(\beta\ell)}\bigg|_{\omega_1} = j\frac{\sec^2 \beta\ell}{\tan^2 \beta\ell} = j\frac{1 + \tan^2 \beta\ell}{\tan^2 \beta\ell} = j\frac{1 + b_c^2}{b_c^2} \simeq \frac{j}{b_c^2},$$

where we have used (6.78) and the assumption that $b_c \ll 1$. Assuming a TEM line, we have $d(\beta\ell)/d\omega = \ell/v_p$, where v_p is the phase velocity of the line. Because $\ell \simeq \pi v_p/\omega_1$, the normalized impedance can be written as

$$z(\omega) \simeq \frac{j\ell(\omega - \omega_1)}{b_c^2 v_p} \simeq \frac{j\pi(\omega - \omega_1)}{\omega_1 b_c^2}. \tag{6.80}$$

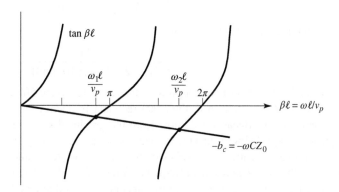

FIGURE 6.17 | Solutions to (6.78) for the resonant frequencies of the gap-coupled microstrip resonator.

So far we have ignored losses, but for a high-Q resonator loss can be included by replacing the resonant frequency, ω_1, with the complex resonant frequency given by $\omega_1(1 + j/2Q_0)$, which follows from (6.10). Applying this procedure to (6.80) gives the input impedance of the gap-coupled lossy resonator as

$$z(\omega) = \frac{\pi}{2Q_0 b_c^2} + j\frac{\pi(\omega - \omega_1)}{\omega_1 b_c^2}. \tag{6.81}$$

Note that an uncoupled $\lambda/2$ open-circuited transmission line resonator looks like a parallel RLC circuit near resonance, but the present case of a capacitive coupled $\lambda/2$ resonator looks like a series RLC circuit near resonance. This is because the series coupling capacitor has the effect of inverting the driving point impedance of the resonator (see the discussion of impedance inverters in Section 8.5).

At resonance the input resistance is $R = Z_0\pi/2Q_0 b_c^2$. For critical coupling we must have $R = Z_0$, or

$$b_c = \sqrt{\frac{\pi}{2Q_0}}. \tag{6.82}$$

The coupling coefficient of (6.76) is found to be

$$g = \frac{Z_0}{R} = \frac{2Q_0 b_c^2}{\pi}. \tag{6.83}$$

If $b_c < \sqrt{\pi/2Q}$, then $g < 1$ and the resonator is undercoupled; if $b_c > \sqrt{\pi/2Q}$, then $g > 1$ and the resonator is overcoupled.

EXAMPLE 6.6 DESIGN OF A GAP-COUPLED MICROSTRIP RESONATOR

A resonator is made from an open-circuited 50 Ω microstrip line and is gap coupled to a 50 Ω feedline, as in Figure 6.13a. The resonator has a length of 2.175 cm, an effective dielectric constant of 1.802, and an attenuation of 0.01 dB/cm near its resonance. Find the value of the coupling capacitor required for critical coupling, and the resulting resonant frequency.

Solution
The first resonant frequency will occur when the resonator is about $\ell = \lambda_g/2$ in length. Ignoring fringing fields, we find that the approximate resonant frequency is

$$f_0 = \frac{v_p}{\lambda_g} = \frac{c}{2\ell\sqrt{\epsilon_e}} = \frac{3 \times 10^8}{2(0.02175)\sqrt{1.802}} = 5.14 \text{ GHz}.$$

This result does not include the effect of the coupling capacitor. From (6.35) the unloaded Q of this resonator is

$$Q_0 = \frac{\beta}{2\alpha} = \frac{\pi}{\lambda_g\alpha} = \frac{\pi}{2\ell\alpha} = \frac{\pi(8.7 \text{ dB/Np})}{2(0.02175 \text{ m})(1 \text{ dB/m})} = 628.$$

From (6.82) the normalized coupling capacitor susceptance is

$$b_c = \sqrt{\frac{\pi}{2Q_0}} = \sqrt{\frac{\pi}{2(628)}} = 0.05,$$

so the coupling capacitor has a value of

$$C = \frac{b_c}{\omega Z_0} = \frac{0.05}{2\pi(5.14 \times 10^9)(50)} = 0.031 \text{ pF},$$

which should provide critical coupling of the resonator to the 50 Ω feedline.

Now that C is determined, the exact resonant frequency can be found by solving the transcendental equation of (6.78). Because we know from the graphical solution of Figure 6.17 that the actual resonant frequency is slightly lower than the unloaded resonant frequency of 5.0 GHz, it is an easy matter to calculate (6.78) for several frequencies in this vicinity, which leads to a value of about 4.918 GHz. This is about 1.6% lower than the unloaded resonant frequency. Figure 6.18 shows a Smith chart plot of the input impedance of the gap-coupled resonator for coupling capacitor values that lead to undercoupled, critically coupled, and overcoupled resonators.

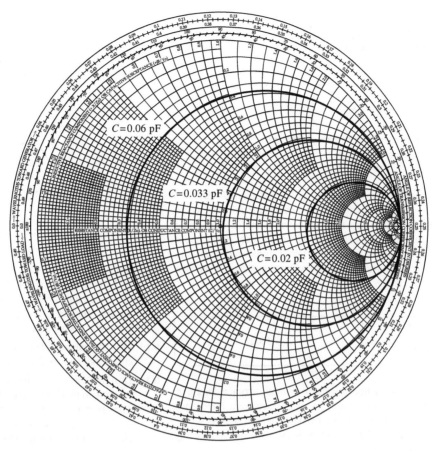

FIGURE 6.18 | Smith chart plot of input impedance of the gap-coupled microstrip resonator of Example 6.6 versus frequency for various values of the coupling capacitor.

REFERENCES

[1] R. E. Collin, *Foundations for Microwave Engineering*, 2nd edition, Wiley–IEEE Press, Hoboken, N.J., 2001.

[2] S. B. Cohn, "Microwave Bandpass Filters Containing High-Q Dielectric Resonators," *IEEE Transactions on Microwave Theory and Techniques*, vol. MTT-16, pp. 218–227, April 1968.

[3] M. W. Pospieszalski, "Cylindrical Dielectric Resonators and Their Applications in TEM Line Microwave Circuits," *IEEE Transactions on Microwave Theory and Techniques*, vol. MTT-27, pp. 233–238, March 1979.

PROBLEMS

6.1 A series *RLC* resonator with an external load is shown below. Find the resonant frequency, the unloaded Q, and the loaded Q.

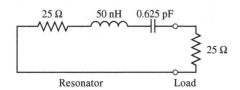

6.2 Derive an expression for the unloaded Q of a transmission line resonator consisting of a short-circuited transmission line 1λ long.

6.3 A transmission line resonator is fabricated from a $\lambda/4$ length of open-circuited line. Find the unloaded Q of this resonator if the complex propagation constant of the line is $\alpha + j\beta$.

6.4 Consider the resonator shown in the accompanying figure, consisting of a $\lambda/2$ length of lossless transmission line shorted at both ends. At an arbitrary point, z, on the line, compute the impedances Z_L and Z_R seen looking to the left and to the right, respectively, and show that $Z_L = Z_R^*$. (This condition holds true for any lossless transmission line resonator and is the basis for the transverse resonance technique discussed in Section 3.9.)

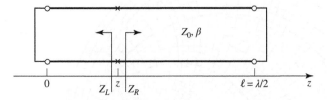

6.5 A resonator is constructed from a 3.5 cm length of 100 Ω air-filled coaxial line, shorted at one end and terminated with a capacitor at the other end, as shown below. (a) Determine the capacitor value to achieve the lowest order resonance at 5.0 GHz. (b) Now assume that loss is introduced by placing a 12,500 Ω resistor in parallel with the capacitor. Calculate the unloaded Q.

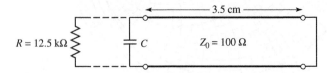

6.6 A transmission line resonator is made from a length ℓ of lossless transmission line of characteristic impedance $Z_0 = 100$ Ω. If the line is terminated at both ends as shown below, find ℓ/λ for the first resonance, and the unloaded Q of this resonator.

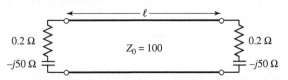

6.7 Write the expressions for the $\bar{E}$ and $\bar{H}$ fields for a short-circuited $\lambda/2$ coaxial line resonator, and show that the time-average stored electric and magnetic energies are equal.

6.8 A series RLC resonant circuit is connected to a length of transmission line that is $\lambda/4$ long at its resonant frequency, as shown below. Show that, in the vicinity of resonance, the input impedance behaves like that of a parallel RLC circuit.

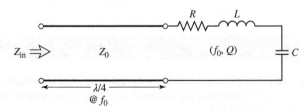

6.9 A rectangular cavity resonator is constructed from a 3.0 cm length of aluminum X-band waveguide. The cavity is air filled. Find the resonant frequency and unloaded Q of the TE_{101} and TE_{102} resonant modes.

6.10 Derive the unloaded Q for the TM_{111} mode of a rectangular cavity, assuming lossy conducting walls and lossless dielectric.

6.11 Consider the rectangular cavity resonator partially filled with dielectric as shown below. Derive a transcendental equation for the resonant frequency of the dominant mode by writing the fields in the air- and dielectric-filled regions in terms of TE_{10} waveguide modes, and enforcing boundary conditions at $z = 0$, $d - t$, and d.

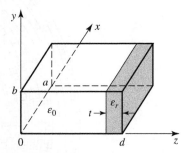

6.12 Determine the resonant frequencies of a rectangular cavity by carrying out a full separation-of-variables solution to the wave equation for E_z (for TM modes) and H_z (for TE modes), subject to the appropriate boundary conditions of the cavity. [Assume a solution of the form $X(x)Y(y)Z(z)$.]

6.13 Find the unloaded Q for the TM_{nm0} resonant mode of a circular cavity. Consider both conductor and dielectric losses.

6.14 Design a circular cavity resonator to operate in the TE_{111} mode with maximum unloaded Q at a frequency of 8 GHz. The cavity is gold plated and filled with a dielectric material having $\epsilon_r = 2.2$ and $\tan \delta = 0.0009$. Find the cavity dimensions and the resulting unloaded Q.

6.15 An air-filled rectangular cavity resonator has its first three resonant modes at the frequencies 5, 6, and 7.5 GHz. Find the dimensions of the cavity.

6.16 Consider the microstrip ring resonator shown below. If the effective dielectric constant of the microstrip line is ϵ_e, find an equation for the frequency of the first resonance. Suggest some methods of coupling to this resonator.

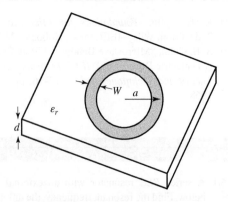

6.17 A circular microstrip disk resonator is shown in the accompanying figure. Solve the wave equation for TM_{nm0} modes for this structure, using the magnetic wall approximation that $H_\varphi = 0$ at $\rho = a$. If fringing fields are neglected, show that the resonant frequency of the dominant mode is given by

$$f_{110} = \frac{1.841c}{2\pi a \sqrt{\epsilon_r}}$$

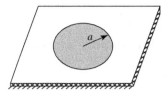

6.18 Compute the resonant frequency of a cylindrical dielectric resonator with $\epsilon_r = 10.8$, $2a = 11.71$ mm, and $L = 2.14$ mm.

6.19 Extend the analysis of Section 6.5 to derive a transcendental equation for the resonant frequency of the next resonant mode of the cylindrical dielectric resonator. (H_z odd in z.)

6.20 Consider the rectangular dielectric resonator shown below. Assume a magnetic wall boundary condition around the edges of the cavity, and allow evanescent fields in the $\pm z$ directions away from the dielectric, similar to the analysis of Section 6.5. Derive a transcendental equation for the resonant frequency.

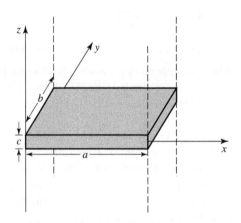

6.21 A high-Q resonator useful at millimeter wave frequencies is the Fabry-Perot resonator, which consists of two parallel metal plates (see the accompanying figure). A plane wave traveling at normal incidence between the two plates will exhibit resonance when the plate separation is equal to a multiple of $\lambda/2$. (a) Derive an expression for the resonant frequency of a Fabry-Perot resonator having a plate separation d and mode number ℓ. (b) If the plates have conductivity σ, derive an expression for the unloaded Q of the resonator. (c) Use these results to find the resonant frequency and unloaded Q of a Fabry-Perot resonator having $d = 5.0$ cm, with copper plates, and with a mode number $\ell = 35$.

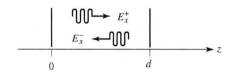

6.22 A parallel *RLC* circuit, with $R = 1.2$ kΩ, $L = 0.89$ nH, $C = 0.79$ pF, is coupled with a series capacitor, C_0, to a 50 Ω transmission line, as shown below. Determine C_0 for critical coupling to the line. What is the resonant frequency?

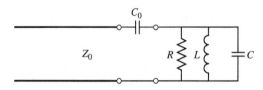

6.23 An aperture-coupled rectangular waveguide cavity has a resonant frequency of 10.0 GHz and an unloaded Q of 11,000. If the waveguide dimensions are $a = 10$ cm and $b = 8$ cm, find the normalized aperture reactance required for critical coupling.

6.24 A microwave resonator is connected as a one-port circuit, and its return loss is measured versus frequency. At resonance the return loss is 14 dB, while at 2.3988 GHz and at 2.4012 GHz the return loss is 11 dB (the half-power points). Determine the unloaded Q of the resonator. Do this for both series and parallel resonators.

6.25 A microwave resonator is measured in a two-port configuration like that shown in Figure 6.21. The minimum insertion loss is measured as 1.54 dB at 3.0000 GHz. The insertion loss is 4.54 dB at 2.9925 GHz and at 3.0075 GHz. What is the unloaded Q of the resonator?

6.26 A thin slab of magnetic material is inserted next to the $z = 0$ wall of the rectangular cavity shown below. If the cavity is operating in the TE_{101} mode, derive a perturbational expression for the change in resonant frequency caused by the magnetic material.

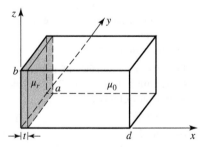

MULTIPLE CHOICE QUESTIONS

6.1 Quality factor Q of a resonant circuit signifies

(a) loss in the resonant circuit
(b) gain in the resonant circuit
(c) magnetic energy stored in the circuit
(d) electric energy stored in the circuit

6.2 The relation between unloaded Q and external Q is

(a) external Q > unloaded Q
(b) unloaded Q > external Q
(c) external Q = unloaded Q
(d) none of the above

6.3 If a parallel RLC circuit is excited with a source of 8 V, 50 Hz and the circuit has an inductor of 1 mH, a capacitor of 1 µF, and a resistor of 50 Ω, then the power loss that occurs in the circuit is

(a) 6.4 mW   (c) 12.8 mW

(b) 3.2 mW   (d) none of the above

6.4 Short-circuited $\lambda/2$ transmission line has a quality factor of

(a) $\dfrac{\beta}{2\alpha}$ (c) $\dfrac{\beta}{\alpha}$

(b) $\dfrac{2\beta}{\alpha}$ (d) $\dfrac{Z_0}{Z_L}$

6.5 An air coaxial cable has attenuation of 0.022 and phase constant of 104.7, then the quality factor of a $\lambda/2$ short-circuited resonator made out of this material is

(a) 1218   (c) 1416

(b) 2380   (d) insufficient data

6.6 A microstrip patch antenna has a width of 5.08 mm and a surface resistivity of 1.84×10^{-2}. Then the attenuation due to conductor loss is

(a) 0.0724   (c) 0.054

(b) 0.034   (d) none of the above

6.7 If the attenuation due to dielectric loss and attenuation due to conductor loss in a microstrip transmission line is 0.024 Np/m and 0.0724 Np/m, respectively, then the unloaded quality factor if the propagation constant is 151 is

(a) 150   (c) 587

(b) 234   (d) 783

6.8 In order to obtain the resonant frequency of a rectangular waveguide, the closed cavity has to satisfy

(a) Gaussian equation   (c) Ampere's law

(b) Helmholtz equation   (d) none of the above

6.9 Unloaded Q of a rectangular waveguide cavity resonator

(a) does not exist

(b) defined as the ratio of length of the waveguide to breadth of the waveguide

(c) defined as the ratio of stored energy to power dissipated in the walls

(d) none of the above

6.10 Find the wave number of a rectangular cavity resonator filled with a dielectric of 2.25 and designed to operate at a frequency of 5 GHz.

(a) 157.08   (c) 345.1

(b) 145.2   (d) 415.08

6.11 The required length of the cavity resonator for $l = 1$ mode $(m = 1, n = 0)$ given that the wave number of the cavity resonator is 157.01 and the broader dimension of the waveguide is 4.755 cm is

(a) 1.10 cm   (c) 2.8 cm

(b) 2.20 cm   (d) 1.8 cm

6.12 If the loss tangent of a rectangular waveguide is 0.0004, then Q due to dielectric loss is

(a) 1250   (c) 2500

(b) 2450   (d) 1800

6.13 A cylindrical cavity resonator can be constructed using a circular waveguide

(a) open at both the ends

(b) shorted at both the ends

(c) matched at both the ends

(d) none of the above

6.14 The mode of a circular cavity resonator used in frequency meters is

(a) TE_{011} mode   (c) TE_{111} mode

(b) TE_{101} mode   (d) TM_{111} mode

6.15 The propagation constant of $TE_{m,n}$ mode of propagation for a cylindrical cavity resonator is

(a) $\sqrt{k^2 - \left(\dfrac{p_{nm}}{a}\right)^2}$ (c) $\sqrt{k^2 + \left(\dfrac{p_{nm}}{a}\right)^2}$

(b) $\sqrt{\dfrac{p_{nm}}{a}}$ (d) none of the above

6.16 A circular cavity resonator is filled with a dielectric of 2.08 and is operating at 5 GHz of frequency. Then the wave number is

(a) 181   (c) 161

(b) 151   (d) 216

6.17 If a dielectric resonator has a dielectric constant of 49, then the reflection coefficient of the dielectric resonator is

(a) 0.5   (c) 0.1

(b) 0.7   (d) 0.75

6.18 The major disadvantage of dielectric resonators is

(a) complex construction

(b) field fringing

(c) requirement of high dielectric constant

(d) none of the above

6.19 When impedance matching is done between a resonator and a feed line, the condition for impedance matching is

(a) $R = Z_0$ (c) $R = 2Z_0$

(b) $R = \dfrac{Z_0}{2}$ (d) $R = \sqrt{Z_0}$

6.20 A measure of the level of coupling between a resonator and a feed is given by

(a) coupling coefficient

(b) power transfer coefficient

(c) voltage coefficient

(d) reflection coefficient

ANSWER KEY

6.1 (a)	**6.5** (b)	**6.9** (c)	**6.13** (b)	**6.17** (d)
6.2 (b)	**6.6** (a)	**6.10** (a)	**6.14** (a)	**6.18** (b)
6.3 (a)	**6.7** (d)	**6.11** (b)	**6.15** (a)	**6.19** (a)
6.4 (a)	**6.8** (b)	**6.12** (c)	**6.16** (b)	**6.20** (a)

Power Dividers and Directional Couplers

Learning Objectives

After completing this chapter, you will be able to

- Understand the concepts of power dividers and couplers, and their properties
- Develop and design T-junction, Wilkinson, and waveguide power dividers

- Do even- and odd-mode analysis of 90° and 180° hybrid couplers

Power dividers and directional couplers are passive microwave components used for power division or power combining, as illustrated in Figure 7.1. In power division, an input signal is divided into two (or more) output signals of lesser power, while a power combiner accepts two or more input signals and combines them at an output port. The coupler or divider may have three ports, four ports, or more, and may be (ideally) lossless. Three-port networks take the form of T-junctions and other power dividers, while four-port networks take the form of directional couplers and hybrids. Power dividers usually provide in-phase output signals with an equal power division ratio (3 dB), but unequal power division ratios are also possible. Directional couplers can be designed for arbitrary power division, while hybrid junctions usually have equal power division. Hybrid junctions have either a 90° or a 180° phase shift between the output ports.

A wide variety of waveguide couplers and power dividers were invented and characterized at the MIT Radiation Laboratory in the 1940s. These included *E*- and *H*-plane waveguide T-junctions, the Bethe hole coupler, multihole directional couplers, the Schwinger coupler, the waveguide magic-T, and various types of couplers using coaxial probes. In the mid-1950s through the 1960s, many of these couplers were reinvented to use stripline or microstrip technology. The increasing use of planar lines also led to the development of new types of couplers and dividers, such as the Wilkinson divider, the branch line hybrid, and the coupled line directional coupler.

We will first discuss some of the general properties of three- and four-port networks, and then treat the analysis and design of several of the most common types of power dividers, couplers, and hybrids.

7.1 BASIC PROPERTIES OF DIVIDERS AND COUPLERS

In this section we will use properties of the scattering matrix developed in Section 4.3 to derive some of the basic characteristics of three- and four-port networks. We will also define isolation, coupling, and directivity, which are important quantities for the characterization of couplers and hybrids.

Three-Port Networks (T-Junctions)

The simplest type of power divider is a *T-junction*, which is a three-port network with two inputs and one output. The scattering matrix of an arbitrary three-port network has nine independent elements:

$$[S] = \begin{bmatrix} S_{11} & S_{12} & S_{13} \\ S_{21} & S_{22} & S_{23} \\ S_{31} & S_{32} & S_{33} \end{bmatrix}. \tag{7.1}$$

If the device is passive and contains no anisotropic materials, then it must be reciprocal and its scattering matrix will be symmetric ($S_{ij} = S_{ji}$). Usually, to avoid power loss, we would like to have a junction that

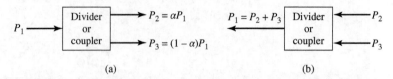

FIGURE 7.1 | Power division and combining. (a) Power division. (b) Power combining.

is lossless and matched at all ports. We can easily show, however, that it is impossible to construct such a three-port lossless reciprocal network that is matched at all ports.

If all ports are matched, then $S_{ii} = 0$, and if the network is reciprocal, the scattering matrix of (7.1) reduces to

$$[S] = \begin{bmatrix} 0 & S_{12} & S_{13} \\ S_{12} & 0 & S_{23} \\ S_{13} & S_{23} & 0 \end{bmatrix}.$$ (7.2)

If the network is also lossless, then energy conservation requires that the scattering matrix satisfy the unitary properties of (4.53), which leads to the following conditions [1, 2]:

$$|S_{12}|^2 + |S_{13}|^2 = 1,$$ (7.3a)

$$|S_{12}|^2 + |S_{23}|^2 = 1,$$ (7.3b)

$$|S_{13}|^2 + |S_{23}|^2 = 1,$$ (7.3c)

$$S_{13}^* S_{23} = 0,$$ (7.3d)

$$S_{23}^* S_{12} = 0,$$ (7.3e)

$$S_{12}^* S_{13} = 0.$$ (7.3f)

Equations (7.3d)–(7.3f) show that at least two of the three parameters (S_{12}, S_{13}, S_{23}) must be zero. However, this condition will always be inconsistent with one of equations (7.3a)–(7.3c), implying that a three-port network cannot be simultaneously lossless, reciprocal, and matched at all ports. If any one of these three conditions is relaxed, then a physically realizable device is possible.

Based on the relaxation of any one condition, three-port networks are categorized as T-junctions, power dividers, and circulators, as shown in Table 7.1. The conditions are as follows:

- If the three-port network satisfies the lossless and reciprocal conditions and fails the matched condition, then it is called a T-junction.
- If the three-port network satisfies the matched and reciprocal conditions and fails the lossless condition, then it is called a power divider.
- If the three-port network satisfies the matched and lossless conditions and fails the reciprocal condition, then it is called a circulator.

If the three-port network is nonreciprocal, then $S_{ij} \neq S_{ji}$, and the conditions of input matching at all ports and energy conservation can be satisfied. Such a device is known as a *circulator*, and generally relies on an anisotropic material, such as ferrite, to achieve nonreciprocal behavior. Ferrite circulators will be discussed in more detail in Chapter 9, but we can demonstrate here that any matched lossless three-port network

TABLE 7.1 | **Categorization of Three-Port Networks**

Three-port Networks			
Matched	Lossless	Reciprocal	Passive Device
✗	✓	✓	T-junctions
✓	✗	✓	Power dividers
✓	✓	✗	Circulators

must be nonreciprocal and, thus, a circulator. The scattering matrix of a matched three-port network has the following form:

$$[S] = \begin{bmatrix} 0 & S_{12} & S_{13} \\ S_{21} & 0 & S_{23} \\ S_{31} & S_{32} & 0 \end{bmatrix}. \tag{7.4}$$

If the network is lossless, $[S]$ must be unitary, which implies the following conditions:

$$S_{31}^* S_{32} = 0, \tag{7.5a}$$

$$S_{21}^* S_{23} = 0, \tag{7.5b}$$

$$S_{12}^* S_{13} = 0, \tag{7.5c}$$

$$|S_{12}|^2 + |S_{13}|^2 = 1, \tag{7.5d}$$

$$|S_{21}|^2 + |S_{23}|^2 = 1, \tag{7.5e}$$

$$|S_{31}|^2 + |S_{32}|^2 = 1. \tag{7.5f}$$

These equations can be satisfied in one of two ways. Either

$$S_{12} = S_{23} = S_{31} = 0, \quad |S_{21}| = |S_{32}| = |S_{13}| = 1, \tag{7.6a}$$

or

$$S_{21} = S_{32} = S_{13} = 0, \quad |S_{12}| = |S_{23}| = |S_{31}| = 1. \tag{7.6b}$$

These results shows that $S_{ij} \neq S_{ji}$ for $i \neq j$, which implies that the device must be nonreciprocal. The scattering matrices for the two solutions of (7.6) are shown in Figure 7.2, together with the symbols for the two possible types of circulators and their scattering parameters representation. The only difference between the two cases is in the direction of power flow between the ports: solution (7.6a) corresponds to a circulator that allows power flow only from port 1 to 2, or port 2 to 3, or port 3 to 1, while solution (7.6b) corresponds to a circulator with the opposite direction of power flow.

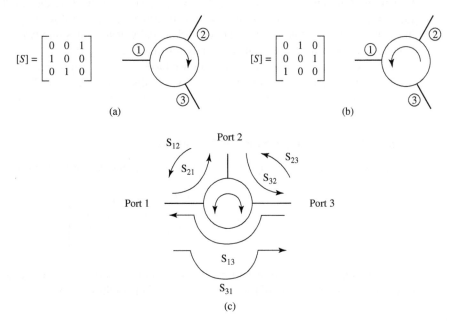

(a)   (b)

(c)

FIGURE 7.2 | Two types of circulators and their scattering matrices. (a) Clockwise circulation. (b) Counterclockwise circulation. (c) Scattering parameters representation. The phase references for the ports are arbitrary.

$$[S] = \begin{bmatrix} 0 & e^{j\theta} & 0 \\ e^{j\theta} & 0 & 0 \\ 0 & 0 & e^{j\phi} \end{bmatrix}$$

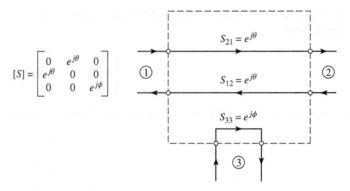

FIGURE 7.3 | A reciprocal lossless three-port network matched at ports 1 and 2.

Alternatively, a lossless and reciprocal three-port network can be physically realized if only two of its ports are matched [1]. If ports 1 and 2 are the matched ports, then the scattering matrix can be written as

$$[S] = \begin{bmatrix} 0 & S_{12} & S_{13} \\ S_{12} & 0 & S_{23} \\ S_{13} & S_{23} & S_{33} \end{bmatrix}. \tag{7.7}$$

To be lossless, the following unitarity conditions must be satisfied:

$$S_{13}^* S_{23} = 0, \tag{7.8a}$$

$$S_{12}^* S_{13} + S_{23}^* S_{33} = 0, \tag{7.8b}$$

$$S_{23}^* S_{12} + S_{33}^* S_{13} = 0, \tag{7.8c}$$

$$|S_{12}|^2 + |S_{13}|^2 = 1, \tag{7.8d}$$

$$|S_{12}|^2 + |S_{23}|^2 = 1, \tag{7.8e}$$

$$|S_{13}|^2 + |S_{23}|^2 + |S_{33}|^2 = 1. \tag{7.8f}$$

Equations (7.8d) and (7.8e) show that $|S_{13}| = |S_{23}|$, so (7.8a) leads to the result that $S_{13} = S_{23} = 0$. Then, $|S_{12}| = |S_{33}| = 1$. The scattering matrix and corresponding signal flow graph for this network are shown in Figure 7.3, where it is seen that the network actually degenerates into two separate components—one a matched two-port line and the other a totally mismatched one-port.

Finally, if the three-port network is allowed to be lossy, it can be reciprocal and matched at all ports; this is the case of the *resistive divider*, which will be discussed in Section 7.2. In addition, a lossy three-port network can be made to have isolation between its output ports (e.g., $S_{23} = S_{32} = 0$).

Four-Port Networks (Directional Couplers)

The scattering matrix of a reciprocal four-port network matched at all ports has the following form:

$$[S] = \begin{bmatrix} 0 & S_{12} & S_{13} & S_{14} \\ S_{12} & 0 & S_{23} & S_{24} \\ S_{13} & S_{23} & 0 & S_{34} \\ S_{14} & S_{24} & S_{34} & 0 \end{bmatrix}. \tag{7.9}$$

If the network is lossless, 10 equations result from the unitarity, or energy conservation, condition [1, 2]. Consider the multiplication of row 1 and row 2, and the multiplication of row 4 and row 3:

$$S_{13}^* S_{23} + S_{14}^* S_{24} = 0, \tag{7.10a}$$

$$S_{14}^* S_{13} + S_{24}^* S_{23} = 0. \tag{7.10b}$$

Multiply (7.10a) by S_{24}^*, and (7.10b) by S_{13}^*, and subtract to obtain

$$S_{14}^*(|S_{13}|^2 - |S_{24}|^2) = 0. \tag{7.11}$$

Similarly, the multiplication of row 1 and row 3, and the multiplication of row 4 and row 2, gives

$$S_{12}^* S_{23} + S_{14}^* S_{34} = 0, \tag{7.12a}$$

$$S_{14}^* S_{12} + S_{34}^* S_{23} = 0. \tag{7.12b}$$

Multiply (7.12a) by S_{12}, and (7.12b) by S_{34}, and subtract to obtain

$$S_{23}(|S_{12}|^2 - |S_{34}|^2) = 0. \tag{7.13}$$

One way for (7.11) and (7.13) to be satisfied is if $S_{14} = S_{23} = 0$, which results in a directional coupler. Then the self-products of the rows of the unitary scattering matrix of (7.9) yield the following equations:

$$|S_{12}|^2 + |S_{13}|^2 = 1, \tag{7.14a}$$

$$|S_{12}|^2 + |S_{24}|^2 = 1, \tag{7.14b}$$

$$|S_{13}|^2 + |S_{34}|^2 = 1, \tag{7.14c}$$

$$|S_{24}|^2 + |S_{34}|^2 = 1, \tag{7.14d}$$

which imply that $|S_{13}| = |S_{24}|$ [using (7.14a) and (7.14b)], and that $|S_{12}| = |S_{34}|$ [using (7.14b) and (7.14d)].

Further simplification can be made by choosing the phase references on three of the four ports. Thus, we choose $S_{12} = S_{34} = \alpha$, $S_{13} = \beta e^{j\theta}$, and $S_{24} = \beta e^{j\phi}$, where α and β are real, and θ and ϕ are phase constants to be determined (one of which we are still free to choose). The dot product of rows 2 and 3 gives

$$S_{12}^* S_{13} + S_{24}^* S_{34} = 0, \tag{7.15}$$

which yields a relation between the remaining phase constants as

$$\theta + \phi = \pi \pm 2n\pi. \tag{7.16}$$

If we ignore integer multiples of 2π, there are two particular choices that commonly occur in practice:

1. *A Symmetric Coupler*: $\theta = \phi = \pi/2$. The phases of the terms having amplitude β are chosen equal. Then the scattering matrix has the following form:

$$[S] = \begin{bmatrix} 0 & \alpha & j\beta & 0 \\ \alpha & 0 & 0 & j\beta \\ j\beta & 0 & 0 & \alpha \\ 0 & j\beta & \alpha & 0 \end{bmatrix}. \tag{7.17}$$

2. *An Antisymmetric Coupler*: $\theta = 0$, $\phi = \pi$. The phases of the terms having amplitude β are chosen to be 180° apart. Then the scattering matrix has the following form:

$$[S] = \begin{bmatrix} 0 & \alpha & \beta & 0 \\ \alpha & 0 & 0 & -\beta \\ \beta & 0 & 0 & \alpha \\ 0 & -\beta & \alpha & 0 \end{bmatrix}. \tag{7.18}$$

Note that these two couplers differ only in the choice of reference planes. In addition, the amplitudes α and β are not independent, as (7.14a) requires that

$$\alpha^2 + \beta^2 = 1. \tag{7.19}$$

Thus, apart from phase references, an ideal four-port directional coupler has only one degree of freedom, leading to two possible configurations.

Another way for (7.11) and (7.13) to be satisfied is if $|S_{13}| = |S_{24}|$ and $|S_{12}| = |S_{34}|$. If we choose phase references, however, such that $S_{13} = S_{24} = \alpha$ and $S_{12} = S_{34} = j\beta$ [which satisfies (7.16)], then (7.10a) yields $\alpha(S_{23} + S_{14}^*) = 0$, and (7.12a) yields $\beta(S_{14}^* - S_{23}) = 0$. These two equations have two possible solutions.

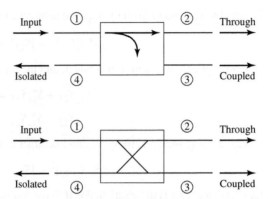

FIGURE 7.4 | Two commonly used symbols for directional couplers, and power flow conventions.

First, $S_{14} = S_{23} = 0$, which is the same as the above solution for the directional coupler. The other solution occurs for $\alpha = \beta = 0$, which implies that $S_{12} = S_{13} = S_{24} = S_{34} = 0$. This is the degenerate case of two decoupled two-port networks (between ports 1 and 4, and ports 2 and 3), which is of trivial interest and will not be considered further. We are thus left with the conclusion that any reciprocal, lossless, matched four-port network is a directional coupler.

The basic operation of a directional coupler can be illustrated with the aid of Figure 7.4, which shows two commonly used symbols for a directional coupler and the port definitions. Power supplied to port 1 is coupled to port 3 (the *coupled* port) with the coupling factor $|S_{13}|^2 = \beta^2$, while the remainder of the input power is delivered to port 2 (the *through* port) with the coefficient $|S_{12}|^2 = \alpha^2 = 1 - \beta^2$. In an ideal directional coupler, no power is delivered to port 4 (the *isolated* port).

The following quantities are commonly used to characterize a directional coupler:

$$\text{Coupling} = C = 10 \log \frac{P_1}{P_3} = -20 \log \beta \text{ dB,} \tag{7.20a}$$

$$\text{Directivity} = D = 10 \log \frac{P_3}{P_4} = 20 \log \frac{\beta}{|S_{14}|} \text{ dB,} \tag{7.20b}$$

$$\text{Isolation} = I = 10 \log \frac{P_1}{P_4} = -20 \log |S_{14}| \text{ dB,} \tag{7.20c}$$

$$\text{Insertion loss } = L = 10 \log \frac{P_1}{P_2} = -20 \log |S_{12}| \text{ dB.} \tag{7.20d}$$

The *coupling factor* indicates the fraction of the input power that is coupled to the output port. The *directivity* is a measure of the coupler's ability to isolate forward and backward waves (or the coupled and uncoupled ports). The *isolation* is a measure of the power delivered to the uncoupled port. These quantities are related as

$$I = D + C \text{ dB.} \tag{7.21}$$

The *insertion loss* accounts for the input power delivered to the through port, diminished by power delivered to the coupled and isolated ports. The ideal coupler has infinite directivity and isolation ($S_{14} = 0$). Then both α and β can be determined from the coupling factor, C.

Hybrid couplers are special cases of directional couplers, where the coupling factor is 3 dB, which implies that $\alpha = \beta = 1/\sqrt{2}$. There are two types of hybrids. The *quadrature hybrid* has a 90° phase shift between ports 2 and 3 ($\theta = \phi = \pi/2$) when fed at port 1, and is an example of a symmetric coupler. Its scattering matrix has the following form:

$$[S] = \frac{1}{\sqrt{2}} \begin{bmatrix} 0 & 1 & j & 0 \\ 1 & 0 & 0 & j \\ j & 0 & 0 & 1 \\ 0 & j & 1 & 0 \end{bmatrix}. \tag{7.22}$$

The *magic-T hybrid* and the *rat-race hybrid* have a 180° phase difference between ports 2 and 3 when fed at port 4, and are examples of an antisymmetric coupler. Its scattering matrix has the following form:

$$[S] = \frac{1}{\sqrt{2}} \begin{bmatrix} 0 & 1 & 1 & 0 \\ 1 & 0 & 0 & -1 \\ 1 & 0 & 0 & 1 \\ 0 & -1 & 1 & 0 \end{bmatrix}. \tag{7.23}$$

POINT OF INTEREST: Measuring Coupler Directivity

The directivity of a directional coupler is a measure of the coupler's ability to separate forward and reverse wave components, and applications of directional couplers often require high (35 dB or greater) directivity. Poor directivity will limit the accuracy of a reflectometer, and can cause variations in the coupled power level from a coupler when there is even a small mismatch on the through line.

The directivity of a coupler generally cannot be measured directly because it involves a low-level signal that can be masked by coupled power from a reflected wave on the through arm. For example, if a coupler has $C = 20$ dB and $D = 35$ dB, with a load having a return loss RL $= 30$ dB, the signal level through the directivity path will be $D + C = 55$ dB below the input power, but the reflected power through the coupled arm will only be RL $+ C = 50$ dB below the input power.

One way to measure coupler directivity uses a sliding matched load, as follows. First, the coupler is connected to a source and a matched load, as shown in the accompanying left-hand figure, and the coupled output power is measured. If we assume an input power P_i, this power will be $P_c = C^2 P_i$, where $C = 10^{(-C\text{dB})/20}$ is the numerical voltage coupling factor of the coupler. Next, the position of the coupler is reversed, and the through line is terminated with a sliding load, as shown in the right-hand figure.

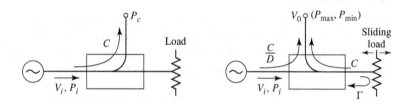

Changing the position of the sliding load introduces a variable phase shift in the signal reflected from the load and coupled to the output port. The voltage at the output port can be written as

$$V_0 = V_i \left(\frac{C}{D} + C|\Gamma| e^{-j\theta} \right),$$

where V_i is the input voltage, $D = 10^{(D\ \text{dB})/20} \geq 1$ is the numerical value of the directivity, $|\Gamma|$ is the reflection coefficient magnitude of the load, and θ is the path length difference between the directivity and reflected signals. Moving the sliding load changes θ, so the two signals will combine to trace out a circular locus, as shown in the following figure.

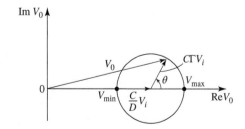

The minimum and maximum output powers are given by

$$P_{\min} = P_i \left(\frac{C}{D} - C|\Gamma| \right)^2, \quad P_{\max} = P_i \left(\frac{C}{D} + C|\Gamma| \right)^2.$$

Let M and m be defined in terms of these powers as follows:

$$M = \frac{P_c}{P_{\text{max}}} = \left(\frac{D}{1 + |\Gamma|D} \right)^2, \quad m = \frac{P_{\text{max}}}{P_{\text{min}}} = \left(\frac{1 + |\Gamma|D}{1 - |\Gamma|D} \right)^2.$$

These ratios can be accurately measured directly by using a variable attenuator between the source and coupler. The coupler directivity (numerical) can then be found as

$$D = M \left(\frac{2m}{m + 1} \right).$$

This method requires that $|\Gamma| < 1/D$ or, in dB, RL > D.

Reference: M. Sucher and J. Fox, eds., *Handbook of Microwave Measurements*, 3rd edition, Volume II, Polytechnic Press, New York, 1963.

7.2 THE T-JUNCTION POWER DIVIDER

The T-junction power divider is a simple three-port network that can be used for power division or power combining, and it can be implemented in virtually any type of transmission line medium. Figure 7.5 shows some commonly used T-junctions in waveguide and microstrip line or stripline form. The junctions shown here are, in the absence of transmission line loss, lossless junctions. Thus, as discussed in the preceding section, such junctions cannot be matched simultaneously at all ports. We will first analyze the T-junction divider, followed by a discussion of the resistive power divider, which can be matched at all ports but is not lossless.

Lossless Divider

The lossless T-junction dividers of Figure 7.5 can all be modeled as a junction of three transmission lines, as shown in Figure 7.6 [3]. In general, there may be fringing fields and higher order modes associated with the discontinuity at such a junction, leading to stored energy that can be accounted for by a lumped susceptance, B. In order for the divider to be matched to the input line of characteristic impedance Z_0, we must have

$$Y_{\text{in}} = jB + \frac{1}{Z_1} + \frac{1}{Z_2} = \frac{1}{Z_0}. \tag{7.24}$$

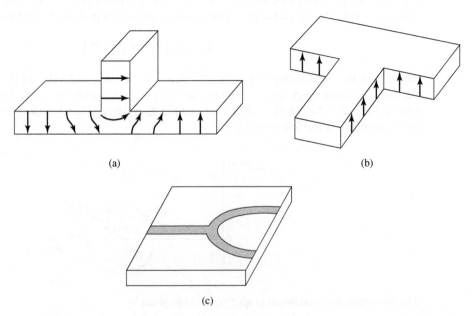

(a) (b)

(c)

FIGURE 7.5 | Various T-junction power dividers. (a) *E*-plane waveguide T. (b) *H*-plane waveguide T. (c) Microstrip line T-junction divider.

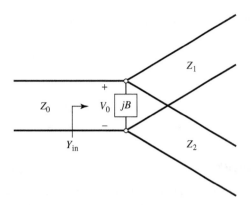

FIGURE 7.6 | Transmission line model of a lossless T-junction divider.

If the transmission lines are assumed to be lossless (or of low loss), then the characteristic impedances are real. If we also assume $B = 0$, then (7.24) reduces to

$$\frac{1}{Z_1} + \frac{1}{Z_2} = \frac{1}{Z_0}. \tag{7.25}$$

In practice, if B is not negligible, some type of discontinuity compensation or a reactive tuning element can usually be used to cancel this susceptance, at least over a narrow frequency range.

The output line impedances, Z_1 and Z_2, can be selected to provide various power division ratios. Thus, for a 50 Ω input line, a 3 dB (equal split) power divider can be made by using two 100 Ω output lines. If necessary, quarter-wave transformers can be used to bring the output line impedances back to the desired levels. If the output lines are matched, then the input line will be matched. There will be no isolation between the two output ports, however, and there will be a mismatch looking into the output ports.

EXAMPLE 7.1   THE T-JUNCTION POWER DIVIDER

A lossless T-junction power divider has a source impedance of 60 Ω. Find the output characteristic impedances so that the output powers are in a 2:1 ratio. Compute the reflection coefficients seen looking into the output ports.

Solution
If the voltage at the junction is V_0, as shown in Figure 7.6, the input power to the matched divider is

$$P_{in} = \frac{1}{2}\frac{V_0^2}{Z_0},$$

while the output powers are

$$P_1 = \frac{1}{2}\frac{V_0^2}{Z_1} = \frac{1}{3}P_{in},$$

$$P_2 = \frac{1}{2}\frac{V_0^2}{Z_2} = \frac{2}{3}P_{in}.$$

These results yield the characteristic impedances as

$$Z_1 = 3Z_0 = 180 \text{ Ω},$$

$$Z_2 = \frac{3Z_0}{2} = 90 \text{ Ω}.$$

The input impedance to the junction is

$$Z_{in} = 90||180 = 60 \text{ Ω},$$

so that the input is matched to the 60 Ω source.

Looking into the 180 Ω output line, we see an impedance of $60 \,||\, 90 = 36\ \Omega$, while at the 90 Ω output line we see an impedance of $60 \,||\, 180 = 45\ \Omega$. The reflection coefficients seen looking into these ports are

$$\Gamma_1 = \frac{36 - 180}{36 + 180} = -0.667,$$

$$\Gamma_2 = \frac{45 - 90}{45 + 90} = -0.333.$$

Resistive Divider

If a three-port divider contains lossy components, it can be made to be matched at all ports, although the two output ports may not be isolated [3]. The circuit for such a divider is illustrated in Figure 7.7, using lumped-element resistors. An equal-split (-3 dB) divider is shown, but unequal power division ratios are also possible.

The resistive divider of Figure 7.7 can easily be analyzed using circuit theory. Assuming that all ports are terminated in the characteristic impedance Z_0, the impedance Z, seen looking into the $Z_0/3$ resistor followed by a terminated output line, is

$$Z = \frac{Z_0}{3} + Z_0 = \frac{4Z_0}{3}. \tag{7.26}$$

Then the input impedance of the divider is

$$Z_{in} = \frac{Z_0}{3} + \frac{2Z_0}{3} = Z_0, \tag{7.27}$$

which shows that the input is matched to the feed line. Because the network is symmetric from all three ports, the output ports are also matched. Thus, $S_{11} = S_{22} = S_{33} = 0$.

If the voltage at port 1 is V_1, then by voltage division the voltage V at the center of the junction is

$$V = V_1 \frac{2Z_0/3}{Z_0/3 + 2Z_0/3} = \frac{2}{3} V_1, \tag{7.28}$$

and the output voltages are, again by voltage division,

$$V_2 = V_3 = V \frac{Z_0}{Z_0 + Z_0/3} = \frac{3}{4} V = \frac{1}{2} V_1. \tag{7.29}$$

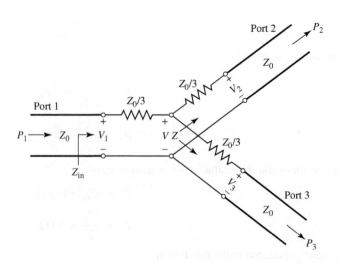

FIGURE 7.7 | An equal-split three-port resistive power divider.

Thus, $S_{21} = S_{31} = S_{23} = 1/2$, so the output powers are 6 dB below the input power level. The network is reciprocal, so the scattering matrix is symmetric, and it can be written as

$$[S] = \frac{1}{2} \begin{bmatrix} 0 & 1 & 1 \\ 1 & 0 & 1 \\ 1 & 1 & 0 \end{bmatrix}. \qquad (7.30)$$

The reader may verify that this is not a unitary matrix.

The power delivered to the input of the divider is

$$P_{in} = \frac{1}{2}\frac{V_1^2}{Z_0}, \qquad (7.31)$$

while the output powers are

$$P_2 = P_3 = \frac{1}{2}\frac{(1/2V_1)^2}{Z_0} = \frac{1}{8}\frac{V_1^2}{Z_0} = \frac{1}{4}P_{in}, \qquad (7.32)$$

which shows that half of the supplied power is dissipated in the resistors.

7.3 THE WILKINSON POWER DIVIDER

The lossless T-junction divider suffers from the disadvantage of not being matched at all ports, and it does not have isolation between output ports. The resistive divider can be matched at all ports, but even though it is not lossless, isolation is still not achieved. From the discussion in Section 7.1, however, we know that a lossy three-port network can be made having all ports matched, with isolation between output ports. The *Wilkinson power divider* [4] is such a network, with the useful property of appearing lossless when the output ports are matched; that is, only reflected power from the output ports is dissipated.

The Wilkinson power divider can be made with arbitrary power division, but we will first consider the equal-split (3 dB) case. This divider is often made in microstrip line or stripline form, as depicted in Figure 7.8a; the corresponding transmission line circuit is given in Figure 7.8b. We will analyze this circuit by reducing it to two simpler circuits driven by symmetric and antisymmetric sources at the output ports. This "even-odd" mode analysis technique [5] will also be useful for other networks that we will study in later sections.

Even-Odd Mode Analysis

For simplicity, we can normalize all impedances to the characteristic impedance Z_0, and redraw the circuit of Figure 7.8b with voltage generators at the output ports as shown in Figure 7.9. This network has been drawn in a form that is symmetric across the midplane; the two source resistors of normalized value 2 combine

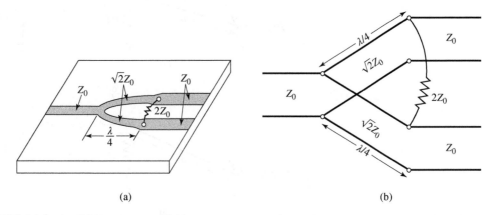

(a)   (b)

FIGURE 7.8 | The Wilkinson power divider. (a) An equal-split Wilkinson power divider in microstrip line form. (b) Equivalent transmission line circuit.

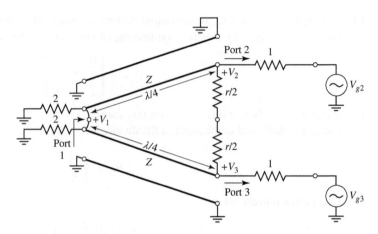

FIGURE 7.9 | The Wilkinson power divider circuit in normalized and symmetric form.

in parallel to give a resistor of normalized value 1, representing the impedance of a matched source. The quarter-wave lines have a normalized characteristic impedance Z, and the shunt resistor has a normalized value of r; we shall show that, for the equal-split power divider, these values should be $Z = \sqrt{2}$ and $r = 2$, as given in Figure 7.8.

Now define two separate modes of excitation for the circuit of Figure 7.9: the *even mode*, where $V_{g2} = V_{g3} = 2V_0$, and the *odd mode*, where $V_{g2} = -V_{g3} = 2V_0$. Superposition of these two modes effectively produces an excitation of $V_{g2} = 4V_0$ and $V_{g3} = 0$, from which we can find the scattering parameters of the network. We now treat these two modes separately.

Even mode: For even-mode excitation, $V_{g2} = V_{g3} = 2V_0$, so $V_2^e = V_3^e$, and therefore no current flows through the $r/2$ resistors or the short circuit between the inputs of the two transmission lines at port 1. We can then bisect the network of Figure 7.9 with open circuits at these points to obtain the network of Figure 7.10a (the grounded side of the $\lambda/4$ line is not shown). Then, looking into port 2, we see an impedance

$$Z_{in}^e = \frac{Z^2}{2}, \tag{7.33}$$

since the transmission line looks like a quarter-wave transformer. Thus, if $Z = \sqrt{2}$, port 2 will be matched for even-mode excitation; then $V_2^e = V_0$ since $Z_{in}^e = 1$. The $r/2$ resistor is superfluous in this case since one

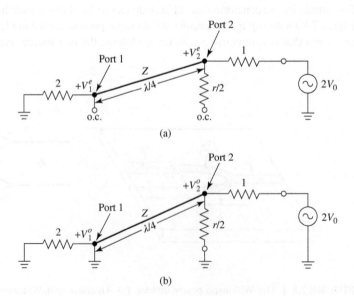

(a)

(b)

FIGURE 7.10 | Bisection of the circuit of Figure 7.9. (a) Even-mode excitation. (b) Odd-mode excitation.

end is open-circuited. Next, we find V_1^e from the transmission line equations. If we let $x = 0$ at port 1 and $x = -\lambda/4$ at port 2, we can write the voltage on the transmission line section as

$$V(x) = V^+(e^{-j\beta x} + \Gamma e^{j\beta x}).$$

Then

$$V_2^e = V(-\lambda/4) = jV^+(1 - \Gamma) = V_0, \tag{7.34a}$$

$$V_1^e = V(0) = V^+(1 + \Gamma) = jV_0\frac{\Gamma + 1}{\Gamma - 1}. \tag{7.34b}$$

The reflection coefficient Γ is that seen at port 1 looking toward the resistor of normalized value 2, so

$$\Gamma = \frac{2 - \sqrt{2}}{2 + \sqrt{2}},$$

and

$$V_1^e = -jV_0\sqrt{2}. \tag{7.35}$$

Odd mode: For odd-mode excitation, $V_{g2} = -V_{g3} = 2V_0$, and so $V_2^o = -V_3^o$, and there is a voltage null along the middle of the circuit in Figure 7.9. We can then bisect this circuit by grounding it at two points on its midplane to give the network of Figure 7.10b. Looking into port 2, we see an impedance of $r/2$ since the parallel-connected transmission line is $\lambda/4$ long and shorted at port 1, and so looks like an open circuit at port 2. Thus, port 2 will be matched for odd-mode excitation if we select $r = 2$. Then $V_2^o = V_0$ and $V_1^o = 0$; for this mode of excitation all power is delivered to the $r/2$ resistors, with none going to port 1.

Finally, we must find the input impedance at port 1 of the Wilkinson divider when ports 2 and 3 are terminated in matched loads. The resulting circuit is shown in Figure 7.11a, where it is seen that this is similar to an even mode of excitation since $V_2 = V_3$. No current flows through the resistor of normalized value 2, so it can be removed, leaving the circuit of Figure 7.11b. We then have the parallel connection of two quarter-wave transformers terminated in loads of unity (normalized). The input impedance is

$$Z_{\text{in}} = \frac{1}{2}(\sqrt{2})^2 = 1. \tag{7.36}$$

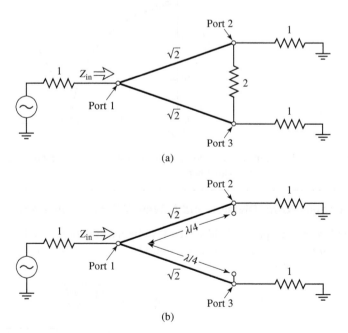

(a)

(b)

FIGURE 7.11 | Analysis of the Wilkinson divider to find S_{11}. (a) The terminated Wilkinson divider. (b) Bisection of the circuit in (a).

In summary, we can establish the following scattering parameters for the Wilkinson divider:

$$S_{11} = 0 \qquad\qquad\qquad\qquad (Z_{\text{in}} = 1 \text{ at port 1})$$

$$S_{22} = S_{33} = 0 \qquad\qquad\qquad \text{(ports 2 and 3 matched for even and odd modes)}$$

$$S_{12} = S_{21} = \frac{V_1^e + V_1^o}{V_2^e + V_2^o} = -j/\sqrt{2} \qquad \text{(symmetry due to reciprocity)}$$

$$S_{13} = S_{31} = -j/\sqrt{2} \qquad\qquad \text{(symmetry of ports 2 and 3)}$$

$$S_{23} = S_{32} = 0 \qquad\qquad\qquad \text{(due to short or open at bisection)}$$

The preceding formula for S_{12} applies because all ports are matched when terminated with matched loads. Note that when the divider is driven at port 1 and the outputs are matched, no power is dissipated in the resistor. Thus the divider is lossless when the outputs are matched; only reflected power from ports 2 or 3 is dissipated in the resistor. Because $S_{23} = S_{32} = 0$, ports 2 and 3 are isolated.

EXAMPLE 7.2   DESIGN AND PERFORMANCE OF A WILKINSON DIVIDER

Design an equal-split Wilkinson power divider for a 75 Ω system impedance at frequency f_0, and plot the return loss (S_{11}), insertion loss ($S_{21} = S_{31}$), and isolation ($S_{23} = S_{32}$) versus frequency from $0.5f_0$ to $1.5f_0$.

Solution
From Figure 7.8 and the above derivation, we have that the quarter-wave transmission lines in the divider should have a characteristic impedance of

$$Z = \sqrt{2}Z_0 = 106.1 \ \Omega,$$

and the shunt resistor a value of

$$R = 2Z_0 = 150 \ \Omega.$$

The transmission lines are $\lambda/4$ long at the frequency f_0. Using a computer-aided design tool for the analysis of microwave circuits, the scattering parameter magnitudes were calculated and plotted in Figure 7.12.

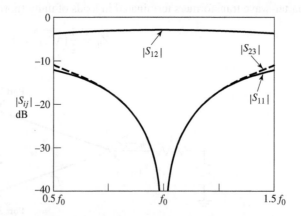

FIGURE 7.12 | Frequency response of an equal-split Wilkinson power divider. Port 1 is the input port; ports 2 and 3 are the output ports.

Unequal Power Division and *N*-Way Wilkinson Dividers

Wilkinson-type power dividers can also be made with unequal power splits; a microstrip line version is shown in Figure 7.13. If the power ratio between ports 2 and 3 is $K^2 = P_3/P_2$, then the following design equations apply:

$$Z_{03} = Z_0 \sqrt{\frac{1 + K^2}{K^3}}, \qquad\qquad\qquad (7.37\text{a})$$

$$Z_{02} = K^2 Z_{03} = Z_0 \sqrt{K(1 + K^2)}, \qquad\qquad (7.37\text{b})$$

$$R = Z_0 \left(K + \frac{1}{K} \right). \qquad\qquad\qquad (7.37\text{c})$$

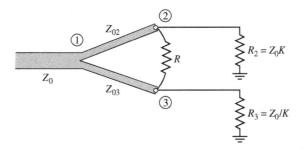

FIGURE 7.13 | A Wilkinson power divider in microstrip form having unequal power division.

Note that the above results reduce to the equal-split case for $K = 1$. Also observe that the output lines are matched to the impedances $R_2 = Z_0 K$ and $R_3 = Z_0/K$, as opposed to the impedance Z_0; matching transformers can be used to transform these output impedances.

The Wilkinson divider can also be generalized to an N-way divider or combiner [4], as shown in Figure 7.14. This circuit can be matched at all ports, with isolation between all ports. A disadvantage, however, is the fact that the divider requires crossovers for the resistors for $N \geq 3$, which makes fabrication difficult in planar form. The Wilkinson divider can also be made with stepped multiple sections, for increased bandwidth. A photograph of a four-way Wilkinson divider network is shown in Figure 7.15.

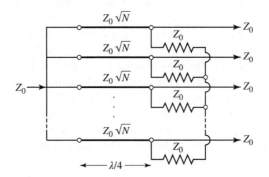

FIGURE 7.14 | An N-way, equal-split Wilkinson power divider.

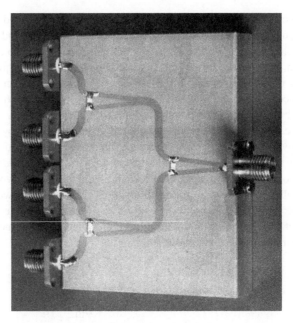

FIGURE 7.15 | Photograph of a four-way corporate power divider network using three microstrip Wilkinson power dividers. Note the isolation chip resistors.

Courtesy of M. D. Abouzahra, MIT Lincoln Laboratory, Lexington, MA.

TABLE 7.2 | Comparison of T-Junction Power Dividers and Wilkinson Power Dividers

Power Divider	Advantages	Disadvantages
Lossless T-junction	Lossless	Not matched at all ports
		No isolation between output ports
Resistive T-junction	Matched at all ports	No isolation between output ports
		Poor power handling, limited by resistor tolerances
		Lossy
Wilkinson	Lossless (if matched at all ports)	Reflected power is dissipated through isolation
	High isolation	resistor if mismatched

Table 7.2 provides a comparison of T-junction power dividers and Wilkinson power dividers.

7.4 WAVEGUIDE DIRECTIONAL COUPLERS

We now turn our attention to directional couplers, which are four-port devices with the characteristics discussed in Section 7.1. To review the basic operation, consider the directional coupler schematic symbols shown in Figure 7.4. Power incident at port 1 will couple to port 2 (the through port) and to port 3 (the coupled port), but not to port 4 (the isolated port). Similarly, power incident in port 2 will couple to ports 1 and 4, but not 3. Thus, ports 1 and 4 are decoupled, as are ports 2 and 3. The fraction of power coupled from port 1 to port 3 is given by C, the coupling coefficient, as defined in (7.20a), and the leakage of power from port 1 to port 4 is given by I, the isolation, as defined in (7.20c). Another quantity that characterizes a coupler is the directivity, $D = I - C$ (dB), which is the ratio of the powers delivered to the coupled port and the isolated port. The ideal coupler is characterized solely by the coupling factor, as the isolation and directivity are infinite. The ideal coupler is also lossless and matched at all ports.

Directional couplers can be made in many different forms. We will first discuss waveguide couplers, followed by hybrid junctions. A hybrid junction is a special case of a directional coupler, where the coupling factor is 3 dB (equal split), and the phase relation between the output ports is either 90° (quadrature hybrid), or 180° (magic-T or rat-race hybrid). Then we will discuss the implementation of directional couplers in coupled transmission line form.

Bethe Hole Coupler

The directional property of all directional couplers is produced through the use of two separate waves or wave components, which add in phase at the coupled port and are canceled at the isolated port. One of the simplest ways of doing this is to couple one waveguide to another through a single small hole in the common broad wall between the two waveguides. Such a coupler is known as a *Bethe hole coupler*, two versions of which are shown in Figure 7.16. We know that an aperture can be replaced with equivalent sources consisting of electric and magnetic dipole moments [6]. The normal electric dipole moment and the axial magnetic dipole moment radiate with even symmetry in the coupled guide, while the transverse magnetic dipole moment radiates with odd symmetry. Thus, by adjusting the relative amplitudes of these two equivalent sources, we can cancel the radiation in the direction of the isolated port, while enhancing the radiation in the direction of the coupled port. Figure 7.16 shows two ways in which these wave amplitudes can be controlled; in the coupler shown in Figure 7.16a, the two waveguides are parallel and the coupling is controlled by s, the aperture offset from the sidewall of the waveguide. For the coupler of Figure 7.16b, the wave amplitudes are controlled by the angle, θ, between the two waveguides.

First consider the configuration of Figure 7.16a, with an incident TE_{10} mode into port 1. These fields can be written as

$$E_y = A \sin \frac{\pi x}{a} e^{-j\beta z}, \tag{7.38a}$$

$$H_x = \frac{-A}{Z_{10}} \sin \frac{\pi x}{a} e^{-j\beta z}, \tag{7.38b}$$

$$H_z = \frac{j\pi A}{\beta a Z_{10}} \cos \frac{\pi x}{a} e^{-j\beta z}, \tag{7.38c}$$

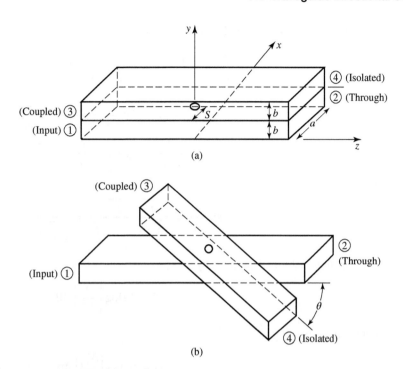

FIGURE 7.16 | Two versions of the Bethe hole directional coupler. (a) Parallel waveguides. (b) Skewed waveguides.

where $Z_{10} = k_0 \eta_0 / \beta$ is the wave impedance of the TE_{10} mode. Then, from (4.124) and (4.125), this incident wave generates the following equivalent polarization currents at the aperture at $x = s, y = b, z = 0$:

$$\bar{P}_e = \epsilon_0 \alpha_e \hat{y} A \sin \frac{\pi s}{a} \delta(x - s)\delta(y - b)\delta(z), \tag{7.39a}$$

$$\bar{P}_m = -\alpha_m A \left[\frac{-\hat{x}}{Z_{10}} \sin \frac{\pi s}{a} + \hat{z}\frac{j\pi}{\beta a Z_{10}} \cos \frac{\pi s}{a} \right] \delta(x - s)\delta(y - b)\delta(z). \tag{7.39b}$$

Using (4.128a) and (4.128b) to relate $\bar{P}_e$ and $\bar{P}_m$ to the currents $\bar{J}$ and $\bar{M}$, and then using (4.118), (4.120), (4.122), and (4.123), gives the amplitudes of the forward and reverse traveling waves in the top guide as

$$A_{10}^+ = \frac{-1}{P_{10}} \int_v \bar{E}_{10}^- \cdot \bar{J}dv + \frac{1}{P_{10}} \int_v \bar{H}_{10}^- \cdot \bar{M}dv$$

$$= \frac{-j\omega A}{P_{10}} \left[\epsilon_0 \alpha_e \sin^2 \frac{\pi s}{a} - \frac{\mu_0 \alpha_m}{Z_{10}^2} \left(\sin^2 \frac{\pi s}{a} + \frac{\pi^2}{\beta^2 a^2} \cos^2 \frac{\pi s}{a} \right) \right], \tag{7.40a}$$

$$A_{10}^- = \frac{-1}{P_{10}} \int_v \bar{E}_{10}^+ \cdot \bar{J}dv + \frac{1}{P_{10}} \int_v \bar{H}_{10}^+ \cdot \bar{M}dv$$

$$= \frac{-j\omega A}{P_{10}} \left[\epsilon_0 \alpha_e \sin^2 \frac{\pi s}{a} + \frac{\mu_0 \alpha_m}{Z_{10}^2} \left(\sin^2 \frac{\pi s}{a} - \frac{\pi^2}{\beta^2 a^2} \cos^2 \frac{\pi s}{a} \right) \right], \tag{7.40b}$$

where $P_{10} = ab/Z_{10}$ is the power normalization constant. Note from (7.40a) and (7.40b) that the amplitude of the wave excited toward port 4 (A_{10}^+) is generally different from that excited toward port 3 (A_{10}^-) (because $H_x^+ = -H_x^-$), so we can cancel the power delivered to port 4 by setting $A_{10}^+ = 0$. If we assume that the aperture is round, then polarizabilities are $\alpha_e = 2r_0^3/3$ and $\alpha_m = 4r_0^3/3$, where r_0 is the radius of the aperture.

Then from (7.40a) we obtain the following condition for $A_{10}^+ = 0$:

$$\left(2\epsilon_0 - \frac{4\mu_0}{Z_{10}^2}\right)\sin^2\frac{\pi s}{a} - \frac{4\pi^2\mu_0}{\beta^2 a^2 Z_{10}^2}\cos^2\frac{\pi s}{a} = 0,$$

$$\left(k_0^2 - 2\beta^2\right)\sin^2\frac{\pi s}{a} = \frac{2\pi^2}{a^2}\cos^2\frac{\pi s}{a},$$

$$\left(\frac{4\pi^2}{a^2} - k_0^2\right)\sin^2\frac{\pi s}{a} = \frac{2\pi^2}{a^2},$$

or

$$\sin\frac{\pi s}{a} = \pi\sqrt{\frac{2}{4\pi^2 - k_0^2 a^2}} = \frac{\lambda_0}{\sqrt{2\left(\lambda_0^2 - a^2\right)}}. \tag{7.41}$$

The coupling factor is then given by

$$C = 20\log\left|\frac{A}{A_{10}^-}\right| \text{ dB} \tag{7.42a}$$

and the directivity by

$$D = 20\log\left|\frac{A_{10}^-}{A_{10}^+}\right| \text{ dB.} \tag{7.42b}$$

Thus, a Bethe hole coupler of the type shown in Figure 7.16a can be designed by first using (7.41) to find s, the position of the aperture, and then using (7.42a) to determine the aperture size, r_0, to give the required coupling factor.

For the skewed geometry of Figure 7.16b, the aperture may be centered at $s = a/2$, and the skew angle θ adjusted for cancellation at port 4. In this case, the normal electric field does not change with θ, but the transverse magnetic field components are reduced by $\cos\theta$. We can account for the skew angle by replacing α_m in the previous derivation by $\alpha_m\cos\theta$. The wave amplitudes of (7.40a) and (7.40b) then become, for $s = a/2$,

$$A_{10}^+ = \frac{-j\omega A}{P_{10}}\left(\epsilon_0\alpha_e - \frac{\mu_0\alpha_m}{Z_{10}^2}\cos\theta\right), \tag{7.43a}$$

$$A_{10}^- = \frac{-j\omega A}{P_{10}}\left(\epsilon_0\alpha_e + \frac{\mu_0\alpha_m}{Z_{10}^2}\cos\theta\right). \tag{7.43b}$$

Setting $A_{10}^+ = 0$ results in the following condition for the angle θ:

$$2\epsilon_0 - \frac{4\mu_0}{Z_{10}^2}\cos\theta = 0,$$

or

$$\cos\theta = \frac{k_0^2}{2\beta^2}. \tag{7.44}$$

The coupling factor then simplifies to

$$C = 20\log\left|\frac{A}{A_{10}^-}\right| = -20\log\frac{4k_0^2 r_0^3}{3ab\beta} \text{ dB.} \tag{7.45}$$

The angular geometry of the skewed Bethe hole coupler is often a disadvantage in terms of fabrication and application. In addition, both coupler designs operate properly only at the design frequency; deviation from this frequency will alter the coupling level and the directivity, as shown in the following example.

EXAMPLE 7.3 BETHE HOLE COUPLER DESIGN AND PERFORMANCE

Design a Bethe hole coupler of the type shown in Figure 7.16a for an X-band waveguide operating at 9 GHz, with a coupling of 20 dB. Calculate and plot the coupling and directivity from 7 to 11 GHz. Assume a round aperture.

Solution
For an X-band waveguide at 9 GHz, we have the following constants:

$$a = 0.02286 \text{ m},$$
$$b = 0.01016 \text{ m},$$
$$\lambda_0 = 0.0333 \text{ m},$$
$$k_0 = 188.5 \text{ m}^{-1},$$
$$\beta = 129.0 \text{ m}^{-1},$$
$$Z_{10} = 550.9 \ \Omega,$$
$$P_{10} = 4.22 \times 10^{-7} \text{ m}^2/\Omega.$$

Equation (7.41) can be used to find the aperture position s:

$$\sin \frac{\pi s}{a} = \frac{\lambda_0}{\sqrt{2 \left(\lambda_0^2 - a^2 \right)}} = 0.972,$$

$$s = \frac{a}{\pi} \sin^{-1} 0.972 = 0.424a = 9.69 \text{ mm}.$$

The coupling is 20 dB, so

$$C = 20 \text{ dB} = 20 \log \left| \frac{A}{A_{10}^-} \right|,$$

or

$$\left| \frac{A}{A_{10}^-} \right| = 10^{20/20} = 10;$$

thus, $|A_{10}^-/A| = 1/10$. Now use (7.40b) to find r_0:

$$\left| \frac{A_{10}^-}{A} \right| = \frac{1}{10} = \frac{\omega}{P_{10}} \left[\left(\epsilon_0 \alpha_e + \frac{\mu_0 \alpha_m}{Z_{10}^2} \right) (0.944) - \frac{\pi^2 \mu_0 \alpha_m}{\beta^2 a^2 Z_{10}^2} (0.056) \right].$$

Because $\alpha_e = 2r_0^3/3$ and $\alpha_m = 4r_0^3/3$, we obtain

$$0.1 = 1.44 \times 10^6 r_0^3,$$

or

$$r_0 = 4.15 \text{ mm}.$$

This completes the design of the Bethe hole coupler. To compute the coupling and directivity versus frequency, we evaluate (7.42a) and (7.42b), using the expressions for A_{10}^- and A_{10}^+ given in (7.40a) and (7.40b). In these expressions the aperture position and size are fixed at $s = 9.69$ mm and $r_0 = 4.15$ mm, and the frequency is varied. A short computer program was used to calculate the data shown in Figure 7.17. Observe that the coupling varies by less than 1 dB over the band. The directivity is very large (>60 dB) at the design frequency but decreases to 15–20 dB at the band edges. The directivity is a more sensitive function of frequency because it depends on the cancellation of two wave components.

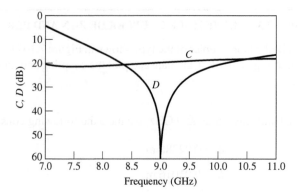

FIGURE 7.17 | Coupling and directivity versus frequency for the Bethe hole coupler of Example 7.3.

Design of Multihole Couplers

As seen from Example 7.3, a single-hole coupler has a relatively narrow bandwidth, at least in terms of its directivity. However, if the coupler is designed with a series of coupling holes, the extra degrees of freedom can be used to increase this bandwidth. The principle of operation and design of such a *multihole waveguide coupler* is very similar to that of the multisection matching transformer.

First let us consider the operation of the two-hole coupler shown in Figure 7.18. Two parallel waveguides sharing a common broad wall are shown, although the same type of structure could be made in microstrip line or stripline form. Two small apertures are spaced $\lambda_g/4$ apart and couple the two guides. A wave entering at port 1 is mostly transmitted through to port 2, but some power is coupled through the two apertures. If a phase reference is taken at the first aperture, then the phase of the wave incident at the second aperture will be $-90°$. Each aperture will radiate a forward wave component and a backward wave component into the upper guide; in general, the forward and backward amplitudes are different. In the direction of port 3, both wave components are in phase because both have traveled $\lambda_g/4$ to the second aperture. However, we obtain a cancellation in the direction of port 4 because the wave coming through the second aperture travels $\lambda_g/2$ further than the wave component coming through the first aperture. Clearly, this cancellation is frequency sensitive, making the directivity a sensitive function of frequency. The coupling is less frequency dependent since the path lengths from port 1 to port 3 are always the same. Thus, in the multihole coupler design, we synthesize the directivity response, as opposed to the coupling response, as a function of frequency.

Now consider the general case of the multihole coupler shown in Figure 7.19, where $N + 1$ equally spaced apertures couple two parallel waveguides. The amplitude of the incident wave in the lower left guide is A and, for small coupling, is essentially the same as the amplitude of the through wave. For instance,

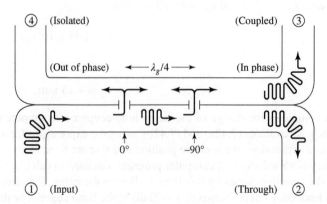

FIGURE 7.18 | Basic operation of a two-hole directional coupler.

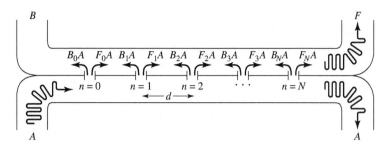

FIGURE 7.19 | Geometry of an $(N + 1)$-hole waveguide directional coupler.

a 20 dB coupler has a power coupling factor of $10^{-20/10} = 0.01$, so the power transmitted through waveguide A is $1 - 0.01 = 0.99$ of the incident power (1% coupled to the upper guide). The voltage (or field) drop in waveguide A is $\sqrt{0.99} = 0.995$, or 0.5%. Thus, the assumption that the amplitude of the incident field is identical at each aperture is a good one. Of course, the phase will change from one aperture to the next.

As we saw in the previous section for the Bethe hole coupler, an aperture generally excites forward and backward traveling waves with different amplitudes. Thus, let

F_n denote the coupling coefficient of the nth aperture in the forward direction.

B_n denote the coupling coefficient of the nth aperture in the backward direction.

Then the amplitude of the forward wave can be written as

$$F = Ae^{-j\beta Nd} \sum_{n=0}^{N} F_n,$$ (7.46)

since all components travel the same path length. The amplitude of the backward wave is

$$B = A \sum_{n=0}^{N} B_n e^{-2j\beta nd},$$ (7.47)

since the path length for the nth component is $2\beta nd$, where d is the spacing between the apertures. In (7.46) and (7.47) the phase reference is taken at the $n = 0$ aperture.

From the definitions in (7.20a) and (7.20b) the coupling and directivity can be computed as

$$C = -20 \log \left| \frac{F}{A} \right| = -20 \log \left| \sum_{n=0}^{N} F_n \right| \text{ dB},$$ (7.48)

$$D = -20 \log \left| \frac{B}{F} \right| = -20 \log \left| \frac{\sum_{n=0}^{N} B_n e^{-2j\beta nd}}{\sum_{n=0}^{N} F_n} \right|$$

$$= -C - 20 \log \left| \sum_{n=0}^{N} B_n e^{-2j\beta nd} \right| \text{ dB}.$$ (7.49)

Now assume that the apertures are round holes with identical positions s relative to the edge of the guide, with r_n being the radius of the nth aperture. Then we know from the preceding section that the coupling coefficients will be proportional to the polarizabilities α_e and α_m of the aperture, and hence proportional to r_n^3. So we can write

$$F_n = K_f r_n^3,$$ (7.50a)

$$B_n = K_b r_n^3,$$ (7.50b)

where K_f and K_b are constants for the forward and backward coupling coefficients that are the same for all apertures, but are functions of frequency. Then (7.48) and (7.49) reduce to

$$C = -20 \log |K_f| - 20 \log \sum_{n=0}^{N} r_n^3 \text{ dB}, \tag{7.51}$$

$$D = -C - 20 \log |K_b| - 20 \log \left| \sum_{n=0}^{N} r_n^3 e^{-2j\beta nd} \right|$$

$$= -C - 20 \log |K_b| - 20 \log S \text{ dB}. \tag{7.52}$$

In (7.51), the second term is constant with frequency. The first term is not affected by the choice of r_n, but is a relatively slowly varying function of frequency. Similarly, in (7.52) the first two terms are slowly varying functions of frequency, representing the directivity of a single aperture, but the last term (S) is a sensitive function of frequency due to phase cancellation in the summation. Thus we can choose the r_n to synthesize a desired frequency response for the directivity, while the coupling should be relatively constant with frequency.

Observe that the last term in (7.52),

$$S = \left| \sum_{n=0}^{N} r_n^3 e^{-2j\beta nd} \right|, \tag{7.53}$$

is very similar in form to the expression obtained in Section 5.5 for multisection quarter-wave matching transformers. As in that case, we can develop coupler designs that yield either a binomial (maximally flat) or a Chebyshev (equal ripple) response for the directivity. Another interpretation of (7.53) may be recognizable to the student familiar with basic antenna theory, as this expression is identical to the array pattern factor of an $(N + 1)$-element array with element weights r_n^3. In that case, too, the pattern may be synthesized in terms of binomial or Chebyshev polynomials.

Binomial response: As in the case of the multisection quarter-wave matching transformers, we can obtain a binomial, or maximally flat, response for the directivity of the multihole coupler by making the coupling coefficients proportional to the binomial coefficients. Thus,

$$r_n^3 = kC_n^N, \tag{7.54}$$

where k is a constant to be determined, and C_n^N is a binomial coefficient given in (5.57). To find k, we evaluate the coupling using (7.51) to give

$$C = -20 \log |K_f| - 20 \log k - 20 \log \sum_{n=0}^{N} C_n^N \text{ dB}. \tag{7.55}$$

Because we know $K_f, N,$ and C, we can solve for k and then find the required aperture radii from (7.54). The spacing, d, should be $\lambda_g/4$ at the center frequency.

Chebyshev response: First assume that N is even (an odd number of holes), and that the coupler is symmetric, so that $r_0 = r_N, r_1 = r_{N-1}$, etc. Then from (7.53) we can write S as

$$S = \left| \sum_{n=0}^{N} r_n^3 e^{-2jn\theta} \right| = 2 \sum_{n=0}^{N/2} r_n^3 \cos(N - 2n)\theta,$$

where $\theta = \beta d$. To achieve a Chebyshev response we equate this to the Chebyshev polynomial of degree N:

$$S = 2 \sum_{n=0}^{N/2} r_n^3 \cos(N - 2n)\theta = k|T_N(\sec \theta_m \cos \theta)|, \tag{7.56}$$

where k and θ_m are constants to be determined. From (7.53) and (7.56), we see that for $\theta = 0, S = \sum_{n=0}^{N} r_n^3 = k|T_N(\sec \theta_m)|$. Using this result in (7.51) gives the coupling as

$$C = -20 \log |K_f| - 20 \log S \Big|_{\theta=0}$$

$$= -20 \log |K_f| - 20 \log k - 20 \log |T_N(\sec \theta_m)| \text{ dB}. \tag{7.57}$$

From (7.52) the directivity is

$$D = -C - 20 \log |K_b| - 20 \log S$$

$$= 20 \log \frac{K_f}{K_b} + 20 \log \frac{T_N(\sec \theta_m)}{T_N(\sec \theta_m \cos \theta)} \text{ dB.}$$ (7.58)

The term $\log K_f/K_b$ is a function of frequency, so D will not have an exact Chebyshev response. This error is usually small, however. We can assume that the smallest value of D will occur when $T_N(\sec \theta_m \cos \theta) = 1$, since $|T_N(\sec \theta_m)| \geq |T_N(\sec \theta_m \cos \theta)|$. So if $D_{\min}$ is the specified minimum value of directivity in the passband, then θ_m can be found from the relation

$$D_{\min} = 20 \log T_N(\sec \theta_m) \text{ dB.}$$ (7.59)

Alternatively, we could specify the bandwidth, which then dictates θ_m and $D_{\min}$. In either case, (7.57) can then be used to find k, and then (7.56) solved for the radii, r_n.

If N is odd (an even number of holes), the results for C, D, and $D_{\min}$ in (7.57), (7.58), and (7.59) still apply, but instead of (7.56), the following relation is used to find the aperture radii:

$$S = 2 \sum_{n=0}^{(N-1)/2} r_n^3 \cos(N - 2n)\theta = k|T_N(\sec \theta_m \cos \theta)|.$$ (7.60)

EXAMPLE 7.4   MULTIHOLE WAVEGUIDE COUPLER DESIGN

Design a four-hole Chebyshev coupler in an X-band waveguide using round apertures located at $s = a/4$. The center frequency is 9 GHz, the coupling is 20 dB, and the minimum directivity is 40 dB. Plot the coupling and directivity response from 7 to 11 GHz.

Solution
For an X-band waveguide at 9 GHz, we have the following constants:

$$a = 0.02286 \text{ m,}$$

$$b = 0.01016 \text{ m,}$$

$$\lambda_0 = 0.0333 \text{ m,}$$

$$k_0 = 188.5 \text{ m}^{-1},$$

$$\beta = 129.0 \text{ m}^{-1},$$

$$Z_{10} = 550.9 \text{ }\Omega,$$

$$P_{10} = 4.22 \times 10^{-7} \text{ m}^2/\Omega.$$

From (7.40a) and (7.40b), we obtain for an aperture at $s = a/4$:

$$|K_f| = \frac{2k_0}{3\eta_0 P_{10}} \left[\sin^2 \frac{\pi s}{a} - \frac{2\beta^2}{k_0^2} \left(\sin^2 \frac{\pi s}{a} + \frac{\pi^2}{\beta^2 a^2} \cos^2 \frac{\pi s}{a} \right) \right] = 3.953 \times 10^5,$$

$$|K_b| = \frac{2k_0}{3\eta_0 P_{10}} \left[\sin^2 \frac{\pi s}{a} + \frac{2\beta^2}{k_0^2} \left(\sin^2 \frac{\pi s}{a} - \frac{\pi^2}{\beta^2 a^2} \cos^2 \frac{\pi s}{a} \right) \right] = 3.454 \times 10^5.$$

For a four-hole coupler, $N = 3$, so (7.59) gives

$$40 = 20 \log T_3(\sec \theta_m) \text{ dB,}$$

$$100 = T_3(\sec \theta_m) = \cosh \left[3 \cosh^{-1}(\sec \theta_m) \right],$$

$$\sec \theta_m = 3.01,$$

where (5.64b) was used. Thus $\theta_m = 70.6°$ and $109.4°$ at the band edges. Then from (7.57) we can solve for k:

$$C = 20 = -20\log(3.953 \times 10^5) - 20\log k - 40 \text{ dB},$$

$$20\log k = -171.94,$$

$$k = 2.53 \times 10^{-9}.$$

Finally, (7.60) and the expansion from (5.66c) for T_3 allow us to solve for the radii as follows:

$$S = 2\left(r_0^3\cos 3\theta + r_1^3\cos\theta\right) = k\left[\sec^3\theta_m(\cos 3\theta + 3\cos\theta) - 3\sec\theta_m\cos\theta\right],$$

$$2r_0^3 = k\sec^3\theta_m \quad \Rightarrow \quad r_0 = r_3 = 3.26 \text{ mm},$$

$$2r_1^3 = 3k(\sec^3\theta_m - \sec\theta_m) \quad \Rightarrow \quad r_1 = r_2 = 4.51 \text{ mm}.$$

The resulting coupling and directivity are plotted in Figure 7.20; note the increased directivity bandwidth compared to that of the Bethe hole coupler of Example 7.3.

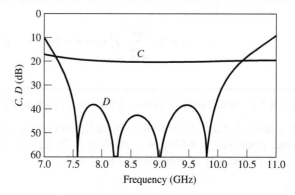

FIGURE 7.20 | Coupling and directivity versus frequency for the four-hole coupler of Example 7.4.

7.5 THE QUADRATURE (90°) HYBRID

Quadrature hybrids are 3 dB directional couplers with a 90° phase difference in the outputs of the through and coupled arms. This type of hybrid is often made in microstrip line or stripline form as shown in Figure 7.21 and is also known as a *branch-line hybrid*. Other 3 dB couplers, such as coupled line couplers or Lange couplers, can also be used as quadrature couplers; these components will be discussed in later sections. Here we will analyze the operation of the quadrature hybrid using an even-odd mode decomposition technique similar to that used for the Wilkinson power divider.

With reference to Figure 7.21, the basic operation of the branch-line coupler is as follows. With all ports matched, power entering port 1 is evenly divided between ports 2 and 3, with a 90° phase shift between these outputs ($S_{21} = jS_{31}$). No power is coupled to port 4 (the isolated port). The scattering matrix has the following form:

$$[S] = \frac{-1}{\sqrt{2}}\begin{bmatrix} 0 & j & 1 & 0 \\ j & 0 & 0 & 1 \\ 1 & 0 & 0 & j \\ 0 & 1 & j & 0 \end{bmatrix}. \tag{7.61}$$

Similarly, the power entering port 4 splits evenly between ports 2 and 3, with a −90° phase difference ($S_{24} = -jS_{34}$). Observe that the branch-line hybrid has a high degree of symmetry, as any port can be used as the input port. The output ports will always be on the opposite side of the junction from the input port, and the isolated port will be the remaining port on the same side as the input port. This symmetry is reflected in the scattering matrix, as each row can be obtained as a transposition of the first row.

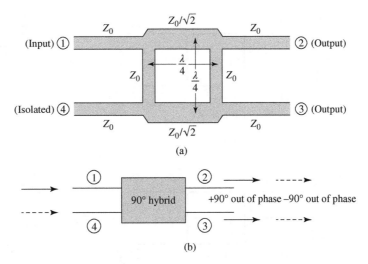

FIGURE 7.21 | (a) Geometry of a branch-line coupler. (b) Working of a branch-line coupler.

Even-Odd Mode Analysis

We first draw the schematic circuit of the branch-line coupler in normalized form, as in Figure 7.22, where it is understood that each line represents a transmission line with indicated characteristic impedance normalized to Z_0. The common ground return for each transmission line is not shown. We assume that a wave of unit amplitude $A_1 = 1$ is incident at port 1.

The circuit of Figure 7.22 can be decomposed into the superposition of an even-mode excitation and an odd-mode excitation [5], as shown in Figure 7.23. Note that superimposing the two sets of excitations produces the original excitation of Figure 7.22, and since the circuit is linear, the actual response (the scattered waves) can be obtained from the sum of the responses to the even and odd excitations.

Because of the symmetry or antisymmetry of the excitation, the four-port network can be decomposed into a set of two decoupled two-port networks, as shown in Figure 7.23. Because the amplitudes of the incident waves for these two-ports are $\pm 1/2$, the amplitudes of the emerging wave at each port of the branch-line hybrid can be expressed as

$$B_1 = \frac{1}{2}\Gamma_e + \frac{1}{2}\Gamma_o, \tag{7.62a}$$

$$B_2 = \frac{1}{2}T_e + \frac{1}{2}T_o, \tag{7.62b}$$

$$B_3 = \frac{1}{2}T_e - \frac{1}{2}T_o, \tag{7.62c}$$

$$B_4 = \frac{1}{2}\Gamma_e - \frac{1}{2}\Gamma_o, \tag{7.62d}$$

where $\Gamma_{e,o}$ and $T_{e,o}$ are the even- and odd-mode reflection and transmission coefficients for the two-port networks of Figure 7.23. First consider the calculation of Γ_e and T_e for the even-mode two-port circuit. This

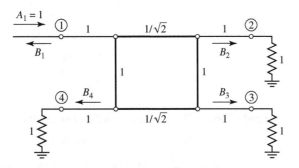

FIGURE 7.22 | Circuit of the branch-line hybrid coupler in normalized form.

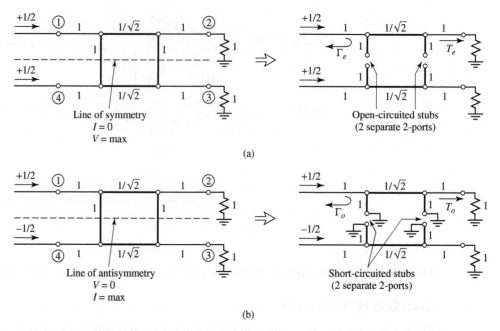

FIGURE 7.23 | Decomposition of the branch-line coupler into even- and odd-mode excitations. (a) Even mode (e). (b) Odd mode (o).

can best be done by multiplying the *ABCD* matrices of each cascade component in that circuit to give

$$
\begin{bmatrix} A & B \\ C & D \end{bmatrix}_e = \underbrace{\begin{bmatrix} 1 & 0 \\ j & 1 \end{bmatrix}}_{\substack{\text{Shunt} \\ Y=j}} \underbrace{\begin{bmatrix} 0 & j/\sqrt{2} \\ j\sqrt{2} & 0 \end{bmatrix}}_{\substack{\lambda/4 \\ \text{Transmission} \\ \text{line}}} \underbrace{\begin{bmatrix} 1 & 0 \\ j & 1 \end{bmatrix}}_{\substack{\text{Shunt} \\ Y=J}} = \frac{1}{\sqrt{2}} \begin{bmatrix} -1 & j \\ j & -1 \end{bmatrix},
$$

(7.63)

where the individual matrices can be found from Table 4.1, and the admittance of the shunt open-circuited $\lambda/8$ stubs is $Y = j\tan\beta\ell = j$. Then Table 4.2 can be used to convert from *ABCD* parameters (defined here with $Z_o = 1$) to S parameters, which are equivalent to the reflection and transmission coefficients. Thus,

$$
\Gamma_e = \frac{A+B-C-D}{A+B+C+D} = \frac{(-1+j-j+1)/\sqrt{2}}{(-1+j+j-1)/\sqrt{2}} = 0,
$$

(7.64a)

$$
T_e = \frac{2}{A+B+C+D} = \frac{2}{(-1+j+j-1)/\sqrt{2}} = \frac{-1}{\sqrt{2}}(1+j).
$$

(7.64b)

Similarly, for the odd mode we obtain

$$
\begin{bmatrix} A & B \\ C & D \end{bmatrix}_o = \frac{1}{\sqrt{2}} \begin{bmatrix} 1 & j \\ j & 1 \end{bmatrix},
$$

(7.65)

which gives the reflection and transmission coefficients as

$$
\Gamma_o = 0,
$$

(7.66a)

$$
T_o = \frac{1}{\sqrt{2}}(1-j).
$$

(7.66b)

Using (7.64) and (7.66) in (7.62) gives the following results:

$$
B_1 = 0 \qquad \text{(port 1 is matched)},
$$

(7.67a)

$$
B_2 = -\frac{j}{\sqrt{2}} \qquad \text{(half-power, } -90° \text{ phase shift from port 1 to 2)},
$$

(7.67b)

FIGURE 7.24 | Photograph of an eight-way microstrip power divider for an array antenna feed network at 1.26 GHz. The circuit uses six quadrature hybrids in a Bailey configuration for unequal power division ratios (see Problem 7.30).

Courtesy of ProSensing, Inc., Amherst, MA.

$$B_3 = -\frac{1}{\sqrt{2}} \quad \text{(half-power, } -180° \text{ phase shift from port 1 to 3),} \tag{7.67c}$$

$$B_4 = 0 \quad \text{(no power to port 4).} \tag{7.67d}$$

These results agree with the first row and column of the scattering matrix given in (7.61); the remaining elements can easily be found by transposition.

In practice, due to the quarter-wave length requirement, the bandwidth of a branch-line hybrid is limited to 10%–20%. However, as with multisection matching transformers and multihole directional couplers, the bandwidth of a branch-line hybrid can be increased to a decade or more by using multiple sections in cascade. In addition, the basic design can be modified for unequal power division and/or different characteristic impedances at the output ports. Another practical point to be aware of is the fact that discontinuity effects at the junctions of the branch-line coupler may require that the shunt arms be lengthened by 10°–20°. Figure 7.24 shows a photograph of a circuit using several quadrature hybrids.

EXAMPLE 7.5   DESIGN AND PERFORMANCE OF A QUADRATURE HYBRID

Design a 75 Ω branch-line quadrature hybrid junction, and plot the scattering parameter magnitudes from $0.5f_0$ to $1.5f_0$, where f_0 is the design frequency.

Solution
After the preceding analysis, the design of a quadrature hybrid is trivial. The lines are $\lambda/4$ at the design frequency f_0, and the branch-line impedances are

$$\frac{Z_0}{\sqrt{2}} = \frac{75}{\sqrt{2}} = 53.03 \ \Omega.$$

The calculated frequency response is plotted in Figure 7.25. Note that we obtain perfect 3 dB power division at ports 2 and 3, and perfect isolation and return loss at ports 4 and 1, respectively, at the design frequency f_0. All of these quantities, however, degrade quickly as the frequency departs from f_0.

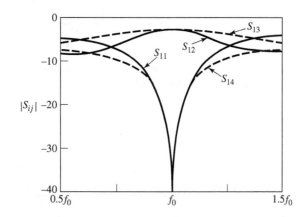

FIGURE 7.25 | Scattering parameter magnitudes versus frequency for the branch-line coupler of Example 7.5.

7.6   COUPLED LINE DIRECTIONAL COUPLERS

When two unshielded transmission lines are in close proximity, power can be coupled from one line to the other due to the interaction of the electromagnetic fields. Such lines are referred to as *coupled transmission lines*, and they usually consist of three conductors in close proximity, although more conductors can be used. Figure 7.26 shows several examples of coupled transmission lines. Coupled transmission lines are sometimes assumed to operate in the TEM mode, which is rigorously valid for coaxial line and stripline structures, but only approximately valid for microstrip line, coplanar waveguide, or slotline structures. Coupled transmission lines can support two distinct propagating modes, and this feature can be used to implement a variety of practical directional couplers, hybrids, and filters.

The coupled lines shown in Figure 7.26 are symmetric, meaning that the two conducting strips have the same width and position relative to ground; this simplifies the analysis of their operation. We will first discuss the basic theory of coupled lines and present some design data for coupled stripline and coupled microstrip line. We will then analyze the operation of a single-section coupled line directional coupler and extend these results to multisection coupled line coupler design.

Coupled Line Theory

The coupled lines of Figure 7.26, or other symmetric three-wire lines, can be represented by the structure and equivalent circuit shown in Figure 7.27. If we assume TEM propagation, then the electrical characteristics of the coupled lines can be completely determined from the effective capacitances between the lines and the velocity of propagation on the line. As depicted in Figure 7.27, C_{12} represents the capacitance between the two strip conductors, and C_{11} and C_{22} represent the capacitance between one strip conductor and ground. Because the strip conductors are identical in size and location relative to the ground conductor, we have $C_{11} = C_{22}$. Note that the designation of "ground" for the third conductor has no special relevance beyond the fact that it is convenient, since in many applications this conductor is the ground plane of a stripline or microstrip circuit.

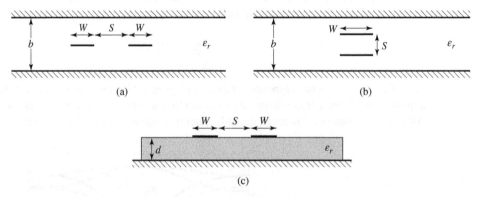

FIGURE 7.26 | Various coupled transmission line geometries. (a) Coupled stripline (planar, or edge-coupled). (b) Coupled stripline (stacked, or broadside-coupled). (c) Coupled microstrip lines.

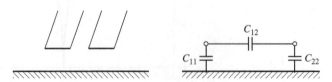

FIGURE 7.27 | A three-wire coupled transmission line and its equivalent capacitance network.

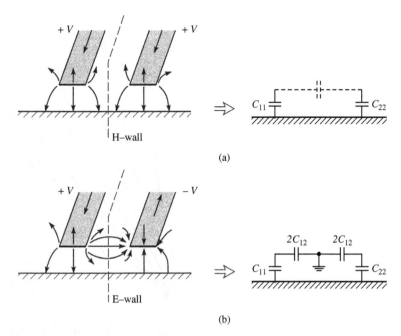

FIGURE 7.28 | Even- and odd-mode excitations for a coupled line, and the resulting equivalent capacitance networks. (a) Even-mode excitation. (b) Odd-mode excitation.

Now consider two special types of excitations for the coupled line: the even mode, where the currents in the strip conductors are equal in amplitude and in the same direction, and the odd mode, where the currents in the strip conductors are equal in amplitude but in opposite directions. The electric field lines for these two cases are sketched in Figure 7.28. Because the line is TEM, the propagation constant and phase velocity are the same for both of these modes: $\beta = \omega / v_p$ and $v_p = c / \sqrt{\epsilon_r}$, where ϵ_r is the relative permittivity of the TEM line.

For the even mode, the electric field has even symmetry about the center line, and no current flows between the two strip conductors. This leads to the equivalent circuit shown, where C_{12} is effectively open-circuited. The resulting capacitance of either line to ground for the even mode is

$$C_e = C_{11} = C_{22},\tag{7.68}$$

assuming that the two strip conductors are identical in size and location. Then the characteristic impedance for the even mode is

$$Z_{0e} = \sqrt{\frac{L_e}{C_e}} = \frac{\sqrt{L_e C_e}}{C_e} = \frac{1}{v_p C_e},\tag{7.69}$$

where $v_p = c / \sqrt{\epsilon_r} = 1 / \sqrt{L_e C_e} = 1 / \sqrt{L_o C_o}$ is the phase velocity of propagation on the line.

For the odd mode, the electric field lines have an odd symmetry about the center line, and a voltage null exists between the two strip conductors. We can imagine this as a ground plane through the middle of C_{12}, which leads to the equivalent circuit shown. In this case the effective capacitance between either strip conductor and ground is

$$C_o = C_{11} + 2C_{12} = C_{22} + 2C_{12},\tag{7.70}$$

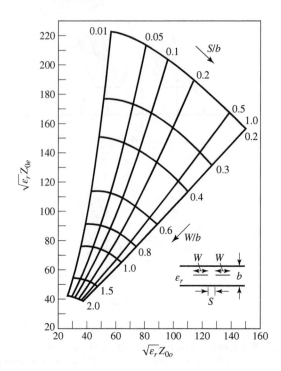

FIGURE 7.29 | Normalized even- and odd-mode characteristic impedance design data for symmetric edge-coupled striplines.

and the characteristic impedance for the odd mode is

$$Z_{0o} = \sqrt{\frac{L_o}{C_o}} = \frac{\sqrt{L_o C_o}}{C_o} = \frac{1}{v_p C_o}. \tag{7.71}$$

In words, Z_{0e} (Z_{0o}) is the characteristic impedance of one of the strip conductors relative to ground when the coupled line is operated in the even (odd) mode. An arbitrary excitation of a coupled line can always be treated as a superposition of appropriate amplitudes of even- and odd-mode excitations. This analysis assumes the lines are symmetric, and that fringing capacitances are identical for even and odd modes.

If the coupled line supports a pure TEM mode, such as coaxial line, parallel plate guide, or stripline, analytical techniques such as conformal mapping [7] can be used to evaluate the capacitances per unit length of line, and the even- and odd-mode characteristic impedances can then be determined. For quasi-TEM lines, such as microstrip line, these results can be obtained numerically or by approximate quasi-static techniques [8]. In either case, such calculations are generally too involved for our consideration, but many commercial microwave CAD packages can provide design data for a variety of coupled lines. Here we will present only graphical design data for two cases of coupled lines.

For a symmetric coupled stripline of the type shown in Figure 7.26a, the design graph in Figure 7.29 can be used to determine the necessary strip widths and spacing for a given set of characteristic impedances, Z_{0e} and Z_{0o}, and the dielectric constant. This graph should cover ranges of parameters for most practical applications, and can be used for any dielectric constant, since the TEM mode of stripline allows scaling by the dielectric constant.

For coupled microstrip lines, the results do not scale with dielectric constant, so design graphs must be made for specific values of dielectric constant. Figure 7.30 shows such a design graph for symmetric coupled microstrip lines on a substrate with $\epsilon_r = 10$. Another difficulty with coupled microstrip lines is the fact that the phase velocity is usually different for the two modes of propagation because the two modes operate with different field configurations in the vicinity of the air–dielectric interface. This can have a degrading effect on coupler directivity.

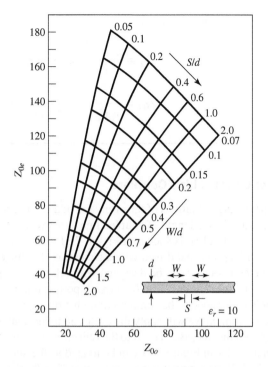

FIGURE 7.30 | Even- and odd-mode characteristic impedance design data for symmetric coupled microstrip lines on a substrate with $\epsilon_r = 10$.

EXAMPLE 7.6   IMPEDANCE OF A SIMPLE COUPLED LINE

For the broadside coupled stripline geometry of Figure 7.26b, assume $W \gg S$ and $W \gg b$, so that fringing fields can be ignored, and determine the even- and odd-mode characteristic impedances.

Solution
We first find the equivalent network capacitances, C_{11} and C_{12} (because the line is symmetric, $C_{22} = C_{11}$). The capacitance per unit length of broadside parallel lines with width W and separation d is

$$\bar{C} = \frac{\epsilon W}{d} \text{ F/m,}$$

where ϵ is the substrate permittivity. This formula ignores fringing fields.

C_{11} is formed by the capacitance of one strip to the ground planes. Thus the capacitance per unit length is

$$\bar{C}_{11} = \frac{2\epsilon_r\epsilon_0 W}{b - s} \text{ F/m.}$$

The capacitance per unit length between the strips is

$$\bar{C}_{12} = \frac{\epsilon_r\epsilon_0 W}{S} \text{ F/m.}$$

Then from (7.68) and (7.70), the even- and odd-mode capacitances are

$$\bar{C}_e = \bar{C}_{11} = \frac{2\epsilon_r\epsilon_0 W}{b - S} \text{ F/m,}$$

$$\bar{C}_o = \bar{C}_{11} + 2\bar{C}_{12} = 2\epsilon_r\epsilon_0 W \left(\frac{1}{b - S} + \frac{1}{S} \right) \text{ F/m.}$$

The phase velocity on the line is $v_p = 1/\sqrt{\epsilon_r \epsilon_0 \mu_0} = c/\sqrt{\epsilon_r}$, so the characteristic impedances are

$$Z_{0e} = \frac{1}{v_p \bar{C}_e} = \eta_0 \frac{b - S}{2W\sqrt{\epsilon_r}},$$

$$Z_{0o} = \frac{1}{v_p \bar{C}_o} = \eta_0 \frac{1}{2W\sqrt{\epsilon_r}[1/(b-S) + 1/S]}.$$

Design of Coupled Line Couplers

With the preceding definitions of the even- and odd-mode characteristic impedances, we can apply an even-odd mode analysis to a length of coupled line to arrive at the design equations for a single-section coupled line coupler. Such a line is shown in Figure 7.31. This four-port network is terminated in the impedance Z_0 at three of its ports, and driven with a generator of voltage $2V_0$ and internal impedance Z_0 at port 1. We will show that a coupler can be designed with arbitrary coupling such that the input (port 1) is matched, while port 4 is isolated. Port 2 is the through port, and port 3 is the coupled port. In Figure 7.31, a ground conductor is understood to be common to both strip conductors.

For this problem we will apply the even-odd mode analysis technique in conjunction with the voltages and currents on the line, as opposed to the reflection and transmission coefficients. So, by superposition, the excitation at port 1 in Figure 7.31 can be treated as the sum of the even- and odd-mode excitations shown in Figure 7.32. From symmetry we can see that $I_1^e = I_3^e$, $I_4^e = I_2^e$, $V_1^e = V_3^e$, and $V_4^e = V_2^e$ for the even mode, while $I_1^o = -I_3^o$, $I_4^o = -I_2^o$, $V_1^o = -V_3^o$, and $V_4^o = -V_2^o$ for the odd mode. The input impedance at port 1 of the coupler of Figure 7.31 can then be expressed as

$$Z_{\text{in}} = \frac{V_1}{I_1} = \frac{V_1^e + V_1^o}{I_1^e + I_1^o}. \tag{7.72}$$

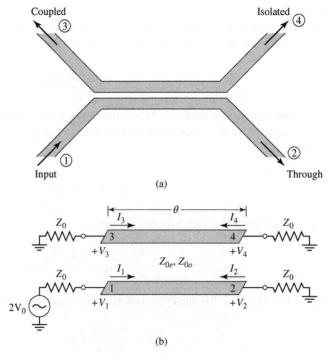

FIGURE 7.31 | A single-section coupled line coupler. (a) Geometry and port designations. (b) The schematic circuit.

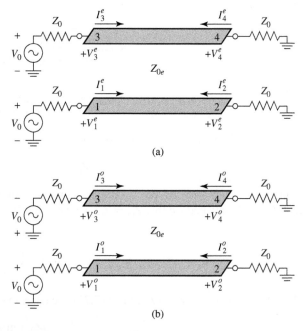

FIGURE 7.32 | Decomposition of the coupled line coupler circuit of Figure 7.31 into even- and odd-mode excitations. (a) Even mode. (b) Odd mode.

If we let Z_{in}^e be the input impedance at port 1 for the even mode, and Z_{in}^o be the input impedance for the odd mode, then we have

$$Z_{in}^e = Z_{0e} \frac{Z_0 + jZ_{0e} \tan \theta}{Z_{0e} + jZ_0 \tan \theta}, \tag{7.73a}$$

$$Z_{in}^o = Z_{0o} \frac{Z_0 + jZ_{0o} \tan \theta}{Z_{0o} + jZ_0 \tan \theta}, \tag{7.73b}$$

since, for each mode, the line looks like a transmission line of characteristic impedance Z_{0e} or Z_{0o}, terminated in a load impedance, Z_0. Then by voltage division,

$$V_1^o = V_0 \frac{Z_{in}^o}{Z_{in}^o + Z_0}, \tag{7.74a}$$

$$V_1^e = V_0 \frac{Z_{in}^e}{Z_{in}^e + Z_0}, \tag{7.74b}$$

and

$$I_1^o = \frac{V_0}{Z_{in}^o + Z_0}, \tag{7.75a}$$

$$I_1^e = \frac{V_0}{Z_{in}^e + Z_0}. \tag{7.75b}$$

Using these results in (7.72) yields

$$Z_{in} = \frac{Z_{in}^o \left(Z_{in}^e + Z_0 \right) + Z_{in}^e \left(Z_{in}^o + Z_0 \right)}{Z_{in}^e + Z_{in}^o + 2Z_0} = Z_0 + \frac{2 \left(Z_{in}^o Z_{in}^e - Z_0^2 \right)}{Z_{in}^e + Z_{in}^o + 2Z_0}. \tag{7.76}$$

Now if we let

$$Z_0 = \sqrt{Z_{0e} Z_{0o}}, \tag{7.77}$$

then (7.73a) and (7.73b) reduce to

$$Z_{in}^e = Z_{0e} \frac{\sqrt{Z_{0o}} + j\sqrt{Z_{0e}}\tan\theta}{\sqrt{Z_{0e}} + j\sqrt{Z_{0o}}\tan\theta},$$

$$Z_{in}^o = Z_{0o} \frac{\sqrt{Z_{0e}} + j\sqrt{Z_{0o}}\tan\theta}{\sqrt{Z_{0o}} + j\sqrt{Z_{0e}}\tan\theta},$$

so that $Z_{in}^e Z_{in}^o = Z_{0e} Z_{0o} = Z_0^2$, and (7.76) reduces to

$$Z_{in} = Z_0. \tag{7.78}$$

Thus, as long as (7.77) is satisfied, port 1 (and, by symmetry, all other ports) will be matched.

Now if (7.77) is satisfied, so that $Z_{in} = Z_0$, we have that $V_1 = V_0$, by voltage division. The voltage at port 3 is

$$V_3 = V_3^e + V_3^o = V_1^e - V_1^o = V_0 \left[\frac{Z_{in}^e}{Z_{in}^e + Z_0} - \frac{Z_{in}^o}{Z_{in}^o + Z_0} \right], \tag{7.79}$$

where (7.74) has been used. From (7.73) and (7.77), we can show that

$$\frac{Z_{in}^e}{Z_{in}^e + Z_0} = \frac{Z_0 + jZ_{0e}\tan\theta}{2Z_0 + j(Z_{0e} + Z_{0o})\tan\theta},$$

$$\frac{Z_{in}^o}{Z_{in}^o + Z_0} = \frac{Z_0 + jZ_{0o}\tan\theta}{2Z_0 + j(Z_{0e} + Z_{0o})\tan\theta},$$

so that (7.79) reduces to

$$V_3 = V_0 \frac{j(Z_{0e} - Z_{0o})\tan\theta}{2Z_0 + j(Z_{0e} + Z_{0o})\tan\theta}. \tag{7.80}$$

Now define the coupling coefficient, C, as

$$C = \frac{Z_{0e} - Z_{0o}}{Z_{0e} + Z_{0o}}, \tag{7.81}$$

which we will soon see is actually the midband voltage coupling coefficient, V_3/V_0. Then,

$$\sqrt{1 - C^2} = \frac{2Z_0}{Z_{0e} + Z_{0o}},$$

so that

$$V_3 = V_0 \frac{jC\tan\theta}{\sqrt{1 - C^2} + j\tan\theta}. \tag{7.82}$$

Similarly, we can show that

$$V_4 = V_4^e + V_4^o = V_2^e - V_2^o = 0, \tag{7.83}$$

$$V_2 = V_2^e + V_2^o = V_0 \frac{\sqrt{1 - C^2}}{\sqrt{1 - C^2}\cos\theta + j\sin\theta}. \tag{7.84}$$

Equations (7.82) and (7.84) can be used to plot the coupled and through port voltages versus frequency, as shown in Figure 7.33. At very low frequencies ($\theta \ll \pi/2$), virtually all power is transmitted through port 2, with none being coupled to port 3. For $\theta = \pi/2$, the coupling to port 3 is at its first maximum; this is where the coupler is generally operated, for small size and minimum line loss. Otherwise, the response is periodic, with maxima in V_3 for $\theta = \pi/2, 3\pi/2, \ldots$

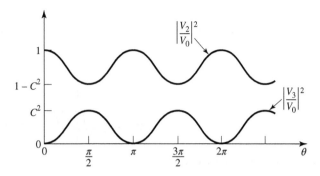

FIGURE 7.33 | Coupled and through port voltages (squared) versus frequency for the coupled line coupler of Figure 7.31.

For $\theta = \pi/2$, the coupler is $\lambda/4$ long, and (7.82) and (7.84) reduce to

$$\frac{V_3}{V_0} = C, \tag{7.85}$$

$$\frac{V_2}{V_0} = -j\sqrt{1 - C^2}, \tag{7.86}$$

which shows that $C < 1$ is the voltage coupling factor at the design frequency, $\theta = \pi/2$. Note that these results satisfy power conservation since $P_{\text{in}} = (1/2)|V_0|^2/Z_0$, while the output powers are $P_2 = (1/2)|V_2|^2/Z_0 = (1/2)(1 - C^2)|V_0|^2/Z_0$, $P_3 = (1/2)|C|^2|V_0|^2/Z_0$, and $P_4 = 0$, so that $P_{\text{in}} = P_2 + P_3 + P_4$. Also observe that there is a 90° phase shift between the two output port voltages; thus this coupler can be used as a quadrature hybrid. In addition, as long as (7.77) is satisfied, the coupler will be matched at the input and have perfect isolation, at any frequency.

Finally, if the characteristic impedance, Z_0, and the voltage coupling coefficient, C, are specified, then the following design equations for the required even- and odd-mode characteristic impedances can be easily derived from (7.77) and (7.81):

$$Z_{0e} = Z_0 \sqrt{\frac{1 + C}{1 - C}}, \tag{7.87a}$$

$$Z_{0o} = Z_0 \sqrt{\frac{1 - C}{1 + C}}. \tag{7.87b}$$

In the above analysis it was assumed that the even and odd modes of the coupled line structure have the same velocities of propagation, so that the line has the same electrical length for both modes. For coupled microstrip lines, or other non-TEM lines, this condition will generally not be satisfied exactly, and the coupler will have poor directivity. The fact that coupled microstrip lines have unequal even- and odd-mode phase velocities can be intuitively explained by considering the field line plots of Figure 7.28, which show that the even mode has less fringing field in the air region than the odd mode. Thus its effective dielectric constant should be higher, indicating a smaller phase velocity for the even mode. Techniques for compensating coupled microstrip lines to achieve equal even- and odd-mode phase velocities include the use of dielectric overlays and anisotropic substrates.

This type of coupler is best suited for weak coupling, as a large coupling factor requires lines that are too close together to be practical, or a combination of even- and odd-mode characteristic impedances that is nonrealizable.

EXAMPLE 7.7   SINGLE-SECTION COUPLER DESIGN AND PERFORMANCE

Design a 20 dB single-section coupled line coupler in stripline with a ground plane spacing of 3.2 mm, a dielectric constant of 2.2, a characteristic impedance of 50 Ω, and a center frequency of 3 GHz. Plot the coupling and directivity from 1 to 5 GHz. Include the effect of losses by assuming a loss tangent of 0.0009 for the dielectric material and copper conductors of 50 μm thickness.

Solution

The voltage coupling factor is $C = 10^{-20/20} = 0.1$. From (7.87), the even- and odd-mode characteristic impedances are

$$Z_{0e} = Z_0 \sqrt{\frac{1+C}{1-C}} = 55.28 \ \Omega,$$

$$Z_{0o} = Z_0 \sqrt{\frac{1-C}{1+C}} = 45.23 \ \Omega.$$

To use Figure 7.29, we have that

$$\sqrt{\epsilon_r} Z_{0e} = 81.99,$$

$$\sqrt{\epsilon_r} Z_{0o} = 67.08,$$

and so $W/b = 0.809$ and $S/b = 0.306$. This gives a conductor width of $W = 2.59$ mm and a conductor separation of $S = 0.098$ cm (these values were actually found using a commercial microwave CAD package).

Figure 7.34 shows the resulting coupling and directivity versus frequency, including the effect of dielectric and conductor losses. Losses have the effect of reducing the directivity, which is typically greater than 70 dB in the absence of loss.

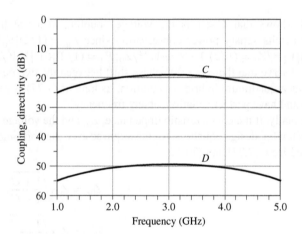

FIGURE 7.34 | Coupling versus frequency for the single-section coupler of Example 7.7.

Design of Multisection Coupled Line Couplers

As Figure 7.33 shows, the coupling of a single-section coupled line coupler is limited in bandwidth due to the $\lambda/4$ length requirement. As in the case of matching transformers and waveguide couplers, bandwidth can be increased by using multiple sections. In fact, there is a very close relationship between multisection coupled line couplers and multisection quarter-wave transformers [9].

Because the phase characteristics are usually better, multisection coupled line couplers are generally made with an odd number of sections, as shown in Figure 7.35. Thus, we will assume that N is odd. We will also assume that the coupling is weak ($C \geq 10$ dB), and that each section is $\lambda/4$ long ($\theta = \pi/2$) at the center frequency.

For a single coupled line section, with $C \ll 1$, (7.82) and (7.84) simplify to

$$\frac{V_3}{V_1} = \frac{jC\tan\theta}{\sqrt{1-C^2}+j\tan\theta} \simeq \frac{jC\tan\theta}{1+j\tan\theta} = jC\sin\theta e^{-j\theta}, \tag{7.88a}$$

$$\frac{V_2}{V_1} = \frac{\sqrt{1-C^2}}{\sqrt{1-C^2}\cos\theta+j\sin\theta} \simeq e^{-j\theta}. \tag{7.88b}$$

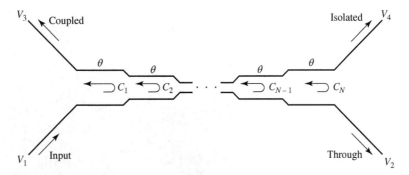

FIGURE 7.35 | An N-section coupled line coupler.

Then for $\theta = \pi/2$ we have that $V_3/V_1 = C$ and $V_2/V_1 = -j$. The above approximation is equivalent to assuming that no power is lost on the through path from one section to the next, and is similar to the approximation used for the multisection waveguide coupler analysis. It is a good assumption for small C, even though power conservation is violated.

Using these results, we can express the total voltage at the coupled port (port 3) of the cascaded coupler in Figure 7.35 as

$$V_3 = \left(jC_1 \sin\theta e^{-j\theta}\right)V_1 + \left(jC_2 \sin\theta e^{-j\theta}\right)V_1 e^{-2j\theta}$$
$$+ \cdots + \left(jC_N \sin\theta e^{-j\theta}\right)V_1 e^{-2j(N-1)\theta}, \tag{7.89}$$

where C_n is the voltage coupling coefficient of the nth section. If we assume that the coupler is symmetric, so that $C_1 = C_N, C_2 = C_{N-1}$, etc., we can simplify (7.89) to

$$V_3 = jV_1 \sin\theta e^{-j\theta}\left[C_1\left(1 + e^{-2j(N-1)\theta}\right) + C_2\left(e^{-2j\theta} + e^{-2j(N-2)\theta}\right) + \cdots + C_M e^{-j(N-1)\theta}\right]$$
$$= 2jV_1 \sin\theta e^{-jN\theta}\left[C_1\cos(N-1)\theta + C_2\cos(N-3)\theta + \cdots + \frac{1}{2}C_M\right], \tag{7.90}$$

where $M = (N+1)/2$.

At the center frequency, we define the voltage coupling factor C_0:

$$C_0 = \left|\frac{V_3}{V_1}\right|_{\theta=\pi/2}. \tag{7.91}$$

Equation (7.90) is in the form of a Fourier series for the coupling as a function of frequency. Thus, we can synthesize a desired coupling response by choosing the coupling coefficients, C_n. Note that in this case we synthesize the coupling response, while in the case of the multihole waveguide coupler we synthesized the directivity response. This is because the path for the uncoupled arm of the multisection coupled line coupler is in the forward direction, and so is less dependent on frequency than the coupled arm path, which is in the reverse direction; this is the opposite situation from the multihole waveguide coupler.

Multisection couplers of this form can achieve decade bandwidths, but coupling levels must be low. Because of the longer electrical length, it is more critical to have equal even- and odd-mode phase velocities than it is for the single-section coupler. This usually means that stripline is the preferred medium for such couplers. Mismatched phase velocities will degrade the coupler directivity, as will junction discontinuities, load mismatches, and fabrication tolerances. A photograph of a coupled line coupler is shown in Figure 7.36.

FIGURE 7.36 | Photograph of a single-section microstrip coupled line coupler.
Courtesy of M. D. Abouzahra, MIT Lincoln Laboratory, Lexington, MA.

EXAMPLE 7.8   MULTISECTION COUPLER DESIGN AND PERFORMANCE

Design a three-section 20 dB coupled line coupler with a binomial (maximally flat) response, a system impedance of 50 Ω, and a center frequency of 3 GHz. Plot the coupling and directivity from 1 to 5 GHz.

Solution
For a maximally flat response for a three-section ($N = 3$) coupler, we require that

$$\frac{d^n}{d\theta^n} C(\theta)\bigg|_{\theta=\pi/2} = 0 \quad \text{for } n = 1, 2.$$

From (7.90),

$$C = \left|\frac{V_3}{V_1}\right| = 2 \sin\theta \left(C_1 \cos 2\theta + \frac{1}{2} C_2\right)$$
$$= C_1(\sin 3\theta - \sin\theta) + C_2 \sin\theta$$
$$= C_1 \sin 3\theta + (C_2 - C_1) \sin\theta,$$

so

$$\frac{dC}{d\theta} = [3C_1 \cos 3\theta + (C_2 - C_1) \cos\theta]\big|_{\pi/2} = 0,$$
$$\frac{d^2C}{d\theta^2} = [-9C_1 \sin 3\theta - (C_2 - C_1) \sin\theta]\big|_{\pi/2} = 10C_1 - C_2 = 0.$$

At midband, $\theta = \pi/2$ and $C_0 = 20$ dB. Thus, $C = 10^{-20/20} = 0.1 = C_2 - 2C_1$. Solving these two equations for C_1 and C_2 gives

$$C_1 = C_3 = 0.0125,$$
$$C_2 = 0.125.$$

From (7.87) the even- and odd-mode characteristic impedances for each section are

$$Z_{0e}^1 = Z_{0e}^3 = 50\sqrt{\frac{1.0125}{0.9875}} = 50.63\ \Omega,$$

$$Z_{0o}^1 = Z_{0o}^3 = 50\sqrt{\frac{0.9875}{1.0125}} = 49.38\ \Omega,$$

$$Z_{0e}^2 = 50\sqrt{\frac{1.125}{0.875}} = 56.69\ \Omega,$$

$$Z_{0o}^2 = 50\sqrt{\frac{0.875}{1.125}} = 44.10\ \Omega.$$

The coupling and directivity for this coupler are plotted in Figure 7.37.

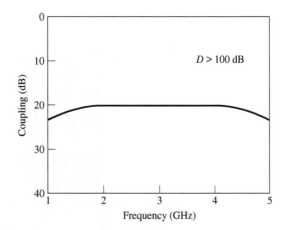

FIGURE 7.37 | Coupling versus frequency for the three-section binomial coupler of Example 7.8.

7.7 THE LANGE COUPLER

Generally the coupling in a coupled line coupler is too loose to achieve coupling factors of 3 or 6 dB. One way to increase the coupling between edge-coupled lines is to use several lines parallel to each other, so that the fringing fields at both edges of a line contribute to the coupling. One of the most practical implementations of this idea is the *Lange coupler* [10], shown in Figure 7.38a, where four parallel coupled lines are used with interconnections to provide tight coupling. This coupler can easily achieve 3 dB coupling ratios, with an octave or more bandwidth. The design tends to compensate for unequal even- and odd-mode phase velocities, which also improves the bandwidth. There is a 90° phase difference between the output lines (ports 2 and 3), so the Lange coupler is a type of quadrature hybrid. The main disadvantage of the Lange coupler is probably practical, as the lines are very narrow and close together, and the required bonding wires across the lines increases complexity. This type of coupled line geometry is also referred to as *interdigitated*; such structures can also be used for filter circuits.

The *unfolded* Lange coupler [11], shown in Figure 7.38b, operates in essentially the same way as the original Lange coupler but is easier to model with an equivalent circuit. Such an equivalent circuit consists of a four-wire coupled line structure, as shown in Figure 7.39a. All of the lines have the same width and spacing. If we make the reasonable assumption that each line couples only to its nearest neighbor, and ignore more-distant couplings, then we effectively have a two-wire coupled line circuit, as shown in Figure 7.39b. Then, if we can derive the even- and odd-mode characteristic impedances, Z_{e4} and Z_{o4}, of the four-wire circuit of Figure 7.39a in terms of Z_{0e} and Z_{0o}, the even- and odd-mode characteristic impedances of any adjacent pair of lines, we can apply the coupled line coupler results of Section 7.6 to analyze the Lange coupler.

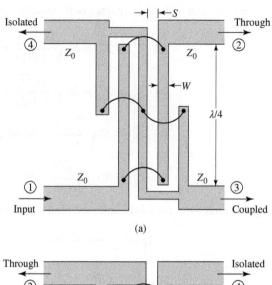

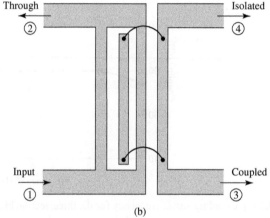

FIGURE 7.38 | The Lange coupler. (a) Layout in microstrip form. (b) The unfolded Lange coupler.

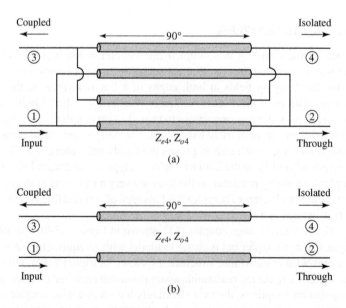

FIGURE 7.39 | Equivalent circuits for the unfolded Lange coupler. (a) Four-wire coupled line model. (b) Approximate two-wire coupled line model.

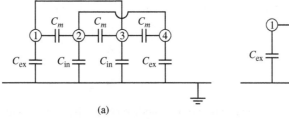

FIGURE 7.40 | Effective capacitance networks for the unfolded Lange coupler equivalent circuits of Figure 7.39. (a) Effective capacitance network for the four-wire model. (b) Effective capacitance network for the two-wire model.

Figure 7.40a shows the effective capacitances between the conductors of the four-wire coupled line of Figure 7.39a. Unlike the two-line case of Section 7.6, the capacitances of the four lines to ground are different depending on whether the line is on the outside (1 and 4), or on the inside (2 and 3). An approximate relation between these capacitances is [12]

$$C_{\text{in}} = C_{\text{ex}} - \frac{C_{\text{ex}} C_m}{C_{\text{ex}} + C_m}. \tag{7.92}$$

For an even-mode excitation, all four conductors in Figure 7.40a are at the same potential, so C_m has no effect, and the total capacitance of any line to ground is

$$C_{e4} = C_{\text{ex}} + C_{\text{in}}. \tag{7.93a}$$

For an odd-mode excitation, electric walls effectively exist through the middle of each C_m, so the capacitance of any line to ground is

$$C_{o4} = C_{\text{ex}} + C_{\text{in}} + 6C_m. \tag{7.93b}$$

The even- and odd-mode characteristic impedances are then

$$Z_{e4} = \frac{1}{v_p C_{e4}}, \tag{7.94a}$$

$$Z_{o4} = \frac{1}{v_p C_{o4}}, \tag{7.94b}$$

where v_p is the phase velocity of propagation on the line.

Now consider any isolated pair of adjacent conductors in the four-line model; the effective capacitances are as shown in Figure 7.40b. The even- and odd-mode capacitances are

$$C_e = C_{\text{ex}}, \tag{7.95a}$$

$$C_o = C_{\text{ex}} + 2C_m. \tag{7.95b}$$

Solving (7.95) for C_{ex} and C_m, and substituting into (7.93) with the aid of (7.92) gives the even-odd mode capacitances of the four-wire line in terms of a two-wire coupled line:

$$C_{e4} = \frac{C_e(3C_e + C_o)}{C_e + C_o}, \tag{7.96a}$$

$$C_{o4} = \frac{C_o(3C_o + C_e)}{C_e + C_o}. \tag{7.96b}$$

Because characteristic impedances are related to capacitance as $Z_0 = 1/v_p C$, we can rewrite (7.96) to give the even/odd mode characteristic impedances of the Lange coupler in terms of the characteristic impedances of a two-conductor line that is identical to any pair of adjacent lines in the coupler:

$$Z_{e4} = \frac{Z_{0o} + Z_{0e}}{3Z_{0o} + Z_{0e}} Z_{0e}, \tag{7.97a}$$

$$Z_{o4} = \frac{Z_{0o} + Z_{0e}}{3Z_{0e} + Z_{0o}} Z_{0o}, \tag{7.97b}$$

where Z_{0e} and Z_{0o} are the even- and odd-mode characteristic impedances of the two-conductor pair.

Now we can apply the results of Section 7.6 to the coupler of Figure 7.39b. From (7.77) the characteristic impedance is

$$Z_0 = \sqrt{Z_{e4} Z_{o4}} = \sqrt{\frac{Z_{0e} Z_{0o} (Z_{0o} + Z_{0e})^2}{(3Z_{0o} + Z_{0e})(3Z_{0e} + Z_{0o})}}, \tag{7.98}$$

and the voltage coupling coefficient is, from (7.81),

$$C = \frac{Z_{e4} - Z_{o4}}{Z_{e4} + Z_{o4}} = \frac{3\left(Z_{0e}^2 - Z_{0o}^2\right)}{3\left(Z_{0e}^2 + Z_{0o}^2\right) + 2Z_{0e} Z_{0o}}, \tag{7.99}$$

where (7.97) was used. For design purposes it is useful to invert these results to give the necessary even- and odd-mode impedances in terms of a desired characteristic impedance and coupling coefficient:

$$Z_{0e} = \frac{4C - 3 + \sqrt{9 - 8C^2}}{2C\sqrt{(1 - C)/(1 + C)}} Z_0, \tag{7.100a}$$

$$Z_{0o} = \frac{4C + 3 - \sqrt{9 - 8C^2}}{2C\sqrt{(1 + C)/(1 - C)}} Z_0. \tag{7.100b}$$

These results are approximate because of the simplifications involved with the application of two-line characteristic impedances to the four-line circuit, and because of the assumption of equal even- and odd-mode phase velocities. In practice, however, these results generally give sufficient accuracy. If necessary, a more complete analysis can be made to directly determine Z_{e4} and Z_{o4} for the four-line circuit, as in reference [13].

7.8 THE 180° HYBRID

The 180° hybrid junction is a four-port network with a 180° phase shift between the two output ports. It can also be operated so that the outputs are in phase. 180° hybrid junction can be used as a power splitter as well as a combiner.

- As a splitter: With reference to Figure 7.41b, a signal applied to port 1 will be evenly split into two in-phase components at ports 2 and 3 $(S_{21} = S_{31})$, and port 4 will be isolated $(S_{41} = 0)$. If the input is applied to port 4, it will be equally split into two components with a 180° phase difference at ports 2 and 3 $(S_{24} = -S_{34})$, and port 1 will be isolated $(S_{14} = 0)$.
- As a combiner: With input signals applied at ports 2 and 3, the sum of the inputs will be formed at port 1, while the difference will be formed at port 4. Hence, ports 1 and 4 are referred to as the sum and difference ports, respectively.

The scattering matrix for the ideal 3 dB 180° hybrid thus has the following form:

$$[S] = \frac{-j}{\sqrt{2}} \begin{bmatrix} 0 & 1 & 1 & 0 \\ 1 & 0 & 0 & -1 \\ 1 & 0 & 0 & 1 \\ 0 & -1 & 1 & 0 \end{bmatrix}. \tag{7.101}$$

The reader may verify that this matrix is unitary and symmetric.

The 180° hybrid can be fabricated in several forms. The ring hybrid, or rat-race, shown in Figures 7.42a and 7.43, can easily be constructed in planar (microstrip or stripline) form, although waveguide versions are also possible. Another type of planar 180° hybrid uses tapered matching sections and coupled lines,

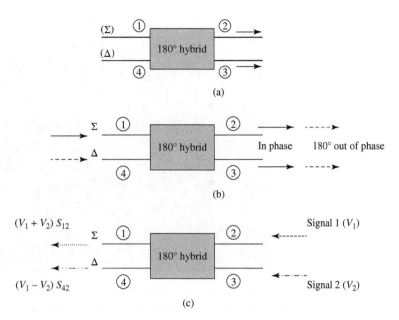

FIGURE 7.41 | (a) Symbol for a 180° hybrid junction. (b) As a splitter. (c) As a combiner.

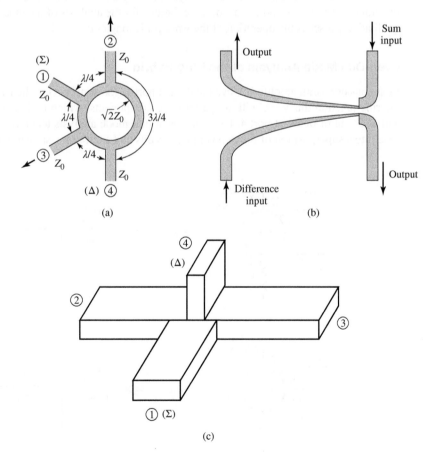

FIGURE 7.42 | Three types of hybrid junctions. (a) A ring hybrid, or *rat-race*, in microstrip line or stripline form. (b) A tapered coupled line hybrid. (c) A waveguide hybrid junction, or *magic-T*.

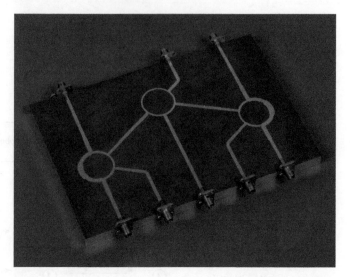

FIGURE 7.43 | Photograph of a microstrip power divider network using three ring hybrids.
Courtesy of M. D. Abouzahra, MIT Lincoln Laboratory, Lexington, MA.

as shown in Figure 7.42b. Yet another type is the hybrid waveguide junction, or magic-T, shown in Figure 7.42c. We will first analyze the ring hybrid, using an even-odd mode analysis similar to that used for the branch-line hybrid, and use a similar technique for the analysis of the tapered line hybrid. Then we will qualitatively discuss the operation of the waveguide magic-T.

Even-Odd Mode Analysis of the Ring Hybrid

First consider a unit amplitude wave incident at port 1 (the sum port) of the ring hybrid of Figure 7.42a. At the ring junction this wave will divide into two components, which both arrive in phase at ports 2 and 3, and 180° out of phase at port 4. Using the even-odd mode analysis technique [5], we can decompose this case into a superposition of the two simpler circuits and excitations shown in Figure 7.44. The amplitudes

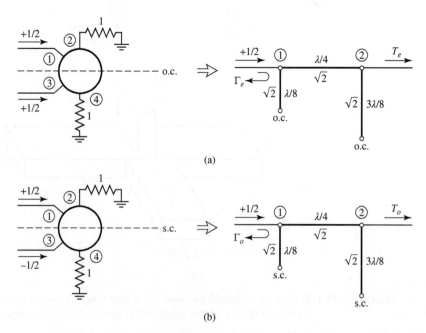

FIGURE 7.44 | Even- and odd-mode decomposition of the ring hybrid when port 1 is excited with a unit amplitude incident wave. (a) Even mode. (b) Odd mode.

of the scattered waves from the ring hybrid will be

$$B_1 = \frac{1}{2}\Gamma_e + \frac{1}{2}\Gamma_o, \tag{7.102a}$$

$$B_2 = \frac{1}{2}T_e + \frac{1}{2}T_o, \tag{7.102b}$$

$$B_3 = \frac{1}{2}\Gamma_e - \frac{1}{2}\Gamma_o, \tag{7.102c}$$

$$B_4 = \frac{1}{2}T_e - \frac{1}{2}T_o. \tag{7.102d}$$

We can evaluate the required reflection and transmission coefficients defined in Figure 7.44 using the *ABCD* matrix for the even- and odd-mode two-port circuits in Figure 7.44. The results are

$$\begin{bmatrix} A & B \\ C & D \end{bmatrix}_e = \begin{bmatrix} 1 & j\sqrt{2} \\ j\sqrt{2} & -1 \end{bmatrix}, \tag{7.103a}$$

$$\begin{bmatrix} A & B \\ C & D \end{bmatrix}_o = \begin{bmatrix} -1 & j\sqrt{2} \\ j\sqrt{2} & 1 \end{bmatrix}. \tag{7.103b}$$

Then with the aid of Table 4.2 we have

$$\Gamma_e = \frac{-j}{\sqrt{2}}, \tag{7.104a}$$

$$T_e = \frac{-j}{\sqrt{2}}, \tag{7.104b}$$

$$\Gamma_o = \frac{j}{\sqrt{2}}, \tag{7.104c}$$

$$T_o = \frac{-j}{\sqrt{2}}. \tag{7.104d}$$

Using these results in (7.102) gives

$$B_1 = 0, \tag{7.105a}$$

$$B_2 = \frac{-j}{\sqrt{2}}, \tag{7.105b}$$

$$B_3 = \frac{-j}{\sqrt{2}}, \tag{7.105c}$$

$$B_4 = 0, \tag{7.105d}$$

which shows that the input port is matched, port 4 is isolated, and the input power is evenly divided and in phase between ports 2 and 3. These results form the first row and column of the scattering matrix in (7.101).

Next consider a unit amplitude wave incident at port 4 (the difference port) of the ring hybrid of Figure 7.42a. The two wave components on the ring will arrive in phase at port 2 and at port 3, with a relative phase difference of 180° between these two ports. The two wave components will be 180° out of phase at port 1. This case can be decomposed into a superposition of the two simpler circuits and excitations shown in Figure 7.45. The amplitudes of the scattered waves will be

$$B_1 = \frac{1}{2}T_e - \frac{1}{2}T_o, \tag{7.106a}$$

$$B_2 = \frac{1}{2}\Gamma_e - \frac{1}{2}\Gamma_o, \tag{7.106b}$$

$$B_3 = \frac{1}{2}T_e + \frac{1}{2}T_o, \tag{7.106c}$$

$$B_4 = \frac{1}{2}\Gamma_e + \frac{1}{2}\Gamma_o. \tag{7.106d}$$

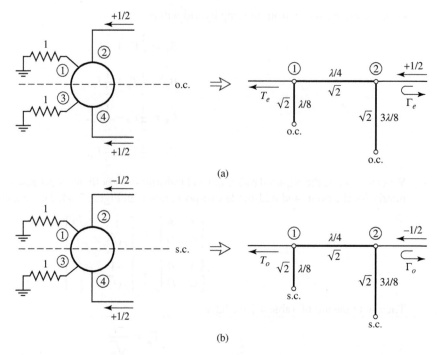

FIGURE 7.45 | Even- and odd-mode decomposition of the ring hybrid when port 4 is excited with a unit amplitude incident wave. (a) Even mode. (b) Odd mode.

The *ABCD* matrices for the even- and odd-mode circuits of Figure 7.45 are

$$
\begin{bmatrix} A & B \\ C & D \end{bmatrix}_e = \begin{bmatrix} -1 & j\sqrt{2} \\ j\sqrt{2} & 1 \end{bmatrix},
\tag{7.107a}
$$

$$
\begin{bmatrix} A & B \\ C & D \end{bmatrix}_o = \begin{bmatrix} 1 & j\sqrt{2} \\ j\sqrt{2} & -1 \end{bmatrix}.
\tag{7.107b}
$$

Then, from Table 4.2, the necessary reflection and transmission coefficients are

$$
\Gamma_e = \frac{j}{\sqrt{2}},
\tag{7.108a}
$$

$$
T_e = \frac{-j}{\sqrt{2}},
\tag{7.108b}
$$

$$
\Gamma_o = \frac{-j}{\sqrt{2}},
\tag{7.108c}
$$

$$
T_o = \frac{-j}{\sqrt{2}}.
\tag{7.108d}
$$

Using these results in (7.106) gives

$$
B_1 = 0,
\tag{7.109a}
$$

$$
B_2 = \frac{j}{\sqrt{2}},
\tag{7.109b}
$$

$$
B_3 = \frac{-j}{\sqrt{2}},
\tag{7.109c}
$$

$$
B_4 = 0,
\tag{7.109d}
$$

which shows that the input port is matched, port 1 is isolated, and the input power is evenly divided into ports 2 and 3 with a 180° phase difference. These results form the fourth row and column of the scattering matrix of (7.101). The remaining elements in this matrix can be found from symmetry considerations.

The bandwidth of the ring hybrid is limited by the frequency dependence of the ring lengths, but is generally on the order of 20%–30%. Increased bandwidth can be obtained by using additional sections, or a symmetric ring circuit, as suggested in reference [14].

EXAMPLE 7.9   DESIGN AND PERFORMANCE OF A RING HYBRID

Design a 180° ring hybrid for a 60 Ω system impedance, and plot the magnitude of the scattering parameters (S_{1j}) from $0.5f_0$ to $1.5f_0$, where f_0 is the design frequency.

Solution
With reference to Figure 7.42a, the characteristic impedance of the ring transmission line is

$$\sqrt{2}Z_0 = 84.9 \text{ Ω,}$$

while the feedline impedances are 60 Ω. The scattering parameter magnitudes are plotted versus frequency in Figure 7.46.

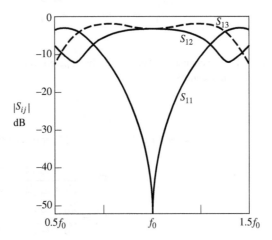

FIGURE 7.46 | Scattering parameter magnitudes versus frequency for the ring hybrid of Example 7.9.

Even-Odd Mode Analysis of the Tapered Coupled Line Hybrid

The tapered coupled line 180° hybrid [15], shown in Figure 7.42b, can provide an arbitrary power division ratio with a bandwidth of a decade or more. This hybrid is also referred to as an asymmetric tapered coupled line coupler.

The schematic circuit of this coupler is shown in Figure 7.47, with the ports numbered to correspond functionally to the ports of the 180° hybrids in Figures 7.41 and 7.42. The coupler consists of two coupled lines with tapering characteristic impedances over the length $0 < z < L$. At $z = 0$ the lines are very weakly coupled, so that $Z_{0e}(0) = Z_{0o}(0) = Z_0$, while at $z = L$ the coupling is such that $Z_{0e}(L) = Z_0/k$ and $Z_{0o}(L) = kZ_0$, where $0 \le k \le 1$ is a coupling factor that will be related to the voltage coupling factor. The even mode of the coupled line thus matches a load impedance of Z_0/k (at $z = L$) to Z_0, while the odd mode matches a load of kZ_0 to Z_0; note that $Z_{0e}(z)Z_{0o}(z) = Z_0^2$ for all z. The Klopfenstein taper is generally used for these tapered matching lines. For $L < z < 2L$ the lines are uncoupled, and both have a characteristic impedance Z_0; these lines are required for phase compensation of the coupled line section. The length of each section, $\theta = \beta L$, must be the same, and they should be electrically long to provide a good impedance match over the desired bandwidth.

First consider an incident voltage wave of amplitude V_0 applied to port 4, the difference input. This excitation can be reduced to the superposition of an even-mode excitation and an odd-mode excitation, as shown in Figures 7.48a and 7.48b, respectively. At the junctions of the coupled and uncoupled lines ($z = L$),

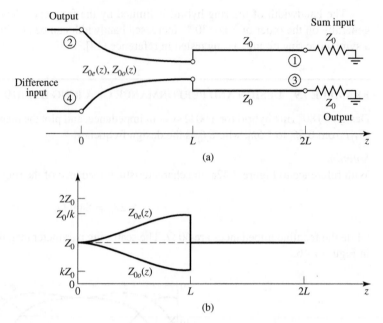

(a)

(b)

FIGURE 7.47 | (a) Schematic diagram of the tapered coupled line hybrid. (b) The variation of characteristic impedances.

the reflection coefficients seen by the even or odd modes of the tapered lines are

$$\Gamma'_e = \frac{Z_0 - Z_0/k}{Z_0 + Z_0/k} = \frac{k - 1}{k + 1}, \tag{7.110a}$$

$$\Gamma'_o = \frac{Z_0 - kZ_0}{Z_0 + kZ_0} = \frac{1 - k}{1 + k}. \tag{7.110b}$$

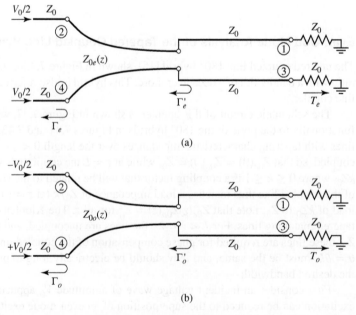

(a)

(b)

FIGURE 7.48 | Excitation of the tapered coupled line hybrid. (a) Even-mode excitation. (b) Odd-mode excitation.

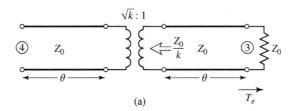

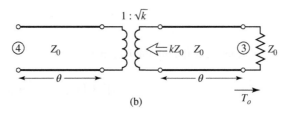

FIGURE 7.49 | Equivalent circuits for the tapered coupled line hybrid, for transmission from port 4 to port 3. (a) Even-mode case. (b) Odd-mode case.

At $z = 0$ these coefficients are transformed to

$$\Gamma_e = \frac{k-1}{k+1}e^{-2j\theta}, \tag{7.111a}$$

$$\Gamma_o = \frac{1-k}{1+k}e^{-2j\theta}. \tag{7.111b}$$

Then by superposition the scattering parameters of ports 2 and 4 are as follows:

$$S_{44} = \frac{1}{2}(\Gamma_e + \Gamma_o) = 0, \tag{7.112a}$$

$$S_{24} = \frac{1}{2}(\Gamma_e - \Gamma_o) = \frac{k-1}{k+1}e^{-2j\theta}. \tag{7.112b}$$

By symmetry, we also have that $S_{22} = 0$ and $S_{42} = S_{24}$.

To evaluate the transmission coefficients into ports 1 and 3, we will use the *ABCD* parameters for the equivalent circuits shown in Figure 7.49, where the tapered matching sections have been assumed to be ideal, and replaced with transformers. The *ABCD* matrix of the transmission line–transformer–transmission line cascade can be found by multiplying the three individual *ABCD* matrices for these components, but it is easier to use the fact that the transmission line sections affect only the phase of the transmission coefficients. The *ABCD* matrix of the transformer is, for the even mode,

$$\begin{bmatrix} \sqrt{k} & 0 \\ 0 & 1/\sqrt{k} \end{bmatrix},$$

and for the odd mode is

$$\begin{bmatrix} 1/\sqrt{k} & 0 \\ 0 & \sqrt{k} \end{bmatrix}.$$

Then the even- and odd-mode transmission coefficients are

$$T_e = T_o = \frac{2\sqrt{k}}{k+1}e^{-2j\theta}, \tag{7.113}$$

since $T = 2/(A + B/Z_0 + CZ_0 + D) = 2\sqrt{k}/(k+1)$ for both modes; the $e^{-2j\theta}$ factor accounts for the phase delay of the two transmission line sections. We can then evaluate the following scattering parameters:

$$S_{34} = \frac{1}{2}(T_e + T_o) = \frac{2\sqrt{k}}{k+1}e^{-2j\theta}, \tag{7.114a}$$

$$S_{14} = \frac{1}{2}(T_e - T_o) = 0. \tag{7.114b}$$

The voltage coupling factor from port 4 to port 3 is

$$\beta = |S_{34}| = \frac{2\sqrt{k}}{k+1}, \quad 0 < \beta < 1, \tag{7.115a}$$

while the voltage coupling factor from port 4 to port 2 is

$$\alpha = |S_{24}| = -\frac{k-1}{k+1}, \quad 0 < \alpha < 1. \tag{7.115b}$$

Power conservation is verified by the fact that

$$|S_{24}|^2 + |S_{34}|^2 = \alpha^2 + \beta^2 = 1.$$

If we now apply even- and odd-mode excitations at ports 1 and 3, so that superposition yields an incident voltage wave at port 1, we can derive the remaining scattering parameters. With a phase reference at the input ports, the even- and odd-mode reflection coefficients at port 1 will be

$$\Gamma_e = \frac{1-k}{1+k}e^{-2j\theta}, \tag{7.116a}$$

$$\Gamma_o = \frac{k-1}{k+1}e^{-2j\theta}. \tag{7.116b}$$

Then we can calculate the following scattering parameters:

$$S_{11} = \frac{1}{2}(\Gamma_e + \Gamma_o) = 0, \tag{7.117a}$$

$$S_{31} = \frac{1}{2}(\Gamma_e - \Gamma_o) = \frac{1-k}{1+k}e^{-2j\theta} = \alpha e^{-2j\theta}. \tag{7.117b}$$

From symmetry we have that $S_{33} = 0, S_{13} = S_{31}$, and $S_{14} = S_{32}, S_{12} = S_{34}$. The tapered coupled line 180° hybrid thus has the following scattering matrix:

$$[S] = \begin{bmatrix} 0 & \beta & \alpha & 0 \\ \beta & 0 & 0 & -\alpha \\ \alpha & 0 & 0 & \beta \\ 0 & -\alpha & \beta & 0 \end{bmatrix} e^{-2j\theta}. \tag{7.118}$$

Waveguide Magic-T

The waveguide magic-T hybrid junction of Figure 7.42c has terminal properties similar to those of the ring hybrid, and a scattering matrix similar in form to (7.101). A rigorous analysis of this junction is too complicated to present here, but we can explain its operation in a qualitative sense by considering the field lines for excitations at the sum and difference ports.

First consider a TE_{10} mode incident at port 1. The resulting E_y field lines are illustrated in Figure 7.50a, where it is seen that there is an odd symmetry about guide 4. Because the field lines of a TE_{10} mode in guide 4 would have even symmetry, there is no coupling between ports 1 and 4. There is identical coupling to ports 2 and 3, however, resulting in an in-phase, equal-split power division.

For a TE_{10} mode incident at port 4, the field lines are as shown in Figure 7.50b. Again ports 1 and 4 are decoupled, due to symmetry (or reciprocity). Ports 2 and 3 are excited equally by the incident wave, but with a 180° phase difference.

In practice, tuning posts or irises are often used for matching; such components must be placed symmetrically to maintain proper operation of the hybrid.

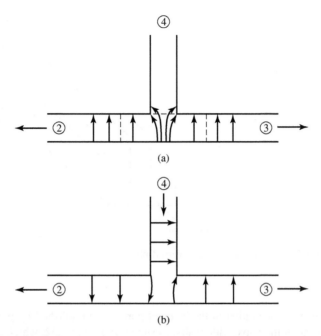

FIGURE 7.50 | Electric field lines for a waveguide hybrid junction. (a) Incident wave at port 1. (b) Incident wave at port 4.

7.9 OTHER COUPLERS

Although we have discussed the general properties of couplers and have analyzed and derived design data for several of the most frequently used couplers, there are many other types that we have not treated in detail. In this section we will briefly describe some of these.

Moreno crossed-guide coupler: This is a waveguide directional coupler consisting of two waveguides at right angles, with coupling provided by two apertures in the common broad wall of the guides. See Figure 7.51. By proper design [16], the two wave components excited by these apertures can be made to cancel in the back direction. The apertures usually consist of crossed slots, in order to couple tightly to the fields of both guides.

Schwinger reversed-phase coupler: This waveguide coupler is designed so that the path lengths for the two coupling apertures are the same for the uncoupled port, so that the directivity is essentially independent of

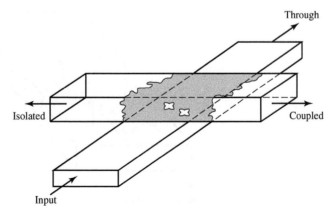

FIGURE 7.51 | The Moreno crossed-guide coupler.

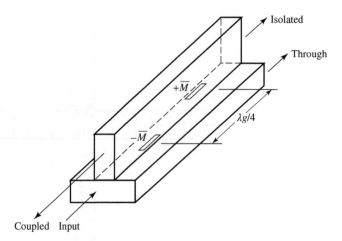

FIGURE 7.52 | The Schwinger reversed-phase coupler.

frequency. Cancellation in the isolated port is accomplished by placing the slots on opposite sides of the centerline of the waveguide walls, as shown in Figure 7.52, which couple to magnetic dipoles with a 180° phase difference. The $\lambda_g/4$ slot spacing leads to in-phase combining at the coupled (backward) port, but this coupling is very frequency sensitive. This is the opposite situation from that of the multihole waveguide coupler discussed in Section 7.4.

Riblet short-slot coupler: This coupler, shown in Figure 7.53, consists of two waveguides with a common sidewall. Coupling takes place in the region where part of the common wall has been removed. In this region both the TE_{10} (even) and the TE_{20} (odd) modes are excited, and by proper design can be made to cause cancellation at the isolated port and addition at the coupled port. The width of the interaction region must be small enough to prevent propagation of the undesired TE_{30} mode. This coupler can usually be made smaller than other waveguide couplers.

Symmetric tapered coupled line coupler: We saw that a continuously tapered transmission line matching transformer was the logical extension of the multisection matching transformer. Similarly, the multisection coupled line coupler can be extended to a continuous taper, yielding a coupled line coupler with good bandwidth characteristics. Such a coupler is shown in Figure 7.54. Generally, both the conductor width and separation can be adjusted to provide a synthesized coupling or directivity response. One way to do this

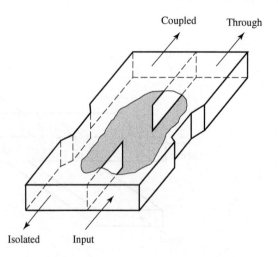

FIGURE 7.53 | The Riblet short-slot coupler.

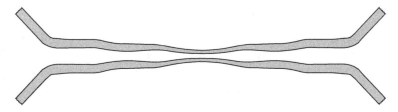

FIGURE 7.54 | A symmetric tapered coupled line coupler.

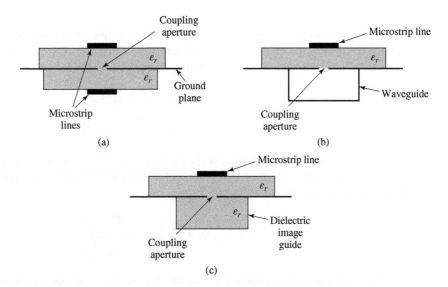

FIGURE 7.55 | Various aperture coupled planar line couplers. (a) Microstrip-to-microstrip coupler. (b) Microstrip-to-waveguide coupler. (c) Microstrip-to-dielectric image line coupler.

involves the computer optimization of a stepped-section approximation to the continuous taper [17]. This coupler provides a 90° phase shift between the outputs.

Couplers with apertures in planar lines: Many of the above-mentioned waveguide couplers can also be fabricated with planar lines such as microstrip line, stripline, dielectric image lines, or various combinations of these. Some possibilities are illustrated in Figure 7.55. In principle, the design of such couplers can be carried out using the small-hole coupling theory and analysis techniques used in this chapter. The evaluation of the fields of planar lines, however, is usually much more complicated than for rectangular waveguides.

POINT OF INTEREST: The Reflectometer

A *reflectometer* is a circuit that uses a directional coupler to isolate and sample the incident and reflected powers from a load. It is a key component in a scalar or vector network analyzer, as it can be used to measure the reflection coefficient of a one-port network and, in a more general configuration, the scattering parameters of a two-port network. It can also be used as an SWR meter, or as a power monitor in systems applications.

The basic reflectometer circuit shown on the left in the accompanying figure can be used to measure the reflection coefficient magnitude of an unknown load. If we assume a reasonably matched coupler with loose coupling ($C \ll 1$), so that $\sqrt{1 - C^2} \simeq 1$, then the circuit can be represented by the signal flow graph shown on the right in the accompanying figure. In operation, the directional coupler provides a sample, V_i, of the incident wave, and a sample, V_r, of the reflected wave. A ratio meter with an appropriately calibrated scale can then measure these voltages and provide a reading in terms of reflection coefficient magnitude, or SWR.

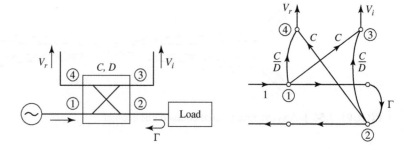

Realistic directional couplers, however, have finite directivity, which means that both incident and reflected powers will contribute to both V_i and V_r, leading to an error. If we assume a unit incident wave from the source, inspection of the signal flow graph leads to the following expressions for V_i and V_r:

$$V_i = C + \frac{C}{D}\Gamma e^{j\theta},$$

$$V_r = \frac{C}{D} + C\Gamma e^{j\phi},$$

where Γ is the reflection coefficient of the load, $D = 10^{(D\,\mathrm{dB}/20)}$ is the numerical directivity of the coupler, and θ, ϕ are the unknown phase delays through the circuit. Then the maximum and minimum values of the magnitude of V_r/V_i can be written as

$$\left|\frac{V_r}{V_i}\right|_{\substack{\max \\ \min}} = \frac{|\Gamma| \pm \dfrac{1}{D}}{1 \mp \dfrac{|\Gamma|}{D}}.$$

For a coupler with infinite directivity this reduces to the desired result of $|\Gamma|$. Otherwise a measurement uncertainty of approximately $\pm 1/D$ is introduced. Good accuracy thus requires a coupler with high directivity, preferably greater than 40 dB.

REFERENCES

[1] A. E. Bailey, ed., *Microwave Measurement*, Peter Peregrinus, London, 1985.

[2] R. E. Collin, *Foundations for Microwave Engineering*, 2nd edition, Wiley–IEEE Press, Hoboken, NJ, 2001.

[3] F. E. Gardiol, *Introduction to Microwaves*, Artech House, Dedham, MA, 1984.

[4] E. Wilkinson, "An *N*-Way Hybrid Power Divider," *IRE Transactions on Microwave Theory and Techniques*, vol. MTT-8, pp. 116–118, January 1960.

[5] J. Reed and G. J. Wheeler, "A Method of Analysis of Symmetrical Four-Port Networks," *IRE Transactions on Microwave Theory and Techniques*, vol. MTT-4, pp. 246–252, October 1956.

[6] C. G. Montgomery, R. H. Dicke, and E. M. Purcell, *Principles of Microwave Circuits*, MIT Radiation Laboratory Series, vol. 8, McGraw-Hill, New York, 1948.

[7] H. Howe, *Stripline Circuit Design*, Artech House, Dedham, MA, 1974.

[8] K. C. Gupta, R. Garg, and I. J. Bahl, *Microstrip Lines and Slot Lines*, Artech House, Dedham, MA, 1979.

[9] L. Young, "The Analytical Equivalence of the TEM-Mode Directional Couplers and Transmission-Line Stepped Impedance Filters," *Proceedings of the IEEE*, vol. 110, pp. 275–281, February 1963.

[10] J. Lange, "Interdigitated Stripline Quadrature Hybrid," *IEEE Transactions on Microwave Theory and Techniques*, vol. MTT-17, pp. 1150–1151, December 1969.

[11] R. Waugh and D. LaCombe, "Unfolding the Lange Coupler," *IEEE Transactions on Microwave Theory and Techniques*, vol. MTT-20, pp. 777–779, November 1972.

[12] W. P. Ou, "Design Equations for an Interdigitated Directional Coupler," *IEEE Transactions on Microwave Theory and Techniques*, vol. MTT-23, pp. 253–255, February 1973.

[13] D. Paolino, "Design More Accurate Interdigitated Couplers," *Microwaves*, vol. 15, pp. 34–38, May 1976.

[14] J. Hughes and K. Wilson, "High Power Multiple IMPATT Amplifiers," in: *Proceedings of the 4th European Microwave Conference*, Montreux, Switzerland, pp. 118–122, 1974.

[15] R. H. DuHamel and M. E. Armstrong, "The Tapered-Line Magic-T," in: *Abstracts of 15th Annual Symposium of the USAF Antenna Research and Development Program*, Monticello, IL, October 12–14, 1965.

[16] T. N. Anderson, "Directional Coupler Design Nomograms," *Microwave Journal*, vol. 2, pp. 34–38, May 1959.

[17] D. W. Kammler, "The Design of Discrete *N*-Section and Continuously Tapered Symmetrical TEM Directional Couplers," *IEEE Transactions on Microwave Theory and Techniques*, vol. MTT-17, pp. 577–590, August 1969.

7.1 Write the scattering matrix for a nonideal symmetric hybrid coupler in terms of the coupler parameters, C, D, I, and L, as defined in (7.20). Repeat for a nonideal antisymmetric hybrid coupler. Assume that the couplers are matched at all ports.

7.2 A 30 dBm power source is connected to the input of a directional coupler having a coupling factor of 25 dB, a directivity of 38 dB, and an insertion loss of 0.5 dB. If all ports are matched, find the output powers (in dBm) at the through, coupled, and isolated ports.

7.3 A directional coupler has the scattering matrix given below. Find the return loss, coupling factor, directivity, and insertion loss. Assume that the ports are terminated in matched loads.

$$[S] = \begin{bmatrix} 0.176\angle45° & 0.944\angle90° & 0.265\angle180° & 0.0053\angle90° \\ 0.944\angle90° & 0.176\angle45° & 0.0053\angle90° & 0.265\angle180° \\ 0.265\angle180° & 0.0053\angle90° & 0.176\angle45° & 0.944\angle90° \\ 0.0053\angle90° & 0.265\angle180° & 0.944\angle90° & 0.176\angle45° \end{bmatrix}$$

7.4 Two identical 90° couplers with $C = 8.34$ dB are connected as shown in the figure. Find the resulting phase and amplitudes at ports 2′ and 3′, relative to port 1.

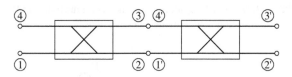

7.5 Consider the T-junction of three lines with characteristic impedances $Z_1, Z_2,$ and Z_3, as shown below. Demonstrate that it is impossible for all three lines to be matched when looking toward the junction.

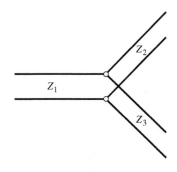

7.6 Design a lossless T-junction divider with a 45 Ω source impedance to give a 3:1 power split. Design quarter-wave matching transformers to convert the impedances of the output lines to 45 Ω. Determine the magnitude of the scattering parameters for this circuit, using a 45 Ω characteristic impedance.

7.7 Consider the T and π resistive attenuator circuits shown in the figure. If the input and output are matched to Z_0, and the ratio of output voltage to input voltage is α, derive the design equations for R_1 and R_2 for each circuit. If $Z_0 = 50$ Ω,

compute R_1 and R_2 for 5, 15, and 22 dB attenuators of each type.

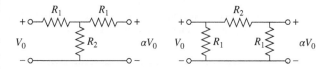

7.8 Design a three-port resistive divider for an equal power split and a 75 Ω system impedance. If port 3 is matched, calculate the change in output power at port 3 (in dB) when port 2 is connected first to a matched load, and then to a load having a mismatch of $\Gamma = 0.4$. See the figure below.

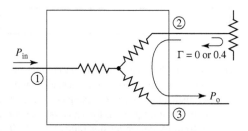

7.9 Consider the general resistive divider shown below. For an arbitrary power division ratio $\alpha = P_2/P_3$, derive expressions for the resistors $R_1, R_2,$ and R_3, and the output characteristic impedances Z_{o2} and Z_{o3} so that all ports are matched, assuming the source impedance is Z_0.

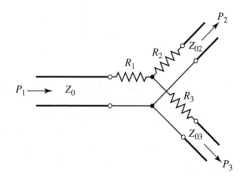

7.10 Design a Wilkinson power divider with a power division ratio of $P_3/P_2 = 1/3$ and a source impedance of 50 Ω.

7.11 Derive the design equations in (7.37a)–(7.37c) for the unequal-split Wilkinson divider.

7.12 For the Bethe hole coupler of the type shown in Figure 7.16a, derive a design for s so that port 3 is the isolated port.

7.13 Design a Bethe hole coupler of the type shown in Figure 7.16a for a Ku-band waveguide operating at 13 GHz. The required coupling is 25 dB.

7.14 Design a Bethe hole coupler of the type shown in Figure 7.16b for a Ku-band waveguide operating at 15 GHz. The required coupling is 28 dB.

7.15 Design a five-hole directional coupler in a Ku-band waveguide with a binomial directivity response. The center frequency is 15 GHz, and the required coupling is 22 dB. Use round apertures centered across the broad wall of the waveguides.

7.16 Develop the necessary equations required to design a two-hole directional coupler using two waveguides with apertures in a common sidewall, as shown below.

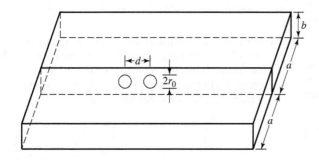

7.17 Consider the general branch-line coupler shown below, with shunt arm characteristic impedances Z_a and series arm characteristic impedances Z_b. Using an even-odd mode analysis, derive design equations for a quadrature hybrid coupler with an arbitrary power division ratio of $\alpha = P_2/P_3$, and with the input port (port 1) matched. Assume all arms are $\lambda/4$ long. Is port 4 isolated, in general?

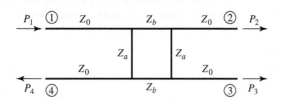

7.18 An edge-coupled stripline with a ground plane spacing of 2.5 mm and a dielectric constant of 4.4 has strip widths of 0.6 mm and a separation of 0.3 mm between the edges of the strips. Use the graph of Figure 7.29 to find the resulting even- and odd-mode characteristic impedances. If possible, compare your results to those obtained from a microwave CAD tool.

7.19 A coupled microstrip line is to be designed for a substrate having a thickness of 2.0 mm and dielectric constant of 10.0. The required even- and odd-mode characteristic impedances are 135 Ω and 75 Ω, respectively. Use the graph of Figure 7.30 to find the required line widths and separation. If possible, compare your results to those obtained from a microwave CAD tool.

7.20 Repeat the derivation in Section 7.6 for the design equations of a single-section coupled line coupler using reflection and transmission coefficients instead of voltages and currents.

7.21 Design a single-section coupled line coupler with a coupling of 18.5 dB, a system impedance of 50 Ω, and a center frequency of 10 GHz. If the coupler is to be made in stripline

(edge-coupled), with $\epsilon_r = 2.2$ and $b = 0.345$ cm, find the necessary strip widths and separation.

7.22 A 19-dB three-section coupled line coupler is required to have a maximally flat coupling response with a center frequency of 5 GHz and $Z_0 = 50$ Ω. (a) Design the coupler and find Z_{0e} and Z_{0o} for each section. Use CAD to plot the resulting coupling (in dB) from 1 to 9 GHz. (b) Lay out the microstrip implementation of the coupler on an FR4 substrate having $\epsilon_r = 4.4$, $d = 1.6$ mm, and $\tan \delta = 0.02$, with copper conductors 0.35 mil thick. Use CAD to plot the insertion loss versus frequency.

7.23 For the Lange coupler, derive the design equations (7.100) for Z_{0e} and Z_{0o} from (7.98) and (7.99).

7.24 Design a 3 dB Lange coupler for operation at 5 GHz. If the coupler is to be fabricated in microstrip on an alumina substrate with $\epsilon_r = 10$ and $d = 1.6$ mm, compute Z_{0e} and Z_{0o} for the two adjacent lines, and find the necessary spacing and widths of the lines.

7.25 An input signal V_1 is applied to the sum port of a 180° hybrid, and another signal V_4 is applied to the difference port. What are the output signals?

7.26 Calculate the even- and odd-mode characteristic impedances for a tapered coupled line 180° hybrid coupler with a 3 dB coupling ratio and a 50 Ω characteristic impedance.

7.27 Find the scattering parameters for the four-port Bagley polygon power divider shown below.

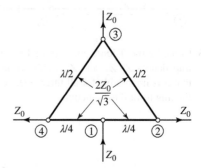

7.28 For the symmetric hybrid shown below, calculate the output voltages if port 1 is fed with an incident wave of $1\angle 0$ V. Assume that the outputs are matched.

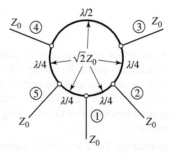

7.29 The Bailey unequal-split power divider uses a 90° hybrid coupler and a T-junction, as shown below. The power division ratio is controlled by adjusting the feed position, a, along the transmission line of length b that connects ports 1 and 4 of the hybrid. A quarter-wave transformer of impedance $Z_0/\sqrt{2}$ is used to match the input of the divider. (a) For $b = \lambda/4$, show that the output power division ratio is given by $P_3/P_2 = \tan^2(\pi a/2b)$. (b) Using a branch-line hybrid with $Z_0 = 50\ \Omega$, design a power divider with a division ratio of $P_3/P_2 = 0.5$, and plot the resulting input return loss and transmission coefficients versus frequency.

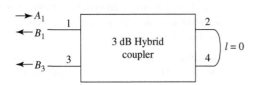

7.30 Derive S_{11}, S_{21}, and S_{31} for the circuit shown below, given $A_1 = 1$ and $A_3 = 0$.

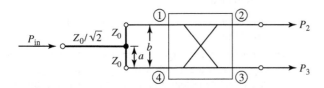

7.31 Derive S_{11} for the circuit shown below, given $A_1 = 1$.

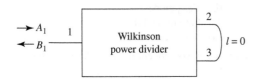

7.32 Consider a 3 dB hybrid coupler and a Wilkinson divider connected with each other as shown below. Calculate the following values

(a) B_3 for $A_1 = A_2 = 1$

(b) B_3 for $A_1 = 1$ and $A_2 = 0$

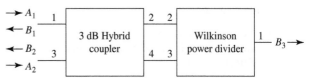

7.33 Consider the 1:4 Wilkinson power divider as shown below. Calculate B_5 and also determine how much power is dissipated in the Wilkinson power divider network. Given $A_1 = 1$, $A_2 = 0.8$, $A_3 = 0.8$, and $A_4 = 0.6$.

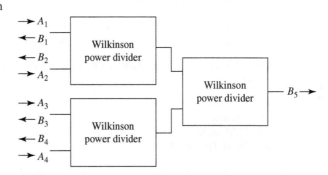

MULTIPLE CHOICE QUESTIONS

7.1 Microwave components such as power dividers and couplers used for power division or power combining are _____ in nature.

(a) linear

(b) nonlinear

(c) active

(d) passive

7.2 If the input power is divided in the ratio of 1:2 in a T-junction coupler and the characteristic impedance of the two output lines is 60 Ω and 120 Ω, then the impedance of the input line is

(a) 50 Ω

(b) 45 Ω

(c) 40 Ω

(d) none of the above

7.3 For an S matrix of a resistive T-junction, the diagonal elements are

(a) 0

(b) 0.5

(c) 1.0

(d) none of the above

7.4 Which of the following is a major disadvantage of the lossless T-junction power dividers?

(a) all the ports are not matched

(b) low-power output

(c) construction is complex

(d) none of the above

7.5 If 12 W is applied to the input port of a standard Wilkinson divider, then the sum of the power measured at both the output ports is

(a) 12 W

(b) 10 W

(c) 7.07 W

(d) 8.49 W

7.6 The analysis of Wilkinson power dividers is done using

(a) symmetry

(b) even-odd mode analysis

(c) S matrix approach

(d) none of the above

7.7 For an equal-split Wilkinson power divider of 75 Ω system impedance, the characteristic impedance of quarter-wave transmission line used is

(a) 50 Ω

(b) 70.7 Ω

(c) 106 Ω

(d) none of the above

7.8 The branch-line impedance for a 60 Ω branch-line quadrature hybrid that has to be designed to operate over a range of frequencies is

(a) 35.35 Ω   (c) 42.43 Ω

(b) 70.70 Ω   (d) 84.85 Ω

7.9 If the branch-line impedance of a coupler designed to operate at 2 GHz is 42.43 Ω, then the characteristic impedance of the material of the arms of the branch-line coupler is

(a) 50 Ω   (c) 100 Ω

(b) 60 Ω   (d) none of the above

7.10 The mode of propagation supported by coupled line couplers is

(a) TE mode

(b) TM mode

(c) TEM mode

(d) Quasi-TEM mode

7.11 For a coupled line coupler, if the characteristic impedance of the microstrip line is 50 Ω and voltage coupling factor is 0.2, then the even-mode characteristic impedance is

(a) 40.82 Ω   (c) 61.24 Ω

(b) 55.24 Ω   (d) 50.82 Ω

7.12 For a coupled line coupler, if the characteristic impedance of the microstrip line is 50 Ω and voltage coupling factor is 0.2, then the odd-mode characteristic impedance is

(a) 40.82 Ω   (c) 61.24 Ω

(b) 55.24 Ω   (d) 50.82 Ω

7.13 For a broadside parallel line with width W and separation d, the capacitance per unit length is

(a) $\epsilon d/W$ (c) dW/ϵ

(b) $\epsilon W/d$ (d) none of the above

7.14 All the four lines of a Lange coupler for even-mode excitation are at

(a) negative potential

(b) zero potential

(c) unequal potential

(d) equal potential

7.15 For a four-wire Lange coupler, if the odd- and even-mode capacitances are 15 pF and 10 pF, respectively, then the equivalent capacitance in terms of two-wired coupled line in even mode is

(a) 18 pF   (c) 10 pF

(b) 15 pF   (d) 33 pF

7.16 For a four-wire Lange coupler, if the odd- and even-mode capacitances are 15 pF and 10 pF, respectively, then the equivalent capacitance in terms of two-wired coupled line in odd mode is

(a) 18 pF   (c) 10 pF

(b) 15 pF   (d) 33 pF

7.17 A 180° ring hybrid with system impedance of 60 Ω has to be designed. The characteristic impedance of the arms of the 180° hybrid is

(a) 50 Ω   (c) 84.85 Ω

(b) 70.70 Ω   (d) 60 Ω

7.18 Which mode of propagation is supported in Riblet short-slot coupler?

(a) TE_{10} mode

(b) TE_{20} mode

(c) both (a) and (b)

(d) none of the above

7.19 A key component in the scalar or vector network analyzer is

(a) frequency meter

(b) reflectometer

(c) radiometer

(d) none of the above

7.20 Which of the following holds correct for a basic reflectometer circuit used to measure the magnitude of the unknown load?

(a) reflection coefficient

(b) transmission coefficient

(c) standing wave ratio

(d) none of the above

ANSWER KEY				
7.1 (d)	**7.5** (d)	**7.9** (b)	**7.13** (b)	**7.17** (c)
7.2 (c)	**7.6** (b)	**7.10** (c)	**7.14** (d)	**7.18** (c)
7.3 (a)	**7.7** (c)	**7.11** (c)	**7.15** (a)	**7.19** (b)
7.4 (a)	**7.8** (c)	**7.12** (a)	**7.16** (d)	**7.20** (a)

Microwave Filters

Learning Objectives

After completing this chapter, you will be able to

- Understand the concept of periodic structures
- Design filters using the image parameter method and the insertion loss method
- Learn about filter transformations and filter implementation
- Understand the concept of stepped-impedance low-pass filters and coupled line filters

$\mathbf{A}$ filter is a two-port network used to control the frequency response at a certain point in an RF or microwave system by providing transmission at frequencies within the passband of the filter and attenuation in the stopband of the filter. Typical frequency responses include low-pass, high-pass, bandpass, and band-reject characteristics. Applications can be found in virtually any type of RF or microwave communication, radar, or test and measurement system.

The development of filter theory and practice began in the years preceding World War II by pioneers such as Mason, Sykes, Darlington, Fano, Lawson, and Richards. The image parameter method of filter design was developed in the late 1930s and was useful for low-frequency filters in radio and telephony. In the early 1950s a group at Stanford Research Institute, consisting of G. Matthaei, L. Young, E. Jones, S. Cohn, and others, became very active in microwave filter and coupler development. A voluminous handbook on filters and couplers resulted from this work and remains a valuable reference [1]. Today, most microwave filter design is done with sophisticated computer-aided design (CAD) packages based on the insertion loss method. Because of continuing advances in network synthesis with distributed elements, the use of low-temperature superconductors and other new materials, and the incorporation of active devices in filter circuits, microwave filter design remains an active research area.

We begin our discussion of filter theory and design with the frequency characteristics of periodic structures, which consist of a transmission line or waveguide periodically loaded with reactive elements. These structures are of interest in themselves because of their application to slow-wave components and traveling-wave amplifier design, and also because they exhibit basic passband-stopband responses that lead to the image parameter method of filter design.

Filters designed using the *image parameter method* consist of a cascade of simpler two-port filter sections to provide the desired cutoff frequencies and attenuation characteristics but do not allow the specification of a particular frequency response over the complete operating range. Thus, although the procedure is relatively simple, the design of filters by the image parameter method often must be iterated many times to achieve the desired results.

A more modern procedure, called the *insertion loss method*, uses network synthesis techniques to design filters with a completely specified frequency response. The design is simplified by beginning with low-pass filter prototypes that are normalized in terms of impedance and frequency. Transformations are then applied to convert the prototype designs to the desired frequency range and impedance level.

Both the image parameter and insertion loss methods of filter design lead to circuits using lumped elements (capacitors and inductors). For microwave applications such designs usually must be modified to employ distributed elements consisting of transmission line sections. The *Richards transformation* and the *Kuroda identities* provide this step. We will also discuss transmission line filters using stepped impedances and coupled lines; filters using coupled resonators will also be briefly described.

The subject of microwave filters is quite extensive due to the importance of these components in practical systems and the wide variety of possible implementations. Here we can treat only the basic principles and some of the more common filter designs, and we refer the reader to references such as [1–4] for further discussion.

8.1   PERIODIC STRUCTURES

An infinite transmission line or waveguide periodically loaded with reactive elements is an example of a *periodic structure*. As shown in Figure 8.1, periodic structures can take various forms, depending on the transmission line media being used. Often the loading elements are formed as discontinuities in the line itself, but in any case they can be modeled as lumped reactances in shunt (or series) on a transmission line, as shown in Figure 8.2. Periodic structures support slow-wave propagation (slower than the phase velocity of the unloaded line), and have passband and stopband characteristics similar to those of filters; they find application in traveling-wave tubes, masers, phase shifters, and antennas.

Analysis of Infinite Periodic Structures

We first consider the propagation characteristics of the infinite loaded line shown in Figure 8.2. Each unit cell of this line consists of a length, d, of transmission line with a shunt susceptance across the midpoint of the line; the susceptance, b, is normalized to the characteristic impedance, Z_0. If we consider the infinite line as being composed of a cascade of identical two-port networks, we can relate the voltages and currents on either side of the nth unit cell using the *ABCD* matrix:

$$\begin{bmatrix} V_n \\ I_n \end{bmatrix} = \begin{bmatrix} A & B \\ C & D \end{bmatrix} \begin{bmatrix} V_{n+1} \\ I_{n+1} \end{bmatrix},$$

(8.1)

where $A, B, C,$ and D are the matrix parameters for a cascade of a transmission line section of length $d/2$, a shunt susceptance b, and another transmission line section of length $d/2$. From Table 4.1 we then have, in

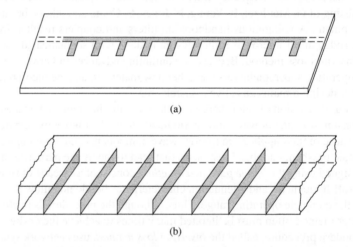

(a)

(b)

FIGURE 8.1 | Examples of periodic structures. (a) Periodic stubs on a microstrip line. (b) Periodic diaphragms in a waveguide.

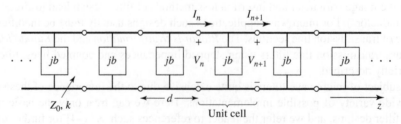

FIGURE 8.2 | Equivalent circuit of a periodically loaded transmission line. The unloaded line has characteristic impedance Z_0 and propagation constant k.

normalized form,

$$
\begin{bmatrix} A & B \\ C & D \end{bmatrix} = \begin{bmatrix} \cos\dfrac{\theta}{2} & j\sin\dfrac{\theta}{2} \\ j\sin\dfrac{\theta}{2} & \cos\dfrac{\theta}{2} \end{bmatrix} \begin{bmatrix} 1 & 0 \\ jb & 1 \end{bmatrix} \begin{bmatrix} \cos\dfrac{\theta}{2} & j\sin\dfrac{\theta}{2} \\ j\sin\dfrac{\theta}{2} & \cos\dfrac{\theta}{2} \end{bmatrix}
$$

$$
= \begin{bmatrix} \left(\cos\theta - \dfrac{b}{2}\sin\theta\right) & j\left(\sin\theta + \dfrac{b}{2}\cos\theta - \dfrac{b}{2}\right) \\ j\left(\sin\theta + \dfrac{b}{2}\cos\theta + \dfrac{b}{2}\right) & \left(\cos\theta - \dfrac{b}{2}\sin\theta\right) \end{bmatrix}, \tag{8.2}
$$

where $\theta = kd$, and k is the propagation constant of the unloaded line. The reader can verify that $AD - BC = 1$, as required for reciprocal networks.

For a wave propagating in the $+z$ direction, we must have

$$
V(z) = V(0)e^{-\gamma z}, \tag{8.3a}
$$

$$
I(z) = I(0)e^{-\gamma z}, \tag{8.3b}
$$

for a phase reference at $z = 0$. Since the structure is infinitely long, the voltage and current at the nth terminals can differ from the voltage and current at the $n + 1$ terminals only by the propagation factor, $e^{-\gamma d}$. Thus,

$$
V_{n+1} = V_n e^{-\gamma d}, \tag{8.4a}
$$

$$
I_{n+1} = I_n e^{-\gamma d}. \tag{8.4b}
$$

Using this result in (8.1) gives the following:

$$
\begin{bmatrix} V_n \\ I_n \end{bmatrix} = \begin{bmatrix} A & B \\ C & D \end{bmatrix} \begin{bmatrix} V_{n+1} \\ I_{n+1} \end{bmatrix} = \begin{bmatrix} V_{n+1}e^{\gamma d} \\ I_{n+1}e^{\gamma d} \end{bmatrix},
$$

or

$$
\begin{bmatrix} A - e^{\gamma d} & B \\ C & D - e^{\gamma d} \end{bmatrix} \begin{bmatrix} V_{n+1} \\ I_{n+1} \end{bmatrix} = 0. \tag{8.5}
$$

For a nontrivial solution, the determinant of the above matrix must vanish:

$$
AD + e^{2\gamma d} - (A + D)e^{\gamma d} - BC = 0, \tag{8.6}
$$

or, since $AD - BC = 1$,

$$
1 + e^{2\gamma d} - (A + D)e^{\gamma d} = 0,
$$

$$
e^{-\gamma d} + e^{\gamma d} = A + D,
$$

$$
\cosh \gamma d = \frac{A + D}{2} = \cos\theta - \frac{b}{2}\sin\theta, \tag{8.7}
$$

where (8.2) was used for the values of A and D. Now, if $\gamma = \alpha + j\beta$, we have that

$$
\cosh \gamma d = \cosh \alpha d \cos \beta d + j\sinh \alpha d \sin \beta d = \cos\theta - \frac{b}{2}\sin\theta. \tag{8.8}
$$

Since the right-hand side of (8.8) is purely real, we must have either $\alpha = 0$ or $\beta = 0$.

Case 1: $\alpha = 0, \beta \neq 0$. This case corresponds to a nonattenuated propagating wave on the periodic structure, and defines the passband of the structure. Equation (8.8) reduces to

$$
\cos \beta d = \cos\theta - \frac{b}{2}\sin\theta, \tag{8.9a}
$$

which can be solved for β if the magnitude of the right-hand side is less than or equal to unity. Note that there are an infinite number of values of β that can satisfy (8.9a).

Case 2: $\alpha \neq 0, \beta = 0, \pi$. In this case the wave does not propagate, but is attenuated along the line; this defines the stopband of the structure. Because the line is lossless, power is not dissipated, but is reflected back to the input of the line. The magnitude of (8.8) reduces to

$$\cosh \alpha d = \left| \cos \theta - \frac{b}{2} \sin \theta \right| \geq 1,$$
(8.9b)

which has only one solution ($\alpha > 0$) for positively traveling waves; $\alpha < 0$ applies for negatively traveling waves. If $\cos \theta - (b/2) \sin \theta \leq -1$, (8.9b) is obtained from (8.8) by letting $\beta = \pi$; then all the lumped loads on the line are $\lambda/2$ apart, yielding an input impedance the same as if $\beta = 0$.

Thus, depending on the frequency and normalized susceptance values, the periodically loaded line will exhibit either passbands or stopbands, and so can be considered as a type of filter. It is important to note that the voltage and current waves defined in (8.3) and (8.4) are meaningful only when measured at the terminals of the unit cells, and do not apply to voltages and currents that may exist at points within a unit cell. These waves are similar to the elastic waves (*Bloch waves*) that propagate through periodic crystal lattices.

Besides the propagation constant of the waves on the periodically loaded line, we will also be interested in the characteristic impedance for these waves. We can define a characteristic impedance at the unit cell terminals as

$$Z_B = Z_0 \frac{V_{n+1}}{I_{n+1}},$$
(8.10)

since V_{n+1} and I_{n+1} in the above derivation were normalized quantities. This impedance is also referred to as the *Bloch impedance*. From (8.5) we have that

$$(A - e^{\gamma d})V_{n+1} + BI_{n+1} = 0,$$

so (8.10) yields

$$Z_B = \frac{-BZ_0}{A - e^{\gamma d}}.$$

From (8.6) we can solve for $e^{\gamma d}$ in terms of A and D as follows:

$$e^{\gamma d} = \frac{(A + D) \pm \sqrt{(A + D)^2 - 4}}{2}.$$

Then the Bloch impedance has two solutions given by

$$Z_B^{\pm} = \frac{-2BZ_0}{A - D \mp \sqrt{(A + D)^2 - 4}}.$$
(8.11)

For symmetric unit cells (as assumed in Figure 8.2) we will always have $A = D$. In this case (8.11) reduces to

$$Z_B^{\pm} = \frac{\pm BZ_0}{\sqrt{A^2 - 1}}.$$
(8.12)

The $\pm$ solutions correspond to the characteristic impedance for positively and negatively traveling waves, respectively. For symmetric networks these impedances are the same except for the sign; the characteristic impedance for a negatively traveling wave is negative because we have defined I_n in Figure 8.2 as always being in the positive direction.

From (8.2) we see that B is always purely imaginary. If $\alpha = 0, \beta \neq 0$ (passband), then (8.7) shows that $\cosh \gamma d = A \leq 1$ (for symmetric networks), and (8.12) shows that Z_B will be real. If $\alpha \neq 0, \beta = 0$ (stopband), then (8.7) shows that $\cosh \gamma d = A \geq 1$, and (8.12) shows that Z_B is imaginary. This situation is similar to that for the wave impedance of a waveguide, which is real for propagating modes and imaginary for cutoff, or evanescent, modes.

Terminated Periodic Structures

Next consider a truncated periodic structure terminated in a load impedance Z_L, as shown in Figure 8.3. At the terminals of an arbitrary unit cell, the incident and reflected voltages and currents can be written as

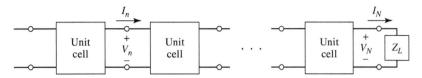

FIGURE 8.3 | A periodic structure terminated in a normalized load impedance Z_L.

(assuming operation in the passband)

$$V_n = V_0^+ e^{-j\beta nd} + V_0^- e^{j\beta nd} \tag{8.13a}$$

$$I_n = I_0^+ e^{-j\beta nd} + I_0^- e^{j\beta nd} = \frac{V_0^+}{Z_B^+} e^{-j\beta nd} + \frac{V_0^-}{Z_B^-} e^{j\beta nd} \tag{8.13b}$$

where we have replaced γz in (8.3) with $j\beta nd$ since we are interested only in terminal quantities.

Now define the following incident and reflected voltages at the nth unit cell:

$$V_n^+ = V_0^+ e^{-j\beta nd} \tag{8.14a}$$

$$V_n^- = V_0^- e^{j\beta nd} \tag{8.14b}$$

Then (8.13) can be written as

$$V_n = V_n^+ + V_n^-, \tag{8.15a}$$

$$I_n = \frac{V_n^+}{Z_B^+} + \frac{V_n^-}{Z_B^-}. \tag{8.15b}$$

At the load, where $n = N$, we have

$$V_N = V_N^+ + V_N^- = Z_L I_N = Z_L \left(\frac{V_N^+}{Z_B^+} + \frac{V_N^-}{Z_B^-} \right), \tag{8.16}$$

and the reflection coefficient at the load can be found as

$$\Gamma = \frac{V_N^-}{V_N^+} = -\frac{Z_L/Z_B^+ - 1}{Z_L/Z_B^- - 1}. \tag{8.17}$$

If the unit cell network is symmetric ($A = D$), then $Z_B^+ = -Z_B^- = Z_B$, which reduces (8.17) to the familiar result

$$\Gamma = \frac{Z_L - Z_B}{Z_L + Z_B}. \tag{8.18}$$

In order to avoid reflections on the terminated periodic structure we must have $Z_L = Z_B$, which is real for a lossless structure operating in a passband. If necessary, a quarter-wave transformer can be used between the periodically loaded line and the load.

k-β Diagrams and Wave Velocities

When studying the passband and stopband characteristics of a periodic structure, it is useful to plot the propagation constant, β, versus the propagation constant of the unloaded line, k (or ω). Such a graph is called a k-β *diagram*, or *Brillouin* diagram, after L. Brillouin, a physicist who studied wave propagation in periodic crystal structures.

The k-β diagram can be plotted from (8.9a), which is the dispersion relation for a general periodic structure. In fact, a k-β diagram can be used to study the dispersion characteristics of many types of microwave components and transmission lines. For instance, consider the dispersion relation for a waveguide mode:

$$\beta = \sqrt{k^2 - k_c^2}, \quad \text{or} \quad k = \sqrt{\beta^2 + k_c^2}, \tag{8.19}$$

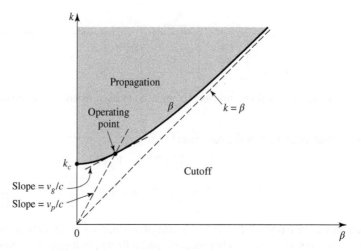

FIGURE 8.4 | k-β diagram for a waveguide mode.

where k_c is the cutoff wave number of the mode, k is the free-space wave number, and β is the propagation constant of the mode. Relation (8.19) is plotted in the k-β diagram of Figure 8.4. For values of $k < k_c$ there is no real solution for β, so the mode is nonpropagating. For $k > k_c$ the mode propagates, and k approaches β for large values of β (TEM propagation).

The k-β diagram is also useful for interpreting the various wave velocities associated with a dispersive structure. The phase velocity is

$$v_p = \frac{\omega}{\beta} = c\frac{k}{\beta}, \tag{8.20}$$

which is seen to be equal to c (speed of light) times the slope of the line from the origin to the operating point on the k-β diagram. The group velocity is

$$v_g = \frac{d\omega}{d\beta} = c\frac{dk}{d\beta}, \tag{8.21}$$

which is the slope of the k-β curve at the operating point. Thus, referring to Figure 8.4, we see that the phase velocity for a propagating waveguide mode is infinite at cutoff and approaches c (from above) as k increases. The group velocity, however, is zero at cutoff and approaches c (from below) as k increases. We finish our discussion of periodic structures with a practical example of a capacitively loaded line.

EXAMPLE 8.1   ANALYSIS OF A PERIODIC STRUCTURE

Consider the periodic capacitively loaded line shown in Figure 8.5 (such a line may be implemented as in Figure 8.1, with short capacitive stubs). If $Z_0 = 50$ Ω, $d = 1.0$ cm, and $C_0 = 2.666$ pF, sketch the k-β diagram and compute the propagation constant, phase velocity, and Bloch impedance at $f = 3.0$ GHz. Assume $k = k_0$.

Solution
We can rewrite the dispersion relation of (8.9a) as

$$\cos \beta d = \cos k_0 d - \left(\frac{C_0 Z_0 c}{2d}\right) k_0 d \sin k_0 d.$$

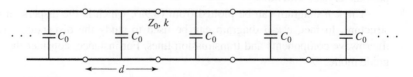

FIGURE 8.5 | A capacitively loaded line.

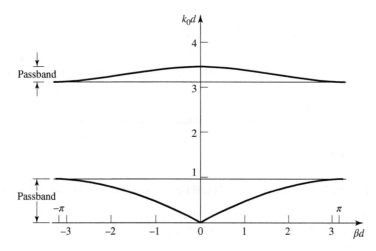

FIGURE 8.6 | k-β diagram for Example 8.1.

Then

$$\frac{C_0 Z_0 c}{2d} = \frac{(2.666 \times 10^{-12})(50)(3 \times 10^8)}{2(0.01)} = 2.0,$$

so we have

$$\cos \beta d = \cos k_0 d - 2k_0 d \sin k_0 d.$$

The most straightforward way to proceed at this point is to numerically evaluate the right-hand side of the above equation for a set of values of $k_0 d$ starting at zero. When the magnitude of the right-hand side is unity or less, we have a passband and can solve for βd. Otherwise we have a stopband. Calculation shows that the first passband exists for $0 \leq k_0 d \leq 0.96$. The second passband does not begin until the $\sin k_0 d$ term changes sign at $k_0 d = \pi$. As $k_0 d$ increases, an infinite number of passbands are possible, but they become narrower. Figure 8.6 shows the k-β diagram for the first two passbands.

At 3.0 GHz, we have

$$k_0 d = \frac{2\pi(3 \times 10^9)}{3 \times 10^8}(0.01) = 0.6283 = 36°,$$

so $\beta d = 1.5$ and the propagation constant is $\beta = 150$ rad/m. The phase velocity is

$$v_p = \frac{k_0 c}{\beta} = \frac{0.6283}{1.5}c = 0.42c,$$

which is much less than the speed of light, indicating that this is a slow-wave structure. To evaluate the Bloch impedance, we use (8.2) and (8.12):

$$\frac{b}{2} = \frac{\omega C_0 Z_0}{2} = 1.256,$$

$$\theta = k_0 d = 36°,$$

$$A = \cos \theta - \frac{b}{2} \sin \theta = 0.0707,$$

$$B = j\left(\sin \theta + \frac{b}{2} \cos \theta - \frac{b}{2}\right) = j0.3479.$$

Then,

$$Z_B = \frac{B Z_0}{\sqrt{A^2 - 1}} = \frac{(j0.3479)(50)}{j\sqrt{1 - (0.0707)^2}} = 17.4 \; \Omega.$$

8.2 FILTER DESIGN BY THE IMAGE PARAMETER METHOD

The image parameter method of filter design involves the specification of passband and stopband characteristics for a cascade of simple two-port networks, and so is related in concept to the periodic structures of Section 8.1. The method is relatively simple but has the disadvantage that an arbitrary frequency response cannot be incorporated into the design. This is in contrast to the insertion loss method, which is the subject of the following section. Nevertheless, the image parameter method is useful for simple filters, and it provides a link between infinite periodic structures and practical filter design. The image parameter method also finds application in solid-state traveling-wave amplifier design.

Image Impedances and Transfer Functions for Two-Port Networks

We begin with definitions of the image impedances and voltage transfer function for an arbitrary reciprocal two-port network; these results are required for the analysis and design of filters by the image parameter method.

Consider the arbitrary two-port network shown in Figure 8.7, where the network is specified by its *ABCD* parameters. Note that the reference direction for the current at port 2 has been chosen according to the convention for *ABCD* parameters. The image impedances, Z_{i1} and Z_{i2}, are defined for this network as follows:

$$Z_{i1} = \text{input impedance at port 1 when port 2 is terminated with } Z_{i2}$$
$$Z_{i2} = \text{input impedance at port 2 when port 1 is terminated with } Z_{i1}.$$

Thus both ports are matched when terminated in their image impedances. We can derive expressions for the image impedances in terms of the *ABCD* parameters of the network.

The port voltages and currents are related as

$$V_1 = AV_2 + BI_2, \tag{8.22a}$$

$$I_1 = CV_2 + DI_2. \tag{8.22b}$$

The input impedance at port 1, with port 2 terminated in Z_{i2}, is

$$Z_{\text{in1}} = \frac{V_1}{I_1} = \frac{AV_2 + BI_2}{CV_2 + DI_2} = \frac{AZ_{i2} + B}{CZ_{i2} + D}, \tag{8.23}$$

since $V_2 = Z_{i2}I_2$.

Now solve (8.22) for V_2, I_2 by inverting the *ABCD* matrix. Since $AD - BC = 1$ for a reciprocal network, we obtain

$$V_2 = DV_1 - BI_1, \tag{8.24a}$$

$$I_2 = -CV_1 + AI_1. \tag{8.24b}$$

Then the input impedance at port 2, with port 1 terminated in Z_{i1}, can be found as

$$Z_{\text{in2}} = \frac{-V_2}{I_2} = -\frac{DV_1 - BI_1}{-CV_1 + AI_1} = \frac{DZ_{i1} + B}{CZ_{i1} + A}, \tag{8.25}$$

since $V_1 = -Z_{i1}I_1$ (circuit of Figure 8.7).

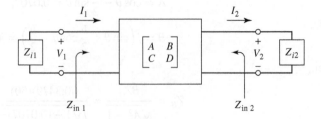

FIGURE 8.7 | A two-port network terminated in its image impedances.

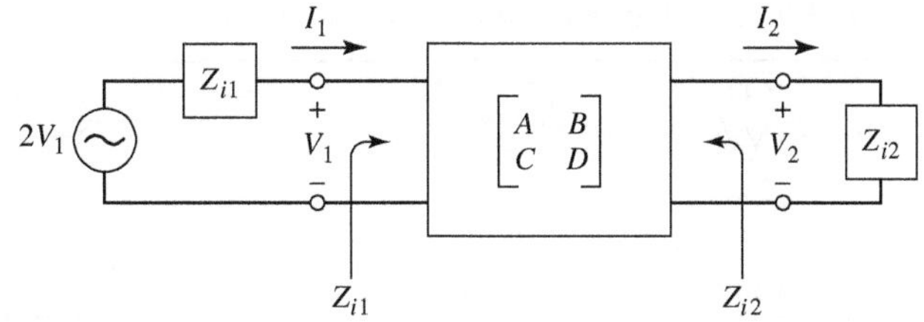

FIGURE 8.8 | A two-port network terminated in its image impedances and driven with a voltage generator.

We desire that $Z_{\text{in}1} = Z_{i1}$ and $Z_{\text{in}2} = Z_{i2}$, so (8.23) and (8.25) give two equations for the image impedances:

$$Z_{i1}(CZ_{i2} + D) = AZ_{i2} + B, \tag{8.26a}$$

$$Z_{i1}D - B = Z_{i2}(A - CZ_{i1}). \tag{8.26b}$$

Solving for Z_{i1} and Z_{i2} gives

$$Z_{i1} = \sqrt{\frac{AB}{CD}}, \tag{8.27a}$$

$$Z_{i2} = \sqrt{\frac{BD}{AC}}, \tag{8.27b}$$

with $Z_{i2} = DZ_{i1}/A$. If the network is symmetric, then $A = D$ and $Z_{i1} = Z_{i2}$ as expected.

Now consider the voltage transfer function for a two-port network terminated in its image impedances. With reference to Figure 8.8 and (8.24a), the output voltage at port 2 can be expressed as

$$V_2 = DV_1 - BI_1 = \left(D - \frac{B}{Z_{i1}}\right)V_1 \tag{8.28}$$

(since we now have $V_1 = I_1 Z_{i1}$), so the voltage ratio is

$$\frac{V_2}{V_1} = D - \frac{B}{Z_{i1}} = D - B\sqrt{\frac{CD}{AB}} = \sqrt{\frac{D}{A}}(\sqrt{AD} - \sqrt{BC}). \tag{8.29a}$$

Similarly, the current ratio is

$$\frac{I_2}{I_1} = -C\frac{V_1}{I_1} + A = -CZ_{i1} + A = \sqrt{\frac{A}{D}}(\sqrt{AD} - \sqrt{BC}). \tag{8.29b}$$

The factor $\sqrt{D/A}$ occurs in reciprocal positions in (8.29a) and (8.29b), and so can be interpreted as a transformer turns ratio. Apart from this factor, we can define a propagation factor for the network as

$$e^{-\gamma} = \sqrt{AD} - \sqrt{BC}, \tag{8.30}$$

with $\gamma = \alpha + j\beta$ as usual. Since

$$e^{\gamma} = 1/(\sqrt{AD} - \sqrt{BC}) = (AD - BC)/(\sqrt{AD} - \sqrt{BC}) = \sqrt{AD} + \sqrt{BC}$$

and $\cosh\gamma = (e^{\gamma} + e^{-\gamma})/2$, we also have that

$$\cosh\gamma = \sqrt{AD}. \tag{8.31}$$

Two important types of two-port networks are the T and π circuits, which can be made in symmetric form. Table 8.1 lists the image impedances and propagation factors, along with other useful parameters, for these two networks.

TABLE 8.1 | Image Parameters for T- and π-Networks

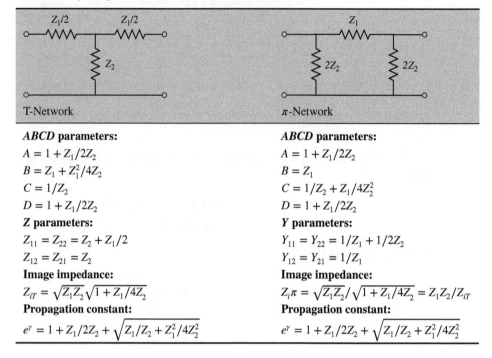

T-Network π-Network

ABCD parameters:

$A = 1 + Z_1/2Z_2$

$B = Z_1 + Z_1^2/4Z_2$

$C = 1/Z_2$

$D = 1 + Z_1/2Z_2$

Z parameters:

$Z_{11} = Z_{22} = Z_2 + Z_1/2$

$Z_{12} = Z_{21} = Z_2$

Image impedance:

$Z_{iT} = \sqrt{Z_1 Z_2}\sqrt{1 + Z_1/4Z_2}$

Propagation constant:

$e^\gamma = 1 + Z_1/2Z_2 + \sqrt{Z_1/Z_2 + Z_1^2/4Z_2^2}$

ABCD parameters:

$A = 1 + Z_1/2Z_2$

$B = Z_1$

$C = 1/Z_2 + Z_1/4Z_2^2$

$D = 1 + Z_1/2Z_2$

Y parameters:

$Y_{11} = Y_{22} = 1/Z_1 + 1/2Z_2$

$Y_{12} = Y_{21} = 1/Z_1$

Image impedance:

$Z_{i\pi} = \sqrt{Z_1 Z_2}/\sqrt{1 + Z_1/4Z_2} = Z_1 Z_2/Z_{iT}$

Propagation constant:

$e^\gamma = 1 + Z_1/2Z_2 + \sqrt{Z_1/Z_2 + Z_1^2/4Z_2^2}$

Constant-k Filter Sections

We can now develop low-pass and high-pass filter sections. First consider the T-network shown in Figure 8.9. Intuitively, we can see that this is a low-pass filter network because the series inductors and shunt capacitor tend to block high-frequency signals while passing low-frequency signals. Comparing with the results given in Table 8.1, we have $Z_1 = j\omega L$ and $Z_2 = 1/j\omega C$, so the image impedance is

$$Z_{iT} = \sqrt{\frac{L}{C}}\sqrt{1 - \frac{\omega^2 LC}{4}}. \tag{8.32}$$

If we define a cutoff frequency, ω_c, as

$$\omega_c = \frac{2}{\sqrt{LC}} \tag{8.33}$$

and a nominal characteristic impedance, R_0, as

$$R_0 = \sqrt{\frac{L}{C}} = k, \tag{8.34}$$

where k is a constant, then we can rewrite (8.32) as

$$Z_{iT} = R_0\sqrt{1 - \frac{\omega^2}{\omega_c^2}}. \tag{8.35}$$

Then $Z_{iT} = R_0$ for $\omega = 0$.

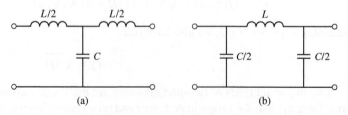

(a)   (b)

FIGURE 8.9 | Low-pass constant-k filter sections in T and π forms. (a) T-section. (b) π-section.

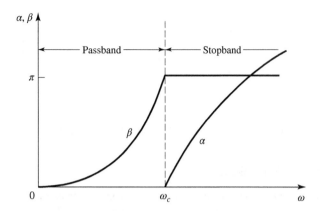

FIGURE 8.10 | Typical passband and stopband characteristics of the low-pass constant-k sections of Figure 8.9.

The propagation factor, also from Table 8.1, is

$$e^{\gamma} = 1 - \frac{2\omega^2}{\omega_c^2} + \frac{2\omega}{\omega_c}\sqrt{\frac{\omega^2}{\omega_c^2} - 1}. \tag{8.36}$$

Now consider two frequency regions:

1. For $\omega < \omega_c$: This is the passband of the filter section. Equation (8.35) shows that Z_{iT} is real, and (8.36) shows that γ is imaginary, since $\omega^2/\omega_c^2 - 1$ is negative and $|e^{\gamma}| = 1$:

$$|e^{\gamma}|^2 = \left(1 - \frac{2\omega^2}{\omega_c^2}\right)^2 + \frac{4\omega^2}{\omega_c^2}\left(1 - \frac{\omega^2}{\omega_c^2}\right) = 1.$$

2. For $\omega > \omega_c$: This is the stopband of the filter section. Equation (8.35) shows that Z_{iT} is imaginary, and (8.36) shows that e^{γ} is real and $-1 < e^{\gamma} < 0$ (as seen from the limits as $\omega \to \omega_c$ and $\omega \to \infty$). The attenuation rate for $\omega \gg \omega_c$ is 40 dB/decade.

Typical phase and attenuation constants are sketched in Figure 8.10. Observe that the attenuation, α, is zero or relatively small near the cutoff frequency, although $\alpha \to \infty$ as $\omega \to \infty$. This type of filter is known as a *constant-k* low-pass prototype. There are only two parameters to choose (L and C), which are determined by ω_c, the cutoff frequency, and R_0, the image impedance at zero frequency.

The above results are valid only when the filter section is terminated in its image impedance at both ports. This is a major weakness of the design because the image impedance is a function of frequency, and is not likely to match a given source or load impedance. This disadvantage, as well as the fact that the attenuation is rather low near cutoff, can be remedied with the modified *m*-derived sections to be discussed shortly.

For the low-pass π-network of Figure 8.9, we have that $Z_1 = j\omega L$ and $Z_2 = 1/j\omega C$, so the propagation factor is the same as that for the low-pass T-network. The cutoff frequency, ω_c, and nominal characteristic impedance, R_0, are the same as the corresponding quantities for the T-network as given in (8.33) and (8.34). At $\omega = 0$ we have that $Z_{iT} = Z_{i\pi} = R_0$, where $Z_{i\pi}$ is the image impedance of the low-pass π-network, but Z_{iT} and $Z_{i\pi}$ are generally not equal at other frequencies.

High-pass constant-k sections are shown in Figure 8.11; we see that the positions of the inductors and capacitors are reversed from those in the low-pass prototype. The design equations are easily shown to be

$$R_0 = \sqrt{\frac{L}{C}}. \tag{8.37}$$

$$\omega_c = \frac{1}{2\sqrt{LC}}. \tag{8.38}$$

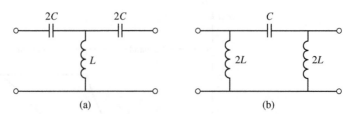

FIGURE 8.11 | High-pass constant-k filter sections in T and π forms. (a) T-section. (b) π-section.

m-Derived Filter Sections

We have seen that the constant-k filter section suffers from the disadvantages of a relatively slow attenuation rate past cutoff, and a nonconstant image impedance. The *m-derived* filter section is a modification of the constant-k section designed to overcome these problems. As shown in Figure 8.12a, b the impedances Z_1 and Z_2 in a constant-k T-section are replaced with Z_1' and Z_2', and we let

$$Z_1' = mZ_1. \tag{8.39}$$

Then we choose Z_2' to obtain the same value of Z_{iT} as for the constant-k section. Thus, from Table 8.1,

$$Z_{iT} = \sqrt{Z_1 Z_2 + \frac{Z_1^2}{4}} = \sqrt{Z_1' Z_2' + \frac{Z_1'^2}{4}} = \sqrt{mZ_1 Z_2' + \frac{m^2 Z_1^2}{4}}. \tag{8.40}$$

Solving for Z_2' gives

$$Z_2' = \frac{Z_2}{m} + \frac{Z_1}{4m} - \frac{mZ_1}{4} = \frac{Z_2}{m} + \frac{(1-m^2)}{4m}Z_1. \tag{8.41}$$

Because the impedances Z_1 and Z_2 represent reactive elements, Z_2' represents two elements in series, as indicated in Figure 8.12c. Note that $m = 1$ reduces to the original constant-k section.

For a low-pass filter, we have $Z_1 = j\omega L$ and $Z_2 = 1/j\omega C$. Then (8.39) and (8.41) give the m-derived components as

$$Z_1' = j\omega L m, \tag{8.42a}$$

$$Z_2' = \frac{1}{j\omega C m} + \frac{(1-m^2)}{4m}j\omega L, \tag{8.42b}$$

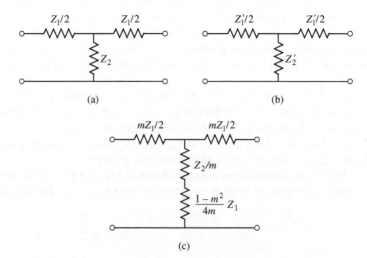

FIGURE 8.12 | Development of an m-derived filter section from a constant-k section. (a) Constant-k section. (b) General m-derived section. (c) Final m-derived section.

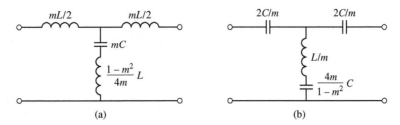

FIGURE 8.13 | *m*-Derived filter sections. (a) Low-pass T-section. (b) High-pass T-section.

which results in the circuit of Figure 8.13. Now consider the propagation factor for the *m*-derived section. From Table 8.1,

$$e^\gamma = 1 + \frac{Z_1'}{2Z_2'} + \sqrt{\frac{Z_1'}{Z_2'}\left(1 + \frac{Z_1'}{4Z_2'}\right)}. \tag{8.43}$$

For the low-pass *m*-derived filter,

$$\frac{Z_1'}{Z_2'} = \frac{j\omega Lm}{(1/j\omega Cm) + j\omega L(1-m^2)/4m} = \frac{-(2\omega m/\omega_c)^2}{1-(1-m^2)(\omega/\omega_c)^2},$$

where $\omega_c = 2/\sqrt{LC}$ as before. Then,

$$1 + \frac{Z_1'}{4Z_2'} = \frac{1-(\omega/\omega_c)^2}{1-(1-m^2)(\omega/\omega_c)^2}.$$

If we restrict $0 < m < 1$, then these results show that e^γ is real and $|e^\gamma| > 1$ for $\omega > \omega_c$. Thus the stopband begins at $\omega = \omega_c$, as for the constant-*k* section. However, when $\omega = \omega_\infty$, where

$$\omega_\infty = \frac{\omega_c}{\sqrt{1-m^2}}, \tag{8.44}$$

the denominators vanish and e^γ becomes infinite, implying infinite attenuation. Physically, this pole in the attenuation characteristic is caused by the resonance of the series *LC* resonator in the shunt arm of the T; this is easily verified by showing that the resonant frequency of this *LC* resonator is ω_∞. Note that (8.44) indicates that $\omega_\infty > \omega_c$, so infinite attenuation occurs after the cutoff frequency, ω_c, as illustrated in Figure 8.14. The position of the pole at ω_∞ can be controlled with the value of *m*.

We now have a very sharp cutoff response, but one problem with the *m*-derived section is that its attenuation decreases for $\omega > \omega_\infty$. Since it is often desirable to have infinite attenuation as $\omega \to \infty$, the

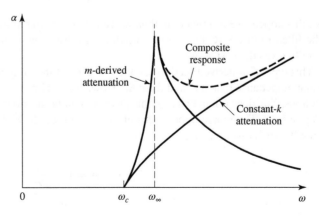

FIGURE 8.14 | Typical attenuation responses for constant-*k*, *m*-derived, and composite filters.

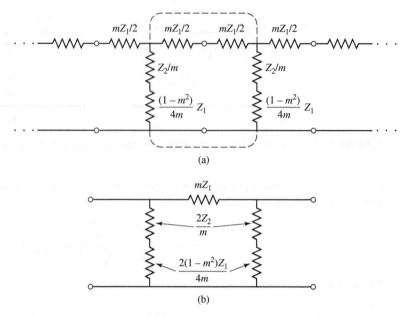

FIGURE 8.15 | Development of an m-derived π-section. (a) Infinite cascade of m-derived T-sections. (b) A de-embedded π-equivalent.

m-derived section can be cascaded with a constant-k section to give the composite attenuation response shown in Figure 8.14.

The m-derived T-section was designed so that its image impedance was identical to that of the constant-k section (independent of m), so we still have the problem of a nonconstant image impedance. However, the image impedance of the π-equivalent will depend on m, and this extra degree of freedom can be used to design an optimum matching section.

The easiest way to obtain the corresponding π-section is to consider it as a piece of an infinite cascade of m-derived T-sections, as shown in Figure 8.15. Then the image impedance of this network is, using the results of Table 8.1 and (8.35),

$$Z_{i\pi} = \frac{Z_1' Z_2'}{Z_{iT}} = \frac{Z_1 Z_2 + Z_1^2(1-m^2)/4}{R_0\sqrt{1-(\omega/\omega_c)^2}}.$$ (8.45)

Now $Z_1 Z_2 = L/C = R_0^2$ and $Z_1^2 = -\omega^2 L^2 = -4R_0^2(\omega/\omega_c)^2$, so (8.45) reduces to

$$Z_{i\pi} = \frac{1-(1-m^2)(\omega/\omega_c)^2}{\sqrt{1-(\omega/\omega_c)^2}}R_0.$$ (8.46)

Since this impedance is a function of m, we can choose m to minimize the variation of $Z_{i\pi}$ over the passband of the filter. Figure 8.16 shows this variation with frequency for several values of m; a value of $m = 0.6$ generally gives the best results.

This type of m-derived section can then be used at the input and output of the filter to provide a nearly constant impedance match to and from R_0. However, the image impedance of the constant-k and m-derived T-sections, Z_{iT}, does not match $Z_{i\pi}$; this problem can be surmounted by bisecting the π-sections, as shown in Figure 8.17. The image impedances of this circuit are $Z_{i1} = Z_{iT}$ and $Z_{i2} = Z_{i\pi}$, which can be shown by finding its *ABCD* parameters:

$$A = 1 + \frac{Z_1'}{4Z_2'},$$ (8.47a)

$$B = \frac{Z_1'}{2},$$ (8.47b)

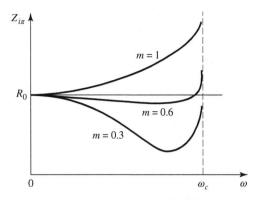

FIGURE 8.16 | Variation of $Z_{i\pi}$ in the passband of a low-pass m-derived section for various values of m.

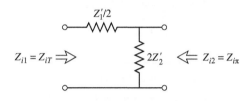

FIGURE 8.17 | A bisected π-section used to match $Z_{i\pi}$ to Z_{iT}.

$$C = \frac{1}{2Z_2'}, \tag{8.47c}$$

$$D = 1, \tag{8.47d}$$

and then using (8.27) for Z_{i1} and Z_{i2}:

$$Z_{i1} = \sqrt{Z_1' Z_2' + \frac{Z_1'^2}{4}} = Z_{iT}, \tag{8.48a}$$

$$Z_{i2} = \sqrt{\frac{Z_1' Z_2'}{1 + Z_1'/4Z_2'}} = \frac{Z_1' Z_2'}{Z_{iT}} = Z_{i\pi}, \tag{8.48b}$$

where (8.40) has been used for Z_{iT}.

Composite Filters

By combining in cascade the constant-k, m-derived sharp cutoff and the m-derived matching sections we can realize a filter with the desired attenuation and matching properties. This type of design is called a *composite filter*, and is shown in Figure 8.18. The sharp-cutoff section, with $m < 0.6$, places an attenuation pole near the cutoff frequency to provide a sharp attenuation response; the constant-k section provides high attenuation further into the stopband. The bisected-π sections at the ends of the filter match the nominal source and load impedance, R_0, to the internal image impedances, Z_{iT}, of the constant-k and m-derived sections.

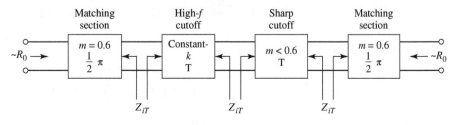

FIGURE 8.18 | The final four-stage composite filter.

TABLE 8.2 | **Summary of Composite Filter Design**

Low-Pass	High-Pass

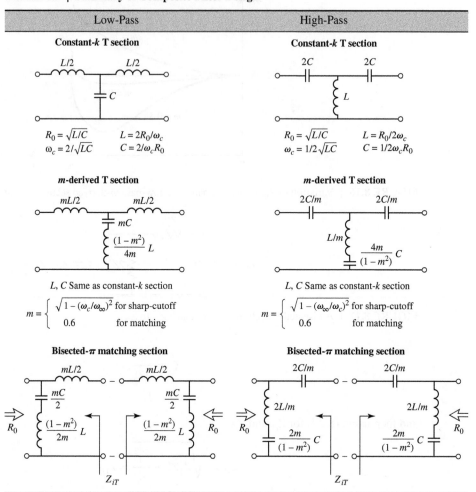

Constant-*k* T section

$$R_0 = \sqrt{L/C} \qquad L = 2R_0/\omega_c$$
$$\omega_c = 2/\sqrt{LC} \qquad C = 2/\omega_c R_0$$

Constant-*k* T section

$$R_0 = \sqrt{L/C} \qquad L = R_0/2\omega_c$$
$$\omega_c = 1/2\sqrt{LC} \qquad C = 1/2\omega_c R_0$$

m-derived T section

L, C Same as constant-*k* section

$$m = \begin{cases} \sqrt{1-(\omega_c/\omega_\infty)^2} & \text{for sharp-cutoff} \\ 0.6 & \text{for matching} \end{cases}$$

m-derived T section

L, C Same as constant-*k* section

$$m = \begin{cases} \sqrt{1-(\omega_\infty/\omega_c)^2} & \text{for sharp-cutoff} \\ 0.6 & \text{for matching} \end{cases}$$

Bisected-π matching section

Bisected-π matching section

Table 8.2 summarizes the design equations for low- and high-pass composite filters; notice that once the cutoff frequency and impedance are specified, there is only one degree of freedom (the value of *m* for the sharp-cutoff section) left to control the filter response. The following example illustrates the design procedure.

EXAMPLE 8.2 LOW-PASS COMPOSITE FILTER DESIGN

Design a low-pass composite filter with a cutoff frequency of 2 MHz and impedance of 50 Ω. Place the infinite attenuation pole at 2.05 MHz, and plot the frequency response from 0 to 4 MHz.

Solution
All of the component values can be found from Table 8.2. For the constant-*k* section

$$L = \frac{2R_0}{\omega_c} = 7.96 \ \mu\text{H}, \quad C = \frac{2}{R_0\omega_c} = 3.18 \ \text{nF}.$$

For the *m*-derived sharp-cutoff section

$$m = \sqrt{1-\left(\frac{f_c}{f_\infty}\right)^2} = 0.2195,$$

$$\frac{mL}{2} = 0.874 \ \mu\text{H},$$

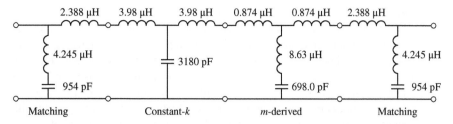

FIGURE 8.19 | Low-pass composite filter for Example 8.2.

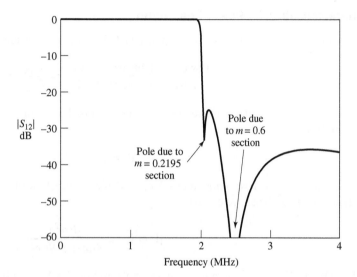

FIGURE 8.20 | Frequency response for the low-pass filter of Example 8.2.

$$mC = 698.0 \text{ pF},$$

$$\frac{1-m^2}{4m}L = 8.63 \text{ μH}.$$

For the $m = 0.6$ matching sections

$$\frac{mL}{2} = 2.388 \text{ μH},$$

$$\frac{mC}{2} = 954 \text{ pF},$$

$$\frac{1-m^2}{2m}L = 4.245 \text{ μH}.$$

The completed filter circuit is shown in Figure 8.19; the series pairs of inductors between the sections have been combined. Figure 8.20 shows the resulting frequency response for $|S_{12}|$. Note the sharp dip at $f = 2.05$ MHz due to the $m = 0.2195$ section, and the pole at 2.50 MHz, which is due to the $m = 0.6$ matching sections.

8.3   FILTER DESIGN BY THE INSERTION LOSS METHOD

A perfect filter would have zero insertion loss in the passband, infinite attenuation in the stopband, and a linear phase response (to avoid signal distortion) in the passband. Of course, such filters do not exist in practice, so compromises must be made; herein lies the art of filter design.

The image parameter method of the previous section may yield a usable filter response for some applications, but there is no methodical way of improving the design. The insertion loss method, however, allows a high degree of control over the passband and stopband amplitude and phase characteristics, with a systematic way to synthesize a desired response. The necessary design trade-offs can be evaluated to best meet the application requirements. If, for example, a minimum insertion loss is most important, a binomial response could be used; a Chebyshev response would satisfy a requirement for the sharpest cutoff. If it is possible to sacrifice the attenuation rate, a better phase response can be obtained by using a linear phase filter design. In addition, in all cases, the insertion loss method allows filter performance to be improved in a straightforward manner, at the expense of a higher order filter. For the filter prototypes to be discussed below, the order of the filter is equal to the number of reactive elements.

Characterization by Power Loss Ratio

In the insertion loss method a filter response is defined by its insertion loss, or *power loss ratio*, P_{LR}:

$$P_{LR} = \frac{\text{Power available from source}}{\text{Power delivered to load}} = \frac{P_{inc}}{P_{load}} = \frac{1}{1 - |\Gamma(\omega)|^2}.$$ (8.49)

Observe that this quantity is the reciprocal of $|S_{12}|^2$ if both load and source are matched. The insertion loss (IL) in dB is

$$IL = 10 \log P_{LR}.$$ (8.50)

From Section 4.1 we know that $|\Gamma(\omega)|^2$ is an even function of ω; therefore it can be expressed as a polynomial in ω^2. Thus we can write

$$|\Gamma(\omega)|^2 = \frac{M(\omega^2)}{M(\omega^2) + N(\omega^2)},$$ (8.51)

where M and N are real polynomials in ω^2. Substituting this form in (8.49) gives the following:

$$P_{LR} = 1 + \frac{M(\omega^2)}{N(\omega^2)}.$$ (8.52)

For a filter to be physically realizable its power loss ratio must be of the form in (8.52). Notice that specifying the power loss ratio simultaneously constrains the magnitude of the reflection coefficient, $|\Gamma(\omega)|$. We now discuss some practical filter responses.

Maximally flat: This characteristic is also called the *binomial* or *Butterworth* response, and is optimum in the sense that it provides the flattest possible passband response for a given filter complexity, or order. For a low-pass filter, it is specified by

$$P_{LR} = 1 + k^2 \left(\frac{\omega}{\omega_c} \right)^{2N},$$ (8.53)

where N is the order of the filter and ω_c is the cutoff frequency. The passband extends from $\omega = 0$ to $\omega = \omega_c$; at the band edge the power loss ratio is $1 + k^2$. If we choose this as the -3 dB point, as is common, we have $k = 1$, which we will assume from now on. For $\omega > \omega_c$, the attenuation increases monotonically with frequency, as shown in Figure 8.21. For $\omega \gg \omega_c, P_{LR} \simeq k^2(\omega/\omega_c)^{2N}$, which shows that the insertion loss increases at the rate of $20N$ dB/decade. Like the binomial response for multisection quarter-wave matching transformers, the first $(2N - 1)$ derivatives of (8.53) are zero at $\omega = 0$.

Equal ripple: If a Chebyshev polynomial is used to specify the insertion loss of an Nth-order low-pass filter as

$$P_{LR} = 1 + k^2 T_N^2 \left(\frac{\omega}{\omega_c} \right),$$ (8.54)

then a sharper cutoff will result, although the passband response will have ripples of amplitude $1 + k^2$, as shown in Figure 8.21, since $T_N(x)$ oscillates between ± 1 for $|x| \, 1$. Thus, k^2 determines the passband ripple

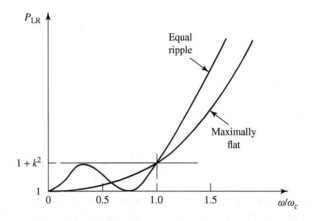

FIGURE 8.21 | Maximally flat and equal-ripple low-pass filter responses ($N = 3$).

level. For large x, $T_N(x) \simeq \frac{1}{2}(2x)^N$, so for $\omega \gg \omega_c$ the insertion loss becomes

$$P_{\text{LR}} \simeq \frac{k^2}{4} \left(\frac{2\omega}{\omega_c} \right)^{2N},$$

which also increases at the rate of $20N$ dB/decade. However, the insertion loss for the Chebyshev case is $(2^{2N})/4$ greater than the binomial response at any given frequency where $\omega \gg \omega_c$.

Elliptical function: The maximally flat and equal-ripple responses both have monotonically increasing attenuation in the stopband. In many applications it is adequate to specify a minimum stopband attenuation, in which case a better cutoff rate can be obtained. Such filters are called *elliptical function filters* [3], and they have equal-ripple responses in the passband as well as in the stopband, as shown in Figure 8.22. The maximum attenuation in the passband, A_{max}, can be specified, as well as the minimum attenuation in the stopband, A_{min}. Elliptical function filters are difficult to synthesize, so we will not consider them further; the interested reader is referred to reference [3].

Linear phase: The above filters specify the amplitude response, but in some applications (such as multi-plexing filters for communication systems) it is important to have a linear phase response in the passband to avoid signal distortion. Since a sharp-cutoff response is generally incompatible with a good phase response, the phase response of a filter must be deliberately synthesized, usually resulting in an inferior attenuation characteristic. A linear phase characteristic can be achieved with the following phase response:

$$\phi(\omega) = A\omega \left[1 + p \left(\frac{\omega}{\omega_c} \right)^{2N} \right], \tag{8.55}$$

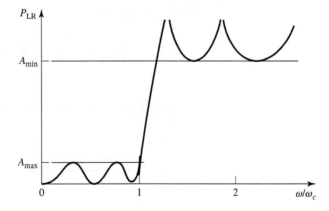

FIGURE 8.22 | Elliptical function low-pass filter response.

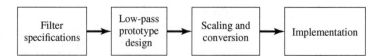

FIGURE 8.23 | The process of filter design by the insertion loss method.

where $\phi(\omega)$ is the phase of the voltage transfer function of the filter, and p is a constant. A related quantity is the *group delay*, defined as

$$\tau_d = \frac{d\phi}{d\omega} = A \left[1 + p(2N + 1) \left(\frac{\omega}{\omega_c} \right)^{2N} \right], \tag{8.56}$$

which shows that the group delay for a linear phase filter is a maximally flat function.

More general filter specifications can be obtained, but the above cases are the most common. We will next discuss the design of *low-pass filter prototypes* that are normalized in terms of impedance and frequency; this normalization simplifies the design of filters for arbitrary frequency, impedance, and type (low-pass, high-pass, bandpass, or bandstop). The low-pass prototypes are then scaled to the desired frequency and impedance, and the lumped-element components replaced with distributed circuit elements for implementation at microwave frequencies. This design process is illustrated in Figure 8.23.

Maximally Flat Low-Pass Filter Prototype

Consider the two-element low-pass filter prototype shown in Figure 8.24; we will derive the normalized element values, L and C, for a maximally flat response. We assume a source impedance of 1 Ω, and a cutoff frequency $\omega_c = 1$ rad/sec. From (8.53), the desired power loss ratio will be, for $N = 2$,

$$P_{\text{LR}} = 1 + \omega^4. \tag{8.57}$$

The input impedance of this filter is

$$Z_{\text{in}} = j\omega L + \frac{R(1 - j\omega RC)}{1 + \omega^2 R^2 C^2}. \tag{8.58}$$

Because

$$\Gamma = \frac{Z_{\text{in}} - 1}{Z_{\text{in}} + 1},$$

the power loss ratio can be written as

$$P_{\text{LR}} = \frac{1}{1 - |\Gamma|^2} = \frac{1}{1 - \left[(Z_{\text{in}} - 1)/(Z_{\text{in}} + 1) \right]\left[(Z_{\text{in}}^* - 1)/(Z_{\text{in}}^* + 1) \right]} = \frac{|Z_{\text{in}} + 1|^2}{2(Z_{\text{in}} + Z_{\text{in}}^*)}. \tag{8.59}$$

Now

$$Z_{\text{in}} + Z_{\text{in}}^* = \frac{2R}{1 + \omega^2 R^2 C^2},$$

$$|Z_{\text{in}} + 1|^2 = \left(\frac{R}{1 + \omega^2 R^2 C^2} + 1 \right)^2 + \left(\omega L - \frac{\omega C R^2}{1 + \omega^2 R^2 C^2} \right)^2,$$

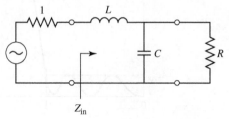

FIGURE 8.24 | Low-pass filter prototype, $N = 2$.

so (8.59) becomes

$$P_{LR} = \frac{1 + \omega^2 R^2 C^2}{4R} \left[\left(\frac{R}{1 + \omega^2 R^2 C^2} + 1 \right)^2 + \left(\omega L - \frac{\omega C R^2}{1 + \omega^2 R^2 C^2} \right)^2 \right]$$

$$= \frac{1}{4R}(R^2 + 2R + 1 + R^2\omega^2 C^2 + \omega^2 L^2 + \omega^4 L^2 C^2 R^2 - 2\omega^2 LCR^2)$$

$$= 1 + \frac{1}{4R}[(1 - R)^2 + (R^2 C^2 + L^2 - 2LCR^2)\omega^2 + L^2 C^2 R^2 \omega^4]. \qquad (8.60)$$

Observe that this expression is a polynomial in ω^2. Comparing to the desired response of (8.57) shows that $R = 1$, since $P_{LR} = 1$ for $\omega = 0$. In addition, the coefficient of ω^2 must vanish, so

$$C^2 + L^2 - 2LC = (C - L)^2 = 0,$$

or $L = C$. Then, for the coefficient of ω^4 to be unity, we must have

$$\frac{1}{4}L^2 C^2 = \frac{1}{4}L^4 = 1,$$

or

$$L = C = \sqrt{2}.$$

In principle, this procedure can be extended to find the element values for filters with an arbitrary number of elements, N, but clearly this is not practical for large N. For a normalized low-pass design, where the source impedance is 1 Ω and the cutoff frequency is $\omega_c = 1$ rad/sec, however, the element values for the ladder-type circuits of Figure 8.25 can be tabulated [1]. Table 8.3 gives such element values for maximally flat low-pass filter prototypes for $N = 1$ to 10. (Notice that the values for $N = 2$ agree with the above analytical solution.) These data can be used with either of the ladder circuits of Figure 8.25 in the following way. The element values are numbered from g_0 at the generator impedance to g_{N+1} at the load impedance for a filter having N reactive elements. The elements alternate between series and shunt connections, and g_k has the following definition:

$$g_0 = \begin{cases} \text{generator resistance (network of Figure 8.25a)} \\ \text{generator conductance (network of Figure 8.25b)} \end{cases}$$

$$\begin{matrix} g_k \\ (k = 1 \text{ to } N) \end{matrix} = \begin{cases} \text{inductance for series inductors} \\ \text{capacitance for shunt capacitors} \end{cases}$$

$$g_{N+1} = \begin{cases} \text{load resistance if } g_N \text{ is a shunt capacitor} \\ \text{load conductance if } g_N \text{ is a series inductor} \end{cases}$$

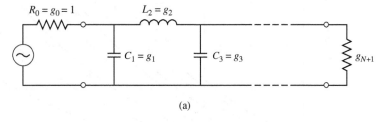

(a)

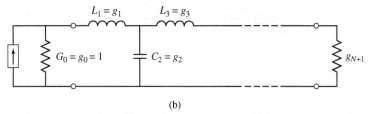

(b)

FIGURE 8.25 | Ladder circuits for low-pass filter prototypes and their element definitions. (a) Prototype beginning with a shunt element. (b) Prototype beginning with a series element.

TABLE 8.3 | Element Values for Maximally Flat Low-Pass Filter Prototypes ($g_0 = 1, \omega_c = 1, N = 1$ to 10)

N	g_1	g_2	g_3	g_4	g_5	g_6	g_7	g_8	g_9	g_{10}	g_{11}
1	2.0000	1.0000									
2	1.4142	1.4142	1.0000								
3	1.0000	2.0000	1.0000	1.0000							
4	0.7654	1.8478	1.8478	0.7654	1.0000						
5	0.6180	1.6180	2.0000	1.6180	0.6180	1.0000					
6	0.5176	1.4142	1.9318	1.9318	1.4142	0.5176	1.0000				
7	0.4450	1.2470	1.8019	2.0000	1.8019	1.2470	0.4450	1.0000			
8	0.3902	1.1111	1.6629	1.9615	1.9615	1.6629	1.1111	0.3902	1.0000		
9	0.3473	1.0000	1.5321	1.8794	2.0000	1.8794	1.5321	1.0000	0.3473	1.0000	
10	0.3129	0.9080	1.4142	1.7820	1.9754	1.9754	1.7820	1.4142	0.9080	0.3129	1.0000

Source: Reprinted from G. L. Matthaei, L. Young, and E. M. T. Jones, *Microwave Filters, Impedance-Matching Networks, and Coupling Structures*, Artech House, Dedham, MA, 1980, with permission.

Then the circuits of Figure 8.25 can be considered as the dual of each other, and both will give the same filter response.

Finally, as a matter of practical design procedure, it will be necessary to determine the size, or order, of the filter. This is usually dictated by a specification on the insertion loss at some frequency in the stopband of the filter. Figure 8.26 shows the attenuation characteristics for various N versus normalized frequency. If a filter with $N > 10$ is required, a good result can usually be obtained by cascading two designs of lower order.

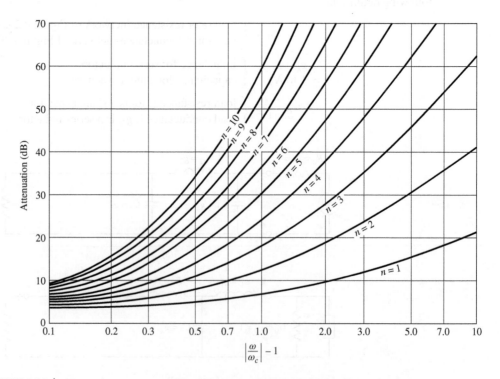

FIGURE 8.26 | Attenuation versus normalized frequency for maximally flat filter prototypes.

Adapted from G. L. Matthaei, L. Young, and E. M. T. Jones, *Microwave Filters, Impedance-Matching Networks, and Coupling Structures*, Artech House, Dedham, MA, 1980, with permission.

Equal-Ripple Low-Pass Filter Prototype

For an equal-ripple low-pass filter with a cutoff frequency $\omega_c = 1$ rad/sec, the power loss ratio from (8.54) is

$$P_{\text{LR}} = 1 + k^2 T_N^2(\omega), \tag{8.61}$$

where $1 + k^2$ is the ripple level in the passband. Since the Chebyshev polynomials have the property that

$$T_N(0) = \begin{cases} 0 & \text{for } N \text{ odd,} \\ 1 & \text{for } N \text{ even,} \end{cases}$$

equation (8.61) shows that the filter will have a unity power loss ratio at $\omega = 0$ for N odd, but a power loss ratio of $1 + k^2$ at $\omega = 0$ for N even. Thus, there are two cases to consider, depending on N.

For the two-element filter of Figure 8.24, the power loss ratio is given in terms of the component values in (8.60). From (5.62b), we see that $T_2(x) = 2x^2 - 1$, so equating (8.61) to (8.60) gives

$$1 + k^2(4\omega^4 - 4\omega^2 + 1) = 1 + \frac{1}{4R}[(1 - R)^2 + (R^2 C^2 + L^2 - 2LCR^2)\omega^2 + L^2 C^2 R^2 \omega^4], \tag{8.62}$$

which can be solved for R, L, and C if the ripple level (as determined by k^2) is known. Thus, at $\omega = 0$ we have that

$$k^2 = \frac{(1 - R)^2}{4R},$$

or

$$R = 1 + 2k^2 \pm 2k\sqrt{1 + k^2} \qquad \text{(for } N \text{ even).} \tag{8.63}$$

Equating coefficients of ω^2 and ω^4 yields the additional relations

$$4k^2 = \frac{1}{4R}L^2 C^2 R^2,$$

$$-4k^2 = \frac{1}{4R}(R^2 C^2 + L^2 - 2LCR^2),$$

which can be used to find L and C. Note that (8.63) gives a value for R that is not unity, so there will be an impedance mismatch if the load has a unity (normalized) impedance; this can be corrected with a quarter-wave transformer, or by using an additional filter element to make N odd. For odd N, it can be shown that $R = 1$. (This is because there is a unity power loss ratio at $\omega = 0$ for N odd.)

Tables exist for designing equal-ripple low-pass filters with a normalized source impedance and cutoff frequency ($\omega_c' = 1$ rad/sec) [1], and these can be applied to either of the ladder circuits of Figure 8.25. This design data depends on the specified passband ripple level. Table 8.4 lists element values for normalized low-pass filter prototypes having 0.5 or 3.0 dB ripple for $N = 1$ to 10. Notice that the load impedance $g_{N+1} \neq 1$ for even N. If the stopband attenuation is specified, the curves in Figure 8.27 can be used to determine the necessary value of N for these ripple values.

Linear Phase Low-Pass Filter Prototypes

Filters having a maximally flat time delay, or a linear phase response, can be designed in the same way, but things are somewhat more complicated because the phase of the voltage transfer function is not as simply expressed as is its amplitude. Design values have been derived for such filters [1], however, again for the ladder circuits of Figure 8.25, and they are given in Table 8.5 for a normalized source impedance and cutoff frequency ($\omega_c' = 1$ rad/sec). The resulting normalized group delay in the passband will be $\tau_d = 1/\omega_c' = 1$ sec.

8.4   FILTER TRANSFORMATIONS

The low-pass filter prototypes of the previous section were normalized designs having a source impedance of $R_s = 1 \ \Omega$ and a cutoff frequency of $\omega_c = 1$ rad/sec. Here we show how these designs can be scaled in

TABLE 8.4 | Element Values for Equal-Ripple Low-Pass Filter Prototypes ($g_0 = 1, \omega_c = 1, N = 1$ to 10, 0.5 dB and 3.0 dB ripple)

N	g_1	g_2	g_3	g_4	g_5	g_6	g_7	g_8	g_9	g_{10}	g_{11}
					0.5 dB Ripple						
1	0.6986	1.0000									
2	1.4029	0.7071	1.9841								
3	1.5963	1.0967	1.5963	1.0000							
4	1.6703	1.1926	2.3661	0.8419	1.9841						
5	1.7058	1.2296	2.5408	1.2296	1.7058	1.0000					
6	1.7254	1.2479	2.6064	1.3137	2.4758	0.8696	1.9841				
7	1.7372	1.2583	2.6381	1.3444	2.6381	1.2583	1.7372	1.0000			
8	1.7451	1.2647	2.6564	1.3590	2.6964	1.3389	2.5093	0.8796	1.9841		
9	1.7504	1.2690	2.6678	1.3673	2.7239	1.3673	2.6678	1.2690	1.7504	1.0000	
10	1.7543	1.2721	2.6754	1.3725	2.7392	1.3806	2.7231	1.3485	2.5239	0.8842	1.9841
					3.0 dB Ripple						
1	1.9953	1.0000									
2	3.1013	0.5339	5.8095								
3	3.3487	0.7117	3.3487	1.0000							
4	3.4389	0.7483	4.3471	0.5920	5.8095						
5	3.4817	0.7618	4.5381	0.7618	3.4817	1.0000					
6	3.5045	0.7685	4.6061	0.7929	4.4641	0.6033	5.8095				
7	3.5182	0.7723	4.6386	0.8039	4.6386	0.7723	3.5182	1.0000			
8	3.5277	0.7745	4.6575	0.8089	4.6990	0.8018	4.4990	0.6073	5.8095		
9	3.5340	0.7760	4.6692	0.8118	4.7272	0.8118	4.6692	0.7760	3.5340	1.0000	
10	3.5384	0.7771	4.6768	0.8136	4.7425	0.8164	4.7260	0.8051	4.5142	0.6091	5.8095

Source: Reprinted from G. L. Matthaei, L. Young, and E. M. T. Jones, *Microwave Filters, Impedance-Matching Networks, and Coupling Structures*, Artech House, Dedham, MA, 1980, with permission.

terms of impedance and frequency, and converted to give high-pass, bandpass, or bandstop characteristics. Several examples will be presented to illustrate the design procedure.

Impedance and Frequency Scaling

Impedance scaling: In the prototype design, the source and load resistances are unity (except for equal-ripple filters with even N, which have nonunity load resistance). A source resistance of R_0 can be obtained by multiplying all the impedances of the prototype design by R_0. Thus, if we let primes denote impedance scaled quantities, the new filter component values are given by

$$L' = R_0 L, \tag{8.64a}$$

$$C' = \frac{C}{R_0}, \tag{8.64b}$$

$$R'_s = R_0, \tag{8.64c}$$

$$R'_L = R_0 R_L, \tag{8.64d}$$

where L, C, and R_L are the component values for the original prototype.

Frequency scaling for low-pass filters: To change the cutoff frequency of a low-pass prototype from unity to ω_c requires that we scale the frequency dependence of the filter by the factor $1/\omega_c$, which is accomplished

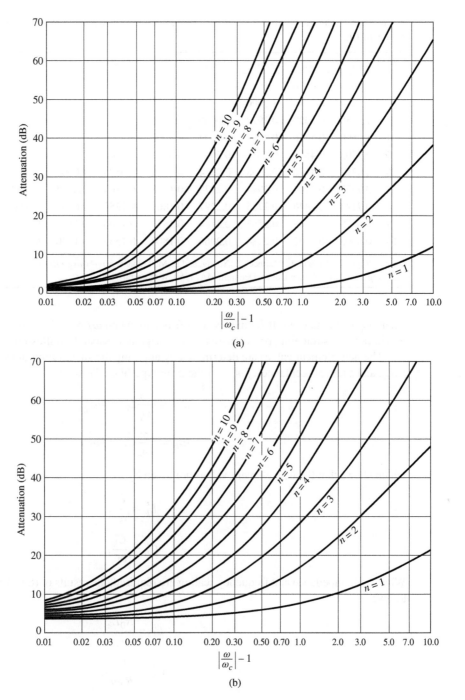

FIGURE 8.27 | Attenuation versus normalized frequency for equal-ripple filter prototypes. (a) 0.5 dB ripple level. (b) 3.0 dB ripple level.

Adapted from G. L. Matthaei, L. Young, and E. M. T. Jones, *Microwave Filters, Impedance-Matching Networks, and Coupling Structures*, Artech House, Dedham, MA, 1980, with permission.

by replacing ω by ω/ω_c:

$$\omega \leftarrow \frac{\omega}{\omega_c}. \tag{8.65}$$

Then the new power loss ratio will be

$$P'_{\mathrm{LR}}(\omega) = P_{\mathrm{LR}}\left(\frac{\omega}{\omega_c}\right),$$

TABLE 8.5 | Element Values for Maximally Flat Time Delay Low-Pass Filter Prototypes ($g_0 = 1$, $\omega_c = 1$, $N = 1$ to 10)

N	g_1	g_2	g_3	g_4	g_5	g_6	g_7	g_8	g_9	g_{10}	g_{11}
1	2.0000	1.0000									
2	1.5774	0.4226	1.0000								
3	1.2550	0.5528	0.1922	1.0000							
4	1.0598	0.5116	0.3181	0.1104	1.0000						
5	0.9303	0.4577	0.3312	0.2090	0.0718	1.0000					
6	0.8377	0.4116	0.3158	0.2364	0.1480	0.0505	1.0000				
7	0.7677	0.3744	0.2944	0.2378	0.1778	0.1104	0.0375	1.0000			
8	0.7125	0.3446	0.2735	0.2297	0.1867	0.1387	0.0855	0.0289	1.0000		
9	0.6678	0.3203	0.2547	0.2184	0.1859	0.1506	0.1111	0.0682	0.0230	1.0000	
10	0.6305	0.3002	0.2384	0.2066	0.1808	0.1539	0.1240	0.0911	0.0557	0.0187	1.0000

Source: Reprinted from G. L. Matthaei, L. Young, and E. M. T. Jones, *Microwave Filters, Impedance-Matching Networks, and Coupling Structures*, Artech House, Dedham, MA, 1980, with permission.

where ω_c is the new cutoff frequency; cutoff occurs when $\omega/\omega_c = 1$, or $\omega = \omega_c$. This transformation can be viewed as a stretching, or expansion, of the original passband, as illustrated in Figure 8.28a, b.

The new element values are determined by applying the substitution of (8.65) to the series reactances, $j\omega L_k$, and shunt susceptances, $j\omega C_k$, of the prototype filter. Thus,

$$jX_k = j\frac{\omega}{\omega_c}L_k = j\omega L'_k,$$

$$jB_k = j\frac{\omega}{\omega_c}C_k = j\omega C'_k,$$

which shows that the new element values are given by

$$L'_k = \frac{L_k}{\omega_c}, \tag{8.66a}$$

$$C'_k = \frac{C_k}{\omega_c}. \tag{8.66b}$$

When both impedance and frequency scaling are required, the results of (8.64) can be combined with (8.66) to give

$$L'_k = \frac{R_0 L_k}{\omega_c}, \tag{8.67a}$$

$$C'_k = \frac{C_k}{R_0 \omega_c}. \tag{8.67b}$$

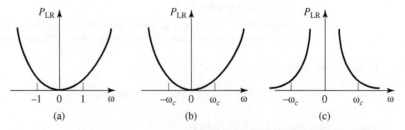

FIGURE 8.28 | Frequency scaling for low-pass filters and transformation to a high-pass response. (a) Low-pass filter prototype response for $\omega_c = 1$ rad/sec. (b) Frequency scaling for low-pass response. (c) Transformation to high-pass response.

Low-pass to high-pass transformation: The frequency substitution

$$\omega \leftarrow -\frac{\omega_c}{\omega} \tag{8.68}$$

can be used to convert a low-pass response to a high-pass response, as shown in Figure 8.28c. This substitution maps $\omega = 0$ to $\omega = \pm\infty$, and vice versa; cutoff occurs when $\omega = \pm\omega_c$. The negative sign is needed to convert inductors (and capacitors) to realizable capacitors (and inductors). Applying (8.68) to the series reactances, $j\omega L_k$, and the shunt susceptances, $j\omega C_k$, of the prototype filter gives

$$jX_k = -j\frac{\omega_c}{\omega}L_k = \frac{1}{j\omega C'_k},$$

$$jB_k = -j\frac{\omega_c}{\omega}C_k = \frac{1}{j\omega L'_k},$$

which shows that series inductors L_k must be replaced with capacitors C'_k, and shunt capacitors C_k must be replaced with inductors L'_k. The new component values are given by

$$C'_k = \frac{1}{\omega_c L_k}, \tag{8.69a}$$

$$L'_k = \frac{1}{\omega_c C_k}. \tag{8.69b}$$

Impedance scaling can be included by using (8.64) to give

$$C'_k = \frac{1}{R_0\omega_c L_k}, \tag{8.70a}$$

$$L'_k = \frac{R_0}{\omega_c C_k}. \tag{8.70b}$$

EXAMPLE 8.3   LOW-PASS FILTER DESIGN COMPARISON

Design a maximally flat low-pass filter with a cutoff frequency of 5 GHz, impedance of 50 Ω, and at least 15 dB insertion loss at 7.5 GHz. Compute and plot the amplitude response and group delay for $f = 0$ to 10 GHz, and compare with an equal-ripple (3.0 dB ripple) and linear phase filter having the same order.

Solution
First find the required order of the maximally flat filter to satisfy the insertion loss specification at 7.5 GHz. We have that $|\omega/\omega_c| - 1 = 0.5$; from Figure 8.26 we see that $N = 5$ will be sufficient. Then Table 8.3 gives the prototype element values as

$$g_1 = 0.618,$$
$$g_2 = 1.618,$$
$$g_3 = 2.000,$$
$$g_4 = 1.618,$$
$$g_5 = 0.618.$$

Then (8.67) can be used to obtain the scaled element values:

$$L'_k = \frac{R_0 L_k}{\omega_c}$$

$$C'_k = \frac{C_k}{R_0\omega_c}$$

$$C'_1 = 0.393 \text{ pF},$$
$$L'_2 = 2.575 \text{ nH},$$
$$C'_3 = 1.273 \text{ pF},$$
$$L'_4 = 2.575 \text{ nH},$$
$$C'_5 = 0.393 \text{ pF}.$$

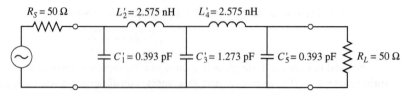

FIGURE 8.29 | Low-pass, maximally flat filter circuit for Example 8.3.

The final filter circuit is shown in Figure 8.29; the ladder circuit of Figure 8.25a was used, but that of Figure 8.25b could have been used just as well.

The component values for the equal-ripple filter and the linear phase filter, for $N = 5$, can be determined from Tables 8.4 and 8.5. The amplitude and group delay results for these three filters are shown in Figure 8.30. These results clearly show the trade-offs involved with the three types of filters. The equal-ripple response has the sharpest cutoff but the worst group delay characteristics. The maximally flat response has a flatter attenuation characteristic in the passband but a slightly lower cutoff rate. The linear phase filter has the worst cutoff rate but a very good group delay characteristic.

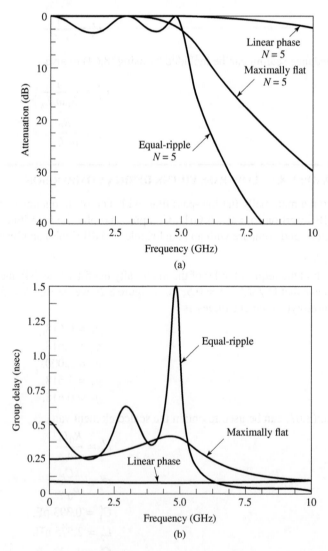

FIGURE 8.30 | Frequency response of the filter design of Example 8.3. (a) Amplitude response. (b) Group delay response.

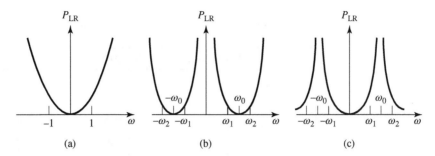

FIGURE 8.31 | Bandpass and bandstop frequency transformations. (a) Low-pass filter prototype response for $\omega_c = 1$. (b) Transformation to bandpass response. (c) Transformation to bandstop response.

Bandpass and Bandstop Transformations

Low-pass prototype filter designs can also be transformed to have the bandpass or bandstop responses illustrated in Figure 8.31. If ω_1 and ω_2 denote the edges of the passband, then a bandpass response can be obtained using the following frequency substitution:

$$\omega \leftarrow \frac{\omega_0}{\omega_2 - \omega_1}\left(\frac{\omega}{\omega_0} - \frac{\omega_0}{\omega}\right) = \frac{1}{\Delta}\left(\frac{\omega}{\omega_0} - \frac{\omega_0}{\omega}\right),\tag{8.71}$$

where

$$\Delta = \frac{\omega_2 - \omega_1}{\omega_0}\tag{8.72}$$

is the fractional bandwidth of the passband. The center frequency, ω_0, could be chosen as the arithmetic mean of ω_1 and ω_2, but the equations are simpler if it is chosen as the geometric mean:

$$\omega_0 = \sqrt{\omega_1 \omega_2}.\tag{8.73}$$

Then the transformation of (8.71) maps the bandpass characteristics of Figure 8.31b to the low-pass response of Figure 8.31a as follows:

$$\text{When } \omega = \omega_0, \quad \frac{1}{\Delta}\left(\frac{\omega}{\omega_0} - \frac{\omega_0}{\omega}\right) = 0.$$

$$\text{When } \omega = \omega_1, \quad \frac{1}{\Delta}\left(\frac{\omega}{\omega_0} - \frac{\omega_0}{\omega}\right) = \frac{1}{\Delta}\left(\frac{\omega_1^2 - \omega_0^2}{\omega_0 \omega_1}\right) = -1.$$

$$\text{When } \omega = \omega_2, \quad \frac{1}{\Delta}\left(\frac{\omega}{\omega_0} - \frac{\omega_0}{\omega}\right) = \frac{1}{\Delta}\left(\frac{\omega_2^2 - \omega_0^2}{\omega_0 \omega_2}\right) = 1.$$

The new filter elements are determined by using (8.71) in the expressions for the series reactance and shunt susceptances. Thus,

$$jX_k = \frac{j}{\Delta}\left(\frac{\omega}{\omega_0} - \frac{\omega_0}{\omega}\right)L_k = j\frac{\omega L_k}{\Delta \omega_0} - j\frac{\omega_0 L_k}{\Delta \omega} = j\omega L_k' - j\frac{1}{\omega C_k'},$$

which shows that a series inductor, L_k, is transformed to a series LC circuit with element values

$$L_k' = \frac{L_k}{\Delta \omega_0},\tag{8.74a}$$

$$C_k' = \frac{\Delta}{\omega_0 L_k}.\tag{8.74b}$$

Similarly,

$$jB_k = \frac{j}{\Delta}\left(\frac{\omega}{\omega_0} - \frac{\omega_0}{\omega}\right)C_k = j\frac{\omega C_k}{\Delta \omega_0} - j\frac{\omega_0 C_k}{\Delta \omega} = j\omega C_k' - j\frac{1}{\omega L_k'},$$

TABLE 8.6 | Summary of Prototype Filter Transformations $\left(\Delta = \dfrac{\omega_2 - \omega_1}{\omega_0}\right)$

Low-Pass	High-Pass	Bandpass	Bandstop

which shows that a shunt capacitor, C_k, is transformed to a shunt LC circuit with element values

$$L'_k = \frac{\Delta}{\omega_0 C_k}, \tag{8.74c}$$

$$C'_k = \frac{C_k}{\Delta \omega_0}. \tag{8.74d}$$

The low-pass filter elements are thus converted to series resonant circuits (having a low impedance at resonance) in the series arms, and to parallel resonant circuits (having a high impedance at resonance) in the shunt arms. Notice that both series and parallel resonator elements have a resonant frequency of ω_0.

The inverse transformation can be used to obtain a bandstop response. Thus,

$$\omega \leftarrow -\Delta \left(\frac{\omega}{\omega_0} - \frac{\omega_0}{\omega}\right)^{-1}, \tag{8.75}$$

where Δ and ω_0 have the same definitions as in (8.72) and (8.73). Then series inductors of the low-pass prototype are converted to parallel LC circuits having element values given by

$$L'_k = \frac{\Delta L_k}{\omega_0}, \tag{8.76a}$$

$$C'_k = \frac{1}{\omega_0 \Delta L_k}. \tag{8.76b}$$

The shunt capacitor of the low-pass prototype is converted to series LC circuits having element values given by

$$L'_k = \frac{1}{\omega_0 \Delta C_k}, \tag{8.76c}$$

$$C'_k = \frac{\Delta C_k}{\omega_0}. \tag{8.76d}$$

The element transformations from a low-pass prototype to a high-pass, bandpass, or bandstop filter are summarized in Table 8.6. These results do not include impedance scaling, which can be made using (8.64).

EXAMPLE 8.4   BANDPASS FILTER DESIGN

Design a bandpass filter having a 0.5 dB equal-ripple response, with $N = 3$. The center frequency is 2 GHz, the bandwidth is 15%, and the impedance is 50 Ω.

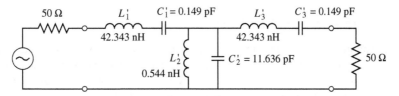

FIGURE 8.32 | Bandpass filter circuit for Example 8.4.

Solution

From Table 8.4 the element values for the low-pass prototype circuit of Figure 8.25b are given as

$$g_1 = 1.5963 = L_1,$$
$$g_2 = 1.0967 = C_2,$$
$$g_3 = 1.5963 = L_3,$$
$$g_4 = 1.000 = R_L.$$

Equations (8.64) and (8.74) give the impedance-scaled and frequency-transformed element values for the circuit of Figure 8.32 as

$$L'_1 = \frac{L_1 R_0}{\omega_0 \Delta} = 42.343 \text{ nH}, \text{ where } \Delta = \frac{\omega_2 - \omega_1}{\omega_0} = \frac{15\% \, \omega_0}{\omega_0} = 0.15$$
$$C'_1 = \frac{\Delta}{\omega_0 L_1 R_0} = 0.149 \text{ pF},$$
$$L'_2 = \frac{\Delta R_0}{\omega_0 C_2} = 0.544 \text{ nH},$$
$$C'_2 = \frac{C_2}{\omega_0 \Delta R_0} = 11.636 \text{ pF},$$
$$L'_3 = \frac{L_3 R_0}{\omega_0 \Delta} = 42.343 \text{ nH},$$
$$C'_3 = \frac{\Delta}{\omega_0 L_3 R_0} = 0.149 \text{ pF}.$$

The resulting amplitude response is shown in Figure 8.33.

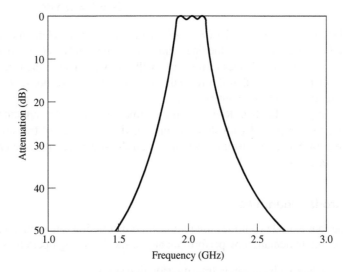

FIGURE 8.33 | Amplitude response for the bandpass filter of Example 8.4.

8.5   FILTER IMPLEMENTATION

The lumped-element filter designs discussed in the previous sections generally work well at low frequencies, but two problems arise at higher RF and microwave frequencies. First, lumped-element inductors and capacitors are generally available only for a limited range of values, and can be difficult to implement at microwave frequencies. Distributed elements, such as open-circuited or short-circuited transmission line stubs, are often used to approximate ideal lumped elements. In addition, at microwave frequencies the distances between filter components is not negligible. The first problem is treated with *Richards' transformation*, which can be used to convert lumped elements to transmission line sections. *Kuroda's identities* can then be used to physically separate filter elements by using transmission line sections. Because such additional transmission line sections do not affect the filter response, this type of design is called *redundant* filter synthesis. It is possible to design microwave filters that take advantage of these sections to improve the filter response [4]; such *nonredundant* synthesis does not have a lumped-element counterpart.

Richards' Transformation

The transformation

$$\Omega = \tan \beta \ell = \tan \left(\frac{\omega \ell}{v_p} \right) \tag{8.77}$$

maps the ω plane to the Ω plane, which repeats with a period of $\omega \ell / v_p = 2\pi$. This transformation was introduced by P. Richards [6] to synthesize an LC network using open- and short-circuited transmission line stubs. Thus, if we replace the frequency variable ω with Ω, we can write the reactance of an inductor as

$$jX_L = j\Omega L = jL \tan \beta \ell, \tag{8.78a}$$

and the susceptance of a capacitor as

$$jB_C = j\Omega C = jC \tan \beta \ell. \tag{8.78b}$$

These results indicate that an inductor can be replaced with a short-circuited stub of length $\beta \ell$ and characteristic impedance L, while a capacitor can be replaced with an open-circuited stub of length $\beta \ell$ and characteristic impedance $1/C$. A unity filter impedance is assumed.

Cutoff occurs at unity frequency for a low-pass filter prototype; to obtain the same cutoff frequency for the Richards'-transformed filter, (8.77) shows that

$$\Omega = 1 = \tan \beta \ell,$$

which gives a stub length of $\ell = \lambda / 8$, where λ is the wavelength of the line at the cutoff frequency, ω_c. At the frequency $\omega_0 = 2\omega_c$, the lines will be $\lambda/4$ long, and an attenuation pole will occur. At frequencies away from ω_c, the impedances of the stubs will no longer match the original lumped-element impedances, and the filter response will differ from the desired prototype response. In addition, the response will be periodic in frequency, repeating every $4\omega_c$.

In principle, then, Richards' transformation allows the inductors and capacitors of a lumped-element filter to be replaced with short-circuited and open-circuited transmission line stubs, as illustrated in Figure 8.34. Since the electrical lengths of all the stubs are the same ($\lambda/8$ at ω_c), these lines are called *commensurate lines*.

Kuroda's Identities

The four Kuroda identities use redundant transmission line sections to achieve a more practical microwave filter implementation by performing any of the following operations:

- Physically separate transmission line stubs
- Transform series stubs into shunt stubs, or vice versa
- Change impractical characteristic impedances into more realizable values

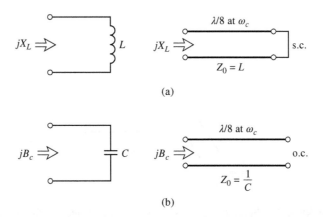

(a)

(b)

FIGURE 8.34 | Richards' transformation. (a) For an inductor to a short-circuited stub. (b) For a capacitor to an open-circuited stub.

The additional transmission line sections are called *unit elements* and are $\lambda/8$ long at ω_c; the unit elements are thus commensurate with the stubs used to implement the inductors and capacitors of the prototype design.

The four Kuroda identities are illustrated in Table 8.7, where each box represents a unit element, or transmission line, of the indicated characteristic impedance and length ($\lambda/8$ at ω_c). The inductors and capacitors represent short-circuit and open-circuit stubs, respectively. We will prove the equivalence of the first case, and then show how to use these identities in Example 8.5.

The two circuits of identity (a) in Table 8.7 can be redrawn as shown in Figure 8.35; we will show that these two networks are equivalent by showing that their *ABCD* matrices are identical. From Table 4.1, the

TABLE 8.7 | The Four Kuroda Identities ($n^2 = 1 + Z_2/Z_1$)

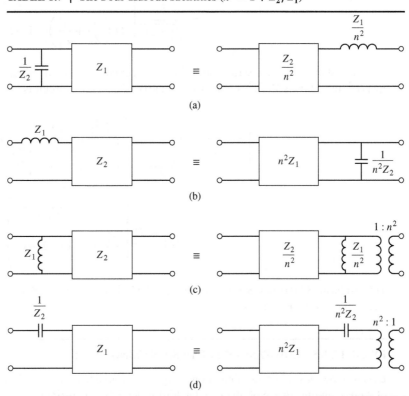

(a)

(b)

(c)

(d)

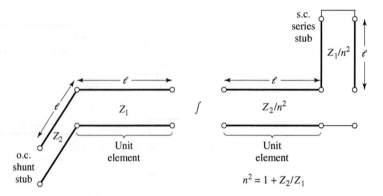

FIGURE 8.35 | Equivalent circuits illustrating Kuroda identity (a) in Table 8.7.

$ABCD$ matrix of a length ℓ of transmission line with characteristic impedance Z_1 is

$$\begin{bmatrix} A & B \\ C & D \end{bmatrix} = \begin{bmatrix} \cos\beta\ell & jZ_1\sin\beta\ell \\ \dfrac{j}{Z_1}\sin\beta\ell & \cos\beta\ell \end{bmatrix} = \frac{1}{\sqrt{1+\Omega^2}}\begin{bmatrix} 1 & j\Omega Z_1 \\ \dfrac{j\Omega}{Z_1} & 1 \end{bmatrix}, \tag{8.79}$$

where $\Omega = \tan\beta\ell$. The open-circuited shunt stub in the first circuit in Figure 8.35 has an impedance of $-jZ_2\cot\beta\ell = -jZ_2/\Omega$, so the $ABCD$ matrix of the entire circuit is

$$\begin{bmatrix} A & B \\ C & D \end{bmatrix}_L = \begin{bmatrix} 1 & 0 \\ \dfrac{j\Omega}{Z_2} & 1 \end{bmatrix}\begin{bmatrix} 1 & j\Omega Z_1 \\ \dfrac{j\Omega}{Z_1} & 1 \end{bmatrix}\frac{1}{\sqrt{1+\Omega^2}}$$

$$= \frac{1}{\sqrt{1+\Omega^2}}\begin{bmatrix} 1 & j\Omega Z_1 \\ j\Omega\left(\dfrac{1}{Z_1} + \dfrac{1}{Z_2}\right) & 1 - \Omega^2\dfrac{Z_1}{Z_2} \end{bmatrix}. \tag{8.80a}$$

The short-circuited series stub in the second circuit in Figure 8.35 has an impedance of $j(Z_1/n^2)\tan\beta\ell = j\Omega Z_1/n^2$, so the $ABCD$ matrix of the entire circuit is

$$\begin{bmatrix} A & B \\ C & D \end{bmatrix}_R = \begin{bmatrix} 1 & j\dfrac{\Omega Z_2}{n^2} \\ \dfrac{j\Omega n^2}{Z_2} & 1 \end{bmatrix}\begin{bmatrix} 1 & \dfrac{j\Omega Z_1}{n^2} \\ 0 & 1 \end{bmatrix}\frac{1}{\sqrt{1+\Omega^2}}$$

$$= \frac{1}{\sqrt{1+\Omega^2}}\begin{bmatrix} 1 & \dfrac{j\Omega}{n^2}(Z_1 + Z_2) \\ \dfrac{j\Omega n^2}{Z_2} & 1 - \Omega^2\dfrac{Z_1}{Z_2} \end{bmatrix}. \tag{8.80b}$$

The results in (8.80a) and (8.80b) are identical if we choose $n^2 = 1 + Z_2/Z_1$. The other identities in Table 8.7 can be proved in the same way.

EXAMPLE 8.5   LOW-PASS FILTER DESIGN USING STUBS

Design a low-pass filter for fabrication using microstrip lines. The specifications include a cutoff frequency of 4 GHz, an impedance of 50 Ω, and a third-order 3 dB equal-ripple passband response.

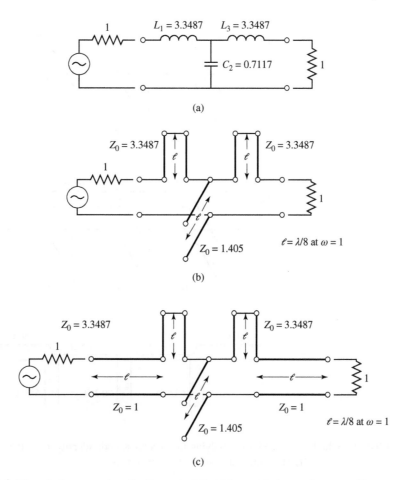

FIGURE 8.36 | Filter design procedure for Example 8.5. (a) Lumped-element low-pass filter prototype. (b) Using Richards' transformations to convert inductors and capacitors to series and shunt stubs. (c) Adding unit elements at the ends of the filter.

Solution

From Table 8.4 the normalized low-pass prototype element values are

$$g_1 = 3.3487 = L_1,$$
$$g_2 = 0.7117 = C_2,$$
$$g_3 = 3.3487 = L_3,$$
$$g_4 = 1.0000 = R_L,$$

with the lumped-element circuit shown in Figure 8.36a.

We now use Richards' transformations to convert series inductors to series stubs, and shunt capacitors to shunt stubs, as shown in Figure 8.36b. According to (8.78), the characteristic impedance of a series stub (inductor) is L, and the characteristic impedance of a shunt stub (capacitor) is $1/C$. For commensurate line synthesis, all stubs are $\lambda/8$ long at $\omega = \omega_c$. (It is usually most convenient to work with normalized quantities until the last step in the design.)

The series stubs of Figure 8.36b would be very difficult to implement in microstrip line form, so we will use one of the Kuroda identities to convert these to shunt stubs. First we add unit elements at either end of the filter, as shown in Figure 8.36c. These redundant elements do not affect filter performance since they are matched to the source and load ($Z_0 = 1$). Then we can apply Kuroda identity (b) from Table 8.7 to both ends of the filter. In both cases we have that

$$n^2 = 1 + \frac{Z_2}{Z_1} = 1 + \frac{1}{3.3487} = 1.299.$$

The result is shown in Figure 8.36d.

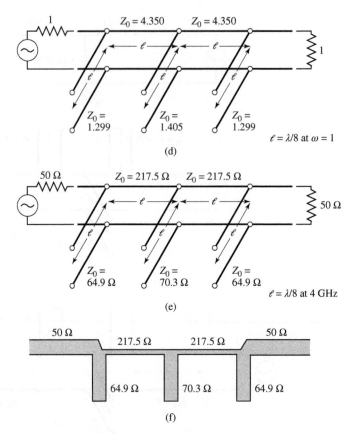

(d)

(e)

(f)

FIGURE 8.36 | Continued. (d) Applying the second Kuroda identity. (e) After impedance and frequency scaling. (f) Microstrip fabrication of the final filter.

Finally, we impedance and frequency scale the circuit, which simply involves multiplying the normalized characteristic impedances by 50 Ω and choosing the line and stub lengths to be $\lambda/8$ at 4 GHz. The final circuit is shown in Figure 8.36e, with a microstrip layout in Figure 8.36f.

The calculated amplitude response of this filter is plotted in Figure 8.37, along with the response of the lumped-element version. Note that the passband characteristics are very similar up to 4 GHz, but the distributed-element filter has a sharper cutoff. Also notice that the distributed-element filter has a response that repeats every 16 GHz, as a result of the periodic nature of Richards' transformation.

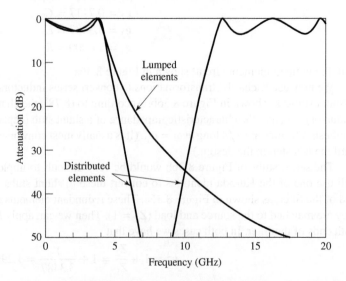

FIGURE 8.37 | Amplitude responses of lumped-element and distributed-element low-pass filter of Example 8.5.

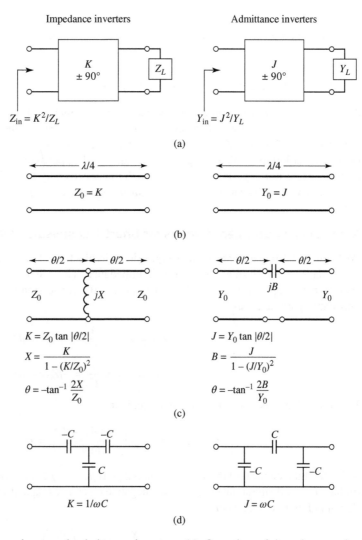

FIGURE 8.38 | Impedance and admittance inverters. (a) Operation of impedance and admittance inverters. (b) Implementation as quarter-wave transformers. (c) Implementation using transmission lines and reactive elements. (d) Implementation using capacitor networks.

Similar procedures can be used for bandstop filters, but the Kuroda identities are not useful for high-pass or bandpass filters.

Impedance and Admittance Inverters

As we have seen, it is often desirable to use only series, or only shunt, elements when implementing a filter with a particular type of transmission line. The Kuroda identities can be used for conversions of this form, but another possibility is to use *impedance* (K) or *admittance* (J) *inverters* [1, 4, 7]. Such inverters are especially useful for bandpass or bandstop filters with narrow (<10%) bandwidths.

The conceptual operation of impedance and admittance inverters is illustrated in Figure 8.38; since these inverters essentially form the inverse of the load impedance or admittance, they can be used to transform series-connected elements to shunt-connected elements, or vice versa. This procedure will be illustrated in later sections for bandpass and bandstop filters.

In its simplest form, an impedance or admittance inverter can be constructed using a quarter-wave transformer of the appropriate characteristic impedance, as shown in Figure 8.38b. This implementation also allows the *ABCD* matrix of the inverter to be easily found from the *ABCD* parameters for a length of transmission line, as given in Table 4.1. Many other types of circuits can also be used as impedance or admittance inverters, with one such alternative being shown in Figure 8.38c. The lengths, $\theta/2$, of the transmission line sections are generally required to be negative for this type of inverter, but this poses no problem if these lines can be absorbed into connecting transmission lines on either side.

8.6   STEPPED-IMPEDANCE LOW-PASS FILTERS

A relatively easy way to implement low-pass filters in microstrip or stripline is to use alternating sections of very high and very low characteristic impedance lines. Such filters are usually referred to as *stepped-impedance*, or hi-Z, low-Z filters, and are popular because they are easier to design and take up less space than a similar low-pass filter using stubs. Because of the approximations involved, however, their electrical performance is not as good, so the use of such filters is usually limited to applications where a sharp cutoff is not required (for instance, in rejecting out-of-band mixer products).

Approximate Equivalent Circuits for Short Transmission Line Sections

We begin by finding the approximate equivalent circuits for a short length of transmission line having either a very large or a very small characteristic impedance. The *ABCD* parameters of a length ℓ of line having characteristic impedance Z_0 are given in Table 4.1; the conversion in Table 4.2 can then be used to find the impedance parameters as

$$Z_{11} = Z_{22} = \frac{A}{C} = -jZ_0 \cot \beta\ell, \tag{8.81a}$$

$$Z_{12} = Z_{21} = \frac{1}{C} = -jZ_0 \csc \beta\ell. \tag{8.81b}$$

The series elements of the T-equivalent circuit are

$$Z_{11} - Z_{12} = -jZ_0 \left(\frac{\cos \beta\ell - 1}{\sin \beta\ell} \right) = jZ_0 \tan \left(\frac{\beta\ell}{2} \right), \tag{8.82}$$

while the shunt element of the T-equivalent is Z_{12}. If $\beta\ell < \pi/2$, the series elements have a positive reactance (inductors), while the shunt element has a negative reactance (capacitor). We thus have the equivalent circuit shown in Figure 8.39a, where

$$\frac{X}{2} = Z_0 \tan \left(\frac{\beta\ell}{2} \right), \tag{8.83a}$$

$$B = \frac{1}{Z_0} \sin \beta\ell. \tag{8.83b}$$

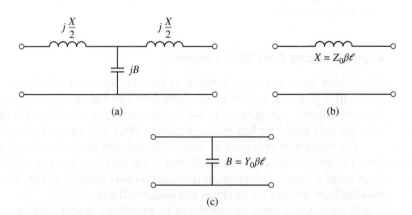

FIGURE 8.39 | Approximate equivalent circuits for short sections of transmission lines. (a) T-equivalent circuit for a transmission line section having $\beta\ell \ll \pi/2$. (b) Equivalent circuit for small $\beta\ell$ and large Z_0. (c) Equivalent circuit for small $\beta\ell$ and small Z_0.

Now assume a short length of line (say $\beta\ell < \pi/4$) and a large characteristic impedance. Then (8.83) approximately reduces to

$$X \simeq Z_0\beta\ell, \tag{8.84a}$$

$$B \simeq 0, \tag{8.84b}$$

which implies the equivalent circuit of Figure 8.39b (a series inductor). For a short length of line and a small characteristic impedance, (8.83) approximately reduces to

$$X \simeq 0, \tag{8.85a}$$

$$B \simeq Y_0\beta\ell, \tag{8.85b}$$

which implies the equivalent circuit of Figure 8.39c (a shunt capacitor). So the series inductors of a low-pass prototype can be replaced with high-impedance line sections ($Z_0 = Z_h$), and the shunt capacitors can be replaced with low-impedance line sections ($Z_0 = Z_\ell$). The ratio Z_h/Z_ℓ should be as large as possible, so the actual values of Z_h and Z_ℓ are usually set to the highest and lowest characteristic impedance that can be practically fabricated. The lengths of the lines can then be determined from (8.84) and (8.85); to get the best response near cutoff, these lengths should be evaluated at $\omega = \omega_c$. Combining the results of (8.84) and (8.85) with the scaling equations of (8.67) allows the electrical lengths of the inductor sections to be calculated as

$$\beta\ell = \frac{LR_0}{Z_h} \quad \text{(inductor)} \tag{8.86a}$$

and the electrical length of the capacitor sections as

$$\beta\ell = \frac{CZ_\ell}{R_0} \quad \text{(capacitor)}, \tag{8.86b}$$

where R_0 is the filter impedance and L and C are the normalized element values (the g_k) of the low-pass prototype.

EXAMPLE 8.6   STEPPED-IMPEDANCE FILTER DESIGN

Design a stepped-impedance low-pass filter having a maximally flat response and a cutoff frequency of 2.5 GHz. It is desired to have more than 20 dB insertion loss at 4 GHz. The filter impedance is 50 Ω; the highest practical line impedance is 120 Ω, and the lowest is 20 Ω. Consider the effect of losses when this filter is implemented with a microstrip substrate having $d = 0.158$ cm, $\epsilon_r = 4.2$, $\tan\delta = 0.02$, and copper conductors of 0.5 mil thickness.

Solution
To use Figure 8.26 we calculate

$$\frac{\omega}{\omega_c} - 1 = \frac{4.0}{2.5} - 1 = 0.6;$$

then the figure indicates $N = 6$ should give the required attenuation at 4.0 GHz. Table 8.3 gives the low-pass prototype values as

$$g_1 = 0.517 = C_1, \quad g_2 = 1.414 = L_2, \quad g_3 = 1.932 = C_3,$$
$$g_4 = 1.932 = L_4, \quad g_5 = 1.414 = C_5, \quad g_6 = 0.517 = L_6.$$

The low-pass prototype filter is shown in Figure 8.40a.

Next, (8.86a) and (8.86b) are used to replace the series inductors and shunt capacitors with sections of low-impedance and high-impedance lines. The required electrical line lengths, $\beta\ell_i$, along with the physical microstrip line widths, W_i, and lengths, ℓ_i, are given in the table.

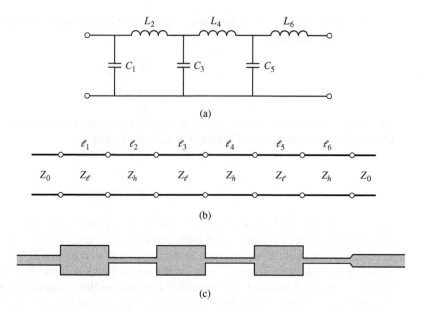

(a)

(b)

(c)

FIGURE 8.40 | Filter design for Example 8.6. (a) Low-pass filter prototype circuit. (b) Stepped-impedance implementation. (c) Microstrip layout of final filter.

Section	$Z_i = Z_\ell$ or $Z_h (\Omega)$	$\beta\ell_i$ (deg)	W_i (mm)	ℓ_i (mm)
1	20	11.8	11.3	2.05
2	120	33.8	0.428	6.63
3	20	44.3	11.3	7.69
4	120	46.1	0.428	9.04
5	20	32.4	11.3	5.63
6	120	12.3	0.428	2.41

The final filter circuit is shown in Figure 8.40b, with $Z_\ell = 20\ \Omega$ and $Z_h = 120\ \Omega$. Note that $\beta\ell < 45°$ for all but one section. The microstrip layout of the filter is shown in Figure 8.40c.

Figure 8.41 shows the calculated amplitude response of the filter, with and without losses. The effect of loss is to increase the passband attenuation to about 1 dB at 2 GHz. The response of the corresponding

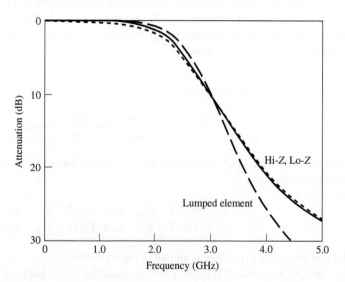

FIGURE 8.41 | Amplitude response of the stepped-impedance low-pass filter of Example 8.6, with (dotted line) and without (solid line) losses. The response of the corresponding lumped-element filter is also shown.

lumped-element filter is also shown in Figure 8.41. The passband characteristic is similar to that of the stepped impedance filter, but the lumped-element filter gives more attenuation at higher frequencies. This is because the stepped-impedance filter elements depart significantly from the lumped-element values at higher frequencies. The stepped-impedance filter may have other passbands at higher frequencies, but the response will not be perfectly periodic because the lines are not commensurate.

Comparison of Richards' Transformation and Stepped-Impedance Method

Table 8.8 shows the comparison of Richards' transformation and stepped-impedance method to design low-pass filters. From the table, it can be observed that the Richards' transformation offers good electrical performance as compared to the stepped-impedance method. Particularly, if we consider the frequency response of low-pass filters at 4 GHz frequency, the third-order low-pass filters designed with Richards' transformation method offer good performance (sharp cutoff) compared to sixth-order low-pass filters designed with the stepped-impedance method.

TABLE 8.8 | Parameters for Richards' Transformation and Stepped-Impedance Methods

Parameter	Richards' Transformation with Kuroda's Identities	Stepped Impedance
Design and implementation	Complex	Easy
Accuracy	High (accurate method)	Less (Approximate method)
Cutoff (for example, the frequency response of low-pass filters)	Sharp	Slow
	For order = 3, f_c = 4 GHz	For order = 6, f_c = 4 GHz
Electrical performance	Good	Poor
Inductor realized as	Short-circuited stub	High-impedance transmission line
Capacitor realized as	Open-circuited stub	Low-impedance transmission line
Impedance	Vary	Fixed as high and low values
Electrical length	Fixed ($\lambda/8$)	Vary
Applications	Wireless communication	Rejecting out-of-band mixer

8.7 COUPLED LINE FILTERS

The parallel coupled transmission lines discussed in Section 7.6 (for directional couplers) can be used to construct many types of filters. Fabrication of multisection bandpass or bandstop coupled line filters is particularly easy in microstrip or stripline form for bandwidths less than about 20%. Wider bandwidth filters generally require very tightly coupled lines, which are difficult to fabricate. We will first study the filter characteristics of a single quarter-wave coupled line section, and then show how these sections can be used to design a bandpass filter [7]. Other filter designs using coupled lines can be found in reference [1].

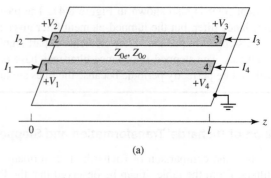

(a)

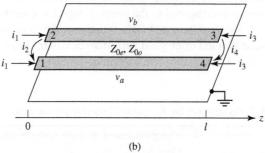

(b)

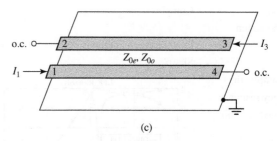

(c)

FIGURE 8.42 | Definitions pertaining to a coupled line filter section. (a) A parallel coupled line section with port voltage and current definitions. (b) A parallel coupled line section with even- and odd-mode current sources. (c) A two-port coupled line section having a bandpass response.

Filter Properties of a Coupled Line Section

A parallel coupled line section is shown in Figure 8.42a, with port voltage and current definitions. We will derive the open-circuit impedance matrix for this four-port network by considering the superposition of even- and odd-mode excitations [8], which are shown in Figure 8.42b. Thus, the current sources i_1 and i_3 drive the line in the even mode, while i_2 and i_4 drive the line in the odd mode. By superposition, we see that the total port currents, I_i, can be expressed in terms of the even- and odd-mode currents as

$$I_1 = i_1 + i_2,$$ (8.87a)

$$I_2 = i_1 - i_2,$$ (8.87b)

$$I_3 = i_3 - i_4,$$ (8.87c)

$$I_4 = i_3 + i_4.$$ (8.87d)

First consider the line as being driven in the even mode by the i_1 current sources. If the other ports are open-circuited, the impedance seen at port 1 or 2 is

$$Z_{in}^e = -jZ_{0e} \cot \beta \ell.$$ (8.88)

The voltage on either conductor can be expressed as

$$v_a^1(z) = v_b^1(z) = V_e^+[e^{-j\beta(z-\ell)} + e^{j\beta(z-\ell)}]$$
$$= 2V_e^+ \cos \beta(\ell - z), \qquad (8.89)$$

so the voltage at port 1 or 2 is

$$v_a^1(0) = v_b^1(0) = 2V_e^+ \cos \beta\ell = i_1 Z_{in}^e.$$

This result and (8.88) can be used to rewrite (8.89) in terms of i_1 as

$$v_a^1(z) = v_b^1(z) = -jZ_{0e}\frac{\cos \beta(\ell - z)}{\sin \beta\ell}i_1. \qquad (8.90)$$

Similarly, the voltages due to current sources i_3 driving the line in the even mode are

$$v_a^3(z) = v_b^3(z) = -jZ_{0e}\frac{\cos \beta z}{\sin \beta\ell}i_3. \qquad (8.91)$$

Now consider the line as being driven in the odd mode by current i_2. If the other ports are open-circuited, the impedance seen at port 1 or 2 is

$$Z_{in}^o = -jZ_{0o} \cot \beta\ell. \qquad (8.92)$$

The voltage on either conductor can be expressed as

$$v_a^2(z) = -v_b^2(z) = V_0^+\left[e^{-j\beta(z-\ell)} + e^{j\beta(z-\ell)}\right] = 2V_0^+ \cos \beta(\ell - z). \qquad (8.93)$$

Then the voltage at port 1 or port 2 is

$$v_a^2(0) = -v_b^2(0) = 2V_0^+ \cos \beta\ell = i_2 Z_{in}^o.$$

This result and (8.92) can be used to rewrite (8.93) in terms of i_2 as

$$v_a^2(z) = -v_b^2(z) = -jZ_{0o}\frac{\cos \beta(\ell - z)}{\sin \beta\ell}i_2. \qquad (8.94)$$

Similarly, the voltages due to current i_4 driving the line in the odd mode are

$$v_a^4(z) = -v_b^4(z) = -jZ_{0o}\frac{\cos \beta z}{\sin \beta\ell}i_4. \qquad (8.95)$$

The total voltage at port 1 is

$$V_1 = v_a^1(0) + v_a^2(0) + v_a^3(0) + v_a^4(0)$$
$$= -j(Z_{0e}i_1 + Z_{0o}i_2) \cot \theta - j(Z_{0e}i_3 + Z_{0o}i_4) \csc \theta, \qquad (8.96)$$

where the results of (8.90), (8.91), (8.94), and (8.95) were used, and $\theta = \beta\ell$. Next, we solve (8.87) for the i_j in terms of the Is:

$$i_1 = \frac{1}{2}(I_1 + I_2), \qquad (8.97a)$$

$$i_2 = \frac{1}{2}(I_1 - I_2), \qquad (8.97b)$$

$$i_3 = \frac{1}{2}(I_3 + I_4), \qquad (8.97c)$$

$$i_4 = \frac{1}{2}(I_4 - I_3), \qquad (8.97d)$$

and use these results in (8.96):

$$V_1 = \frac{-j}{2}(Z_{0e}I_1 + Z_{0e}I_2 + Z_{0o}I_1 - Z_{0o}I_2)\cot\theta$$

$$\frac{-j}{2}(Z_{0e}I_3 + Z_{0e}I_4 + Z_{0o}I_4 - Z_{0o}I_3)\csc\theta. \tag{8.98}$$

This result yields the top row of the open-circuit impedance matrix [Z] that describes the coupled line section. From symmetry, all other matrix elements can be found once the first row is known. The matrix elements are then

$$Z_{11} = Z_{22} = Z_{33} = Z_{44} = \frac{-j}{2}\left(Z_{0e} + Z_{0o}\right)\cot\theta \tag{8.99a}$$

$$Z_{12} = Z_{21} = Z_{34} = Z_{43} = \frac{-j}{2}\left(Z_{0e} - Z_{0o}\right)\cot\theta \tag{8.99b}$$

$$Z_{13} = Z_{31} = Z_{24} = Z_{42} = \frac{-j}{2}\left(Z_{0e} - Z_{0o}\right)\csc\theta \tag{8.99c}$$

$$Z_{14} = Z_{41} = Z_{23} = Z_{32} = \frac{-j}{2}\left(Z_{0e} + Z_{0o}\right)\csc\theta \tag{8.99d}$$

A two-port network can be formed from a coupled line section by terminating two of the four ports with either open or short circuits, or by connecting two ends; there are 10 possible combinations, as illustrated in Table 8.9. As indicated in the table, the various circuits have different frequency responses, including low-pass, bandpass, all pass, and all stop. For bandpass filters, we are most interested in the case shown in Figure 8.42c, as open circuits are easier to fabricate in microstrip than are short circuits. In this case, $I_2 = I_4 = 0$, so the four-port impedance matrix equations reduce to

$$V_1 = Z_{11}I_1 + Z_{13}I_3, \tag{8.100a}$$

$$V_3 = Z_{31}I_1 + Z_{33}I_3, \tag{8.100b}$$

where Z_{ij} is given in (8.99).

We can analyze the filter characteristics of this circuit by calculating the image impedance (which is the same at ports 1 and 3), and the propagation constant. From Table 8.1, the image impedance in terms of the impedance parameters is

$$Z_i = \sqrt{Z_{11}^2 - \frac{Z_{11}Z_{13}^2}{Z_{33}}}$$

$$= \frac{1}{2}\sqrt{(Z_{0e} - Z_{0o})^2\csc^2\theta - (Z_{0e} + Z_{0o})^2\cot^2\theta}. \tag{8.101}$$

When the coupled line section is $\lambda/4$ long ($\theta = \pi/2$), the image impedance reduces to

$$Z_i = \frac{1}{2}(Z_{0e} - Z_{0o}), \tag{8.102}$$

which is real and positive since $Z_{0e} > Z_{0o}$. However, when $\theta \to 0$ or π, $Z_i \to \pm j\infty$, indicating a stopband. The real part of the image impedance is sketched in Figure 8.43, where the cutoff frequencies can be found from (8.101) as

$$\cos\theta_1 = -\cos\theta_2 = \frac{Z_{0e} - Z_{0o}}{Z_{0e} + Z_{0o}}.$$

The propagation constant can also be calculated from the results of Table 8.1 as

$$\cos\beta = \sqrt{\frac{Z_{11}Z_{33}}{Z_{13}^2}} = \frac{Z_{11}}{Z_{13}} = \frac{Z_{0e} + Z_{0o}}{Z_{0e} - Z_{0o}}\cos\theta, \tag{8.103}$$

which shows β is real for $\theta_1 < \theta < \theta_2 = \pi - \theta_1$, where $\cos\theta_1 = (Z_{0e} - Z_{0o})/(Z_{0e} + Z_{0o})$.

TABLE 8.9 | Ten Canonical Coupled Line Circuits

Circuit	Image Impedance	Response
	$Z_{i1} = \dfrac{2Z_{0e}Z_{0o}\cos\theta}{\sqrt{(Z_{0e}+Z_{0o})^2\cos^2\theta-(Z_{0e}-Z_{0o})^2}}$ $Z_{i2} = \dfrac{Z_{0e}Z_{0o}}{Z_{i1}}$	Low-pass
	$Z_{i1} = \dfrac{2Z_{0e}Z_{0o}\sin\theta}{\sqrt{(Z_{0e}-Z_{0o})^2-(Z_{0e}+Z_{0o})^2\cos^2\theta}}$	Bandpass
	$Z_{i1} = \dfrac{\sqrt{(Z_{0e}-Z_{0o})^2-(Z_{0e}+Z_{0o})^2\cos^2\theta}}{2\sin\theta}$	Bandpass
	$Z_{i1} = \dfrac{\sqrt{Z_{0e}Z_{0o}}\,\sqrt{(Z_{0e}-Z_{0o})^2-(Z_{0e}+Z_{0o})^2\cos^2\theta}}{(Z_{0e}+Z_{0o})\sin\theta}$ $Z_{i2} = \dfrac{Z_{0e}Z_{0o}}{Z_{i1}}$	Bandpass
	$Z_{i1} = \dfrac{Z_{0e}+Z_{0o}}{2}$	All pass
	$Z_{i1} = \dfrac{2Z_{0e}Z_{0o}}{Z_{0e}+Z_{0o}}$	All pass
	$Z_{i1} = \sqrt{Z_{0e}Z_{0o}}$	All pass
	$Z_{i1} = -j\,\dfrac{2Z_{0e}Z_{0o}}{Z_{0e}+Z_{0o}}\cot\theta$ $Z_{i2} = \dfrac{Z_{0e}Z_{0o}}{Z_{i1}}$	All stop
	$Z_{i1} = j\,\sqrt{Z_{0e}Z_{0o}}\,\tan\theta$	All stop
	$Z_{i1} = -j\,\sqrt{Z_{0e}Z_{0o}}\,\cot\theta$	All stop

Design of Coupled Line Bandpass Filters

Narrowband bandpass filters can be made with cascaded coupled line sections of the form shown in Figure 8.42c. To derive the design equations for filters of this type, we first show that a single coupled line section can be approximately modeled by the equivalent circuit shown in Figure 8.44. We will do this by calculating the image impedance and propagation constant of the equivalent circuit and showing that they are approximately equal to those of the coupled line section for $\theta = \pi/2$, which will correspond to the center frequency of the bandpass response.

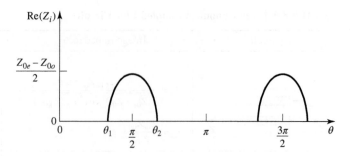

FIGURE 8.43 | The real part of the image impedance of the bandpass network of Figure 8.42c.

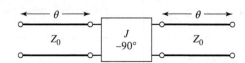

FIGURE 8.44 | Equivalent circuit of the coupled line section of Figure 8.42c.

The *ABCD* parameters of the equivalent circuit can be computed using the *ABCD* matrices for transmission lines from Table 4.1:

$$
\begin{bmatrix} A & B \\ C & D \end{bmatrix} = \begin{bmatrix} \cos\theta & jZ_0\sin\theta \\ \dfrac{j\sin\theta}{Z_0} & \cos\theta \end{bmatrix} \begin{bmatrix} 0 & -j/J \\ -jJ & 0 \end{bmatrix} \begin{bmatrix} \cos\theta & jZ_0\sin\theta \\ \dfrac{j\sin\theta}{Z_0} & \cos\theta \end{bmatrix}
$$

$$
= \begin{bmatrix} \left(JZ_0 + \dfrac{1}{JZ_0}\right)\sin\theta\cos\theta & j\left(JZ_0^2\sin^2\theta - \dfrac{\cos^2\theta}{J}\right) \\ j\left(\dfrac{1}{JZ_0^2}\sin^2\theta - J\cos^2\theta\right) & \left(JZ_0 + \dfrac{1}{JZ_0}\right)\sin\theta\cos\theta \end{bmatrix}. \tag{8.104}
$$

The *ABCD* parameters of the admittance inverter were obtained by considering it as a quarter-wave length of transmission of characteristic impedance, $1/J$. From (8.27) the image impedance of the equivalent circuit is

$$
Z_i = \sqrt{\frac{B}{C}} = \sqrt{\frac{JZ_0^2\sin^2\theta - (1/J)\cos^2\theta}{(1/JZ_0^2)\sin^2\theta - J\cos^2\theta}}, \tag{8.105}
$$

which reduces to the following value at the center frequency, $\theta = \pi/2$:

$$
Z_i = JZ_0^2. \tag{8.106}
$$

From (8.31) the propagation constant is

$$
\cos\beta = A = \left(JZ_0 + \frac{1}{JZ_0}\right)\sin\theta\cos\theta. \tag{8.107}
$$

Equating the image impedances in (8.102) and (8.106), and the propagation constants of (8.103) and (8.107), yields the following equations:

$$
\frac{1}{2}(Z_{0e} - Z_{0o}) = JZ_0^2,
$$

$$
\frac{Z_{0e} + Z_{0o}}{Z_{0e} - Z_{0o}} = JZ_0 + \frac{1}{JZ_0},
$$

where we have assumed $\sin\theta \simeq 1$ for θ near $\pi/2$. These equations can be solved for the even- and odd-mode line impedances to give

$$Z_{0e} = Z_0[1 + JZ_0 + (JZ_0)^2], \tag{8.108a}$$

$$Z_{0o} = Z_0[1 - JZ_0 + (JZ_0)^2]. \tag{8.108b}$$

Now consider a bandpass filter composed of a cascade of $N + 1$ coupled line sections, as shown in Figure 8.45a. The sections are numbered from left to right, with the load on the right, but the filter can be reversed without affecting the response. Since each coupled line section has an equivalent circuit of the form shown in Figure 8.44, the equivalent circuit of the cascade is as shown in Figure 8.45b. Between any two consecutive inverters we have a transmission line section that is effectively 2θ in length. This line is approximately $\lambda/2$ long in the vicinity of the bandpass region of the filter, and has an approximate equivalent circuit that consists of a shunt parallel LC resonator, as in Figure 8.45c.

The first step in establishing this equivalence is to find the parameters for the T-equivalent and ideal transformer circuit of Figure 8.45c (an exact equivalent). The $ABCD$ matrix for this circuit can be calculated using the results in Table 4.1 for a T-circuit and an ideal transformer:

$$\begin{bmatrix} A & B \\ C & D \end{bmatrix} = \begin{bmatrix} \dfrac{Z_{11}}{Z_{12}} & \dfrac{Z_{11}^2 - Z_{12}^2}{Z_{12}} \\ \dfrac{1}{Z_{12}} & \dfrac{Z_{11}}{Z_{12}} \end{bmatrix} \begin{bmatrix} -1 & 0 \\ 0 & -1 \end{bmatrix} = \begin{bmatrix} \dfrac{-Z_{11}}{Z_{12}} & \dfrac{Z_{12}^2 - Z_{11}^2}{Z_{12}} \\ \dfrac{-1}{Z_{12}} & \dfrac{-Z_{11}}{Z_{12}} \end{bmatrix}. \tag{8.109}$$

Equating this result to the $ABCD$ parameters for a transmission line of length 2θ and characteristic impedance Z_0 gives the parameters of the equivalent circuit as

$$Z_{12} = \frac{-1}{C} = \frac{jZ_0}{\sin 2\theta}, \tag{8.110a}$$

$$Z_{11} = Z_{22} = -Z_{12}A = -jZ_0 \cot 2\theta. \tag{8.110b}$$

Then the series arm impedance is

$$Z_{11} - Z_{12} = -jZ_0 \frac{\cos 2\theta + 1}{\sin 2\theta} = -jZ_0 \cot \theta. \tag{8.111}$$

The $1: -1$ transformer provides a $180°$ phase shift, which cannot be obtained with the T-network alone; since this does not affect the amplitude response of the filter, it can be discarded. For $\theta \sim \pi/2$ the series arm impedances of (8.111) are near zero and can also be ignored. The shunt impedance Z_{12}, however, looks like the impedance of a parallel resonant circuit for $\theta \sim \pi/2$. If we let $\omega = \omega_0 + \Delta\omega$, where $\theta = \pi/2$ at the center frequency ω_0, then we have $2\theta = \beta\ell = \omega\ell/v_p = (\omega_0 + \Delta\omega)\pi/\omega_0 = \pi(1 + \Delta\omega/\omega_0)$, so (8.110a) can be written for small $\Delta\omega$ as

$$Z_{12} = \frac{jZ_0}{\sin \pi(1 + \Delta\omega/\omega_0)} \simeq \frac{-jZ_0\omega_0}{\pi(\omega - \omega_0)}. \tag{8.112}$$

From Section 6.1 the impedance near resonance of a parallel LC circuit is

$$Z = \frac{-jL\omega_0^2}{2(\omega - \omega_0)}, \tag{8.113}$$

with $\omega_0^2 = 1/LC$. Equating this to (8.112) gives the equivalent inductor and capacitor values as

$$L = \frac{2Z_0}{\pi\omega_0}, \tag{8.114a}$$

$$C = \frac{1}{\omega_0^2 L} = \frac{\pi}{2Z_0\omega_0}. \tag{8.114b}$$

The end sections of the circuit of Figure 8.45b require a different treatment. The lines of length θ on either end of the filter are matched to Z_0 and so can be ignored. The end inverters, J_1 and J_{N+1}, can each

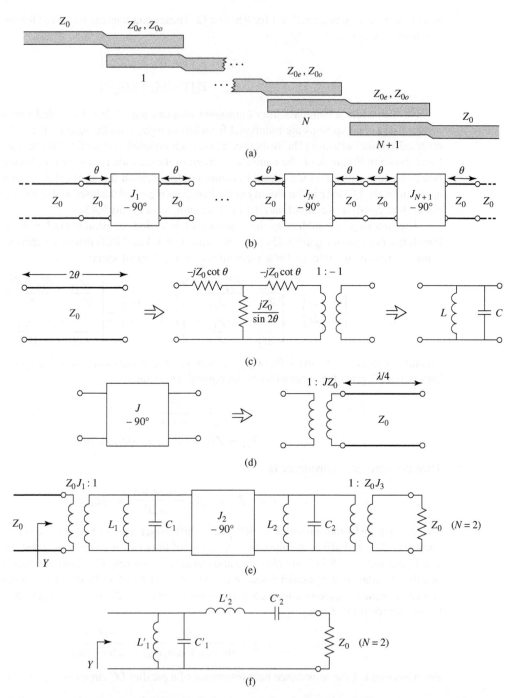

FIGURE 8.45 | Development of an equivalent circuit for derivation of design equations for a coupled line bandpass filter. (a) Layout of an $(N + 1)$-section coupled line bandpass filter. (b) Using the equivalent circuit of Figure 8.44 for each coupled line section. (c) Equivalent circuit for transmission lines of length 2θ. (d) Equivalent circuit of the admittance inverters. (e) Using results of (c) and (d) for the $N = 2$ case. (f) Lumped-element circuit for a bandpass filter for $N = 2$.

be represented as a transformer followed by a $\lambda/4$ section of line, as shown in Figure 8.45d. The *ABCD* matrix of a transformer with a turns ratio N in cascade with a quarter-wave line is

$$\begin{bmatrix} A & B \\ C & D \end{bmatrix} = \begin{bmatrix} \dfrac{1}{N} & 0 \\ 0 & N \end{bmatrix} \begin{bmatrix} 0 & -jZ_0 \\ \dfrac{-j}{Z_0} & 0 \end{bmatrix} = \begin{bmatrix} 0 & \dfrac{-jZ_0}{N} \\ \dfrac{-jN}{Z_0} & 0 \end{bmatrix}. \qquad (8.115)$$

Comparing this to the *ABCD* matrix of an admittance inverter [part of (8.104)] shows that the necessary turns ratio is $N = JZ_0$. The $\lambda/4$ line merely produces a phase shift and so can be ignored.

Using these results for the interior and end sections allows the circuit of Figure 8.45b to be transformed into the circuit of Figure 8.45e, which is specialized to the $N = 2$ case. We see that each pair of coupled line sections leads to an equivalent shunt *LC* resonator, and an admittance inverter occurs between each pair of *LC* resonators. Next, we show that the admittance inverters have the effect of transforming a shunt *LC* resonator into a series *LC* resonator, leading to the final equivalent circuit of Figure 8.45f (shown for $N = 2$). This will then allow the admittance inverter constants, J_n, to be determined from the element values of a low-pass prototype. We will demonstrate this for the $N = 2$ case.

With reference to Figure 8.45e, the admittance just to the right of the J_2 inverter is

$$j\omega C_2 + \frac{1}{j\omega L_2} + Z_0 J_3^2 = j\sqrt{\frac{C_2}{L_2}}\left(\frac{\omega}{\omega_0} - \frac{\omega_0}{\omega}\right) + Z_0 J_3^2,$$

since the transformer scales the load admittance by the square of the turns ratio. Then the admittance seen at the input of the filter is

$$
\begin{aligned}
Y &= \frac{1}{J_1^2 Z_0^2}\left\{ j\omega C_1 + \frac{1}{j\omega L_1} + \frac{J_2^2}{j\sqrt{C_2/L_2}\left[(\omega/\omega_0) - (\omega_0/\omega)\right] + Z_0 J_3^2} \right\} \\
&= \frac{1}{J_1^2 Z_0^2}\left\{ j\sqrt{\frac{C_1}{L_1}}\left(\frac{\omega}{\omega_0} - \frac{\omega_0}{\omega}\right) + \frac{J_2^2}{j\sqrt{C_2/L_2}\left[(\omega/\omega_0) - (\omega_0/\omega)\right] + Z_0 J_3^2} \right\}.
\end{aligned}
\tag{8.116}
$$

These results also use the fact, from (8.114), that $L_n C_n = 1/\omega_0^2$ for all *LC* resonators. Now the admittance seen looking into the circuit of Figure 8.45f is

$$
\begin{aligned}
Y &= j\omega C_1' + \frac{1}{j\omega L_1'} + \frac{1}{j\omega L_2' + 1/j\omega C_2' + Z_0} \\
&= j\sqrt{\frac{C_1'}{L_1'}}\left(\frac{\omega}{\omega_0} - \frac{\omega_0}{\omega}\right) + \frac{1}{j\sqrt{L_2'/C_2'}\left[(\omega/\omega_0) - (\omega_0/\omega)\right] + Z_0},
\end{aligned}
\tag{8.117}
$$

which is identical in form to (8.116). Thus, the two circuits will be equivalent if the following conditions are met:

$$\frac{1}{J_1^2 Z_0^2}\sqrt{\frac{C_1}{L_1}} = \sqrt{\frac{C_1'}{L_1'}}, \tag{8.118a}$$

$$\frac{J_1^2 Z_0^2}{J_2^2}\sqrt{\frac{C_2}{L_2}} = \sqrt{\frac{L_2'}{C_2'}}, \tag{8.118b}$$

$$\frac{J_1^2 Z_0^3 J_3^2}{J_2^2} = Z_0. \tag{8.118c}$$

We know L_n and C_n from (8.114); L_n' and C_n' are determined from the element values of a lumped-element low-pass prototype that has been impedance scaled and frequency transformed to a bandpass filter. Using the results in Table 8.6 and the impedance scaling formulas of (8.64) allows the L_n' and C_n' values to be written as

$$L_1' = \frac{\Delta Z_0}{\omega_0 g_1}, \tag{8.119a}$$

$$C_1' = \frac{g_1}{\Delta \omega_0 Z_0}, \tag{8.119b}$$

$$L_2' = \frac{g_2 Z_0}{\Delta \omega_0}, \tag{8.119c}$$

$$C_2' = \frac{\Delta}{\omega_0 g_2 Z_0}, \tag{8.119d}$$

where $\Delta = (\omega_2 - \omega_1)/\omega_0$ is the fractional bandwidth of the filter. Then (8.118) can be solved for the inverter constants with the following results (for $N = 2$):

$$J_1 Z_0 = \left(\frac{C_1 L_1'}{L_1 C_1'} \right)^{1/4} = \sqrt{\frac{\pi \Delta}{2g_1}}, \tag{8.120a}$$

$$J_2 Z_0 = J_1 Z_0^2 \left(\frac{C_2 C_2'}{L_2 L_2'} \right)^{1/4} = \frac{\pi \Delta}{2\sqrt{g_1 g_2}}, \tag{8.120b}$$

$$J_3 Z_0 = \frac{J_2}{J_1} = \sqrt{\frac{\pi \Delta}{2g_2}}. \tag{8.120c}$$

After the J_n are found, Z_{0e} and Z_{0o} for each coupled line section can be calculated from (8.108).

The above results were derived for the special case of $N = 2$ (three coupled line sections), but more general results can be derived for any number of sections, and for the case where $Z_L \neq Z_0$ (or $g_{N+1} \neq 1$, as in the case of an equal-ripple response with N even). Thus, the design equations for a bandpass filter with $N + 1$ coupled line sections are

$$Z_0 J_1 = \sqrt{\frac{\pi \Delta}{2g_1}}, \tag{8.121a}$$

$$Z_0 J_n = \frac{\pi \Delta}{2\sqrt{g_{n-1} g_n}} \quad \text{for } n = 2, 3, \dots, N, \tag{8.121b}$$

$$Z_0 J_{N+1} = \sqrt{\frac{\pi \Delta}{2g_N g_{N+1}}}. \tag{8.121c}$$

The even- and odd-mode characteristic impedances for each section are found from (8.108).

EXAMPLE 8.7   COUPLED LINE BANDPASS FILTER DESIGN

Design a coupled line bandpass filter with $N = 3$ and a 0.5 dB equal-ripple response. The center frequency is 2.4 GHz, the bandwidth is 10%, and $Z_0 = 75\ \Omega$. What is the attenuation at 2.1 GHz?

Solution
The fractional bandwidth is $\Delta = 0.1$. We can use Figure 8.27a to obtain the attenuation at 2.1 GHz, but first we must use (8.71) to convert this frequency to the normalized low-pass form ($\omega_c = 1$):

$$\omega \leftarrow \frac{1}{\Delta} \left(\frac{\omega}{\omega_0} - \frac{\omega_0}{\omega} \right) = \frac{1}{0.1} \left(\frac{2.1}{2.4} - \frac{2.4}{2.1} \right) = -2.67.$$

Then the value on the horizontal scale of Figure 8.27a is

$$\left| \frac{\omega}{\omega_c} \right| - 1 = |-2.67| - 1 = 1.67,$$

which indicates an attenuation of about 20 dB for $N = 3$.

The low-pass prototype values, g_n, are given in Table 8.4; then (8.121) can be used to calculate the admittance inverter constants, J_n. Finally, the even- and odd-mode characteristic impedances can be found from (8.108). These results are summarized in the following table:

n	g_n	$Z_0 J_n$	$Z_{0e}(\Omega)$	$Z_{0o}(\Omega)$
1	1.5963	0.3137	105.9	58.85
2	1.0967	0.1187	84.96	67.15
3	1.5963	0.1187	84.96	67.15
4	1.0000	0.3137	105.9	58.85

Note that the filter sections are symmetric about the midpoint. The calculated response of this filter is shown in Figure 8.46; passbands also occur at 7 GHz, 11.5 GHz, etc.

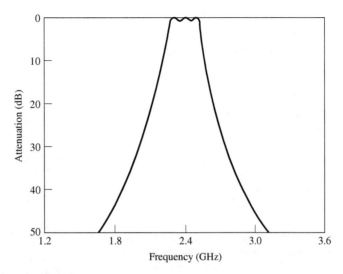

FIGURE 8.46 | Amplitude response of the coupled line bandpass filter of Example 8.7.

Many other types of filters can be constructed using coupled line sections; most of these are of the bandpass or bandstop variety. One particularly compact design is the interdigitated filter, which can be obtained from a coupled line filter by folding the lines at their midpoints; see references [1] and [3] for details.

REFERENCES

[1] G. L. Matthaei, L. Young, and E. M. T. Jones, *Microwave Filters, Impedance-Matching Networks, and Coupling Structures*, Artech House, Dedham, MA, 1980.

[2] R. E. Collin, *Foundations for Microwave Engineering*, 2nd edition, Wiley-IEEE Press, Hoboken, NJ, 2001.

[3] J. A. G. Malherbe, *Microwave Transmission Line Filters*, Artech House, Dedham, MA, 1979.

[4] W. A. Davis, *Microwave Semiconductor Circuit Design*, Van Nostrand Reinhold, New York, 1984.

[5] R. F. Harrington, *Time-Harmonic Electromagnetic Fields*, McGraw-Hill, New York, 1961.

[6] P. I. Richards, "Resistor-Transmission Line Circuits," *Proceedings of the IRE*, vol. 36, pp. 217–220, February 1948.

[7] S. B. Cohn, "Parallel-Coupled Transmission-Line-Resonator Filters," *IRE Transactions on Microwave Theory and Techniques*, vol. MTT-6, pp. 223–231, April 1958.

[8] E. M. T. Jones and J. T. Bolljahn, "Coupled-Strip-Transmission Line Filters and Directional Couplers," *IRE Transactions on Microwave Theory and Techniques*, vol. MTT-4, pp. 78–81, April 1956.

PROBLEMS

8.1 Sketch the k-β diagram for the infinite periodic structure shown below. Assume $Z_0 = 75\ \Omega, d = 1.0$ cm, $k = k_0$, and $L_0 = 1.25$ nH.

8.3 Compute the image impedances and propagation factor for the network shown below.

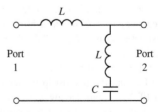

8.2 Verify the expression for the image impedance of a π-network given in Table 8.1.

8.4 Design a composite low-pass filter by the image parameter method with the following specifications: $R_0 = 75\ \Omega$,

$f_c = 60$ MHz, and $f_\infty = 64$ MHz. Use CAD to plot the insertion loss versus frequency.

8.5 Design a composite high-pass filter by the image parameter method with the following specifications: $R_0 = 50\ \Omega$, $f_c = 50$ MHz, and $f_\infty = 48$ MHz. Use CAD to plot the insertion loss versus frequency.

8.6 Solve the design equations in Section 8.3 for the elements of an $N = 2$ equal-ripple filter if the ripple specification is 1.2 dB.

8.7 Design a low-pass, maximally flat lumped-element filter having a passband of 0–2.5 GHz, and an attenuation of at least 15 dB at 4.4 GHz. The characteristic impedance is 50 Ω. Use CAD to plot the insertion loss versus frequency.

8.8 Design a high-pass lumped-element filter with a 3 dB equal-ripple response, a cutoff frequency of 4 GHz, and at least 25 dB insertion loss at 2 GHz. The characteristic impedance is 75 Ω. Use CAD to plot the insertion loss versus frequency.

8.9 Design a four-section bandpass lumped-element filter having a maximally flat group delay response. The bandwidth should be 5% with a center frequency of 2 GHz. The impedance is 50 Ω. Use CAD to plot the insertion loss versus frequency.

8.10 Design a three-section bandstop lumped-element filter with a 0.5 dB equal-ripple response, a bandwidth of 15% centered at 4 GHz, and an impedance of 50 Ω. What is the resulting attenuation at 4.1 GHz? Use CAD to plot the insertion loss versus frequency.

8.11 Verify the second Kuroda identity in Table 8.7 by calculating the *ABCD* matrices for both circuits.

8.12 Design a low-pass, third-order, maximally flat filter using only series stubs. The cutoff frequency is 5 GHz and the impedance is 75 Ω. Use CAD to plot the insertion loss versus frequency.

8.13 Design a low-pass, fourth-order, maximally flat filter using only shunt stubs. The cutoff frequency is 10 GHz and the impedance is 75 Ω. Use CAD to plot the insertion loss versus frequency.

8.14 Verify the operation of the admittance inverter of Figure 8.38c by calculating its *ABCD* matrix and comparing it to the *ABCD* matrix of the admittance inverter made from a quarter-wave line.

8.15 Show that the π-equivalent circuit for a short length of transmission line leads to equivalent circuits identical to those in Figures 8.39b and 8.39c for large and small characteristic impedance, respectively.

8.16 Design a stepped-impedance low-pass filter having a cutoff frequency of 5 GHz and a fifth-order 0.5 dB equal-ripple response. Assume $R_0 = 50\ \Omega$, $Z_\ell = 10\ \Omega$, and $Z_h = 150\ \Omega$. (a) Find the required electrical lengths of the five sections, and use CAD to plot the insertion loss from 0 to 9 GHz. (b) Lay out the microstrip implementation of the filter on an FR4 substrate having $\epsilon_r = 4.4$, $d = 0.079$ cm, and tan $\delta = 0.02$, and with copper conductors 2 mil thick. Use CAD

to plot the insertion loss versus frequency in the passband of the filter, and compare with the lossless case.

8.17 Design a stepped-impedance low-pass filter with $f_c = 3.5$ GHz and $R_0 = 60\ \Omega$, using the exact transmission line equivalent circuit of Figure 8.39a. Assume a maximally flat $N = 5$ response, and solve for the necessary line lengths and impedances if $Z_\ell = 15\ \Omega$ and $Z_h = 180\ \Omega$. Use CAD to plot the insertion loss versus frequency.

8.18 Design a four-section coupled line bandpass filter with a 0.5 dB equal ripple response. The center frequency is 2.45 GHz, the bandwidth is 10%, and the impedance is 50 Ω. (a) Find the required even- and odd-mode impedances of the coupled line sections, and calculate the expected attenuation at 2.1 GHz. Use CAD to plot the insertion loss from 1.55 to 3.35 GHz. (b) Lay out the microstrip implementation of the filter on an FR4 substrate having $\epsilon_r = 4.2$, $d = 0.158$ cm, and tan $\delta = 0.01$, and with copper conductors 0.5 mil thick. Use CAD to plot the insertion loss versus frequency in the passband of the filter, and compare with the lossless case.

8.19 The *Schiffman phase shifter*, shown below, can produce a 90° differential phase shift over a relatively broad frequency range. It consists of a coupled line section of length θ, and a transmission line section of length 3θ; $\theta = \pi/2$ at midband. The characteristic impedance of the transmission line is $Z_0 = \sqrt{Z_{0e} Z_{0o}}$, where Z_{0e} and Z_{0o} are the even- and odd-mode impedances of the coupled line section. Use the analysis of Section 8.7 to find the phase shift through the coupled line section, and then find the differential phase shift between the two outputs. Plot the differential phase shift for $\theta = 0$ to π for $Z_{0e}/Z_{0o} = 2.7$ and determine the bandwidth for which the phase shift is 90° $\pm 2.5°$.

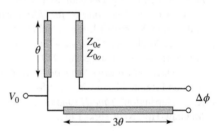

8.20 Design a four-section bandpass filter having maximally flat group delay response. The bandwidth should be 5% with a center frequency of 2 GHz. The impedance is 50 Ω. Consider $g_0 = \omega_c = g_5 = 1$, $g_1 = 0.7654 = g_4$, $g_2 = 1.8478 = g_3$.

8.21 Design a low-pass maximally flat lumped-element filter having a 3 dB cutoff frequency of 3 GHz with an attenuation of 20 dB at 5 GHz. The characteristic impedance is 75 Ω.

8.22 Design a five-section high-pass filter with 3 dB equiripple response, a cutoff frequency of 1 GHz, and an impedance of 50 Ω. The first element of the filter should start with a capacitor. Consider $g_1 = g_3 = 3.4817$, $g_2 = g_4 = 0.7618$, $g_3 = 4.5381$.

8.1 What kind of circuit is a filter?

(a) frequency-damping circuit
(b) amplitude-selective circuit
(c) frequency-selective circuit
(d) amplitude-damping circuit

8.2 What type of filter is created by using capacitors and resistors or inductors and capacitors?

(a) differential filters
(b) passive filters
(c) continuous filters
(d) active filters

8.3 Which of the following equations defines a characteristic impedance at the unit cell terminal?

(a) $Z_B = Z_0 \dfrac{V_{n+1}}{I_{n+1}}$

(c) $Z_B = Z_0 \dfrac{V_{n-1}}{I_{n-1}}$

(b) $Z_0 = Z_B \dfrac{V_{n+1}}{I_{n+1}}$

(d) $Z_0 = Z_B \dfrac{V_{n-1}}{I_{n-1}}$

8.4 Consider a periodic capacitively loaded line with $Z_0 = 50\ \Omega$, $f = 3.0$ GHz, $K_0^* d = 0.65$ rad, $d = 1.0$ cm, and $\beta = 160$ rad/m. Determine the phase velocity. (Note: c is velocity of light in m/sec.)

(a) 0.41c
(b) 0.51c
(c) 0.61c
(d) 0.81c

8.5 Determine the value of L in design of a low-pass composite filter with a cutoff frequency of 3 MHz and an impedance of 75 Ω.

(a) 11.94 µH
(b) 7.96 µH
(c) 2.21 nH
(d) 1.42 nH

8.6 Determine the value of C in design of a low-pass composite filter with a cutoff frequency of 3 MHz and an impedance of 75 Ω.

(a) 11.94 µF
(b) 7.96 µF
(c) 2.21 nF
(d) 1.42 nF

8.7 A filter response in the insertion loss method defined by its insertion loss, or power loss ratio, is given by

(a) $\dfrac{1}{1 + \Gamma(\omega)}$

(c) $\dfrac{1}{1 - |\Gamma(\omega)|^2}$

(b) $\dfrac{1}{1 - \Gamma(\omega)}$

(d) $\dfrac{1}{1 + |\Gamma(\omega)|^2}$

8.8 $|\Gamma(\omega)|^2$ is an even function of ω. Therefore, it can be expressed as a polynomial in ω^2 if M and N are real polynomials in ω^2 as

(a) $|\Gamma(\omega)|^2 = \dfrac{M(\omega^2)}{M(\omega^2) + N(\omega^2)}$

(b) $|\Gamma(\omega)|^2 = \dfrac{M(\omega^2)}{M(\omega^2) - N(\omega^2)}$

(c) $|\Gamma(\omega)|^2 = \dfrac{N(\omega^2)}{M(\omega^2) + N(\omega^2)}$

(d) $|\Gamma(\omega)|^2 = \dfrac{N(\omega^2)}{M(\omega^2) - N(\omega^2)}$

8.9 If ω_1 and ω_2 denote the edges of the passband, then for a low-pass prototype filter design that can also be transformed to have the bandpass or bandstop responses, the value of fractional bandwidth of the passband Δ is:

(a) $\Delta = \dfrac{\omega_1 + \omega_2}{\omega_0}$

(c) $\Delta = \dfrac{\omega_2 - \omega_1}{\omega_0}$

(b) $\Delta = \dfrac{\omega_1 - \omega_2}{\omega_0}$

(d) $\Delta = \dfrac{\omega_2 + \omega_1}{\omega_0}$

8.10 Which of the following definitions is Richards' transformation to synthesize an LC network using open- and short-circuited transmission line stubs?

(a) $\Omega = 1 + j\tan(\beta \ell) = 1 + j\tan\left(\dfrac{\omega \ell}{v_p}\right)$

(b) $\Omega = \tan(\beta \ell) = \tan\left(\dfrac{\omega \ell}{v_p}\right)$

(c) $\Omega = 1 - j\tan(\beta \ell) = 1 - j\tan\left(\dfrac{\omega \ell}{v_p}\right)$

(d) $\Omega = -\tan(\beta \ell) = -\tan\left(\dfrac{\omega \ell}{v_p}\right)$

8.11 By performing which of the following operations, the four Kuroda identities use redundant transmission line sections to achieve a more practical microwave filter implementation

(a) physically separate transmission line stubs
(b) transform series stubs into shunt stubs, or vice versa
(c) change impractical characteristic impedances into more realizable values
(d) all of the above

8.12 The way to implement low-pass filters in stripline or microstrip is to use alternating sections of very high and very low characteristic impedance lines. Such filters are usually referred to as

(a) stepped-impedance filters
(b) hi-Z filters
(c) low-Z filters
(d) all of the above

8.13 For a stepped-impedance low-pass filter, the values of $\beta \ell$ for inductor and capacitor are given as

(a) $\beta \ell = \dfrac{LR_0}{Z_h}$ and $\beta \ell = \dfrac{CZ_\ell}{R_0}$

(b) $\beta \ell = \dfrac{LR_0}{Z_h}$ and $\beta \ell = \dfrac{Z_\ell}{CR_0}$

(c) $\beta \ell = \dfrac{R_0}{LZ_h}$ and $\beta \ell = \dfrac{CZ_\ell}{R_0}$

(d) $\beta \ell = \dfrac{R_0}{LZ_h}$ and $\beta \ell = \dfrac{CR_0}{Z_\ell}$

8.14 To design a coupled line bandpass filter with $N = 3$ and a 0.5 dB equal-ripple response, the center frequency is

2.0 GHz, the bandwidth is 10%, and $Z_0 = 50$ Ω. At 1.8 GHz convert frequency to the normalized low-pass form ($\omega_c = 1$)

(a) −1.11   (c) 2.11

(b) −2.11   (d) 1.11

8.15 For a bandpass filter using capacitively coupled shunt resonator, the admittance inverter constant is

(a) $Z_0 J_{N,N+1} = \sqrt{\dfrac{\pi \Delta}{4 g_N g_{N+1}}}$

(b) $Z_0 J_{N,N+1} = \dfrac{1}{4} \sqrt{\dfrac{\pi \Delta}{g_N g_{N+1}}}$

(c) $Z_0 J_{N,N-1} = \sqrt{\dfrac{\pi \Delta}{4 g_N g_{N+1}}}$

(d) $Z_0 J_{N,N-1} = \dfrac{1}{4} \sqrt{\dfrac{\pi \Delta}{g_N g_{N+1}}}$

8.16 What is the value of inductor for a coupled line bandpass filter design at resonance at 2.4 GHz if $C = 1.89$ pF?

(a) 2.3 nH   (c) 5.3 nH

(b) 4.6 nH   (d) 1.8 nH

8.17 What is the value of inductor for a coupled line bandpass filter design at resonance at 2.4 GHz if the characteristic impedance given is $Z_0 = 50$ Ω?

(a) 5.1 nH   (c) 3.1 nH

(b) 4.1 nH   (d) 2.1 nH

8.18 What is the value of capacitor for a coupled line bandpass filter design at resonance at 2.4 GHz if $L = 2.2$ nH and $Z_0 = 50$ Ω?

(a) 2.1 pF   (c) 4.1 pF

(b) 3.1 pF   (d) 5.1 pF

8.19 Which of the following filters performs exactly opposite to a bandpass filter?

(a) band-stop filter   (c) band-elimination filter

(b) band-reject filter   (d) all of the above

ANSWER KEY

8.1 (c)	**8.5** (b)	**8.9** (c)	**8.13** (a)	**8.17** (d)
8.2 (b)	**8.6** (d)	**8.10** (b)	**8.14** (b)	**8.18** (a)
8.3 (a)	**8.7** (c)	**8.11** (d)	**8.15** (a)	**8.19** (d)
8.4 (a)	**8.8** (a)	**8.12** (d)	**8.16** (a)	

Theory and Design of Ferrimagnetic Components

Learning Objectives

After completing this chapter, you will be able to

- Understand the basic properties of ferrimagnetic materials
- Learn about plane wave propagation in a ferrite medium

- Understand the concepts of different modes of propagation in a ferrite-loaded rectangular waveguide
- Design ferrite isolators, ferrite phase shifters, and ferrite circulators

The components and networks discussed up to this point have all been reciprocal. That is, the response between any two ports i and j of a component did not depend on the direction of signal flow (thus, $S_{ij} = S_{ji}$). This will always be the case when the component is passive and consists of only isotropic materials, but if the component contains either active devices or anisotropic material, nonreciprocal behavior can be obtained. In some cases nonreciprocity is a useful property (e.g., circulators, isolators), while in other cases nonreciprocity is an ancillary property (e.g., transistor amplifiers, ferrite phase shifters).

In Chapter 1 we discussed materials having electric anisotropy (tensor permittivity) and magnetic anisotropy (tensor permeability). Some of the most practical anisotropic materials for microwave applications are *ferrimagnetic compounds*, also known as *ferrites*, such as yttrium iron garnet (YIG) and materials composed of iron oxides and various other elements such as aluminum, cobalt, manganese, and nickel. In contrast to ferromagnetic materials (e.g., iron, steel), ferrimagnetic compounds have high resistivity and a significant amount of anisotropy at microwave frequencies. As we will see, the magnetic anisotropy of a ferrimagnetic material is actually induced by applying a DC magnetic bias field. This field aligns the magnetic dipoles in the ferrite material to produce a net (nonzero) magnetic dipole moment, and causes the magnetic dipoles to precess at a frequency controlled by the strength of the bias field. A microwave signal circularly polarized in the same direction as this precession will interact strongly with the dipole moments, while an oppositely polarized field will interact less strongly. Since, for a given direction of rotation, the sense of polarization changes with the direction of propagation, a microwave signal will propagate through a magnetically biased ferrite differently in different directions. This effect can be utilized to fabricate directional devices such as isolators, circulators, and gyrators. Another useful characteristic of ferrimagnetic materials is that the interaction with an applied microwave signal can be controlled by adjusting the strength of the bias field. This effect leads to a variety of control devices such as phase shifters, switches, and tunable resonators and filters.

It is interesting to compare ferrimagnetic materials to *paraelectric* materials, which are almost the dual of ferrimagnetic materials. Certain ceramic compounds, such as lithium niobate and barium titanate, have the property that their dielectric permittivity can be controlled with the application of a DC bias electric field. Paraelectric materials can therefore be used for variable phase shifters and other control components. Unlike ferrimagnetic materials, paraelectric materials are isotropic, and therefore paraelectric devices are reciprocal. Paraelectric materials typically have very high dielectric constants and loss tangents when used in bulk form, so modern applications generally use thin films of paraelectric material layered on a substrate. An important advantage of paraelectric devices over ferrite devices is that the need for a large and heavy magnet or biasing coil is eliminated.

We will begin by considering the microscopic behavior of a ferrimagnetic material and its interaction with a microwave signal to derive the permeability tensor. This macroscopic description of the material can then be used with Maxwell's equations to analyze wave propagation in an infinite ferrite medium and in a ferrite-loaded waveguide. These canonical problems will illustrate the nonreciprocal propagation properties of ferrimagnetic materials, including Faraday rotation and birefringence effects, and will be used in later sections when we discuss the operation and design of waveguide phase shifters and isolators.

9.1   BASIC PROPERTIES OF FERRIMAGNETIC MATERIALS

In this section we will show how the permeability tensor for a ferrimagnetic material can be deduced from a relatively simple microscopic view of the atom. We will also discuss how loss affects the permeability tensor and the demagnetization field inside a finite-sized piece of ferrite.

The Permeability Tensor

The magnetic properties of a material are due to the existence of magnetic dipole moments, which arise primarily from electron spin. From quantum mechanical considerations [1], the magnetic dipole moment of an electron due to its spin is given by

$$m = \frac{q\hbar}{2m_e} = 9.27 \times 10^{-24} \text{ A-m}^2, \tag{9.1}$$

where $\hbar$ is Planck's constant divided by 2π, q is the electron charge, and m_e is the mass of the electron. An electron in orbit around a nucleus gives rise to an effective current loop and thus an additional magnetic moment, but this effect is generally insignificant compared to the magnetic moment due to spin. The *Landé g factor* is a measure of the relative contributions of the orbital moment and the spin moment to the total magnetic moment; $g = 1$ when the moment is due only to orbital motion, and $g = 2$ when the moment is due only to spin. For most microwave ferrite materials, g is in the range 1.98–2.01, so $g = 2$ is a good approximation.

In most solids, electron spins occur in pairs with opposite signs, so the overall magnetic moment is negligible. In a magnetic material, however, a large fraction of the electron spins are unpaired (more left-hand spins than right-hand spins, or vice versa), but are generally oriented in random directions so that the net magnetic moment is still small. An external magnetic field, however, can cause the dipole moments to align in the same direction to produce a large overall magnetic moment. The existence of exchange forces can keep adjacent electron spins aligned after the external field is removed; the material is then said to be permanently magnetized.

An electron has a spin angular momentum given in terms of Planck's constant as [1, 2]

$$s = \frac{\hbar}{2}. \tag{9.2}$$

The vector direction of this momentum is opposite the direction of the spin magnetic dipole moment, as indicated in Figure 9.1. The ratio of the spin magnetic moment to the spin angular momentum is a constant called the *gyromagnetic ratio*:

$$\gamma = \frac{m}{s} = \frac{q}{m_e} = 1.759 \times 10^{11} \text{ C/kg}, \tag{9.3}$$

where (9.1) and (9.2) have been used. Then we can write the following vector relation between the magnetic moment and the angular momentum:

$$\bar{m} = -\gamma \bar{s}, \tag{9.4}$$

where the negative sign is due to the fact that these vectors are oppositely directed.

When a magnetic bias field $\bar{H}_0 = \hat{z}H_0$ is present, a torque will be exerted on the magnetic dipole:

$$\bar{T} = \bar{m} \times \bar{B}_0 = \mu_0 \bar{m} \times \bar{H}_0 = -\mu_0 \gamma \bar{s} \times \bar{H}_0. \tag{9.5}$$

Since torque is equal to the time rate of change of angular momentum, we have

$$\frac{d\bar{s}}{dt} = \frac{-1}{\gamma} \frac{d\bar{m}}{dt} = \bar{T} = \mu_0 \bar{m} \times \bar{H}_0,$$

or

$$\frac{d\bar{m}}{dt} = -\mu_0 \gamma \bar{m} \times \bar{H}_0. \tag{9.6}$$

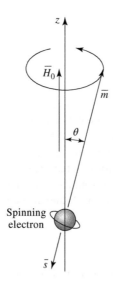

FIGURE 9.1 | Spin magnetic dipole moment and angular momentum vectors for a spinning electron.

This is the equation of motion for the magnetic dipole moment, $\bar{m}$. We will solve this equation to show that the magnetic dipole precesses around the H_0-field vector, similar to a spinning top precessing around a vertical axis.

Writing (9.6) in terms of its three vector components gives

$$\frac{dm_x}{dt} = -\mu_0 \gamma m_y H_0, \tag{9.7a}$$

$$\frac{dm_y}{dt} = \mu_0 \gamma m_x H_0, \tag{9.7b}$$

$$\frac{dm_z}{dt} = 0. \tag{9.7c}$$

Now use (9.7a) and (9.7b) to obtain two equations for m_x and m_y:

$$\frac{d^2 m_x}{dt^2} + \omega_0^2 m_x = 0, \tag{9.8a}$$

$$\frac{d^2 m_y}{dt^2} + \omega_0^2 m_y = 0, \tag{9.8b}$$

where

$$\omega_0 = \mu_0 \gamma H_0 \tag{9.9}$$

is called the *Larmor*, or *precession*, frequency. One solution to (9.8) that is compatible with (9.7a) and (9.7b) is given by

$$m_x = A \cos \omega_0 t, \tag{9.10a}$$

$$m_y = A \sin \omega_0 t. \tag{9.10b}$$

Equation (9.7c) shows that m_z is a constant, and (9.1) shows that the magnitude of $\bar{m}$ is also a constant, so we have the relation that

$$|\bar{m}|^2 = \left(\frac{q\hbar}{2m_e}\right)^2 = m_x^2 + m_y^2 + m_z^2 = A^2 + m_z^2. \tag{9.11}$$

Thus the precession angle, θ, between $\bar{m}$ and $\bar{H}_0$ (the z-axis) is given by

$$\sin\theta = \frac{\sqrt{m_x^2 + m_y^2}}{|\bar{m}|} = \frac{A}{|\bar{m}|}. \tag{9.12}$$

The projection of $\bar{m}$ on the xy plane is given by (9.10), which shows that $\bar{m}$ traces a circular path in this plane. The position of this projection at time t is given by $\phi = \omega_0 t$, so the angular rate of rotation is $d\phi/dt = \omega_0$, the precession frequency. In the absence of any damping forces, the actual precession angle will be determined by the initial position of the magnetic dipole, and the dipole will precess about $\bar{H}_0$ at this angle indefinitely (free precession). In reality, however, the existence of damping forces will cause the magnetic dipole moment to spiral in from its initial angle until $\bar{m}$ is aligned with $\bar{H}_0$ ($\theta = 0$).

Now assume that there are N unbalanced electron spins (magnetic dipoles) per unit volume, so that the total magnetization is

$$\bar{M} = N\bar{m}, \tag{9.13}$$

and the equation of motion in (9.6) becomes

$$\frac{d\bar{M}}{dt} = -\mu_0 \gamma \bar{M} \times \bar{H}, \tag{9.14}$$

where $\bar{H}$ is the internal applied field. (Note: In Chapter 1 we used $\bar{P}_m$ for magnetization and $\bar{M}$ for magnetic currents; here we use $\bar{M}$ for magnetization, as this is common practice in ferrimagnetics work. Since we will not be using magnetic currents in this chapter, there should be no confusion.) As the strength of the bias field H_0 is increased, more magnetic dipole moments will align with H_0 until all are aligned, and $\bar{M}$ reaches an upper limit. See Figure 9.2. The material is then said to be *magnetically saturated*, and M_s is denoted as the *saturation magnetization*. M_s is thus a physical property of the ferrite material, and it typically ranges from $4\pi M_s = 300$–5000 G. (Appendix I lists the saturation magnetization and other physical properties of several types of microwave ferrite materials.) Below saturation, ferrite materials can be very lossy at microwave frequencies, and the RF interaction is reduced. For this reason ferrites are usually operated in the saturated state, and this assumption is made for the remainder of this chapter.

The saturation magnetization of a material is a strong function of temperature, decreasing as temperature increases. This effect can be understood by noting that the vibrational energy of an atom increases with temperature, making it more difficult to align all the magnetic dipoles. At a high enough temperature the thermal energy is greater than the energy supplied by the internal magnetic field, and a zero net magnetization results. This temperature is called the *Curie temperature*, T_C.

Now consider the interaction of a small AC (microwave) magnetic field with a magnetically saturated ferrite material. Such a field will cause a forced precession of the dipole moments around the $\bar{H}_0(\hat{z})$ axis at the frequency of the applied AC field, much like the operation of an AC synchronous motor. The small-signal approximation will apply to all the ferrite components of interest to us, but there are applications where high-power signals can be used to obtain useful nonlinear effects.

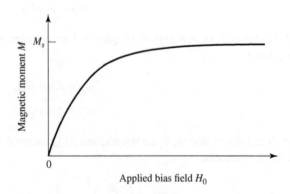

FIGURE 9.2 | Magnetic moment of a ferrimagnetic material versus bias field, H_0.

If $\bar{H}$ is the applied AC field, the total magnetic field is

$$\bar{H}_t = H_0\hat{z} + \bar{H}, \tag{9.15}$$

where we assume that $|\bar{H}| \ll H_0$. This field produces a total magnetization in the ferrite material given by

$$\bar{M}_t = M_s\hat{z} + \bar{M}, \tag{9.16}$$

where M_s is the (DC) saturation magnetization and $\bar{M}$ is the additional (AC) magnetization (in the xy plane) caused by $\bar{H}$. Substituting (9.16) and (9.15) into (9.14) gives the following component equations of motion:

$$\frac{dM_x}{dt} = -\mu_0\gamma M_y(H_0 + H_z) + \mu_0\gamma(M_s + M_z)H_y, \tag{9.17a}$$

$$\frac{dM_y}{dt} = \mu_0\gamma M_x(H_0 + H_z) - \mu_0\gamma(M_s + M_z)H_x, \tag{9.17b}$$

$$\frac{dM_z}{dt} = -\mu_0\gamma M_x H_y + \mu_0\gamma M_y H_x, \tag{9.17c}$$

since $dM_s/dt = 0$. Since $|\bar{H}| \ll H_0$, we have $|\bar{M}||\bar{H}| \ll |\bar{M}|H_0$ and $|\bar{M}||\bar{H}| \ll M_s|\bar{H}|$, so we can ignore MH products. Then (9.17) reduces to

$$\frac{dM_x}{dt} = -\omega_0 M_y + \omega_m H_y, \tag{9.18a}$$

$$\frac{dM_y}{dt} = \omega_0 M_x - \omega_m H_x, \tag{9.18b}$$

$$\frac{dM_z}{dt} = 0, \tag{9.18c}$$

where $\omega_0 = \mu_0\gamma H_0$ and $\omega_m = \mu_0\gamma M_s$. Solving (9.18a) and (9.18b) for M_x and M_y gives the following equations:

$$\frac{d^2M_x}{dt^2} + \omega_0^2 M_x = \omega_m\frac{dH_y}{dt} + \omega_0\omega_m H_x, \tag{9.19a}$$

$$\frac{d^2M_y}{dt^2} + \omega_0^2 M_y = -\omega_m\frac{dH_x}{dt} + \omega_0\omega_m H_y. \tag{9.19b}$$

These are the equations of motion for the forced precession of the magnetic dipoles, assuming small-signal conditions. It is now an easy step to arrive at the permeability tensor for ferrites; after doing this, we will try to gain some physical insight into the magnetic interaction process by considering circularly polarized AC fields.

If the AC $\bar{H}$ field has an $e^{j\omega t}$ time-harmonic dependence, the AC steady-state form of (9.19) reduces to the following phasor equations:

$$(\omega_0^2 - \omega^2)M_x = \omega_0\omega_m H_x + j\omega\omega_m H_y, \tag{9.20a}$$

$$(\omega_0^2 - \omega^2)M_y = -j\omega\omega_m H_x + \omega_0\omega_m H_y, \tag{9.20b}$$

which shows the linear relationship between $\bar{H}$ and $\bar{M}$. As in (1.24), (9.20) can be written with a tensor susceptibility, $[\chi]$, to relate $\bar{H}$ and $\bar{M}$:

$$\bar{M} = [\chi]\bar{H} = \begin{bmatrix} \chi_{xx} & \chi_{xy} & 0 \\ \chi_{yx} & \chi_{yy} & 0 \\ 0 & 0 & 0 \end{bmatrix}\bar{H}, \tag{9.21}$$

where the elements of $[\chi]$ are given by

$$\chi_{xx} = \chi_{yy} = \frac{\omega_0\omega_m}{\omega_0^2 - \omega^2}, \tag{9.22a}$$

$$\chi_{xy} = -\chi_{yx} = \frac{j\omega\omega_m}{\omega_0^2 - \omega^2}. \tag{9.22b}$$

The $\hat{z}$ component of $\bar{H}$ does not affect the magnetic moment of the material under the above assumptions.

To relate $\bar{B}$ and $\bar{H}$, we have from (1.23) that

$$\bar{B} = \mu_0(\bar{M} + \bar{H}) = [\mu]\bar{H}, \tag{9.23}$$

where the tensor permeability $[\mu]$ is given by

$$[\mu] = \mu_0([U] + [\chi]) = \begin{bmatrix} \mu & j\kappa & 0 \\ -j\kappa & \mu & 0 \\ 0 & 0 & \mu_0 \end{bmatrix} \quad (\hat{z} \text{ bias}). \tag{9.24}$$

The elements of the permeability tensor are then

$$\mu = \mu_0(1 + \chi_{xx}) = \mu_0(1 + \chi_{yy}) = \mu_0\left(1 + \frac{\omega_0\omega_m}{\omega_0^2 - \omega^2}\right), \tag{9.25a}$$

$$\kappa = -j\mu_0\chi_{xy} = j\mu_0\chi_{yx} = \mu_0\frac{\omega\omega_m}{\omega_0^2 - \omega^2}. \tag{9.25b}$$

A material having a permeability tensor of this form is called *gyrotropic*; note that an $\hat{x}$ (or $\hat{y}$) component of $\bar{H}$ gives rise to both $\hat{x}$ and $\hat{y}$ components of $\bar{B}$, with a 90° phase shift between them.

If the direction of bias is reversed, both H_0 and M_s will change signs, so ω_0 and ω_m will change signs. Equation (9.25) then shows that μ will be unchanged, but κ will change sign. If the bias field is suddenly removed ($H_0 = 0$), the ferrite will generally remain magnetized ($0 < |M| < M_s$); only by demagnetizing the ferrite (e.g., with a decreasing AC bias field) can $M = 0$ be obtained. Since the results of (9.22) and (9.25) assume a saturated ferrite sample, both M_s and H_0 should be set to zero for the unbiased, demagnetized case. Then $\omega_0 = \omega_m = 0$ and (9.25) show that $\mu = \mu_0$ and $\kappa = 0$, as expected for a nonmagnetic material.

The tensor results of (9.24) assume magnetic bias in the $\hat{z}$ direction. If the ferrite is biased in a different direction, the permeability tensor will be transformed according to the change in coordinates. Thus, if $\bar{H}_0 = \hat{x}H_0$, the permeability tensor will be

$$[\mu] = \begin{bmatrix} \mu_0 & 0 & 0 \\ 0 & \mu & j\kappa \\ 0 & -j\kappa & \mu \end{bmatrix} \quad (\hat{x} \text{ bias}), \tag{9.26}$$

while if $\bar{H}_0 = \hat{y}H_0$ the permeability tensor will be

$$[\mu] = \begin{bmatrix} \mu & 0 & -j\kappa \\ 0 & \mu_0 & 0 \\ j\kappa & 0 & \mu \end{bmatrix} \quad (\hat{y} \text{ bias}). \tag{9.27}$$

A comment must be made about units. By tradition most practical work in magnetics is done with CGS units, with magnetization measured in gauss (1 gauss [G] = 10^{-4} weber/m^2), and field strength measured in oersteds ($4\pi \times 10^{-3}$ oersted [Oe] = 1 A/m). Thus, $\mu_0 = 1$ G/Oe in CGS units, implying that B and H have the same numerical values in a nonmagnetic material. Saturation magnetization is usually expressed as $4\pi M_s$ gauss; the corresponding MKS value is then $\mu_0 M_s$ weber/m^2 = 10^{-4} ($4\pi M_s$ gauss). In CGS units, the Larmor frequency can be expressed as $f_0 = \omega_0/2\pi = \mu_0\gamma H_0/2\pi = (2.8 \text{ MHz/Oe}) \times (H_0 \text{ oersted})$, and $f_m = \omega_m/2\pi = \mu_0\gamma M_s/2\pi = (2.8 \text{ MHz/Oe}) \times (4\pi M_s \text{ gauss})$. In practice, these units are convenient and easy to use.

Circularly Polarized Fields

To get a better physical understanding of the interaction of an AC signal with a saturated ferrimagnetic material we will consider circularly polarized fields. As discussed in Section 1.5, a right-hand circularly polarized (RHCP) field can be expressed in phasor form as

$$\bar{H}^+ = H^+(\hat{x} - j\hat{y}), \tag{9.28a}$$

and in time domain form as

$$\bar{H}^+ = \text{Re}\{\bar{H}^+ e^{j\omega t}\} = H^+(\hat{x}\cos\omega t + \hat{y}\sin\omega t), \tag{9.28b}$$

where we have assumed the amplitude H^+ as real. This latter form shows that $\bar{\mathcal{H}}^+$ is a vector that rotates with time, such that at time t it is oriented at the angle ωt from the x-axis; thus its angular velocity is ω. (Also note that $|\bar{\mathcal{H}}^+| = H^+ \neq |\bar{H}^+|$.) Applying the RHCP field of (9.28a) to (9.20) gives the magnetization components as

$$M_x^+ = \frac{\omega_m}{\omega_0 - \omega} H^+,$$

$$M_y^+ = \frac{-j\omega_m}{\omega_0 - \omega} H^+,$$

so the magnetization vector resulting from $\bar{H}^+$ can be written as

$$\bar{M}^+ = M_x^+ \hat{x} + M_y^+ \hat{y} = \frac{\omega_m}{\omega_0 - \omega} H^+ (\hat{x} - j\hat{y}), \tag{9.29}$$

which shows that the magnetization is also RHCP, and so it rotates with angular velocity ω in synchronism with the driving field, $\bar{H}^+$. Since $\bar{M}^+$ and $\bar{H}^+$ are vectors in the same direction, we can write $\bar{B}^+ = \mu_0(\bar{M}^+ + \bar{H}^+) = \mu^+ \bar{H}^+$, where μ^+ is the effective permeability for an RHCP wave given by

$$\mu^+ = \mu_0 \left(1 + \frac{\omega_m}{\omega_0 - \omega} \right) = \mu + \kappa. \tag{9.30}$$

The angle, θ_M, between M^+ and the z-axis is given by

$$\tan \theta_M = \frac{|\mathcal{M}^+|}{M_s} = \frac{\omega_m H^+}{(\omega_0 - \omega) M_s} = \frac{\omega_0 H^+}{(\omega_0 - \omega) H_0}, \tag{9.31}$$

while the angle, θ_H, between $\bar{H}^+$ and the z-axis, is given by

$$\tan \theta_H = \frac{|\mathcal{H}^+|}{H_0} = \frac{H^+}{H_0}. \tag{9.32}$$

For frequencies such that $\omega < 2\omega_0$, (9.31) and (9.32) show that $\theta_M > \theta_H$, as illustrated in Figure 9.3a. In this case the magnetic dipole is precessing in the same direction as it would freely precess in the absence of $\bar{H}^+$.

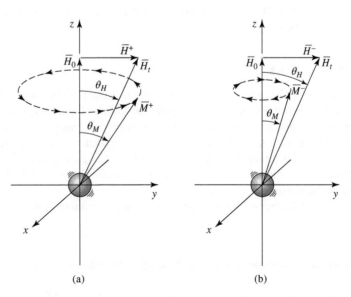

(a)   (b)

FIGURE 9.3 | Forced precession of a magnetic dipole with circularly polarized fields. (a) RHCP, $\theta_M > \theta_H$. (b) LHCP, $\theta_M < \theta_H$.

Now consider a left-hand circularly polarized field (LHCP), expressed in phasor form as

$$\bar{H}^- = H^-(\hat{x} + j\hat{y}), \tag{9.33a}$$

and in time domain form as

$$\bar{H}^- = \text{Re}\{\bar{H}^- e^{j\omega t}\} = H^-(\hat{x}\cos\omega t - \hat{y}\sin\omega t). \tag{9.33b}$$

Equation (9.33b) shows that $\bar{H}^-$ is a vector rotating in the $-\omega$ (left-hand) direction. Applying the LHCP field of (9.33a) to (9.20) gives the magnetization components as

$$M_x^- = \frac{\omega_m}{\omega_0 + \omega} H^-,$$

$$M_y^- = \frac{j\omega_m}{\omega_0 + \omega} H^-,$$

so the vector magnetization can be written as

$$\bar{M}^- = M_x^- \hat{x} + M_y^- \hat{y} = \frac{\omega_m}{\omega_0 + \omega} H^-(\hat{x} + j\hat{y}), \tag{9.34}$$

which shows that the magnetization is LHCP, rotating in synchronism with $\bar{H}^-$. Writing $\bar{B}^- = \mu_0(\bar{M}^- + \bar{H}^-) = \mu^- \bar{H}^-$ gives the effective permeability for an LHCP wave as

$$\mu^- = \mu_0\left(1 + \frac{\omega_m}{\omega_0 + \omega}\right) = \mu - \kappa. \tag{9.35}$$

The angle, θ_M, between $\bar{M}^-$ and the z-axis is given by

$$\tan\theta_M = \frac{|\bar{\mathcal{M}}^-|}{M_s} = \frac{\omega_m H^-}{(\omega_0 + \omega)M_s} = \frac{\omega_0 H^-}{(\omega_0 + \omega)H_0}, \tag{9.36}$$

which is seen to be less than θ_H of (9.32), as shown in Figure 9.3b. In this case the magnetic dipole is precessing in the opposite direction of its free precession.

Thus we see that the interaction of a circularly polarized wave with a magnetically biased ferrite depends on the sense of the polarization (RHCP or LHCP). This is because the bias field sets up a preferential precession direction coinciding with the direction of forced precession for an RHCP wave, but opposite to that of an LHCP wave. As we will see in Section 9.2, this effect leads to nonreciprocal propagation characteristics.

Effect of Loss

Equations (9.22) and (9.25) show that the elements of the susceptibility or permeability tensors become infinite when the frequency, ω, equals the Larmor frequency, ω_0. This effect is known as *gyromagnetic resonance*, and it occurs when the forced precession frequency is equal to the free precession frequency. In the absence of loss the response may be unbounded, in the same way that the response of an *LC* resonant circuit will be unbounded when driven with an AC signal having a frequency equal to the resonant frequency of the *LC* circuit. All real ferrite materials, however, have various magnetic loss mechanisms that damp out such singularities.

As with other resonant systems, loss can be accounted for by making the resonant frequency complex:

$$\omega_0 \leftarrow \omega_0 + j\alpha\omega, \tag{9.37}$$

where α is a damping factor. Substituting (9.37) into (9.22) makes the susceptibilities complex:

$$\chi_{xx} = \chi_{xx}' - j\chi_{xx}'' \tag{9.38a}$$

$$\chi_{xy} = \chi_{xy}'' + j\chi_{xy}' \tag{9.38b}$$

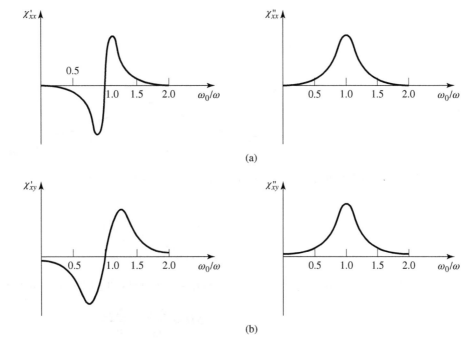

FIGURE 9.4 | Complex susceptibilities for a typical ferrite. (a) Real and imaginary parts of χ_{xx}. (b) Real and imaginary parts of χ_{xy}.

where the real and imaginary parts are given by

$$\chi'_{xx} = \frac{\omega_0 \omega_m (\omega_0^2 - \omega^2) + \omega_0 \omega_m \omega^2 \alpha^2}{\left[\omega_0^2 - \omega^2(1 + \alpha^2)\right]^2 + 4\omega_0^2 \omega^2 \alpha^2}, \tag{9.39a}$$

$$\chi''_{xx} = \frac{\alpha \omega \omega_m \left[\omega_0^2 + \omega^2(1 + \alpha^2)\right]}{\left[\omega_0^2 - \omega^2(1 + \alpha^2)\right]^2 + 4\omega_0^2 \omega^2 \alpha^2}, \tag{9.39b}$$

$$\chi'_{xy} = \frac{\omega \omega_m \left[\omega_0^2 - \omega^2(1 + \alpha^2)\right]}{\left[\omega_0^2 - \omega^2(1 + \alpha^2)\right]^2 + 4\omega_0^2 \omega^2 \alpha^2}, \tag{9.39c}$$

$$\chi''_{xy} = \frac{2\omega_0 \omega_m \omega^2 \alpha}{\left[\omega_0^2 - \omega^2(1 + \alpha^2)\right]^2 + 4\omega_0^2 \omega^2 \alpha^2}. \tag{9.39d}$$

Equation (9.37) can also be applied to (9.25) to give a complex $\mu = \mu' - j\mu''$, and $\kappa = \kappa' - j\kappa''$; this is why (9.38b) appears to define χ'_{xy} and χ''_{xy} backward, as $\chi_{xy} = j\kappa/\mu_0$. For most ferrite materials the loss is small, so $\alpha \ll 1$, and the $(1 + \alpha^2)$ terms in (9.39) can be approximated as unity. The real and imaginary parts of the susceptibilities of (9.39) are sketched in Figure 9.4 for a typical ferrite.

The damping factor, α, is related to the *linewidth*, ΔH, of the susceptibility curve near resonance. Consider the plot of χ''_{xx} versus bias field, H_0, shown in Figure 9.5. For a fixed frequency ω, resonance occurs when $H_0 = H_r$, such that $\omega_0 = \mu_0 \gamma H_r$. The linewidth, ΔH, is defined as the width of the curve of χ''_{xx} versus H_0 where χ''_{xx} has decreased to half its peak value. If we assume $(1 + \alpha^2) \simeq 1$, (9.39b) shows that the maximum value of χ''_{xx} is $\omega_m/2\alpha\omega$, and occurs when $\omega = \omega_0$. Now let ω_{02} be the Larmor frequency for which $H_0 = H_2$, where χ''_{xx} has decreased to half its maximum value. Then we can solve (9.39b) for α in terms of ω_{02}:

$$\frac{\alpha \omega \omega_m (\omega_{02}^2 + \omega^2)}{(\omega_{02}^2 - \omega^2)^2 + 4\omega_{02}^2 \omega^2 \alpha^2} = \frac{\omega_m}{4\alpha\omega},$$

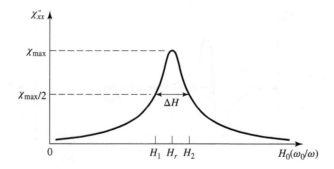

FIGURE 9.5 | Definition of the linewidth, ΔH, of the gyromagnetic resonance.

$$4\alpha^2\omega^4 = \left(\omega_{02}^2 - \omega^2\right)^2,$$

$$\omega_{02} = \omega\sqrt{1 + 2\alpha} \simeq \omega(1 + \alpha).$$

Then $\Delta\omega_0 = 2(\omega_{02} - \omega_0) \simeq 2[\omega(1 + \alpha) - \omega] = 2\alpha\omega$, and using (9.9) gives the linewidth as

$$\Delta H = \frac{\Delta\omega_0}{\mu_0\gamma} = \frac{2\alpha\omega}{\mu_0\gamma}. \tag{9.40}$$

Typical linewidths range from less than 100 Oe (for YIG) to 100–500 Oe (for ferrites); single-crystal YIG can have a linewidth as low as 0.3 Oe. Also note that this loss is separate from the dielectric loss that a ferrimagnetic material may have.

Demagnetization Factors

The DC bias field, H_0, internal to a ferrite sample is generally different from the externally applied field, H_a, because of the boundary conditions at the surface of the ferrite. To illustrate this effect, consider a thin ferrite plate, as shown in Figure 9.6. When the applied field is normal to the plate, continuity of B_n at the surface of the plate gives

$$B_n = \mu_0 H_a = \mu_0(M_s + H_0),$$

so the internal magnetic bias field is

$$H_0 = H_a - M_s.$$

This shows that the internal field is less than the applied field by an amount equal to the saturation magnetization. When the applied field is parallel to the ferrite plate, continuity of H_t at the surfaces of the plate gives

$$H_t = H_a = H_0.$$

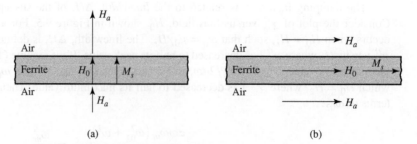

(a)   (b)

FIGURE 9.6 | Internal and external fields for a thin ferrite plate. (a) Normal bias. (b) Tangential bias.

TABLE 9.1 | **Demagnetization Factors for Some Simple Shapes**

Shape		N_x	N_y	N_z
Thin disk or plate		0	0	1
Thin rod		$\dfrac{1}{2}$	$\dfrac{1}{2}$	0
Sphere		$\dfrac{1}{3}$	$\dfrac{1}{3}$	$\dfrac{1}{3}$

In this case the internal field is not reduced. In general, the internal field (AC or DC), $\bar{H}$, is affected by the shape of the ferrite sample and its orientation with respect to the external field, $\bar{H}_e$, and can be expressed as

$$\bar{H} = \bar{H}_e - N\bar{M}, \tag{9.41}$$

where $N = N_x, N_y$, or N_z is called the *demagnetization factor* for that direction of the external field. Different shapes have different demagnetization factors, which depend on the direction of the applied field. Table 9.1 lists the demagnetization factors for a few simple shapes. The demagnetization factors are defined such that $N_x + N_y + N_z = 1$.

The demagnetization factors can also be used to relate the internal and external RF fields near the boundary of a ferrite sample. For a z-biased ferrite with transverse RF fields, (9.41) reduces to

$$H_x = H_{xe} - N_x M_x, \tag{9.42a}$$

$$H_y = H_{ye} - N_y M_y, \tag{9.42b}$$

$$H_z = H_a - N_z M_s, \tag{9.42c}$$

where H_{xe}, H_{ye} are the RF fields external to the ferrite, and H_a is the externally applied bias field. Equation (9.21) relates the internal transverse RF fields and magnetization as

$$M_x = \chi_{xx} H_x + \chi_{xy} H_y,$$

$$M_y = \chi_{yx} H_x + \chi_{yy} H_y.$$

Using (9.42a, b) to eliminate H_x and H_y gives

$$M_x = \chi_{xx} H_{xe} + \chi_{xy} H_{ye} - \chi_{xx} N_x M_x - \chi_{xy} N_y M_y,$$

$$M_y = \chi_{yx} H_{xe} + \chi_{yy} H_{ye} - \chi_{yx} N_x M_x - \chi_{yy} N_y M_y.$$

These equations can be solved for M_x, M_y to give

$$M_x = \frac{\chi_{xx}(1 + \chi_{yy} N_y) - \chi_{xy} \chi_{yx} N_y}{D} H_{xe} + \frac{\chi_{xy}}{D} H_{ye}, \tag{9.43a}$$

$$M_y = \frac{\chi_{yx}}{D} H_{xe} + \frac{\chi_{yy}(1 + \chi_{xx} N_x) - \chi_{yx} \chi_{xy} N_x}{D} H_{ye}, \tag{9.43b}$$

where

$$D = (1 + \chi_{xx}N_x)(1 + \chi_{yy}N_y) - \chi_{yx}\chi_{xy}N_xN_y. \tag{9.44}$$

This result is of the form $\bar{M} = [\chi_e]\bar{H}$, where the coefficients of H_{xe} and H_{ye} in (9.43) can be defined as "external" susceptibilities since they relate magnetization to the external RF fields.

For an infinite ferrite medium gyromagnetic resonance occurs when the denominator of the susceptibilities of (9.22) vanishes, at the frequency $\omega_r = \omega = \omega_0$. However, for a finite-sized ferrite sample the gyromagnetic resonance frequency is altered by the demagnetization factors, and is given by the condition that $D = 0$ in (9.43). Using the expressions in (9.22) for the susceptibilities in (9.44), and setting the result equal to zero, gives

$$\left(1 + \frac{\omega_0\omega_m N_x}{\omega_0^2 - \omega^2}\right)\left(1 + \frac{\omega_0\omega_m N_y}{\omega_0^2 - \omega^2}\right) - \frac{\omega^2\omega_m^2}{\left(\omega_0^2 - \omega^2\right)^2}N_xN_y = 0.$$

After some algebraic manipulations this result can be reduced to give the resonance frequency, ω_r, as

$$\omega_r = \omega = \sqrt{(\omega_0 + \omega_m N_x)(\omega_0 + \omega_m N_y)}. \tag{9.45}$$

Since $\omega_0 = \mu_0\gamma H_0 = \mu_0\gamma(H_a - N_zM_s)$, and $\omega_m = \mu_0\gamma M_s$, (9.45) can be rewritten in terms of the applied bias field strength and saturation magnetization as

$$\omega_r = \mu_0\gamma\sqrt{[H_a + (N_x - N_z)M_s][H_a + (N_y - N_z)M_s]}. \tag{9.46}$$

This result is known as *Kittel's equation* [4].

POINT OF INTEREST: Permanent Magnets

Since ferrite components such as isolators, gyrators, and circulators generally use permanent magnets to supply the required DC bias field, it may be useful to discuss some of the important characteristics of permanent magnets.

A permanent magnet is made by placing the magnetic material in a strong magnetic field, and then removing the field to leave the material magnetized in a remanent state. Unless the magnet shape forms a closed path (like a toroid), the demagnetization factors at the magnet ends will cause a slightly negative H field to be induced in the magnet. Thus the "operating point" of a permanent magnet will be in the second quadrant of the B–H hysteresis curve for the magnet material. This portion of the curve is called the *demagnetization curve*. A typical example is shown below.

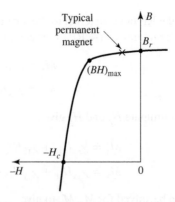

The residual magnetization, for $H = 0$, is called the *remanence*, B_r, of the material. This quantity characterizes the strength of the magnet, so generally a magnet material is chosen to have a large remanence. Another important parameter is the *coercivity*, H_c, which is the value of the negative H field required to reduce the magnetization to zero. A good permanent magnet should have a high coercivity to reduce the effects of vibration, temperature changes, and external fields, which can lead to a loss of magnetization. An overall figure of merit for a permanent magnet is sometimes given as the maximum value of the BH product, $(BH)_{\text{max}}$, on the demagnetization curve. This quantity is essentially the maximum magnetic energy density that can be stored by the magnet, and can be useful in electromechanical applications. The following table lists the remanence, coercivity, and $(BH)_{\text{max}}$ for some of the most common permanent magnet materials.

Material	Composition	B_r (Oe)	H_c (G)	$(BH)_{\text{max}}$ (G-Oe)$\times 10^6$
ALNICO 5	Al, Ni, Co, Cu	12,000	720	5.0
ALNICO 8	Al, Ni, Co, Cu, Ti	7100	2000	5.5
ALNICO 9	Al, Ni, Co, Cu, Ti	10,400	1600	8.5
Remalloy	Mo, Co, Fe	10,500	250	1.1
Platinum cobalt	Pt, Co	6450	4300	9.5
Ceramic	$BaO_6Fe_2O_3$	3950	2400	3.5
Cobalt samarium	Co, Sm	8400	7000	16.0

9.2 PLANE WAVE PROPAGATION IN A FERRITE MEDIUM

The previous section gives an explanation of the microscopic phenomena that occur inside a biased ferrite material to produce a tensor permeability of the form given in (9.24) [or in (9.26) or (9.27), depending on the bias direction]. Once we have this macroscopic description of the ferrite material, we can solve Maxwell's equations for wave propagation in various geometries involving ferrite materials. We begin with plane wave propagation in an infinite ferrite medium, for propagation either in the direction of bias, or transverse to the bias field. These problems will illustrate the important effects of Faraday rotation and birefringence.

Propagation in Direction of Bias (Faraday Rotation)

Consider an infinite ferrite-filled region with a DC magnetic bias field given by $\bar{H}_0 = \hat{z}H_0$, and a tensor permittivity $[\mu]$ given by (9.24). Maxwell's equations can be written as

$$\nabla \times \bar{E} = -j\omega[\mu]\bar{H}, \tag{9.47a}$$

$$\nabla \times \bar{H} = j\omega\epsilon\bar{E}, \tag{9.47b}$$

$$\nabla \cdot \bar{D} = 0, \tag{9.47c}$$

$$\nabla \cdot \bar{B} = 0. \tag{9.47d}$$

Now assume plane wave propagation in the z direction, with $\partial/\partial x = \partial/\partial y = 0$. Then the electric and magnetic fields will have the following form:

$$\bar{E} = \bar{E}_0 e^{-j\beta z}, \tag{9.48a}$$

$$\bar{H} = \bar{H}_0 e^{-j\beta z}. \tag{9.48b}$$

The two curl equations of (9.47a) and (9.47b) reduce to the following, after using (9.24):

$$j\beta E_y = -j\omega(\mu H_x + j\kappa H_y), \tag{9.49a}$$

$$-j\beta E_x = -j\omega(-j\kappa H_x + \mu H_y), \tag{9.49b}$$

$$0 = -j\omega\mu_0 H_z, \tag{9.49c}$$

$$j\beta H_y = j\omega\epsilon E_x, \tag{9.49d}$$

$$-j\beta H_x = j\omega\epsilon E_y, \tag{9.49e}$$

$$0 = j\omega\epsilon E_z. \tag{9.49f}$$

Equations (9.49c) and (9.49f) show that $E_z = H_z = 0$, as expected for TEM plane waves. We also have $\nabla \cdot \bar{D} = \nabla \cdot \bar{B} = 0$ since $\partial/\partial x = \partial/\partial y = 0$. Equations (9.49d) and (9.49e) give relations between the transverse field components as

$$Y = \frac{H_y}{E_x} = \frac{-H_x}{E_y} = \frac{\omega \epsilon}{\beta}, \tag{9.50}$$

where Y is the wave admittance. Using (9.50) in (9.49a) and (9.49b) to eliminate H_x and H_y gives the following results:

$$j\omega^2 \epsilon \kappa E_x + (\beta^2 - \omega^2 \mu \epsilon)E_y = 0, \tag{9.51a}$$

$$(\beta^2 - \omega^2 \mu \epsilon)E_x - j\omega^2 \epsilon \kappa E_y = 0. \tag{9.51b}$$

For a nontrivial solution for E_x and E_y the determinant of this set of equations must vanish:

$$\omega^4 \epsilon^2 \kappa^2 - (\beta^2 - \omega^2 \mu \epsilon)^2 = 0,$$

or

$$\beta_\pm = \omega \sqrt{\epsilon(\mu \pm \kappa)}. \tag{9.52}$$

There are two possible propagation constants, β_+ and β_-.

First consider the fields associated with β_+, which can be found by substituting β_+ into (9.51a) or (9.51b):

$$j\omega^2 \epsilon \kappa E_x + \omega^2 \epsilon \kappa E_y = 0,$$

or

$$E_y = -jE_x.$$

Then the electric field of (9.48a) must have the following form:

$$\bar{E}_+ = E_0(\hat{x} - j\hat{y})e^{-j\beta_+ z}, \tag{9.53a}$$

which is seen to be a right-hand circularly polarized plane wave. Using (9.50) gives the associated magnetic field as

$$\bar{H}_+ = E_0 Y_+ (j\hat{x} + \hat{y})e^{-j\beta_+ z}, \tag{9.53b}$$

where Y_+ is the wave admittance for this wave:

$$Y_+ = \frac{\omega \epsilon}{\beta_+} = \sqrt{\frac{\epsilon}{\mu + \kappa}}. \tag{9.53c}$$

Similarly, the fields associated with β_- are left-hand circularly polarized:

$$\bar{E}_- = E_0(\hat{x} + j\hat{y})e^{-j\beta_- z}, \tag{9.54a}$$

$$\bar{H}_- = E_0 Y_- (-j\hat{x} + \hat{y})e^{-j\beta_- z}, \tag{9.54b}$$

where Y_- is the wave admittance for this wave:

$$Y_- = \frac{\omega \epsilon}{\beta_-} = \sqrt{\frac{\epsilon}{\mu - \kappa}}. \tag{9.54c}$$

Thus we see that RHCP and LHCP plane waves are the source-free modes of the $\hat{z}$-biased ferrite medium, and these waves propagate through the ferrite medium with different propagation constants. As discussed in the previous section, the physical explanation for this effect is that the magnetic bias field creates a preferred direction for magnetic dipole precession, and one sense of circular polarization causes precession in this preferred direction, while the other sense of polarization causes precession in the opposite direction. Also note that for an RHCP wave, the ferrite material can be represented with an effective permeability of $\mu + \kappa$, while for an LHCP wave the effective permeability is $\mu - \kappa$. In mathematical terms, we can state that $(\mu + \kappa)$ and $(\mu - \kappa)$, or β_+ and β_-, are the *eigenvalues* of the system of equations in (9.51), and that $\bar{E}_+$ and $\bar{E}_-$ are the associated *eigenvectors*. When losses are present, the attenuation constants for RHCP and LHCP waves will also be different.

Now consider a linearly polarized electric field at $z = 0$, represented as the sum of an RHCP and an LHCP wave:

$$\bar{E}|_{z=0} = \hat{x}E_0 = \frac{E_0}{2}(\hat{x} - j\hat{y}) + \frac{E_0}{2}(\hat{x} + j\hat{y}). \tag{9.55}$$

The RHCP component will propagate in the z direction as $e^{-j\beta_+ z}$, and the LHCP component will propagate as $e^{-j\beta_- z}$, so the total field of (9.55) will propagate as

$$\begin{aligned}
\bar{E} &= \frac{E_0}{2}(\hat{x} - j\hat{y})e^{-j\beta_+ z} + \frac{E_0}{2}(\hat{x} + j\hat{y})e^{-j\beta_- z} \\
&= \frac{E_0}{2}\hat{x}(e^{-j\beta_+ z} + e^{-j\beta_- z}) - j\frac{E_0}{2}\hat{y}(e^{-j\beta_+ z} - e^{-j\beta_- z}) \\
&= E_0\left[\hat{x}\cos\left(\frac{\beta_+ - \beta_-}{2}\right)z - \hat{y}\sin\left(\frac{\beta_+ - \beta_-}{2}\right)z\right]e^{-j(\beta_+ + \beta_-)z/2}.
\end{aligned} \tag{9.56}$$

This is still a linearly polarized wave, but one whose direction of polarization rotates as the wave propagates along the z-axis. At a given point along the z-axis the polarization direction measured from the x-axis is given by

$$\phi = \tan^{-1}\frac{E_y}{E_x} = \tan^{-1}\left[-\tan\left(\frac{\beta_+ - \beta_-}{2}\right)z\right] = -\left(\frac{\beta_+ - \beta_-}{2}\right)z. \tag{9.57}$$

This effect is called *Faraday rotation*, after Michael Faraday, who first observed this phenomenon during his study of the propagation of light through liquids that had magnetic properties. Note that for a fixed position on the z-axis, the polarization angle is fixed, unlike the case for a circularly polarized wave, where the polarization direction rotates with time.

For $\omega < \omega_0$, μ and κ are positive and $\mu > \kappa$. Then $\beta_+ > \beta_-$, and (9.57) shows that ϕ becomes more negative as z increases, meaning that the polarization (direction of $\bar{E}$) rotates counterclockwise as we look in the $+z$ direction. Reversing the bias direction (sign of H_0 and M_s) changes the sign of κ, which changes the direction of rotation to clockwise. Similarly, for $+z$ bias, a wave traveling in the $-z$ direction will rotate its polarization clockwise as we look in the direction of propagation ($-z$); if we were looking in the $+z$ direction, however, the direction of rotation would be counterclockwise (same as a wave propagating in the $+z$ direction). Thus, a wave that travels from $z = 0$ to $z = L$ and back again to $z = 0$ undergoes a total polarization rotation of 2ϕ, where ϕ is given in (9.57) with $z = L$. So, unlike the situation of a screw being driven into a block of wood and then backed out, the polarization does not "unwind" when the direction of propagation is reversed. Faraday rotation is thus seen to be a nonreciprocal effect.

EXAMPLE 9.1 PLANE WAVE PROPAGATION IN A FERRITE MEDIUM

Consider an infinite ferrite medium with $4\pi M_s = 1800$ G, $\Delta H = 75$ Oe, $\epsilon_r = 14$, and $\tan\delta = 0.001$. If the bias field strength is $H_0 = 3570$ Oe, calculate and plot the phase and attenuation constants for RHCP and LHCP plane waves versus frequency for $f = 0$ to 20 GHz.

Solution
The Larmor precession frequency is

$$f_0 = \frac{\omega_0}{2\pi} = (2.8 \text{ MHz/Oe})(3570 \text{ Oe}) = 10.0 \text{ GHz},$$

and

$$f_m = \frac{\omega_m}{2\pi} = (2.8 \text{ MHz/Oe})(1800 \text{ G}) = 5.04 \text{ GHz}.$$

At each frequency we can compute the complex propagation constant as

$$\gamma_\pm = \alpha_\pm + j\beta_\pm = j\omega\sqrt{\epsilon(\mu \pm \kappa)},$$

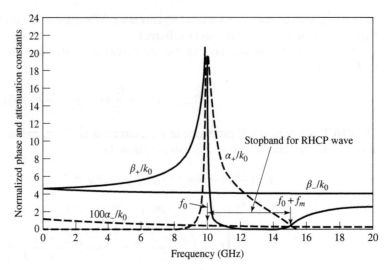

FIGURE 9.7 | Normalized phase and attenuation constants for circularly polarized plane waves in the ferrite medium of Example 9.1.

where $\epsilon = \epsilon_0 \epsilon_r (1 - j \tan \delta)$ is the complex permittivity, and μ, κ are given by (9.25). The following substitution for ω_0 is used to account for ferrimagnetic loss:

$$\omega_0 \leftarrow \omega_0 + j \frac{\mu_0 \gamma \Delta H}{2},$$

or

$$f_0 \leftarrow f_0 + j \frac{(2.8 \text{ MHz/Oe})(75 \text{ Oe})}{2} = (10.0 + j0.105) \text{ GHz},$$

which is derived from (9.37) and (9.40). The quantities $\mu \pm \kappa$ can be simplified to the following, by using (9.25):

$$\mu + \kappa = \mu_0 \left(1 + \frac{\omega_m}{\omega_0 - \omega} \right),$$

$$\mu - \kappa = \mu_0 \left(1 + \frac{\omega_m}{\omega_0 + \omega} \right).$$

The phase and attenuation constants are plotted in Figure 9.7, normalized to the free-space wave number, k_0.

Observe that β_+ and α_+ (for an RHCP wave) show a resonance near $f = f_0 = 10 \text{ GHz}$; β_- and α_- (for an LHCP wave) do not, however, because the singularities in μ and κ cancel in the $(\mu - \kappa)$ term contained in γ_-. Also note from Figure 9.7 that a stopband (β_+ near zero, large α_+) exists for RHCP waves for frequencies between f_0 and $f_0 + f_m$ (between ω_0 and $\omega_0 + \omega_m$). For frequencies in this range, the above expression for $(\mu + \kappa)$ shows that this quantity is negative, and $\beta_+ = 0$ (in the absence of loss), so an RHCP wave incident on such a ferrite medium would be totally reflected.

Propagation Transverse to Bias (Birefringence)

Next consider the case where an infinite ferrite region is biased in the $\hat{x}$ direction, transverse to the direction of propagation; the permeability tensor is given in (9.26). For plane wave fields of the form in (9.48), Maxwell's curl equations reduce to

$$j\beta E_y = -j\omega \mu_0 H_x, \tag{9.58a}$$

$$-j\beta E_x = -j\omega(\mu H_y + j\kappa H_z), \tag{9.58b}$$

$$0 = -j\omega(-j\kappa H_y + \mu H_z), \tag{9.58c}$$

$$j\beta H_y = j\omega\epsilon E_x, \tag{9.58d}$$

$$-j\beta H_x = j\omega\epsilon E_y, \tag{9.58e}$$

$$0 = j\omega\epsilon E_z. \tag{9.58f}$$

Then $E_z = 0$, and $\nabla \cdot \bar{D} = 0$ since $\partial/\partial x = \partial/\partial y = 0$. Equations (9.58d) and (9.58e) give an admittance relation between the transverse field components:

$$Y = \frac{H_y}{E_x} = \frac{-H_x}{E_y} = \frac{\omega\epsilon}{\beta}. \tag{9.59}$$

Using (9.59) in (9.58a) and (9.58b) to eliminate H_x and H_y, and using (9.58c) in (9.58b) to eliminate H_z, gives the following results:

$$\beta^2 E_y = \omega^2\mu_0\epsilon E_y, \tag{9.60a}$$

$$\mu(\beta^2 - \omega^2\mu\epsilon)E_x = -\omega^2\epsilon\kappa^2 E_x. \tag{9.60b}$$

One solution to (9.60) occurs for

$$\beta_o = \omega\sqrt{\mu_0\epsilon}, \tag{9.61}$$

with $E_x = 0$. Then the complete fields are

$$\bar{E}_o = \hat{y}E_0 e^{-j\beta_o z}, \tag{9.62a}$$

$$\bar{H}_o = -\hat{x}E_0 Y_o e^{-j\beta_o z}, \tag{9.62b}$$

since (9.59) shows that $H_y = 0$ when $E_x = 0$, and (9.58c) shows that $H_z = 0$ when $H_y = 0$. The admittance is

$$Y_o = \frac{\omega\epsilon}{\beta_o} = \sqrt{\frac{\epsilon}{\mu_0}}. \tag{9.63}$$

This wave is called the *ordinary wave* because it is unaffected by the magnetic properties of the ferrite. This happens whenever the magnetic field components transverse to the bias direction are zero ($H_y = H_z = 0$). The wave propagates in either the $+z$ or $-z$ direction with the same propagation constant, which is independent of H_0.

Another solution to (9.60) occurs for

$$\beta_e = \omega\sqrt{\mu_e\epsilon}, \tag{9.64}$$

with $E_y = 0$, where μ_e is an effective permeability given by

$$\mu_e = \frac{\mu^2 - \kappa^2}{\mu}. \tag{9.65}$$

This wave is called the *extraordinary wave* and is affected by the ferrite magnetization. Note that the effective permeability may be negative for certain values of ω, ω_0. The electric field is

$$\bar{E}_e = \hat{x}E_0 e^{-j\beta_e z}. \tag{9.66a}$$

Since $E_y = 0$, (9.58e) shows that $H_x = 0$. H_y can be found from (9.58d), and H_z from (9.58c), giving the complete magnetic field as

$$\bar{H}_e = E_0 Y_e \left(\hat{y} + \hat{z}\frac{j\kappa}{\mu}\right) e^{-j\beta_e z}, \tag{9.66b}$$

where

$$Y_e = \frac{\omega\epsilon}{\beta_e} = \sqrt{\frac{\epsilon}{\mu_e}}. \tag{9.67}$$

These fields constitute a linearly polarized wave, but note that the magnetic field has a component in the direction of propagation. Except for the existence of H_z, the extraordinary wave has electric and magnetic

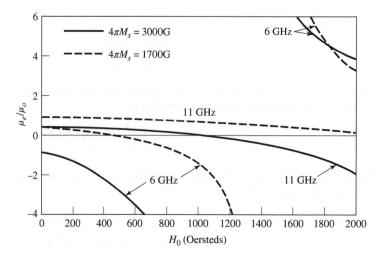

FIGURE 9.8 | Effective permeability, μ_e, versus bias field, H_0, for various saturation magnetizations and frequencies.

fields that are perpendicular to the corresponding fields of the ordinary wave. Thus, a wave polarized in the y direction will have a propagation constant β_o (ordinary wave), but a wave polarized in the x direction will have a propagation constant β_e (extraordinary wave). This effect, where the propagation constant depends on the polarization direction, is called *birefringence* [2]. Birefringence often occurs in optics work, where the index of refraction can have different values depending on the polarization. The double image seen through a calcite crystal is an example of this effect.

From (9.65) we can see that μ_e, the effective permeability for the extraordinary wave, can be negative if $\kappa^2 > \mu^2$. This condition depends on the values of ω, ω_0, and ω_m, or f, H_0, and M_s, but for a fixed frequency and saturation magnetization there will always be some range of bias field for which $\mu_e < 0$ (ignoring loss). When this occurs β_e will become imaginary, as seen from (9.64), which implies that the wave will be cut off, or evanescent. An $\hat{x}$-polarized plane wave incident at the interface of such a ferrite region would be totally reflected. The effective permeability is plotted versus bias field strength in Figure 9.8 for several values of frequency and saturation magnetization.

9.3 PROPAGATION IN A FERRITE-LOADED RECTANGULAR WAVEGUIDE

In the previous section we introduced the effects of a ferrite material on electromagnetic fields by considering the propagation of plane waves in an infinite ferrite medium. In practice, however, most ferrite components use waveguide or other types of transmission lines loaded with ferrite material. Many of these geometries are very difficult to treat without the use of complex numerical methods, but it is possible to analyze some of the simpler cases involving ferrite-loaded rectangular waveguides. This will allow us to quantitatively demonstrate the operation and design of several types of practical ferrite components.

TE$_{m0}$ Modes of Waveguide with a Single Ferrite Slab

We first consider the geometry shown in Figure 9.9, where a rectangular waveguide is loaded with a vertical slab of ferrite material biased in the $\hat{y}$ direction. This geometry and its analysis will be used in later sections to treat the operation and design of resonance isolators, field displacement isolators, and remanent (nonreciprocal) phase shifters.

In the ferrite slab, Maxwell's equations can be written as

$$\nabla \times \bar{E} = -j\omega[\mu]\bar{H}, \tag{9.68a}$$

$$\nabla \times \bar{H} = j\omega\epsilon\bar{E}, \tag{9.68b}$$

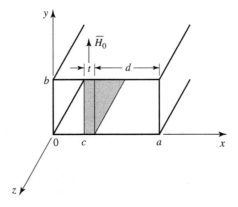

FIGURE 9.9 | Geometry of a rectangular waveguide loaded with a transversely biased ferrite slab.

where $[\mu]$ is the permeability tensor for $\hat{y}$ bias, as given in (9.27). If we let $\bar{E}(x, y, z) = [\bar{e}(x, y) + \hat{z}e_z(x, y)]e^{-j\beta z}$ and $\bar{H}(x, y, z) = [\bar{h}(x, y) + \hat{z}h_z(x, y)]e^{-j\beta z}$, (9.68) reduces to

$$\frac{\partial e_z}{\partial y} + j\beta e_y = -j\omega(\mu h_x - j\kappa h_z),\tag{9.69a}$$

$$-j\beta e_x - \frac{\partial e_z}{\partial x} = -j\omega\mu_0 h_y,\tag{9.69b}$$

$$\frac{\partial e_y}{\partial x} - \frac{\partial e_x}{\partial y} = -j\omega(j\kappa h_x + \mu h_z),\tag{9.69c}$$

$$\frac{\partial h_z}{\partial y} + j\beta h_y = j\omega\epsilon e_x,\tag{9.69d}$$

$$-j\beta h_x - \frac{\partial h_z}{\partial x} = j\omega\epsilon e_y,\tag{9.69e}$$

$$\frac{\partial h_y}{\partial x} - \frac{\partial h_x}{\partial y} = j\omega\epsilon e_z.\tag{9.69f}$$

For TE$_{m0}$ modes, we know that $E_z = 0$ and $\partial/\partial y = 0$. Then (9.69b) and (9.69d) imply that $e_x = h_y = 0$ (since $\beta^2 \neq \omega^2\mu_0\epsilon$ for a waveguide mode) and so (9.69) reduces to three equations:

$$j\beta e_y = -j\omega(\mu h_x - j\kappa h_z),\tag{9.70a}$$

$$\frac{\partial e_y}{\partial x} = -j\omega(j\kappa h_x + \mu h_z),\tag{9.70b}$$

$$j\omega\epsilon e_y = -j\beta h_x - \frac{\partial h_z}{\partial x}.\tag{9.70c}$$

We can solve (9.70a) and (9.70b) for h_x and h_z as follows. Multiply (9.70a) by μ and (9.70b) by $j\kappa$, then add to obtain

$$h_x = \frac{1}{\omega\mu\mu_e}\left(-\mu\beta e_y - \kappa\frac{\partial e_y}{\partial x}\right).\tag{9.71a}$$

Now multiply (9.70a) by $j\kappa$ and (9.71a) by μ, then add to obtain

$$h_z = \frac{j}{\omega\mu\mu_e}\left(\kappa\beta e_y + \mu\frac{\partial e_y}{\partial x}\right),\tag{9.71b}$$

where $\mu_e = (\mu^2 - \kappa^2)/\mu$. Substituting (9.71) into (9.70c) gives a wave equation for e_y:

$$j\omega\epsilon e_y = \frac{-j\beta}{\omega\mu\mu_e}\left(-\mu\beta e_y - \kappa\frac{\partial e_y}{\partial x}\right) - \frac{j}{\omega\mu\mu_e}\left(\kappa\beta\frac{\partial e_y}{\partial x} + \mu\frac{\partial^2 e_y}{\partial x^2}\right),$$

or

$$\left(\frac{\partial^2}{\partial x^2} + k_f^2\right) e_y = 0,$$

(9.72)

where k_f is defined as a cutoff wave number for the ferrite:

$$k_f^2 = \omega^2 \mu_e \epsilon - \beta^2.$$

(9.73)

We can obtain the corresponding results for the air regions by letting $\mu = \mu_0, \kappa = 0$, and $\epsilon_r = 1$, to obtain

$$\left(\frac{\partial^2}{\partial x^2} + k_a^2\right) e_y = 0,$$

(9.74)

where k_a is the cutoff wave number for the air regions:

$$k_a^2 = k_0^2 - \beta^2.$$

(9.75)

The magnetic field in the air region is given by

$$h_x = \frac{-\beta}{\omega \mu_0} e_y = \frac{-1}{Z_w} e_y,$$

(9.76a)

$$h_z = \frac{j}{\omega \mu_0} \frac{\partial e_y}{\partial x}.$$

(9.76b)

The general solutions for e_y in the air-ferrite-air regions of the waveguide are then

$$e_y = \begin{cases} A \sin k_a x & \text{for } 0 < x < c, \\ B \sin k_f(x-c) + C \sin k_f(c+t-x) & \text{for } c < x < c+t, \\ D \sin k_a(a-x) & \text{for } c+t < x < a, \end{cases}$$

(9.77a)

which have been constructed to facilitate the enforcement of boundary conditions at $x = 0, c, c+t$, and a [3]. We will also need h_z, which can be found from (9.77a), (9.71b), and (9.76b):

$$h_z = \begin{cases} (jk_a A/\omega \mu_0) \cos k_a x & \text{for } 0 < x < c, \\ (j/\omega \mu \mu_e)\{\kappa \beta [B \sin k_f(x-c) + C \sin k_f(c+t-x)] \\ \quad + \mu k_f [B \cos k_f(x-c) - C \cos k_f(c+t-x)]\} & \text{for } c < x < c+t, \\ (-jk_a D/\omega \mu_0) \cos k_a(a-x) & \text{for } c+t < x < a. \end{cases}$$

(9.77b)

Matching e_y and h_z at $x = c$ and $x = c+t = a-d$ gives four equations for the constants A, B, C, D:

$$A \sin k_a c = C \sin k_f t,$$

(9.78a)

$$B \sin k_f t = D \sin k_a d,$$

(9.78b)

$$A \frac{k_a}{\mu_0} \cos k_a c = B \frac{k_f}{\mu_e} - C \frac{1}{\mu \mu_e}(-\kappa \beta \sin k_f t + \mu k_f \cos k_f t),$$

(9.78c)

$$B \frac{1}{\mu \mu_e}(\kappa \beta \sin k_f t + \mu k_f \cos k_f t) - C \frac{k_f}{\mu_e} = -D \frac{k_a}{\mu_0} \cos k_a d.$$

(9.78d)

Solving (9.78a) and (9.78b) for C and D, substituting into (9.78c) and (9.78d), and then eliminating A or B gives the following transcendental equation for the propagation constant, β:

$$\left(\frac{k_f}{\mu_e}\right)^2 + \left(\frac{\kappa \beta}{\mu \mu_e}\right)^2 - k_a \cot k_a c \left(\frac{k_f}{\mu_0 \mu_e} \cot k_f t + \frac{\kappa \beta}{\mu_0 \mu \mu_e}\right) - \left(\frac{k_a}{\mu_0}\right)^2$$

$$\times \cot k_a c \cot k_a d - k_a \cot k_a d \left(\frac{k_f}{\mu_0 \mu_e} \cot k_f t - \frac{\kappa \beta}{\mu_0 \mu \mu_e}\right) = 0.$$

(9.79)

After using (9.73) and (9.75) to express the cutoff wave numbers k_f and k_a in terms of β, we can solve (9.79) numerically. The fact that (9.79) contains terms that are odd in $\kappa\beta$ indicates that the resulting wave propagation will be nonreciprocal since changing the direction of the bias field (which is equivalent to changing the direction of propagation) changes the sign of κ, which leads to a different solution for β. We will identify these two solutions as β_+ and β_- for positive bias and propagation in the $+z$ direction (positive κ) and in the $-z$ direction (negative κ), respectively. The effects of magnetic loss can easily be included by allowing ω_o to be complex, as in (9.37).

In later sections we will also need to evaluate the electric field in the guide, as given in (9.77a). If we choose the arbitrary amplitude constant as A, then B, C, and D can be found in terms of A by using (9.78a), (9.78b), and (9.78c). Note from (9.75) that if $\beta > k_o$, then k_a will be imaginary. In this case, the $\sin k_a x$ function of (9.77a) becomes $j \sinh |k_a| x$, indicating an almost exponential variation in the field distribution.

A useful approximate result can be obtained for the differential phase shift, $\beta_+ - \beta_-$, by expanding β in (9.79) in a Taylor series about $t = 0$. This can be accomplished with implicit differentiation after using (9.73) and (9.75) to express k_f and k_a in terms of β [4]. The result is

$$\beta_+ - \beta_- \simeq \frac{2k_c t\kappa}{a\mu} \sin 2k_c c = 2k_c \frac{\kappa}{\mu} \frac{\Delta S}{S} \sin 2k_c c, \tag{9.80}$$

where $k_c = \pi/a$ is the cutoff frequency of the empty guide, and $\Delta S/S = t/a$ is the *filling factor*, or ratio of slab cross-sectional area to waveguide cross-sectional area. Thus, this formula can be applied to other geometries, such as waveguides loaded with small ferrite strips or rods, although the appropriate demagnetization factors may be required for some ferrite shapes. The result in (9.80) is accurate, however, only for very small ferrite cross sections, typically for $\Delta S/S < 0.01$.

This same technique can be used to obtain an approximate expression for the forward and reverse attenuation constants in terms of the imaginary parts of the susceptibilities defined in (9.39):

$$\alpha_\pm \simeq \frac{\Delta S}{S\beta_0} \left(\beta_0^2 \chi''_{xx} \sin^2 k_c x + k_c^2 \chi''_{zz} \cos^2 k_c x \mp \chi''_{xy} k_c \beta_0 \sin 2k_c x \right), \tag{9.81}$$

where $\beta_0 = \sqrt{k_0^2 - k_c^2}$ is the propagation constant of the empty guide. This result will be useful in the design of resonance isolators. Both (9.80) and (9.81) can also be derived using a perturbation method with the empty waveguide fields [4], and so are usually referred to as the *perturbation theory* results.

TE$_{m0}$ Modes of Waveguide with Two Symmetric Ferrite Slabs

A related geometry is a rectangular waveguide loaded with two symmetrically placed ferrite slabs, as shown in Figure 9.10. With equal but opposite $\hat{y}$-directed bias fields on the ferrite slabs, this configuration provides a useful model for the nonreciprocal remanent phase shifter, which will be discussed in Section 9.5. Its analysis is very similar to that of the single-slab geometry.

Since the h_y and h_z fields (including the bias fields) are antisymmetric about the midplane of the waveguide at $x = a/2$, a magnetic wall can be placed at this point. Then we only need to consider the region for

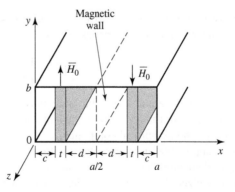

FIGURE 9.10 | Geometry of a rectangular waveguide loaded with two symmetric ferrite slabs.

$0 < x < a/2$. The electric field in this region can be written as

$$e_y = \begin{cases} A \sin k_a x, & 0 < x < c, \\ B \sin k_f(x-c) + C \sin k_f(c+t-x), & c < x < c+t, \\ D \cos k_a(a/2-x), & c+t < x < a/2, \end{cases} \tag{9.82a}$$

which is similar in form to (9.77a), except that the expression for $c + t < x < a/2$ was constructed to have a maximum at $x = a/2$ (since h_z must be zero at $x = a/2$). The cutoff wave numbers k_f and k_a are defined in (9.73) and (9.75).

Using (9.71) and (9.76) gives the h_z field as

$$h_z = \begin{cases} (jk_aA/\omega\mu_0)\cos k_a x, & 0 < x < c, \\ (j/\omega\mu\mu_e)\{-\kappa\beta[B\sin k_f(x-c) + C\sin k_f(c+t-x)] \\ \quad +\mu k_f[B\cos k_f(x-c) - C\cos k_f(c+t-x)]\}, & c < x < c+t, \\ (jk_aD/\omega\mu_0)\sin k_a(a/2-x), & c+t < x < a/2. \end{cases} \tag{9.82b}$$

Matching e_y and h_z at $x = c$ and $x = c+t = a/2 - d$ gives four equations for the constants A, B, C, D:

$$A \sin k_a c = C \sin k_f t, \tag{9.83a}$$

$$B \sin k_f t = D \cos k_a d, \tag{9.83b}$$

$$A \frac{k_a}{\mu_0} \cos k_a c = B \frac{k_f}{\mu_e} - C \frac{1}{\mu\mu_e}(-\kappa\beta\sin k_f t + \mu k_f\cos k_f t), \tag{9.83c}$$

$$\frac{B}{\mu\mu_e}(\kappa\beta\sin k_f t + \mu k_f\cos k_f t) - C\frac{k_f}{\mu_e} = D\frac{k_a}{\mu_0}\sin k_a d. \tag{9.83d}$$

Reducing these results gives a transcendental equation for the propagation constant, β:

$$\left(\frac{k_f}{\mu_e}\right)^2 + \left(\frac{\kappa\beta}{\mu\mu_e}\right)^2 - k_a\cot k_a c\left(\frac{k_f}{\mu_0\mu_e}\cot k_f t + \frac{\kappa\beta}{\mu_0\mu\mu_e}\right) + \left(\frac{k_a}{\mu_0}\right)^2$$

$$\times \cot k_a c\tan k_a d + k_a\tan k_a d\left(\frac{k_f}{\mu_0\mu_e}\cot k_f t - \frac{\kappa\beta}{\mu_0\mu\mu_e}\right) = 0. \tag{9.84}$$

This equation can be solved numerically for β. As in (9.79) for the single-slab case, κ and β appear in (9.84) only as $\kappa\beta$, κ^2, or β^2, which implies nonreciprocal propagation since changing the sign of κ (or bias fields) necessitates a change in sign for β (propagation direction) for the same root. At first glance it may seem that, for the same waveguide and slab dimensions and parameters, two slabs would give twice the phase shift of one slab, but this is generally untrue because the fields are highly concentrated in the ferrite regions.

9.4   FERRITE ISOLATORS

One of the most useful microwave ferrite components is the *isolator*, which is a two-port device having unidirectional transmission characteristics. The scattering matrix for an ideal isolator has the form

$$[S] = \begin{bmatrix} 0 & 0 \\ 1 & 0 \end{bmatrix}, \tag{9.85}$$

indicating that both ports are matched, but transmission occurs only in the direction from port 1 to port 2. Since the scattering matrix is not unitary, the isolator must be lossy. And, of course, $[S]$ is not symmetric, since an isolator is a nonreciprocal component.

A common application uses an isolator between a high-power source and a load to prevent possible reflections from damaging the source. An isolator can be used in place of a matching or tuning network, but it should be realized that any power reflected from the load will be absorbed by the isolator, as opposed to being reflected back to the load, which is the case when a matching network is used.

Although there are several types of ferrite isolators, we will concentrate on the resonance isolator and the field displacement isolator. These devices are of practical importance, and can be analyzed and designed using the results for the ferrite slab-loaded waveguide of the previous section.

Resonance Isolators

We have seen that a circularly polarized plane wave rotating in the same direction as the precessing magnetic dipoles of a ferrite medium will have a strong interaction with the material, while a circularly polarized wave rotating in the opposite direction will have a weaker interaction. Such a result was illustrated in Example 9.1, where the attenuation of a circularly polarized wave was very large near the gyromagnetic resonance of the ferrite, while the attenuation of a wave propagating in the opposite direction was very small. This effect can be used to construct an isolator; such isolators must operate near gyromagnetic resonance and so are called *resonance isolators*. Resonance isolators usually consist of a ferrite slab or strip mounted at a certain point in a waveguide. We will discuss the two isolator geometries shown in Figure 9.11.

Ideally, the RF fields inside the ferrite material should be circularly polarized. In an empty rectangular waveguide the magnetic fields of the TE_{10} mode can be written as

$$H_x = \frac{j\beta_0}{k_c} A \sin k_c x e^{-j\beta_0 z},$$

$$H_z = A \cos k_c x e^{-j\beta_0 z},$$

where $k_c = \pi/a$ is the cutoff wave number and $\beta_0 = \sqrt{k_0^2 - k_c^2}$ is the propagation constant of the empty guide. Since a circularly polarized wave must satisfy the condition that $H_x/H_z = \pm j$, the location x of the point of circular polarization in the empty guide is given by

$$\tan k_c x = \pm \frac{k_c}{\beta_0}. \tag{9.86}$$

Ferrite loading, however, may perturb the fields so that (9.86) may not give the actual optimum position, or it may prevent the internal fields from being circularly polarized for any position.

First consider the full-height E-plane slab geometry of Figure 9.11a; we can analyze this case using the exact results from the previous section. Alternatively, we could use the perturbation result of (9.81), but this would require the use of a demagnetization factor for h_x, and would be less accurate than the exact results. Thus, for a given set of parameters, (9.79) can be solved numerically for the complex propagation constants of the forward and reverse waves of the ferrite-loaded guide. It is necessary to include the effect of magnetic loss, which can be done by using (9.37) for the complex resonant frequency, ω_0, in the expressions for μ and κ. The imaginary part of ω_0 can be related to the linewidth, ΔH, of the ferrite through (9.40). Usually the waveguide width, a, frequency, ω, and ferrite parameters, $4\pi M_s$ and ϵ_r, will be fixed, and the bias field and slab position and thickness will be determined to give the optimum design.

Ideally, the forward attenuation constant (α_+) would be zero, with a nonzero attenuation constant (α_-) in the reverse direction. However, for the E-plane ferrite slab there is no position $x = c$ where the fields are

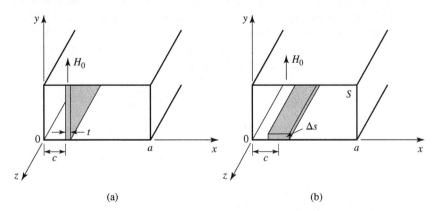

(a)   (b)

FIGURE 9.11 | Two resonance isolator geometries. (a) E-plane, full-height slab. (b) H-plane, partial-height slab.

perfectly circularly polarized in the ferrite (this is because the demagnetization factor $N_x \simeq 1$ [4]). Hence, the forward and reverse waves both contain an RHCP component and an LHCP component, so ideal attenuation characteristics cannot be obtained. The optimum design, then, generally minimizes the forward attenuation, which determines the slab position. Alternatively, it may be desired to maximize the ratio of the reverse to forward attenuations. Since the maximum reverse attenuation generally does not occur at the same slab position as the minimum forward attenuation, such a design will involve a trade-off of the forward loss.

For a long, thin slab, the demagnetization factors are approximately those of a thin disk: $N_x \simeq 1, N_y = N_z = 0$. It can then be shown via the Kittel equation of (9.45) that the gyromagnetic resonance frequency of the slab is given by

$$\omega = \sqrt{\omega_0(\omega_0 + \omega_m)}, \tag{9.87}$$

which determines H_0, given the operating frequency and saturation magnetization. This is an approximate result; the transcendental equation of (9.79) accounts for demagnetization exactly, so the actual internal bias field, H_0, can be found by numerically solving (9.79) for the attenuation constants for values of H_0 near the approximate value given by (9.87).

Once the slab position, c, and bias field, H_0, have been found, the slab length, L, can be chosen to give the desired total reverse attenuation (or isolation) as $(\alpha_-)L$. The slab thickness can also be used to adjust this value. Typical numerical results are given in Example 9.2.

One advantage of this geometry is that the full-height slab is easy to bias with an external C-shaped permanent magnet, with no demagnetization factor. However, it suffers from several disadvantages:

- Zero forward attenuation cannot be obtained because the internal magnetic field is not truly circularly polarized.
- The bandwidth of the isolator is relatively narrow, dictated essentially by the linewidth, ΔH, of the ferrite.
- The geometry is not well suited for high-power applications because of poor heat transfer from the middle of the slab, and an increase in temperature will cause a change in M_s, which will degrade performance.

The first two problems noted above can be remedied to a significant degree with the addition of a dielectric loading slab; see reference [5] for details.

EXAMPLE 9.2   FERRITE RESONANCE ISOLATOR DESIGN

Design an E-plane resonance isolator in an X-band waveguide to operate at 10 GHz with a minimum forward insertion loss and 30 dB reverse attenuation. Use a 0.5 mm thick ferrite slab with $4\pi M_s = 1700$ G, $\Delta H = 200$ Oe, and $\epsilon_r = 13$. Determine the bandwidth for which the reverse attenuation is at least 27 dB.

Solution
The complex roots of (9.79) were found numerically using an interval-halving routine followed by a Newton–Raphson iteration. The approximate bias field, H_0, given by (9.87) is 2820 Oe, but numerical results indicate the actual field to be closer to 2840 Oe for resonance at 10 GHz. Figure 9.12a shows the calculated forward (α_+) and reverse (α_-) attenuation constants at 10 GHz versus slab position, and it can be seen that the minimum forward attenuation occurs for $c/a = 0.125$; the reverse attenuation at this point is $\alpha_- = 12.4$ dB/cm. Figure 9.12b shows the attenuation constants versus frequency for this slab position. For a total reverse attenuation of 30 dB, the length of the slab must be

$$L = \frac{30 \text{ dB}}{12.4 \text{ dB/cm}} = 2.42 \text{ cm}.$$

For the total reverse attenuation to be at least 27 dB, we must have

$$\alpha_- > \frac{27 \text{ dB}}{2.42 \text{ cm}} = 11.16 \text{ dB/cm}.$$

The bandwidth according to the above definition is, from the data of Figure 9.12b, less than 2%. This result could be improved by using a ferrite with a larger linewidth, at the expense of a longer or thicker slab and a higher forward attenuation.

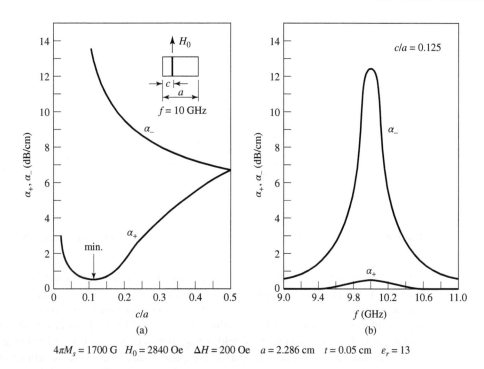

$$4\pi M_s = 1700\ \text{G} \quad H_0 = 2840\ \text{Oe} \quad \Delta H = 200\ \text{Oe} \quad a = 2.286\ \text{cm} \quad t = 0.05\ \text{cm} \quad \varepsilon_r = 13$$

FIGURE 9.12 | Forward and reverse attenuation constants for the resonance isolator of Example 9.2. (a) Versus slab position. (b) Versus frequency.

Next we consider a resonance isolator using the H-plane slab geometry of Figure 9.11b. If the slab is much thinner than it is wide, the demagnetization factors will be approximately $N_x = N_z = 0, N_y = 1$. This means that a stronger applied bias field will be required to produce the internal field H_0 in the y direction. However, the RF magnetic field components, h_x and h_z, will not be affected by the air–ferrite boundary since $N_x = N_z = 0$, and perfect circular polarized fields will exist in the ferrite when it is positioned at the circular polarization point of the empty guide, as given by (9.86). Another advantage of this geometry is that it has better thermal properties than the E-plane version since the ferrite slab has a large surface area in contact with the waveguide wall for heat dissipation.

Unlike the full-height E-plane slab case, the H-plane geometry of Figure 9.11b cannot be analyzed exactly. However, if the slab occupies only a very small fraction of the total guide cross section ($\Delta S/S \ll 1$, where ΔS and S are the cross-sectional areas of the slab and waveguide, respectively), the perturbational result for α_+ in (9.81) can be used with reasonable results. This expression is given in terms of the susceptibilities $\chi_{xx} = \chi'_{xx} - j\chi''_{xx}, \chi_{zz} = \chi'_{zz} - j\chi''_{zz}$, and $\chi_{xy} = \chi''_{xy} + j\chi'_{xy}$, as defined for a $\hat{y}$-biased ferrite in a manner similar to (9.22). For ferrite shapes other than a thin H-plane slab, these susceptibilities would have to be modified with the appropriate demagnetization factors, as in (9.43) [4].

As seen from the susceptibility expressions of (9.22), gyromagnetic resonance for this geometry will occur when $\omega = \omega_0$, which determines the internal bias field, H_0. The center of the slab is positioned at the circular polarization point of the empty guide, as given by (9.86). This should result in a near-zero forward attenuation constant. The total reverse attenuation, or isolation, can be controlled with either the length, L, of the ferrite slab or its cross section, ΔS, since (9.81) shows $\alpha_\pm$ is proportional to $\Delta S/S$. If $\Delta S/S$ is too large, however, the purity of circular polarization over the slab cross section will be degraded, and forward loss will increase. One practical alternative is to use a second, identical ferrite slab on the top wall of the guide to double $\Delta S/S$ without significantly degrading polarization purity.

The Field Displacement Isolator

Another type of isolator uses the fact that the electric field distributions of the forward and reverse waves in a ferrite slab-loaded waveguide can be quite different. As illustrated in Figure 9.13, the electric field for the forward wave can be made to vanish at the side of the ferrite slab at $x = c + t$, while the electric field of the

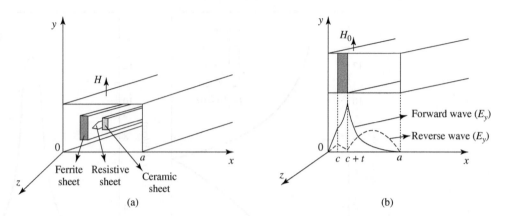

FIGURE 9.13 | (a) Geometry of a field displacement isolator. (b) Electric fields of a field displacement isolator.

reverse wave can be quite large at this same plane. If a thin resistive sheet is placed in this position the forward wave will be essentially unaffected, but the reverse wave will be attenuated. Such an isolator is called a *field displacement isolator*; high values of isolation with a relatively compact device can be obtained with bandwidths on the order of 10%. Another advantage of the field displacement isolator over the resonance isolator is that a much smaller bias field is required since it operates well below gyromagnetic resonance.

The main problem in designing a field displacement isolator is to determine the design parameters that produce field distributions like those shown in Figure 9.13. The general form of the electric field is given in (9.77a), from the analysis of the ferrite slab-loaded waveguide. This shows that for the electric field of the forward wave to have a sinusoidal dependence for $c + t < x < a$, and to vanish at $x = c + t$, the cutoff wave number k_a^+ must be real and satisfy the condition that

$$k_a^+ = \frac{\pi}{d}, \tag{9.88}$$

where $d = a - c - t$. In addition, the electric field of the reverse wave should have a hyperbolic dependence for $c + t < x < a$, which implies that k_a^- must be imaginary. Since from (9.75), $k_a^2 = k_0^2 - \beta^2$, the above conditions imply that $\beta^+ < k_0$ and $\beta^- > k_0$, where $k_0 = \omega\sqrt{\mu_0\epsilon_0}$. These conditions on $\beta_\pm$ depend critically on the slab position, which must be determined by numerically solving (9.79) for the propagation constants. The slab thickness also affects this result, but less critically; a typical value is $t = a/10$.

It also turns out that in order to satisfy (9.88), to force $E_y = 0$ at $x = c + t$, $\mu_e = (\mu^2 - \kappa^2)/\mu$ must be negative. This requirement can be intuitively understood by thinking of the waveguide mode for $c + t < x < a$ as a superposition of two obliquely traveling plane waves. The magnetic field components H_x and H_z of these waves are both perpendicular to the bias field, a situation that is similar to the extraordinary plane waves discussed in Section 9.2, where it was seen that propagation would not occur for $\mu_e < 0$. Applying this cutoff condition to the ferrite-loaded waveguide will allow a null in E_y for the forward wave to be formed at $x = c + t$.

The condition that μ_e be negative depends on the frequency, saturation magnetization, and bias field. Figure 9.8 shows the dependence of μ_e versus bias field for several frequencies and saturation magnetization. This type of data can be used to select the saturation magnetization and bias field to give $\mu_e < 0$ at the design frequency. Observe that higher frequencies will require a ferrite with higher saturation magnetization and a higher bias field, but $\mu_e < 0$ always occurs before the resonance in μ_e at $\sqrt{\omega_0(\omega_0 + \omega_m)}$. Further design details will be given in the following example.

EXAMPLE 9.3   FIELD DISPLACEMENT ISOLATOR DESIGN

Design a field displacement isolator in an X-band waveguide to operate at 11 GHz. The ferrite has $4\pi M_s = 3000$ G and $\epsilon_r = 13$. Ferrite loss can be ignored.

Solution
We first determine the internal bias field, H_0, such that $\mu_e < 0$. This can be found from Figure 9.8, which shows μ_e/μ_o versus H_0 for $4\pi M_s = 3000$ G at 11 GHz. We see that $H_0 = 1200$ Oe should be sufficient.

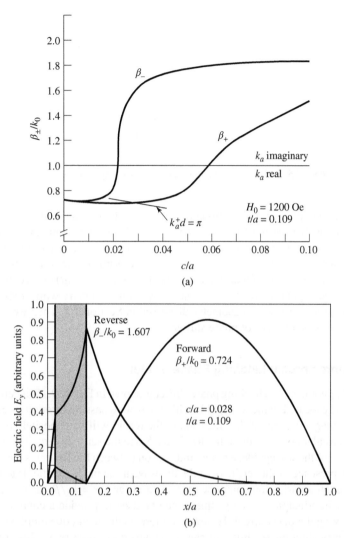

FIGURE 9.14 | Propagation constants and electric field distribution for the field displacement isolator of Example 9.3. (a) Forward and reverse propagation constants versus slab position. (b) Electric field amplitudes for the forward and reverse waves.

Also note from this figure that a ferrite with a smaller saturation magnetization would require a much larger bias field.

Next we determine the slab position, c/a, by numerically solving (9.79) for the propagation constants, $\beta_{\pm}$, as a function of c/a. The slab thickness was set to $t = 0.25$ cm, which is approximately $a/10$. Figure 9.14a shows the resulting propagation constants, as well as the locus of points where β_+ and c/a satisfy the condition of (9.88). The intersection of β_+ with this locus will ensure that $E_y = 0$ at $x = c + t$ for the forward wave; this intersection occurs for a slab position of $c/a = 0.028$. The resulting propagation constants are $\beta_+ = 0.724k_0 < k_0$ and $\beta_- = 1.607k_0 > k_0$.

The electric fields are plotted in Figure 9.14b. Note that the forward wave has a null at the face of the ferrite slab, while the reverse wave has a peak (the relative amplitudes of these fields are arbitrary). A resistive sheet can be placed at this point to attenuate the reverse wave. The actual isolation will depend on the resistivity of this sheet; a value of 75 Ω per square is typical.

9.5 FERRITE PHASE SHIFTERS

Another important application of ferrite materials is in *phase shifters*, which are two-port components that provide variable phase shift by changing the bias field of the ferrite. (Microwave diodes and FETs can also be

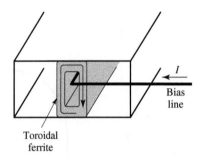

FIGURE 9.15 | Geometry of a nonreciprocal latching phase shifter using a ferrite toroid.

used to implement phase shifters; see Chapter 11.) Phase shifters find application in test and measurements systems, but the most significant use is in phased array antennas where the antenna beam can be steered in space by electronically controlled phase shifters. Because of this demand, many different types of phase shifters have been developed, both reciprocal (same phase shift in either direction) and nonreciprocal [2, 6]. One of the most useful designs is the latching (or *remanent*) nonreciprocal phase shifter using a ferrite toroid in a rectangular waveguide; we can analyze this geometry with a reasonable degree of approximation using the double ferrite slab geometry discussed in Section 9.3. Then we will qualitatively discuss the operation of a few other types of phase shifters.

Nonreciprocal Latching Phase Shifter

The geometry of a latching phase shifter is shown in Figure 9.15; it consists of a toroidal ferrite core symmetrically located in the waveguide with a bias wire passing through its center. When the ferrite is magnetized, the magnetization of the sidewalls of the toroid will be oppositely directed and perpendicular to the plane of circular polarization of the RF fields. Since the sense of circular polarization is also opposite on opposite sides of the waveguide, a strong interaction between the RF fields and the ferrite can be obtained. Of course, the presence of the ferrite perturbs the waveguide fields (the fields tend to concentrate in the ferrite), so the circular polarization point does not occur at $\tan k_c x = k_c / \beta_0$, as it does for an empty guide.

In principle, such a geometry can be used to provide a continuously variable (analog) phase shift by varying the bias current. However, a more useful technique employs the magnetic hysteresis of the ferrite to provide a phase shift that can be switched between two values (digital). A typical hysteresis curve is shown in Figure 9.16, showing the variation in magnetization, M, with bias field, H_0. When the ferrite is initially demagnetized and the bias field is off, both M and H_0 are zero. As the bias field is increased, the magnetization increases along the dashed-line path until the ferrite is magnetically saturated, and $M = M_s$. If the bias field is now reduced to zero, the magnetization will decrease to a remanent condition (like a permanent magnet), where $M = M_r$. A bias field in the opposite direction will saturate the ferrite with $M = -M_s$, whereupon removal of the bias field will leave the ferrite in a remanent state with $M = -M_r$. Thus we can "latch" the ferrite magnetization in one of two states, where $M = \pm M_r$, giving a digital phase shift.

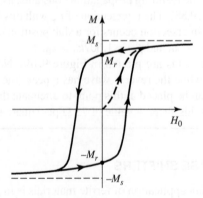

FIGURE 9.16 | A hysteresis curve for a ferrite toroid.

The amount of differential phase shift between these two states is controlled by the length of the ferrite toroid. In practice, several sections having individual bias lines and decreasing lengths are used in series to give binary differential phase shifts of 180°, 90°, 45°, etc., to as fine a resolution as desired (or can be afforded). An important advantage of the latching mode of operation is that the bias current does not have to be continuously applied, but only pulsed with one polarity or the other to change the polarity of the remanent magnetization; switching speeds can be on the order of a few microseconds. The bias wire can be oriented perpendicular to the electric field in the guide, with a negligible perturbing effect. The top and bottom walls of the ferrite toroid have very little magnetic interaction with the RF fields because the magnetization is not perpendicular to the plane of circular polarization, and the top and bottom magnetizations are oppositely directed. These walls provide mainly a dielectric loading effect, and the essential operating features of the remanent phase shifters can be obtained by considering the simpler dual ferrite slab geometry of Section 9.3.

For a given operating frequency and waveguide size, the design of a remanent dual-slab phase shifter mainly involves the determination of the slab thickness, t, the spacing between the slabs, $s = 2d = a - 2c - 2t$ (see Figure 9.10), and the length of the slabs for the desired phase shift. This requires the propagation constants, $\beta_{\pm}$, for the dual-slab geometry, which can be numerically evaluated from the transcendental equation of (9.84). This equation requires values for μ and κ, which can be determined from (9.25) for the remanent state by setting $H_0 = 0$ ($\omega_0 = 0$) and $M_s = M_r$ ($\omega_m = \mu_0 \gamma M_r$):

$$\mu = \mu_0, \tag{9.89a}$$

$$\kappa = -\mu_0 \frac{\omega_m}{\omega}. \tag{9.89b}$$

The differential phase shift, $\beta_+ - \beta_-$, is linearly proportional to κ for κ/μ_0 up to about 0.5. Then, since κ is proportional to M_r, as seen by (9.89b), it follows that a shorter length of ferrite can be used to provide a given phase shift if a ferrite with a higher remanent magnetization is selected. The insertion loss of the phase shifter increases with length but is a function of the ferrite linewidth, ΔH. A figure of merit commonly used to characterize phase shifters is the ratio of phase shift to insertion loss, measured in degrees/dB.

EXAMPLE 9.4   REMANENT PHASE SHIFTER DESIGN

Design a two-slab remanent phase shifter at 10 GHz using an X-band waveguide with ferrite having $4\pi M_r = 1786$ G and $\epsilon_r = 13$. Assume that the ferrite slabs are spaced 2 mm apart. Determine the slab thickness for maximum differential phase shift, and the lengths of the slabs for 180° and 90° phase shifter sections.

Solution
From (9.89) we have that

$$\frac{\mu}{\mu_0} = 1,$$

$$\frac{\kappa}{\mu_0} = \pm \frac{\omega_m}{\omega} = \pm \frac{(2.8 \text{ MHz/Oe})(1786 \text{ G})}{10,000 \text{ MHz}} = \pm 0.5.$$

Using a numerical root-finding technique, such as interval halving, we can solve (9.84) for the propagation constants β_+ and β_- by using positive and negative values of κ. Figure 9.17 shows the resulting differential phase shift, $(\beta_+ - \beta_-)/k_0$, versus slab thickness, t, for several slab spacings. Observe that the phase shift increases as the spacing, s, between the slabs decreases and as the slab thickness increases, for t/a up to about 0.12.

From the curve in Figure 9.17 for $s = 1$ mm, we see that the optimum slab thickness for maximum phase shift is $t/a = 0.12$, or $t = 2.74$ mm, since $a = 2.286$ cm for an X-band guide. The corresponding normalized differential phase shift is 0.39, so

$$\beta_+ - \beta_- = 0.39 k_0 = 0.39 \left(\frac{2.09 \text{ rad}}{\text{cm}} \right)$$

$$= 0.8151 \text{ rad/cm} = 46.7°/\text{cm}.$$

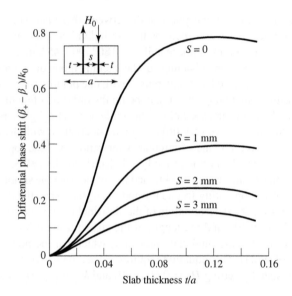

FIGURE 9.17 | Differential phase shift for the two-slab remanent phase shifter of Example 9.4.

The ferrite length required for the 180° phase shift section is then

$$L = \frac{180°}{46.7°/\text{cm}} = 3.85 \text{ cm,}$$

while the length required for a 90° section is

$$L = \frac{90°}{46.7°/\text{cm}} = 1.93 \text{ cm.}$$

Other Types of Ferrite Phase Shifters

Many other types of ferrite phase shifters have been developed, with various combinations of rectangular or circular waveguide, transverse or longitudinal biasing, latching or continuous phase variation, and reciprocal or nonreciprocal operation. Phase shifters using printed transmission lines have also been proposed. Even though PIN diode and FET circuits offer a less bulky and more integratable alternative to ferrite components, ferrite phase shifters often have advantages in terms of cost, power-handling capacity, and power requirements. However, there is still a great need for a low-cost, compact phase shifter, primarily for phased array antenna systems.

Several waveguide phase shifter designs are derived from the nonreciprocal Faraday rotation phase shifter shown in Figure 9.18. In operation, a rectangular waveguide TE_{10} mode entering at the left is converted to a TE_{11} circular waveguide mode with a short transition section. Then a quarter-wave dielectric plate, oriented 45° from the electric field vector, converts the wave to an RHCP wave by providing a 90° phase difference between the field components that are parallel and perpendicular to the plate. In the ferrite-loaded region the phase delay is $\beta_+ z$, which can be controlled with the bias field strength. A second quarter-wave plate converts the wave back to a linearly polarized field. The operation is similar for a wave entering at the right, except now the phase delay is $\beta_- z$; the phase shift is thus nonreciprocal. The ferrite rod is biased longitudinally, in the direction of propagation, with a solenoid coil. This type of phase shifter can be made reciprocal by using nonreciprocal quarter-wave plates to convert a linearly polarized wave to the same sense of circular polarization for either propagation direction.

The *Reggia-Spencer* phase shifter, shown in Figure 9.19, is a popular reciprocal phase shifter. In either rectangular or circular waveguide form, a longitudinally biased ferrite rod is centered in the guide. When the diameter of the rod is greater than a certain critical size, the fields become tightly bound to the ferrite and are circularly polarized. A large reciprocal phase shift can be obtained over relatively short lengths, although the phase shift is rather frequency sensitive.

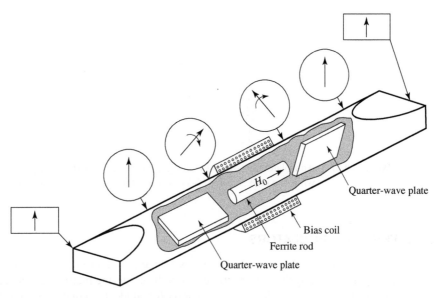

FIGURE 9.18 | Nonreciprocal Faraday rotation phase shifter.

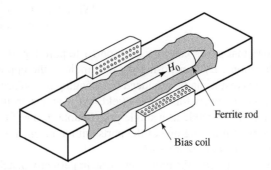

FIGURE 9.19 | Reggia-Spencer reciprocal phase shifter.

The Gyrator

An important canonical nonreciprocal component is the *gyrator*, which is a two-port device having a 180° differential phase shift. The schematic symbol for a gyrator is shown in Figure 9.20, and the scattering matrix for an ideal gyrator is

$$[S] = \begin{bmatrix} 0 & 1 \\ -1 & 0 \end{bmatrix}, \tag{9.90}$$

which shows that it is lossless, matched, and nonreciprocal. Use of the gyrator as a basic nonreciprocal building block in combination with reciprocal dividers and couplers can lead to useful equivalent circuits for nonreciprocal components such as isolators and circulators. Figure 9.21, for example, shows an equivalent circuit for an isolator using a gyrator and two quadrature hybrids.

The gyrator can be implemented as a phase shifter with a 180° differential phase shift; bias can be provided with a permanent magnet, making the gyrator a passive device.

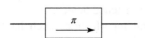

FIGURE 9.20 | Symbol for a gyrator, which has a differential phase shift of 180°.

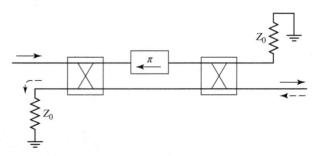

FIGURE 9.21 | An isolator constructed with a gyrator and two quadrature hybrids. The forward wave ($\rightarrow$) is passed, while the reverse wave ($\leftarrow$) is absorbed in the matched load of the first hybrid.

9.6 FERRITE CIRCULATORS

As discussed in Section 7.1, a *circulator* is a three-port microwave device that can be lossless and matched at all ports; by using the unitary properties of the scattering matrix we were able to show that such a device must be nonreciprocal. The scattering matrix for an ideal circulator thus has the following form:

$$[S] = \begin{bmatrix} 0 & 0 & 1 \\ 1 & 0 & 0 \\ 0 & 1 & 0 \end{bmatrix}, \tag{9.91}$$

which shows that power can flow from port 1 to port 2, port 2 to port 3, and port 3 to port 1, but not in the reverse directions. By transposing the port indices, the opposite circularity can be obtained. For a ferrite circulator, this result can be produced by changing the polarity of the magnetic bias field. Most circulators use permanent magnets for the bias field, but if an electromagnet is used the circulator can operate in a latching (remanent) mode as a single-pole double-throw (SPDT) switch. A circulator can also be used as an isolator by terminating one of the ports with a matched load. A photograph of a disassembled stripline circulator is shown in Figure 9.22.

We will first discuss the properties of an imperfectly matched circulator in terms of its scattering matrix, and then we will analyze the operation of the stripline junction circulator. The operation of waveguide circulators is similar in principle.

Properties of a Mismatched Circulator

If we assume that a circulator has circular symmetry around its three ports and is lossless, but not perfectly matched, we can write its scattering matrix as

$$[S] = \begin{bmatrix} \Gamma & \beta & \alpha \\ \alpha & \Gamma & \beta \\ \beta & \alpha & \Gamma \end{bmatrix}. \tag{9.92}$$

Since the circulator is assumed lossless, the scattering matrix must be unitary, which implies the following two conditions:

$$|\Gamma|^2 + |\beta|^2 + |\alpha|^2 = 1, \tag{9.93a}$$

$$\Gamma\beta^* + \alpha\Gamma^* + \beta\alpha^* = 0. \tag{9.93b}$$

If the circulator were matched ($\Gamma = 0$), then (9.93) shows that either $\alpha = 0$ and $|\beta| = 1$, or $\beta = 0$ and $|\alpha| = 1$; this describes the ideal circulator with its two possible circularity states. Observe that this condition depends only on a lossless and matched device.

Now assume small imperfections, such that $|\Gamma| \ll 1$. To be specific, consider the circularity state where power flows primarily in the 1-2-3 direction, so that $|\alpha|$ is close to unity and $|\beta|$ is small. Then $\beta\Gamma \sim 0$, and (9.93b) shows that $\alpha\Gamma^* + \beta\alpha^* \simeq 0$, so $|\Gamma| \simeq |\beta|$. Then (9.93a) shows that $|\alpha|^2 \simeq 1 - 2|\beta|^2 \simeq 1 - 2|\Gamma|^2$, or

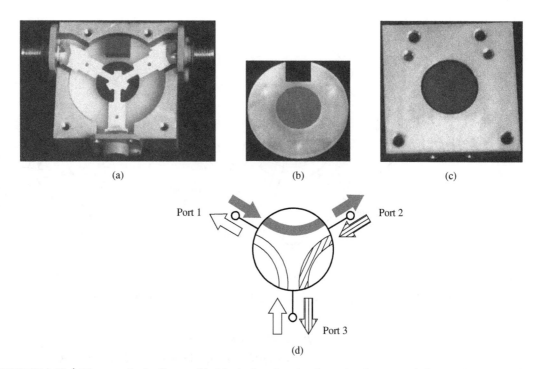

FIGURE 9.22 | Photograph of a disassembled ferrite junction circulator, showing (a) a stripline conductor, (b) a ferrite disk, and (c) a bias magnet. (d) Power flow between the ports. The middle port of the circulator is terminated with a matched load, so this circulator is actually configured as an isolator. Note the change in the width of the stripline conductors due to the different dielectric constants of the ferrite and the surrounding plastic material.

$|\alpha| \simeq 1 - |\Gamma|^2$. Then the scattering matrix of (9.92) can be written as

$$[S] = \begin{bmatrix} \Gamma & \Gamma & 1-\Gamma^2 \\ 1-\Gamma^2 & \Gamma & \Gamma \\ \Gamma & 1-\Gamma^2 & \Gamma \end{bmatrix}, \tag{9.94}$$

ignoring phase factors. This result shows that circulator isolation, $\beta \simeq \Gamma$, and transmission, $\alpha \simeq 1 - \Gamma^2$, both deteriorate as the input ports become mismatched.

Junction Circulator

The geometry of a stripline junction circulator is shown in Figure 9.23, and in the photograph of Figure 9.22. Two ferrite disks fill the spaces between the center metallic disk and the ground planes of the stripline. Three stripline conductors are attached to the periphery of the center disk at 120° intervals, forming the three ports of the circulator. The DC bias field is applied normal to the ground planes.

In operation, the ferrite disks form a dielectric resonator; in the absence of a bias field, this resonator has a single lowest order resonant mode with a $\cos\phi$ (or $\sin\phi$) dependence. When the ferrite is magnetically biased this mode breaks into two resonant modes having slightly different resonant frequencies. The operating frequency of the circulator can then be chosen so that the superposition of these two modes adds at the output port and cancels at the isolated port.

We can analyze the junction circulator by treating it as a thin cavity resonator with electric walls on the top and bottom, and an approximate magnetic wall on the side. Then $E_\rho = E_\phi \simeq 0$, and $\partial/\partial z = 0$, so we have TM modes. Since E_z on either side of the center conducting disk is antisymmetric, we need only consider the solution for one of the ferrite disks [7].

We first transform (9.23), $\bar{B} = [\mu]\bar{H}$, from rectangular to cylindrical coordinates:

$$\begin{aligned} B_\rho &= B_x \cos\phi + B_y \sin\phi \\ &= (\mu H_x + j\kappa H_y)\cos\phi + (-j\kappa H_x + \mu H_y)\sin\phi \\ &= \mu H_\rho + j\kappa H_\phi, \end{aligned} \tag{9.95a}$$

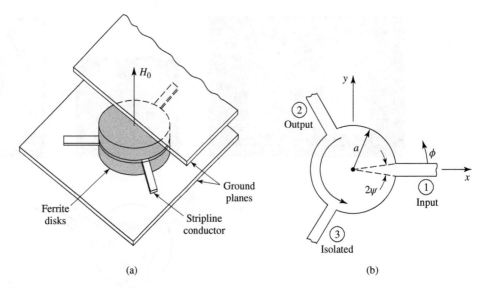

FIGURE 9.23 | A stripline junction circulator. (a) Pictorial view. (b) Geometry.

$$B_\phi = -B_x \sin \phi + B_y \cos \phi$$
$$= -(\mu H_x + j\kappa H_y) \sin \phi + (-j\kappa H_x + \mu H_y) \cos \phi$$
$$= -j\kappa H_\rho + \mu H_\phi. \tag{9.95b}$$

So we have that

$$\begin{bmatrix} B_\rho \\ B_\phi \\ B_z \end{bmatrix} = [\mu] \begin{bmatrix} H_\rho \\ H_\phi \\ H_z \end{bmatrix}, \tag{9.96}$$

where $[\mu]$ is the same matrix as for rectangular coordinates, as given in (9.24).

In cylindrical coordinates, with $\partial/\partial z = 0$, Maxwell's curl equations reduce to the following:

$$\frac{1}{\rho} \frac{\partial E_z}{\partial \phi} = -j\omega(\mu H_\rho + j\kappa H_\phi), \tag{9.97a}$$

$$-\frac{\partial E_z}{\partial \rho} = -j\omega(-j\kappa H_\rho + \mu H_\phi), \tag{9.97b}$$

$$\frac{1}{\rho} \left[\frac{\partial(\rho H_\phi)}{\partial \rho} - \frac{\partial H_\rho}{\partial \phi} \right] = j\omega\epsilon E_z. \tag{9.97c}$$

Solving (9.97a) and (9.97b) for H_ρ and H_ϕ in terms of E_z gives

$$H_\rho = \frac{jY}{k\mu} \left(\frac{\mu}{\rho} \frac{\partial E_z}{\partial \phi} + j\kappa \frac{\partial E_z}{\partial \rho} \right), \tag{9.98a}$$

$$H_\phi = \frac{-jY}{k\mu} \left(\frac{-j\kappa}{\rho} \frac{\partial E_z}{\partial \phi} + \mu \frac{\partial E_z}{\partial \rho} \right), \tag{9.98b}$$

where $k^2 = \omega^2 \epsilon (\mu^2 - \kappa^2)/\mu = \omega^2 \epsilon \mu_e$ is an effective wave number, and $Y = \sqrt{\epsilon/\mu_e}$ is an effective admittance. Using (9.98) to eliminate H_ρ and H_ϕ in (9.97c) gives a wave equation for E_z:

$$\frac{\partial^2 E_z}{\partial \rho^2} + \frac{1}{\rho} \frac{\partial E_z}{\partial \rho} + \frac{1}{\rho^2} \frac{\partial^2 E_z}{\partial \phi^2} + k^2 E_z = 0. \tag{9.99}$$

This equation is identical in form to the equation for E_z for the TM mode of a circular waveguide, so the general solution can be written as

$$E_{zn} = \left[A_{+n} e^{jn\phi} + A_{-n} e^{-jn\phi} \right] J_n(k\rho), \tag{9.100a}$$

where we have excluded the solution with $Y_n(k\rho)$ because E_z must be finite at $\rho = 0$. We will also need $H_{\phi n}$, which can be found using (9.98b):

$$H_{\phi n} = -jY \left\{ A_{+n}e^{jn\phi} \left[J_n'(k\rho) + \frac{n\kappa}{k\rho\mu}J_n(k\rho) \right] \right.$$

$$\left. + A_{-n}e^{-jn\phi} \left[J_n'(k\rho) - \frac{n\kappa}{k\rho\mu}J_n(k\rho) \right] \right\}. \tag{9.100b}$$

The resonant modes can now be found by enforcing the boundary condition that $H_\phi = 0$ at $\rho = a$.

If the ferrite is not magnetized, then $H_0 = M_s = 0$ and $\omega_0 = \omega_m = 0$, so that $\kappa = 0$ and $\mu = \mu_e = \mu_0$, and resonance occurs when

$$J_n'(ka) = 0,$$

or $ka = x_0 = p_{11}' = 1.841$. Define this frequency as ω_0 (not to be confused with $\omega_0 = \gamma\mu_0 H_0$):

$$\omega_0 = \frac{x_0}{a\sqrt{\epsilon\mu_e}} = \frac{1.841}{a\sqrt{\epsilon\mu_0}}. \tag{9.101}$$

When the ferrite is magnetized there are two possible resonant modes for each value of n, as associated with either a $e^{jn\phi}$ variation or a $e^{-jn\phi}$ variation. The resonance condition for the two $n = 1$ modes is

$$\frac{\kappa}{\mu x}J_1(x) \pm J_1'(x) = 0, \tag{9.102}$$

where $x = ka$. This result shows the nonreciprocal property of the circulator, since changing the sign of κ (the polarity of the bias field) in (9.102) leads to the other root and propagation in the opposite direction in ϕ.

If we let x_+ and x_- be the two roots of (9.102), then we can express the resonant frequencies for these two $n = 1$ modes as

$$\omega_\pm = \frac{x_\pm}{a\sqrt{\epsilon\mu_e}}. \tag{9.103}$$

We can develop an approximate result for $\omega_\pm$ if we assume that κ/μ is small, so that $\omega_\pm$ will be close to ω_0 of (9.101). Using a Taylor series about x_0 for the two terms in (9.102) gives the following results, since $J_1'(x_0) = 0$:

$$J_1(x) \simeq J_1(x_0) + (x - x_0)J_1'(x_0) = J_1(x_0),$$

$$J_1'(x) \simeq J_1'(x_0) + (x - x_0)J_1''(x_0)$$

$$= -(x - x_0)\left(1 - \frac{1}{x_0^2}\right)J_1(x_0).$$

Then (9.102) becomes

$$\frac{\kappa}{\mu x_0} \mp (x_\pm - x_0)\left(1 - \frac{1}{x_0^2}\right) = 0,$$

or

$$x_\pm \simeq x_0\left(1 \pm 0.418\frac{\kappa}{\mu}\right), \tag{9.104}$$

since $x_0 = 1.841$. This result gives the resonant frequencies as

$$\omega_\pm \simeq \omega_0\left(1 \pm 0.418\frac{\kappa}{\mu}\right). \tag{9.105}$$

Note that $\omega_\pm$ approaches ω_0 as $\kappa \to 0$, and that

$$\omega_- \leq \omega_0 \leq \omega_+.$$

We can use a superposition of these two modes to design a circulator. The amplitudes of these modes give two degrees of freedom that can be used to provide coupling from the input to the output port, and to provide cancellation at the isolated port. It will turn out that ω_0 will be the operating frequency, which will be between the resonances of the $\omega_\pm$ modes. Thus, $H_\phi \neq 0$ over the periphery of the ferrite disks since $\omega \neq \omega_\pm$. If we select port 1 as the input port, port 2 as the output port, and port 3 as the isolated port, as in Figure 9.23, we can assume the following E_z field at the ports at $\rho = a$:

$$
E_z(\rho = a, \phi) = \begin{cases} E_0, & \text{for } \phi = 0 \quad \text{(port 1)}, \\ -E_0, & \text{for } \phi = 120° \text{ (port 2)}, \\ 0, & \text{for } \phi = 240° \text{ (port 3)}. \end{cases} \tag{9.106a}
$$

If the feedlines are narrow, the E_z field will be relatively constant across their width. The corresponding H_ϕ field should be

$$
H_\phi(\rho = a, \phi) = \begin{cases} H_0 & \text{for } -\psi < \phi < \psi, \\ H_0 & \text{for } 120° - \psi < \phi < 120° + \psi, \\ 0 & \text{elsewhere}. \end{cases} \tag{9.106b}
$$

Equating (9.106a) to E_z of (9.100a) gives the mode amplitude constants as

$$
A_{+1} = \frac{E_0(1 + j/\sqrt{3})}{2J_1(ka)}, \tag{9.107a}
$$

$$
A_{-1} = \frac{E_0(1 - j/\sqrt{3})}{2J_1(ka)}. \tag{9.107b}
$$

Then (9.100a) and (9.100b) can be reduced to give the electric and magnetic fields as

$$
\begin{aligned}
E_{z1} &= \frac{E_0 J_1(k\rho)}{2J_1(ka)}\left[\left(1 + \frac{j}{\sqrt{3}}\right)e^{j\phi} + \left(1 - \frac{j}{\sqrt{3}}\right)e^{-j\phi}\right] \\
&= \frac{E_0 J_1(k\rho)}{J_1(ka)}\left(\cos\phi - \frac{\sin\phi}{\sqrt{3}}\right),
\end{aligned} \tag{9.108a}
$$

$$
\begin{aligned}
H_{\phi 1} = \frac{-jYE_0}{2J_1(ka)}&\left\{\left(1 + \frac{j}{\sqrt{3}}\right)\left[J_1'(k\rho) + \frac{\kappa}{k\rho\mu}J_1(k\rho)\right]e^{j\phi}\right. \\
&\left. + \left(1 - \frac{j}{\sqrt{3}}\right)\left[J_1'(k\rho) - \frac{\kappa}{k\rho\mu}J_1(k\rho)\right]e^{-j\phi}\right\}.
\end{aligned} \tag{9.108b}
$$

To approximately equate $H_{\phi 1}$ to H_ϕ in (9.106b) requires that H_ϕ be expanded in a Fourier series:

$$
\begin{aligned}
H_\phi(\rho = a, \phi) &= \sum_{n=-\infty}^{\infty} C_n e^{jn\phi} = \frac{2H_0\psi}{\pi} \\
&+ \frac{H_0}{\pi}\sum_{n=1}^{\infty}\left[(1 + e^{-j2\pi n/3})e^{jn\phi} + (1 + e^{j2\pi n/3})e^{-jn\phi}\right] \\
&\times \frac{\sin n\psi}{n}.
\end{aligned} \tag{9.109}
$$

The $n = 1$ term of this result is

$$
H_{\phi 1}(\rho = a, \phi) = \frac{-j\sqrt{3}H_0\sin\psi}{2\pi}\left[\left(1 + \frac{j}{\sqrt{3}}\right)e^{j\phi} - \left(1 - \frac{j}{\sqrt{3}}\right)e^{-j\phi}\right],
$$

which can be equated to (9.108b) for $\rho = a$. Equivalence can be obtained if two conditions are met:

$$
J_1'(ka) = 0,
$$

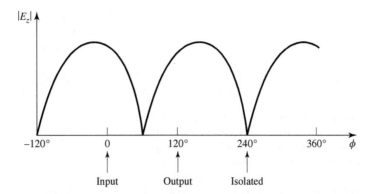

FIGURE 9.24 | Magnitude of the electric field around the periphery of the junction circulator.

and

$$\frac{YE_0\kappa}{ka\mu} = \frac{\sqrt{3}H_0\sin\psi}{\pi}.$$

The first condition is identical to the condition for resonance in the absence of bias, which implies that the operating frequency is ω_0, as given by (9.101). For a given operating frequency, (9.101) can then be used to find the disk radius, a. The second condition can be related to the wave impedance at port 1 or 2:

$$Z_w = \frac{E_0}{H_0} = \frac{\sqrt{3}ka\mu\sin\psi}{\pi Y\kappa} \simeq \frac{\mu\sin\psi}{\kappa Y}, \tag{9.110}$$

since $\sqrt{3}ka/\pi = \sqrt{3}(1.841)/\pi \simeq 1.0$. Thus, Z_w can be controlled for impedance matching by adjusting κ/μ via the bias field.

We can compute power flows at the three ports as follows:

$$P_{\text{in}} = P_1 = -\hat{\rho} \cdot \bar{E} \times \bar{H}^* = E_z H_\phi \Big|_{\phi=0} = \frac{E_0 H_0 \sin\psi}{\pi} = \frac{E_0^2 \kappa Y}{\pi\mu}, \tag{9.111a}$$

$$P_{\text{out}} = P_2 = \hat{\rho} \cdot \bar{E} \times \bar{H}^* = -E_z H_\phi \Big|_{\phi=120°} = \frac{E_0 H_0 \sin\psi}{\pi} = \frac{E_0^2 \kappa Y}{\pi\mu}, \tag{9.111b}$$

$$P_{\text{iso}} = P_3 = \hat{\rho} \cdot \bar{E} \times \bar{H}^* = -E_z H_\phi \Big|_{\phi=240°} = 0. \tag{9.111c}$$

These results show that power flow occurs from port 1 to port 2, but not from port 1 to port 3. By the azimuthal symmetry of the circulator, this also implies that power can be coupled from port 2 to port 3, or from port 3 to port 1, but not in the reverse directions.

The electric field of (9.108a) is sketched in Figure 9.24 along the periphery of the circulator, showing that the amplitudes and phases of the $e^{\pm j\phi}$ modes are such that their superposition gives a null at the isolated port, with equal voltages at the input and output ports. This result ignores the loading effect of the input and output lines, which will distort the field from that shown in Figure 9.24. This design is narrowband, but bandwidth can be improved using dielectric loading; the analysis then requires consideration of higher order modes.

REFERENCES

[1] R. F. Soohoo, *Microwave Magnetics*, Harper and Row, New York, 1985.

[2] A. J. Baden Fuller, *Ferrites at Microwave Frequencies*, Peter Peregrinus, London, 1987.

[3] R. E. Collin, *Field Theory of Guided Waves*, McGraw-Hill, New York, 1960.

[4] B. Lax and K. J. Button, *Microwave Ferrites and Ferrimagnetics*, McGraw-Hill, New York, 1962.

[5] F. E. Gardiol and A. S. Vander Vorst, "Computer Analysis of E-Plane Resonance Isolators," *IEEE Transactions on Microwave Theory and Techniques*, vol. MTT-19, pp. 315–322, March 1971.

[6] G. P. Rodrigue, "A Generation of Microwave Ferrite Devices," *Proceedings of the IEEE*, vol. 76, pp. 121–137, February 1988.

[7] C. E. Fay and R. L. Comstock, "Operation of the Ferrite Junction Circulator," *IEEE Transactions on Microwave Theory and Techniques*, vol. MTT-13, pp. 15–27, January 1965.

PROBLEMS

9.1 An LHCP RF magnetic field of $\bar{H} = (0.6\hat{x} + j0.5\hat{y})$ A/m is applied to a calcium vanadium garnet (CVG) ferrite medium having a saturation magnetization of $4\pi M_s = 900$ G. Ignoring loss, calculate the resulting magnetic flux density $\bar{B}$ at $f = 2$ GHz for two cases: (a) no magnetic bias field and ferrite demagnetized ($M_s = H_0 = 0$), and (b) a z-directed magnetic bias field of 750 Oe.

9.2 Consider the following field transformations from linearly polarized to circular polarized components:

$$B^+ = (B_x + jB_y)/2, \quad H^+ = (H_x + jH_y)/2, \quad \text{(RHCP)}$$
$$B^- = (B_x - jB_y)/2, \quad H^- = (H_x - jH_y)/2, \quad \text{(LHCP)}$$
$$B_z = B_z, \quad\quad H_z = H_z.$$

For a z-biased ferrite medium, show that the relation between $\bar{B}$ and $\bar{H}$ can be expressed in terms of a diagonal tensor permeability as follows:

$$\begin{bmatrix} B^+ \\ B^- \\ B_z \end{bmatrix} = \begin{bmatrix} (\mu + \kappa) & 0 & 0 \\ 0 & (\mu - \kappa) & 0 \\ 0 & 0 & \mu_0 \end{bmatrix} \begin{bmatrix} H^+ \\ H^- \\ H_z \end{bmatrix}.$$

9.3 A tunable oscillator uses a YIG sphere with $4\pi M_s = 1700$ G and requires a magnetic bias field of 800 Oe internal to the sphere. What is the external applied magnetic field strength that must be applied to produce this internal field?

9.4 A thin ferrite rod with $4\pi M_s = 900$ G is magnetically biased along its axis. Find the external bias field strength required to produce a gyromagnetic resonance at 2.5 GHz.

9.5 An infinite lossless ferrite medium with a saturation magnetization of $4\pi M_s = 1250$ G and a dielectric constant of 10 is biased to a field strength of 600 Oe. At 8 GHz, calculate the differential phase shift per meter between an RHCP and an LHCP plane wave propagating in the direction of bias. If a linearly polarized wave is propagating in this material, what is the distance it must travel in order that its polarization is rotated 90°?

9.6 An infinite lossless ferrite medium with a saturation magnetization of $4\pi M_s = 1780$ G and a dielectric constant of 13 is biased in the $\hat{x}$ direction with a field strength of 2100 Oe. At 5 GHz, two plane waves propagate in the $+z$ direction, one linearly polarized in x and the other linearly polarized in y. What is the distance these two waves must travel so that the differential phase shift between them is 180°?

9.7 Consider a circularly polarized plane wave normally incident on an infinite ferrite medium, as shown in the accompanying figure. Calculate the reflection and transmission coefficients for an RHCP (Γ^+, T^+) and an LHCP (Γ^-, T^-) incident wave. HINT: The transmitted wave will be polarized in the same sense as the incident wave, but the reflected wave will be oppositely polarized.

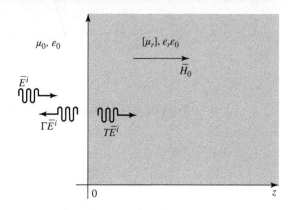

9.8 An infinite lossless ferrite material with $4\pi M_s = 1200$ G is biased in the $\hat{y}$ direction with $\bar{H}_0 = H_0\hat{y}$. Determine the range of H_0, in oersteds, where an extraordinary wave (polarized in $\hat{y}$, propagating in $\hat{x}$) will be cut off. The frequency is 4 GHz.

9.9 Find the forward and reverse propagation constants for a waveguide half-filled with a transversely biased ferrite. (The geometry of Figure 9.9 with $c = 0$ and $t = a/2$.) Assume $a = 1.0$ cm, $f = 10$ GHz, $4\pi M_s = 1700$ G, and $\epsilon_r = 13$. Plot versus $H_0 = 0$ to 1500 Oe. Ignore loss and the fact that the ferrite may not be saturated for small H_0.

9.10 Find the forward and reverse propagation constants for a waveguide filled with two pieces of oppositely biased ferrite. (The geometry of Figure 9.10 with $c = 0$ and $t = a/2$.) Assume $a = 1.0$ cm, $f = 10$ GHz, $4\pi M_s = 1700$ G, and $\epsilon_r = 13$. Plot versus $H_0 = 0$ to 1500 Oe. Ignore loss and the fact that the ferrite may not be saturated for small H_0.

9.11 Consider a wide, thin ferrite slab in a rectangular X-band waveguide, as shown in Figure 9.11b. If $f = 10$ GHz, $4\pi M_s = 1700$ G, $c = a/4$, and $\Delta S = 2$ mm², use the perturbation formula of (9.80) to plot the differential phase shift, $(\beta_+ - \beta_-)/k_0$, versus the bias field for $H_0 = 0$ to 1200 Oe. Ignore loss.

9.12 An E-plane resonance isolator with the geometry of Figure 9.11a is to be designed to operate at 10 GHz, with a ferrite having a saturation magnetization of $4\pi M_s = 1200$ G. (a) What is the approximate bias field, H_0, required for resonance? (b) What is the required bias field if the H-plane geometry of Figure 9.11b is used?

9.13 Design a resonance isolator using the H-plane ferrite slab geometry of Figure 9.11b in an X-band waveguide. The isolator should have minimum forward insertion loss, and a reverse attenuation of 30 dB at 10 GHz. Use a ferrite slab having $\Delta S/S = 0.01$, $4\pi M_s = 1700$ G, and $\Delta H = 200$ Oe.

9.14 Calculate and plot the two normalized positions, x/a, where the magnetic fields of the TE_{10} mode of an empty rectangular waveguide are circularly polarized, for $k_0 = k_c$ to $2k_c$.

9.15 A conceptual latching ferrite phase shifter using the bire-fringence effect is shown below. In state 1, the ferrite is magnetized so that $H_0 = 0$ and $\bar{M} = M_r\hat{x}$. In state 2, the ferrite is magnetized so that $H_0 = 0$ and $\bar{M} = M_r\hat{y}$. If $f = 10$ GHz, $\epsilon_r = 13$, and $4\pi M_r = 1780$ G, find the required length L to achieve a differential phase shift of 90°. Assume the incident plane wave is $\hat{x}$ polarized for both states, and ignore reflections.

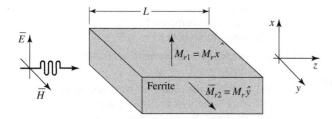

9.16 Rework Example 9.4 with a slab spacing of $s = 2$ mm and a remanent magnetization of 1000 G. (Assume all other parameters as unchanged and that the differential phase shift is linearly proportional to κ.)

9.17 Consider a latching phase shifter constructed with a wide, thin H-plane ferrite slab in an X-band waveguide, as shown in Figure 9.11b. If $f = 10$ GHz, $4\pi M_r = 1200$ G, $c = a/4$, and $\Delta S = 2$ mm², use the perturbation formula of (9.80) to calculate the required length for a differential phase shift of 22.5°.

9.18 Design a gyrator using the twin H-plane ferrite slab geometry shown below. The frequency is 10 GHz, and the saturation magnetization is $4\pi M_s = 1720$ G. The cross-sectional area of each slab is 3.0 mm² and the guide is X-band waveguide. The permanent magnet has a field strength of $H_a = 4000$ Oe. Determine the internal field in the ferrite, H_0, and use the perturbation formula of (9.80) to determine the optimum location of the slabs and the length, L, to give the necessary 180° differential phase shift.

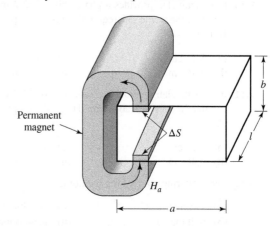

9.19 Draw an equivalent circuit for a circulator using a gyrator and two couplers.

9.20 A certain lossless circulator has a return loss of 15 dB. What is the isolation? What is the isolation if the return loss is 30 dB?

MULTIPLE CHOICE QUESTIONS

9.1 Which of the following is a nonreciprocal device?

(a) branch-line coupler
(b) circulator
(c) magic-T
(d) Wilkinson power divider

9.2 Magnetic properties exist in a material due to the existence of which of the following

(a) magnetic dipole moment
(b) electric dipole moment
(c) electrons in atoms
(d) none of the above

9.3 Which of the following is a measure of the relative contribution of the spin moment and the orbital moment to the total magnetic moment?

(a) Newton's ratio   (c) Landé's factor
(b) Gibbs factor   (d) none of the above

9.4 In which quadrant, does the operating point of a permanent magnetic lie?

(a) first quadrant   (c) third quadrant
(b) second quadrant   (d) fourth quadrant

9.5 How many ports does a microwave device need to work as a ferrite isolator?

(a) 4   (c) 2
(b) 3   (d) 1

9.6 Which of the following has a very high attenuation near the gyromagnetic resonance of ferrites?

(a) linearly polarized wave
(b) left polarized wave
(c) circularly polarized wave
(d) right polarized wave

9.7 Using ferrite materials, the isolators constructed must operate at

(a) magnetic resonance
(b) gyromagnetic resonance
(c) isolator resonance
(d) none of the above

9.8 The length of a ferrite slab for E-plane resonance isolator operating at 8 GHz with a minimum forward insertion loss and 28 dB reverse attenuation and reverse attenuation at this point being $\alpha_- = 11.5$ dB/cm is

(a) 2.4 cm   (c) 4.2 cm
(b) 3.2 cm   (d) 3.8 cm

9.9 The electric field distribution of the reverse and forward waves in a ferrite slab-loaded waveguide is quite different. This property is used in

(a) field displacement isolator
(b) waveguide isolator
(c) resonance isolator
(d) none of the above

9.10 If a ferrite slab provides a phase shift of 52°/cm, then the length of the ferrite slab required to produce a phase shift of 180° is

(a) 1.73 cm   (c) 5.19 cm
(b) 3.46 cm   (d) 6.92 cm

9.11 If a ferrite slab provides a phase shift of 52°/cm, then the length of the ferrite slab required to produce a phase shift of 90° is

(a) 1.73 cm   (c) 5.19 cm
(b) 3.46 cm   (d) 6.92 cm

9.12 The scattering matrix for a gyrator is

(a) symmetric   (c) identity matrix
(b) null matrix   (d) skew symmetric

9.13 If a ferrite slab produces a phase shift of 0.504 rad/cm, then the length of the slab required to produce a phase shift of 150° is

(a) 1.73 cm   (c) 5.19 cm
(b) 3.46 cm   (d) 6.92 cm

9.14 Around the periphery of the junction circulator, the curve in the plot of the magnitude of an electric field has

(a) one peak   (c) three peaks
(b) two peaks   (d) four peaks

9.15 In which form is the ferrite material present in a stripline junction circulator?

(a) ferrite cubes
(b) ferrite disk
(c) ferrite material is not used in a microstrip circulator
(d) slab

9.16 The diagonal elements of the matrix in the scattering matrix representation of a nonideal circulator are

(a) one
(b) zero
(c) reflection coefficient Γ
(d) none of the above

9.17 In a practical ferrite circulator, opposite circulatory can be obtained by

(a) changing the polarity of the magnetic bias field
(b) impedance matching the input ports
(c) changing the order of port operation
(d) none of the above

9.18 In an ideal three-port circulator the total number of 1's in the scattering matrix is

(a) 2   (c) 4
(b) 3   (d) 6

9.19 The residual magnetization retained in a magnetic material after demagnetization of the material is called

(a) retardation   (c) remanence
(b) residue   (d) none of the above

9.20 Which of the following is the gyromagnetic resonance frequency of a slab?

(a) $\omega = \sqrt{\omega_0(\omega_0 + \omega_m)}$
(b) $\omega = \sqrt{\omega_0(\omega_m - \omega_0)}$
(c) $\omega = \sqrt{\omega_m(\omega_m - \omega_0)}$
(d) none of the above

ANSWER KEY

9.1 (b)	**9.5** (c)	**9.9** (a)	**9.13** (c)	**9.17** (a)
9.2 (a)	**9.6** (c)	**9.10** (b)	**9.14** (c)	**9.18** (b)
9.3 (c)	**9.7** (b)	**9.11** (a)	**9.15** (b)	**9.19** (c)
9.4 (b)	**9.8** (a)	**9.12** (d)	**9.16** (c)	**9.20** (a)

Noise and Nonlinear Distortion

Learning Objectives

After completing this chapter, you will be able to

- Describe the concept of noise in microwave circuits
- Understand the concept of noise figure of a passive two-port network, mismatched lossy line, and mismatched amplifier

- Apply the concept of nonlinear distortion for harmonic and intermodulation distortion and third-order intercept point
- Learn about linear and spurious free dynamic range

The effect of noise is critical to the performance of most RF and microwave communication, radar, and remote sensing systems because noise ultimately determines the threshold for the minimum signal that can be reliably detected by a receiver. Noise power in a receiver will be introduced from the external environment through the receiving antenna, as well as generated internally by the receiver circuitry. Here we will study the sources of noise in RF and microwave systems, and the characterization of components in terms of noise temperature and noise figure, including the effect of impedance mismatch. The additional noise-related topics of transistor amplifier noise figure, oscillator phase noise, and antenna noise temperature will be discussed in later chapters.

We will also discuss the related topics of compression, harmonic distortion, intermodulation distortion, and dynamic range. These can have important limiting effects when large signal levels are present in mixers, amplifiers, and other components that use nonlinear devices such as diodes and transistors.

10.1 NOISE IN MICROWAVE CIRCUITS

Noise power is a result of random processes such as the flow of charges or holes in an electron tube or solid-state device, propagation through the ionosphere or other ionized gas, or, most basic of all, the thermal vibrations in any component at a temperature above absolute zero. Noise can be passed into a microwave system from external sources, or generated within the system itself. In either case the noise level of a system sets the lower limit on the strength of a signal that can be detected in the presence of the noise. Thus, it is generally desired to minimize the residual noise level of a radar or communication receiver to achieve the best performance. In some cases, such as radiometers or radio astronomy systems, the desired signal is actually the noise power received by an antenna, and it is necessary to distinguish between the received noise power and the undesired noise generated by the receiver system itself.

Dynamic Range and Sources of Noise

In previous chapters we have implicitly assumed that all components were *linear* (meaning that the output signal level is directly proportional to the input signal level), and *deterministic* (meaning that the output signal is predictable from the input signal). In reality no component can perform in this way over an unlimited range of input/output signal levels. In practice, however, there is usually a range of signal levels over which such assumptions are approximately valid; this range is called the *dynamic range* of the component.

As an example, consider a realistic microwave transistor amplifier having a power gain G, as shown in Figure 10.1. If the amplifier were ideal, the output power would be related to the input power as $P_{\text{out}} = GP_{\text{in}}$, and this relation would hold true for any value of P_{in}. Thus, if $P_{\text{in}} = 0$, we would have $P_{\text{out}} = 0$, and if $P_{\text{in}} = 10^6$ W and $G = 10$ dB, we would have $P_{\text{out}} = 10^7$ W. Neither of these results would actually occur

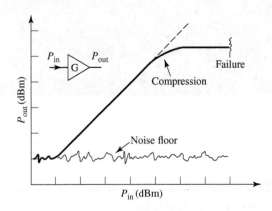

FIGURE 10.1 | Illustrating the dynamic range of a realistic amplifier.

in practice, however. Because of noise generated by the amplifier itself, some nonzero noise power will always be delivered by the amplifier, even when the input power is zero. At the other extreme, very high input power will cause the amplifier to fail. Thus, the actual relation between the output and input power will be as shown in Figure 10.1. At very low input power levels, the output will be dominated by the noise generated by the amplifier. This level is often called the *noise floor* of the component or system; typical values may range from −80 to −140 dBm over the bandwidth of the system, with the lowest values being obtained with thermally cooled components. Above the noise floor, the amplifier will have a range of input power for which $P_{out} = GP_{in}$ is closely approximated. This is the usable *dynamic range* of the component. At the upper end of this range, the output will begin to saturate, meaning that the output power no longer increases linearly as the input power increases. Excessive input power will lead to failure of the amplifier.

Noise that is generated internally in a device or component is usually caused by random motions of charges or charge carriers in devices and materials. Such motions may be due to any of several mechanisms, leading to various types of noise:

- *Thermal noise* is the most basic type of noise, being caused by thermal vibration of bound charges. It is also known as *Johnson* or *Nyquist* noise.
- *Shot noise* is due to random fluctuations of charge carriers in an electron tube or solid-state device.
- *Flicker noise* occurs in solid-state components and vacuum tubes. Flicker noise power varies inversely with frequency, and so is often called $1/f$-noise.
- *Plasma noise* is caused by random motion of charges in an ionized gas, such as a plasma, the ionosphere, or sparking electrical contacts.
- *Quantum noise* results from the quantized nature of charge carriers and photons; it is often insignificant relative to other noise sources.

External noise may be introduced into a system either by a receiving antenna or by electromagnetic coupling. Some sources of external RF noise include the following:

- Thermal noise from the ground
- Cosmic background noise from the sky
- Noise from stars (including the sun)
- Lightning
- Gas discharge lamps
- Radio, TV, and cellular stations
- Wireless devices
- Microwave ovens
- Deliberate jamming devices

The characterization of noise effects in RF and microwave systems in terms of noise temperature and noise figure will apply to all types of noise, regardless of the source, as long as the spectrum of the noise is relatively flat over the bandwidth of the system. Noise with a flat frequency spectrum is called *white noise*.

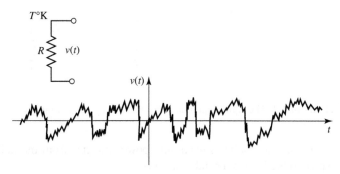

FIGURE 10.2 | A random voltage generated by a noisy resistor.

Noise Power and Equivalent Noise Temperature

Consider a resistor at a physical temperature of T degrees kelvin (K), as depicted in Figure 10.2. The electrons in the resistor are in random motion, with a kinetic energy that is proportional to the temperature. These random motions produce small, random voltage fluctuations at the resistor terminals, as illustrated in Figure 10.2. This voltage has a zero average value but a nonzero root mean square (rms) value given by Planck's blackbody radiation law,

$$V_n = \sqrt{\frac{4hf\, BR}{e^{hf/kT} - 1}}, \tag{10.1}$$

where

$h = 6.626 \times 10^{-34}$ J-sec is Planck's constant.
$k = 1.380 \times 10^{-23}$ J/K is Boltzmann's constant.
$T = $ the temperature in degrees kelvin (K).
$B = $ the bandwidth of the system in Hz.
$f = $ the center frequency of the bandwidth in Hz.
$R = $ the resistance in Ω.

This result comes from quantum mechanical considerations, and is valid for any frequency f. At microwave frequencies the above result can be simplified by making use of the fact that $hf \ll kT$. (As a worst-case example, let $f = 100$ GHz and $T = 100$ K. Then $hf = 6.6 \times 10^{-23} \ll kT = 1.4 \times 10^{-21}$.) Using the first two terms of a Taylor series expansion for the exponential in (10.1) gives

$$e^{hf/kt} - 1 \simeq \frac{hf}{kT},$$

so that (10.1) reduces to

$$V_n = \sqrt{4kTBR}. \tag{10.2}$$

This is the *Rayleigh–Jeans approximation,* and is the result that is most commonly used in microwave work [1]. For very high frequencies or very low temperatures, however, this approximation may be invalid, in which case (10.1) should be used.

The noisy resistor of Figure 10.2 can be replaced with a Thevenin equivalent circuit consisting of a noiseless resistor and a generator with a voltage given by (10.2), as shown in Figure 10.3. Connecting

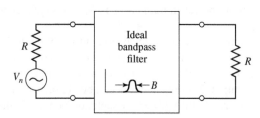

FIGURE 10.3 | Equivalent circuit of a noisy resistor delivering maximum power to a load resistor through an ideal bandpass filter.

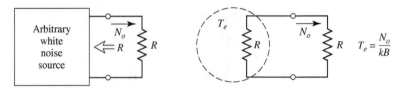

FIGURE 10.4 | The equivalent noise temperature, T_e, of an arbitrary white noise source.

a load resistor R results in maximum power transfer from the noisy resistor, with the result that power delivered to the load in a bandwidth B is

$$P_n = \left(\frac{V_n}{2R}\right)^2 R = \frac{V_n^2}{4R} = kTB, \tag{10.3}$$

since V_n is an rms voltage. This important result gives the maximum available noise power from the noisy resistor at temperature T. Note that this noise power is independent of frequency; such a noise source has a power spectral density that is constant with frequency, and is an example of a white noise source. The noise power is directly proportional to the bandwidth, which in practice is usually limited by the passband of the RF or microwave system. Independent white noise sources can be treated as Gaussian-distributed random variables, so the noise powers (variances) of independent noise sources are additive.

The following trends can be observed from (10.3):

- As $B \to 0, P_n \to 0$. This means that systems with smaller bandwidths collect less noise power.
- As $T \to 0, P_n \to 0$. This means that cooler devices and components generate less noise power.
- As $B \to \infty, P_n \to \infty$. This is the so-called *ultraviolet catastrophe*, which does not occur in reality because (10.2)–(10.3) are not valid as f (or B) $\to \infty$; (10.1) must be used in this case.

If an arbitrary source of noise (thermal or nonthermal) is "white," so that the noise power is not a strong function of frequency over the bandwidth of interest, it can be modeled as an equivalent thermal noise source, and characterized with an *equivalent noise temperature*. Thus, consider the arbitrary white noise source of Figure 10.4, which has a driving-point impedance of R and delivers a noise power N_o to a load resistor R. This noise source can be replaced by a noisy resistor of value R at temperature T_e, where T_e is an equivalent temperature selected so that the same noise power is delivered to the load. That is,

$$T_e = \frac{N_o}{kB}. \tag{10.4}$$

Components and systems can then be characterized by saying that they have an equivalent noise temperature T_e; this implies some fixed bandwidth B, which is generally the operational bandwidth of the component or system.

For example, consider a noisy amplifier with a bandwidth B and gain G. Let the amplifier be matched to noiseless source and load resistors, as shown in Figure 10.5. If the source resistor is at a (hypothetical) temperature of $T_s = 0\,K$, then the input power to the amplifier will be $N_i = 0$, and the output noise power N_o will be due only to the noise generated by the amplifier itself. We can obtain the same load noise power by driving an ideal noiseless amplifier with a resistor at the temperature

$$T_e = \frac{N_o}{GkB}, \tag{10.5}$$

so that the output power in both cases is $N_o = GkT_eB$. Then T_e is the equivalent noise temperature of the amplifier.

It is sometimes useful for measurement purposes to have a calibrated noise source. A passive noise source may simply consist of a resistor held at a constant temperature, either in a temperature-controlled oven, or in a cryogenic flask. Active noise sources may use a diode, transistor, or tube to provide a calibrated noise power output. Noise generators can be characterized by an equivalent noise temperature, but a more common measure of noise power for such components is the *excess noise ratio* (ENR), defined as

$$\text{ENR (dB)} = 10 \log \frac{N_g - N_o}{N_o} = 10 \log \frac{T_g - T_0}{T_0}, \tag{10.6}$$

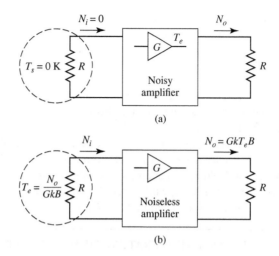

FIGURE 10.5 | Defining the equivalent noise temperature of a noisy amplifier. (a) Noisy amplifier. (b) Noiseless amplifier.

where N_g and T_g are the noise power and equivalent noise temperature of the generator, and N_o and T_0 are the noise power and temperature associated with a room-temperature ($T_0 = 290$ K) passive source (a matched load). Solid-state noise generators typically have ENRs ranging from 20 to 40 dB.

Measurement of Noise Temperature

In principle, the equivalent noise temperature of a component can be determined by measuring the output power when a matched load at 0 K is connected at the input of the component. In practice, of course, a 0 K source temperature cannot be obtained, so a different method must be used. If two matched loads at significantly different temperatures are available, then the *Y-factor method* can be applied.

This technique is illustrated in Figure 10.6, where the amplifier (or other component) under test is connected to one of two matched loads at different temperatures, and the output power is measured for each case. Let T_1 be the temperature of the hot load and T_2 the temperature of the cold load ($T_1 > T_2$), and let P_1 and P_2 be the respective powers measured at the amplifier output. The output noise power consists of noise power generated by the amplifier as well as noise power from the source resistor. Thus we have

$$N_1 = GkT_1B + GkT_eB, \tag{10.7a}$$

$$N_2 = GkT_2B + GkT_eB, \tag{10.7b}$$

which are two equations for the two unknowns, T_e and GB (the gain–bandwidth product of the amplifier). Define the Y-factor as

$$Y = \frac{N_1}{N_2} = \frac{T_1 + T_e}{T_2 + T_e} > 1, \tag{10.8}$$

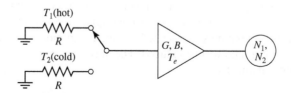

FIGURE 10.6 | The Y-factor method for measuring the equivalent noise temperature of an amplifier.

which is determined as the ratio of the output power measurements. Then (10.7) can be solved for the equivalent noise temperature of the device under test as

$$T_e = \frac{T_1 - YT_2}{Y - 1},$$
(10.9)

in terms of the load temperatures and the Y-factor.

Note that to obtain accurate results from this method, the two source temperatures must not be too close together. If they are, N_1 will be close to N_2, Y will be close to unity, and the evaluation of (10.9) will involve the subtractions of numbers close to each other, resulting in a loss of accuracy. In practice, one noise source is usually a load resistor at room temperature ($T_0 = 290$ K), while the other noise source is either "hotter" or "colder," depending on whether T_e is greater or less than T_0. An active noise generator can be used as a "hot" source, while a "cold" source can be obtained by immersing a load resistor in liquid nitrogen ($T = 77$ K) or liquid helium ($T = 4$ K).

EXAMPLE 10.1   NOISE TEMPERATURE MEASUREMENT

An X-band amplifier has a gain of 20 dB and a 1.5 GHz bandwidth. Its equivalent noise temperature is to be measured via the Y-factor method. The following data are obtained:

$$\text{For } T_1 = 290 \text{ K}, \qquad N_1 = -62.0 \text{ dBm.}$$
$$\text{For } T_2 = 77 \text{ K}, \qquad N_2 = -64.7 \text{ dBm.}$$

Determine the equivalent noise temperature of the amplifier. If the amplifier is used with a source having an equivalent noise temperature of $T_s = 500$ K, what is the output noise power from the amplifier, in dBm?

Solution
From (10.8), the Y-factor in dB is

$$Y = (N_1 - N_2) \text{ dB} = (-62.0) - (-64.7) = 2.7 \text{ dB},$$

which is a numeric value of $Y = 1.86$. Using (10.9) gives the equivalent noise temperature as

$$T_e = \frac{T_1 - YT_2}{Y - 1} = \frac{290 - (1.86)(77)}{1.86 - 1} = 170 \text{ K.}$$

If a source with an equivalent noise temperature of $T_s = 500$ K drives the amplifier, the noise power into the amplifier will be kT_sB. The total noise power out of the amplifier will be

$$N_o = GkT_sB + GkT_eB = 100(1.38 \times 10^{-23})(1.5 \times 10^9)(500 + 170)$$
$$= 1.39 \times 10^{-9} \text{ W} = -58.58 \text{ dBm.}$$

10.2   NOISE FIGURE

Definition of Noise Figure

We have seen that a noisy microwave component can be characterized by an equivalent noise temperature. An alternative characterization is the *noise figure* of the component, which is a measure of the degradation in the signal-to-noise ratio between the input and output of the component. The *signal-to-noise ratio* is the ratio of desired signal power to undesired noise power, and so is dependent on the signal power. When noise and a desired signal are applied to the input of a noiseless network, both noise and signal will be attenuated or amplified by the same factor, so that the signal-to-noise ratio will be unchanged. However, if the network is noisy, the output noise power will be increased more than the output signal power, so that the output signal-to-noise ratio will be reduced. The noise figure, F, is a measure of this reduction in signal-to-noise ratio, and is defined as

$$F = \frac{S_i/N_i}{S_o/N_o} \geq 1,$$
(10.10)

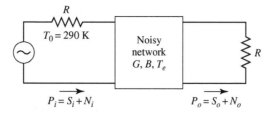

FIGURE 10.7 | Determining the noise figure of a noisy network.

where S_i, N_i are the input signal and noise powers, and S_o, N_o are the output signal and noise powers. By definition, the input noise power is assumed to be the noise power resulting from a matched resistor at $T_0 = 290$ K; that is, $N_i = kT_0B$.

Consider Figure 10.7, which shows noise power N_i and signal power S_i being fed into a noisy two-port network. The network is characterized by a gain, G, a bandwidth, B, and an equivalent noise temperature, T_e. The input noise power is $N_i = kT_0B$, and the output noise power is a sum of the amplified input noise and the internally generated noise: $N_o = kGB(T_0 + T_e)$. The output signal power is $S_o = GS_i$. Using these results in (10.10) gives the noise figure as

$$F = \frac{S_i}{kT_0B} \frac{kGB(T_0 + T_e)}{GS_i} = 1 + \frac{T_e}{T_0} \geq 1. \tag{10.11}$$

In dB, $F = 10\log(1 + T_e/T_0)$ dB ≥ 0. If the network were noiseless, T_e would be zero, giving $F = 1$, or 0 dB. Solving (10.11) for T_e gives

$$T_e = (F - 1)T_0. \tag{10.12}$$

It is important to keep in mind two things concerning the definition of noise figure: noise figure is defined for a matched input source, and for a noise source equivalent to a matched load at temperature $T_0 = 290$ K. Noise figure and equivalent noise temperatures are interchangeable characterizations of the noise properties of a component.

An important special case occurs in practice for a two-port network consisting of a passive, lossy component, such as an attenuator or lossy transmission line, held at a physical temperature T. Consider such a network with a matched source resistor that is also at temperature T, as shown in Figure 10.8. The power gain, G, of a lossy network is less than unity; the loss factor, L, can be defined as $L = 1/G > 1$. Because the entire system is in thermal equilibrium at the temperature T, and has a driving point impedance of R, the output noise power must be $N_o = kTB$. However, we can also think of this power as coming from the source resistor (attenuated by the lossy line), and from the noise generated by the line itself. Thus we also have that

$$N_o = kTB = GkTB + GN_{\text{added}}, \tag{10.13}$$

where N_{added} is the noise generated by the line, as if it appeared at the input terminals of the line. Solving (10.13) for this power gives

$$N_{\text{added}} = \frac{1 - G}{G}kTB = (L - 1)kTB. \tag{10.14}$$

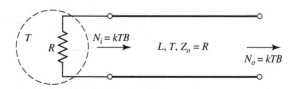

FIGURE 10.8 | Determining the noise figure of a lossy line or attenuator with loss L and temperature T.

Then (10.4) shows that the lossy line has an equivalent noise temperature (referred to the input) given by

$$T_e = \frac{1-G}{G}T = (L-1)T.$$ (10.15)

From (10.11) the noise figure is

$$F = 1 + (L-1)\frac{T}{T_0}.$$ (10.16)

If the line is at temperature T_0, then $F = L$. For instance, a 6 dB attenuator at room temperature has a noise figure of $F = 6$ dB.

Noise Figure of a Cascaded System

In a typical microwave system the input signal travels through a cascade of many different components, each of which may degrade the signal-to-noise ratio to some degree. If we know the noise figure (or noise temperature) of the individual stages, we can determine the noise figure (or noise temperature) of the cascade connection of stages. We will see that the noise performance of the first stage is usually the most critical, an interesting result that is very important in practice.

Consider the cascade of two components, having gains G_1, G_2, noise figures F_1, F_2, and equivalent noise temperatures T_{e1}, T_{e2}, as shown in Figure 10.9. We wish to find the overall noise figure and equivalent noise temperature of the cascade, as if it were a single component. The overall gain of the cascade is $G_1 G_2$.

Using noise temperatures, we can write the noise power at the output of the first stage as

$$N_1 = G_1 k T_0 B + G_1 k T_{e1} B,$$ (10.17)

since $N_i = k T_0 B$ for noise figure calculations. The noise power at the output of the second stage is

$$N_o = G_2 N_1 + G_2 k T_{e2} B$$
$$= G_1 G_2 k B \left(T_0 + T_{e1} + \frac{1}{G_1} T_{e2} \right).$$ (10.18)

For the equivalent system we have

$$N_o = G_1 G_2 k B (T_{cas} + T_0),$$ (10.19)

so comparison with (10.18) gives the noise temperature of the cascaded system as

$$T_{cas} = T_{e1} + \frac{1}{G_1} T_{e2}.$$ (10.20)

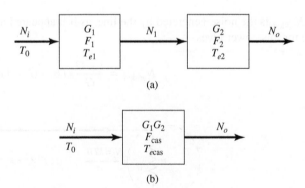

(a)

(b)

FIGURE 10.9 | Noise figure and equivalent noise temperature of a cascaded system. (a) Two cascaded networks. (b) Equivalent network.

Using (10.12) to convert the temperatures in (10.20) to noise figures yields the noise figure of the cascaded system as

$$F_{\text{cas}} = F_1 + \frac{1}{G_1}(F_2 - 1). \tag{10.21}$$

Equations (10.20) and (10.21) show that the noise characteristics of a cascaded system are dominated by the characteristics of the first stage since the effect of the second stage is reduced by the gain of the first (assuming $G_1 > 1$). Thus, for the best overall system noise performance, the first stage should have a low noise figure and at least moderate gain. Expense and effort should be devoted primarily to the first stage, as opposed to later stages, since later stages have a diminished impact on the overall noise performance.

Equations (10.20) and (10.21) can be generalized to an arbitrary number of stages, as follows:

$$T_{\text{cas}} = T_{e1} + \frac{T_{e2}}{G_1} + \frac{T_{e3}}{G_1 G_2} + \cdots, \tag{10.22}$$

$$F_{\text{cas}} = F_1 + \frac{F_2 - 1}{G_1} + \frac{F_3 - 1}{G_1 G_2} + \cdots. \tag{10.23}$$

EXAMPLE 10.2   NOISE ANALYSIS OF A WIRELESS RECEIVER

The block diagram of a wireless receiver front-end is shown in Figure 10.10. Compute the overall noise figure of this subsystem. If the input noise power from a feeding antenna is $N_i = kT_A B$, where $T_A = 150$ K, find the output noise power in dBm. If we require a minimum signal-to-noise ratio (SNR) of 20 dB at the output of the receiver, what is the minimum signal voltage that should be applied at the receiver input? Assume the system is at temperature T_0, with a characteristic impedance of 50 Ω, and an IF bandwidth of 10 MHz.

Solution
We first perform the required conversions from dB to numerical values:

$$G_a = 10 \text{ dB} = 10 \qquad G_f = -1.0 \text{ dB} = 0.79 \qquad G_m = -3.0 \text{ dB} = 0.5$$
$$F_a = 2 \text{ dB} = 1.58 \qquad F_f = 1 \text{ dB} = 1.26 \qquad F_m = 4 \text{ dB} = 2.51$$

Next, use (10.23) to find the overall noise figure of the system:

$$F = F_a + \frac{F_f - 1}{G_a} + \frac{F_m - 1}{G_a G_f} = 1.58 + \frac{(1.26 - 1)}{10} + \frac{(2.51 - 1)}{(10)(0.79)}$$
$$= 1.80 = 2.55 \text{ dB}.$$

The best way to compute the output noise power is to use noise temperatures. From (10.12), the equivalent noise temperature of the overall system is

$$T_e = (F - 1)T_0 = (1.80 - 1)(290) = 232 \text{ K}.$$

The overall gain of the system is $G = (10)(0.79)(0.5) = 3.95$. Then we can find the output noise power as

$$N_o = k(T_A + T_e)BG = (1.38 \times 10^{-23})(150 + 232)(10 \times 10^6)(3.95)$$
$$= 2.08 \times 10^{-13} \text{ W} = -96.8 \text{ dBm}.$$

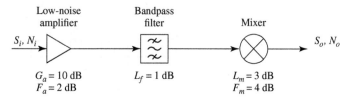

FIGURE 10.10 | Block diagram of a wireless receiver front-end for Example 10.2.

For an output SNR of 20 dB = 100, the input signal power must be

$$S_i = \frac{S_o}{G} = \frac{S_o}{N_o}\frac{N_o}{G} = 100\frac{2.08 \times 10^{-13}}{3.95} = 5.27 \times 10^{-12} \text{ W} = -82.8 \text{ dBm.}$$

For a 50 Ω system impedance, this corresponds to an input signal voltage of

$$V_i = \sqrt{Z_o S_i} = \sqrt{(50)(5.27 \times 10^{-12})} = 1.62 \times 10^{-5} \text{ V} = 16.2 \ \mu\text{V (rms).}$$

Note: It may be tempting to compute the output noise power from the definition of the noise figure, as

$$N_o = N_i F\left(\frac{S_o}{S_i}\right) = N_i FG = kT_A BFG$$

$$= (1.38 \times 10^{-23})(150)(10 \times 10^6)(1.8)(3.95) = 1.47 \times 10^{-13} \text{ W.}$$

This is an *incorrect* result! The reason for the disparity with the earlier result is that the definition of noise figure assumes an input noise level of $kT_0 B$, while this problem involves an input noise of $kT_A B$, with $T_A = 150$ K $\neq T_0$. This is a common error, and suggests that when computing absolute noise power it is often safer to use noise temperatures to avoid this confusion.

Noise Figure of a Passive Two-Port Network

We previously derived the noise figure for a matched lossy line or attenuator by using a thermodynamic argument. Here we generalize that technique to evaluate the noise figure of general passive networks (networks that do not contain active devices such as diodes or transistors, which generate nonthermal noise). In addition, this method will account for the change in noise figure that occurs when a component is impedance mismatched at either its input or output port. Generally it is easier and more accurate to find the noise characteristics of an active device, such as a diode or transistor, by direct measurement than by calculation from first principles.

Figure 10.11 shows an arbitrary passive two-port network, with a generator at port 1 and a load at port 2. The network is characterized by its scattering matrix, $[S]$. In the general case, impedance mismatches may exist at each port, and we define these mismatches in terms of the following reflection coefficients:

$$\Gamma_s = \text{reflection coefficient looking toward generator,}$$
$$\Gamma_{in} = \text{reflection coefficient looking toward port 1 of network,}$$
$$\Gamma_{out} = \text{reflection coefficient looking toward port 2 of network,}$$
$$\Gamma_L = \text{reflection coefficient looking toward load.}$$

If we assume the network is at temperature T, and that an available input noise power of $N_1 = kTB$ is applied to the input of the network, we can write the available output noise power at port 2 as

$$N_2 = G_{21}kTB + G_{21}N_{\text{added}}, \tag{10.24}$$

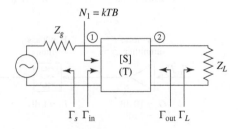

FIGURE 10.11 | A passive two-port network with impedance mismatches. The network is at physical temperature T.

where N_{added} is the noise power generated internally by the network (referenced to port 1), and G_{21} is the *available power gain* of the network from port 1 to port 2. The available power gain can be expressed in terms of the scattering parameters of the network and the port mismatches as (also see Section 12.1),

$$G_{21} = \frac{\text{power available from network}}{\text{power available from source}} = \frac{|S_{21}|^2(1 - |\Gamma_S|^2)}{|1 - S_{11}\Gamma_S|^2(1 - |\Gamma_{out}|^2)}. \tag{10.25}$$

As derived in Example 4.8, the output port mismatch is given by

$$\Gamma_{out} = S_{22} + \frac{S_{12}S_{21}\Gamma_S}{1 - S_{11}\Gamma_S}. \tag{10.26}$$

Observe that when the network is matched to its external circuitry, so that $\Gamma_s = 0$ and $S_{22} = 0$, we have $\Gamma_{out} = 0$ and $G_{21} = |S_{21}|^2$, which is the gain of the network when it is matched. Also observe that the available gain of the network does not depend on the load mismatch, Γ_L. This is because available gain is defined in terms of the maximum power that is available from the network, which occurs when the load impedance is conjugately matched to the output impedance of the network.

Since the input noise power is kTB, and the network is passive and at temperature T, the network is in thermodynamic equilibrium, and so the available output noise power must be $N_2 = kTB$. Then we can solve for N_{added} from (10.24) to give

$$N_{added} = \frac{1 - G_{21}}{G_{21}}kTB. \tag{10.27}$$

Then the equivalent noise temperature of the network is

$$T_e = \frac{N_{added}}{kB} = \frac{1 - G_{21}}{G_{21}}T, \tag{10.28}$$

and the noise figure of the network is

$$F = 1 + \frac{T_e}{T_0} = 1 + \frac{1 - G_{21}}{G_{21}}\frac{T}{T_0}. \tag{10.29}$$

Note the similarity of (10.27)–(10.29) to the results in (10.14)–(10.16) for the lossy line—the essential difference is that here we are using the available gain of the network, which accounts for impedance mismatches between the network and the external circuit. We can illustrate the use of this result with some applications to problems of practical interest.

Noise Figure of a Mismatched Lossy Line

Earlier we found the noise figure of a lossy transmission line under the assumption that it was matched to its input and output circuits. Now we consider the case where the line is mismatched to its input circuit. Figure 10.12 shows a transmission line of length ℓ at temperature T, with a power loss factor $L = 1/G$, and an impedance mismatch between the line and the generator. Thus, $Z_g \neq Z_0$, and the reflection coefficient looking toward the generator is

$$\Gamma_s = \frac{Z_g - Z_0}{Z_g + Z_0} \neq 0.$$

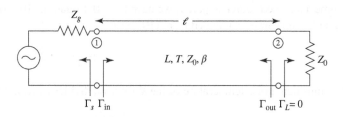

FIGURE 10.12 | A lossy transmission line at temperature T with an impedance mismatch at its input port.

The scattering matrix of the lossy line of characteristic impedance Z_0 can be written as

$$[S] = \begin{bmatrix} 0 & 1 \\ 1 & 0 \end{bmatrix} \frac{e^{-j\beta\ell}}{\sqrt{L}}, \tag{10.30}$$

where β is the propagation constant of the line. Using the elements of (10.30) in (10.26) gives the reflection coefficient looking into port 2 of the line as

$$\Gamma_{out} = S_{22} + \frac{S_{12}S_{21}\Gamma_s}{1 - S_{11}\Gamma_s} = \frac{\Gamma_s}{L}e^{-2j\beta\ell}. \tag{10.31}$$

Then the available gain, from (10.25), is

$$G_{21} = \frac{\frac{1}{L}(1 - |\Gamma_s|^2)}{1 - |\Gamma_{out}|^2} = \frac{L(1 - |\Gamma_s|^2)}{L^2 - |\Gamma_s|^2}. \tag{10.32}$$

We can verify two limiting cases of (10.32): when $L = 1$ we have $G_{21} = 1$, and when $\Gamma_s = 0$ we have $G_{21} = 1/L$. Using (10.32) in (10.28) gives the equivalent noise temperature of the mismatched lossy line as

$$T_e = \frac{1 - G_{21}}{G_{21}}T = \frac{(L - 1)(L + |\Gamma_s|^2)}{L(1 - |\Gamma_s|^2)}T. \tag{10.33}$$

The corresponding noise figure can then be evaluated using (10.11). Observe that when the line is matched, $\Gamma_s = 0$, and (10.33) reduces to $T_e = (L - 1)T$, in agreement with the result for the matched lossy line given by (10.15). If the line is lossless, then $L = 1$, and (10.33) reduces to $T_e = 0$ regardless of mismatch, as expected. However, when the line is lossy and mismatched, so that $L > 1$ and $|\Gamma_s| > 0$, then the noise temperature given by (10.33) is greater than $T_e = (L - 1)T$, the noise temperature of the matched lossy line. The reason for this increase is that the lossy line actually delivers noise power out of both its ports, but when the input port is mismatched some of the available noise power at port 1 is reflected from the source back into port 1 and appears at port 2. When the generator is matched to port 1, none of the available power from port 1 is reflected back into the line, so the noise power available at port 2 is a minimum. This result implies that impedance matching is important in minimizing noise temperature and noise figure.

EXAMPLE 10.3   APPLICATION TO A WILKINSON POWER DIVIDER

Find the noise figure of a Wilkinson power divider when one of the output ports is terminated in a matched load. Assume an insertion loss factor of L from the input to either output port.

Solution
From Chapter 7 the scattering matrix of a Wilkinson divider is given as

$$[S] = \frac{-j}{\sqrt{2L}} \begin{bmatrix} 0 & 1 & 1 \\ 1 & 0 & 0 \\ 1 & 0 & 0 \end{bmatrix},$$

where the factor $L \geq 1$ accounts for the dissipative loss from port 1 to port 2 or 3 (note that dissipative loss is distinct from the -3 dB power division ratio). To evaluate the noise figure of the Wilkinson divider, we first terminate port 3 with a matched load; this converts the three-port device into a two-port device. If we assume a matched source at port 1, we have $\Gamma_s = 0$. Equation (10.26) then gives $\Gamma_{out} = S_{22} = 0$, and so the available gain can be calculated from (10.25) as

$$G_{21} = |S_{21}|^2 = \frac{1}{2L}.$$

The equivalent noise temperature of the Wilkinson divider is, from (10.28),

$$T_e = \frac{1 - G_{21}}{G_{21}}T = (2L - 1)T,$$

where T is the physical temperature of the divider. Using (10.11) gives the noise figure as

$$F = 1 + \frac{T_e}{T_0} = 1 + (2L - 1)\frac{T}{T_0}.$$

Observe that if the divider is at room temperature, then $T = T_0$ and the above reduces to $F = 2L$. If the divider is at room temperature and lossless, this reduces to $F = 2 = 3$ dB. In this case the source of the noise power is the isolation resistor contained in the Wilkinson divider circuit.

Because the network is matched at its input and output, it is easy to obtain these same results using the thermodynamic argument directly. Thus, if we apply an input noise power of kTB to port 1 of the matched divider at temperature T, the system will be in thermal equilibrium and the output noise power must be kTB. We can also express the output noise power as the sum of the input power times the gain of the divider, and N_{added}, the noise power added by the divider itself (referenced to the input to the divider):

$$kTB = \frac{kTB}{2L} + \frac{N_{\text{added}}}{2L}.$$

Solving for N_{added} gives $N_{\text{added}} = kTB(2L - 1)$, so the equivalent noise temperature is

$$T_e = \frac{N_{\text{added}}}{kB} = (2L - 1)T,$$

in agreement with the above.

Noise Figure of a Mismatched Amplifier

Finally, consider the effect of an input impedance mismatch on the noise figure of an amplifier. As shown in Figure 10.13, the amplifier, when matched, has a gain G, a noise figure F, and a bandwidth B. The amplifier output is matched, but there is an impedance mismatch at the input represented by the reflection coefficient, Γ. Our previous results involving the effect of mismatch on noise figure made use of (10.29), but that was derived for a passive network and so cannot be directly used in this case. Instead we will use noise temperatures.

Since we are dealing with noise figure, let the input noise power to the amplifier be $N_i = kT_0 B$. Then the output noise power from the amplifier (referenced to the input) is given by

$$N_o = kT_0 GB(1 - |\Gamma|^2) + kT_0(F - 1)GB \tag{10.34}$$

where the first term is due to the input noise power, decreased by the reflection at the input, and the second term is the noise power due to the amplifier itself, based on the equivalent noise temperature as given by (10.12). For an applied signal power S_i, the output signal power is

$$S_o = G(1 - |\Gamma|^2)S_i. \tag{10.35}$$

The overall noise figure, F_m, of the mismatched amplifier can be found from (10.10) as

$$F_m = \frac{S_i N_o}{S_o N_i} = 1 + \frac{F - 1}{1 - |\Gamma|^2}. \tag{10.36}$$

Observe from (10.36) the limiting case that $F_m = F$ when $|\Gamma| = 0$ (no mismatch), and that this is the minimum noise figure that can be achieved since F_m increases as the mismatch increases. This result

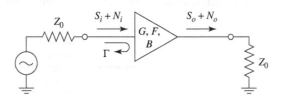

FIGURE 10.13 | A noisy amplifier with an impedance mismatch at its input.

demonstrates that good noise figure requires good impedance matching. This problem would be more complicated if a mismatch also existed at the output of the amplifier, particularly if the amplifier is not unilateral.

10.3 NONLINEAR DISTORTION

We have seen that thermal noise is generated by any lossy component. Since all realistic components have at least a small loss, the ideal linear component does not exist in practice because all realistic devices are nonlinear at very low signal levels due to noise effects. In addition, practical components may also become nonlinear at high signal levels. In the case of active devices, such as diodes and transistors, this may be due to effects such as gain compression or the generation of spurious frequency components due to device nonlinearities, but all devices ultimately fail at very high power levels. In either case, these effects set a minimum and maximum realistic power range, or *dynamic range*, over which a given component or network will operate as desired. In this section we will study the response of nonlinear devices in general, and two definitions of dynamic range. These results will be useful for our later discussions of amplifiers (Chapter 12), mixers (Chapter 13), and wireless receivers (Chapter 14).

Devices such as diodes and transistors have nonlinear characteristics, and it is this nonlinearity that is of great utility for desirable functions such as amplification, detection, and frequency conversion [2]. Nonlinear device characteristics, however, can also lead to undesirable effects such as gain compression and the generation of spurious frequency components. These effects may lead to increased losses, signal distortion, and possible interference with other radio channels or services. Some of the many possible effects of nonlinearity in RF and microwave circuits are listed below [3]:

- Harmonic generation (multiples of a fundamental signal)
- Saturation (gain reduction in an amplifier)
- Intermodulation distortion (products of a two-tone input signal)
- Cross-modulation (modulation transfer from one signal to another)
- AM-PM conversion (amplitude variation causes phase shift)
- Spectral regrowth (intermodulation with many closely spaced signals)

Figure 10.14 shows a general nonlinear network, having an input voltage v_i and an output voltage v_o. In the most general sense, the output response of a nonlinear circuit can be modeled as a Taylor series in terms of the input signal voltage:

$$v_o = a_0 + a_1 v_i + a_2 v_i^2 + a_3 v_i^3 + \cdots, \tag{10.37}$$

where the Taylor coefficients are defined as

$$a_0 = v_o(0) \qquad \text{(DC output)} \tag{10.38a}$$

$$a_1 = \left.\frac{dv_o}{dv_i}\right|_{v_i=0} \qquad \text{(linear output)} \tag{10.38b}$$

$$a_2 = \left.\frac{d^2 v_o}{dv_i^2}\right|_{v_i=0} \qquad \text{(squared output)} \tag{10.38c}$$

and higher order terms. Different functions can be obtained from the nonlinear network depending on the dominance of particular terms in the expansion. The constant term, with coefficient a_0, in (10.37) leads to rectification, converting an AC input signal to DC. The linear term, with coefficient a_1, models a linear attenuator ($a_1 < 1$) or amplifier ($a_1 > 1$). The second-order term, with coefficient a_2, can be used for mixing

FIGURE 10.14 | A general nonlinear device or network.

and other frequency conversion functions. Practical nonlinear devices usually have a series expansion containing many nonzero terms, and a combination of several of the above effects will occur. We will consider some important special cases below.

Gain Compression

First consider the case where a single-frequency sinusoid is applied to the input of a general nonlinear network, such as an amplifier:

$$v_i = V_0 \cos \omega_0 t. \tag{10.39}$$

Equation (10.37) gives the output voltage as

$$
\begin{aligned}
v_o &= a_0 + a_1 V_0 \cos \omega_0 t + a_2 V_0^2 \cos^2 \omega_0 t + a_3 V_0^3 \cos^3 \omega_0 t + \cdots \\
&= \left(a_0 + \frac{1}{2} a_2 V_0^2 \right) + \left(a_1 V_0 + \frac{3}{4} a_3 V_0^3 \right) \cos \omega_0 t + \frac{1}{2} a_2 V_0^2 \cos 2\omega_0 t \\
&\quad + \frac{1}{4} a_3 V_0^3 \cos 3\omega_0 t + \cdots .
\end{aligned}
\tag{10.40}
$$

This result leads to the voltage gain of the signal component at frequency ω_0:

$$G_v = \frac{v_o^{(\omega_0)}}{v_i^{(\omega_0)}} = \frac{a_1 V_0 + \frac{3}{4} a_3 V_0^3}{V_0} = a_1 + \frac{3}{4} a_3 V_0^2, \tag{10.41}$$

where we have retained only terms through the third order.

The result of (10.41) shows that the voltage gain is equal to a_1, the coefficient of the linear term, as expected, but with an additional term proportional to the square of the input voltage amplitude. In most practical amplifiers a_3 typically has the opposite sign of a_1, so that the output of the amplifier tends to be reduced from the expected linear dependence for large values of V_0. This effect is called *gain compression*, or *saturation*. Physically, this is usually due to the fact that the instantaneous output voltage of an amplifier is limited by the power supply voltage used to bias the active device.

A typical amplifier response is shown in Figure 10.15. For an ideal linear amplifier a plot of the output power versus input power would be a straight line with a slope of unity, and the power gain of the amplifier given by the ratio of the output power to the input power. The amplifier response of Figure 10.15 tracks the ideal response over a limited range, then begins to saturate, resulting in reduced gain. To quantify the linear operating range of the amplifier, we define the *1 dB compression point* as the power level for which the output power has decreased by 1 dB from the ideal linear characteristic. This power level is usually denoted by P_{1dB}, and can be stated in terms of either input power (IP_{1dB}) or output power (OP_{1dB}).

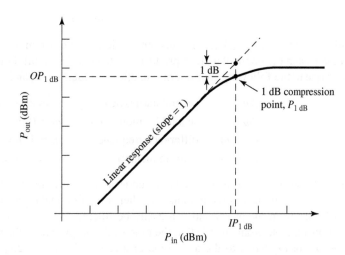

FIGURE 10.15 | Definition of the 1 dB compression point for a nonlinear amplifier.

The 1 dB compression point is typically given as the larger of these two options, so for amplifiers $P_{1\text{dB}}$ is usually specified as an output power, while for mixers $P_{1\text{dB}}$ is usually specified in terms of input power. The relation between a compression point referenced at the input versus the output is given as, in dB, $OP_{1\text{dB}} = IP_{1\text{dB}} + G - 1$ dB [4, 5].

Harmonic and Intermodulation Distortion

Observe from the expansion of (10.40) that a portion of the input signal at frequency ω_0 is converted to other frequency components. For example, the first term of (10.40) represents a DC voltage, which would be a useful response in a rectifier application. The voltage components at frequencies $2\omega_0$ or $3\omega_0$ can be useful for frequency multiplier circuits. In amplifiers, however, the presence of other frequency components will lead to signal distortion if those components are in the passband of the amplifier.

For a single input frequency, or *tone*, ω_0, the output will in general consist of harmonics of the input frequency of the form $n\omega_0$, for $n = 0, 1, 2, \ldots$. Often these harmonics lie outside the passband of the amplifier and so do not interfere with the desired signal at frequency ω_0. The situation is different, however, when the input signal consists of two closely spaced frequencies.

Consider a *two-tone* input voltage, consisting of two closely spaced frequencies ω_1 and ω_2:

$$v_i = V_0(\cos \omega_1 t + \cos \omega_2 t). \tag{10.42}$$

From (10.37) the output is

$$
\begin{aligned}
v_o &= a_0 + a_1 V_0(\cos \omega_1 t + \cos \omega_2 t) + a_2 V_0^2(\cos \omega_1 t + \cos \omega_2 t)^2 \\
&\quad + a_3 V_0^3(\cos \omega_1 t + \cos \omega_2 t)^3 + \cdots \\
&= a_0 + a_1 V_0 \cos \omega_1 t + a_1 V_0 \cos \omega_2 t + \frac{1}{2} a_2 V_0^2(1 + \cos 2\omega_1 t) + \frac{1}{2} a_2 V_0^2(1 + \cos 2\omega_2 t) \\
&\quad + a_2 V_0^2 \cos(\omega_1 - \omega_2)t + a_2 V_0^2 \cos(\omega_1 + \omega_2)t \\
&\quad + a_3 V_0^3 \left(\frac{3}{4} \cos \omega_1 t + \frac{1}{4} \cos 3\omega_1 t \right) + a_3 V_0^3 \left(\frac{3}{4} \cos \omega_2 t + \frac{1}{4} \cos 3\omega_2 t \right) \\
&\quad + a_3 V_0^3 \left[\frac{3}{2} \cos \omega_2 t + \frac{3}{4} \cos(2\omega_1 - \omega_2)t + \frac{3}{4} \cos(2\omega_1 + \omega_2)t \right] \\
&\quad + a_3 V_0^3 \left[\frac{3}{2} \cos \omega_1 t + \frac{3}{4} \cos(2\omega_2 - \omega_1)t + \frac{3}{4} \cos(2\omega_2 + \omega_1)t \right] + \cdots. \tag{10.43}
\end{aligned}
$$

where standard trigonometric identities have been used to expand the initial expression. We see that the output spectrum consists of harmonics of the form

$$m\omega_1 + n\omega_2, \tag{10.44}$$

with $m, n = 0, \pm 1, \pm 2, \pm 3, \ldots$. These combinations of the two input frequencies are called *intermodulation products*, and the *order* of a given product is defined as $|m| + |n|$. For example, the squared term of (10.43) gives rise to the following four intermodulation products of second order:

$$
\begin{array}{llll}
2\omega_1 & \text{(second harmonic of } \omega_1) & m = 2 \ \ n = 0 & \text{order} = 2, \\
2\omega_2 & \text{(second harmonic of } \omega_2) & m = 0 \ \ n = 2 & \text{order} = 2, \\
\omega_1 - \omega_2 & \text{(difference frequency)} & m = 1 \ \ n = -1 & \text{order} = 2, \\
\omega_1 + \omega_2 & \text{(sum frequency)} & m = 1 \ \ n = 1 & \text{order} = 2.
\end{array}
$$

All of these second-order products are undesired in an amplifier, but in a mixer the sum or difference frequencies form the desired outputs. In either case, if ω_1 and ω_2 are close, all of the second-order products will be far from ω_1 or ω_2 and can easily be filtered (either passed or rejected) from the output of the component. Note from (10.43) that the ratio of the amplitude of the second-order intermodulation product $\omega_1 - \omega_2$ (or $\omega_1 + \omega_2$) to the amplitude of a second harmonic $2\omega_1$ (or $2\omega_2$) is 2.0, so the second-order harmonic power will be 6 dB less than the power in the second-order sum or difference terms.

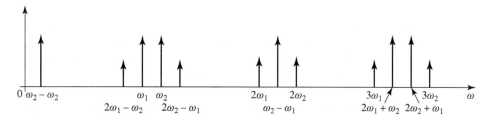

FIGURE 10.16 | Output spectrum of second- and third-order two-tone intermodulation products, assuming $\omega_1 < \omega_2$.

The cubed term of (10.43) leads to six third-order intermodulation products: $3\omega_1, 3\omega_2, 2\omega_1 + \omega_2, 2\omega_2 + \omega_1, 2\omega_1 - \omega_2$, and $2\omega_2 - \omega_1$. The first four of these will again be located far from ω_1 or ω_2, and will typically be outside the passband of the component. However, the two difference terms produce products located near the original input signals at ω_1 and ω_2, and so cannot be easily filtered from the passband of an amplifier. Figure 10.16 shows a typical spectrum of the second- and third-order two-tone intermodulation products. For an arbitrary input signal consisting of many frequencies of varying amplitude and phase, the resulting in-band intermodulation products will cause distortion of the output signal. This effect is called *third-order intermodulation distortion.*

It can be seen from (10.43) that the ratio of the amplitude of the third-order intermodulation product $2\omega_1 - \omega_2$ (or $2\omega_2 - \omega_1$) to the amplitude of the third harmonic $3\omega_1$ (or $3\omega_2$) is 3.0, so the third-order harmonic power will be 9.54 dB less than the power in the third-order intermodulation terms.

Third-Order Intercept Point

Equation (10.43) shows that as the input voltage V_0 increases, the voltage associated with the third-order products increases as V_0^3. Since power is proportional to the square of voltage, we can also say that the output power of third-order products must increase as the cube of the input power. So for small input powers the third-order intermodulation products will be very small, but will increase quickly as input power increases. We can view this effect graphically by plotting the output power for the first- and third-order products versus input power on log-log scales (or in dB), as shown in Figure 10.17.

The output power of the first-order, or linear, product is proportional to the input power, and so the line describing this response has a slope of unity (before the onset of compression). The line describing the response of the third-order products has a slope of 3. (The second-order products would have a slope of 2, but since these products are generally not in the passband of the component, we have not plotted their response in Figure 10.17.) Both the linear and third-order responses will exhibit compression at high input

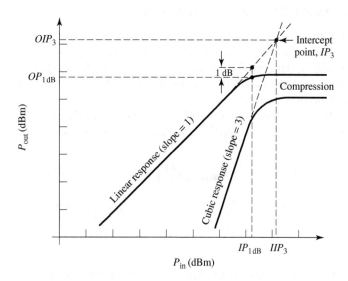

FIGURE 10.17 | Third-order intercept diagram for a nonlinear component.

powers, so we show the extension of their idealized responses with dotted lines. Since these two lines have different slopes, they will intersect, typically at a point above the onset of compression, as shown in the figure. This hypothetical intersection point where the first-order and third-order powers would be equal is called the *third-order intercept point*, denoted as IP_3; it may be specified as either an input power level (IIP_3), or an output power level (OIP_3). The relation between an intercept point referenced at the input versus the output is simply $OIP_3 = G\left(IIP_3\right)$. As with the 1 dB compression point, the reference for IP_3 is typically chosen to result in the largest value, so IP_3 is usually referenced at the output for amplifiers and at the input for mixers. As depicted in Figure 10.17, IP_3 generally occurs at a higher power level than $P_{1\,\mathrm{dB}}$, the 1 dB compression point. Many practical components follow the approximate rule that IP_3 is 10–15 dB greater than $P_{1\,\mathrm{dB}}$, assuming these powers are referenced at the same point.

We can express IP_3 in terms of the Taylor coefficients of the expansion of (10.43) as follows. Define P_{ω_1} as the output power of the desired signal at frequency ω_1. Then from (10.43) we have

$$P_{\omega_1} = \frac{1}{2}a_1^2 V_0^2. \tag{10.45}$$

Similarly, define $P_{2\omega_1 - \omega_2}$ as the output power of the intermodulation product of frequency $2\omega_1 - \omega_2$. Then from (10.43) we have

$$P_{2\omega_1 - \omega_2} = \frac{1}{2}\left(\frac{3}{4}a_3 V_0^3\right)^2 = \frac{9}{32}a_3^2 V_0^6. \tag{10.46}$$

By definition, these two powers are equal at the third-order intercept point. If we define the input signal voltage at the intercept point as V_{IP}, then equating (10.45) and (10.46) gives

$$\frac{1}{2}a_1^2 V_{IP}^2 = \frac{9}{32}a_3^2 V_{IP}^6.$$

Solving for V_{IP} yields

$$V_{IP} = \sqrt{\frac{4a_1}{3a_3}}. \tag{10.47}$$

Since OIP_3 is equal to the linear response of P_{ω_1} at the intercept point, we have from (10.45) and (10.47) that

$$OIP_3 = P_{\omega_1}\Big|_{V_0 = V_{IP}} = \frac{1}{2}a_1^2 V_{IP}^2 = \frac{2a_1^3}{3a_3}, \tag{10.48}$$

where IP_3 in this case is referred to the output port. These expressions will be useful in the following sections.

Intercept Point of a Cascaded System

As in the case of noise figure, a cascade connection of components usually has the effect of degrading (lowering) the third-order intercept point. Unlike noise powers, however, intermodulation products in a cascaded system are deterministic and may be in phase coherence, in which case we cannot simply add powers but must deal with voltages [5]. We will first consider the coherent (in-phase) cascade case, then the noncoherent case.

With reference to Figure 10.18, G_1 and OIP_3' are the power gain and third-order intercept point for the first stage, and G_2 and OIP_3'' are the corresponding values for the second stage. Let P_{ω_1}' be the first-stage

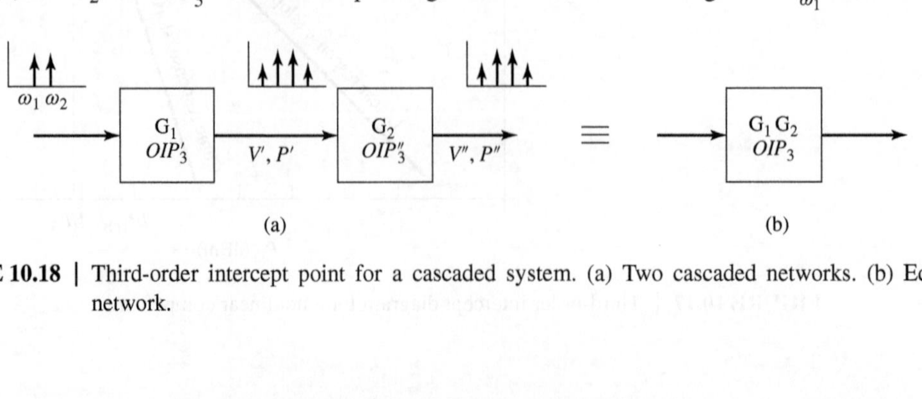

(a) (b)

FIGURE 10.18 | Third-order intercept point for a cascaded system. (a) Two cascaded networks. (b) Equivalent network.

output power of the desired signal at frequency ω_1, and let $P'_{2\omega_1-\omega_2}$ be the first-stage output power at the third-order intermodulation product. From (10.46), $P'_{2\omega_1-\omega_2}$ can be rewritten in terms of P'_{ω_1} and OIP'_3, using (10.45) and (10.48), as follows:

$$P'_{2\omega_1-\omega_2} = \frac{9a_3^2 V_0^6}{32} = \frac{\frac{1}{8}a_1^6 V_0^6}{\frac{4a_1^6}{9a_3^2}} = \frac{(P'_{\omega_1})^3}{(OIP'_3)^2}. \tag{10.49}$$

The first-stage output voltage associated with this power is

$$V'_{2\omega_1-\omega_2} = \sqrt{P'_{2\omega_1-\omega_2} Z_0} = \frac{\sqrt{(P'_{\omega_1})^3 Z_0}}{OIP'_3}, \tag{10.50}$$

where Z_0 is the system impedance.

For coherent intermodulation products, the total third-order distortion voltage at the output of the second stage is the sum of the above voltage times the voltage gain of the second stage, and the distortion voltage generated by the second stage. This is because these voltages are deterministic and phase related, unlike uncorrelated noise powers that arise in cascaded components. Adding these voltages gives the worst-case result for the overall distortion level because there may be phase delays within the stages that could cause partial cancellation. Thus we can write the worst-case total distortion voltage at the output of the second stage as

$$V''_{2\omega_1-\omega_2} = \frac{\sqrt{G_2 (P'_{\omega_1})^3 Z_0}}{OIP'_3} + \frac{\sqrt{(P''_{\omega_1})^3 Z_0}}{OIP''_3}.$$

Since $P''_{\omega_1} = G_2 P'_{\omega_1}$, we have

$$V''_{2\omega_1-\omega_2} = \left(\frac{1}{G_2 (OIP'_3)} + \frac{1}{OIP''_3} \right) \sqrt{(P''_{\omega_1})^3 Z_0}. \tag{10.51}$$

The total output distortion power is

$$P''_{2\omega_1-\omega_2} = \frac{(V''_{2\omega_1-\omega_2})^2}{Z_0} = \left(\frac{1}{G_2 (OIP'_3)} + \frac{1}{OIP''_3} \right)^2 (P''_{\omega_1})^3 = \frac{(P''_{\omega_1})^3}{(OIP_3)^2}. \tag{10.52}$$

Thus the third-order intercept point of the cascaded system with coherent products is

$$OIP_3 = \left(\frac{1}{G_2 (OIP'_3)} + \frac{1}{OIP''_3} \right)^{-1}. \tag{10.53}$$

Note that $OIP_3 = G_2 (OIP'_3)$ for $OIP''_3 \to \infty$, which is the limiting case when the second stage has no third-order distortion.

If the intermodulation products from each stage have relatively random phases, which may occur when the intermodulation products are not very close to the fundamental signals, it may be proper to treat the individual contributions as incoherent, allowing us to add powers. It is straightforward to show that the overall intercept point in this case is given by

$$OIP_3 = \left(\frac{1}{G_2^2 (OIP'_3)^2} + \frac{1}{(OIP''_3)^2} \right)^{-1/2}. \tag{10.54}$$

EXAMPLE 10.4   CALCULATION OF CASCADE INTERCEPT POINT

A low-noise amplifier and mixer are shown in Figure 10.19. The amplifier has a gain of 20 dB and a third-order intercept point of 24 dBm (referenced at output), and the mixer has a conversion loss of 6.5 dB and a third-order intercept point of 13 dBm (referenced at input). Find the intercept points of the cascade network for both a phase coherence assumption and a random-phase (noncoherence) assumption.

Solution
First we transfer the reference of IP_3 for the mixer from its input to its output:

$$OIP_3'' = \left(IIP_3''\right) G_2 = 13 \text{ dBm} - 6.5 \text{ dB} = 6.5 \text{ dBm}.$$

Converting the necessary dB values to numerical values yields:

$$OIP_3' = 24 \text{ dBm} = 251.2 \text{ mW} \qquad \text{(for amplifier)},$$
$$OIP_3'' = 6.5 \text{ dBm} = 4.5 \text{ mW} \qquad \text{(for mixer)},$$
$$G_2 = -6.5 \text{ dB} = 0.224 \qquad \text{(for mixer)}.$$

Assuming coherence, Equation (10.53) gives the intercept point of the cascade as

$$OIP_3 = \left(\frac{1}{G_2\left(OIP_3'\right)} + \frac{1}{OIP_3''}\right)^{-1} = \left(\frac{1}{(0.224)(251 \times 10^{-3})} + \frac{1}{4.5 \times 10^{-3}}\right)^{-1}$$

$$= 4.2 \text{ mW} = 6.23 \text{ dBm},$$

which is seen to be lower than the minimum IP_3 of the individual components. Equation (10.54) gives the results for the noncoherent case as

$$OIP_3 = \left(\frac{1}{G_2^2\left(OIP_3'\right)^2} + \frac{1}{\left(OIP_3''\right)^2}\right)^{-1/2} = \left(\frac{1}{(0.224)^2\left(251 \times 10^{-3}\right)^2} + \frac{1}{\left(4.5 \times 10^{-3}\right)^2}\right)^{-1/2}$$

$$= 4.49 \text{ mW} = 6.52 \text{ dBm}.$$

As expected, the noncoherent case results in a slightly higher intercept point.

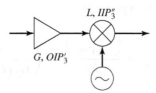

FIGURE 10.19 | System for Example 10.4.

Passive Intermodulation

The above discussion of intermodulation distortion was in the context of active circuits involving diodes and transistors, but it is also possible for intermodulation products to be generated by passive nonlinear effects in connectors, cables, antennas, or almost any component where there is a metal-to-metal contact. This effect is called *passive intermodulation* (PIM) and, as in the case of intermodulation in amplifiers and mixers, it occurs when signals at two or more closely spaced frequencies mix to produce spurious products.

Passive intermodulation can be caused by a number of factors, such as poor mechanical contact, oxidation of junctions between ferrous-based metals, contamination of conducting surfaces at RF junctions, or the use of nonlinear materials such as carbon fiber composites or ferromagnetic materials. In addition, when high powers are involved, thermal effects may contribute to the overall nonlinearity of a junction. It is very difficult to predict PIM levels from first principles, so measurement techniques must usually be used.

Because of the third-power dependence of the third-order intermodulation products with input power, passive intermodulation is usually only significant when input signal powers are relatively large. This is frequently the case in cellular telephone base station transmitters, which may operate with powers of 30–40 dBm, with many closely spaced RF channels. It is often desired to maintain the PIM level below -125 dBm, with two 40 dBm transmit signals. This is a very wide dynamic range, and requires careful selection of components used in the high-power portions of the transmitter, including cables, connectors, and antenna components. Because these components are often exposed to the weather, deterioration due to oxidation, vibration, and sunlight must be offset by a careful maintenance program. Communication satellites often face similar problems with passive intermodulation. Passive intermodulation is generally not a problem in receiver systems due to the much lower power levels.

10.4   DYNAMIC RANGE

Linear and Spurious Free Dynamic Range

We can define *dynamic range* in a general sense as the operating range for which a component or system has desirable characteristics. For a power amplifier this may be the power range that is limited at the low end by noise and at the high end by the compression point. This is essentially the linear operating range for the amplifier, and is called the *linear dynamic range* (LDR). For low-noise amplifiers or mixers, operation may be limited by noise at the low end and the maximum power level for which intermodulation distortion becomes unacceptable. This is effectively the operating range for which spurious responses are minimal, and it is called the *spurious-free dynamic range* (SFDR).

We can find the linear dynamic range LDR as the ratio of $P_{1\,\text{dB}}$, the 1 dB compression point, to the noise level of the component, as shown in Figure 10.20. In dB, this can be written in terms of output powers as

$$\text{LDR (dB)} = OP_{1\,\text{dB}} - N_o, \tag{10.55}$$

for $OP_{1\,\text{dB}}$ and N_o expressed in dBm. Note that some authors prefer to define the linear dynamic range in terms of a minimum detectable power level. This definition is more appropriate for a receiver system rather than an individual component, as it depends on factors external to the component itself, such as the type of modulation used, the recommended system SNR, effects of error-correcting coding, and related factors.

The spurious free dynamic range is defined as the maximum output signal power for which the power of the third-order intermodulation product is equal to the noise level of the component, divided by the output noise level. This situation is shown in Figure 10.20. If P_{ω_1} is the output power of the desired signal at

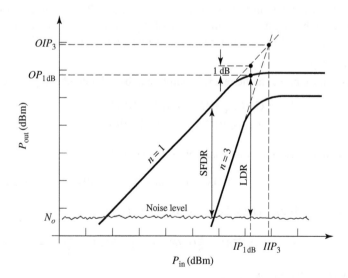

FIGURE 10.20 | Illustrating linear dynamic range (LDR) and spurious free dynamic range (SFDR).

frequency ω_1, and $P_{2\omega_1-\omega_2}$ is the output power of the third-order intermodulation product, then the spurious free dynamic range can be expressed as

$$\text{SFDR} = \frac{P_{\omega_1}}{P_{2\omega_1-\omega_2}}, \tag{10.56}$$

with $P_{2\omega_1-\omega_2}$ taken equal to the noise level of the component. As in (10.49), $P_{2\omega_1-\omega_2}$ can be written in terms of OIP_3 and P_{ω_1} as

$$P_{2\omega_1-\omega_2} = \frac{(P_{\omega_1})^3}{(OIP_3)^2}. \tag{10.57}$$

Observe that this result clearly shows that the third-order intermodulation power increases as the cube of the input signal power. Solving (10.57) for P_{ω_1} and applying the result to (10.56) gives the spurious free dynamic range in terms of OIP_3 and N_o, the output noise power of the component:

$$\text{SFDR} = \frac{P_{\omega_1}}{P_{2\omega_1-\omega_2}}\bigg|_{P_{2\omega_1-\omega_2}=N_o} = \left(\frac{OIP_3}{N_o}\right)^{2/3}. \tag{10.58}$$

This result can be written in terms of dB as

$$\text{SFDR (dB)} = \frac{2}{3}(OIP_3 - N_o), \tag{10.59}$$

for OIP_3 and N_o expressed in dBm. Although this result was derived for the $2\omega_1 - \omega_2$ product, the same result applies for the $2\omega_2 - \omega_1$ product.

In a receiver it may be required to have a minimum detectable signal level, or minimum SNR, in order to achieve a specified performance level. This requires an increase in the input signal level, resulting in a corresponding decrease in dynamic range, since the spurious power level is still equal to the noise power. In this case, the spurious free dynamic range of (10.59) would be modified as [5, 6]:

$$\text{SFDR (dB)} = \frac{2}{3}(OIP_3 - N_o) - \text{SNR}. \tag{10.60}$$

EXAMPLE 10.5   DYNAMIC RANGES

A receiver has a noise figure of 7 dB, a 1 dB compression point of 25 dBm (referenced to output), a gain of 40 dB, and a third-order intercept point of 34 dBm (referenced to output). If the receiver is fed with an antenna having a noise temperature of $T_A = 180$ K, and the desired output SNR is 12 dB, find the linear and spurious free dynamic ranges. Assume a receiver bandwidth of 120 MHz.

Solution
The noise power at the receiver output can be calculated using noise temperatures as

$$N_o = GkB[T_A + (F-1)T_0] = 10^4(1.38 \times 10^{-23})(1.2 \times 10^8)[180 + (5.01 - 1)(290)]$$
$$= 2.22 \times 10^{-8} \text{ W} = -46.53 \text{ dBm}.$$

The linear dynamic range is, from (10.55), in dB,

$$\text{LDR} = OP_{1\text{dB}} - N_o = 25 \text{ dBm} + 46.53 \text{ dBm} = 71.53 \text{ dB}.$$

Equation (10.60) gives the spurious free dynamic range as

$$\text{SFDR} = \frac{2}{3}(OIP_3 - N_o) - \text{SNR} = \frac{2}{3}(34 + 46.53) - 12 = 41.69 \text{ dB}.$$

Observe that SFDR $\ll$ LDR.

REFERENCES

[1] F. T. Ulaby, R. K. Moore, and A. K. Fung, *Microwave Remote Sensing: Active and Passive, Volume I, Microwave Remote Sensing, Fundamentals and Radiometry.* Addison-Wesley, Reading, MA, 1981.

[2] M. E. Hines, "The Virtues of Nonlinearity—Detection, Frequency Conversion, Parametric Amplification and Harmonic Generation," *IEEE Transactions on Microwave Theory and Techniques,* vol. MTT-32, pp. 1097–1104, September 1984.

[3] S. A. Maas, *Nonlinear Microwave and RF Circuits*, 2nd ed., Artech House, Norwood, MA, 2003.

[4] K. Chang, *RF and Microwave Wireless Systems*, John Wiley & Sons, New York, 2000.

[5] W. Egan, *Practical RF System Design*, John Wiley & Sons, Hoboken, NJ, 2003.

[6] M. Steer, *Microwave and RF Design: A Systems Approach*, SciTech, Raleigh, NC, 2010.

PROBLEMS

10.1 The noise figure of a microwave receiver front-end is measured using the Y-factor method. A noise source having an ENR of 25 dB, and a liquid nitrogen cold load (80 K) are used, resulting in a measured Y-factor ratio of 15.83 dB. What is the noise figure of the receiver?

10.2 Assume that measurement error introduces an uncertainty of ΔY into the measurement of Y in a Y-factor measurement. Derive an expression for the normalized error, $\Delta T_e/T_e$, of the equivalent noise temperature in terms of $\Delta Y/Y$ and the temperatures T_1, T_2, and T_e. Minimize this result with respect to T_e to obtain an expression for T_e in terms of T_1 and T_2 that will result in minimum error.

10.3 A lossy transmission line has a noise figure of F_0 at temperature $T_0 = 290$ K. Calculate and plot the noise figure of this line as its physical temperature ranges from $T = 0$ K to 1500 K, for $F_0 = 1$ dB and for $F_0 = 3$ dB.

10.4 An amplifier with a gain of 20 dB, a bandwidth of 180 MHz, and a noise figure of 3.5 dB feeds a receiver with a noise temperature of 900 K. Find the noise figure of the overall system.

10.5 A cellular telephone receiver front-end circuit is shown below. The operating frequency is 1805–1880 MHz, and the physical temperature of the system is 300 K. A noise source with $N_i = -90$ dBm is applied to the receiver input. (a) What is the equivalent noise temperature of the source over the operating bandwidth? (b) What is the noise figure (in dB) of the amplifier? (c) What is the noise figure (in dB) of the cascaded transmission line and amplifier? (d) What is the total noise power output (in dBm) of the receiver over the operating bandwidth?

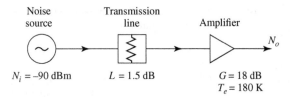

Noise source Transmission line Amplifier

$N_i = -90$ dBm $L = 1.5$ dB $G = 18$ dB
$T_e = 180$ K

10.6 Consider the wireless local area network (WLAN) receiver front-end shown in the accompanying figure, where the bandwidth of the bandpass filter is 100 MHz centered at 2.4 GHz. If the system is at room temperature, find the noise figure of the overall system. What is the resulting signal-to-noise ratio at the output if the input signal power level is −95 dBm? Can the components be rearranged to give a better noise figure?

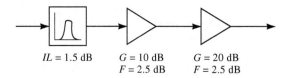

$IL = 1.5$ dB $G = 10$ dB $G = 20$ dB
$F = 2.5$ dB $F = 2.5$ dB

10.7 A two-way power divider has one output port terminated in a matched load, as shown below. Find the noise figure of the resulting two-port network if the divider is (a) an equal-split two-way resistive divider, (b) a two-way Wilkinson divider, and (c) a 3 dB quadrature hybrid. Assume the divider in each case is matched, and at room temperature.

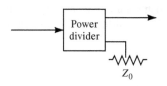

Power divider

Z_0

10.8 Show that, for fixed loss $L > 1$, the equivalent noise temperature of a mismatched lossy line given in (10.33) is minimized when $|\Gamma_s| = 0$.

10.9 Consider the mismatched amplifier of Figure 10.13, having a noise figure F when matched at its input. Calculate and plot the resulting noise figure as the input reflection coefficient magnitude, $|\Gamma|$, varies from 0 to 0.9 for $F = 1$, 3, and 10 dB.

10.10 A lossy line at temperature T feeds an amplifier with noise figure F, as shown below. If an impedance mismatch Γ is present at the input of the amplifier, find the overall noise figure of the system.

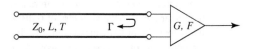

Z_0, L, T Γ G, F

10.11 A balanced amplifier circuit is shown in the accompanying figure. The two amplifiers are identical, each with power gain G and noise figure F. The two quadrature hybrids are also identical, with an insertion loss from the input to either output of $L > 1$ (not including the 3 dB power division factor). Derive an expression for the overall noise figure of the balanced amplifier. What does this result reduce to when the hybrids are lossless?

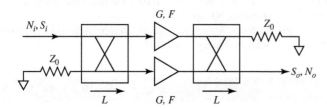

10.12 Show that the following relations involving the third-order intercept point of a two-port nonlinear network are valid. $P^i_{\omega_1}$ and $P^o_{\omega_1}$ are the input and output power levels of an applied two-tone signal, and $P^i_{2\omega_1-\omega_2}$ and $P^o_{2\omega_1-\omega_2}$ are the power levels of the third-order products referenced to the input and output.

$$\frac{OIP_3 - P^o_{\omega_1}}{IIP_3 - P^i_{\omega_1}} = 1, \qquad \frac{OIP_3 - P^o_{2\omega_1-\omega_2}}{IIP_3 - P^i_{2\omega_1-\omega_2}} = 3.$$

10.13 In practice, the third-order intercept point is extrapolated from measured data taken at input power levels well below IP_3. For the spectrum analyzer display shown below, where ΔP is the difference in power between P_{ω_1} and $P_{2\omega_1-\omega_2}$, show that the third-order intercept point is given by $OIP_3 = P_{\omega_1} + (1/2)\Delta P$. Calculate the input and output third-order intercept points for the following data: $P_{\omega_1} = 5$ dBm, $P_{2\omega_1-\omega_2} = -27$ dBm, $P_{in} = -4$ dBm.

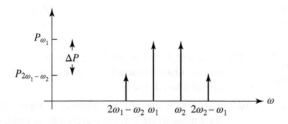

10.14 A two-tone input with a 7 dB difference in the two signal levels is applied to a nonlinear component. What is the relative power ratio of the resulting two third-order intermodulation products $2\omega_1 - \omega_2$ and $2\omega_2 - \omega_1$, if ω_1 and ω_2 are close together?

10.15 For a low-noise amplifier and mixer shown in the accompanying figure, the amplifier has a gain of 30 dB and a third-order intercept point of 22 dBm (referenced at output), and the mixer has a conversion loss of 6 dB and a third-order intercept point of 13 dBm (referenced at input). Find the intercept points of the cascade network for both a phase coherence assumption and a random-phase (noncoherence) assumption.

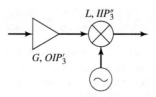

10.16 It is possible to approximately relate the 1 dB compression point to the third-order intercept point. For a single-tone input, use (10.40) to find the amplitudes of the fundamental and third harmonic terms, and assume that a_3 is of opposite sign to a_1. Let V_0 be the voltage where the third-order term reduces the first-order power by 1 dB, and solve for $|a_3/a_1|$. For a two-tone input, use (10.43) to find the amplitude of the third-order intermodulation product, then use (10.44) to relate OP_{1dB} to OIP_3.

10.17 An amplifier with a bandwidth of 1 GHz has a gain of 18 dB and a noise temperature of 350 K. If the 1 dB compression point occurs for an output power level of 5 dBm, what is the linear dynamic range of the amplifier?

10.18 A receiver subsystem has a noise figure of 7 dB, a 1 dB compression point of 22 dBm (referenced to output), a gain of 30 dB, and a third-order intercept point of 33 dBm (referenced to output). If the subsystem is fed with a noise source with $N_i = -108$ dBm and the desired output SNR is 10 dB, find the linear and spurious free dynamic ranges of the subsystem. Assume a system bandwidth of 20 MHz.

10.19 Consider the receiver chain shown in the figure below:

(a) Determine the LNA gain so that the system noise figure is < 4 dB.

(b) For a system noise figure of 5 dB (at the antenna), a signal level of -90 dBm in a 100 MHz bandwidth and an LNA gain of 20 dB, determine the signal and noise voltage at ADC (assume $R_{ADC} = 100\ \Omega$). What is the SNR at ADC?

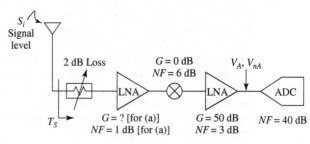

10.20 Calculate the overall noise figure and IIP_3 of the system shown in the figure below:

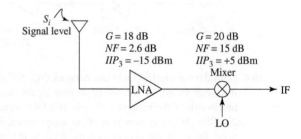

10.1 Which of the following noise is caused by vibration of bound charges?

(a) thermal noise
(b) flicker noise
(c) shot noise
(d) quantum noise

10.2 Which of the following noise is caused by random motion of charges in ionized gas?

(a) flicker noise
(b) quantum noise
(c) thermal noise
(d) plasma noise

10.3 An X-band amplifier has a bandwidth of 1 GHz and a gain of 20 dB. Noise figure of the amplifier is −61 dBm at 290 K and −63.9 dBm at 77 K. The Y-factor of the amplifier is

(a) 3.2 dB
(b) 2.9 dB
(c) 2.5 dB
(d) 5.8 dB

10.4 The Y-factor of an amplifier obtained by measuring the noise figure at temperatures 77 K and 290 K is 2.9 dB. The equivalent noise temperature of the amplifier is

(a) 147 K
(b) 150 K
(c) 170 K
(d) 210 K

10.5 An amplifier with 1 GHz bandwidth and 22 dB gain has a noise equivalent temperature of 170 K. If the amplifier is used with a source having an equivalent noise temperature of $T_S = 450$ K, then the output noise power of the amplifier is

(a) −50 dBm
(b) −58.7 dBm
(c) −60 dBm
(d) −68.7 dBm

10.6 If the noise figure of the first stage of a two-stage cascade network is 9 dB and the noise figure of the second stage is 8 dB and the gain of the first stage is 12, then the noise figure of the cascade is

(a) 5.2 dB
(b) 7.2 dB
(c) 8.3 dB
(d) 9.2 dB

10.7 For a two-stage cascade network the noise equivalent temperature is given by

(a) $T_{e1} + \dfrac{T_{e2}}{G_1}$

(b) $T_{e1} + T_{e2}$

(c) $\dfrac{T_{e1}}{T_{e2}}$

(d) $T_{e1} - \dfrac{T_{e2}}{G_1}$

10.8 For a power divider with insertion loss L and the coupler matched to an external circuitry, the gain of the coupler in terms of insertion loss is

(a) $2L$
(b) $1/2L$
(c) L
(d) $1/L$

10.9 The overall noise figure expression for a mismatched amplifier is

(a) $1 + (F - 1)/(1 - |\Gamma|^2)$
(b) $1 + (F - 1)$
(c) 1
(d) $(F - 1)/(1 - |\Gamma|^2)$

10.10 Active devices such as transistors and diodes become non-linear at high power levels due to

(a) thermal noise
(b) instability of transistor
(c) gain compression
(d) none of the above

10.11 For a single input frequency in a nonlinear device, the output will consist of

(a) harmonics of the input frequency
(b) single frequency component
(c) constant gain
(d) none of the above

10.12 When more than one closely spaced frequency is present in the input frequency, it results in

(a) intermodal distortion
(b) signal fading
(c) gain compression
(d) signal attenuation

10.13 Between the intercept points referenced at the input versus the output, the relation is given by

(a) $OIP_3 = G\,(IIP_3)^2$
(b) $(IIP_3) = G\,(OIP_3)$
(c) $OIP_3 = G\,(IIP_3)$
(d) none of the above

10.14 For a two-port network modeled as Thevenin equivalent, the associated noise power is

(a) k/TB
(b) kTB
(c) TB/k
(d) none of the above

10.15 For an arbitrary white noise source, the equivalent noise temperature is

(a) N_o/kB
(b) N_o
(c) N_o/GkB
(d) none of the above

10.16 For a Wilkinson coupler having a gain of $1/2L$, the equivalent noise temperature is

(a) $T\,(2L + 1)$
(b) $T\,(2L - 1)$
(c) $T\,(2L \times 1)$
(d) $T/(2L - 1)$

10.17 The gain of a two-port network when a network is matched to its external circuitry is given by

(a) $|S_{21}|^2$
(b) $|S_{22}|^2$
(c) $|S_{12}|^2$
(d) $|S_{11}|^2$

10.18 For a two-port network considering other lossy components and the noise due to transmission lines, the expression for noise is

(a) $GkTB - GN_{added}$
(b) $GkTB$
(c) GN_{added}
(d) $GkTB + GN_{added}$

10.19 The ratio of desired signal power to undesired noise power is

   (a) noise temperature
   (b) noise-to-signal ratio
   (c) noise figure
   (d) signal-to-noise ratio

10.20 For a nonlinear circuit, the expression for the output response is given by

   (a) $V_0 = a_0 + a_1 V_{i1}$
   (b) $V_0 = a_0 + a_1 V_{i1} + a_2 V_{i2} + a_3 V_{i3} + ...$
   (c) $V_0 = a_2 V_{i2} + a_3 V_{i3}$
   (d) $V_0 = a_0$

ANSWER KEY

10.1 (a)	**10.5** (b)	**10.9** (a)	**10.13** (c)	**10.17** (a)
10.2 (d)	**10.6** (d)	**10.10** (c)	**10.14** (b)	**10.18** (d)
10.3 (b)	**10.7** (a)	**10.11** (a)	**10.15** (c)	**10.19** (d)
10.4 (a)	**10.8** (b)	**10.12** (a)	**10.16** (b)	**10.20** (b)

11

Learning Objectives

After completing this chapter, you will be able to

- Learn about diodes and diode circuits
- Understand the concepts of bipolar junction transistors and field effect transistors
- Learn about microwave integrated circuits
- Describe the concepts of microwave tubes

Active devices include diodes, transistors, and electron tubes, which can be used for signal detection, mixing, amplification, frequency multiplication, and switching, and as sources of RF and microwave signals. We will discuss some of the basic characteristics of such devices in this chapter. We will avoid detailed discussion of the physics of active devices (see references [1–5] for such material) since for our purposes it will be adequate to work with the terminal characteristics of diodes and transistors using equivalent circuits or scattering parameters. These results will be used to study some basic diode detector and control circuits, and in later chapters for the design of amplifier, mixer, and oscillator circuits using diodes and transistors. We will conclude this chapter with an overview of microwave integrated circuits (MICs) and a brief discussion of some microwave tubes.

Historically, the development of useful RF and microwave active devices has been a long and slow process. The first detector diode was probably the "cat-whisker" crystal detector used in early radio work of the nineteenth century. The advent of electron tubes used as detectors and amplifiers later eliminated this component in most radio systems, but crystal diodes were used by Southworth in his 1930s experiments with waveguides since tube detectors could not operate at such high frequencies. Frequency conversion and heterodyning were also first developed for radio applications in the 1920s. These same techniques were later applied to microwave radars at the MIT Radiation Laboratory during World War II (using crystal diodes as mixers) [1], but it was not until the 1960s that the subject of microwave semiconductor devices saw significant progress. The invention of the transistor led to advances in the theory of solid-state materials and devices, as well as the availability of new semiconductor materials. This led to the development of many new types of diodes and transistors for high-frequency applications. The invention of the gallium arsenide field effect transistor (FET) in the late 1960s [2] was one of the most far-reaching developments in modern microwave engineering. RF and microwave transistors are critical components in wireless systems, finding application as amplifiers, oscillators, switches, phase shifters, mixers, and active filters.

Following the lead from integrated circuitry at lower frequencies, monolithic microwave integrated circuits (MMICs) combine transmission lines, active devices, and other components on a semiconductor substrate. The first single-function MMICs were developed in the late 1960s, but more sophisticated circuits and subsystems, such as multistage FET amplifiers, transmit/receive radar modules, front ends for wireless products, and many other circuits, are now being fabricated as MMICs [2]. The trend is toward MMICs having higher performance, lower power requirements, greater complexity, and lower cost.

11.1 DIODES AND DIODE CIRCUITS

We begin our discussion of active devices with some of the major types of diodes used in RF and microwave circuits. A diode is a two-terminal semiconductor device having a nonlinear V–I relationship. This nonlinearity can be exploited for the useful functions of signal detection, demodulation, switching, frequency multiplication, and oscillation [1]. RF and microwave diodes can be packaged as axial or beam lead components or as surface-mountable chips, or be monolithically integrated with other components on a single

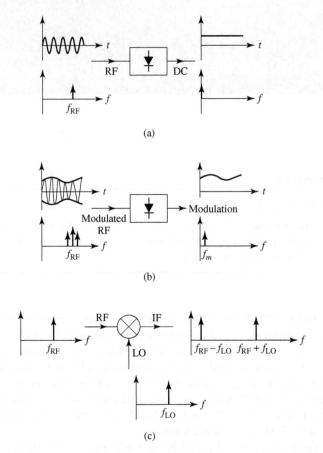

(a)

(b)

(c)

FIGURE 11.1 | Basic frequency conversion operations of rectification, detection, and mixing. (a) Diode rectifier. (b) Diode detector. (c) Mixer.

semiconductor substrate. We first consider detector diodes and circuits, then discuss PIN diodes and control circuits, varactor diodes, and a summary of other types of diodes.

Schottky Diodes and Detectors

The classical *pn* junction diode commonly used at low frequencies has a relatively large junction capacitance that makes it unsuitable for high frequency application. The Schottky barrier diode, however, relies on a semiconductor–metal junction that results in a much lower junction capacitance [3, 4], allowing operation at higher frequencies. Commercially available microwave Schottky diodes generally use *n*-type gallium arsenide (GaAs) material, while lower frequency versions may use *n*-type silicon. Schottky diodes are often biased with a small DC forward current, but can be used without bias.

The primary application of Schottky diodes is in frequency conversion of an input signal. Figure 11.1 illustrates the three basic frequency conversion operations of *rectification* (conversion to DC), *detection* (demodulation of an amplitude-modulated signal), and *mixing* (frequency shifting).

A junction diode can be modeled as a nonlinear resistor, with a small-signal *V–I* relationship expressed as

$$I(V) = I_s(e^{\alpha V} - 1), \tag{11.1}$$

where $\alpha = q/nkT$, and q is the charge of an electron, k is Boltzmann's constant, T is temperature, n is the ideality factor, and I_s is the saturation current [3–5]. Typically, I_s is between 10^{-6} and 10^{-15} A, and $\alpha = q/nkT$ is approximately $1/(25$ mV) for $T = 290$ K. The ideality factor, n, depends on the structure of the diode, and can vary from about 1.05 for Schottky barrier diodes to about 2.0 for point-contact silicon diodes. Figure 11.2 shows a typical diode *V–I* characteristic for a Schottky diode.

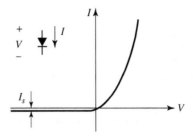

FIGURE 11.2 | *V–I* characteristics of a Schottky diode.

Small-signal approximation: Let the diode voltage be expressed as

$$V = V_0 + v, \tag{11.2}$$

where V_0 is a DC bias voltage and v is a small AC signal voltage. Then (11.1) can be expanded in a Taylor series about V_0 as follows:

$$I(V) = I_0 + v\frac{dI}{dV}\bigg|_{V_0} + \frac{1}{2}v^2\frac{d^2I}{dV^2}\bigg|_{V_0} + \cdots, \tag{11.3}$$

where $I_0 = I(V_0)$ is the DC bias current. The first derivative can be evaluated as

$$\frac{dI}{dV}\bigg|_{V_0} = \alpha I_s e^{\alpha V_0} = \alpha(I_0 + I_s) = G_d = \frac{1}{R_j}, \tag{11.4}$$

which defines R_j, the junction resistance of the diode, and $G_d = 1/R_j$, which is called the dynamic conductance of the diode. The second derivative is

$$\frac{d^2I}{dV^2}\bigg|_{V_0} = \frac{dG_d}{dV}\bigg|_{V_0} = \alpha^2 I_s e^{\alpha V_0} = \alpha^2(I_0 + I_s) = \alpha G_d = G_d'. \tag{11.5}$$

Then (11.3) can be rewritten as the sum of the DC bias current, I_0, and an AC current, i:

$$I(V) = I_0 + i = I_0 + vG_d + \frac{v^2}{2}G_d' + \cdots \tag{11.6}$$

The three-term approximation for the diode current in (11.6) is called the *small-signal approximation*, and will be adequate for most of our purposes.

The small-signal approximation is based on the DC voltage–current relationship of (11.1), and shows that the equivalent circuit of a diode will involve a nonlinear resistance. In practice, however, the AC characteristics of a diode also involve reactive effects due to the structure and packaging of the diode. A typical equivalent circuit for an RF diode is shown in Figure 11.3. The leads or contacts of the diode package are modeled as a series inductance, L_s, and shunt capacitance, C_p. The series resistor, R_s, accounts for contact and current-spreading resistance. The junction capacitance, C_j, and the junction resistance, R_j, are bias dependent. Table 11.1 lists some parameters for a few commercially available Schottky diodes.

Diode rectifiers and detectors: In a rectifier application, a diode is used to convert a fraction of an RF input signal to DC power. Rectification is a very common function and is used for power monitors, automatic gain control circuits, and signal strength indicators. If the total diode voltage consists of a DC bias voltage

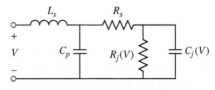

FIGURE 11.3 | Equivalent AC circuit model for a Schottky diode.

TABLE 11.1 | **Parameters for Some Commercial Schottky Diodes**

Schottky Diode	I_s (A)	R_s (Ω)	C_j (pF)	L_s (nH)	C_p (pF)
Skyworks SMS1546	3×10^{-7}	4	0.38	1.0	0.07
Skyworks SMS7630	5×10^{-6}	20	0.14	0.05	0.005
Avago HSMS2800	3×10^{-8}	30	1.6	—	—
Macom MA4E2054	3×10^{-8}	11	0.1	—	0.11

and a small-signal RF voltage,

$$V = V_0 + v_0 \cos \omega_0 t, \tag{11.7}$$

then (11.6) shows that the diode current will be

$$
\begin{aligned}
I &= I_0 + v_0 G_d \cos \omega_0 t + \frac{v_0^2}{2} G_d' \cos^2 \omega_0 t \\
&= I_0 + \frac{v_0^2}{4} G_d' + v_0 G_d \cos \omega_0 t + \frac{v_0^2}{4} G_d' \cos 2\omega_0 t,
\end{aligned}
\tag{11.8}
$$

where I_0 is the bias current and $v_0^2 G_d'/4$ is the DC rectified current. The output also contains AC signals of frequency ω_0 and $2\omega_0$ (as well as higher order harmonics), which are usually filtered out with a simple low-pass filter. A *current sensitivity*, β_i, can be defined as a measure of the change in the DC output current for a given RF input power. From (11.6) the RF input power is $v_0^2 G_d/2$ (using only the first term), while (11.8) shows the change in DC current is $v_0^2 G_d'/4$. The current sensitivity is then

$$\beta_i = \frac{\Delta I_{\text{dc}}}{P_{\text{in}}} = \frac{G_d'}{2G_d} \text{ A/W}. \tag{11.9}$$

An open-circuit *voltage sensitivity*, β_v, can be defined in terms of the voltage drop across the junction resistance when the diode is open circuited. Thus,

$$\beta_v = \beta_i R_j. \tag{11.10}$$

Typical values for the voltage sensitivity of an RF diode range from 400 to 1500 mV/mW.

In a detector application the nonlinearity of a diode is used to demodulate an amplitude-modulated (AM) RF carrier. In this case, the diode voltage can be expressed as

$$v(t) = v_0(1 + m \cos \omega_m t) \cos \omega_0 t, \tag{11.11}$$

where ω_m is the modulation frequency, ω_0 is the RF carrier frequency ($\omega_0 \gg \omega_m$), and m is defined as the *modulation index* ($0 \leq m \leq 1$). Using (11.11) in (11.6) gives the diode current:

$$
\begin{aligned}
i(t) &= v_0 G_d(1 + m \cos \omega_m t) \cos \omega_0 t + \frac{v_0^2}{2} G_d'(1 + m \cos \omega_m t)^2 \cos^2 \omega_0 t \\
&= v_0 G_d \left[\cos \omega_0 t + \frac{m}{2} \cos(\omega_0 + \omega_m)t + \frac{m}{2} \cos(\omega_0 - \omega_m)t \right] \\
&\quad + \frac{v_0^2}{4} G_d' \left[1 + \frac{m^2}{2} + 2m \cos \omega_m t + \frac{m^2}{2} \cos 2\omega_m t + \cos 2\omega_0 t \right. \\
&\quad + m \cos(2\omega_0 + \omega_m)t + m \cos(2\omega_0 - \omega_m)t + \frac{m^2}{2} \cos 2\omega_0 t \\
&\quad \left. + \frac{m^2}{4} \cos 2(\omega_0 + \omega_m)t + \frac{m^2}{4} \cos 2(\omega_0 - \omega_m)t \right].
\end{aligned}
\tag{11.12}
$$

The frequency spectrum of this output is shown in Figure 11.4. The output current terms that are linear in the diode voltage (terms multiplying $v_0 G_d$) have frequencies of ω_0 and $\omega_0 \pm \omega_m$, while the terms that are proportional to the square of the diode voltage (terms multiplying $v_0^2 G_d'/2$) include the frequencies and relative amplitudes listed in Table 11.2.

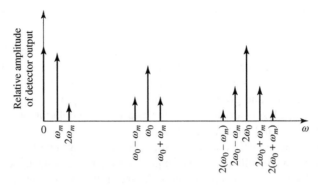

FIGURE 11.4 | Output spectrum of a detected AM signal.

TABLE 11.2 | **Frequencies and Relative Amplitudes of the Square-Law Output of a Detected AM Signal**

Frequency	Relative Amplitude
0	$1 + m^2/2$
ω_m	$2m$
$2\omega_m$	$m^2/2$
$2\omega_0$	$1 + m^2/2$
$2\omega_0 \pm \omega_m$	m
$2(\omega_0 \pm \omega_m)$	$m^2/4$

The desired demodulated output of frequency ω_m is easily separated from the undesired frequency components with a low-pass filter. Observe that the amplitude of this current is $mv_0^2 G_d'/2$, which is proportional to the square of the input signal voltage, and hence the input signal power. This *square-law* behavior is the usual operating condition for detector diodes, but it can be obtained only over a restricted range of input power. If the input power is too large, small-signal conditions will not apply, and the output will become saturated and approach a linear, and then a constant, i versus P characteristic. At very low signal levels the input signal will be lost in the noise floor of the device. Figure 11.5 shows a typical v_{out} versus P_{in} characteristic, where the output voltage can be considered as the voltage drop across a resistor in series with the diode. Square-law operation is particularly important for applications where power levels are inferred from detector voltage, as in SWR indicators and signal level indicators. Detectors may be DC biased to an operating point that provides the best sensitivity.

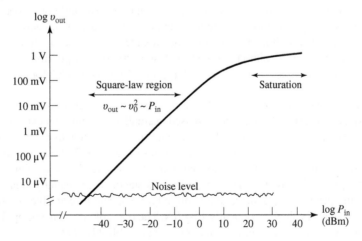

FIGURE 11.5 | Square-law region for a typical diode detector.

POINT OF INTEREST: The Spectrum Analyzer

A spectrum analyzer gives a frequency domain representation of a signal, displaying the average power density versus frequency. Thus, its function is dual to that of an oscilloscope, which displays a time domain representation of a signal. A spectrum analyzer is basically a sensitive receiver that tunes over a specified frequency band and gives a video output that is proportional to the signal power in a narrow bandwidth. Spectrum analyzers are invaluable for measuring modulation products, harmonic and intermodulation distortion, noise, and interference effects.

The diagram below shows a simplified block diagram of a spectrum analyzer. A microwave spectrum analyzer can typically cover any frequency band in the range of several hundred megahertz to tens of gigahertz. The frequency resolution is set by the IF bandwidth, and is typically adjustable from about 100 Hz to 1 MHz. A sweep generator is used to repetitively scan the receiver over the desired frequency band by adjusting the local oscillator frequency, and to provide horizontal deflection of the display. An important part of a modern spectrum analyzer is the YIG-tuned bandpass filter at the input to the mixer. This filter is tuned along with the local oscillator, and acts as a preselector to reduce spurious intermodulation products. An IF amplifier with a logarithmic response is generally used to accommodate a wide dynamic range. Modern spectrum analyzers usually contain a computer to control the system and the measurement process. This improves performance and makes the analyzer more versatile, but can be a disadvantage in that the computer can sometimes remove the user from the physical reality of the measurement.

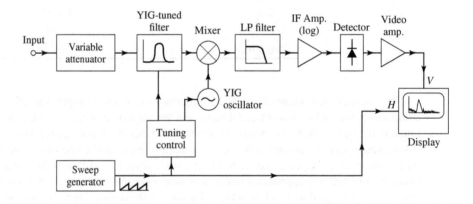

PIN Diodes and Control Circuits

Switches are used extensively in microwave systems for directing signal or power flow between components. Switches can also be used to construct other types of control circuits, such as phase shifters and attenuators. Mechanical switches can be made in waveguide or coaxial form, and can handle high powers but are bulky and slow. PIN diodes, however, can be used to construct an electronic switching element easily integrated with planar circuitry and capable of high-speed operation. Switching speeds typically range from 1 to 10 μsec, although speeds as fast as 20 nsec are possible with careful design of the diode driving circuit. PIN diodes can also be used as power limiters, modulators, and variable attenuators.

PIN diode characteristics: A PIN diode contains an intrinsic (lightly doped) layer between the p and n semiconductor layers. When reverse biased, a small series junction capacitance leads to a relatively high diode impedance, while a forward bias current removes the junction capacitance and leaves the diode in a low-impedance state [3, 4]. These characteristics make the PIN diode a useful RF switching element. Equivalent circuits for the forward- and reverse-biased states are shown in Figure 11.6. The parasitic inductance, L_i, is typically less than 1 nH. The reverse resistance, R_r, is usually small relative to the series reactance due to the junction capacitance and is often ignored. The forward bias current is typically 10–30 mA, and the reverse bias voltage is typically 10–60 V. The bias voltages must be applied to the diode with RF chokes and DC blocks for isolation from the RF signal. Table 11.3 lists parameters for some commercially available PIN diodes.

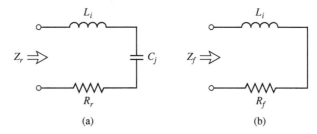

FIGURE 11.6 | Equivalent circuits for the reverse- and forward-biased states of a PIN diode. (a) Reverse bias state. (b) Forward bias state.

TABLE 11.3 | **Parameters for Some Commercial PIN Diodes**

PIN Diode	R_f (Ω)	C_j (pF)
ASI 8001	3.0	0.03
Skyworks DSG9500	4.0	0.025
Infineon BA592	0.36	1.4
Microsemi UM9605	1.5	0.5

Single-pole PIN diode switches: A PIN diode can be used in either a series or a shunt configuration to form a single-pole, single-throw RF switch. These circuits are shown in Figure 11.7, along with the required bias networks. In the series configuration of Figure 11.7a, the switch is ON when the diode is forward biased, while in the shunt configuration the switch is ON when the diode is reverse biased. In both cases, input power is reflected when the switch is in the OFF state. The DC blocking capacitors should have a relatively low impedance at the RF operating frequency, while the RF choke inductors should have a relatively high RF impedance. In some designs, high-impedance quarter-wavelength lines can be used in place of the chokes, to provide RF blocking.

An ideal switch would have zero insertion loss in the ON state, and infinite attenuation in the OFF state. Realistic switching elements, of course, result in some insertion loss for the ON state and finite attenuation for the OFF state. Knowing the diode parameters for the equivalent circuits of Figure 11.6 allows the insertion loss for the ON and OFF states to be calculated for the series and shunt switches. With reference to the simplified switch circuits of Figure 11.8, we can define the insertion loss in terms of the actual load voltage, V_L, and V_0, which is the load voltage that would appear if the switch (Z_d) were absent:

$$\text{IL} = -20 \log \left| \frac{V_L}{V_0} \right|. \tag{11.13}$$

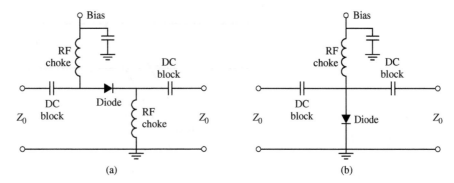

FIGURE 11.7 | Single-pole PIN diode switches. (a) Series configuration. (b) Shunt configuration.

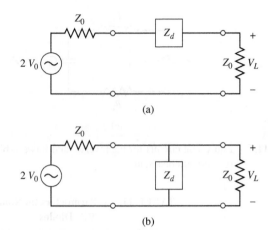

(a)

(b)

FIGURE 11.8 | Simplified equivalent circuits for the series and shunt single-pole PIN diode switches. (a) Series switch. (b) Shunt switch.

Simple circuit analysis applied to the two cases of Figure 11.8 gives the following results:

$$\text{IL} = -20 \log \left| \frac{2Z_0}{2Z_0 + Z_d} \right| \quad \text{(series switch)}, \tag{11.14a}$$

$$\text{IL} = -20 \log \left| \frac{2Z_d}{2Z_d + Z_0} \right| \quad \text{(shunt switch)}. \tag{11.14b}$$

In both cases, Z_d is the diode impedance for either the reverse or forward bias state. Thus,

$$Z_d = \begin{cases} Z_r = R_r + j(\omega L_i - 1/\omega C_j) & \text{for reverse bias} \\ Z_f = R_f + j\omega L_i & \text{for forward bias}. \end{cases} \tag{11.15}$$

The ON-state or OFF-state insertion loss of a switch can usually be improved by adding an external reactance in series or in parallel with the diode, to compensate for the diode reactance. This technique usually reduces the bandwidth, however.

Several single-throw switches can be combined to form a variety of multiple-pole and/or multiple-throw configurations. Figure 11.9 shows series and shunt circuits for a single-pole, double-throw switch; such a switch requires at least two switching elements. In operation, one diode is forward biased in the low-impedance state, with the other diode reverse biased in the high-impedance state. The input signal is switched from one output to the other by reversing the diode bias states. The quarter-wave lines of the shunt circuit limit the bandwidth of this configuration. A photograph of a PIN SP3T switch is shown in Figure 11.10.

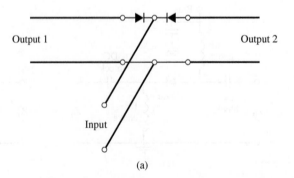

(a)

FIGURE 11.9 | Circuits for single-pole, double-throw PIN diode switches. (a) Series.

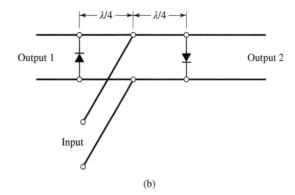

(b)

FIGURE 11.9 | Continued. (b) Shunt.

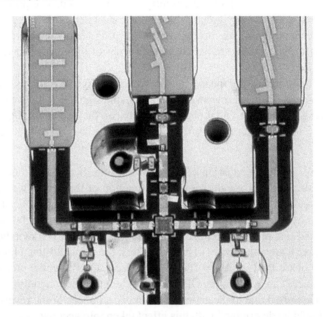

FIGURE 11.10 | Photograph of an SP3T GaAs PIN diode switch, operating from 6 to 27 GHz. The diode chips are 15 mils square.

Courtesy of LNX Corporation, Salem, NH.

EXAMPLE 11.1   SINGLE-POLE PIN DIODE SWITCH

A single-pole switch operating at 1.5 GHz is to be constructed using an Infineon BA592 PIN diode with $C_j = 1.4$ pF and $R_f = 0.36$ Ω. What switch circuit (series or shunt) should be used to obtain the greatest ratio of off-to-on attenuation? Assume that $L_i = 0.5$ nH, $R_r = 0.5$ Ω, and $Z_0 = 50$ Ω.

Solution
First use (11.15) to compute the diode impedance for the reverse and forward bias states:

$$Z_d = \begin{cases} Z_r = R_r + j\left(\omega L_i - \dfrac{1}{\omega C_j}\right) = 0.5 - j71.08 \ \Omega \\ Z_f = R_f + j\omega L_i \qquad\qquad = 0.36 + j4.71 \ \Omega \end{cases}$$

Then (11.14) gives the insertion losses for the ON and OFF states of the series and shunt switches as follows:
 For the series circuit,

$$\text{IL}_\text{on} = -20 \log\left|\frac{2Z_0}{2Z_0 + Z_f}\right| = 0.041 \text{ dB}$$

$$\text{IL}_\text{off} = -20 \log\left|\frac{2Z_0}{2Z_0 + Z_r}\right| = 1.8 \text{ dB}$$

For the shunt circuit,

$$IL_{on} = -20 \log \left| \frac{2Z_r}{2Z_r + Z_0} \right| = 0.528 \text{ dB}$$

$$IL_{off} = -20 \log \left| \frac{2Z_f}{2Z_f + Z_0} \right| = 14.74 \text{ dB}$$

The shunt configuration has the greatest difference in attenuation between the ON and OFF states and series circuit has the lowest ON insertion loss.

PIN diode phase shifters: Several types of microwave phase shifters can be constructed with PIN diode switching elements. Compared with ferrite phase shifters, diode phase shifters have the advantages of small size, integrability with planar circuitry, and high speed. The power requirements for diode phase shifters, however, are generally greater than those for a latching ferrite phase shifter (Section 9.5) because diodes require continuous bias current, while a latching ferrite device requires only a pulsed current to change its magnetic state. There are basically three types of PIN diode phase shifters: *switched line, loaded line*, and *reflection*.

The switched-line phase shifter, shown in Figure 11.11, is the most straightforward type, using two single-pole, double-throw switches to route the signal flow between one of two transmission lines of different length. The differential phase shift between the two paths is given by

$$\Delta\phi = \beta(\ell_2 - \ell_1), \tag{11.16}$$

where β is the propagation constant of the line. If the transmission lines are TEM (or quasi-TEM, like microstrip), this phase shift is a linear function of frequency, which implies a true time delay between the input and output ports. This is a useful feature in wideband systems. This type of phase shifter is also inherently reciprocal, and so it can be used for both receive and transmit functions. The insertion loss of the switched-line phase shifter is equal to the loss of the SPDT switches plus line losses.

Like many other types of phase shifters, the switched-line phase shifter is usually designed for discrete binary phase shifts of $\Delta\phi = 180°, 90°, 45°$, etc. One potential problem with this type of phase shifter is that resonances can occur in the OFF line if its length is near a multiple of $\lambda/2$. The resonant frequency will be slightly shifted due to the series junction capacitances of the reversed biased diodes, so the lengths ℓ_1 and ℓ_2 should be determined with this effect taken into account.

A design that is useful for small amounts of phase shift (generally 45°, or less) is the loaded-line phase shifter. The basic principle of this type of phase shifter can be illustrated with the circuit of Figure 11.12a, which shows a transmission line loaded with a shunt susceptance, jB. The reflection and transmission coefficients can be written as

$$\Gamma = \frac{1 - (1 + jb)}{1 + (1 + jb)} = \frac{-jb}{2 + jb}, \tag{11.17a}$$

$$T = 1 + \Gamma = \frac{2}{2 + jb}, \tag{11.17b}$$

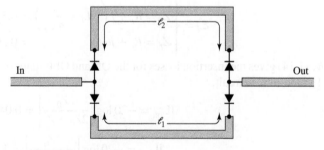

FIGURE 11.11 | A switched-line phase shifter.

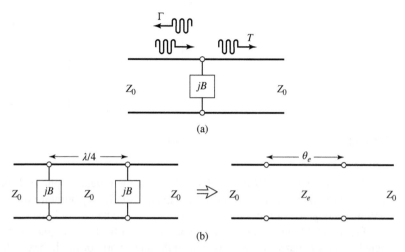

FIGURE 11.12 | Loaded-line phase shifters. (a) Basic circuit. (b) Practical loaded-line phase shifter and its equivalent circuit.

where $b = BZ_0$ is the normalized susceptance. Thus the phase shift in the transmitted wave introduced by the load is

$$\Delta\phi = \tan^{-1}\frac{b}{2}, \tag{11.18}$$

which can be made positive or negative, depending on the sign of b. A disadvantage is the insertion loss that is inherently present due to the reflection from the shunt load. In addition, increasing b to obtain a larger $\Delta\phi$ entails a greater insertion loss, as seen from (11.17b).

The reflections from the shunt susceptance can be reduced by using the circuit of Figure 11.12b, where two shunt loads are separated by a $\lambda/4$ length of line. Then the partial reflection from the second load will be 180° out of phase with the partial reflection from the first load, leading to cancellation. We can analyze this circuit by calculating its $ABCD$ matrix and comparing it to the $ABCD$ matrix of an equivalent line having a length θ_e and characteristic impedance Z_e. Thus, for the loaded line,

$$\begin{bmatrix} A & B \\ C & D \end{bmatrix} = \begin{bmatrix} 1 & 0 \\ jB & 1 \end{bmatrix}\begin{bmatrix} 0 & jZ_0 \\ j/Z_0 & 0 \end{bmatrix}\begin{bmatrix} 1 & 0 \\ jB & 1 \end{bmatrix}$$

$$= \begin{bmatrix} -BZ_0 & jZ_0 \\ j(1/Z_0 - B^2 Z_0) & -BZ_0 \end{bmatrix}, \tag{11.19a}$$

while the equivalent transmission line has an $ABCD$ matrix given by

$$\begin{bmatrix} A & B \\ C & D \end{bmatrix} = \begin{bmatrix} \cos\theta_e & jZ_e\sin\theta_e \\ j\sin\theta_e/Z_e & \cos\theta_e \end{bmatrix}. \tag{11.19b}$$

Then we have that

$$\cos\theta_e = -BZ_0 = -b, \tag{11.20a}$$

$$Z_e = Z_0\cos\theta_e = \frac{Z_0}{\sqrt{1-b^2}}. \tag{11.20b}$$

For small values of b, θ_e will be close to $\pi/2$, and these results will reduce to

$$\theta_e \simeq \frac{\pi}{2} + b, \tag{11.21a}$$

$$Z_e \simeq Z_0\left(1 + \frac{b}{2}\right). \tag{11.21b}$$

The susceptance, B, can be implemented with a lumped inductor or capacitor, or with a stub, and switched between two states with an SPST diode switch.

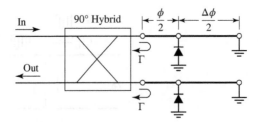

FIGURE 11.13 | A reflection phase shifter using a quadrature hybrid.

The third type of PIN diode phase shifter is the reflection phase shifter, which uses an SPST switch to control the path length of a reflected signal. Usually a quadrature hybrid is used to provide a two-port circuit, although other types of hybrids, or even a circulator, could be used for this purpose.

Figure 11.13 shows a reflection-type phase shifter using a quadrature hybrid. In operation, an input signal divides equally between the two right-hand ports of the hybrid. The diodes are both biased in the same state (forward or reverse biased), so the waves reflected from the two terminations will add in phase at the indicated output port. Turning the diodes on or off changes the total path length for both reflected waves by $\Delta\phi$, producing a phase shift of $\Delta\phi$ at the output. Ideally, the diodes would look like short circuits in their ON state, and open circuits in their OFF state, so that the reflection coefficients at the right side of the hybrid can be written as $\Gamma = e^{-j(\phi+\pi)}$ for the diodes in their ON state, and $\Gamma = e^{-j(\phi+\Delta\phi)}$ for the diodes in their OFF state. There is an infinite number of choices of line lengths that give the desired $\Delta\phi$ (i.e., the value of $\phi/2$ is a degree of freedom), but it can be shown that bandwidth is optimized if the reflection coefficients for the two states are phase conjugates. Thus, if $\Delta\phi = 90°$, the best bandwidth will be obtained for $\phi = 45°$.

A good input match for the reflection-type phase shifter requires that the diodes be well matched. The insertion loss is limited by the loss of the hybrid, as well as by the forward and reverse resistances of the diodes. Impedance transformation sections can be used to improve performance in this regard.

Varactor Diodes

We have seen that a PIN diode has a junction capacitance that can be switched on or off with bias voltage. This effect can be enhanced by tailoring the size and doping profile of the intrinsic layer of the diode to provide a desired junction capacitance versus junction voltage (C vs. V) behavior when reverse biased. Such a device is called a *varactor diode*, and it produces a junction capacitance that varies smoothly with bias voltage, thus providing an electrically adjustable reactive circuit element. One of the most common applications of varactor diodes is to provide electronic frequency tuning of the local oscillator in a multichannel receiver, such as those used in cellular telephones, wireless local area network radios, and television receivers. This is accomplished by using a varactor diode in the resonant circuit of a transistor oscillator, and controlling the DC reverse bias voltage applied to the diode. The nonlinearity of varactor diodes also makes them useful for frequency multipliers (discussed in Chapter 13). Varactor diodes are generally made from silicon for RF applications, and gallium arsenide for microwave applications.

A simplified equivalent circuit for a reverse-biased varactor diode is shown in Figure 11.14. The junction capacitance is dependent on the (negative) junction bias voltage, V, according to

$$C_j(V) = \frac{C_0}{(1 - V/V_0)^\gamma},$$ (11.22)

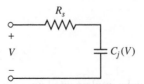

FIGURE 11.14 | Equivalent circuit of a reverse-biased varactor diode.

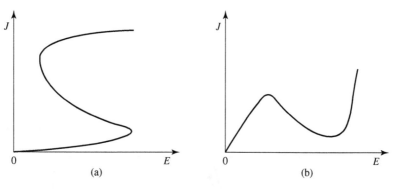

FIGURE 11.15 | (a) Current-controlled NDR. (b) Voltage-controlled NDR.

where C_0 is the junction capacitance with no bias; $V_0 = 0.5$ V for silicon diodes, and $V_0 = 1.3$ V for GaAs diodes. The exponent γ depends on the doping profile of the intrinsic layer of the diode. An ideal hyperabrupt varactor diode has $\gamma = 0.5$; many practical diodes have an exponent of about $\gamma = 0.47$, although the value can be as high as 1.5 or 2.0 for some diodes. In the equivalent circuit, R_s is the series junction and contact resistance, typically on the order of a few ohms. A typical GaAs varactor diode may have $C_0 = 0.5$–2.0 pF, and a junction capacitance that varies from about 0.1 to 2.0 pF as the bias voltage ranges from -20 to 0 V. Parasitic reactances due to the diode package should be included in a realistic design.

Ridley–Watkins–Hilsum (RWH) Theory

To explain the phenomenon of negative differential resistance (NDR) in certain bulk solid-state compounds, Ridley and Watkins proposed a theory known as RWH theory. The negative differential resistance is developed in devices when either current or voltage is applied to the terminals of the sample. This theory explains the operation of many microwave semiconductor devices such as Gunn diodes, which are used to generate microwave power. Negative differential resistance devices are operated in two modes, namely current-controlled mode and voltage-controlled mode, as illustrated in Figure 11.15.

In the current-controlled mode the voltage is multivalued, and in the voltage-controlled mode the current is multivalued. The effect of a negative differential resistance region in current density field curve is to make the sample electrically unstable. As a result, the homogeneous sample becomes electrically heterogeneous to obtain stability. It results in high- and low-current filaments in current-controlled mode, and high- and low-field domains in voltage-controlled mode, as shown in Figure 11.16.

Two-Valley Model Theory

Two-valley model theory explains the effect of generation of negative resistance in certain types of bulk semiconductor materials. Based on the energy band theory, in n-type GaAs, a high-mobility lower valley is separated from a low-mobility upper valley by an energy difference of 0.36 eV, as shown in Figure 11.17. Under equilibrium conditions, electron densities in the lower and upper valleys remain the same. When an electric field is applied, a transfer of electrons takes place between the upper and lower valleys, as shown in Figure 11.18, based on the applied field.

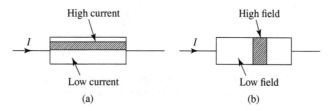

FIGURE 11.16 | (a) High-current filament. (b) High-field domain.

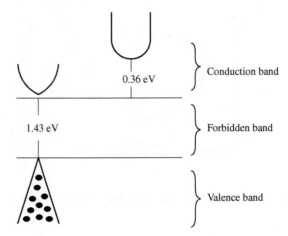

FIGURE 11.17 | Two-valley model of *n*-type GaAs.

When $E < E_l$, no electrons will transfer to the upper valley, as shown in Figure 11.18a. When $E_l < E < E_u$, electrons will begin transfer to the upper valley as shown in Figure 11.18b. And when $E_u < E$, all electrons will transfer to the upper valley as shown in Figure 11.18c.

In order to exhibit negative resistance, the band structure of a semiconductor must satisfy the following conditions:

1. The difference between the energy levels at upper and lower valleys must be several times greater than the thermal energy at room temperature (about 0.026 eV).
2. The difference between the energy levels at the two valleys must be smaller than the energy gap between the valence and conduction bands. If not, then the semiconductor will break down and because of the electron–hole pair formation, the semiconductor will become highly conductive before the electrons start to transfer to upper valleys.
3. The electrons in the upper valley must have large effective mass, low mobility, and high density of state, whereas the lower valley electrons must have small effective mass, high mobility, and low density of state. In other words, the electron velocity must be much larger in the lower valleys than in the upper valleys.

Semiconductors such as germanium and silicon do not satisfy all these criteria, whereas compound semiconductors such as gallium arsenide (GaAs), cadmium telluride (CdTe), and indium phosphide (InP) satisfy these criteria. Other semiconductors such as gallium phosphide (GaP), indium arsenide (InAs), and indium antimonide also do not satisfy the above criteria.

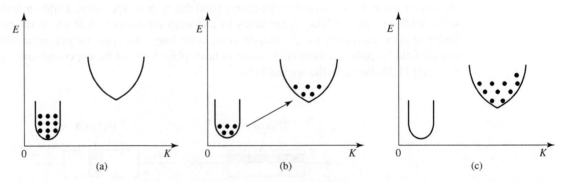

FIGURE 11.18 | Transfer of electron densities for various conditions: (a) $E < E_l$; (b) $E_l < E < E_u$; (c) $E_u < E$, where E is the applied electric field, E_l is the electric field of the lower valley, and E_u is the electric field of the upper valley.

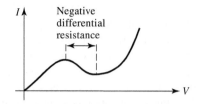

FIGURE 11.19 | The DC $I–V$ characteristic of a Gunn diode, showing the region of negative differential resistance. Other negative resistance devices, such as impact avalanche and transit time (IMPATT) and tunnel diodes, have similar $I–V$ characteristics.

Other Diodes

Here we briefly summarize the characteristics of several other diode devices that are commonly used in microwave circuits. Historically, diode devices were developed long before high-frequency three-terminal devices (e.g., junction and field effect transistors), and for many years diodes provided the only means of microwave power generation and amplification without using electron tubes. Today, many of these devices are most useful at millimeter wave frequencies, since there are now many types of transistors that offer better performance and more design flexibility at RF and microwave frequencies. Further information on diode devices can be found in the literature.

Gunn diodes: The operation of a *Gunn diode* is based on the transferred electron effect (also known as the Gunn effect), which was discovered by J. B. Gunn in 1963. Practical Gunn diodes typically use GaAs or InP materials in a specially doped bulk form, as opposed to a traditional *pn* junction. The Gunn diode has an $I–V$ characteristic that exhibits a negative differential resistance (negative slope) that can be used to generate RF power directly from a DC source when properly biased. Figure 11.19 shows a DC $I–V$ curve that is characteristic of Gunn diodes, where the region of negative differential resistance (negative slope) corresponds to the operating point of the device. Gunn diodes can produce continuous power of up to several hundred milliwatts, at frequencies from 1 to 200 GHz, with efficiencies ranging from 5% to 15%. Oscillator circuits using Gunn diodes require a high-Q resonant circuit or cavity, which is often tuned mechanically. Electronic tuning by bias adjustment is limited to 1% or less, but varactor diodes are sometimes included in the resonant circuit to provide a greater range of electronic tuning. Gunn diode sources are used extensively in low-cost applications such as traffic radars, motion detectors for door openers and security alarms, and test and measurement systems. A photograph of a W-band Gunn diode oscillator is shown in Figure 11.20.

IMPATT diodes: An *impact avalanche and transit time* (IMPATT) diode has a physical structure similar to a PIN diode, but is operated with a relatively high voltage (70–100 V) to produce a reverse-biased avalanche breakdown current. It exhibits a negative resistance over a broad frequency band that can extend into the submillimeter range, and it can be used to directly convert DC to RF power. IMPATT sources are generally noisier than Gunn diodes but are capable of higher powers and higher DC-to-RF conversion efficiencies.

FIGURE 11.20 | A W-band Gunn diode oscillator. The output power is 16 dBm, and the source is mechanically tunable over a frequency range of 4 GHz.

Courtesy of Millitech, Inc., Northampton, MA.

IMPATTs also have better temperature stability than Gunn diodes. Typical IMPATTs operate at frequencies from 10 to 300 GHz, with efficiencies ranging up to 15%. IMPATT diodes are among the few practical solid-state devices that can provide fundamental frequency power above 100 GHz. IMPATT devices can also be used for frequency multiplication and amplification.

Silicon IMPATT diodes can provide CW power ranging from 10 W at 10 GHz to 1 W at 94 GHz, with efficiencies typically below 10%. GaAs IMPATTs can provide CW power ranging from 20 W at 10 GHz to 5 mW at 130 GHz. Pulsed operation generally results in higher powers and higher efficiencies. Because of the low efficiency of these devices, thermal considerations are a limiting factor for both CW and pulsed operation. IMPATT oscillators can be mechanically or electrically tuned. A disadvantage of IMPATT oscillators is that their AM noise level is generally higher than that of other sources.

Tunnel diodes: The *tunnel diode*, invented by L. Esaki in 1957, is a *pn* junction diode with a doping profile that allows electron tunneling through a narrow energy band gap, leading to negative resistance at high frequencies. Tunnel diodes can be used for oscillators as well as amplifiers. Before high-frequency transistors were available, tunnel diodes provided the only means of high-frequency amplification with a solid-state device. Such an amplifier employs the diode in a one-port reflection circuit, where the negative RF resistance of the device produces a reflection coefficient with a magnitude greater than unity, and therefore amplification of an incident signal. Such amplifiers have been made obsolete by modern RF and microwave transistors, but tunnel diodes are still used in some applications today.

BARITT diodes: A *barrier injection transit time* (*BARITT*) diode has a structure similar to a junction transistor without a base contact. Like the IMPATT diode, it is a transit time device. It generally has a lower power capability than the IMPATT diode, but the advantage of lower AM noise. This makes it useful for local oscillator applications at frequencies up to 94 GHz. BARITT diodes are also useful for detector and mixer applications.

TRAPATT diodes: A *trapped plasma avalanche triggered transit* (*TRAPATT*) diode is a microwave generator which is used both as an amplifier and as an oscillator because of its high efficiency and capability to operate from several hundreds of MHz to several GHz. It is a semiconductor *pn* junction diode with high-peak-power diode structure (p^+nn^+). The *n*-type depletion region width varies from 2.5 to 12.5 μm and the p^+ region width varies from 2.5 to 7.5 μm. At lower frequencies the TRAPATT diode's diameter lies between 50 and 750 μm. TRAPATT diodes have low power dissipation and can be operated in the range of 3 GHz to 50 GHz. They are highly suitable for pulsed operations. A disadvantage is that they operate below the millimeter wave region.

Power Combining

In many applications RF power requirement exceeds the power capacity of a single solid-state source; this is especially common at millimeter wave frequencies. Because of the many advantages offered by solid-state sources compared to electron tubes, substantial effort has been directed toward increasing output power through the use of various power combining techniques. Thus, the outputs of two or more sources are combined, effectively multiplying the output power of a single source by the number of individual sources being used. It is important that the individual sources to be combined are coherent and in phase. In principle, an unlimited amount of RF power can be generated in this manner; in practice, however, factors such as high-order modes and combiner losses limit the multiplication factor to about 10–20 dB.

Power combining can be done by combining at the device level or at the circuit level. In addition, in some applications power can be combined spatially by using an array of antennas, where each radiating element is fed with a separate source (the sources must have phase coherence, perhaps by using injection-locked oscillators). At the device level, several diode (or transistor) junctions are essentially connected in parallel over an electrically small region and used as a single device. This technique is limited to a relatively few device junctions. At the circuit level, the power output from N devices can be combined with an N-way combiner. The combining circuit may be an N-way Wilkinson-type network or a similar type of planar combining network. Resonant cavities can also be used for this purpose; cavity combiners often have the advantage of providing self-locking of individual oscillator phases. These various power combining techniques all have their own advantages and disadvantages in terms of efficiency, bandwidth, isolation between sources, and circuit complexity.

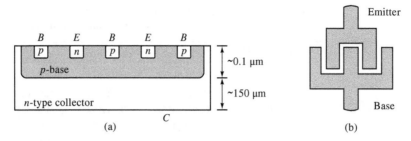

FIGURE 11.21 | (a) Cross section of an interdigitated microwave bipolar junction transistor. (b) Top view, showing base and emitter contacts.

11.2 BIPOLAR JUNCTION TRANSISTORS

Transistors are three-terminal semiconductor devices, and can be categorized as either *junction transistors* or *field effect transistors* [3–6]. Junction transistors include *bipolar junction transistors* (BJTs) that use a single semiconductor material (usually silicon), and *heterojunction bipolar transistors* (HBTs) that use compound semiconductors. Both *npn* and *pnp* configurations are possible, but most RF junction transistors are usually of the *npn* type due to higher electron mobility at higher frequencies

Bipolar Junction Transistor

RF bipolar junction transistors (BJTs) are usually made using silicon (Si), and this transistor is one of the oldest and most popular active RF devices in use today because of its low cost and good operating performance in terms of frequency range, power capacity, and noise characteristics. Silicon junction transistors are useful for amplifiers up to the range of 2–10 GHz, and in oscillators up to about 20 GHz. Bipolar transistors typically have very low $1/f$-noise characteristics, making them well suited for oscillators with low-phase noise.

Bipolar junction transistors are sometimes preferred over FETs at frequencies below about 2–4 GHz because of higher gain and lower cost, and the possibility of biasing with a single power supply. Bipolar transistors are subject to shot noise as well as thermal noise effects, so their noise figure is not as good as that of FETs. Figure 11.21 shows the construction of a typical silicon bipolar transistor having multiple fingers for the base and emitter electrodes. The BJT is current driven, with the base current modulating the collector current. The upper frequency limit of the bipolar transistor is controlled primarily by the base length, which is typically on the order of 0.1 μm.

A small-signal equivalent circuit model for an RF bipolar transistor is shown in Figure 11.22 for a common emitter configuration. This model, known as the hybrid-π model, is popular because of its similarity to the equivalent circuit of a FET, and because of its utility in circuit analysis. This model does not include parasitic resistances and inductances due to the base and emitter leads. More sophisticated equivalent circuits may be advantageous for computer-aided design and modeling over wide frequency ranges. The Gummel–Poon model [6], for example, is used extensively in computer modeling with the SPICE circuit analysis software package, and can include parasitic effects.

In many cases the capacitor, C_c, between the base and collector in the hybrid-π model, has a relatively small value and may be ignored. This has the effect of making $S_{12} = 0$, implying that power only flows in one

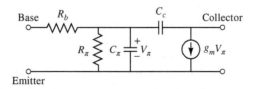

FIGURE 11.22 | Simplified hybrid-π equivalent circuit for a microwave bipolar junction transistor in the common emitter configuration.

TABLE 11.4 | Scattering Parameters for an NPN Silicon BJT (NEC NE 58219, $V_{ce} = 5.0$ V, $I_c = 5.0$ mA, common emitter)

Frequency (GHz)	S_{11}	S_{12}	S_{21}	S_{22}
0.1	$0.78\angle-33°$	$0.03\angle71°$	$12.7\angle155°$	$0.93\angle-17°$
0.5	$0.46\angle-113°$	$0.08\angle52°$	$6.3\angle104°$	$0.53\angle-38°$
1.0	$0.38\angle-158°$	$0.11\angle54°$	$3.5\angle80°$	$0.40\angle-43°$
2.0	$0.40\angle157°$	$0.19\angle56°$	$1.9\angle52°$	$0.33\angle-63°$
4.0	$0.52\angle117°$	$0.38\angle45°$	$1.1\angle14°$	$0.33\angle-127°$

direction through the device (from port 1 to port 2); such a device is called *unilateral*. This approximation is often used to simplify analysis.

The hybrid-π model is roughly based on the physics of the junction transistor, and can be useful under circumstances where the element values of the model are fairly constant over a range of operating bias conditions, load conditions, and frequency. Otherwise, the element values become frequency, bias, or load dependent, in which case the hybrid-π model (or any other equivalent circuit model) becomes much less useful. In this case, it is simpler to treat the transistor as a two-port network, characterized by two-port parameters. In practice, scattering parameters, measured under typical operating conditions, are usually used for this purpose and are supplied by the device manufacturer. Table 11.4 shows scattering parameters for a typical RF silicon junction transistor in a common emitter configuration. Note that there are relatively large mismatches at the base (port 1) and the collector (port 2), and that the gain (given roughly by $|S_{21}|$) drops quickly with an increase in frequency. Also note that $|S_{12}|$ is relatively small (particularly at low frequencies), making the device approximately unilateral.

The equivalent circuit of Figure 11.22 can be used to estimate the upper frequency limit, f_T, defined as the threshold frequency where the short-circuit current gain of the transistor is unity. If we assume an input current I_{in} at the base, and ignore the series base resistance, R_b (typically small), and the shunt resistance, R_π (typically large), then the voltage across the capacitor C_π is $V_\pi = I_{in}/j\omega C_\pi$. The output short-circuit current at the collector is $I_{out} = g_m V_\pi$, so the short-circuit current gain is

$$G_I^{sc} = \left| \frac{I_{out}}{I_{in}} \right| = \frac{g_m}{\omega C_\pi}.$$

The current gain is seen to decrease with frequency, and is unity at the threshold frequency,

$$f_T = \frac{g_m}{2\pi C_\pi}. \tag{11.23}$$

Figure 11.23a shows typical DC operating characteristics for a BJT. The biasing point for the transistor depends on the application and type of device, with low collector currents generally giving the best noise figure, and higher collector currents giving the best power gain. Figure 11.23b shows a typical bias and decoupling circuit for a bipolar transistor in a common emitter configuration.

Heterojunction Bipolar Transistor

The operation of a heterojunction bipolar transistor (HBT) is essentially the same as that of a BJT, but an HBT has a base-emitter junction made from a compound semiconductor material such as GaAs, indium phosphide (InP), or silicon germanium (SiGe), often in conjunction with thin layers of other materials (e.g., aluminum). This structure offers much improved performance at high frequencies. Some HBTs can operate at frequencies exceeding 100 GHz, and recent developments with HBTs using SiGe have demonstrated that these devices are useful in low-cost circuits operating at frequencies of 60 GHz or higher.

Since the HBT is similar in structure and operation to the BJT, the equivalent circuit model of Figure 11.22 can be used for both transistor types. As with BJTs, equivalent circuit models may have limited applicability when attempting to model HBTs over a range of operating conditions, so scattering parameter data, measured for a particular bias point, may be more useful. Table 11.5 gives the scattering parameters at several frequencies for a popular microwave HBT. Observe that $|S_{21}|$ decreases much less rapidly with

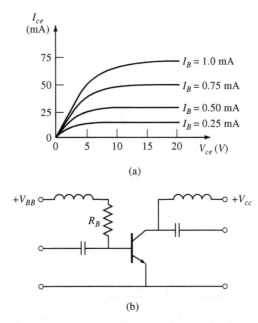

FIGURE 11.23 | (a) DC characteristics of an *npn* BJT. (b) Biasing and decoupling circuit for an *npn* BJT.

TABLE 11.5 | **Scattering Parameters for a SiGe HBT (Infineon BFP640F, $V_{ce} = 2.0$ V, $I_c = 1.2$ mA, common emitter)**

Frequency (GHz)	S_{11}	S_{12}	S_{21}	S_{22}
1.0	$0.91\angle-44°$	$0.06\angle68°$	$3.92\angle149°$	$0.93\angle-17°$
2.0	$0.75\angle-86°$	$0.10\angle46°$	$3.39\angle120°$	$0.79\angle-31°$
4.0	$0.59\angle-144°$	$0.11\angle29°$	$2.18\angle82°$	$0.64\angle-43°$
6.0	$0.54\angle176°$	$0.11\angle34°$	$1.64\angle57°$	$0.58\angle-53°$

frequency when compared with the BJT of Table 11.4. The device also is seen to be approximately unilateral, as $|S_{12}|$ is relatively small.

High levels of monolithic integration are easy and inexpensive with SiGe HBTs, so this technology is proving to be very useful for low-cost millimeter wave circuits for both defense and commercial applications.

11.3 FIELD EFFECT TRANSISTORS

In contrast to BJTs, field effect transistors (FETs) are *monopolar*, with only one carrier type (holes or electrons) providing current flow through the device: *n*-channel FETs employ electrons, while *p*-channel devices use holes. In addition, while a BJT is a current-controlled device, a FET is a voltage-controlled device, having a source-to-drain characteristic that is similar to that of a voltage-dependent variable resistor.

Field effect transistors can take many forms, including the MESFET (metal semiconductor FET), the MOSFET (metal oxide semiconductor FET), the HEMT (high electron mobility transistor), and the PHEMT (pseudomorphic HEMT). FET transistor technology has been under continuous development for more than 50 years—the first junction FETs were developed in the 1950s, while the HEMT was proposed in the early 1980s. GaAs MESFETs are among the most commonly used transistors for microwave and millimeter wave applications, being usable at frequencies up to 60 GHz or more. Even higher operating frequencies can be obtained with GaAs HEMTs. GaAs MESFETs and HEMTs are especially useful for low-noise amplifiers since these transistors have lower noise figures than any other active devices. Recently developed gallium nitride (GaN) HEMTs are very useful for high power RF and microwave amplifiers. CMOS FETs are increasingly being used for RF integrated circuits, offering high levels of integration at low cost and

TABLE 11.6 | Performance Characteristics of Microwave Transistors

Device	BJT	HBT	CMOS	MESFET	HEMT	HEMT
Semiconductor	Si	SiGe	Si	GaAs	GaAs	GaN
Frequency range (GHz)	10	30	20	60	100	10
Typical gain (dB)	10–15	10–15	10–20	5–20	10–20	10–15
Noise figure (dB)	2.0	0.6	1.0	1.0	0.5	1.6
(frequency, GHz)	(2)	(8)	(4)	(10)	(12)	(6)
Power capacity	High	Medium	Low	Medium	Medium	High
Cost	Low	Medium	Low	Medium	High	Medium
Single-polarity supply	Yes	Yes	Yes	No	No	No

low power requirements, for commercial wireless applications. Table 11.6 summarizes the performance characteristics of some of the most popular microwave transistors.

Metal Semiconductor Field Effect Transistor

One of the most important developments in microwave technology has been the GaAs metal semiconductor field effect transistor (MESFET), as this device permitted the first practical solid-state implementation of amplifiers, oscillators, and mixers at microwave frequencies, leading to key applications in radar, GPS, remote sensing, and wireless communication. GaAs MESFETs can be used at frequencies well into the millimeter wave range, with high gain and low noise figure, often making them the device of choice for hybrid and monolithic integrated circuits at frequencies above 10 GHz.

Figure 11.24 shows the cross section of a typical n-channel GaAs MESFET. The gate junction is formed as a Schottky barrier. The desirable gain and noise features of this transistor are a result of the higher electron mobility of GaAs compared to silicon, and the absence of shot noise. The device is biased with a drain-to-source voltage, V_{ds}, and a gate-to-source voltage, V_{gs}. In operation, electrons are drawn from the source to the drain by the positive V_{ds} supply voltage. An applied signal voltage on the gate then modulates these majority electron carriers, producing voltage amplification. The maximum frequency of operation is limited by the gate length; present FETs have gate lengths on the order of 0.2–0.6 μm, with corresponding upper frequency limits of 100 to 50 GHz.

A small-signal equivalent circuit for a microwave MESFET is shown in Figure 11.25 for a common-source configuration. The components and some typical values for this model are listed below:

$$
\begin{aligned}
R_i \text{ (series gate resistance)} &= 7\ \Omega \\
R_{ds} \text{ (drain-to-source resistance)} &= 400\ \Omega \\
C_{gs} \text{ (gate-to-source capacitance)} &= 0.3\ \text{pF} \\
C_{ds} \text{ (drain-to-source capacitance)} &= 0.12\ \text{pF} \\
C_{gd} \text{ (gate-to-drain capacitance)} &= 0.01\ \text{pF} \\
g_m \text{ (transconductance)} &= 40\ \text{mS}
\end{aligned}
$$

This model does not include package parasitics, which typically introduce small series resistances and inductances at the three terminals due to ohmic contacts and bonding leads. The dependent current generator $g_m V_c$ depends on the voltage across the gate-to-source capacitor C_{gs}, leading to a value of $|S_{21}| > 1$ under normal operating conditions (where port 1 is at the gate, and port 2 is at the drain). The reverse signal path,

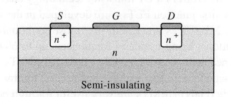

FIGURE 11.24 | Cross section of an n-channel GaAs MESFET.

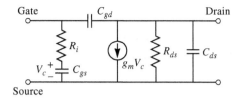

FIGURE 11.25 | Small-signal equivalent circuit for a microwave FET in the common-source configuration.

given by S_{12}, is due solely to the capacitance C_{gd}. As seen from the above data, this is typically a very small capacitor, which can often be ignored in practice. In this case, $S_{12} = 0$, and the device is unilateral. The scattering parameters for a typical GaAs MESFET are given in Table 11.7.

As we did for the BJT, we can use the equivalent circuit model of Figure 11.25 to determine the upper frequency of operation for a MESFET. For a FET, the short-circuit current gain, G_I^{sc}, is defined as the ratio of drain current to gate current when the output is short circuited. For the unilateral case, where $C_{gd} = 0$, the short circuit current gain is

$$G_I^{sc} = \left| \frac{I_d}{I_g} \right| = \left| \frac{g_m V_c}{I_g} \right| = \frac{g_m}{\omega C_{gs}}.$$

The upper frequency threshold, f_T, where the short-circuit current gain is unity, is then given by

$$f_T = \frac{g_m}{2\pi C_{gs}}, \tag{11.24}$$

a result that is equivalent to (11.23) for a bipolar junction transistor.

For proper operation, the transistor must be biased at an appropriate operating point. This depends on the application (low noise, high gain, high power), the class of the amplifier (class A, class AB, class B), and the transistor. Figure 11.26a shows a typical family of DC I_{ds} versus V_{ds} curves for a GaAs MESFET. For low-noise design, the drain current is generally chosen to be about 15% of I_{dss} (the saturated drain-to-source current). High-power circuits generally use higher values of drain current. DC bias voltage must be applied to both the gate and drain, without disturbing the RF signal paths. This can be done with biasing and decoupling circuitry for a dual-polarity supply, as shown in Figure 11.26b. The RF chokes provide a very low DC resistance for biasing, and a very high impedance at RF frequencies to isolate the signal from the bias supply. Similarly, the input and output decoupling capacitors block DC from the input and output lines while allowing passage of RF signals. More sophisticated bias circuits can provide compensation for temperature and device variations, and may work with single-polarity power supplies.

Metal Oxide Semiconductor Field Effect Transistor

The silicon metal oxide semiconductor field effect transistor (MOSFET) is the most common type of FET, being used extensively in analog and digital integrated circuits. Figure 11.27 shows a cross section of an n-channel MOSFET. It consists of a lightly doped p substrate, and differs from a MESFET by having a thin insulating layer (SiO$_2$) between the gate contact and the channel region. Because the gate is insulated, it does not conduct DC bias current.

TABLE 11.7 | Scattering Parameters for an n-Channel GaAs MESFET (NEC NE76184A, $V_{ds} = 3.0$ V, $I_D = 10.0$ mA, Common Source)

Frequency (GHz)	S_{11}	S_{12}	S_{21}	S_{22}
1.0	$0.97\angle-28°$	$0.04\angle72°$	$3.82\angle154°$	$0.70\angle-19°$
2.0	$0.90\angle-55°$	$0.08\angle54°$	$3.56\angle129°$	$0.65\angle-37°$
4.0	$0.72\angle-103°$	$0.12\angle28°$	$2.91\angle86°$	$0.53\angle-68°$
8.0	$0.52\angle179°$	$0.14\angle-1°$	$2.0\angle20°$	$0.42\angle-129°$
12.0	$0.49\angle103°$	$0.17\angle-19°$	$1.5\angle-38°$	$0.44\angle170°$

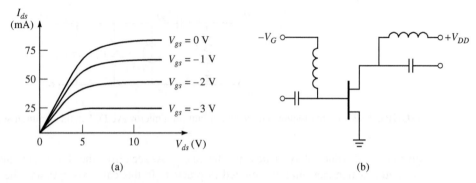

FIGURE 11.26 | (a) DC characteristics of an *n*-channel GaAs MESFET. (b) Biasing and decoupling circuitry.

MOSFETs can be used at frequencies into the UHF range, and can provide powers of several hundred watts when devices are packaged in parallel. Laterally diffused MOSFETs (LDMOS) have direct grounding of the source, and can operate at low microwave frequencies with high powers. These devices are commonly used for high-power transmitters for cellular base stations at 900 and 1900 MHz.

High-density integrated circuits typically use complementary MOS (CMOS), where both *n*-channel and *p*-channel devices are used. This technology is very mature, and has the advantages of low power requirements and low unit cost. Most RF and microwave MOSFETs use *n*-channel silicon devices, although GaN devices are possible.

The small-signal equivalent circuit for a MOSFET is the same as that of the MESFET, given in Figure 11.25. Scattering parameters are available for most nMOS devices intended for high-frequency applications.

High Electron Mobility Transistor

The high electron mobility transistor (HEMT) is a heterojunction FET, meaning that it does not use a single semiconductor material, but instead is constructed with several layers of compound semiconductor materials. These may include transitions between gallium aluminum arsenide (GaAlAs), GaAs, gallium indium arsenide (GaInAs), and similar compounds. These structures result in high carrier mobility—about twice that found in a standard MESFET. GaAs HEMTs can operate at frequencies above 100 GHz.

Figure 11.28 shows the cross section of a HEMT device. It consists of semi-insulating GaAs substrate, followed by an undoped GaAs layer, and then a very thin undoped GaAlAs layer. This is topped with an

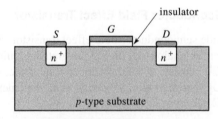

FIGURE 11.27 | Cross section of an *n*-channel MOSFET.

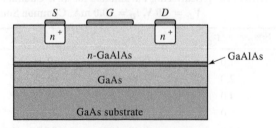

FIGURE 11.28 | Cross section of an *n*-channel HEMT.

TABLE 11.8 | **Scattering Parameters for a GaN HEMT (Cree CGH21120, V_{DD} = 328 V, I_D = 500 mA, Common Source)**

Frequency (GHz)	S_{11}	S_{12}	S_{21}	S_{22}
0.5	0.96∠180°	0.007∠−16°	3.67∠68°	0.72∠−174°
1.0	0.95∠172°	0.008∠−35°	2.03∠44°	0.78∠−172°
2.0	0.78∠153°	0.014∠−83°	2.09∠−17°	0.91∠−174°
4.0	0.88∠−51°	0.008∠79°	0.84∠88°	0.88∠171°

n-doped GaAlAs layer. To reduce thermal and mechanical stress the layers usually have matched crystal lattices. Several variations on this device are possible, including the use of different compound semiconductors, and the pseudomorphic HEMT, which uses a lattice mismatch between the layers. The relatively complicated structure of the HEMT requires sophisticated fabrication techniques, leading to a relatively high cost. The HEMT is also referred to in the literature as a MODFET (modulation-doped FET), a TEGFET (two-dimensional electron gas FET), and an SDFET (selectively doped FET).

A relatively new type of HEMT uses GaN and aluminum gallium nitride (AlGaN) on a silicon or SiC substrate. GaN HEMTs operate with drain voltages in the range of 20–40 V, and can deliver powers up to 100 W at frequencies in the low microwave range, making these devices popular for high-power transmitters.

The equivalent circuit model of Figure 11.25 can also be used for HEMTs, and the DC bias characteristics of a HEMT are similar to those of the MESFET. Table 11.8 gives the scattering parameters for a medium power GaN HEMT.

11.4   MICROWAVE INTEGRATED CIRCUITS

The trend of any maturing electronic technology is toward smaller size, lighter weight, lower power requirements, lower cost, and increased complexity. Microwave technology has been moving in this direction for the last 10–30 years with the development of microwave integrated circuits (MICs) [2]. This technology strives to replace bulky and expensive waveguide and coaxial components with small and inexpensive planar components, and is analogous to the digital integrated circuitry that has led to the rapid increase in sophistication of computer systems. Microwave integrated circuitry can incorporate transmission lines, discrete resistors, capacitors, and inductors, as well as active devices such as diodes and transistors. MIC technology has advanced to the point where complete microwave subsystems, such as receiver front ends and radar transmit/receive modules, can be integrated on a chip that is only a few square millimeters in size.

There are two distinct types of microwave integrated circuits. *Hybrid* MICs have one layer of metallization for conductors and transmission lines, with discrete components (resistors, capacitors, integrated circuit chips, transistors, diodes, etc.) bonded to the substrate. In a thin-film hybrid MIC, some of the simpler components are deposited on the substrate. Hybrid MICs were first developed in the 1960s, and still provide a very flexible and cost-effective means for circuit implementation. *Monolithic* microwave integrated circuits (MMICs) are a more recent development, where the active and passive circuit elements are grown on the substrate. The substrate is a semiconductor material, and several layers of metal, dielectric, and resistive films are used. Below we will briefly describe these two types of MICs in terms of the materials and fabrication processes that are required and the relative merits of each type of circuitry.

Hybrid Microwave Integrated Circuits

Material selection is an important consideration for a hybrid integrated circuit; characteristics such as electrical conductivity, dielectric constant, loss tangent, thermal transfer, mechanical strength, and manufacturing compatibility must be evaluated. Generally the substrate material is of primary importance. For hybrid circuits, alumina, quartz, and Teflon fiber are commonly used for substrates. Alumina is a rigid, ceramic-like material with a dielectric constant of about 9–10. A high dielectric constant is often desirable for lower frequency circuits because it results in a smaller circuit size. At higher frequencies, however, the substrate

FIGURE 11.29 | Photograph of one of the 25,344 hybrid integrated T/R modules used in Raytheon's Ground Based Radar system. This X-band module contains phase shifters, amplifiers, switches, couplers, a ferrite circulator, and associated control and bias circuitry.
Courtesy of Raytheon Company, Waltham, MA.

thickness must be decreased to prevent radiation loss and other spurious effects; then the transmission lines (typically microstrip, slotline, or coplanar waveguide) can become too narrow to be practical. Quartz has a lower dielectric constant (~4), which, with its rigidity, makes it useful for higher frequency (>20 GHz) circuits. Teflon and similar types of soft plastic substrates have dielectric constants ranging from 2 to 10, and can provide a large substrate area at a low cost as long as rigidity and good thermal transfer are not required. Transmission line conductors for hybrid integrated circuits are typically copper or gold.

Computer-aided design (CAD) tools are used extensively for MIC design, optimization, layout, and mask generation. Commonly used software packages include CADENCE (Cadence Design Systems), ADS (Agilent Technologies), Microwave Office (Applied Wave Research), and DESIGNER (Ansoft). The mask itself may be made on Rubylith (a soft Mylar film), usually at a magnified scale (2×, 5×, 10×, etc.) for high accuracy. Then an actual-size mask is made on a thin sheet of glass or quartz. The metallized substrate is coated with photoresist, covered with the mask, and exposed to a light source. The substrate can then be etched to remove the unwanted areas of metal. Plated-through, or via, holes can be made by evaporating a layer of metal inside a hole that has been drilled in the substrate. Finally, the discrete components are soldered or wire bonded to the conductors. This may be done manually, but today the process is usually automated using computer-controlled pick-and-place machines. The fabricated MIC can then be tested. Often provision is made for variations in component values and other circuit tolerances by providing tuning or trim stubs that can be manually trimmed for each circuit. This increases circuit yield but also increases cost since trimming involves labor at a highly skilled level. A photograph of a hybrid MIC module is shown in Figure 11.29.

Monolithic Microwave Integrated Circuits

Progress in GaAs and related semiconductor material processing and device development since the late 1970s has led to the feasibility of the MMIC, where all passive and active components required for a given circuit can be grown or implanted in the substrate. Potentially, an MMIC can be made at low cost because the labor involved with fabricating hybrid MICs is reduced. In addition, a single wafer can contain a large number of circuits, all of which can be processed and fabricated simultaneously.

The substrate of an MMIC must be a semiconductor material to accommodate the fabrication of active devices; the type of devices and the frequency range dictate the type of substrate material. The GaAs MESFET is a very versatile device, finding applications in low-noise amplifiers, high-gain amplifiers, broadband amplifiers, mixers, oscillators, phase shifters, and switches. Thus, GaAs is one of the most common substrates for MMICs, but silicon, silicon-on-sapphire (SoS), silicon carbide (SiC), and InP are also used.

Transmission lines and other conductors are usually made with gold metallization. To improve adhesion of the gold to the substrate, a thin layer of chromium or titanium may be deposited first. These metals are relatively lossy, so the gold layer must be made at least several skin depths thick to reduce attenuation. Capacitors and overlaying lines require insulating dielectric films, such as SiO, SiO_2, Si_2N_4, and Ta_2O_5. These materials have high dielectric constants and low loss, and are compatible with integrated circuit processing. Resistors require the deposition of lossy films; NiCr, Ta, Ti, and doped GaAs are commonly used.

Designing an MMIC requires extensive use of CAD software for circuit design and optimization, as well as for mask generation. Careful consideration must be given to the circuit design to allow for component variations and tolerances, and the fact that circuit trimming after fabrication will be difficult or impossible (and defeats the goal of low-cost production). Thus, effects such as transmission line discontinuities, bias networks, spurious coupling, and package resonances must be taken into account.

After the circuit design has been finalized, the masks can be generated. One or more masks are generally required for each processing step. Processing begins by forming an active layer in the semiconductor substrate for the necessary active devices; this can be done by ion implantation or by epitaxial techniques. Then, active areas are isolated by etching or additional implantation, leaving mesas for the active devices. Next, ohmic contacts are made to the active device areas by alloying a gold or gold/germanium layer onto the substrate. FET gates are then formed with a titanium/platinum/gold compound deposited between the source and drain areas. At this time, the active device processing has been essentially completed, and intermediate tests may be made to evaluate the wafer. If it meets specifications, the next step is to deposit the first layer of metallization for contacts, transmission lines, inductors, and other conducting areas. Then, resistors are formed by depositing resistive films, and the dielectric films required for capacitors and overlays are deposited. A second layer of metallization completes the formation of capacitors and any remaining interconnections. The final processing steps involve the bottom, or back, of the substrate. First it is lapped to the required thickness, and then via holes are formed by etching and plating. Via holes provide ground connections to the circuitry on the top side of the substrate, and a heat dissipation path from the active devices to the ground plane. After processing has been completed, the individual circuits can be cut from the wafer and tested. Figure 11.30 shows the structure of a typical MMIC, and Figure 11.31 shows a photograph of an X-band monolithic integrated GaAs FET amplifier.

Monolithic microwave integrated circuits are not without some disadvantages when compared with hybrid MICs or other types of circuitry. First, MMICs tend to waste large areas of relatively expensive semiconductor substrate for components such as transmission lines and hybrids. In addition, the processing steps and required tolerances for an MMIC are very critical, often resulting in low yields. These factors tend to make MMICs expensive, especially when made in small quantities (less than several hundred). MMICs generally require a more thorough design procedure to include effects such as component tolerances and

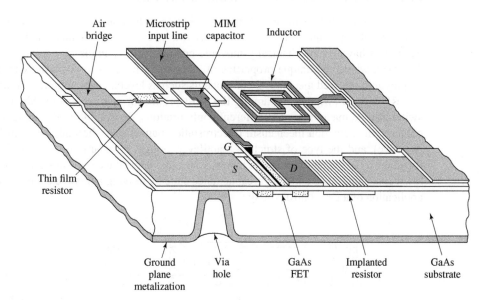

FIGURE 11.30 | Layout of a monolithic microwave integrated circuit.

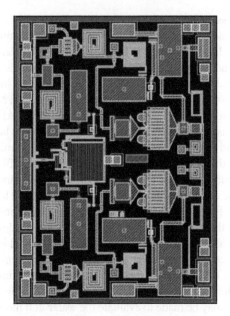

FIGURE 11.31 | Photograph of a typical GaAs X-band MMIC incorporating a pair of two-stage amplifiers.
Courtesy of M. Adlerstein, Raytheon Company, Waltham, MA.

discontinuities, and debugging, tuning, or trimming after fabrication is difficult. Because their small size limits heat dissipation, MMICs cannot be used for circuits requiring more than moderate power levels. In addition, high-Q resonators and filters are difficult to implement in MMIC form because of the inherent resistive losses in MMIC materials.

Besides the obvious features of small size and weight, MMICs have some unique advantages over other types of circuits. Since it is very easy to fabricate additional FETs in an MMIC design, circuit flexibility and performance can often be enhanced with little additional cost. In addition, monolithically integrated devices have much less parasitic reactance than discrete packaged devices, so MMIC circuits can often be made with broader bandwidth than hybrid circuits. MMICs generally give very reproducible results, especially for circuits from the same wafer.

POINT OF INTEREST: RF MEMS Switch Technology

An exciting new field is the use of micromachining techniques to form suspended or movable structures in a silicon substrate that can be used for microwave resonators, antennas, and switches. A micromachined RF switch having a mechanically movable contact is an example of a micro-electro-mechanical system (MEMS), where the unique properties of silicon can be used to construct extremely small devices that employ miniaturized mechanical components such as levers, gears, motors, and actuators.

RF MEMS switches are among the most promising applications of this new technology. A MEMS switch can be made in several different configurations, depending on the signal path (capacitive or direct contact), the actuation mechanism (electrostatic, magnetic, or thermal), the pull-back mechanism (spring or active), and the type of structure (cantilever, bridge, lever arm, or rotary). One popular configuration for microwave switches is shown below, where the capacitance of the signal path is switched between a low-capacitance state and a high-capacitance state by moving a flexible conductive membrane through the application of a DC control voltage.

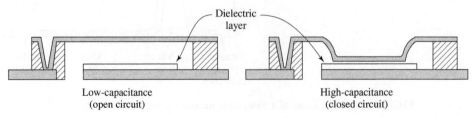

Low-capacitance
(open circuit)

High-capacitance
(closed circuit)

MEMS switches have very good loss characteristics, very low power consumption, and wide bandwidth, and (unlike diode or transistor switches) they exhibit virtually no intermodulation distortion or other nonlinear effects. The table below compares some of the key parameters of MEMS switches with those of popular solid-state switch technology over the 10–20 GHz band.

Switch Technology	Insertion Loss (dB)	Isolation (dB)	Switching Power	DC Voltage (V)	Switching Speed
PIN diode	0.1–0.8	25–45	1–5 mW	1–10	1–5 nsec
FET	0.5–1.0	20–50	1–5 mW	1–10	2–10 nsec
MEMS	0.1–1.0	25–60	1 μW	10–20	>30 μsec

Probably the most significant drawbacks of RF MEMS switches are the relatively slow switching time and potential lifetime limitations; both of these are a result of the mechanical nature of the device. One of the most important applications foreseen for MEMS switches is for low-cost switched-line-length phase shifters, which are required in large numbers for phased array antennas.

11.5   MICROWAVE TUBES

Although solid-state sources of RF and microwave power are preferred due to size, weight, power, and cost considerations, electron tubes provided the first practical sources of high-frequency power, and for several decades they were the only sources available at these frequencies. Today, solid-state diode or transistor sources are used in the majority of RF or microwave applications, and progress in solid-state technology is steadily improving the power versus frequency performance of solid-state sources. There are, however, still some systems that are best served by electron tubes. These are generally microwave or millimeter wave applications that require very high powers and/or very high frequencies.

Radar systems generally require a relatively high-power source, sometimes as high as 1–10 kW, for the transmitter (in addition to one or more low-power sources for local oscillator and down conversion functions in the receiver). Radar transmitters are often operated in a pulsed mode, and peak powers that are much greater than the continuous power rating of a given source can then be attained. Electronic warfare systems use sources with powers in the range of 100 W to 1 kW, with the additional requirement for tunability over a wide bandwidth. In addition, the microwave oven, that most common of all microwave systems, requires a single-frequency, high-power source in the range of 700 W. Usually the most practical way to meet the power requirements of these systems is with electron tubes.

As frequency increases into the millimeter and submillimeter ranges it becomes increasingly difficult to produce even moderate power with solid-state devices, so tubes become more useful at these frequencies. Generally, the division is between solid-state sources for low to moderate powers at low to moderate frequencies, and tubes for high powers and/or high frequencies. Figure 11.32 illustrates the power versus frequency performance for solid-state and tube sources. Solid-state sources have the advantages of small size, ruggedness, low cost, and compatibility with microwave integrated circuits, and so they are usually preferred whenever they can meet the necessary power and frequency requirements. However, very high power applications are dominated by microwave tubes, and even though the power and frequency performance of solid-state sources is steadily improving, it appears that the need for microwave tubes will not be eliminated any time soon.

The first practical microwave source was the *magnetron tube*, developed in England in the 1930s, which later provided the impetus for the development of microwave radar during World War II. Since then, a large variety of microwave tubes have been designed for the generation and amplification of microwave power. Although solid-state devices have been progressively filling roles that once could only be met by microwave tubes, tubes are still essential for the generation of very high powers (10 kW and higher), and at the higher millimeter wave frequencies (100 GHz and higher). Here we will provide a brief overview of some of the most common microwave tubes and their basic characteristics. Several of these tubes are not actually sources by themselves but are high-power amplifiers. Such tubes can used in conjunction with lower power sources (such as solid-state sources), and the combination is referred to as a microwave power module (MPM).

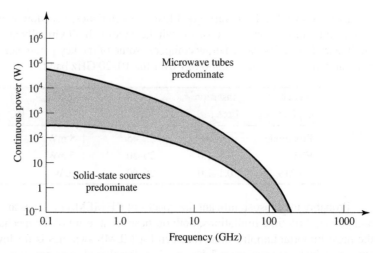

FIGURE 11.32 | Power versus frequency performance of solid-state sources and microwave tubes.

There is a wide variety of microwave tube geometries, as well as a wide variety of principles on which tube operation is based, but all tubes have several common features. First, all tubes involve the interaction of an electron beam with an electromagnetic field, inside a glass or metal vacuum envelope. Thus, a way must be provided for RF energy to be coupled outside the envelope; this is usually accomplished with transparent windows or coaxial coupling probes or loops. Next, a hot cathode is used to generate a stream of electrons by thermionic emission. Cathodes are usually fabricated from a barium oxide–coated metal surface, or an impregnated tungsten surface. The electron stream is then focused into a narrow beam by a focusing anode with a high voltage bias. Alternatively, a solenoidal electromagnet can be used to focus the electron beam. For pulsed operation, a beam modulating electrode is used between the cathode and anode. A positive bias voltage will attract electrons from the cathode and turn the beam on, while a negative bias will turn the beam off. After the electron beam leaves the region of the tube where the desired interaction with the RF field takes place, a collector element is used to provide a complete current path back to the cathode power supply. The assembly of the cathode, focusing anode, and modulating electrode is called the *electron gun*. Because of the requirement for a high vacuum and the need to dissipate large amounts of heat, microwave tubes are generally very large and bulky. In addition, tubes often require large, heavy biasing magnets and high voltage power supplies. Factors to consider when choosing a particular type of tube include power output, frequency, bandwidth, tuning range, and noise.

Microwave tubes can be grouped into two categories, depending on the type of electron beam–field interaction. In *linear-beam*, or "O," type tubes the electron beam traverses the length of the tube and is parallel to the electric field. In the *crossed-field*, or "m," type tube the focusing field is perpendicular to the accelerating electric field. Microwave tubes can also be classified as either oscillators or amplifiers. A more detailed classification of microwave tubes into different categories is shown in Figure 11.33.

Klystron

The klystron is a linear-beam tube that can be used either as an amplifier or as an oscillator. In a klystron amplifier, the electron beam passes through two or more resonant cavities. A single-cavity klystron (reflex klystron) is used to generate microwave signals, whereas two-cavity or multicavity klystrons are used to amplify microwave signals.

Reflex klystron: The reflex klystron is a single-cavity klystron tube that operates as an oscillator by using a reflector electrode after the cavity to provide positive feedback via the electron beam. It can be tuned by mechanically adjusting the cavity size. The cathode injects the electron beam, and the direction of emitted electron beams is controlled with the help of anode connected to the positive DC supply voltage. A single cavity in the reflex klystron acts as a buncher and catcher cavity for forward and reverse moving electrons. The structure of a reflex Klystron is shown in Figure 11.34.

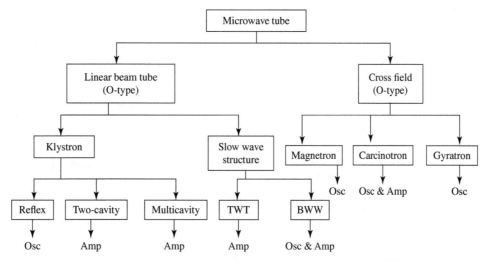

FIGURE 11.33 | Types of microwave tubes.

The velocity of electrons fluctuates according to the cavity gap voltage. That is, the electrons are accelerated and decelerated based on the positive and negative gap voltage, respectively. When an electron moves from the cavity gap to the repeller space, it travels more distance for higher velocity and less distance for lower velocity. In bunching cavity, all the electrons are grouped, and the energy of electrons transferred as the RF output. Reflex klystrons are used in various applications such as a signal source in microwave generators, pump and local oscillators for parametric amplifiers and microwave receivers, and radio and radar receivers. The major disadvantage of klystrons is their narrow bandwidth, which is a result of the high-Q cavities required for electron bunching. Klystrons have very low AM and FM noise levels.

Two-cavity klystron: The two-cavity klystron amplifies a microwave signal by using the principle of velocity and current modulation. It consists of two cavities; a buncher or an input cavity, and a catcher or an output cavity. The region between these two cavities is known as microwave interaction region. An RF signal is inputted at the buncher cavity, and an output signal is collected from the catcher cavity. The electrons emitted from the electron gun or cathode reach the buncher cavity with uniform velocity. In this buncher cavity, the velocity of electrons is modulated by the input RF signal, which is called as velocity modulation. This causes bunching of electrons and the density of electrons varies passing through the catcher cavity,

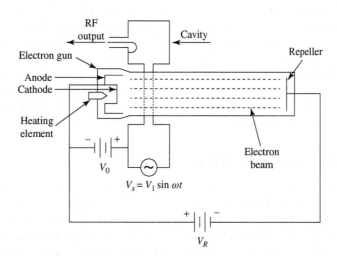

FIGURE 11.34 | A single-cavity reflex klystron.

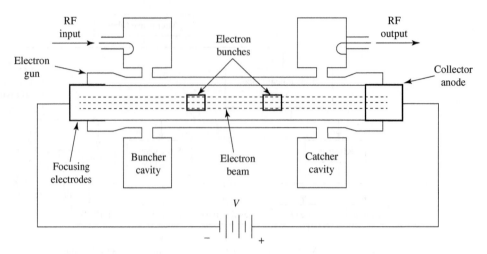

FIGURE 11.35 | A two-cavity reflex klystron.

causing current modulation in the catcher cavity. The maximum bunching of electrons occurs, and the electrons' kinetic energy is transferred to the microwave field. The microwave signal gets amplified and collected by collectors. The structure of a two-cavity klystron is shown in Figure 11.35. The efficiency, output power, and power gain of two-cavity klystrons are 40%, 30 MW at 10 GHz frequency, and 30 dB, respectively.

Multicavity klystron: In a multicavity klystron, the electron beam passes through multiple resonant cavities. The first cavity accepts an RF input and modulates the electron beam by bunching it into high-density and low-density regions. The bunched beam then travels to the next cavity, which accentuates the bunching effect. At the final cavity the RF power is extracted, at a highly amplified level. Two cavities can produce up to about 20 dB of gain, while using four cavities (about the practical limit) can give 80–90 dB of gain. Klystrons are capable of peak powers in the megawatt range, with RF output/DC input power conversion efficiencies of 30%–50%.

Traveling Wave Tube

The narrow bandwidth of the klystron amplifier is overcome in the *traveling wave tube* (TWT). The TWT is a linear-beam amplifier that uses an electron gun and a focusing magnet to accelerate a beam of electrons through an interaction region. Usually the interaction region consists of a slow-wave helix structure, with an RF input at the electron gun end and an RF output at the collector end. The helical structure slows down the propagating RF wave so that it travels at the same velocity as the wave and beam travel along the interaction region, and amplification is achieved. Then the amplified signal is coupled from the end of the helix. The TWT has the highest bandwidth of any amplifier tube, ranging from 30% to 120%; this makes it very useful for electronic warfare systems, which require high power over broad bandwidths. It has a power rating of several hundred watts (typically), but this can be increased to several kilowatts by using an interaction region consisting of a set of coupled cavities; the bandwidth will be reduced, however. The efficiency of the TWT is relatively small, typically ranging from 20% to 40%.

Backward Wave Oscillator

A variation of the TWT is the *backward wave oscillator* (BWO). The difference between a TWT and the BWO is that in a BWO, the RF wave travels along the helix from the collector toward the electron gun. Thus the signal for amplification is provided by the bunched electron beam itself, and oscillation occurs. A very useful feature of the BWO is that its output frequency can be tuned by varying the DC voltage between the cathode and the helix; tuning ranges of an octave or more can be achieved. The power output of the BWO, however, is relatively low (typically less than 1 W), so these tubes are generally being replaced with solid-state sources.

Extended Interaction Oscillator

Another type of linear-beam oscillator tube is the *extended interaction oscillator* (EIO). The EIO is very similar to a klystron, and uses an interaction region consisting of several cavities coupled together, with positive feedback to support oscillation. It has a narrow tuning bandwidth and a moderate efficiency, but it can supply high powers at frequencies up to several hundred GHz. Only the gyratron can deliver more power.

Crossed-field tubes include the *magnetron*, the *crossed-field amplifier*, and the *gyratron*. As previously mentioned, the magnetron was the first high-power microwave source. It consists of a cylindrical cathode surrounded by a cylindrical anode with several cavity resonators along the inside of its periphery. A magnetic bias field is applied parallel to the cathode–anode axis. In operation, a cloud of electrons is formed that rotates around the cathode in the interaction region. As with linear-beam devices, electron bunching occurs, and energy is transferred from the electron beam to the RF wave. RF power can be coupled out of the tube with a probe, loop, or aperture window.

Magnetrons

Magnetrons are capable of very high power outputs, on the order of several kilowatts, and with efficiencies of 80% or more. A significant disadvantage, however, is that they are very noisy and cannot maintain frequency or phase coherence when operated in a pulsed mode. These factors are important for high-performance pulsed radars, where processing techniques operate on a sequence of returned pulses. (Modern radars of this type today generally use a stable low-noise solid-state source, followed by a TWT for power amplification.) The main application of magnetrons today is primarily for microwave cooking.

Cross-Field Amplifier

The *crossed-field amplifier* (CFA) has a geometry similar to that of a TWT, but employs a crossed-field interaction region that is similar to that of the magnetron. The RF input is applied to a slow-wave structure in the interaction region of the CFA, but the electron beam is deflected by a negatively biased electrode to force the beam perpendicular to the slow-wave structure. In addition, a magnetic bias field is applied perpendicular to this electric field, and perpendicular to the electron beam direction. The magnetic field exerts a force on the electron beam that counteracts the field from the sole. In the absence of an RF input, the electric and magnetic fields are adjusted so that their effects on the electron beam cancel, leaving the beam to travel parallel to the slow-wave structure. Applying an RF field causes velocity modulation of the beam, and bunching occurs. The beam is also periodically deflected toward the slow-wave circuit, producing an amplified signal. Crossed-field amplifiers have very good efficiencies—up to 80%, but the gain is limited to 10–15 dB. In addition, the CFA has a noisier output than either a klystron amplifier or TWT. Its bandwidth can be up to 40%.

Gyratron

Another crossed-field tube is the *gyratron*, which can be used as an amplifier or an oscillator. This tube consists of an electron gun with input and output cavities along the axis of the electron beam, similar to a klystron amplifier. However, the gyratron also has a solenoidal bias magnet that provides an axial magnetic field. This field forces the electrons to travel in tight spirals down the length of the tube. The electron velocity is high enough that relativistic effects are important. Bunching occurs, and energy from the transverse component of the electron velocity is coupled to the RF field. A significant feature of the gyratron is that the frequency of operation is determined by the bias field strength and the electron velocity, as opposed to the dimensions of the tube itself. This makes the gyratron especially useful for millimeter wave frequencies; it offers the highest output power (10–100 kW) of any tube in this frequency range. It also has a high efficiency for tubes in the millimeter wave range. The gyratron is a relatively new type of tube, but it is rapidly replacing tubes such as reflex klystrons and EIOs as sources of millimeter wave power.

Figures 11.36 and 11.37 summarize the power versus frequency performance of microwave tube oscillators and amplifiers.

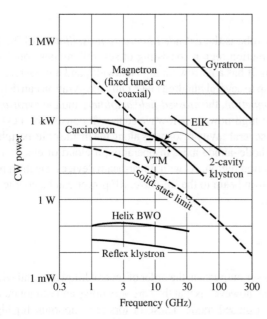

FIGURE 11.36 | Power versus frequency performance of microwave oscillator tubes.

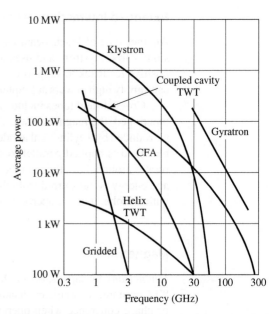

FIGURE 11.37 | Power versus frequency performance of microwave amplifier tubes.

REFERENCES

[1] M. E. Hines, "The Virtues of Nonlinearity—Detection, Frequency Conversion, Parametric Amplification and Harmonic Generation," *IEEE Transactions on Microwave Theory and Techniques,* vol. MTT-32, pp. 1097–1104, September 1984.

[2] D. N. McQuiddy, Jr., J. W. Wassel, J. B. Lagrange, and W. R. Wisseman, "Monolithic Microwave Integrated Circuits: An Historical Perspective," *IEEE Transactions on Microwave Theory and Techniques,* vol. MTT-32, pp. 997–1008, September 1984.

[3] S. Yngvesson, *Microwave Semiconductor Devices*, Kluwer, Norwell, MA, 1991.

[4] R. Ludwig and P. Bretchko, *RF Circuit Design: Theory and Applications*, Prentice-Hall, Upper Saddle River, NJ, 2000.

[5] S. A. Maas, *Nonlinear Microwave and RF Circuits*, 2nd edition, Artech House, Norwood, MA, 2003.

[6] M. Steer, *Microwave and RF Design: A Systems Approach*, SciTech, Raleigh, NC, 2010.

PROBLEMS

11.1 The Skyworks SMS1546 Schottky diode has the following parameters: $C_j = 0.38$ pF, $R_s = 4$ Ω, $I_s = 0.3$ μA, and $L_p = C_p \simeq 0$. Compute the open-circuit voltage sensitivity at 6 GHz for $I_0 = 0, 20$, and 50 μA. Assume $\alpha = 1/(25$ mV), and neglect the effect of bias current on the junction capacitance.

11.2 A single-pole, single-throw switch uses an Infineon BA592 PIN diode in a shunt configuration. The operating frequency is 10 GHz, $Z_0 = 75$ Ω, and the diode parameters are $C_j = 1.4$ pF, $R_r = 0.5$ Ω, $R_f = 0.36$ Ω, and $L_i = 0.5$ nH. Find the electrical length of an open-circuited shunt stub placed across the diode to minimize the insertion loss for the ON state of the switch. Calculate the resulting insertion losses for the ON and OFF states.

11.3 A single-pole, single-throw switch is constructed using two identical PIN diodes in the arrangement shown in the accompanying figure. In the ON state, the series diode is forward biased and the shunt diode is reversed biased; and vice versa for the OFF state. If $f = 8$ GHz, $Z_0 = 50$ Ω, $C_j = 0.1$ pF, $R_r = 0.5$ Ω, $R_f = 0.3$ Ω, and $L_i = 0.4$ nH, determine the insertion losses for the ON and OFF states.

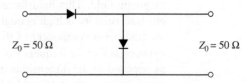

11.4 Consider the loaded-line phase shifter shown in the accompanying figure. If $Z_0 = 50$ Ω, find the necessary stub lengths for a differential phase shift of 45°, and calculate the resulting insertion loss for both states of the phase shifter. Assume all lines are lossless and that the diodes can be approximated as ideal shorts or opens.

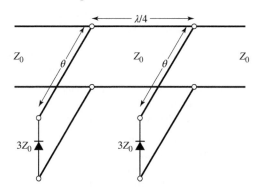

11.5 Use the equivalent circuit of Figure 11.22 to derive the expression for the short-circuit current gain of a bipolar transistor. Assume a unilateral device, where $C_c = 0$.

11.6 Show that the scattering parameters for a FET can be expressed in terms of the parameters of the equivalent circuit of Figure 11.25 as given below. Assume the device is unilateral.

$$S_{11} = \frac{Z_{11} - Z_0}{Z_{11} + Z_0}, \quad S_{12} = 0, \quad S_{21} = \frac{2jZ_0 g_m / \omega C_{gs}}{(Z_{11} + Z_0)(Z_{22} + Z_0)},$$

$$S_{22} = \frac{Z_{22} - Z_0}{Z_{22} + Z_0},$$

where $Z_{11} = R_i - j/\omega C_{gs}$ and $Z_{22} = (1/R_{ds} + j\omega C_{ds})^{-1}$.

11.7 Given the scattering parameters for a FET, derive expressions for the parameters of the equivalent circuit model of Figure 11.25, assuming a unilateral device. Use these results to find the equivalent circuit parameters for the HEMT device whose scattering parameters are given in Table 11.7, at a frequency of 1 GHz. Ignore S_{12}.

11.8 A single-pole switch operating at 1.5 GHz is to be constructed using a Microsemi UM9605 PIN diode with $C_j = 0.5$ pF and $R_f = 1.5\ \Omega$. What switch circuit (series or shunt) should be used to obtain the greatest ratio of off-to-on attenuation? Assume that $L_i = 0.5$ nH, $R_r = 2.0\ \Omega$, and $Z_0 = 50\ \Omega$.

MULTIPLE CHOICE QUESTIONS

11.1 For a shunt configuration, the terminated line characteristic impedance is 50 Ω and the diode impedance is 35 Ω. Then the insertion loss is

(a) 2.34 dB   (c) 5.21 dB
(b) 4.68 dB   (d) 2.61 dB

11.2 For a series configuration, the terminated line characteristic impedance is 50 Ω and the diode impedance is 35 Ω. Then the insertion loss is

(a) 2.34 dB   (c) 5.21 dB
(b) 4.68 dB   (d) 2.61 dB

11.3 Which of the following can be varied as a function of reverse voltage for a varactor diode?

(a) junction capacitance
(b) junction resistance
(c) junction impedance
(d) none of the above

11.4 In which region are the varactor diodes operated to achieve maximum efficiency?

(a) active region
(b) saturation region
(c) reverse saturation region
(d) cutoff region

11.5 Gunn diode electrodes are made up of

(a) molybdenum   (c) GaAs
(b) gold   (d) copper

11.6 How many modes of operation does n-type GaAs has?

(a) 1   (c) 3
(b) 2   (d) 4

11.7 Determine the intrinsic frequency of a Gunn diode oscillator, where the electron drift velocity is 10^5 m/sec and the effective length is 25 microns.

(a) 2 GHz   (c) 5 GHz
(b) 4 GHz   (d) 6 GHz

11.8 For an IMPATT diode when a reverse bias voltage exceeds the breakdown voltage it results in

(a) avalanche multiplication
(b) high reverse saturation current
(c) breakdown of depletion region
(d) none of the above

11.9 Determine the nominal frequency for an IMPATT diode, where the drift velocity is 10^5 m/sec and the length of the intrinsic region is 2.5 microns.

(a) 5 GHz   (c) 15 GHz
(b) 10 GHz   (d) 20 GHz

11.10 Which of the following is a major disadvantage of TRAPATT diodes?

(a) low operational bandwidth
(b) fabrication is costly
(c) low gain
(d) high noise figure

11.11 For RF applications the bipolar junction transistors are suitable because of their

(a) power capacity
(b) good performance in terms of frequency
(c) noise characteristics
(d) all of the above

11.12 Which of the following is suitable for S_{22} parameter of a BJT when the operating frequency of the transistor increases?

(a) decreases
(b) increases

(c) remains constant
(d) none of the above

11.13 Determine the threshold frequency of operation of a transistor if its capacitance measured in the hybrid-π model is 50 pF and it has a short-circuit current gain of 22.

(a) 44 MHz
(b) 65 GHz

(c) 70 GHz
(d) 74 GHz

11.14 For a field effect transistor the frequency of operation is limited by

(a) effective area of the FET
(b) gate length
(c) gate-to-source voltage
(d) drain-to-source voltage

11.15 For a field effect transistor the expression for short-circuit current gain is given by

(a) $\omega C_{gs}/g_m$
(b) $I_g/g_m V_c$

(c) $g_m/\omega C_{gs}$
(d) none of the above

11.16 Using which of the following packing methods are the devices packed such that MOSFETs can provide a power of several hundred watts?

(a) diagonal
(b) parallel

(c) series
(d) none of the above

11.17 Which of the following is an important consideration for a hybrid integrated circuit?

(a) design complexity
(b) processing units

(c) material selection
(d) active sources

11.18 Which of the following characteristics of a material is chosen in order to fabricate a low-frequency circuit using the hybrid microwave IC technology?

(a) high resistivity
(b) low dielectric constant
(c) low resistivity
(d) high dielectric constant

11.19 Which of the following is used in processing in MMICs?

(a) floor planning
(b) net list generation

(c) ion implantation
(d) none of the above

11.20 Which of the following type of beam amplifier is used in a klystron amplifier?

(a) linear beam
(b) parallel field
(c) crossed field
(d) none of the above

ANSWER KEY				
11.1 (b)	**11.5** (a)	**11.9** (d)	**11.13** (c)	**11.17** (c)
11.2 (d)	**11.6** (d)	**11.10** (d)	**11.14** (b)	**11.18** (d)
11.3 (a)	**11.7** (b)	**11.11** (d)	**11.15** (c)	**11.19** (c)
11.4 (c)	**11.8** (a)	**11.12** (a)	**11.16** (b)	**11.20** (a)

12

Microwave Amplifier Design

Learning Objectives

After completing this chapter, you will be able to

- Understand the concept of two-port power gain and stability
- Learn about power amplifiers

- Develop and design single-stage and broadband transistor amplifiers

Signal amplification is one of the most basic and prevalent circuit functions in modern RF and microwave systems. Early microwave amplifiers relied on tubes, such as klystrons and traveling-wave tubes, or solid-state reflection amplifiers based on the negative resistance characteristics of tunnel or varactor diodes. However, due to the dramatic improvements and innovations in solid-state technology that have occurred since the 1970s, most RF and microwave amplifiers today use transistor devices such as Si BJTs, GaAs or SiGe HBTs, Si MOSFETs, GaAs MESFETs, or GaAs or GaN HEMTs [1–5]. Microwave transistor amplifiers are rugged, low-cost, and reliable and can be easily integrated in both hybrid and monolithic integrated circuitry. Transistor amplifiers can be used at frequencies in excess of 100 GHz in a wide range of applications requiring small size, low noise figure, broad bandwidth, and medium to high power capacity. Although microwave tubes are still useful for very high power and/or very high frequency applications, continuing improvement in the performance of microwave transistors is steadily reducing the need for microwave tubes.

Our discussion of transistor amplifier design will primarily rely on the terminal characteristics of the transistor, as represented by either scattering parameters or one of the equivalent circuit models introduced in the previous chapter. We will begin with some general definitions of two-port power gains that are useful for amplifier design and then discuss the subject of stability. These results will then be applied to single-stage transistor amplifiers, including designs for maximum gain, specified gain, and low noise figure. Broadband balanced and distributed amplifiers are discussed in Section 12.4. We conclude with a brief treatment of transistor power amplifiers.

12.1 TWO-PORT POWER GAINS

In this section we develop several expressions for the gain and stability of a general two-port amplifier circuit in terms of the scattering parameters of the transistor. These results will be used in the following sections for amplifier design and in Chapter 13 for oscillator design.

Definitions of Two-Port Power Gains

Consider an arbitrary two-port network, characterized by its scattering matrix $[S]$, connected to source and load impedances Z_S and Z_L, respectively, as shown in Figure 12.1. We will derive expressions for three types of power gain in terms of the scattering parameters of the two-port network and the reflection coefficients, Γ_S and Γ_L, of the source and load.

- *Power gain* $= G = P_L/P_{\text{in}}$ is the ratio of power dissipated in the load Z_L to the power delivered to the input of the two-port network. This gain is independent of Z_S, although the characteristics of some active devices may be dependent on Z_S.

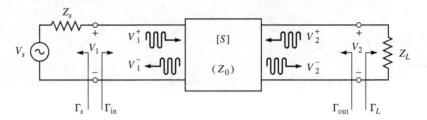

FIGURE 12.1 | A two-port network with arbitrary source and load impedances.

- *Available power gain* $= G_A = P_{avn}/P_{avs}$ is the ratio of the power available from the two-port network to the power available from the source. This assumes conjugate matching of both the source and the load, and depends on Z_S, but not Z_L.
- *Transducer power gain* $= G_T = P_L/P_{avs}$ is the ratio of the power delivered to the load to the power available from the source. This depends on both Z_S and Z_L.

These definitions differ primarily in the way the source and load are matched to the two-port device; if the input and output are both conjugately matched to the two-port device, then the gain is maximized and $G = G_A = G_T$.

With reference to Figure 12.1, the reflection coefficient seen looking toward the load is

$$\Gamma_L = \frac{Z_L - Z_0}{Z_L + Z_0},\tag{12.1a}$$

while the reflection coefficient seen looking toward the source is

$$\Gamma_S = \frac{Z_S - Z_0}{Z_S + Z_0},\tag{12.1b}$$

where Z_0 is the characteristic impedance reference for the scattering parameters of the two-port network.

In general, the input impedance of the terminated two-port network will be mismatched with a reflection coefficient given by Γ_{in}, which can be determined using a signal flow graph (see Example 4.8) or by the following analysis. From the definition of the scattering parameters, and the fact that $V_2^+ = \Gamma_L V_2^-$, we have

$$V_1^- = S_{11} V_1^+ + S_{12} V_2^+ = S_{11} V_1^+ + S_{12} \Gamma_L V_2^-,\tag{12.2a}$$

$$V_2^- = S_{21} V_1^+ + S_{22} V_2^+ = S_{21} V_1^+ + S_{22} \Gamma_L V_2^-.\tag{12.2b}$$

Eliminating V_2^- from (12.2a) and solving for V_1^-/V_1^+ gives

$$\Gamma_{in} = \frac{V_1^-}{V_1^+} = S_{11} + \frac{S_{12} S_{21} \Gamma_L}{1 - S_{22} \Gamma_L} = \frac{Z_{in} - Z_0}{Z_{in} + Z_0},\tag{12.3a}$$

where Z_{in} is the impedance seen looking into port 1 of the terminated network. Similarly, the reflection coefficient seen looking into port 2 of the network when port 1 is terminated by Z_S is

$$\Gamma_{out} = \frac{V_2^-}{V_2^+} = S_{22} + \frac{S_{12} S_{21} \Gamma_S}{1 - S_{11} \Gamma_S}.\tag{12.3b}$$

By voltage division,

$$V_1 = V_S \frac{Z_{in}}{Z_S + Z_{in}} = V_1^+ + V_1^- = V_1^+ \left(1 + \Gamma_{in}\right).$$

Using

$$Z_{in} = Z_0 \frac{1 + \Gamma_{in}}{1 - \Gamma_{in}}$$

from (12.3a) and solving for V_1^+ in terms of V_S gives

$$V_1^+ = \frac{V_S}{2} \frac{(1 - \Gamma_S)}{(1 - \Gamma_S \Gamma_{in})}.$$ (12.4)

If peak values are assumed for all voltages, the average power delivered to the network is

$$P_{in} = \frac{1}{2Z_0} \left| V_1^+ \right|^2 \left(1 - |\Gamma_{in}|^2 \right) = \frac{|V_S|^2}{8Z_0} \frac{|1 - \Gamma_S|^2}{|1 - \Gamma_S \Gamma_{in}|^2} \left(1 - |\Gamma_{in}|^2 \right),$$ (12.5)

where (12.4) was used. The power delivered to the load is

$$P_L = \frac{\left| V_2^- \right|^2}{2Z_0} \left(1 - |\Gamma_L|^2 \right).$$ (12.6)

Solving for V_2^- from (12.2b), substituting into (12.6), and using (12.4) gives

$$P_L = \frac{\left| V_1^+ \right|^2}{2Z_0} \frac{|S_{21}|^2 \left(1 - |\Gamma_L|^2 \right)}{|1 - S_{22} \Gamma_L|^2} = \frac{|V_S|^2}{8Z_0} \frac{|S_{21}|^2 \left(1 - |\Gamma_L|^2 \right) |1 - \Gamma_S|^2}{|1 - S_{22} \Gamma_L|^2 |1 - \Gamma_S \Gamma_{in}|^2}.$$ (12.7)

The power gain can then be expressed as

$$G = \frac{P_L}{P_{in}} = \frac{|S_{21}|^2 \left(1 - |\Gamma_L|^2 \right)}{\left(1 - |\Gamma_{in}|^2 \right) |1 - S_{22} \Gamma_L|^2}.$$ (12.8)

The power available from the source, P_{avs}, is the maximum power that can be delivered to the network. This occurs when the input impedance of the terminated network is conjugately matched to the source impedance, as discussed in Section 2.5. Thus, from (12.5),

$$P_{avs} = P_{in} \Big|_{\Gamma_{in} = \Gamma_S^*} = \frac{|V_S|^2}{8Z_0} \frac{|1 - \Gamma_S|^2}{\left(1 - |\Gamma_S|^2 \right)}.$$ (12.9)

Similarly, the power available from the network, P_{avn}, is the maximum power that can be delivered to the load. Thus, from (12.7),

$$P_{avn} = P_L \Big|_{\Gamma_L = \Gamma_{out}^*} = \frac{|V_S|^2}{8Z_0} \frac{|S_{21}|^2 \left(1 - |\Gamma_{out}|^2 \right) |1 - \Gamma_S|^2}{|1 - S_{22} \Gamma_{out}^*|^2 |1 - \Gamma_S \Gamma_{in}|^2} \Bigg|_{\Gamma_L = \Gamma_{out}^*}.$$ (12.10)

In (12.10), Γ_{in} must be evaluated for $\Gamma_L = \Gamma_{out}^*$. From (12.3a), it can be shown that

$$|1 - \Gamma_S \Gamma_{in}|^2 \Big|_{\Gamma_L = \Gamma_{out}^*} = \frac{|1 - S_{11} \Gamma_S|^2 \left(1 - |\Gamma_{out}|^2 \right)^2}{|1 - S_{22} \Gamma_{out}^*|^2},$$

which reduces (12.10) to

$$P_{avn} = \frac{|V_S|^2}{8Z_0} \frac{|S_{21}|^2 |1 - \Gamma_S|^2}{|1 - S_{11} \Gamma_S|^2 \left(1 - |\Gamma_{out}|^2 \right)}.$$ (12.11)

Observe that P_{avs} and P_{avn} have been expressed in terms of the source voltage, V_S, which is independent of the input or load impedances. There would be confusion if these quantities were expressed in terms of V_1^+ since V_1^+ is different for each of the calculations of P_L, P_{avs}, and P_{avn}.

Using (12.11) and (12.9), we obtain the available power gain as

$$G_A = \frac{P_{\text{avn}}}{P_{\text{avs}}} = \frac{|S_{21}|^2 \left(1 - |\Gamma_S|^2\right)}{|1 - S_{11}\Gamma_S|^2 \left(1 - |\Gamma_{\text{out}}|^2\right)}. \tag{12.12}$$

From (12.7) and (12.9), the transducer power gain is

$$G_T = \frac{P_L}{P_{\text{avs}}} = \frac{|S_{21}|^2 \left(1 - |\Gamma_S|^2\right)\left(1 - |\Gamma_L|^2\right)}{|1 - \Gamma_S\Gamma_{\text{in}}|^2 \, |1 - S_{22}\Gamma_L|^2}. \tag{12.13}$$

A special case of the transducer power gain occurs when both the input and output are matched for zero reflection (in contrast to conjugate matching). Then $\Gamma_L = \Gamma_S = 0$, and (12.13) reduces to

$$G_T = |S_{21}|^2. \tag{12.14}$$

Another special case is the *unilateral transducer power gain*, G_{TU}, where $S_{12} = 0$ (or is negligibly small). This nonreciprocal characteristic is approximately true for many transistors devices. From (12.3a), $\Gamma_{\text{in}} = S_{11}$ when $S_{12} = 0$, so (12.13) gives the unilateral transducer power gain as

$$G_{TU} = \frac{|S_{21}|^2 \left(1 - |\Gamma_S|^2\right)\left(1 - |\Gamma_L|^2\right)}{|1 - S_{11}\Gamma_S|^2 \, |1 - S_{22}\Gamma_L|^2}. \tag{12.15}$$

EXAMPLE 12.1   COMPARISON OF POWER GAIN DEFINITIONS

AT-41485 silicon bipolar junction transistor has the following scattering parameters at 1.0 GHz, with a 50 Ω reference impedance:

$$\begin{aligned} S_{11} &= 0.57\angle{-161°} \\ S_{12} &= 0.04\angle{38°} \\ S_{21} &= 6.8\angle{87°} \\ S_{22} &= 0.46\angle{-33°} \end{aligned}$$

The source impedance is $Z_S = 25\ \Omega$ and the load impedance is $Z_L = 35\ \Omega$. Compute the power gain, the available power gain, and the transducer power gain.

Solution
From (12.1a) and (12.1b) the reflection coefficients at the source and load are

$$\Gamma_S = \frac{Z_S - Z_0}{Z_S + Z_0} = \frac{25 - 50}{25 + 50} = -0.333,$$

$$\Gamma_L = \frac{Z_L - Z_0}{Z_L + Z_0} = \frac{35 - 50}{35 + 50} = -0.176.$$

From (12.3a) and (12.3b) the reflection coefficients seen looking at the input and output of the terminated network are

$$\Gamma_{\text{in}} = S_{11} + \frac{S_{12}S_{21}\Gamma_L}{1 - S_{22}\Gamma_L} = 0.558\angle{-156.6°},$$

$$\Gamma_{\text{out}} = S_{22} + \frac{S_{12}S_{21}\Gamma_S}{1 - S_{11}\Gamma_S} = 0.566\angle{-36.39°}.$$

Then from (12.8) the power gain is

$$G = \frac{|S_{21}|^2 \left(1 - |\Gamma_L|^2\right)}{\left(1 - |\Gamma_{\text{in}}|^2\right) |1 - S_{22}\Gamma_L|^2} = 56.96.$$

From (12.12) the available power gain is

$$G_A = \frac{|S_{21}|^2 \left(1 - |\Gamma_S|^2\right)}{|1 - S_{11}\Gamma_S|^2 \left(1 - |\Gamma_{\text{out}}|^2\right)} = 89.34.$$

From (12.13) the transducer power gain is

$$G_T = \frac{|S_{21}|^2 \left(1 - |\Gamma_S|^2\right)\left(1 - |\Gamma_L|^2\right)}{|1 - \Gamma_S\Gamma_{\text{in}}|^2 \, |1 - S_{22}\Gamma_L|^2} = 39.35.$$

Further Discussion of Two-Port Power Gains

A single-stage microwave transistor amplifier can be modeled by the circuit of Figure 12.2, where matching networks are used on both sides of the transistor to transform the input and output impedance Z_0 to the source and load impedances Z_S and Z_L. The most useful gain definition for amplifier design is the transducer power gain of (12.13), which accounts for both source and load mismatch. From (12.13) we can define separate effective gain factors for the input (source) matching network, the transistor itself, and the output (load) matching network as follows:

$$G_S = \frac{1 - |\Gamma_S|^2}{|1 - \Gamma_{\text{in}}\Gamma_S|^2}, \tag{12.16a}$$

$$G_0 = |S_{21}|^2, \tag{12.16b}$$

$$G_L = \frac{1 - |\Gamma_L|^2}{|1 - S_{22}\Gamma_L|^2}. \tag{12.16c}$$

The overall transducer gain is then $G_T = G_S G_0 G_L$. The effective gains G_S and G_L of the matching networks may be greater than unity. This is because the unmatched transistor would incur power loss due to reflections at the input and output of the transistor, and the matching sections can reduce these losses.

If the transistor is unilateral, so that $S_{12} = 0$ (or is small enough to be ignored), then (12.3) reduces to $\Gamma_{\text{in}} = S_{11}, \Gamma_{\text{out}} = S_{22}$, and the unilateral transducer gain reduces to $G_{TU} = G_S G_0 G_L$, where

$$G_S = \frac{1 - |\Gamma_S|^2}{|1 - S_{11}\Gamma_S|^2}, \tag{12.17a}$$

$$G_0 = |S_{21}|^2, \tag{12.17b}$$

$$G_L = \frac{1 - |\Gamma_L|^2}{|1 - S_{22}\Gamma_L|^2}. \tag{12.17c}$$

The above results have been derived using the scattering parameters of the transistor, but it is possible to obtain alternative expressions for gain in terms of the equivalent circuit parameters of the transistor. As an example, consider the evaluation of the unilateral transducer gain for a conjugately matched FET using

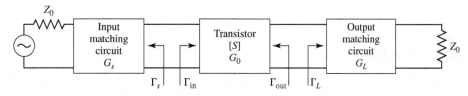

FIGURE 12.2 | The general transistor amplifier circuit.

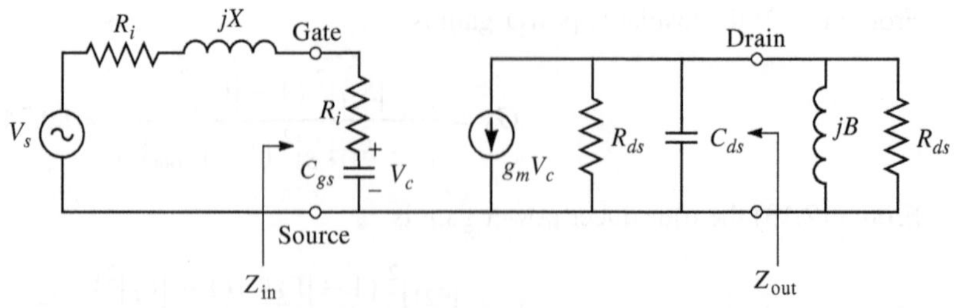

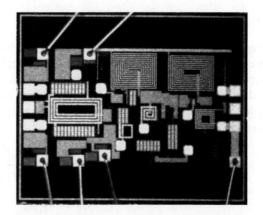

FIGURE 12.3 | Unilateral FET equivalent circuit and source and load terminations for the calculation of unilateral transducer power gain.

FIGURE 12.4 | Photograph of a low-noise MMIC amplifier that is switchable between 2.4, 3.6, and 5.8 GHz. The amplifier uses pHEMTs in a cascode configuration with source inductance, followed by a common source stage with feedback. Gain is approximately 13 dB in each band. Chip dimensions are 1.85 mm by 1 mm.

Courtesy of J. Shatzman and R. W. Jackson of the University of Massachusetts at Amherst and H. Yu of TriQuint, Lowell, MA.

the equivalent circuit of Figure 11.21 (with $C_{gd} = 0$). To conjugately match the transistor we choose source and load impedances as shown in Figure 12.3. Setting the series source inductive reactance $X = 1/\omega C_{gs}$ will make $Z_{\text{in}} = Z_S^*$, and setting the shunt load inductive susceptance $B = -\omega C_{ds}$ will make $Z_{\text{out}} = Z_L^*$; this effectively eliminates the reactive elements from the transistor equivalent circuit. Then by voltage division $V_c = V_S/2j\omega R_i C_{gs}$, and the gain can be easily evaluated as

$$G_{TU} = \frac{P_L}{P_{\text{avs}}} = \frac{\frac{1}{8}|g_m V_c|^2 R_{ds}}{\frac{1}{8}|V_S|^2/R_i} = \frac{g_m^2 R_{ds}}{4\omega^2 R_i C_{gs}^2} = \frac{R_{ds}}{4R_i}\left(\frac{f_T}{f}\right)^2. \tag{12.18}$$

where the last step has been written in terms of the cutoff frequency, f_T, from (11.24). This shows the interesting result that the gain of a conjugately matched FET amplifier drops off as $1/f^2$, or 6 dB per octave. A photograph of a low-noise MMIC amplifier is shown in Figure 12.4.

12.2 STABILITY

We now discuss the necessary conditions for a transistor amplifier to be stable. In the circuit of Figure 12.2, oscillation is possible if either the input or output port impedance has a negative real part; this would then imply that $|\Gamma_{\text{in}}| > 1$ or $|\Gamma_{\text{out}}| > 1$. Because Γ_{in} and Γ_{out} depend on the source and load matching networks, the stability of the amplifier depends on Γ_S and Γ_L as presented by the matching networks. Thus, we define two types of stability:

- *Unconditional stability*: The network is unconditionally stable if $|\Gamma_{in}| < 1$ and $|\Gamma_{out}| < 1$ for all passive source and load impedances (i.e., $|\Gamma_S| < 1$ and $|\Gamma_L| < 1$).
- *Conditional stability*: The network is conditionally stable if $|\Gamma_{in}| < 1$ and $|\Gamma_{out}| < 1$ only for a certain range of passive source and load impedances. This case is also referred to as *potentially unstable*.

Note that the stability condition of an amplifier circuit is usually frequency dependent since the input and output matching networks generally depend on frequency. It is therefore possible for an amplifier to be stable at its design frequency but unstable at other frequencies. Careful amplifier design should consider this possibility. We must also point out that the following discussion of stability is limited to two-port amplifier circuits of the type shown in Figure 12.2, and where the scattering parameters of the active device can be measured without oscillations over the frequency band of interest. The rigorous general treatment of stability requires that the network scattering parameters (or other network parameters) have no poles in the right-half complex frequency plane, in addition to the conditions that $|\Gamma_{in}| < 1$ and $|\Gamma_{out}| < 1$ [6]. This can be a difficult assessment in practice, but for the special case considered here, where the scattering parameters are known to be pole free (as confirmed by measurability), the following stability conditions are adequate.

Stability Circles

Applying the above requirements for unconditional stability to (12.3) gives the following conditions that must be satisfied by Γ_S and Γ_L if the amplifier is to be unconditionally stable:

$$|\Gamma_{in}| = \left| S_{11} + \frac{S_{12}S_{21}\Gamma_L}{1 - S_{22}\Gamma_L} \right| < 1, \tag{12.19a}$$

$$|\Gamma_{out}| = \left| S_{22} + \frac{S_{12}S_{21}\Gamma_S}{1 - S_{11}\Gamma_S} \right| < 1. \tag{12.19b}$$

If the device is unilateral ($S_{12} = 0$), these conditions reduce to the simple results that $|S_{11}| < 1$ and $|S_{22}| < 1$ are sufficient for unconditional stability. Otherwise, the inequalities of (12.19) define a range of values for Γ_S and Γ_L where the amplifier will be stable. Finding this range for Γ_S and Γ_L can be facilitated by using the Smith chart and plotting the input and output *stability circles*. The stability circles are defined as the loci in the Γ_L (or Γ_S) plane for which $|\Gamma_{in}| = 1$ (or $|\Gamma_{out}| = 1$). The stability circles then define the boundaries between stable and potentially unstable regions of Γ_S and Γ_L. Γ_S and Γ_L must lie on the Smith chart ($|\Gamma_S| < 1$, $|\Gamma_L| < 1$ for passive matching networks).

We can derive the equation for the output stability circle as follows. First use (12.19a) to express the condition that $|\Gamma_{in}| = 1$ as

$$\left| S_{11} + \frac{S_{12}S_{21}\Gamma_L}{1 - S_{22}\Gamma_L} \right| = 1, \tag{12.20}$$

or

$$|S_{11}(1 - S_{22}\Gamma_L) + S_{12}S_{21}\Gamma_L| = |1 - S_{22}\Gamma_L|.$$

Now define Δ as the determinant of the scattering matrix:

$$\Delta = S_{11}S_{22} - S_{12}S_{21}. \tag{12.21}$$

Then we can write the above result as

$$|S_{11} - \Delta\Gamma_L| = |1 - S_{22}\Gamma_L|. \tag{12.22}$$

Now square both sides and simplify to obtain

$$|S_{11}|^2 + |\Delta|^2|\Gamma_L|^2 - \left(\Delta\Gamma_L S_{11}^* + \Delta^*\Gamma_L^* S_{11}\right) = 1 + |S_{22}|^2|\Gamma_L|^2 - \left(S_{22}^*\Gamma_L^* + S_{22}\Gamma_L\right)$$

$$\left(|S_{22}|^2 - |\Delta|^2\right)\Gamma_L\Gamma_L^* - \left(S_{22} - \Delta S_{11}^*\right)\Gamma_L - \left(S_{22}^* - \Delta^* S_{11}\right)\Gamma_L^* = |S_{11}|^2 - 1$$

$$\Gamma_L\Gamma_L^* - \frac{\left(S_{22} - \Delta S_{11}^*\right)\Gamma_L + \left(S_{22}^* - \Delta^* S_{11}\right)\Gamma_L^*}{|S_{22}|^2 - |\Delta|^2} = \frac{|S_{11}|^2 - 1}{|S_{22}|^2 - |\Delta|^2}. \tag{12.23}$$

Next, complete the square by adding $\left|S_{22} - \Delta S_{11}^*\right|^2 / \left(|S_{22}|^2 - |\Delta|^2\right)^2$ to both sides:

$$\left|\Gamma_L - \frac{\left(S_{22} - \Delta S_{11}^*\right)^*}{|S_{22}|^2 - |\Delta|^2}\right|^2 = \frac{\left|S_{11}^2\right| - 1}{|S_{22}|^2 - |\Delta|^2} + \frac{\left|S_{22} - \Delta S_{11}^*\right|^2}{\left(|S_{22}|^2 - |\Delta|^2\right)^2},$$

or

$$\left|\Gamma_L - \frac{\left(S_{22} - \Delta S_{11}^*\right)^*}{|S_{22}|^2 - |\Delta|^2}\right| = \left|\frac{S_{12}S_{21}}{|S_{22}|^2 - |\Delta|^2}\right|. \tag{12.24}$$

In the complex Γ plane, an equation of the form $|\Gamma - C| = R$ represents a circle having a center at C (a complex number) and a radius R (a real number). Thus, (12.24) defines the output stability circle with a center C_L and radius R_L, where

$$C_L = \frac{\left(S_{22} - \Delta S_{11}^*\right)^*}{|S_{22}|^2 - |\Delta|^2} \qquad \text{(center)}, \tag{12.25a}$$

$$R_L = \left|\frac{S_{12}S_{21}}{|S_{22}|^2 - |\Delta|^2}\right| \qquad \text{(radius)}. \tag{12.25b}$$

Similar results can be obtained for the input stability circle by interchanging S_{11} and S_{22}:

$$C_S = \frac{\left(S_{11} - \Delta S_{22}^*\right)^*}{|S_{11}|^2 - |\Delta|^2} \qquad \text{(center)}, \tag{12.26a}$$

$$R_S = \left|\frac{S_{12}S_{21}}{|S_{11}|^2 - |\Delta|^2}\right| \qquad \text{(radius)}. \tag{12.26b}$$

Given the scattering parameters of the transistor, we can plot the input and output stability circles to define where $|\Gamma_{in}| = 1$ and $|\Gamma_{out}| = 1$. On one side of the input stability circle we will have $|\Gamma_{out}| < 1$, while on the other side we will have $|\Gamma_{out}| > 1$. Similarly, we will have $|\Gamma_{in}| < 1$ on one side of the output stability circle, and $|\Gamma_{in}| > 1$ on the other side. We need to determine which areas on the Smith chart represent the stable region, for which $|\Gamma_{in}| < 1$ and $|\Gamma_{out}| < 1$.

Consider the output stability circles plotted in the Γ_L plane for $|S_{11}| < 1$ and $|S_{11}| > 1$, as shown in Figure 12.5. If we set $Z_L = Z_0$, then $\Gamma_L = 0$, and (12.19a) shows that $|\Gamma_{in}| = |S_{11}|$. Now if $|S_{11}| < 1$, then $|\Gamma_{in}| < 1$, so $\Gamma_L = 0$ must be in a stable region. This means that the center of the Smith chart ($\Gamma_L = 0$) is in the stable region, so all of the Smith chart ($|\Gamma_L| < 1$) that is exterior to the stability circle defines the stable range for Γ_L. This region is shaded in Figure 12.5a. Alternatively, if we set $Z_L = Z_0$ but have $|S_{11}| > 1$, then

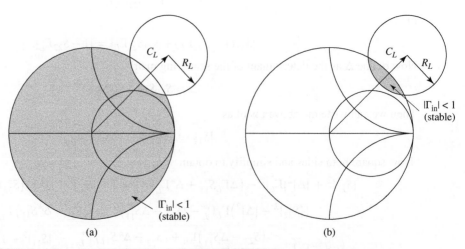

FIGURE 12.5 | Output stability circles for a conditionally stable device. (a) $|S_{11}| < 1$. (b) $|S_{11}| > 1$.

$|\Gamma_{\text{in}}| > 1$ for $\Gamma_L = 0$, and the center of the Smith chart must be in an unstable region. In this case the stable region is the inside region of the stability circle that intersects the Smith chart, as illustrated in Figure 12.5b. Similar results apply to the input stability circle.

If the device is unconditionally stable, the stability circles must be completely outside (or totally enclose) the Smith chart. We can state this result mathematically as

$$||C_L| - R_L| > 1 \quad \text{for} \quad |S_{11}| < 1, \tag{12.27a}$$

$$||C_S| - R_S| > 1 \quad \text{for} \quad |S_{22}| < 1. \tag{12.27b}$$

If $|S_{11}| > 1$ or $|S_{22}| > 1$, the amplifier cannot be unconditionally stable because we can always have a source or load impedance of Z_0 leading to $\Gamma_S = 0$ or $\Gamma_L = 0$, thus causing $|\Gamma_{\text{in}}| > 1$ or $|\Gamma_{\text{out}}| > 1$. If the device is only conditionally stable, operating points for Γ_S and Γ_L must be chosen in stable regions, and it is good practice to check stability at several frequencies over the range where the device operates. Also note that the scattering parameters of a transistor depend on the bias conditions, and so stability will also depend on bias conditions. If it is possible to accept a design with less than maximum gain, a transistor can usually be made to be unconditionally stable by using resistive loading.

Tests for Unconditional Stability

The stability circles discussed above can be used to determine regions for Γ_S and Γ_L where the amplifier circuit will be conditionally stable, but simpler tests can be used to determine unconditional stability. One of these is the $K - \Delta$ *test*, where it can be shown that a device will be unconditionally stable if *Rollet's condition*, defined as

$$K = \frac{1 - |S_{11}|^2 - |S_{22}|^2 + |\Delta|^2}{2|S_{12}S_{21}|} > 1, \tag{12.28}$$

along with the auxiliary condition that

$$|\Delta| = |S_{11}S_{22} - S_{12}S_{21}| < 1, \tag{12.29}$$

are simultaneously satisfied. These two conditions are necessary and sufficient for unconditional stability, and are easily evaluated. If the device scattering parameters do not satisfy the $K - \Delta$ test, the device is not unconditionally stable, and stability circles must be used to determine if there are values of Γ_S and Γ_L for which the device will be conditionally stable. Also recall that we must have $|S_{11}| < 1$ and $|S_{22}| < 1$ if the device is to be unconditionally stable.

While the $K - \Delta$ test of (12.28)–(12.29) is a mathematically rigorous condition for unconditional stability, it cannot be used to compare the relative stability of two or more devices because it involves constraints on two separate parameters. A criterion that combines the scattering parameters in a test involving only a single parameter, μ, defined as [7]

$$\mu = \frac{1 - |S_{11}|^2}{\left|S_{22} - \Delta S_{11}^*\right| + |S_{12}S_{21}|} > 1. \tag{12.30}$$

Thus, if $\mu > 1$, the device is unconditionally stable. In addition, it can be said that larger values of μ imply greater stability.

We can derive the μ-test of (12.30) by starting with the expression from (12.3b) for Γ_{out}:

$$\Gamma_{\text{out}} = S_{22} + \frac{S_{12}S_{21}\Gamma_S}{1 - S_{11}\Gamma_S} = \frac{S_{22} - \Delta\Gamma_S}{1 - S_{11}\Gamma_S}, \tag{12.31}$$

where Δ is the determinant of the scattering matrix defined in (12.21). Unconditional stability implies that $|\Gamma_{\text{out}}| < 1$ for any passive source termination, Γ_S. The reflection coefficient for a passive source impedance must lie within the unit circle on a Smith chart, and the outer boundary of this circle can be written as $\Gamma_S = e^{j\phi}$. The expression given in (12.31) maps this circle into another circle in the Γ_{out} plane. We can show this by substituting $\Gamma_S = e^{j\phi}$ into (12.31) and solving for $e^{j\phi}$:

$$e^{j\phi} = \frac{S_{22} - \Gamma_{\text{out}}}{\Delta - S_{11}\Gamma_{\text{out}}}.$$

Taking the magnitude of both sides gives

$$\left| \frac{S_{22} - \Gamma_{\text{out}}}{\Delta - S_{11}\Gamma_{\text{out}}} \right| = 1.$$

Squaring both sides and expanding gives

$$|\Gamma_{\text{out}}|^2 \left(1 - |S_{11}|^2\right) + \Gamma_{\text{out}}\left(\Delta^* S_{11} - S_{22}^*\right) + \Gamma_{\text{out}}^*\left(\Delta S_{11}^* - S_{22}\right) = |\Delta|^2 - |S_{22}|^2.$$

Now divide by $1 - |S_{11}|^2$ to obtain

$$|\Gamma_{\text{out}}|^2 + \frac{\left(\Delta^* S_{11} - S_{22}^*\right)\Gamma_{\text{out}} + \left(\Delta S_{11}^* - S_{22}\right)\Gamma_{\text{out}}^*}{1 - |S_{11}|^2} = \frac{|\Delta|^2 - |S_{22}|^2}{1 - |S_{11}|^2}.$$

Complete the square by adding $\dfrac{\left|\Delta^* S_{11} - S_{22}^*\right|^2}{\left(1 - |S_{11}|^2\right)^2}$ to both sides:

$$\left| \Gamma_{\text{out}} + \frac{\Delta S_{11}^* - S_{22}}{1 - |S_{11}|^2} \right|^2 = \frac{|\Delta|^2 - |S_{22}|^2}{1 - |S_{11}|^2} + \frac{\left|\Delta^* S_{11} - S_{22}^*\right|^2}{\left(1 - |S_{11}|^2\right)^2} = \frac{|S_{12}S_{21}|^2}{\left(1 - |S_{11}|^2\right)^2}. \tag{12.32}$$

This equation is of the form $|\Gamma_{\text{out}} - C| = R$, representing a circle with center C and radius R in the Γ_{out} plane. Thus the center and radius of the mapped $|\Gamma_S| = 1$ circle are given by

$$C = \frac{S_{22} - \Delta S_{11}^*}{1 - |S_{11}|^2}, \tag{12.33a}$$

$$R = \frac{|S_{12}S_{21}|}{1 - |S_{11}|^2}. \tag{12.33b}$$

If points within this circular region are to satisfy $|\Gamma_{\text{out}}| < 1$, then we must have that

$$|C| + R < 1. \tag{12.34}$$

Substituting (12.33) into (12.34) gives

$$\left| S_{22} - \Delta S_{11}^* \right| + |S_{12}S_{21}| < 1 - |S_{11}|^2,$$

which after rearranging yields the μ-test of (12.30):

$$\frac{1 - |S_{11}|^2}{\left| S_{22} - \Delta S_{11}^* \right| + |S_{12}S_{21}|} > 1.$$

The $K - \Delta$ test of (12.28)–(12.29) can be derived from a similar starting point, or more simply from the μ-test of (12.30). Rearranging (12.30) and squaring gives

$$\left| S_{22} - \Delta S_{11}^* \right|^2 < \left(1 - |S_{11}|^2 - |S_{12}S_{21}|\right)^2. \tag{12.35}$$

It can be verified by direct expansion that

$$\left| S_{22} - \Delta S_{11}^* \right|^2 = |S_{12}S_{21}|^2 + \left(1 - |S_{11}|^2\right)\left(|S_{22}|^2 - |\Delta|^2\right),$$

so (12.35) expands to

$$|S_{12}S_{21}|^2 + \left(1 - |S_{11}|^2\right)\left(|S_{22}|^2 - |\Delta|^2\right) < \left(1 - |S_{11}|^2\right)\left(1 - |S_{11}|^2 - 2|S_{12}S_{21}|\right) + |S_{12}S_{21}|^2.$$

Simplifying gives

$$|S_{22}|^2 - |\Delta|^2 < 1 - |S_{11}|^2 - 2|S_{12}S_{21}|,$$

which yields Rollet's condition of (12.28) after rearranging:

$$\frac{1 - |S_{11}|^2 - |S_{22}|^2 + |\Delta|^2}{2|S_{12}S_{21}|} = K > 1.$$

In addition to (12.28), the $K - \Delta$ test also requires the auxiliary condition of (12.29) to guarantee unconditional stability. Although we derived Rollet's condition from the necessary and sufficient result of the μ-test, the squaring step used in (12.35) introduces an ambiguity in the sign of the right-hand side, thus requiring the additional condition. This can be derived by requiring that the right-hand side of (12.35) be positive before squaring. Thus,

$$|S_{12}S_{21}| < 1 - |S_{11}|^2.$$

Because similar conditions can be derived for the input side of the circuit, we can interchange S_{11} and S_{22} to obtain the analogous condition that

$$|S_{12}S_{21}| < 1 - |S_{22}|^2.$$

Adding these two inequalities gives

$$2|S_{12}S_{21}| < 2 - |S_{11}|^2 - |S_{22}|^2.$$

From the triangle inequality we know that

$$|\Delta| = |S_{11}S_{22} - S_{12}S_{21}| \le |S_{11}S_{22}| + |S_{12}S_{21}|,$$

so we have that

$$|\Delta| < |S_{11}||S_{22}| + 1 - \frac{1}{2}|S_{11}|^2 - \frac{1}{2}|S_{22}|^2 < 1 - \frac{1}{2}\left(|S_{11}|^2 - |S_{22}|^2\right) < 1,$$

which is identical to (12.29).

EXAMPLE 12.2   TRANSISTOR STABILITY

The Triquint T1G6000528 GaN HEMT has the following scattering parameters at 2 GHz ($Z_0 = 50 \ \Omega$):

$$S_{11} = 0.867\angle-160.5°,$$
$$S_{12} = 0.03\angle-9.5°,$$
$$S_{21} = 4.05\angle60.3°,$$
$$S_{22} = 0.51\angle-119.2°.$$

Determine the stability of this transistor by using the $K - \Delta$ test and the μ-test, and plot the stability circles on a Smith chart.

Solution
From (12.28) and (12.29) we compute K and $|\Delta|$ as

$$|\Delta| = |S_{11}S_{22} - S_{12}S_{21}| = 0.341,$$

$$K = \frac{1 - |S_{11}|^2 - |S_{22}|^2 + |\Delta|^2}{2|S_{12}S_{21}|} = 0.43.$$

Thus we have $|\Delta| < 1$ but not $K > 1$, so the unconditional stability criteria of (12.28)–(12.29) are not satisfied, and the device is potentially unstable. The stability of this device can also be evaluated using the μ-test, for which (12.30) gives $\mu = 0.716$, again indicating potential instability.

The centers and radii of the stability circles are given by (12.25) and (12.26):

$$C_L = \frac{\left(S_{22} - \Delta S_{11}^*\right)^*}{|S_{22}|^2 - |\Delta|^2} = 1.56\angle132.49°,$$

$$R_L = \frac{|S_{12}S_{21}|}{|S_{22}|^2 - |\Delta|^2} = 0.845,$$

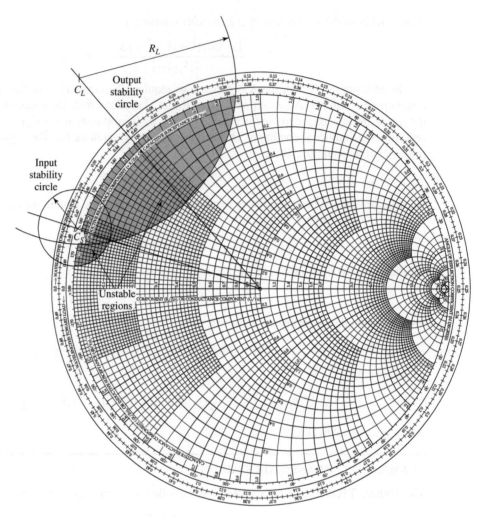

FIGURE 12.6 | Stability circles for Example 12.2.

$$C_S = \frac{\left(S_{11} - \Delta S_{22}^*\right)^*}{|S_{11}|^2 - |\Delta|^2} = 1.105\angle 163°,$$

$$R_S = \frac{|S_{12}S_{21}|}{|S_{11}|^2 - |\Delta|^2} = 0.193.$$

These data can be used to plot the input and output stability circles, as shown in Figure 12.6. Since $|S_{11}| < 1$ and $|S_{22}| < 1$, the central part of the Smith chart represents the stable operating region for Γ_S and Γ_L. The unstable regions are shaded.

12.3 SINGLE-STAGE TRANSISTOR AMPLIFIER DESIGN

Design for Maximum Gain (Conjugate Matching)

After the stability of the transistor has been determined and the stable regions for Γ_S and Γ_L have been located on the Smith chart, the input and output matching sections can be designed. Since G_0 of (12.16b) is fixed for a given transistor, the overall transducer gain of the amplifier will be controlled by the gains,

G_S and G_L, of the matching sections. Maximum gain will be realized when these sections provide a conjugate match between the amplifier source or load impedance and the transistor. Because most transistors exhibit a significant impedance mismatch (large $|S_{11}|$ and $|S_{22}|$), the resulting frequency response may be narrowband. In the following section we will discuss how to design for less than maximum gain, with a corresponding improvement in bandwidth. Broadband amplifier design will be discussed in Section 12.4.

With reference to Figure 12.2 and our discussion in Section 2.5 on conjugate impedance matching, we know that maximum power transfer from the input matching network to the transistor will occur when

$$\Gamma_{\text{in}} = \Gamma_S^*, \tag{12.36a}$$

and that maximum power transfer from the transistor to the output matching network will occur when

$$\Gamma_{\text{out}} = \Gamma_L^*. \tag{12.36b}$$

With the assumption of lossless matching sections, these conditions will maximize the overall transducer gain. From (12.13), this maximum gain will be given by

$$G_{T_{\max}} = \frac{1}{1 - |\Gamma_S|^2}|S_{21}|^2 \frac{1 - |\Gamma_L|^2}{|1 - S_{22}\Gamma_L|^2}. \tag{12.37}$$

In addition, with conjugate matching and lossless matching sections, the input and output ports of the amplifier will be matched to Z_0.

In the general case with a *bilateral* ($S_{12} \neq 0$) transistor, Γ_{in} is affected by Γ_{out} and vice versa, so the input and output sections must be matched simultaneously. Using (12.36) in (12.3) gives the necessary equations:

$$\Gamma_S^* = S_{11} + \frac{S_{12}S_{21}\Gamma_L}{1 - S_{22}\Gamma_L}, \tag{12.38a}$$

$$\Gamma_L^* = S_{22} + \frac{S_{12}S_{21}\Gamma_S}{1 - S_{11}\Gamma_S}. \tag{12.38b}$$

We can solve for Γ_S by first rewriting these equations as follows:

$$\Gamma_S = S_{11}^* + \frac{S_{12}^* S_{21}^*}{1/\Gamma_L^* - S_{22}^*},$$

$$\Gamma_L^* = \frac{S_{22} - \Delta\Gamma_S}{1 - S_{11}\Gamma_S},$$

where $\Delta = S_{11}S_{22} - S_{12}S_{21}$. Substituting the expression for Γ_L^* into the expression for Γ_S and expanding gives

$$\Gamma_S(1 - |S_{22}|^2) + \Gamma_S^2(\Delta S_{22}^* - S_{11}) = \Gamma_S(\Delta S_{11}^* S_{22}^* - |S_{11}|^2 - \Delta S_{12}^* S_{21}^*)$$
$$+ S_{11}^*(1 - |S_{22}|^2) + S_{12}^* S_{21}^* S_{22}.$$

Using the result that $\Delta(S_{11}^* S_{22}^* - S_{12}^* S_{21}^*) = |\Delta|^2$ allows this to be rewritten as a quadratic equation for Γ_S:

$$(S_{11} - \Delta S_{22}^*)\Gamma_S^2 + (|\Delta|^2 - |S_{11}|^2 + |S_{22}|^2 - 1)\Gamma_S + (S_{11}^* - \Delta^* S_{22}) = 0. \tag{12.39}$$

The solution is

$$\Gamma_S = \frac{B_1 \pm \sqrt{B_1^2 - 4|C_1|^2}}{2C_1}. \tag{12.40a}$$

Similarly, the solution for Γ_L can be written as

$$\Gamma_L = \frac{B_2 \pm \sqrt{B_2^2 - 4|C_2|^2}}{2C_2}. \tag{12.40b}$$

The variables B_1, C_1, B_2, C_2 are defined as

$$B_1 = 1 + |S_{11}|^2 - |S_{22}|^2 - |\Delta|^2, \tag{12.41a}$$

$$B_2 = 1 + |S_{22}|^2 - |S_{11}|^2 - |\Delta|^2, \tag{12.41b}$$

$$C_1 = S_{11} - \Delta S_{22}^*, \tag{12.41c}$$

$$C_2 = S_{22} - \Delta S_{11}^*. \tag{12.41d}$$

Solutions to (12.40) are only possible if the quantity within the square root is positive, and it can be shown that this is equivalent to requiring $K > 1$. Thus, unconditionally stable devices can always be conjugately matched for maximum gain, and potentially unstable devices can be conjugately matched if $K > 1$ and $|\Delta| < 1$. The results are much simpler for the unilateral case. When $S_{12} = 0$, (12.38) shows that $\Gamma_S = S_{11}^*$ and $\Gamma_L = S_{22}^*$, and then maximum transducer gain of (12.37) reduces to

$$G_{TU_{\max}} = \frac{1}{1 - |S_{11}|^2} |S_{21}|^2 \frac{1}{1 - |S_{22}|^2}. \tag{12.42}$$

The maximum transducer power gain given by (12.37) occurs when the source and load are conjugately matched to the transistor, as given by the conditions of (12.36). If the transistor is unconditionally stable, so that $K > 1$, the maximum transducer power gain of (12.37) can be simply rewritten as follows:

$$G_{T_{\max}} = \frac{|S_{21}|}{|S_{12}|} \left(K - \sqrt{K^2 - 1} \right). \tag{12.43}$$

This result can be obtained by substituting (12.40) and (12.41) for Γ_S and Γ_L into (12.37) and simplifying. The maximum transducer power gain is also sometimes referred to as the *matched gain*. The maximum gain does not provide a meaningful result if the device is only conditionally stable since simultaneous conjugate matching of the source and load is not possible if $K < 1$ (see Problem 12.8). In this case a useful figure of merit is the *maximum stable gain*, defined as the maximum transducer power gain of (12.43) with $K = 1$. Thus,

$$G_{\text{msg}} = \frac{|S_{21}|}{|S_{12}|}. \tag{12.44}$$

The maximum stable gain is easy to compute and offers a convenient way to compare the gain of various devices under stable operating conditions.

EXAMPLE 12.3   CONJUGATELY MATCHED AMPLIFIER DESIGN

Design an amplifier for maximum gain at 4 GHz using single-stub matching sections. Calculate and plot the input return loss and the gain from 3 to 5 GHz. The transistor is a GaAs MESFET with the following scattering parameters ($Z_0 = 50\ \Omega$):

f(GHz)	S_{11}	S_{12}	S_{21}	S_{22}
3.0	0.80∠−89°	0.03∠56°	2.86∠99°	0.76∠−41°
4.0	0.72∠−116°	0.03∠57°	2.60∠76°	0.73∠−54°
5.0	0.66∠−142°	0.03∠62°	2.39∠54°	0.72∠−68°

Solution

In practice, scattering parameters are usually provided by the manufacturer over a wide frequency range, and it is prudent to check stability over the entire range. Here we have limited the data to three frequencies to illustrate the point without undue computational burden. Using (12.28) and (12.29) to calculate K and Δ from the scattering parameters at each frequency in the above table gives the following results:

f(GHz)	K	Δ
3.0	0.77	0.592
4.0	1.19	0.487
5.0	1.53	0.418

We see that $K > 1$ and $|\Delta| < 1$ at 4 and 5 GHz, so the transistor is unconditionally stable at these frequencies, but it is only conditionally stable at 3 GHz. We can proceed with the design at 4 GHz, but should check stability at 3 GHz after we find the matching networks (which determine Γ_S and Γ_L).

For maximum gain, we should design the matching sections for a conjugate match to the transistor. Thus, $\Gamma_S = \Gamma_{in}^*$ and $\Gamma_L = \Gamma_{out}^*$, and Γ_S, Γ_L can be determined from (12.40):

$$\Gamma_S = \frac{B_1 \pm \sqrt{B_1^2 - 4|C_1|^2}}{2C_1} = 0.872\angle 123°,$$

$$\Gamma_L = \frac{B_2 \pm \sqrt{B_2^2 - 4|C_2|^2}}{2C_2} = 0.876\angle 61°.$$

The effective gain factors of (12.16) can be calculated as

$$G_S = \frac{1}{1 - |\Gamma_S|^2} = 4.17 = 6.20 \text{ dB},$$

$$G_0 = |S_{21}|^2 = 6.76 = 8.30 \text{ dB},$$

$$G_L = \frac{1 - |\Gamma_L|^2}{|1 - S_{22}\Gamma_L|^2} = 1.67 = 2.22 \text{ dB}.$$

Then the overall transducer gain is

$$G_{T_{max}} = 6.20 + 8.30 + 2.22 = 16.7 \text{ dB}.$$

The matching networks can easily be determined using the Smith chart. For the input matching section, first plot Γ_S, as shown in Figure 12.7c. The impedance, Z_S, represented by this reflection coefficient is the impedance seen looking into the matching section toward the source impedance, Z_0. Thus, the matching section must transform Z_0 to the impedance Z_S. There are several ways of doing this, but we will use an open-circuited shunt stub followed by a length of line. We convert to the normalized admittance y_s, and work backward (toward the load on the Smith chart) to find that a line of length 0.120λ will bring us to the $1 + jb$ circle. Then we see that the required stub admittance is $+j3.5$, for an open-circuited stub length of 0.206λ. A similar procedure gives a line length of 0.206λ and a stub length of 0.206λ for the output matching circuit.

The final amplifier circuit is shown in Figure 12.7a. This circuit only shows the RF components; the amplifier will also require bias circuitry. The return loss and gain were calculated using a CAD package, interpolating the necessary scattering parameters from the data given above. The results are plotted in Figure 12.7b, and show the expected gain of 16.7 dB at 4 GHz, with a very good return loss. The bandwidth where the gain drops by 1 dB is about 2.5%.

With regard to the potential instability at 3 GHz, we leave it to the reader to show that the designed matching sections present source and load impedances that lie within the stable regions of the appropriate stability circles. Note that the matching sections are frequency dependent, so the impedances and reflection coefficients are different at 3 GHz than their design values at 4 GHz. The fact that CAD simulation did not show any indication of instability over the frequency range of 3–5 GHz is evidence that the circuit is stable over this frequency range.

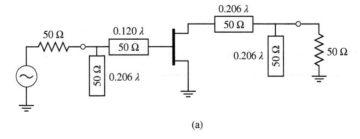

(a)

FIGURE 12.7 | Circuit design and frequency response for the transistor amplifier of Example 12.3. (a) RF circuit.

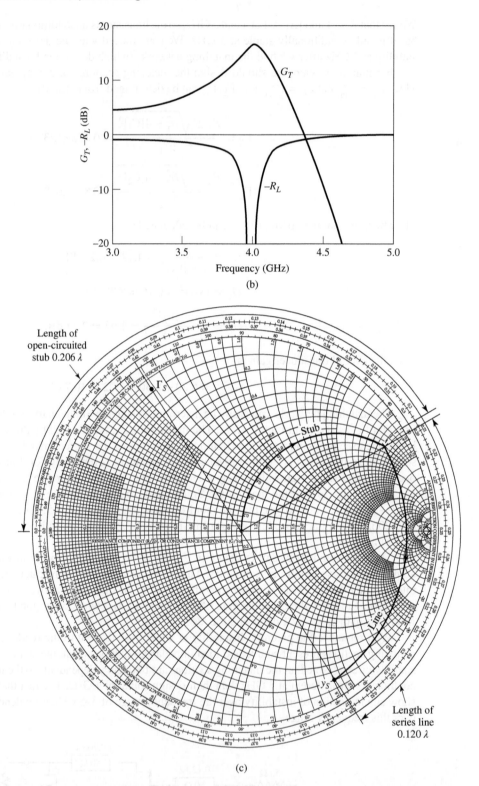

(b)

(c)

FIGURE 12.7 | Continued. (b) Frequency response. (c) Smith chart for the design of the input matching network.

Constant-Gain Circles and Design for Specified Gain

In many cases it is preferable to design for less than the maximum obtainable gain, to improve bandwidth or to obtain a specific value of amplifier gain. This can be done by designing the input and output matching sections to have less than maximum gains; in other words, mismatches are purposely introduced to reduce the overall gain. The design procedure is facilitated by plotting *constant-gain circles* on the Smith chart to represent loci of Γ_S and Γ_L that give fixed values of gain (G_S and G_L). To simplify our discussion, we will only treat the case of a unilateral device; the more general case of a bilateral device must sometimes be considered in practice, and is discussed in detail in references [1–2].

For many transistors $|S_{12}|$ is small enough to be ignored, and the device can be assumed to be unilateral. This greatly simplifies the design procedure. The error in the transducer gain caused by approximating $|S_{12}|$ as zero is given by the ratio G_T/G_{TU}. It can be shown that this ratio is bounded by

$$\frac{1}{(1+U)^2} < \frac{G_T}{G_{TU}} < \frac{1}{(1-U)^2}, \tag{12.45}$$

where U is defined as the *unilateral figure of merit*,

$$U = \frac{|S_{12}||S_{21}||S_{11}||S_{22}|}{\left(1-|S_{11}|^2\right)\left(1-|S_{22}|^2\right)}. \tag{12.46}$$

Usually an error of a few tenths of a dB or less justifies the unilateral assumption.

The expression for G_S and G_L for the unilateral case are given by (12.17a) and (12.17c):

$$G_S = \frac{1-|\Gamma_S|^2}{|1-S_{11}\Gamma_S|^2},$$

$$G_L = \frac{1-|\Gamma_L|^2}{|1-S_{22}\Gamma_L|^2}.$$

These gains are maximized when $\Gamma_S = S_{11}^*$ and $\Gamma_L = S_{22}^*$, resulting in the maximum values given by

$$G_{S_{max}} = \frac{1}{1-|S_{11}|^2}, \tag{12.47a}$$

$$G_{L_{max}} = \frac{1}{1-|S_{22}|^2}. \tag{12.47b}$$

Define normalized gain factors g_S and g_L as

$$g_S = \frac{G_S}{G_{S_{max}}} = \frac{1-|\Gamma_S|^2}{|1-S_{11}\Gamma_S|^2}\left(1-|S_{11}|^2\right), \tag{12.48a}$$

$$g_L = \frac{G_L}{G_{L_{max}}} = \frac{1-|\Gamma_L|^2}{|1-S_{22}\Gamma_L|^2}\left(1-|S_{22}|^2\right). \tag{12.48b}$$

Then we have that $0 \le g_S \le 1$ and $0 \le g_L \le 1$.

For fixed values of g_S and g_L, (12.48) represents circles in the Γ_S or Γ_L plane. To show this, consider (12.48a), which can be expanded to give

$$g_S|1-S_{11}\Gamma_S|^2 = \left(1-|\Gamma_S|^2\right)\left(1-|S_{11}|^2\right),$$

$$\left(g_S|S_{11}|^2 + 1 - |S_{11}|^2\right)|\Gamma_S|^2 - g_S\left(S_{11}\Gamma_S + S_{11}^*\Gamma_S^*\right) = 1 - |S_{11}|^2 - g_S,$$

$$\Gamma_S\Gamma_S^* - \frac{g_S\left(S_{11}\Gamma_S + S_{11}^*\Gamma_S^*\right)}{1-(1-g_S)|S_{11}|^2} = \frac{1-|S_{11}|^2 - g_S}{1-(1-g_S)|S_{11}|^2}. \tag{12.49}$$

Now add $\left(g_S^2|S_{11}|^2\right)/\left[1-(1-g_S)|S_{11}|^2\right]^2$ to both sides to complete the square:

$$\left|\Gamma_S - \frac{g_S S_{11}^*}{1-(1-g_S)|S_{11}|^2}\right|^2 = \frac{\left(1-|S_{11}|^2-g_S\right)\left[1-(1-g_S)|S_{11}|^2\right]+g_S^2|S_{11}|^2}{\left[1-(1-g_S)|S_{11}|^2\right]^2}.$$

Simplifying gives

$$\left|\Gamma_S - \frac{g_S S_{11}^*}{1-(1-g_S)|S_{11}|^2}\right| = \frac{\sqrt{1-g_S}\left(1-|S_{11}|^2\right)}{1-(1-g_S)|S_{11}|^2}, \tag{12.50}$$

which is the equation of a circle with its center and radius given by

$$C_S = \frac{g_S S_{11}^*}{1-(1-g_S)|S_{11}|^2}, \tag{12.51a}$$

$$R_S = \frac{\sqrt{1-g_S}\left(1-|S_{11}|^2\right)}{1-(1-g_S)|S_{11}|^2}. \tag{12.51b}$$

The results for the constant gain circles of the output section can be shown to be

$$C_L = \frac{g_L S_{22}^*}{1-(1-g_L)|S_{22}|^2}, \tag{12.52a}$$

$$R_L = \frac{\sqrt{1-g_L}\left(1-|S_{22}|^2\right)}{1-(1-g_L)|S_{22}|^2}. \tag{12.52b}$$

The centers of each family of circles lie along straight lines given by the angle of S_{11}^* or S_{22}^*. Note that when g_S (or g_L) $= 1$ (maximum gain), the radius R_S (or R_L) $= 0$, and the center reduces to S_{11}^* (or S_{22}^*), as expected. In addition, it can be shown that the 0 dB gain circles ($G_S = 1$ or $G_L = 1$) will always pass through the center of the Smith chart. These results can be used to plot a family of circles of constant gain for the input and output sections. Then Γ_S and Γ_L can be chosen along these circles to provide the desired gains. The choices for Γ_S and Γ_L are not unique, but it makes sense to choose points close to the center of the Smith chart to minimize mismatch, and thus maximize bandwidth. Alternatively, as we will see in the next section, the input network mismatch can be chosen to provide a low-noise design.

EXAMPLE 12.4   AMPLIFIER DESIGN FOR SPECIFIED GAIN

Design an amplifier to have a gain of 11 dB at 4.0 GHz. Plot constant-gain circles for $G_S = 2$ and 3 dB, and $G_L = 0$ and 1 dB. Calculate and plot the input return loss and overall amplifier gain from 3 to 5 GHz. The transistor has the following scattering parameters ($Z_0 = 50\ \Omega$):

f(GHz)	S_{11}	S_{12}	S_{21}	S_{22}
3	0.80∠−90°	0	2.8∠100°	0.66∠−50°
4	0.75∠−120°	0	2.5∠80°	0.60∠−70°
5	0.71∠−140°	0	2.3∠60°	0.58∠−85°

Solution
Since $S_{12} = 0$ and $|S_{11}| < 1$ and $|S_{22}| < 1$, the transistor is unilateral and unconditionally stable at each frequency in the above table. From (12.47) we calculate the maximum matching section gains as

$$G_{S_{\max}} = \frac{1}{1-|S_{11}|^2} = 2.29 = 3.6\ \text{dB},$$

$$G_{L_{\max}} = \frac{1}{1-|S_{22}|^2} = 1.56 = 1.9\ \text{dB}.$$

The gain of the mismatched transistor is

$$G_0 = |S_{21}|^2 = 6.25 = 8.0 \text{ dB},$$

so the maximum unilateral transducer gain is

$$G_{TU_{max}} = 3.6 + 1.9 + 8.0 = 13.5 \text{ dB}.$$

We therefore have 2.5 dB more available gain than is required by the specifications.

Next, use (12.48), (12.51), and (12.52) to calculate the following data for the constant-gain circles:

$$
\begin{aligned}
G_S = 3 \text{ dB} \quad & g_S = 0.875 \quad && C_S = 0.706\angle120° \quad && R_S = 0.166 \\
G_S = 2 \text{ dB} \quad & g_S = 0.691 \quad && C_S = 0.627\angle120° \quad && R_S = 0.294 \\
G_L = 1 \text{ dB} \quad & g_L = 0.806 \quad && C_L = 0.520\angle70° \quad && R_L = 0.303 \\
G_L = 0 \text{ dB} \quad & g_L = 0.640 \quad && C_L = 0.440\angle70° \quad && R_L = 0.440
\end{aligned}
$$

The constant-gain circles are shown in Figure 12.8a. We choose $G_S = 2$ dB and $G_L = 1$ dB, for an overall amplifier gain of 11 dB. Then we select Γ_S and Γ_L along these circles as shown, to minimize the distance from the center of the chart (this places Γ_S and Γ_L along the radial lines at 120° and 70°, respectively). Thus, $\Gamma_S = 0.33\angle120°$ and $\Gamma_L = 0.22\angle70°$, and the matching networks can be designed using shunt stubs as in Example 12.3.

The final amplifier circuit is shown in Figure 12.8b. The response was calculated using CAD software, with interpolation of the given scattering parameter data. The results are shown in Figure 12.8c, where it is seen the desired gain of 11 dB is achieved at 4.0 GHz. The bandwidth over which the gain varies by ±1 dB

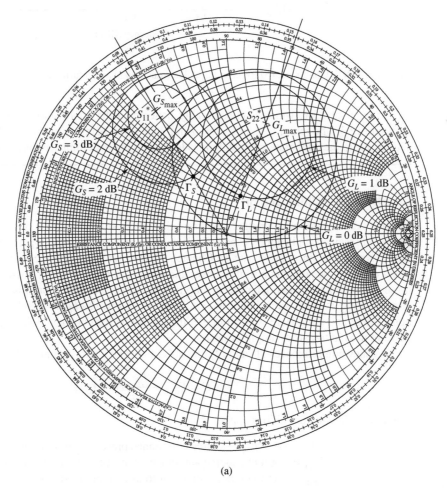

(a)

FIGURE 12.8 | Circuit design and frequency response for the transistor amplifier of Example 12.4. (a) Constant-gain circles.

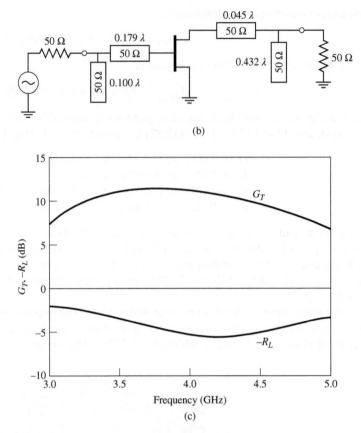

(b)

(c)

FIGURE 12.8 | Continued. (b) RF circuit. (c) Transducer gain and return loss.

or less is about 25%, which is considerably better than the bandwidth of the maximum gain design in Example 12.3. The return loss, however, is not very good, being only about 5 dB at the design frequency. This is due to the deliberate mismatch introduced into the matching sections to achieve the specified gain.

Low-Noise Amplifier Design

Besides stability and gain, another important design consideration for a microwave amplifier is its noise figure. In receiver applications especially it is often required to have a preamplifier with as low a noise figure as possible since, as we saw in Chapter 10, the first stage of a receiver front end has the dominant effect on the noise performance of the overall system. Generally it is not possible to obtain both minimum noise figure and maximum gain for an amplifier, so some sort of compromise must be made. This can be done by using constant-gain circles and *circles of constant noise figure* to select a usable trade-off between noise figure and gain. Here we will derive the equations for constant–noise figure circles and show how they are used in transistor amplifier design.

As shown in references [1] and [2], the noise figure of a two-port amplifier can be expressed as

$$F = F_{\min} + \frac{R_N}{G_S}|Y_S - Y_{\mathrm{opt}}|^2,$$ (12.53)

where the following definitions apply:

$Y_S = G_S + jB_S$ = source admittance presented to transistor.
Y_{opt} = optimum source admittance that results in minimum noise figure.
$F_{\min}$ = minimum noise figure of transistor, attained when $Y_S = Y_{\mathrm{opt}}$.
R_N = equivalent noise resistance of transistor.
G_S = real part of source admittance.

Instead of the admittance Y_S and Y_{opt}, we can use the reflection coefficients Γ_S and Γ_{opt}, where

$$Y_S = \frac{1}{Z_0} \frac{1 - \Gamma_S}{1 + \Gamma_S}, \tag{12.54a}$$

$$Y_{opt} = \frac{1}{Z_0} \frac{1 - \Gamma_{opt}}{1 + \Gamma_{opt}}. \tag{12.54b}$$

Γ_S is the source reflection coefficient defined in Figure 12.1. The quantities F_{min}, Γ_{opt}, and R_N are characteristics of the particular transistor being used, and are called the *noise parameters* of the device; they may be given by the manufacturer or measured.

Using (12.54), we can express the quantity $|Y_S - Y_{opt}|^2$ in terms of Γ_S and Γ_{opt}:

$$|Y_S - Y_{opt}|^2 = \frac{4}{Z_0^2} \frac{|\Gamma_S - \Gamma_{opt}|^2}{|1 + \Gamma_S|^2 |1 + \Gamma_{opt}|^2}. \tag{12.55}$$

In addition,

$$G_S = \text{Re}\{Y_S\} = \frac{1}{2Z_0} \left(\frac{1 - \Gamma_S}{1 + \Gamma_S} + \frac{1 - \Gamma_S^*}{1 + \Gamma_S^*} \right) = \frac{1}{Z_0} \frac{1 - |\Gamma_S|^2}{|1 + \Gamma_S|^2}. \tag{12.56}$$

Using these results in (12.53) gives the noise figure as

$$F = F_{min} + \frac{4R_N}{Z_0} \frac{|\Gamma_S - \Gamma_{opt}|^2}{\left(1 - |\Gamma_S|^2\right)|1 + \Gamma_{opt}|^2}. \tag{12.57}$$

For a fixed noise figure F we can show that this result defines a circle in the Γ_S plane. First define the *noise figure parameter*, N, as

$$N = \frac{|\Gamma_S - \Gamma_{opt}|^2}{1 - |\Gamma_S|^2} = \frac{F - F_{min}}{4R_N/Z_0} |1 + \Gamma_{opt}|^2, \tag{12.58}$$

which is a constant for a given noise figure and set of noise parameters. Then rewrite (12.58) as

$$(\Gamma_S - \Gamma_{opt})(\Gamma_S^* - \Gamma_{opt}^*) = N(1 - |\Gamma_S|^2),$$

$$\Gamma_S \Gamma_S^* - (\Gamma_S \Gamma_{opt}^* + \Gamma_S^* \Gamma_{opt}) + \Gamma_{opt} \Gamma_{opt}^* = N - N|\Gamma_S|^2,$$

$$\Gamma_S \Gamma_S^* - \frac{(\Gamma_S \Gamma_{opt}^* + \Gamma_S^* \Gamma_{opt})}{N + 1} = \frac{N - |\Gamma_{opt}|^2}{N + 1}.$$

Add $|\Gamma_{opt}|^2/(N + 1)^2$ to both sides to complete the square to obtain

$$\left| \Gamma_S - \frac{\Gamma_{opt}}{N + 1} \right| = \frac{\sqrt{N\left(N + 1 - |\Gamma_{opt}|^2\right)}}{(N + 1)}. \tag{12.59}$$

This result defines circles of constant noise figure with centers at

$$C_F = \frac{\Gamma_{opt}}{N + 1}, \tag{12.60a}$$

and radii of

$$R_F = \frac{\sqrt{N\left(N + 1 - |\Gamma_{opt}|^2\right)}}{N + 1}. \tag{12.60b}$$

EXAMPLE 12.5 LOW-NOISE AMPLIFIER DESIGN

A GaAs MESFET is biased for minimum noise figure, with the following scattering parameters and noise parameters at 4 GHz ($Z_0 = 50\ \Omega$): $S_{11} = 0.6\angle-60°$, $S_{12} = 0.05\angle26°$, $S_{21} = 1.9\angle81°$, $S_{22} = 0.5\angle-60°$, $F_{min} = 1.6$ dB, $\Gamma_{opt} = 0.62\angle100°$, and $R_N = 20\ \Omega$. For design purposes, assume the device is unilateral, and calculate the maximum error in G_T resulting from this assumption. Then design an amplifier having a 2.0 dB noise figure with the maximum gain that is compatible with this noise figure.

Solution

We first calculate that $K = 2.78$ and $\Delta = 0.37$, so the device is unconditionally stable even without the approximation of a unilateral device. Next, compute the unilateral figure of merit from (12.46):

$$U = \frac{|S_{12}S_{21}S_{11}S_{22}|}{\left(1 - |S_{11}|^2\right)\left(1 - |S_{22}|^2\right)} = 0.059.$$

From (12.45) the ratio G_T/G_{TU} is bounded as

$$\frac{1}{(1+U)^2} < \frac{G_T}{G_{TU}} < \frac{1}{(1-U)^2},$$

or

$$0.891 < \frac{G_T}{G_{TU}} < 1.130.$$

In dB, this is

$$-0.50 < G_T - G_{TU} < 0.53 \text{ dB},$$

where G_T and G_{TU} are now in dB. Thus, we should expect less than about ±0.5 dB error in gain.

Now use (12.58) and (12.60) to compute the center and radius of the 2 dB noise figure circle:

$$N = \frac{F - F_{min}}{4R_N/Z_0}|1 + \Gamma_{opt}|^2 = \frac{1.58 - 1.445}{4(20/50)}|1 + 0.62\angle100°|^2$$

$$= 0.0986,$$

$$C_F = \frac{\Gamma_{opt}}{N + 1} = 0.56\angle100°,$$

$$R_F = \frac{\sqrt{N\left(N + 1 - |\Gamma_{opt}|^2\right)}}{N + 1} = 0.24.$$

This noise figure circle is plotted in Figure 12.9a. Minimum noise figure ($F_{min} = 1.6$ dB) occurs for $\Gamma_S = \Gamma_{opt} = 0.62\angle100°$.

Next we calculate data for several input section constant-gain circles. From (12.51), we have the following results:

G_S (dB)	g_S	C_S	R_S
1.0	0.805	0.52∠60°	0.300
1.5	0.904	0.56∠60°	0.205
1.7	0.946	0.58∠60°	0.150

These circles are plotted in Figure 12.9a. We see that the $G_S = 1.7$ dB gain circle just intersects the $F = 2$ dB noise figure circle, and that any higher gain will result in a worse noise figure. From the Smith chart the optimum solution is $\Gamma_S = 0.53\angle75°$, yielding $G_S = 1.7$ dB and $F = 2.0$ dB.

For the output section we choose $\Gamma_L = S_{22}^* = 0.5\angle60°$ for a maximum G_L of

$$G_L = \frac{1}{1 - |S_{22}|^2} = 1.33 = 1.25 \text{ dB}.$$

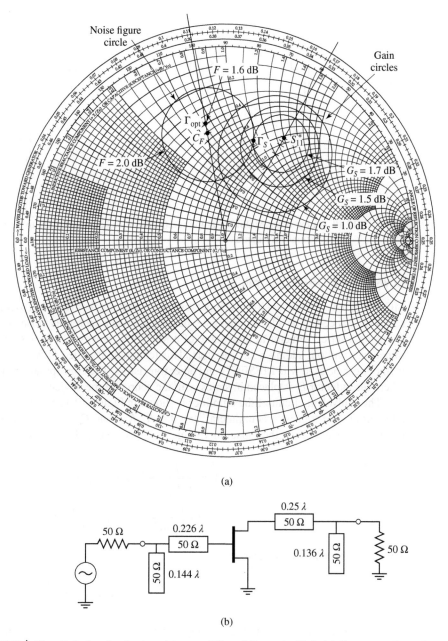

(a)

(b)

FIGURE 12.9 | Circuit design for the transistor amplifier of Example 12.5. (a) Constant-gain and constant–noise figure circles. (b) RF circuit.

The transistor gain is

$$G_0 = |S_{21}|^2 = 3.61 = 5.58 \text{ dB},$$

so the overall transducer gain will be

$$G_{TU} = G_S + G_0 + G_L = 8.53 \text{ dB}.$$

A complete AC circuit for the amplifier, using open-circuited shunt stubs in the matching sections, is shown in Figure 12.9b. A computer analysis of the circuit gives a gain of 8.36 dB.

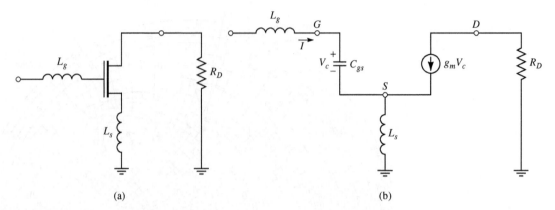

FIGURE 12.10 | Low-noise MOSFET amplifier. (a) Basic AC circuit. (b) Equivalent circuit using a simplified unilateral FET model.

Low-Noise MOSFET Amplifier

MOSFETs have a relatively low AC input resistance, making them difficult to impedance match. An external series resistance can be added to the gate, but this approach increases noise power and degrades efficiency. By using a series inductor at the source of a MOSFET, however, it is possible to create a resistive input impedance without adding noisy resistors. This technique is called *inductive source degeneration*; similar methods can be used with MESFETs and other transistors. The conceptual circuit is shown in Figure 12.10a, where the inductor L_s is placed in series with the source of the device.

The equivalent circuit of the amplifier is shown in Figure 12.10b, where we have simplified the model by assuming the transistor is unilateral, and that R_i, R_{ds}, and C_{ds} can be ignored. For an input current I at the gate of the transistor, the capacitor voltage is $V_c = I/j\omega C_{gs}$. The gate voltage, relative to ground, is then

$$V = \frac{I}{j\omega C_{gs}} + j\omega L_s \left(I + g_m V_c \right)$$

$$= I \left(\frac{1}{j\omega C_{gs}} + j\omega L_s + \frac{g_m L_s}{C_{gs}} \right). \tag{12.61}$$

The input impedance at the gate is

$$Z = \frac{V}{I} = \frac{g_m L_s}{C_{gs}} + j \left(\omega L_s - \frac{1}{\omega C_{gs}} \right), \tag{12.62}$$

showing that the circuit has produced an input resistance of $g_m L_s / C_{gs}$. The series inductor, L_s, can be chosen to match the input resistance of the amplifier to a source impedance, Z_0. The inductor at the gate, L_g, can then be chosen to cancel the residual input reactance, which is usually capacitive. The combination of the series matching inductor, the gate capacitance, and the effective input resistance forms a series RLC resonator. The Q of this resonator is

$$Q = \frac{\omega L_g C_{gs}}{g_m L_s}. \tag{12.63}$$

The bandwidth of this circuit may be relatively narrow if this Q is high.

EXAMPLE 12.6   LOW-NOISE MOSFET AMPLIFIER DESIGN

An Infineon BF1005 n-channel MOSFET transistor having $C_{gs} = 2.2$ pF and $g_m = 24$ mS is used in a 800 MHz low-noise amplifier with inductive source degeneration, as shown in Figure 12.10. Determine the source and gate inductors, and estimate the bandwidth of the amplifier. Assume a source impedance of $Z_0 = 50$ Ω.

Solution

From (12.62), matching the input resistance to Z_0 determines the source inductor as

$$L_s = \frac{Z_0 C_{gs}}{g_m} = \frac{(50)(2.2 \times 10^{-12})}{0.024} = 4.58 \text{ nH.}$$

The net reactance at the input is $jX = j\left(\omega L_s - \dfrac{1}{\omega C_{gs}}\right) = -j67.4 \ \Omega$, so the required series inductance for matching is

$$L_g = \frac{-X}{\omega} = \frac{67.4}{2\pi\left(800 \times 10^6\right)} = 13.41 \text{ nH.}$$

From (12.63) we can estimate the Q as

$$Q = \frac{\omega L_g C_{gs}}{g_m L_s} = 1.35,$$

so the bandwidth of the amplifier could be as high as 80%. This value is probably higher than what would be obtained in practice, due to the approximations that have been made in our analysis.

12.4   BROADBAND TRANSISTOR AMPLIFIER DESIGN

The ideal amplifier would have constant gain and good input matching over the desired frequency bandwidth. As the examples of the last section have shown, conjugate matching will give maximum gain only over a relatively narrow bandwidth, while designing for less than maximum gain will improve the gain bandwidth, but the input and output ports of the amplifier will be poorly matched. Another consideration, as shown earlier in this chapter, is that $|S_{21}|$ decreases with frequency at the rate of 6 dB/octave. For these reasons, special consideration must be given to the problem of designing broadband amplifiers. Some of the common approaches to this problem are listed below; note in each case that an improvement in bandwidth is achieved only at the expense of gain, complexity, or similar factors.

- *Compensated matching networks*: Input and output matching sections can be designed to compensate for the gain rolloff in $|S_{21}|$, but generally at the expense of the input and output matching.
- *Resistive matching networks*: Good input and output matching can be obtained by using resistive matching networks, with a corresponding loss in gain and increase in noise figure.
- *Negative feedback*: Negative feedback can be used to flatten the gain response of the transistor, improve the input and output match, and improve the stability of the device. Amplifier bandwidths in excess of a decade are possible with this method, at the expense of gain and noise figure.
- *Balanced amplifiers*: Two amplifiers having 90° couplers at their input and output can provide good matching over an octave bandwidth, or more. The gain is equal to that of a single amplifier, however, and the design requires two transistors and twice the DC power.
- *Distributed amplifiers*: Several transistors are cascaded together along a transmission line, giving good gain, matching, and noise figure over a wide bandwidth. The circuit is large, and does not give as much gain as a cascade amplifier with the same number of stages.
- *Differential amplifiers*: Driving two devices in a differential mode, with input signals of opposite polarity, results in an effective series connection of device capacitance, thus roughly doubling f_T. Differential amplifiers can also provide a larger output voltage swing than a single device, and common mode noise rejection.

Below we discuss in detail the operation of balanced, differential, and distributed amplifiers.

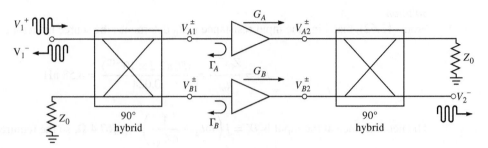

FIGURE 12.11 | A balanced amplifier using 90° hybrid couplers.

Balanced Amplifiers

As we saw in Example 12.4, a fairly flat gain response can be obtained if the amplifier is designed for less than maximum gain, but the input and output matching will be poor. The balanced amplifier circuit solves this problem by using two 90° couplers to cancel input and output reflections from two identical amplifiers. The basic circuit of a balanced amplifier is shown in Figure 12.11. The first 90° hybrid coupler divides the input signal into two equal-amplitude components, with a 90° phase difference, which drive the two amplifiers. The second coupler recombines the amplifier outputs. Because of the phasing properties of the hybrid coupler, reflections from the amplifier inputs cancel at the input to the hybrid, resulting in an improved impedance match; a similar effect occurs at the output of the balanced amplifier. The gain bandwidth is not improved over that of the single amplifier sections. This type of circuit is more complex than a single-stage amplifier since it requires two hybrid couplers and two separate amplifier sections, but it has a number of interesting advantages:

- The individual amplifier stages can be optimized for gain flatness or noise figure, without concern for input and output matching.
- Reflections are absorbed in the coupler terminations, improving input/output matching, as well as the stability of the individual amplifiers.
- The circuit provides a graceful degradation of a 6 dB loss in gain if a single amplifier section fails.
- Bandwidth can be an octave or more, primarily limited by the bandwidth of the couplers.

In practice, balanced MMIC amplifiers often use Lange couplers, which are broadband and very compact, but quadrature hybrids and Wilkinson power dividers (with an extra 90° line on one arm) can also be used.

If we assume ideal hybrid couplers, then, with reference to Figure 12.11, the voltages incident at the amplifiers can be written as

$$V_{A1}^+ = \frac{1}{\sqrt{2}} V_1^+, \tag{12.64a}$$

$$V_{B1}^+ = \frac{-j}{\sqrt{2}} V_1^+, \tag{12.64b}$$

where V_1^+ is the incident input voltage. The output voltage can be found as

$$V_2^- = \frac{-j}{\sqrt{2}} V_{A2}^+ + \frac{1}{\sqrt{2}} V_{B2}^+ = \frac{-j}{\sqrt{2}} G_A V_{A1}^+ + \frac{1}{\sqrt{2}} G_B V_{B1}^+ = \frac{-j}{2} V_1^+ \left(G_A + G_B \right), \tag{12.65}$$

where G_A and G_B are the voltage gains of the amplifiers. Then we can write S_{21} as

$$S_{21} = \frac{V_2^-}{V_1^+} = \frac{-j}{2} \left(G_A + G_B \right), \tag{12.66}$$

which shows that the overall gain of the balanced amplifier is the average of the individual amplifier voltage gains.

The total reflected voltage at the input can be expressed as

$$V_1^- = \frac{1}{\sqrt{2}}V_{A1}^- + \frac{-j}{\sqrt{2}}V_{B1}^- = \frac{1}{\sqrt{2}}\Gamma_A V_{A1}^+ + \frac{-j}{\sqrt{2}}\Gamma_B V_{B1}^+ = \frac{1}{2}V_1^+\left(\Gamma_A - \Gamma_B\right). \tag{12.67}$$

Then we can write S_{11} as

$$S_{11} = \frac{V_1^-}{V_1^+} = \frac{1}{2}\left(\Gamma_A - \Gamma_B\right). \tag{12.68}$$

If the amplifiers are identical, then $G_A = G_B$ and $\Gamma_A = \Gamma_B$, and (12.68) shows that $S_{11} = 0$, and (12.66) shows that the gain of the balanced amplifier will be the same as the gain of an individual amplifier. If one amplifier fails, the overall gain will drop by 6 dB, with the remaining power lost in the coupler terminations. It can also be shown that the noise figure of the balanced amplifier is $F = (F_A + F_B)/2$, where F_A and F_B are the noise figures of the individual amplifiers.

EXAMPLE 12.7 PERFORMANCE AND OPTIMIZATION OF A BALANCED AMPLIFIER

Use the amplifier of Example 12.4 in a balanced configuration operating from 3 to 5 GHz. Use quadrature hybrids, and plot the gain and return loss over this frequency range. Using microwave CAD software, optimize the amplifier matching networks to give 10 dB gain over this band.

Solution
The amplifier of Example 12.4 was designed for a gain of 11 dB at 4 GHz. As seen from Figure 12.8c, the gain varies by a few dB from 3 to 5 GHz, and the return loss is no better than 5 dB. We can design a quadrature hybrid, according to the discussion in Chapter 7, to have a center frequency of 4 GHz. Then the balanced amplifier configuration of Figure 12.11 can be modeled using a microwave CAD package, with the results shown in Figure 12.12. Note the dramatic improvement in return loss over the band as compared with the result for the original amplifier in Figure 12.8c. The input matching is best at 4 GHz since this was the design frequency of the coupler; a coupler with better bandwidth will give improved results at the band edges. Also observe that the gain at 4 GHz is still 11 dB, and that it drops by a few dB at the band edges.

Most modern microwave CAD software packages have an optimization feature with which a small set of design variables can be adjusted to optimize a particular performance variable. In the present example, we will reduce the gain specification to 10 dB, and allow the CAD software to adjust the four transmission line stub and line lengths in the amplifier circuit of Figure 12.8b to give the best fit to this gain over the frequency range 3–5 GHz. Both amplifiers in the balanced circuit remain identical, so we should still see the improved input matching.

The results of this optimization are shown in Figure 12.12, where it can be seen that the gain response is much flatter over the operating band. The input match is still very good in the vicinity of the center

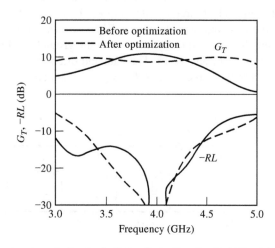

FIGURE 12.12 | Gain and return loss, before and after optimization, for the balanced amplifier of Example 12.7.

frequency, with a slightly worse result at the low-frequency end. The optimized stub and line lengths for the amplifier matching networks are listed below:

Matching Network Parameter	Before Optimization	After Optimization
Input section stub length	0.100 λ	0.109 λ
Input section line length	0.179 λ	0.113 λ
Output section line length	0.045 λ	0.134 λ
Output section stub length	0.432 λ	0.461 λ

These represent fairly small deviations from the lengths in the original matching networks.

Distributed Amplifiers

The concept of the distributed amplifier dates back to the 1940s, when it was used in the design of broadband vacuum tube amplifiers. With recent advances in microwave integrated circuit and device processing technology, the distributed amplifier has found new applications in broadband microwave amplifiers. Bandwidths in excess of a decade are possible, with good input and output matching. Distributed amplifiers are not capable of very high gains or very low noise figure, however, and generally are larger than an amplifier having comparable gain over a narrower bandwidth.

The basic configuration of a microwave distributed amplifier is shown in Figure 12.13. A cascade of N identical FETs have their gates connected to a transmission line having a characteristic impedance Z_g, with a spacing of ℓ_g, while the drains are connected to a transmission line of characteristic impedance Z_d, with a spacing ℓ_d. The operation of the distributed amplifier is very similar to that of the multihole waveguide coupler discussed in Section 7.4. The input signal propagates down the gate line, with each FET tapping off some of the input power. The amplified output signals from the FETs form a traveling wave on the drain line. The propagation constants and lengths of the gate and drain lines are chosen for constructive phasing of the output signals, and the termination impedances on the lines serve to absorb waves traveling in the reverse directions. The gate and drain capacitances of the FET effectively become part of the gate and drain transmission lines, while the gate and drain resistances introduce loss on these lines. This type of circuit is also known as a *traveling wave amplifier*.

Here we will analyze the distributed amplifier in terms of the loaded gate and drain transmission lines [8], although it is also possible to apply the concept of image parameters [9], or to simply model using CAD software. An analytical treatment has the advantage of illustrating the underlying principles of operation of the amplifier, while the numerical CAD approach is recommended for better accuracy and optimization capabilities.

The first step in the analysis of the distributed amplifier is to employ the unilateral ($C_{gd} = 0$) version of the FET equivalent circuit to decompose the circuit of Figure 12.13 into separate loaded transmission lines for the gate and drain terminals. These are shown in Figures 12.14 and 12.15. The gate and drain transmission lines are typically microstrip; the ground conductors are not shown in Figure 12.13, but they are in Figures 12.14 and 12.15. The gate and drain lines are isolated except for the coupling through the dependent current sources, where $I_{dn} = g_m V_{cn}$, and are matched at both ends. Figures 12.14b and 12.15b

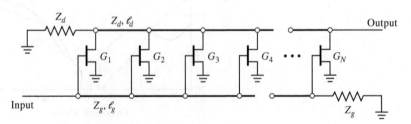

FIGURE 12.13 | Configuration of an N-stage distributed amplifier.

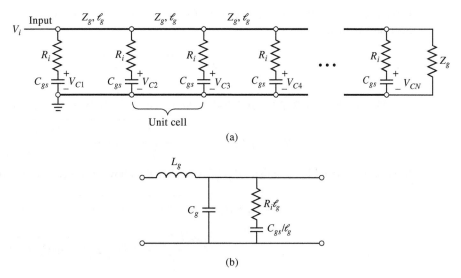

FIGURE 12.14 | (a) Transmission line circuit for the gate line of the distributed amplifier. (b) Equivalent circuit of a single unit cell of the gate line.

show the equivalent circuits for a single unit cell from the gate and drain lines, respectively. L_g and C_g are the inductance and capacitance per unit length of the gate transmission line, while $R_i \ell_g$ and C_{gs}/ℓ_g represent the equivalent per-unit-length loading due to the FET input resistance R_i and gate-to-source capacitance C_{gs}. Similar definitions apply to the quantities $L_d, C_d, R_{ds}\ell_d$, and C_{ds}/ℓ_d for the drain line. Thus we have taken the lumped loading of each FET and distributed its circuit parameters over the transmission lines of each unit cell. This approximation is generally valid when the electrical lengths of the unit cells are small.

We can now use basic transmission line theory to find the effective characteristic impedance and propagation constants of the gate and drain lines. For the gate line, the series impedance and shunt admittance per unit length can be written as

$$Z = j\omega L_g, \tag{12.69a}$$

$$Y = j\omega C_g + \frac{j\omega C_{gs}/\ell_g}{1 + j\omega R_i C_{gs}}. \tag{12.69b}$$

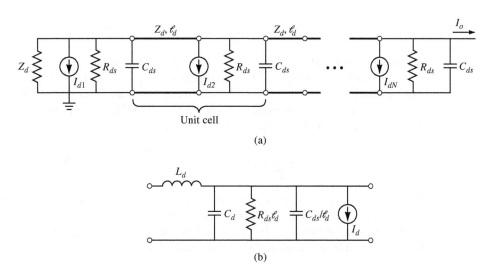

FIGURE 12.15 | (a) Transmission line circuit for the drain line of the distributed amplifier. (b) Equivalent circuit of a single unit cell of the drain line.

If we assume that loss can be neglected for the calculation of characteristic impedance, as discussed in Section 2.6, then we have

$$Z_g = \sqrt{\frac{Z}{Y}} = \sqrt{\frac{L_g}{C_g + C_{gs}/\ell_g}}. \tag{12.70}$$

For the calculation of the propagation constant we retain the resistive term since this will lead to attenuation:

$$\gamma_g = \alpha_g + j\beta_g = \sqrt{ZY} = \sqrt{j\omega L_g\left(j\omega C_g + \frac{j\omega C_{gs}/\ell_g}{1 + j\omega R_i C_{gs}}\right)}.$$

If we assume small loss, so that $\omega R_i C_{gs} \ll 1$, then the above result can be simplified as follows:

$$\gamma_g = \alpha_g + j\beta_g \cong \sqrt{-\omega^2 L_g \left[C_g + C_{gs}\left(1 - j\omega R_i C_{gs}\right)/\ell_g\right]}$$

$$\cong \frac{\omega^2 R_i C_{gs}^2 Z_g}{2\ell_g} + j\omega\sqrt{L_g\left(C_g + C_{gs}/\ell_g\right)}. \tag{12.71}$$

For the drain line, the series impedance and shunt admittance per unit length are

$$Z = j\omega L_d, \tag{12.72a}$$

$$Y = \frac{1}{R_{ds}\ell_d} + j\omega\left(C_d + C_{ds}/\ell_d\right). \tag{12.72b}$$

The characteristic impedance of the drain line can be written as

$$Z_d = \sqrt{\frac{Z}{Y}} = \sqrt{\frac{L_d}{C_d + C_{ds}/\ell_d}}, \tag{12.73}$$

and the propagation constant can be simplified using the small-loss approximation as

$$\gamma_d = \alpha_d + j\beta_d = \sqrt{ZY} = \sqrt{j\omega L_d\left[\frac{1}{R_{ds}\ell_d} + j\omega\left(C_d + C_{ds}/\ell_d\right)\right]}$$

$$\cong \frac{Z_d}{2R_{ds}\ell_d} + j\omega\sqrt{L_d\left(C_d + C_{ds}/\ell_d\right)}. \tag{12.74}$$

For an incident input voltage, V_i, the voltage on the gate-to-source capacitance of the nth FET can be written as

$$V_{cn} = V_i e^{-(n-1)\gamma_g\ell_g}\left(\frac{1}{1 + j\omega R_i C_{gs}}\right), \tag{12.75}$$

for a phase reference at the first transistor. The factor in parentheses in (12.75) accounts for voltage division between R_i and C_{gs}; for typical FET parameters $\omega R_i C_{gs} \ll 1$, so this factor can be approximated as unity over the bandwidth of the amplifier. The output current on the drain line can be found by recognizing that each current generator contributes waves of the form $(-1/2)I_{dn}e^{\pm\gamma_d z}$ in each direction. Since $I_{dn} = g_m V_{cn}$, the total output current at the Nth terminal of the drain line is

$$I_o = -\frac{1}{2}\sum_{n=1}^{n} I_{dn}e^{-(N-n)\gamma_d\ell_d} = -\frac{g_m V_i}{2}e^{-N\gamma_d\ell_d}e^{\gamma_g\ell_g}\sum_{n=1}^{N}e^{-n\left(\gamma_g\ell_g - \gamma_d\ell_d\right)}. \tag{12.76}$$

The terms in the summation will add in phase only when $\beta_g\ell_g = \beta_d\ell_d$, so that the phase delays on the gate and drain lines are synchronized. There is also a backward traveling wave component on the drain line, but the individual contributions to this wave will not be in phase, and therefore they at least partially cancel; the residual will be absorbed in the termination Z_d. Use of the summation formula

$$\sum_{n=1}^{N} x^n = \frac{x^{N+1} - x}{x - 1}$$

allows (12.76) to be simplified as follows:

$$I_o = -\frac{g_m V_i}{2} \frac{e^{\gamma_d \ell_d}\left(e^{-N\gamma_g \ell_g} - e^{-N\gamma_d \ell_d}\right)}{e^{-(\gamma_g \ell_g - \gamma_d \ell_d)} - 1} = -\frac{g_m V_i}{2} \frac{e^{-N\gamma_g \ell_g} - e^{-N\gamma_d \ell_d}}{e^{-\gamma_g \ell_g} - e^{-\gamma_d \ell_d}}. \tag{12.77}$$

For matched input and output ports, the amplifier gain can be calculated as

$$G = \frac{P_{\text{out}}}{P_{\text{in}}} = \frac{\frac{1}{2}|I_o|^2 Z_d}{\frac{1}{2}|V_i|^2/Z_g} = \frac{g_m^2 Z_d Z_g}{4}\left|\frac{e^{-N\gamma_g \ell_g} - e^{-N\gamma_d \ell_d}}{e^{-\gamma_g \ell_g} - e^{-\gamma_d \ell_d}}\right|^2. \tag{12.78}$$

Applying the synchronization condition that $\beta_g \ell_g = \beta_d \ell_d$ allows this result to be further simplified to

$$G = \frac{g_m^2 Z_d Z_g}{4} \frac{\left(e^{-N\alpha_g \ell_g} - e^{-N\alpha_d \ell_d}\right)^2}{\left(e^{-\alpha_g \ell_g} - e^{-\alpha_d \ell_d}\right)^2}. \tag{12.79}$$

If the losses are small, the denominator in (12.79) can be approximated as $(\alpha_g \ell_g - \alpha_d \ell_d)$.

Several interesting aspects of the distributed amplifier can be deduced from the gain expression of (12.79). For the ideal case of a lossless amplifier ($\alpha_g = \alpha_d = 0$), the gain reduces to

$$G = \frac{g_m^2 Z_d Z_g N^2}{4},$$

showing that gain increases as N^2. This is in contrast to the gain of a cascade of N amplifier stages, which increases as $(G_0)^N$. When loss is present, (12.79) shows that the gain of a distributed amplifier approaches zero as $N \to \infty$. This surprising behavior is explained by the fact that the input voltage on the gate line decays exponentially, so the FETs at the end of the amplifier receive no input signal; similarly, the amplified signals from the FETs near the beginning of the amplifier are attenuated along the drain line. The multiplicative increase in gain with N is not enough to compensate for an exponential decay for large N. This implies that, for a given set of transistor parameters, there will be an optimum value of N that maximizes the gain of a distributed amplifier. This can be found by differentiating (12.79) with respect to N and setting the result to zero to obtain

$$N_{\text{opt}} = \frac{\ln\left(\alpha_g \ell_g / \alpha_d \ell_d\right)}{\alpha_g \ell_g - \alpha_d \ell_d}. \tag{12.80}$$

This result depends on frequency, the device parameters, and the line lengths through the attenuation constants given in (12.71) and (12.74).

EXAMPLE 12.8   DISTRIBUTED AMPLIFIER PERFORMANCE

Use (12.79) to calculate the gain of a distributed amplifier from 0 to 20 GHz for $N = 4$, 8, and 16 stages. Assume $Z_d = Z_g = Z_0 = 50\ \Omega$ and the following FET parameters: $R_i = 5\ \Omega$, $R_{ds} = 200\ \Omega$, $C_{gs} = 0.30$ pF, and $g_m = 40$ mS. Find the optimum value of N that will give maximum gain at 16 GHz.

Solution

We use (12.71) and (12.74) to evaluate the attenuation constants α_g and α_d, and then compute the gain versus frequency and N using (12.79). Note that the products $\alpha_g \ell_g$ and $\alpha_d \ell_d$ are independent of ℓ_g and ℓ_d:

$$\alpha_g \ell_g = \frac{\omega^2 R_i C_{gs}^2 Z_0}{2} = 0.1137$$

$$\alpha_d \ell_d = \frac{Z_0}{2R_{ds}} = 0.125$$

The results are shown in Figure 12.16. Observe that the gain drops off with frequency faster for larger N, and that at high frequencies the gain for $N = 16$ is less than the gain for smaller N. The optimum size

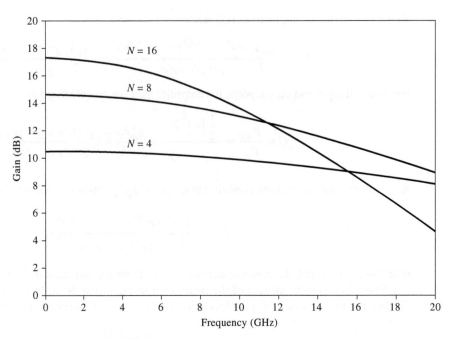

FIGURE 12.16 | Gain versus frequency for the distributed amplifier of Example 12.8.

for maximum gain at 16 GHz can be calculated using (12.80). At 16 GHz we have $\alpha_g \ell_g = 0.1137$ and $\alpha_d \ell_d = 0.125$. The optimum size is then

$$N_{\text{opt}} = \frac{\ln(\alpha_g \ell_g / \alpha_d \ell_d)}{\alpha_g \ell_g - \alpha_d \ell_d} = \frac{\ln(0.1137/0.125)}{0.1137 - 0.125} = 8.4,$$

or about nine stages. Finally, note that $\omega R_i C_{gs} = 0.19$ at 20 GHz, justifying the approximation of unity for the voltage divider factor of (12.75).

Differential Amplifiers

The amplifiers considered above are *single-ended* circuits, meaning that the input and output signals are referenced to a common ground. In contrast, a *differential amplifier* uses balanced inputs and outputs, meaning that there are two signal lines, with opposite polarities, at each port. Figure 12.17 shows the symbols commonly used for single-ended and differential amplifiers. Differential circuits have several advantages over single-ended circuits, including cancellation of interference that is common to both signal lines. Such *common mode interference* is frequently a problem with sensitive receiver circuitry on highly integrated monolithic circuits, and for this reason many of the circuits used in modern RFICs use differential topologies. Another advantage of differential amplifiers is that they can provide output voltage swings that are approximately double that obtained with a single-ended amplifier. A disadvantage of differential circuits is that they use roughly twice the device count as the single-ended equivalents, and more associated bias power.

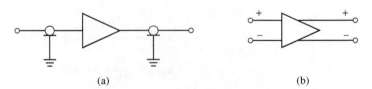

(a)   (b)

FIGURE 12.17 | (a) Single-ended amplifier, with symbols denoting unbalanced input and output lines. (b) Differential amplifier, having balanced input and output lines.

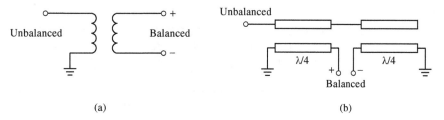

(a)   (b)

FIGURE 12.18 | Balun circuits. (a) A transformer balun. (b) The Marchand balun.

A differential amplifier can be constructed using two single-ended amplifiers and 180° hybrids at the input and output to split and then recombine the signals (similar to the balanced amplifier of Section 12.4 that used 90° hybrids). In this case the initial input and final output signals at the hybrids would be single ended (referenced to ground). Such amplifiers are sometimes referred to as *pseudo-differential* [5], in contrast to fully differential amplifiers, which have balanced input and output signals. In general, a *balun* (balanced-to-unbalanced) circuit is used to transition from an unbalanced signal to a balanced signal (or vice versa). At low frequencies a simple transformer can be used as a balun, as shown in Figure 12.18a. At higher frequencies, a 180° hybrid coupler can be used as a balun, with the unbalanced port at the difference input port of the hybrid, and the two output ports providing the balanced port. Various types of coupled line circuits can also provide a balun function, with one of the most popular being the *Marchand balun*, shown in Figure 12.18b.

Figure 12.19 shows the AC circuit of a differential amplifier using two FETs. The balanced input signal is applied to the gates of the devices, and the balanced output signal is formed across the drains. In practice, an additional transistor is often used at the device sources to provide a current source; this is modeled by the resistor R_s. Usually, the desired input to a differential amplifier consists of equal-amplitude signals with opposite polarities at the two gates, forming an odd-mode excitation. An interference signal, however, will usually appear as equal-amplitude signals with the same polarity at the inputs, forming an even-mode excitation. These modes are also referred to as the *differential mode* and the *common mode*, respectively. We can analyze the differential amplifier by decomposing an arbitrary input into the superposition of an odd mode and an even mode, similar to the analysis we used previously for the quadrature hybrid and other symmetric circuits.

First consider differential (odd) mode excitation, which corresponds to the usual mode of operation for the amplifier. The equivalent circuit is shown in Figure 12.20a, where the unilateral FET model has been used. The input signals in this case are $V_i^+ = V_i$ and $V_i^- = -V_i$; this antisymmetry establishes a zero potential at the midplane of the circuit, so there is a virtual ground at the sources, and the resistor R_s can be removed. The voltages on the capacitors are

$$V_c^\pm = \frac{\pm V_i}{1 + j\omega R_i C_{gs}}, \tag{12.81}$$

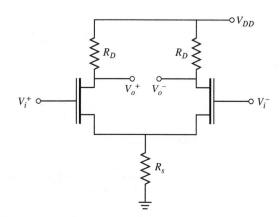

FIGURE 12.19 | Differential amplifier circuit using two FETs. The source resistance R_s may model a current source.

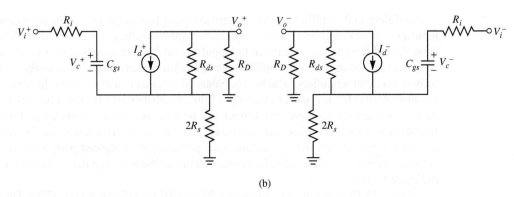

and the output voltages on the drains are

$$V_o^\pm = -I_d^\pm \frac{R_D R_{ds}}{R_D + R_{ds}} = \mp V_i \frac{g_m R_D R_{ds}}{\left(1 + j\omega R_i C_{gs}\right)\left(R_D + R_{ds}\right)}. \tag{12.82}$$

The voltage gain for the differential (odd) mode is then

$$A_d = \frac{V_o^+ - V_o^-}{V_i^+ - V_i^-} = \frac{-g_m R_D R_{ds}}{\left(1 + j\omega R_i C_{ds}\right)\left(R_D + R_{ds}\right)}. \tag{12.83}$$

For the common (even) mode, the input signals are $V_i^+ = V_i^- = V_i$, and the equivalent circuit is as shown in Figure 12.20b. Due to the symmetry of the excitation, no current flows between the sources of the two devices, so the circuit can be bisected as shown, with the original resistor R_s being split into two resistors of $2R_s$. The voltages on the capacitors are

$$V_c^\pm = \frac{V_i}{1 + j\omega C_{gs}\left(R_i + 2R_s\right)}. \tag{12.84}$$

The voltage across either current source is

$$V_I = -I_d \frac{R_{ds}\left(R_D + 2R_s\right)}{R_{ds} + R_D + 2R_s},$$

so the output voltages on the drains are

$$V_o^\pm = V_I \frac{R_D}{R_D + 2R_s} = -I_d \frac{R_{ds} R_D}{R_{ds} + R_D + 2R_s}$$

$$= -V_i \frac{g_m R_{ds} R_D}{\left[1 + j\omega C_{gs}\left(R_i + 2R_s\right)\right]\left(R_{ds} + R_D + 2R_s\right)}. \tag{12.85}$$

The voltage gain for the common (even) mode is then

$$A_c = \frac{V_o^+ + V_o^-}{V_i^+ + V_i^-} = \frac{-g_m R_{ds} R_D}{[1 + j\omega C_{gs}(R_i + 2R_s)](R_{ds} + R_D + 2R_s)}. \tag{12.86}$$

The *common mode rejection ratio* (CMRR) of an amplifier is defined as the ratio of the differential voltage gain to the common mode voltage gain, and is a measure of how well a differential amplifier can provide cancellation of a common mode interference signal. For the differential amplifier considered here, the common mode rejection ratio is

$$\text{CMRR} = \frac{A_d}{A_c} = \frac{(R_{ds} + R_D + 2R_s)}{(R_{ds} + R_D)} \frac{1 + j\omega C_{gs}(R_i + 2R_s)}{1 + j\omega C_{gs} R_i}$$

$$= \left(1 + \frac{2R_s}{R_{ds} + R_D}\right)\left(1 + \frac{2j\omega C_{gs} R_s}{1 + j\omega C_{gs} R_i}\right). \tag{12.87}$$

From this result we see that if $R_s = 0$ we have CMRR $= 1$, which provides no common mode rejection. This is because the two circuits of Figure 12.20 are identical when $R_s = 0$. When $R_s \to \infty$, however (which is the case for an ideal current source feeding the FET sources), we have CMRR $\to \infty$, providing cancellation of the common mode signal.

12.5 POWER AMPLIFIERS

Power amplifiers are used in the final stages of radar and radio transmitters to increase the radiated power level. Typical output powers may be on the order of 100–500 mW for mobile voice or data communication systems, or in the range of 1–100 W for radar or fixed point radio systems. Important considerations for RF and microwave power amplifiers are efficiency, gain, intermodulation distortion, and thermal effects. Single transistors can provide output powers of 10–100 W at UHF frequencies, while devices at higher frequencies are generally limited to output powers less than 10 W. Various power-combining techniques can be used in conjunction with multiple transistors if higher output powers are required.

So far we have considered only *small-signal amplifiers*, where the input signal power is low enough that the transistor can be assumed to operate as a linear device. The scattering parameters of linear devices are well defined and do not depend on the input power level or output load impedance, a fact that greatly simplifies the design of fixed-gain and low-noise amplifiers. For high input powers (e.g., in the range of the 1 dB compression point or third-order intercept point), transistors do not behave linearly. In this case the impedances seen at the input and output of the transistor will depend on the input power level, and this greatly complicates the design of power amplifiers.

Characteristics of Power Amplifiers and Amplifier Classes

The power amplifier is usually the primary consumer of DC power in most hand-held wireless devices, so amplifier efficiency is an important consideration. One measure of amplifier efficiency is the ratio of RF output power to DC input power:

$$\eta = \frac{P_{\text{out}}}{P_{\text{DC}}}. \tag{12.88}$$

This quantity is sometimes referred to as *drain efficiency* (or *collector efficiency*). One drawback of this definition is that it does not account for the RF power delivered at the input to the amplifier. Since most power amplifiers have relatively low gains, the efficiency of (12.88) tends to overrate the actual efficiency. A better measure that includes the effect of input power is the *power added efficiency*, defined as

$$\eta_{PAE} = \text{PAE} = \frac{P_{\text{out}} - P_{\text{in}}}{P_{\text{DC}}} = \left(1 - \frac{1}{G}\right)\frac{P_{\text{out}}}{P_{\text{DC}}} = \left(1 - \frac{1}{G}\right)\eta, \tag{12.89}$$

where G is the power gain of the amplifier. Silicon bipolar junction transistor amplifiers in the cellular telephone band of 800–900 MHz band have power added efficiencies on the order of 80%, but efficiency drops quickly with increasing frequency. Power amplifiers are often designed to provide the best efficiency, even if this means that the resulting gain is less than the maximum possible.

Another useful parameter for power amplifiers is the *compressed gain*, G_1, defined as the gain of the amplifier at the 1 dB compression point. Thus, if G_0 is the small-signal (linear) power gain, we have

$$G_1(\text{dB}) = G_0(\text{dB}) - 1. \tag{12.90}$$

As we have seen in Chapter 10, nonlinearities can lead to the generation of spurious frequencies and intermodulation distortion. This can be a serious issue in wireless transmitters, especially in a multicarrier system, where spurious signals may appear in adjacent channels. Linearity is also critical for nonconstant envelope modulations, such as amplitude shift keying and higher order quadrature amplitude modulation methods.

Class A amplifiers are inherently linear circuits, where the transistor is biased to conduct over the entire range of the input signal cycle. Because of this, class A amplifiers have a theoretical maximum efficiency of 50%. Most small-signal and low-noise amplifiers operate as class A circuits. In contrast, the transistor in a class B amplifier is biased to conduct only during one-half of the input signal cycle. Usually two complementary transistors are operated in a class B push-pull amplifier to provide amplification over the entire cycle. The theoretical efficiency of a class B amplifier is 78%. Class C amplifiers are operated with the transistor near cutoff for more than half of the input signal cycle, and generally use a resonant circuit in the output stage to recover the fundamental. Class C amplifiers can achieve efficiencies near 100% but can only be used with constant envelope modulations. Higher classes, such as class D, E, F, and S, use the transistor as a switch to pump a highly resonant tank circuit, and may achieve very high efficiencies. The majority of communication transmitters operating at UHF frequencies or above rely on class A, AB, or B power amplifiers because of the need for low distortion products.

Large-Signal Characterization of Transistors

A transistor behaves linearly for signal powers well below the 1 dB compression point ($\text{IP}_{1\,\text{dB}}$), and so the small-signal scattering parameters should not depend on either the input power level or the output termination impedance. However, for power levels comparable to or greater than $\text{IP}_{1\,\text{dB}}$, where the nonlinearity of the transistor becomes apparent, the measured scattering parameters will depend on input power level and the output termination impedance (as well as frequency, bias conditions, and temperature). Thus large-signal scattering parameters are not uniquely defined and do not satisfy linearity, and cannot be used in place of small-signal parameters. (For device stability calculations, however, small-signal scattering parameters can generally be used with good results.)

A more useful way to characterize transistors under large-signal operating conditions is to measure the gain and output power as a function of source and load impedances. One way of doing this is to determine the large-signal source and load reflection coefficients, Γ_{SP} and Γ_{LP}, (or impedances, Z_{SP} and Z_{LP}) that maximize power gain for a particular output power (often chosen as $\text{OP}_{1\,\text{dB}}$), and versus frequency. Table 12.1 shows typical large-signal source and load reflection coefficients for an *npn* silicon bipolar power transistor, along with the small-signal scattering parameters.

Another way of characterizing the large-signal behavior of a transistor is to plot contours of constant power output on a Smith chart as a function of the load reflection coefficient, Γ_{LP}, with the transistor conjugately matched at its input. These are called *load-pull contours*, and they can be obtained using an automated measurement set-up with computer-controlled electromechanical stub tuners. A typical set of load-pull

TABLE 12.1 | Small-Signal Scattering Parameters and Large-Signal Reflection Coefficients (Silicon Bipolar Junction Power Transistor)

f(MHz)	S_{11}	S_{12}	S_{21}	S_{22}	Γ_{SP}	Γ_{LP}	G (dB)
800	$0.76 \angle 176°$	$4.10 \angle 76°$	$0.065 \angle 49°$	$0.35 \angle -163°$	$0.856 \angle -167°$	$0.455 \angle 129°$	13.5
900	$0.76 \angle 172°$	$3.42 \angle 72°$	$0.073 \angle 52°$	$0.35 \angle -167°$	$0.747 \angle -177°$	$0.478 \angle 161°$	12.0
1000	$0.76 \angle 169°$	$3.08 \angle 69°$	$0.079 \angle 53°$	$0.36 \angle -169°$	$0.797 \angle -187°$	$0.491 \angle 185°$	10.0

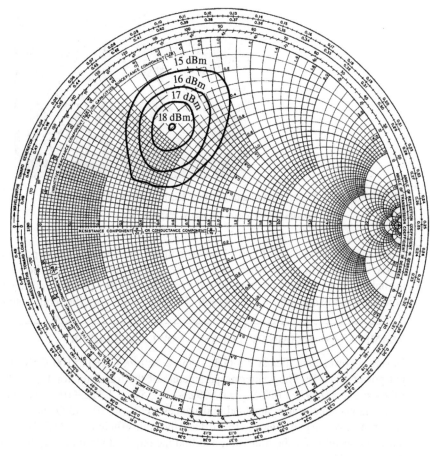

FIGURE 12.21 | Constant–output power contours versus load impedance for a typical power FET.

contours is shown in Figure 12.21. Load-pull contours are similar in function to the constant-gain contours of Section 12.3, but are not perfect circles due to the nonlinearities of the device.

Nonlinear equivalent circuit models can also be developed and used to predict the large-signal performance of FETs and BJTs [10]. The dominant nonlinear parameters for a microwave FET are C_{gs}, g_m, C_{gd}, and R_{ds}. An important consideration in modeling large-signal transistors is the fact that most parameters are dependent on temperature, which of course increases with output power. Equivalent circuit models can be very useful when combined with computer-aided design software.

Design of Class A Power Amplifiers

In this section we will discuss the use of large-signal parameters for the design of class A amplifiers. Since class A amplifiers are ideally linear, it is sometimes possible to use small-signal scattering parameters for design, but better results are usually obtained if large-signal parameters are available. As with small-signal amplifier design, the first step is to check the stability of the device. Since instabilities begin at low signal levels, small-signal scattering parameters can be used for this purpose. Stability is especially important for power amplifiers, as high-power oscillations can easily damage active devices and related circuitry.

The transistor should be chosen on the basis of frequency range and power output, ideally with about 20% more power capacity than is required by the design. Silicon bipolar transistors have higher power outputs than GaAs FETs at frequencies up to a few GHz, and are generally cheaper; GaN HBTs are becoming very popular for high-power applications at RF and low microwave frequencies. Good thermal contact of the transistor package to a heat sink is essential for any amplifier with more than a few tenths of a watt power output. Input matching networks may be designed for maximum power transfer (conjugate matching), while output matching networks are designed for maximum output power (as derived from Γ_{LP}). The optimum values of source and load reflection coefficients are different from those obtained from small-signal

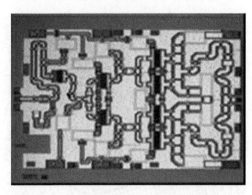

FIGURE 12.22 | Photograph of a three-stage Ku-band GaN MMIC amplifier.
Courtesy of Raytheon Company, Waltham, MA.

scattering parameters via (12.40). Low-loss matching elements are important for good efficiency, particularly in the output stage, where currents are highest. Internally matched chip transistors are sometimes available and have the advantage of reducing the effect of parasitic package reactances, thus improving efficiency and bandwidth. A photograph of a GaN power amplifier chip is shown in Figure 12.22.

EXAMPLE 12.9   DESIGN OF A CLASS A POWER AMPLIFIER

Design a power amplifier at 2.3 GHz using a Nitronex NPT25100 GaN HEMT transistor, with an output power of 10 W. The scattering parameters of the transistor for $V_{DS} = 28$ V and $I_D = 600$ mA are as follows: $S_{11} = 0.593\angle178°$, $S_{12} = 0.009\angle-127°$, $S_{21} = 1.77\angle-106°$, and $S_{22} = 0.958\angle175°$, and the optimum large-signal source and load impedances are $Z_{SP} = 10 - j3$ Ω and $Z_{LP} = 2.5 - j2.3$ Ω. For an output power of 10 W, the power gain is 16.4 dB and the drain efficiency is 26%. Design input and output impedance matching sections for the transistor, and find the required input power, the required DC drain current, and the power added efficiency.

Solution
First establish the stability of the device. Using the small-signal scattering parameters in (12.28) and (12.29) gives

$$|\Delta| = |S_{11}S_{22} - S_{12}S_{21}| = 0.579 < 1,$$

$$K = \frac{1 - |S_{11}|^2 - |S_{22}|^2 + |\Delta|^2}{2|S_{12}S_{21}|} = 2.08 > 1,$$

showing that the device is unconditionally stable.

Converting the large-signal source and load impedances to reflection coefficients gives

$$\Gamma_{SP} = 0.668\angle187°,$$

$$\Gamma_{LP} = 0.905\angle-175°.$$

For comparison, using the small-signal scattering parameters in (12.40) to find the source and load reflection coefficients for conjugate matching gives

$$\Gamma_S = \frac{B_1 \pm \sqrt{B_1^2 - 4|C_1|^2}}{2C_1} = 0.508\angle166°,$$

$$\Gamma_L = \frac{B_2 \pm \sqrt{B_2^2 - 4|C_2|^2}}{2C_2} = 0.954\angle-176°.$$

Note that these values are approximately equal to the large-signal values Γ_{SP} and Γ_{LP}, but not exactly, due to the fact that the scattering parameters used to calculate Γ_S and Γ_L do not apply for large power levels. We should use the large-signal reflection coefficients to design the input and output matching networks. The AC amplifier circuit is shown in Figure 12.23.

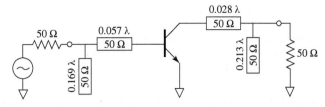

FIGURE 12.23 | RF circuit for the amplifier of Example 12.9.

For an output power of 10 W, the required input drive power is

$$P_{\text{in}} = P_{\text{out}}(\text{dBm}) - G(\text{dB}) = 10\log(10{,}000) - 16.4 = 23.6 \text{ dBm} = 229 \text{ mW}.$$

The DC input power can be found from the drain efficiency as $P_{DC} = P_{\text{out}}/\eta = 38.5$ W, so the DC drain current is $I_D = P_{DC}/V_{DS} = 1.37$ A. The power added efficiency of the amplifier can be found from (12.89) to be

$$\eta_{PAE} = \frac{P_{\text{out}} - P_{\text{in}}}{P_{DC}} = \frac{10.0 - 0.229}{38.5} = 25\%.$$

REFERENCES

[1] G. D. Vendelin, A. M. Pavio, and U. L. Rohde, *Microwave Circuit Design Using Linear and Nonlinear Techniques*, John Wiley & Sons, New York, 1990.

[2] G. Gonzalez, *Microwave Transistor Amplifiers: Analysis and Design*, 2nd edition, Prentice-Hall, Upper Saddle River, NJ, 1997.

[3] R. Ludwig and P. Bretchko, *RF Circuit Design: Theory and Applications*, Prentice-Hall, Upper Saddle River, NJ, 2000.

[4] T. H. Lee, *The Design of CMOS Radio-Frequency Integrated Circuits*, 2nd edition, Cambridge University Press, Cambridge, 2004.

[5] M. Steer, *Microwave and RF Design: A Systems Approach*, SciTech, Raleigh, NC, 2010.

[6] M. Ohtomo, "Proviso on the Unconditional Stability Criteria for Linear Twoports," *IEEE Transactions on Microwave Theory and Techniques*, vol. MTT-43, pp. 1197–1200, May 1995.

[7] M. L. Edwards and J. H. Sinksy, "A New Criteria for Linear 2-Port Stability Using a Single Geometrically Derived Parameter," *IEEE Transactions on Microwave Theory and Techniques*, vol. MTT-40, pp. 2803–2811, December 1992.

[8] Y. Ayasli, R. L. Mozzi, J. L. Vorhous, L. D. Reynolds, and R. A. Pucel, "A Monolithic GaAs 1–13 GHz Traveling-Wave Amplifier," *IEEE Transactions on Microwave Theory and Techniques*, vol. MTT-30, pp. 976–981, July 1982.

[9] J. B. Beyer, S. N. Prasad, R. C. Becker, J. E. Nordman, and G. K. Hohenwarter, "MESFET Distributed Amplifier Design Guidelines," *IEEE Transactions on Microwave Theory and Techniques*, vol. MTT-32, pp. 268–275, March 1984.

[10] W. R. Curtice and M. Ettenberg, "A Nonlinear GaAs FET Model for Use in the Design of Output Circuits for Power Amplifiers," *IEEE Transactions on Microwave Theory and Techniques*, vol. MTT-33, pp. 1383–1394, December 1985.

PROBLEMS

12.1 Consider the microwave network shown below, consisting of a 50 Ω source, a 50 Ω, 3 dB matched attenuator, and a 50 Ω load. (a) Compute the power gain, the available power gain, and the transducer power gain. (b) How do these gains change if the load is changed to 25 Ω? (c) How do these gains change if the source impedance is changed to 25 Ω?

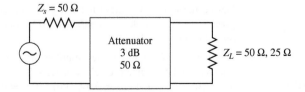

12.2 The Infineon BFP640F SiGe HBT has the following scattering parameters at 900 MHz ($Z_0 = 50$ Ω): $S_{11} = 0.89\angle{-42°}$, $S_{12} = 0.04\angle72°$, $S_{21} = 4.22\angle68°$, and $S_{22} = 0.89\angle{-15°}$. For the transistor in the configuration of Figure 12.1, with no matching networks: (a) Compute the power gain, available power gain, and transducer gain for $Z_S = Z_L = 50$ Ω. (b) Can you find Z_S and Z_L (or Γ_S and Γ_L) to maximize each of these gains (for this case, assume the device is unilateral, with $S_{12} = 0$)?

12.3 An amplifier uses a GaAs HBT-NE52118 device having the following scattering parameters ($Z_0 = 50$ Ω) at 1.5 GHz: $S_{11} = 0.679\angle{-89°}$, $S_{12} = 0.119\angle29°$, $S_{21} = 6.155\angle101.8°$, and $S_{22} = 0.687\angle{-66°}$. The input of the transistor is connected to a source with $V_s = 2$ V (peak) and

$Z_S = 25\ \Omega$, and the output of the transistor is connected to a load of $Z_L = 100\ \Omega$. (a) What are the power gain, the available power gain, the transducer power gain, and the unilateral transducer power gain? (b) Compute the available power from the source, and the power delivered to the load.

12.4 A transistor device has the following scattering parameters at 2 GHz: $S_{11} = 0.894\angle-60.6°$, $S_{12} = 0.02\angle62.4°$, $S_{21} = 3.122\angle123.6°$, and $S_{22} = 0.781\angle-27.6°$. Determine the stability of the device, and plot the stability circles if the device is potentially unstable.

12.5 The scattering parameters of a GaN HEMT device are given at four frequencies in Table 11.8. Use the $K - \Delta$ test to determine the stability of this transistor at each frequency.

12.6 Use the μ-parameter test to determine which of the following devices are unconditionally stable and, of those, which has the greatest stability:

Device	S_{11}	S_{12}	S_{21}	S_{22}
A	$0.36\angle-175°$	$0.06\angle75°$	$4.5\angle85°$	$0.56\angle-25°$
B	$0.86\angle-65°$	$0.3\angle75°$	$5.6\angle85°$	$0.56\angle68°$
C	$0.56\angle-145°$	$0.05\angle65°$	$2.6\angle60°$	$0.66\angle-75°$

12.7 Show that for a unilateral device, where $S_{12} = 0$, the μ-parameter test of (12.30) implies that $|S_{11}| < 1$ and $|S_{22}| < 1$ for unconditional stability.

12.8 Prove that the condition for a positive discriminant in (12.40a), that is, $B_1^2 > 4|C_1|^2$, is equivalent to the condition that $K^2 > 1$.

12.9 Using the scattering parameter data for the GaAs MESFET given in Table 11.7, design an amplifier for maximum gain at 12 GHz. Design matching sections using open-circuited shunt stubs, and compute the gain. Check the stability of the resulting design using CAD modeling over the frequency range of 1–15 GHz, or by using stability circles at a few frequencies.

12.10 Consider the impedance matching network shown at left below, where a load $Z_L(\Gamma_L)$ at port 2 is matched to a source impedance Z_0 at port 1. Show that the same network will present the impedance $Z = Z_L^*$ at port 2 when port 1 is terminated with Z_0, as shown in the figure below at right. Assume the matching network is reciprocal and lossless. This relationship allows the impedance tuning techniques of Chapter 5 to be used to design the input and output matching networks for an amplifier.

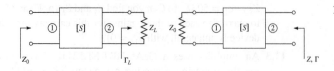

12.11 Design an amplifier with maximum G_{TU} using a transistor with the following scattering parameters ($Z_0 = 50\ \Omega$) at

5.0 GHz: $S_{11} = 0.504\angle-125°$, $S_{12} = 0$, $S_{21} = 4.499\angle72°$, and $S_{22} = 0.471\angle-87.6°$. Design L-section matching sections using lumped elements.

12.12 Design an amplifier to have a gain of 15 dB at 5.0 GHz, using a transistor with the following scattering parameters ($Z_0 = 50\ \Omega$): $S_{11} = 0.448\angle-153.3°$, $S_{12} = 0$, $S_{21} = 2.49\angle7.1°$, and $S_{22} = 0.25\angle-154°$. Plot (and use) constant-gain circles for $G_S = 1$ dB and $G_L = 2$ dB. Use matching sections with open-circuited shunt stubs.

12.13 Compute the unilateral figure of merit for the transistor of Problem 12.4. What is the maximum error in the transducer gain if an amplifier is designed assuming the device is unilateral?

12.14 Show that the 0 dB gain circle for G_S ($G_S = 1$), defined by (12.51), will pass through the center of the Smith chart.

12.15 A GaAs FET-NE52118 has the following scattering and noise parameters at 8 GHz ($Z_0 = 50\ \Omega$), $I_c = 5$ mA: $S_{11} = 0.605\angle-93.5°$, $S_{12} = 0.175\angle51°$, $S_{21} = 1.69\angle-57.6°$, $S_{22} = 0.344\angle-109.3°$, $F_{min} = 2.6$ dB, $\Gamma_{opt} = 0.74\angle124°$, and $R_N = 15\ \Omega$. Design an amplifier with minimum noise figure and maximum possible gain. Use open-circuited shunt stubs in the matching sections.

12.16 A GaAs FET has the following scattering and noise parameters at 6 GHz ($Z_0 = 50\ \Omega$): $S_{11} = 0.484\angle-60°$, $S_{12} = 0$, $S_{21} = 2.104\angle-15.7°$, $S_{22} = 0.247\angle-60°$, $F_{min} = 2.0$ dB, $\Gamma_{opt} = 0.64\angle100°$, and $R_N = 25\ \Omega$. Design an amplifier to have a gain of 6 dB and the minimum noise figure possible with this gain. Use open-circuited shunt stubs in the matching sections.

12.17 Repeat the analysis of the balanced amplifier of Example 12.7 using a 3 dB coupled line hybrid coupler. Use CAD software to optimize the input and output matching networks of the amplifiers to obtain a flat 10 dB gain response from 3 to 5 GHz, and compare the results with those obtained using the quadrature hybrid.

12.18 If the individual amplifier stages in a balanced amplifier have mismatches of Γ_A and Γ_B at their output ports, show that the output mismatch of the balanced amplifier is $S_{22} = (\Gamma_A - \Gamma_B)/2$.

12.19 Derive the result for the optimum size of a distributed amplifier given in (12.80).

12.20 Consider a distributed amplifier using GaAs MESFETs with the following parameters: $R_i = 5\ \Omega$, $R_{ds} = 250\ \Omega$, $C_{gs} = 0.3$ pF, and $g_m = 30$ mS. Calculate and plot the gain from 1 to 18 GHz for $N = 2, 4, 8$, and 16 sections. Find the optimum value of N that will give maximum gain at 16 GHz. Assume $Z_d = Z_g = Z_0 = 50\ \Omega$.

12.21 Use the transistor data given in Table 12.1 to design a power amplifier at 900 MHz with a power output of 1 W. Design the input and output matching circuits using the given large-signal reflection coefficients. Compute the required input power level.

12.22 Consider a 1 GHz low-noise amplifier with the S parameters and Γ_{opt} given as: $S_{11} = 0.7\angle-30°$, $S_{12} = 0.02$, $S_{21} = 10.1\angle60°$, and $S_{22} = 0.6\angle-110°$, $\Gamma_{opt} = 0.5\angle20°$.

The source and load impedances are 30 Ω and 40 Ω, respectively. Comment on the stability of the problem. Compute the power gain, the available power gain, and the transducer power gain.

12.23 A silicon transistor has following S-matrix at 5 GHz with a 50 Ω reference impedance,

$$[S] = \begin{bmatrix} 0.63\angle 115° & 0.138\angle 54° \\ 1.5\angle 15° & 0.44\angle -73° \end{bmatrix}$$

The source and load impedances are 25 Ω and 40 Ω, respectively. Compute the unilateral power gain, unilateral available power gain, and unilateral transducer power gain.

12.24 Determine the length of short-circuited shunt stubs (ℓ_1 and ℓ_4) and the length of the transmission line (ℓ_2 and ℓ_3) at 3 GHz for the network shown below. The S-matrix of the network is.

$$[S] = \begin{bmatrix} 0.63\angle 145° & 0.03\angle 64° \\ 2.4\angle 75° & 0.74\angle -68° \end{bmatrix}$$

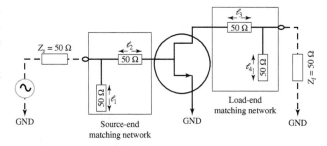

12.25 The Triquint T1G6000528 GaN HEMT has the following scattering parameters at 5 GHz ($Z_0 = 50$ Ω):

$$S_{11} = 0.836\angle 144.5°$$
$$S_{12} = 0.007\angle 164.1°$$
$$S_{21} = 2.82\angle 143.3°$$
$$S_{22} = 0.57\angle -165.89°$$

Determine the stability of this transistor by using the $K - \Delta$ test and the μ-test, and plot the stability circles on a Smith chart.

MULTIPLE CHOICE QUESTIONS

12.1 The voltage reflection coefficient for a two-port network seen looking towards the load, Γ_S, is

(a) $\dfrac{Z_S - Z_0}{Z_S + Z_0}$

(b) $\dfrac{Z_S + Z_0}{Z_S - Z_0}$

(c) $\dfrac{2Z_S}{Z_S + Z_0}$

(d) $\dfrac{2Z_0}{Z_S - Z_0}$

12.2 The S parameter that is zero for a unilateral transistor is

(a) S_{22}
(b) S_{21}
(c) S_{12}
(d) S_{11}

12.3 When both output and input of an amplifier are matched to zero reflection (in contrast to conjugate matching), the transducer power gain is

(a) $|S_{22}|^2$
(b) $|S_{21}|^2$
(c) $|S_{12}|^2$
(d) $|S_{11}|^2$

12.4 For conditional stability, a network should satisfy which of the following:

(a) $|\Gamma_{in}| > 1, |\Gamma_{out}| > 1$
(b) $|\Gamma_{in}| < 1, |\Gamma_{out}| > 1$
(c) $|\Gamma_{in}| > 1, |\Gamma_{out}| < 1$
(d) $|\Gamma_{in}| < 1, |\Gamma_{out}| < 1$

12.5 In terms of S parameters for a unilateral device, condition for unconditional stability is

(a) $|S_{11}| < 1, |S_{22}| < 1$
(b) $|S_{11}| > 1, |S_{22}| > 1$
(c) $|S_{11}| > 1, |S_{22}| < 1$
(d) $|S_{11}| < 1, |S_{22}| > 1$

12.6 If the S parameters of a transistor are given as below

$$S_{11} = 0.869\angle -159°$$
$$S_{12} = 0.031\angle -9°$$
$$S_{21} = 4.25\angle 61°$$
$$S_{22} = 0.507\angle -117°$$

Then Δ for the given transistor is

(a) 0.225
(b) 0.325
(c) 0.336
(d) 0.456

12.7 For an amplifier the frequency response is

(a) passband
(b) narrowband
(c) wideband
(d) none of the above

12.8 From the input matching port to the transistor, the maximum power transfer will occur on which of the following conditions:

(a) $\Gamma_{in} = \Gamma_S^*$
(b) $\Gamma_{in} = \Gamma_S$
(c) $\Gamma_{in} = \Gamma_S e^{-j\omega}$
(d) none of the above

12.9 For maximum gain, unconditionally stable devices can always be

(a) forward biased
(b) conjugate matched
(c) lossless matched
(d) driven with high current

12.10 Using BJT amplifiers, high gain is not achievable at microwave frequencies because

(a) at high frequencies ports are not matched
(b) complex architecture
(c) device construction
(d) none of the above

12.11 The gain response of a transistor can be flattened by

(a) increasing the biasing current
(b) increasing the input signal level
(c) increasing the operational bandwidth
(d) giving negative feedback to the amplifier

12.12 A flat gain response is achieved at the cost of reduced gain in conventional amplifiers. But this drawback can be overcome by using

(a) balanced amplifiers
(b) distributed amplifiers
(c) differential amplifiers
(d) none of the above

12.13 To achieve the required performance, coupler that is mostly used in balanced amplifiers is

(a) Wilkinson coupler   (c) lange coupler
(b) branch-line coupler   (d) waveguide coupler

12.14 Distributed amplifiers offer very high

(a) gain   (c) attenuation
(b) intensity   (d) bandwidth

12.15 Which of the following amplifiers uses balanced input and output, meaning that there are two signal lines, with opposite polarity at each port?

(a) differential amplifier   (c) distributed amplifier
(b) balanced amplifier   (d) none of the above

12.16 For differential amplifiers the major advantage is

(a) low input impedance
(b) high gain
(c) higher output voltage swing
(d) none of the above

12.17 With increase in operating frequency, the gain of power amplifiers

(a) increases   (c) increases exponentially
(b) decreases   (d) decreases exponentially

12.18 In order to conduct the input signal over the entire cycle, a class B amplifier consists of _____ transistors.

(a) 4   (c) 2
(b) 3   (d) 1

12.19 If the input power supplied to an amplifier is 0.15 V and the output power of the amplifier is 12 V, given that the DC voltage used is 33.8 V, determine the efficiency of the power amplifier.

(a) 25%   (c) 50%
(b) 35%   (d) 75%

12.20 If a power amplifier has a gain of 15.8 dB and an output power of 12 W, then the input drive power is

(a) 631.3 mW   (c) 315.6 mW
(b) 250.8 mW   (d) 236.6 mW

ANSWER KEY

12.1 (a)	**12.5** (a)	**12.9** (b)	**12.13** (c)	**12.17** (b)
12.2 (c)	**12.6** (c)	**12.10** (a)	**12.14** (d)	**12.18** (c)
12.3 (b)	**12.7** (b)	**12.11** (d)	**12.15** (a)	**12.19** (b)
12.4 (d)	**12.8** (a)	**12.12** (a)	**12.16** (c)	**12.20** (c)

Learning Objectives

After completing this chapter, you will be able to

- Understand the concepts of RF and microwave oscillators
- Learn about the representation and modeling of the oscillator phase noise
- Understand the concept of frequency multipliers
- Learn about the design of different types of mixers

RF and microwave oscillators are found in all modern wireless communication, radar, and remote sensing systems to provide signal sources for frequency conversion and carrier generation. A solid-state oscillator uses an active nonlinear device, such as a diode or transistor, in conjunction with a passive circuit to convert DC to a sinusoidal steady-state RF signal. Basic transistor oscillator circuits can generally be used at low frequencies, often with crystal resonators to provide improved frequency stability and low noise performance. At higher frequencies, diodes or transistors biased to a negative resistance operating point can be used with cavity, transmission line, or dielectric resonators to produce fundamental frequency oscillations up to 100 GHz. Alternatively, frequency multipliers, in conjunction with a lower frequency source, can be used to produce power at millimeter wave frequencies. Because of the requirement of a nonlinear active device, the rigorous analysis and design of oscillator circuits can be difficult, and is usually carried out today with sophisticated CAD tools.

In this chapter we begin with an overview of low-frequency transistor oscillator circuits, including the well-known Hartley and Colpitts configurations, as well as crystal-controlled oscillators. Next we consider oscillators for use at microwave frequencies, which differ from their lower frequency counterparts primarily due to different transistor characteristics and the ability to make practical use of negative resistance devices and high-Q microwave resonators. We also discuss the important topic of oscillator phase noise. Finally, an introduction to frequency multiplication techniques is given. A related topic is that of frequency conversion, or *mixing*, so we also discuss in this chapter the fundamental operations of frequency up-conversion and down-conversion. Detectors and single-ended mixers using both diodes and transistors are discussed, along with some specialized mixer circuits.

Important considerations for oscillators used in RF and microwave systems include the following:

- Tuning range (specified in MHz/V for voltage-tuned oscillators)
- Frequency stability (specified in PPM/°C)
- AM and FM noise (specified in dBc/Hz below carrier, offset from carrier)
- Harmonics (specified in dBc below carrier)

Typical frequency stability requirements can range from 2 to 0.5 PPM/°C, while phase noise requirements may range from −80 to −110 dBc/Hz at a 10 kHz offset from the carrier.

Transistor oscillators generally have lower frequency and power capabilities than diode sources (e.g., tunnel, Gunn, or IMPATT diodes), but offer several advantages over diodes. First, oscillators using transistors are readily compatible with monolithic integrated circuitry, allowing easy integration with transistor amplifiers and mixers, while diode devices are often less compatible. In addition, a transistor oscillator circuit is much more flexible than a diode source. This is because the negative resistance oscillation mechanism of a diode is determined and limited by the physical characteristics of the device itself, while the operating characteristics of a transistor can be adjusted to a greater degree by the bias point, as well as the source or load impedances presented to the device. Transistor oscillators usually allow more control of the frequency of oscillation, temperature stability, and output noise than do diode sources. Transistor oscillator circuits also lend themselves well to frequency tuning, phase or injection locking, and various modulation requirements. Transistor sources are relatively efficient but usually are not capable of very high power outputs.

Tunable sources are necessary in many types of electronic warfare systems, frequency-hopping radar and communication systems, and test systems. Transistor oscillators can be made tunable by using an adjustable element in the resonant load, such as a varactor diode or a magnetically biased YIG sphere. Thus, a *voltage-controlled oscillator* (VCO) can be made by using a reverse-biased varactor diode in the tank circuit of a transistor oscillator. In a YIG-tuned oscillator (YTO), a single-crystal YIG sphere is used to control the inductance of a coil in the tank circuit of the oscillator. Since YIG is a ferrimagnetic material, its effective permeability can be controlled with an external DC magnetic bias field, thus controlling the oscillator frequency. YIG oscillators can be made to tune over a decade or more of bandwidth, while varactor-tuned oscillators are limited to a tuning range of about an octave. YIG-tuned oscillators, however, cannot be tuned as fast as varactor oscillators.

13.1 RF OSCILLATORS

In the most general sense, an oscillator is a nonlinear circuit that converts DC power to an AC waveform. Most RF oscillators provide sinusoidal outputs, which minimizes undesired harmonics and noise sidebands. The basic conceptual operation of a sinusoidal oscillator can be described with the linear feedback circuit shown in Figure 13.1. An amplifier with voltage gain A has an output voltage V_o. This voltage passes through a feedback network with a frequency-dependent transfer function $H(\omega)$, and is added to the input V_i of the circuit. The output voltage can be expressed as

$$V_o(\omega) = AV_i(\omega) + H(\omega)AV_o(\omega), \tag{13.1}$$

which can be solved to yield the output voltage in terms of the input voltage as

$$V_o(\omega) = \frac{A}{1 - AH(\omega)}V_i(\omega). \tag{13.2}$$

If the denominator of (13.2) becomes zero at a particular frequency, it is possible to achieve a nonzero output voltage for a zero input voltage, thus forming an oscillator. This is known as the *Nyquist criterion*, or the *Barkhausen criterion*. In contrast to the design of an amplifier, where we design to achieve at least conditional stability, oscillator design depends on an unstable circuit.

The oscillator circuit of Figure 13.1 is useful conceptually but provides little helpful information for the design of practical transistor oscillators. Thus we consider next a general analysis of transistor oscillator circuits.

General Analysis

There are a large number of possible RF oscillator circuits using bipolar or field effect transistors in either common emitter/source, base/gate, or collector/drain configurations. Various types of feedback networks lead to the well-known *Hartley, Colpitts, Clapp*, and *Pierce* oscillator circuits [1–3]. All of these variations can be represented by the general oscillator circuit shown in Figure 13.2.

The equivalent circuit on the right-hand side of Figure 13.2 is used to model either a bipolar or a field effect transistor. We have assumed here a unilateral transistor, which is usually a good approximation in practice. We can simplify the analysis by assuming real input and output admittances of the transistor, defined as G_i and G_o, respectively, with a transistor transconductance g_m. The feedback network on the left side of the circuit is formed from three admittances in a bridged-T configuration. These components

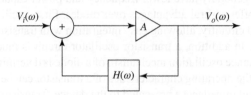

FIGURE 13.1 | Block diagram of a sinusoidal oscillator using an amplifier with a frequency-dependent feedback path.

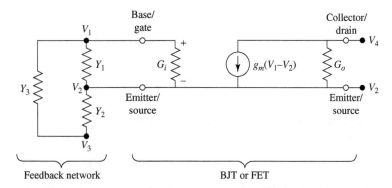

FIGURE 13.2 | General circuit for a transistor oscillator. The transistor may be either a bipolar junction transistor or a field effect transistor. This circuit can be used for common emitter/source, base/gate, or collector/drain configurations by grounding V_2, V_1, or V_4, respectively. Feedback is provided by connecting node V_3 to V_4.

are usually reactive elements (capacitors or inductors) in order to provide a frequency-selective transfer function with high Q. A common emitter/source configuration can be obtained by setting $V_2 = 0$, while common base/gate or common collector/drain configurations can be modeled by setting either $V_1 = 0$ or $V_4 = 0$, respectively. As shown, the circuit of Figure 13.2 does not include a completed feedback path—this can be achieved by connecting node V_3 to node V_4.

Writing Kirchhoff's equation for the four voltage nodes of the circuit of Figure 13.2 gives the following matrix equation:

$$
\begin{bmatrix}
(Y_1 + Y_3 + G_i) & -(Y_1 + G_i) & -Y_3 & 0 \\
-(Y_1 + G_i + g_m) & (Y_1 + Y_2 + G_i + G_o + g_m) & -Y_2 & -G_o \\
-Y_3 & -Y_2 & (Y_2 + Y_3) & 0 \\
g_m & -(G_o + g_m) & 0 & G_o
\end{bmatrix}
\begin{bmatrix}
V_1 \\
V_2 \\
V_3 \\
V_4
\end{bmatrix} = 0 \tag{13.3}
$$

Recall from circuit analysis that if the ith node of the circuit is grounded, so that $V_i = 0$, the matrix of (13.3) will be modified by eliminating the ith row and column, reducing the order of the matrix by one. In addition, if two nodes are connected together, the matrix is modified by adding the corresponding rows and columns.

Oscillators Using a Common Emitter BJT

As a specific example, consider an oscillator using a bipolar junction transistor in a common emitter configuration. In this case we have $V_2 = 0$, with feedback provided from the collector, so that $V_3 = V_4$. In addition, the output admittance of the transistor is negligible, so we set $G_o = 0$. These conditions serve to reduce the matrix of (13.3) to the following:

$$
\begin{bmatrix}
(Y_1 + Y_3 + G_i) & -Y_3 \\
(g_m - Y_3) & (Y_2 + Y_3)
\end{bmatrix}
\begin{bmatrix}
V_1 \\
V
\end{bmatrix} = 0, \tag{13.4}
$$

where $V = V_3 = V_4$.

If the circuit is to operate as an oscillator, then (13.4) must be satisfied for nonzero values of V_1 and V, so the determinant of the matrix must be zero. If the feedback network consists only of lossless capacitors and inductors, then Y_1, Y_2, and Y_3 must be imaginary, so we let $Y_1 = jB_1$, $Y_2 = jB_2$, and $Y_3 = jB_3$. Also recall that the transconductance g_m, and transistor input conductance G_i, are real. The determinant of (13.4) then simplifies to

$$
\begin{vmatrix}
G_i + j(B_1 + B_3) & -jB_3 \\
g_m - jB_3 & j(B_2 + B_3)
\end{vmatrix} = 0. \tag{13.5}
$$

Separately equating the real and imaginary parts of the determinant to zero gives two equations:

$$\frac{1}{B_1} + \frac{1}{B_2} + \frac{1}{B_3} = 0,$$ (13.6a)

$$\frac{1}{B_3} + \left(1 + \frac{g_m}{G_i}\right)\frac{1}{B_2} = 0.$$ (13.6b)

If we convert susceptances to reactances, and let $X_1 = 1/B_1$, $X_2 = 1/B_2$, and $X_3 = 1/B_3$, then we can write (13.6a) as

$$X_1 + X_2 + X_3 = 0.$$ (13.7a)

Using (13.6a) to eliminate B_3 from (13.6b) reduces that equation to the following:

$$X_1 = \frac{g_m}{G_i}X_2.$$ (13.7b)

Since g_m and G_i are positive, (13.7b) implies that X_1 and X_2 have the same sign, and therefore are either both capacitors or both inductors. Equation (13.7a) then shows that X_3 must be opposite in sign from X_1 and X_2, and therefore the opposite type of reactive component. This conclusion leads to two of the most commonly used oscillator circuits.

If X_1 and X_2 are capacitors and X_3 is an inductor, we have a *Colpitts* oscillator. Let $X_1 = -1/\omega_0 C_1$, $X_2 = -1/\omega_0 C_2$, and $X_3 = \omega_0 L_3$. Then (13.7a) becomes

$$\frac{-1}{\omega_0}\left(\frac{1}{C_1} + \frac{1}{C_2}\right) + \omega_0 L_3 = 0,$$

which can be solved for the frequency of oscillation, ω_0, as

$$\omega_0 = \sqrt{\frac{1}{L_3}\left(\frac{C_1 + C_2}{C_1 C_2}\right)}.$$ (13.8)

Using these same substitutions in (13.7b) gives a necessary condition for oscillation of the Colpitts circuit as

$$\frac{C_2}{C_1} = \frac{g_m}{G_i}.$$ (13.9)

The resulting common emitter Colpitts oscillator circuit is shown in Figure 13.3a.

Alternatively, if we choose X_1 and X_2 to be inductors and X_3 to be a capacitor, then we have a *Hartley* oscillator. Let $X_1 = \omega_0 L_1$, $X_2 = \omega_0 L_2$, and $X_3 = -1/\omega_0 C_3$. Then (13.7a) becomes

$$\omega_0(L_1 + L_2) - \frac{1}{\omega_0 C_3} = 0,$$

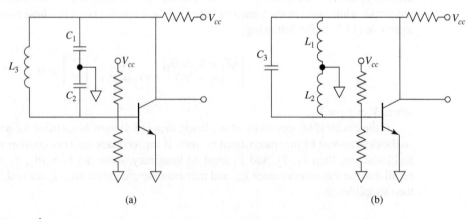

(a)   (b)

FIGURE 13.3 | Transistor oscillator circuits using a common emitter BJT. (a) Colpitts oscillator. (b) Hartley oscillator.

which can be solved for ω_0 to give

$$\omega_0 = \sqrt{\frac{1}{C_3(L_1 + L_2)}}. \tag{13.10}$$

These same substitutions used in (13.7b) gives a necessary condition for oscillation of the Hartley circuit as

$$\frac{L_1}{L_2} = \frac{g_m}{G_i}. \tag{13.11}$$

The resulting common emitter Hartley oscillator circuit is shown in Figure 13.3b.

Oscillators Using a Common Gate FET

Next consider an oscillator using a FET in a common gate configuration. In this case $V_1 = 0$, and again $V_3 = V_4$ provides the feedback path. For a FET the input admittance can be neglected, so we set $G_i = 0$. Then the matrix of (13.3) reduces to

$$\begin{bmatrix} (Y_1 + Y_2 + g_m + G_o) & -(Y_2 + G_o) \\ -(G_o + g_m + Y_2) & (Y_2 + Y_3 + G_o) \end{bmatrix} \begin{bmatrix} V_2 \\ V \end{bmatrix} = 0, \tag{13.12}$$

where $V = V_3 = V_4$. Again we assume the feedback network is composed of lossless reactive elements, so that Y_1, Y_2, and Y_3 can be replaced with their susceptances. Setting the determinant of (13.12) to zero then gives

$$\begin{vmatrix} (g_m + G_o) + j(B_1 + B_2) & -G_o - jB_2 \\ -(G_o + g_m) - jB_2 & G_o + j(B_2 + B_3) \end{vmatrix} = 0. \tag{13.13}$$

Equating the real and imaginary parts to zero gives two equations:

$$\frac{1}{B_1} + \frac{1}{B_2} + \frac{1}{B_3} = 0, \tag{13.14a}$$

$$\frac{G_o}{B_3} + \frac{g_m}{B_1} + \frac{G_o}{B_1} = 0. \tag{13.14b}$$

As before, let X_1, X_2, and X_3 be the reciprocals of the corresponding susceptances. Then (13.14a) can be rewritten as

$$X_1 + X_2 + X_3 = 0. \tag{13.15a}$$

Using (13.14a) to eliminate B_3 from (13.14b) reduces that equation to

$$\frac{X_2}{X_1} = \frac{g_m}{G_o}. \tag{13.15b}$$

Since g_m and G_o are positive, (13.15b) shows that X_1 and X_2 must have the same sign, while (13.15a) indicates that X_3 must have the opposite sign. If X_1 and X_2 are chosen to be negative, then these elements will be capacitive and X_3 will be inductive. This corresponds to a Colpitts oscillator. Since (13.15a) is identical to (13.7a), its solution gives the result for the resonant frequency for the common gate Colpitts oscillator as

$$\omega_0 = \sqrt{\frac{1}{L_3}\left(\frac{C_1 + C_2}{C_1 C_2}\right)}, \tag{13.16}$$

which is identical to the result obtained in (13.8) for the common emitter Colpitts oscillator. This is because the resonant frequency is determined by the feedback network, which is identical in both cases. The further condition for oscillation given by (13.15b) reduces to

$$\frac{C_1}{C_2} = \frac{g_m}{G_o}. \tag{13.17}$$

If we choose X_1 and X_2 to be positive (inductive), then X_3 will be capacitive, and we have a Hartley oscillator. The resonant frequency of the common gate Hartley oscillator is given by

$$\omega_0 = \sqrt{\frac{1}{C_3(L_1 + L_2)}}, \tag{13.18}$$

which is identical to the result of (13.10) for the common emitter Hartley oscillator. Equation (13.15b) reduces to

$$\frac{L_2}{L_1} = \frac{g_m}{G_o}. \tag{13.19}$$

The circuits for common gate Colpitts and Hartley oscillators are similar to the circuits shown in Figure 13.3 if the BJT is replaced with a FET device.

Practical Considerations

It must be emphasized that the above analysis is based on very idealized assumptions, and in practice successful oscillator design requires attention to factors such as the reactances associated with the input and output transistor ports, the variation of transistor properties with temperature, transistor bias and decoupling circuitry, and the effect of inductor losses. For these purposes computer-aided design software can be very helpful.

The above analysis can be extended to account for more realistic feedback network inductors having series resistance, which invariably occurs in practice. For example, consider the case of a common emitter BJT Colpitts oscillator, with the impedance of the inductor given by $Z_3 = 1/Y_3 = R + j\omega L_3$. Substituting into (13.4) and setting the real and imaginary parts of the determinant to zero gives the following result for resonant frequency:

$$\omega_0 = \sqrt{\frac{1}{L_3}\left(\frac{1}{C_1} + \frac{1}{C_2} + \frac{G_i R}{C_1}\right)} = \sqrt{\frac{1}{L_3}\left(\frac{1}{C_1'} + \frac{1}{C_2}\right)}. \tag{13.20}$$

This equation is similar to the result of (13.8) for the lossless inductor, except that C_1' is defined as

$$C_1' = \frac{C_1}{1 + RG_i}. \tag{13.21}$$

The corresponding condition for oscillation is

$$\frac{R}{G_i} = \frac{1 + g_m/G_i}{\omega_0^2 C_1 C_2} - \frac{L_3}{C_1}. \tag{13.22}$$

This result sets the maximum value of the series resistance R; the left-hand side of (13.22) should generally be chosen to be less than the right-hand side to ensure oscillation.

EXAMPLE 13.1   COLPITTS OSCILLATOR DESIGN

Design a 200 MHz Colpitts oscillator using a bipolar junction transistor in a common emitter configuration with $\beta = g_m/G_i = 30$, and a transistor input resistance of $R_i = 1/G_i = 1200\ \Omega$. Use an inductor with $L_3 = 15$ nH and an unloaded Q of 50. What is the minimum Q of the inductor for which oscillation will be sustained?

Solution
From (13.20) the series combination of C_1' and C_2 is found to be

$$\frac{C_1' C_2}{C_1' + C_2} = \frac{1}{\omega_0^2 L_3} = \frac{1}{(2\pi)^2 (200 \times 10^6)^2 (15 \times 10^{-9})} = 42.2 \text{ pF}.$$

This value can be obtained in several ways, but here we will choose $C_1' = C_2 = 84.4$ pF.

From Chapter 6 we know that the unloaded Q of an inductor is related to its series resistance by $Q_0 = \omega L/R$, so the series resistance of the 15 nH inductor is

$$R = \frac{\omega_0 L_3}{Q_0} = \frac{(2\pi)(200 \times 10^6)(15 \times 10^{-9})}{50} = 0.38 \ \Omega.$$

Then (13.21) gives C_1 as

$$C_1 = C_1'(1 + RG_i) = (84.4 \times 10^{-12})\left(1 + \frac{0.38}{1200}\right) = 84.4 \text{ pF},$$

which we see is essentially unchanged from the value found by neglecting the inductor loss. Using (13.22) with the above values gives

$$\frac{R}{G_i} = \frac{1 + \beta}{\omega_0^2 C_1 C_2} - \frac{L_3}{C_1}$$

$$(0.38)(1200) < \frac{1 + 30}{(2\pi)^2(200 \times 10^6)^2(84.4 \times 10^{-12})^2} - \frac{15 \times 10^{-9}}{84.4 \times 10^{-12}}$$

$$456 < 2755.85 - 177.73 = 2578$$

which indicates that the condition for oscillation will be satisfied. This condition can be used to find the minimum unloaded inductor Q by first solving for the maximum value of series resistance R:

$$R_{\text{max}} = \frac{1}{R_i}\left(\frac{1 + \beta}{\omega_0^2 C_1 C_2} - \frac{L_3}{C_1}\right) = \frac{2578}{1200} = 2.15 \ \Omega.$$

So the minimum unloaded Q is

$$Q_{\text{min}} = \frac{\omega_0 L_3}{R_{\text{max}}} = \frac{(2\pi)(200 \times 10^6)(15 \times 10^{-9})}{2.15} = 8.8.$$

Crystal Oscillators

As we have seen from the above analysis, the resonant frequency of an oscillator is determined from the condition that a 180° phase shift occurs between the input and output of the transistor. If the resonant feedback circuit has a high Q, so that there is a very rapid change in the phase shift with frequency, the oscillator will have good frequency stability. Quartz crystals are useful for this purpose, especially at frequencies below a few hundred MHz, where LC resonators seldom have unloaded Qs greater than a few hundred. Quartz crystals may have unloaded Qs as high as 100,000 and temperature drift less than 0.001%/C°. Crystal-controlled oscillators therefore find extensive use as stable frequency sources in RF systems; further stability can be obtained by controlling the temperature of the quartz crystal.

A quartz crystal resonator consists of a small, thin sheet of quartz mounted between two metallic plates. Mechanical oscillations can be excited in the crystal through the piezoelectric effect. The equivalent circuit of a quartz crystal near its lowest resonant mode is shown in Figure 13.4a. This circuit has series and parallel

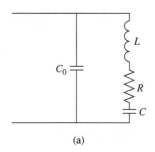

(a)

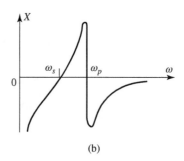

(b)

FIGURE 13.4 | (a) Equivalent circuit of a crystal resonator. (b) Input reactance of a crystal resonator.

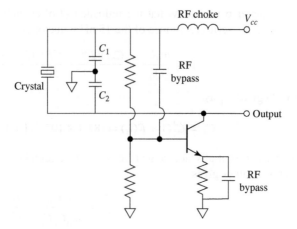

FIGURE 13.5 | Pierce crystal oscillator circuit.

resonant frequencies, ω_s and ω_p, given by

$$\omega_s = \frac{1}{\sqrt{LC}}, \tag{13.23a}$$

$$\omega_p = \frac{1}{\sqrt{L\left(\dfrac{C_0 C}{C_0 + C}\right)}}. \tag{13.23b}$$

The reactance of the circuit of Figure 13.4a is plotted in Figure 13.4b, where we see that the reactance is inductive in the frequency range between the series and parallel resonances. This is the usual operating point of the crystal, so that the crystal may be used in place of the inductor in a Colpitts or Pierce oscillator. A typical crystal oscillator circuit is shown in Figure 13.5.

13.2   MICROWAVE OSCILLATORS

In this section we focus on oscillator circuits that are useful at microwave frequencies, primarily employing negative resistance diodes or transistors.

Figure 13.6 shows the canonical RF circuit for a one-port negative resistance oscillator, where $Z_{in} = R_{in} + jX_{in}$ is the input impedance of the active device (e.g., a biased diode or transistor). In general, this impedance is current (or voltage) dependent, as well as frequency dependent, which we indicate by writing $Z_{in}(I, j\omega) = R_{in}(I, j\omega) + jX_{in}(I, j\omega)$. The device is terminated with a passive load impedance, $Z_L = R_L + jX_L$. Applying Kirchhoff's voltage law gives

$$(Z_L + Z_{in})I = 0. \tag{13.24}$$

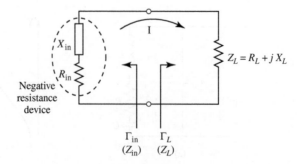

FIGURE 13.6 | Circuit for a one-port negative resistance oscillator.

If oscillation is occurring, such that the RF current I is nonzero, then the following two conditions must be satisfied:

$$R_L + R_{in} = 0, \tag{13.25a}$$

$$X_L + X_{in} = 0. \tag{13.25b}$$

Since the load is passive, $R_L > 0$, and (13.25a) implies that $R_{in} < 0$. Thus, while a positive resistance implies energy dissipation, a negative resistance implies an energy source. The condition of (13.25b) controls the frequency of oscillation. The condition in (13.24), that $Z_L = -Z_{in}$ for steady-state oscillation, implies that the reflection coefficients Γ_L and Γ_{in} are related as

$$\Gamma_L = \frac{Z_L - Z_0}{Z_L + Z_0} = \frac{-Z_{in} - Z_0}{-Z_{in} + Z_0} = \frac{Z_{in} + Z_0}{Z_{in} - Z_0} = \frac{1}{\Gamma_{in}}. \tag{13.26}$$

The process of oscillation is critically dependent on the nonlinear behavior of Z_{in}, as follows. Initially, it is necessary for the overall circuit to be unstable at a certain frequency, that is, $R_{in}(I, j\omega) + R_L < 0$. Then any transient excitation or noise will cause an oscillation to build up at the frequency ω. As I increases, $R_{in}(I, j\omega)$ must become less negative until the current I_0 is reached such that $R_{in}(I_0, j\omega_0) + R_L = 0$, and $X_{in}(I_0, j\omega_0) + X_L(j\omega_0) = 0$. At this point the oscillator can run in a stable state. The final frequency, ω_0, generally differs from the startup frequency because X_{in} is current dependent, so that $X_{in}(I, j\omega) \neq X_{in}(I_0, j\omega_0)$.

Thus we see that the conditions of (13.25) are not enough to guarantee a stable state of oscillation. In particular, stability requires that any perturbation in current or frequency will be damped out, allowing the oscillator to return to its original state. This condition can be quantified by considering the effect of a small change, δI, in the current, and a small change, δs, in the complex frequency $s = \alpha + j\omega$. If we let $Z_T(I, s) = Z_{in}(I, s) + Z_L(s)$, then we can write a Taylor series for $Z_T(I, s)$ about the stable operating point I_0, ω_0 as

$$Z_T(I, s) = Z_T(I_0, s_0) + \left.\frac{\partial Z_T}{\partial s}\right|_{s_0, I_0} \delta s + \left.\frac{\partial Z_T}{\partial I}\right|_{s_0, I_0} \delta I = 0, \tag{13.27}$$

since $Z_T(I, s)$ must still equal zero if oscillation is occurring. In (13.27), $s_0 = j\omega_0$ is the complex frequency at the original operating point. Now use the fact that $Z_T(I_0, s_0) = 0$, and that $\partial Z_T/\partial s = -j(\partial Z_T/\partial \omega)$, to solve (13.27) for $\delta s = \delta \alpha + j\delta \omega$:

$$\delta s = \delta \alpha + j\delta \omega = \left.\frac{-\partial Z_T/\partial I}{\partial Z_T/\partial s}\right|_{s_0, I_0} \delta I = \frac{-j\left(\partial Z_T/\partial I\right)\left(\partial Z_T^*/\partial \omega\right)}{\left|\partial Z_T/\partial \omega\right|^2} \delta I. \tag{13.28}$$

If the transient caused by δI and $\delta \omega$ is to decay, we must have $\delta \alpha < 0$ when $\delta I > 0$. Equation (13.28) then implies that

$$\mathrm{Im}\left\{\frac{\partial Z_T}{\partial I}\frac{\partial Z_T^*}{\partial \omega}\right\} < 0,$$

or

$$\frac{\partial R_T}{\partial I}\frac{\partial X_T}{\partial \omega} - \frac{\partial X_T}{\partial I}\frac{\partial R_T}{\partial \omega} > 0. \tag{13.29}$$

This relation is sometimes known as *Kurokawa's condition*. For a passive load, $\partial R_L/\partial I = \partial X_L/\partial I = \partial R_L/\partial \omega = 0$, so (13.29) reduces to

$$\frac{\partial R_{in}}{\partial I}\frac{\partial}{\partial \omega}(X_L + X_{in}) - \frac{\partial X_{in}}{\partial I}\frac{\partial R_{in}}{\partial \omega} > 0. \tag{13.30}$$

As discussed above, we usually have that $\partial R_{in}/\partial I > 0$, so (13.30) can be satisfied if $\partial(X_L + X_{in})/\partial \omega \gg 0$. This implies that a high-Q circuit will result in maximum oscillator stability. Cavity and dielectric resonators are often used for this purpose.

Effective oscillator design requires the consideration of several other issues, such as the selection of an operating point for stable operation and maximum power output, frequency pulling, large-signal effects, and noise characteristics. We leave these topics to more advanced texts [4, 5].

EXAMPLE 13.2 NEGATIVE RESISTANCE OSCILLATOR DESIGN

A one-port oscillator uses a negative resistance diode having $\Gamma_{in} = 1.25\angle 40°$ ($Z_0 = 50 \ \Omega$) at its desired operating point, for $f = 6$ GHz. Design a load matching network for a 50 Ω load impedance.

Solution
From either the Smith chart (see Problem 13.5) or by direct calculation, we find the input impedance of the diode as

$$Z_{in} = Z_0 \frac{1 + \Gamma_{in}}{1 - \Gamma_{in}} = -43.4 + j124.1 \ \Omega.$$

Then, by (13.25), the load impedance must be

$$Z_L = -Z_{in} = 43.4 - j124.1 \ \Omega.$$

A shunt stub and series section of line can be used to convert 50 Ω to Z_L, as shown in the circuit of Figure 13.7.

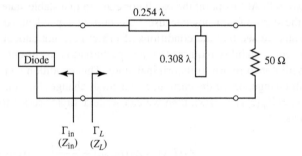

FIGURE 13.7 | Load matching circuit for the one-port oscillator of Example 13.2.

Transistor Oscillators

In a transistor oscillator, a negative resistance one-port network is effectively created by terminating a potentially unstable transistor with an impedance designed to drive the device in an unstable region. The circuit model of a transistor oscillator is shown in Figure 13.8. In this circuit, the RF output port is part of the load network on the output side of the transistor, but it is also possible to use the terminating network to the left of the transistor as the output port. In the case of an amplifier, we preferred a device with a high degree of stability—ideally, an unconditionally stable device. For an oscillator, we require a device with a high degree of instability. Typically, common source or common gate FET configurations are used (common emitter or common base for bipolar junction devices), often with positive feedback to enhance the instability of the device. After the transistor configuration is selected, the output stability circle can be drawn in the Γ_L plane, and Γ_L selected to produce a large value of negative resistance at the input to the transistor. Then the terminating impedance $Z_S = R_S + jX_S$ can be chosen to match Z_{in}. Because such a design often relies on the small-signal scattering parameters, and because R_{in} will become less negative as the oscillator

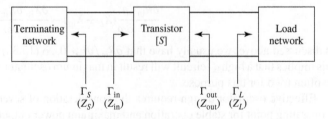

FIGURE 13.8 | Circuit for a two-port transistor oscillator.

power builds up, it is often necessary to choose R_S so that $R_S + R_{in} < 0$. Otherwise, oscillation may cease if increasing RF power increases R_{in} to the point where $R_S + R_{in} > 0$. In practice, a value of

$$R_S = \frac{-R_{in}}{3} \tag{13.31a}$$

is often used. The reactive part of Z_S is chosen to resonate the circuit,

$$X_S = -X_{in}. \tag{13.31b}$$

When oscillation occurs between the termination network and the transistor, oscillation will simultaneously occur at the output port, which we can show as follows. For steady-state oscillation at the input port, we must have $\Gamma_S \Gamma_{in} = 1$, analogous to the condition of (13.26). Then from (12.3a) we have

$$\frac{1}{\Gamma_S} = \Gamma_{in} = S_{11} + \frac{S_{12}S_{21}\Gamma_L}{1 - S_{22}\Gamma_L} = \frac{S_{11} - \Delta\Gamma_L}{1 - S_{22}\Gamma_L}, \tag{13.32}$$

where $\Delta = S_{11}S_{22} - S_{12}S_{21}$. Solving for Γ_L gives

$$\Gamma_L = \frac{1 - S_{11}\Gamma_S}{S_{22} - \Delta\Gamma_S}. \tag{13.33}$$

From (12.3b) we have that

$$\Gamma_{out} = S_{22} + \frac{S_{12}S_{21}\Gamma_S}{1 - S_{11}\Gamma_S} = \frac{S_{22} - \Delta\Gamma_S}{1 - S_{11}\Gamma_S}, \tag{13.34}$$

which shows that $\Gamma_L \Gamma_{out} = 1$, and hence $Z_L = -Z_{out}$. Thus, the condition for oscillation at the load network is satisfied. Note that it is preferable to use the large-signal scattering parameters of the transistor in the above development.

EXAMPLE 13.3   TRANSISTOR OSCILLATOR DESIGN

Design a transistor oscillator at 4 GHz using a GaAs MESFET in a common gate configuration, with a 5 nH inductor in series with the gate to increase the instability. Choose a load network to match to a 50 Ω load, and an appropriate terminating network at the input to the transistor. The scattering parameters of the transistor in a common source configuration are ($Z_0 = 50\ \Omega$): $S_{11} = 0.72\angle-116°$, $S_{12} = 0.03\angle57°$, $S_{21} = 2.60\angle76°$, and $S_{22} = 0.73\angle-54°$.

Solution
The first step is to convert the common source scattering parameters to the scattering parameters that apply to the transistor in a common gate configuration with a series inductor. (See Figure 13.9a.) This is most easily done using a microwave CAD package. The new scattering parameters are

$$S'_{11} = 2.18\angle-35°,$$
$$S'_{12} = 1.26\angle18°,$$
$$S'_{21} = 2.75\angle96°,$$
$$S'_{22} = 0.52\angle155°.$$

Note that $|S'_{11}|$ is significantly greater than $|S_{11}|$, which suggests that the configuration of Figure 13.9a is more unstable than the common source configuration. Calculating the output stability circle (Γ_L plane) parameters gives

$$C_L = \frac{\left(S'_{22} - \Delta' S'^{*}_{11}\right)^*}{\left|S'_{22}\right|^2 - |\Delta'|^2} = 1.08\angle33°,$$

$$R_L = \left|\frac{S'_{12}S'_{21}}{\left|S'_{22}\right|^2 - |\Delta'|^2}\right| = 0.665.$$

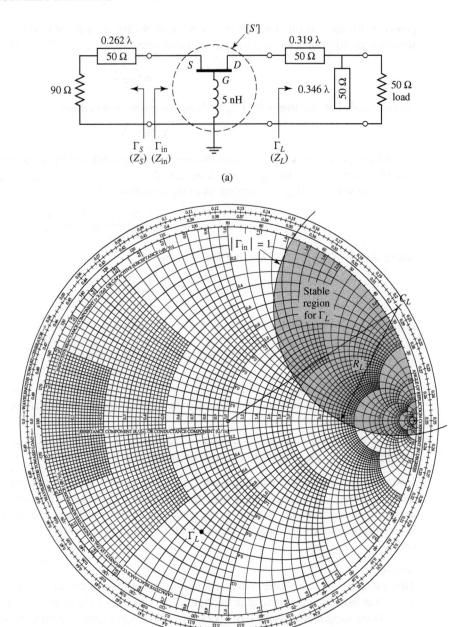

FIGURE 13.9 | Circuit design for the transistor oscillator of Example 13.3. (a) Oscillator circuit. (b) Smith chart for determining Γ_L.

Since $\left| S'_{11} \right| = 2.18 > 1$, the stable region is inside this circle, as shown in the Smith chart in Figure 13.9b.

There is a great amount of freedom in our choice for Γ_L, but one objective is to make $|\Gamma_{in}|$ large. We therefore try several values of Γ_L located on the opposite side of the chart from the stability circle, and select $\Gamma_L = 0.59\angle{-104°}$. Then we can design a single-stub matching network to convert a 50 Ω load to $Z_L = 20 - j35$ Ω, as shown in Figure 13.9a.

For the given value of Γ_L, we calculate Γ_{in} as

$$\Gamma_{in} = S'_{11} + \frac{S'_{12}S'_{21}\Gamma_L}{1 - S'_{22}\Gamma_L} = 3.96\angle{-2.4°}$$

or $Z_{in} = -84 - j1.9\ \Omega$. Then, from (13.31), we find Z_S as

$$Z_S = \frac{-R_{in}}{3} - jX_{in} = 28 + j1.9\ \Omega.$$

Using $R_{in}/3$ should ensure enough instability for the startup of oscillation. The easiest way to implement the impedance Z_S is to use a 90 Ω load with a short length of line, as shown in the figure. It is likely that the steady-state oscillation frequency will differ from 4 GHz because of the nonlinearity of the transistor parameters.

Dielectric Resonator Oscillators

As we saw from the result of (13.30), oscillator stability is enhanced with the use of a high-Q tuning network. The unloaded Q of a resonant network using lumped elements or microstrip lines and stubs is typically limited to a few hundred (see Chapter 6), and while waveguide cavity resonators can have unloaded Qs of 10^4 or more, they are not well suited for integration in miniature microwave integrated circuitry. Another disadvantage of metal cavities is the significant frequency drift caused by dimensional expansion due to temperature variations. The dielectric cavity resonator discussed in Section 6.5 overcomes most of these disadvantages, as it can have an unloaded Q as high as several thousand, it is compact and easily integrated with planar circuitry, and it can be made from ceramic materials that have excellent temperature stability. For these reasons, transistor *dielectric resonator oscillators* (DROs) are in common use over the entire microwave, and lower millimeter wave, frequency range.

A dielectric resonator is usually coupled to an oscillator circuit by positioning it in close proximity to a microstrip line, as shown in Figure 13.10a. The resonator operates in the $TE_{01\delta}$ mode, and couples to the fringing magnetic field of the microstrip line. The strength of coupling is determined by the spacing, d, between the resonator and microstrip line. Because coupling is via the magnetic field, the resonator appears as a series load on the microstrip line, as shown in the equivalent circuit of Figure 13.10b. The resonator is modeled as a parallel RLC circuit, and the coupling to the feedline is modeled by the turns ratio, N, of the transformer. Using the result of (6.19) for the impedance of a parallel RLC resonator, we can express the equivalent series impedance, Z, seen by the microstrip line as

$$Z = \frac{N^2 R}{1 + j2Q_0 \Delta\omega/\omega_0}, \tag{13.35}$$

where $Q_0 = R/\omega_0 L$ is the unloaded resonator Q, $\omega_0 = 1/\sqrt{LC}$ is the resonant frequency, and $\Delta\omega = \omega - \omega_0$. The coupling factor, defined in (6.76), between the resonator and the feedline is the ratio of the unloaded to external Q, and can be found as

$$g = \frac{Q_0}{Q_e} = \frac{R/\omega_0 L}{R_L/N^2 \omega_0 L} = \frac{N^2 R}{2Z_0}, \tag{13.36}$$

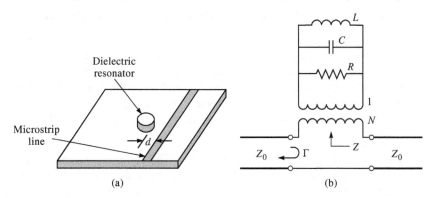

(a)   (b)

FIGURE 13.10 | (a) Geometry of a dielectric resonator coupled to a microstrip line. (b) Equivalent circuit.

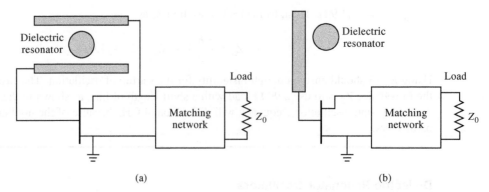

(a)   (b)

FIGURE 13.11 | (a) Dielectric resonator oscillator using parallel feedback. (b) Dielectric resonator oscillator using series feedback.

where $R_L = 2Z_0$ is the load resistance for a feedline with source and termination resistances Z_0. In some cases the feedline is terminated with an open-circuit $\lambda/4$ from the resonator to maximize the magnetic field at that point; in this case $R_L = Z_0$, and the coupling factor is twice the value given in (13.36).

The reflection coefficient seen on the terminated microstrip line looking toward the resonator can be written as

$$\Gamma = \frac{(Z_0 + N^2R) - Z_0}{(Z_0 + N^2R) + Z_0} = \frac{N^2R}{2Z_0 + N^2R} = \frac{g}{1 + g}. \tag{13.37}$$

This allows the coupling coefficient to be found from $g = \Gamma/(1 - \Gamma)$ after the simple procedure of measuring Γ at resonance; the resonant frequency and Q can also be found by measurement. Alternatively, these quantities can be calculated using approximate analytical solutions [6]. Note that this procedure leaves a degree of freedom between N and R since only the product N^2R is uniquely determined.

There are many oscillator configurations using common source (emitter), common gate (base), or common drain (collector) connections of either BJTs or FETs, in addition to the optional use of series or shunt elements to increase the instability of the device [4, 5]. A dielectric resonator can be incorporated into the circuit to provide frequency stability using either the parallel feedback arrangement of Figure 13.11a, or the series feedback technique shown in Figure 13.11b. The parallel configuration uses a resonator coupled to two microstrip lines, functioning as a high-Q bandpass filter that couples a portion of the transistor output back to its input. The amount of coupling is controlled by the spacing between the resonator and the lines, and the phase is controlled by the length of the lines. The series feedback configuration is simpler, using only a single microstrip feedline, but typically does not have a tuning range as wide as that obtained with parallel feedback. Design of an oscillator using parallel feedback is most conveniently done using microwave CAD software, but a dielectric resonator oscillator using series feedback can be designed using the same procedure that was discussed in the previous section on two-port oscillators.

EXAMPLE 13.4   DIELECTRIC RESONATOR OSCILLATOR DESIGN

A wireless local area network application requires a local oscillator operating at 2.4 GHz. Design a dielectric resonator oscillator using the series feedback circuit of Figure 13.11b with a bipolar transistor having the following scattering parameters ($Z_0 = 50\ \Omega$): $S_{11} = 1.8\angle130°$, $S_{12} = 0.4\angle45°$, $S_{21} = 3.8\angle36°$, and $S_{22} = 0.7\angle-63°$. Determine the required coupling coefficient for the dielectric resonator, and the required microstrip matching network for the load. Plot the magnitude of Γ_{out} versus $\Delta f/f_0$ for small variations in frequency about the design value, assuming an unloaded resonator Q of 1000.

Solution
The DRO circuit is shown in Figure 13.12a. The dielectric resonator is placed $\lambda/4$ from the open end of the microstrip line; the line length ℓ_r can be adjusted to match the phase of the required value of Γ_S. In contrast to the oscillator of the previous example, the output load impedance for this circuit is part of the terminating network.

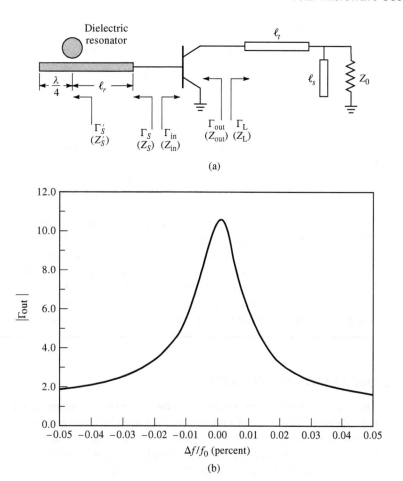

FIGURE 13.12 | (a) Circuit for the dielectric resonator of Example 13.4. (b) $|\Gamma_{out}|$ vs. frequency in Example 13.4.

The stability circles for the transistor can be plotted if desired, but are not necessary for the design since we can begin by choosing Γ_S to provide a large value of $|\Gamma_{out}|$. From (13.34) we have

$$\Gamma_{out} = S_{22} + \frac{S_{12}S_{21}\Gamma_S}{1 - S_{11}\Gamma_S},$$

which indicates that we can maximize Γ_{out} by making $1 - S_{11}\Gamma_S$ close to zero. Thus we choose $\Gamma_S = 0.6\angle-130°$, which gives $\Gamma_{out} = 10.7\angle132°$. This corresponds to an impedance

$$Z_{out} = Z_0\frac{1 + \Gamma_{out}}{1 - \Gamma_{out}} = 50\frac{1 + 10.7\angle132°}{1 - 10.7\angle132°} = -43.7 + j6.1 \ \Omega.$$

Applying the analogous startup condition of (13.31) for the output side gives the required termination impedance as

$$Z_L = \frac{-R_{out}}{3} - jX_{out} = 5.5 - j6.1 \ \Omega.$$

The matching network can now be designed using a Smith chart. The shortest transmission line length for matching Z_L to the load impedance Z_0 is $\ell_t = 0.481\lambda$, and the required open-circuit stub length is $\ell_s = 0.307\lambda$.

Next we match Γ_S to the resonator network. From (13.35) we know that the equivalent impedance of the resonator seen by the microstrip line is real at the resonant frequency, so the phase angle of the reflection coefficient at this point, Γ'_S, must be either zero or 180°. For an undercoupled parallel RLC resonator, $R < Z_0$, so the proper phase will be 180°, which can be achieved by transformation through the line length ℓ_r. The magnitude of the reflection coefficient is unchanged, so we have the relation

$$\Gamma'_S = \Gamma_S e^{2j\beta\ell_r} = (0.6\angle-130°)e^{2j\beta\ell_r} = 0.6\angle180°,$$

which gives $\ell_r = 0.431\lambda$. The equivalent impedance of the resonator at resonance is then

$$Z'_S = Z_0 \frac{1+\Gamma'_S}{1-\Gamma'_S} = 12.5 \ \Omega.$$

The coupling coefficient can be found using (13.36), with a factor of two to account for the $\lambda/4$ stub termination, as

$$g = \frac{N^2 R}{Z_0} = \frac{12.5}{50} = 0.25.$$

The variation of $|\Gamma_{out}|$ with frequency will give an indication of the frequency stability of the oscillator. We can calculate Γ_{out} from (13.34), after first using (13.35) to compute Z'_S, Γ'_S, and then transforming down the line of length ℓ_r to obtain Γ_S. The electrical line length can be approximated as constant for the small changes in frequency associated with this calculation. A short computer program, or microwave CAD software, can be used to generate data for $-0.01 < \Delta f/f_0 < 0.01$, which is shown in the graph of Figure 13.12b. Observe that $|\Gamma_{out}|$ decreases rapidly with a change in frequency as small as a few hundredths of a percent, demonstrating the sharp selectivity that can be obtained with a dielectric resonator.

13.3 OSCILLATOR PHASE NOISE

The noise produced by an oscillator or other signal source is important in practice because it may severely degrade the performance of a communication or radar receiver system. Besides adding to the noise level of the receiver, a noisy local oscillator will lead to down-conversion of undesired nearby signals, thus limiting the selectivity of the receiver and how closely adjacent channels may be spaced. *Phase noise* refers to the short-term random fluctuation in the frequency (or phase) of an oscillator signal. Phase noise also introduces uncertainty during the detection of digitally modulated signals.

An ideal oscillator would have a frequency spectrum consisting of a single delta function at its operating frequency, but a realistic oscillator will have a spectrum more like that shown in Figure 13.13. Spurious signals due to oscillator harmonics or intermodulation products appear as discrete spikes in the spectrum. Phase noise, due to random fluctuations caused by thermal and other noise sources, appears as a broad, continuous distribution localized about the output signal. Phase noise is defined as the ratio of power in one phase modulation sideband to the total signal power per unit bandwidth (1 Hz) at a particular offset, f_m, from the signal frequency, and is denoted as $\mathscr{L}(f_m)$. It is usually expressed in decibels relative to the carrier power per hertz of bandwidth (dBc/Hz). A typical oscillator phase noise specification for a cellular radio, for example, may be -110 dBc/Hz at 25 kHz from the carrier. In the following sections we show how phase noise may be represented, and present a widely used model for characterizing the phase noise of an oscillator.

Representation of Phase Noise

In general, the output voltage of an oscillator or synthesizer can be written as

$$v_o(t) = V_0[1 + A(t)] \cos[\omega_o t + \theta(t)], \tag{13.38}$$

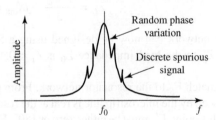

FIGURE 13.13 | Output spectrum of a typical RF oscillator.

where $A(t)$ represents the amplitude fluctuations of the output, and $\theta(t)$ represents the phase variation of the output waveform. Of these, amplitude variations can usually be well controlled, and generally have less impact on system performance. Phase variations may be discrete (due to deterministic spurious mixer products or harmonics), or random in nature (due to thermal or other random noise sources). Note from (13.38) that an instantaneous phase variation is indistinguishable from a variation in frequency.

Small changes in the oscillator frequency can be represented as a frequency modulation of the carrier by letting

$$\theta(t) = \frac{\Delta f}{f_m} \sin \omega_m t = \theta_p \sin \omega_m t, \tag{13.39}$$

where $f_m = \omega_m / 2\pi$ is the modulating frequency. The peak phase deviation is $\theta_p = \Delta f / f_m$ (also called the *modulation index*). Substituting (13.39) into (13.38) and expanding gives

$$v_o(t) = V_o[\cos \omega_o t \cos(\theta_p \sin \omega_m t) - \sin \omega_o t \sin(\theta_p \sin \omega_m t)], \tag{13.40}$$

where we set $A(t) = 0$ to ignore amplitude fluctuations. Assuming the phase deviations are small, so that $\theta_p \ll 1$, we can use the small-argument expressions that $\sin x \simeq x$ and $\cos x \simeq 1$ to simplify (13.40) to

$$v_o(t) = V_o\left(\cos \omega_o t - \theta_p \sin \omega_m t \sin \omega_o t\right)$$
$$= V_o\left\{\cos \omega_o t - \frac{\theta_p}{2}[\cos(\omega_o + \omega_m)t - \cos(\omega_o - \omega_m)t]\right\}. \tag{13.41}$$

This expression shows that small phase or frequency deviations in the output of an oscillator result in modulation sidebands at $\omega_o \pm \omega_m$, located on either side of the carrier signal at ω_o. When these deviations are due to random changes in temperature or device noise, the output spectrum of the oscillator will take the form shown in Figure 13.13.

According to the definition of phase noise as the ratio of noise power in a single sideband to the carrier power, the waveform of (13.41) has a corresponding phase noise of

$$\mathcal{L}(f) = \frac{P_n}{P_c} = \frac{\dfrac{1}{2}\left(\dfrac{V_o \theta_p}{2}\right)^2}{\dfrac{1}{2}V_o^2} = \frac{\theta_p^2}{4} = \frac{\theta_{\text{rms}}^2}{2}, \tag{13.42}$$

where $\theta_{\text{rms}} = \theta_p / \sqrt{2}$ is the rms value of the phase deviation. The two-sided power spectral density associated with phase noise includes power in both sidebands:

$$S_\theta(f_m) = 2\mathcal{L}(f_m) = \frac{\theta_p^2}{2} = \theta_{\text{rms}}^2. \tag{13.43}$$

White noise generated by passive or active devices can be interpreted in terms of phase noise by using the same definition. From Chapter 10 we know that the noise power at the output of a noisy two-port network is $kT_0 BFG$, where $T_0 = 290$ K, B is the measurement bandwidth, F is the noise figure of the network, and G is the gain of the network. For a 1 Hz bandwidth, the ratio of output noise power density to output signal power gives the power spectral density as

$$S_\theta(f_m) = \frac{kT_0 F}{P_c}, \tag{13.44}$$

where P_c is the input signal (carrier) power. Note that the gain of the network cancels in this expression.

Leeson's Model for Oscillator Phase Noise

In this section we present Leeson's model for characterizing the power spectral density of oscillator phase noise [2, 7]. As in Section 13.1, we will model the oscillator as an amplifier with a feedback path, as shown in Figure 13.14. If the voltage gain of the amplifier is included in the feedback transfer function $H(\omega)$, then

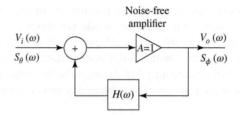

FIGURE 13.14 | Feedback amplifier model for characterizing oscillator phase noise.

the voltage transfer function for the oscillator circuit is

$$V_o(\omega) = \frac{V_i(\omega)}{1 - H(\omega)}. \tag{13.45}$$

If we consider oscillators that use a high-Q resonant circuit in the feedback loop (e.g., Colpitts, Hartley, Clapp, and similar oscillators), then $H(\omega)$ can be represented as the voltage transfer function of a parallel *RLC* resonator:

$$H(\omega) = \frac{1}{1 + jQ_0\left(\dfrac{\omega}{\omega_0} - \dfrac{\omega_0}{\omega}\right)} = \frac{1}{1 + 2jQ_0\Delta\omega/\omega_0}, \tag{13.46}$$

where ω_0 is the resonant frequency of the oscillator, and $\Delta\omega = \omega - \omega_0$ is the frequency offset relative to the resonant frequency.

Since the input and output power spectral densities are related by the square of the magnitude of the voltage transfer function [8], we can use (13.45)–(13.46) to write

$$S_\phi(\omega) = \left|\frac{1}{1 - H(\omega)}\right|^2 S_\theta(\omega) = \frac{1 + 4Q_0^2\Delta\omega^2/\omega_0^2}{4Q_0^2\Delta\omega^2/\omega_0^2}S_\theta(\omega)$$

$$= \left(1 + \frac{\omega_0^2}{4Q_0^2\Delta\omega_0^2}\right)S_\theta(\omega) = \left(1 + \frac{\omega_h^2}{\Delta\omega_0^2}\right)S_\theta(\omega), \tag{13.47}$$

where $S_\theta(\omega)$ is the input power spectral density, and $S_\phi(\omega)$ is the output power spectral density. In (13.47) we have also defined $\omega_h = \omega_0/2Q_0$ as the half-power (3 dB) bandwidth of the resonator.

The noise spectrum of a typical transistor amplifier with an applied sinusoidal signal at f_0 is shown in Figure 13.15. Besides *kTB* thermal noise, transistors generate additional noise that varies as $1/f$ at frequencies below the frequency f_α. This $1/f$, or *flicker*, noise is likely caused by random fluctuations of the carrier density in the active device. Due to the nonlinearity of the transistor, the $1/f$ noise will modulate the applied signal at f_0, and appear as $1/f$ noise sidebands around f_0. Since the $1/f$ noise component dominates the phase noise power at frequencies close to the carrier, it is important to include it in our model. Thus we consider an input power spectral density as shown in Figure 13.16, where $K/\Delta f$ represents the $1/f$ noise component around the carrier, and kT_0F/P_0 represents the thermal noise. Thus the power spectral density

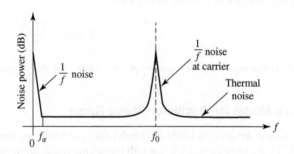

FIGURE 13.15 | Noise power versus frequency for an amplifier with an applied input signal.

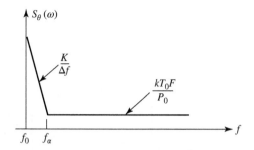

FIGURE 13.16 | Idealized power spectral density of amplifier noise, including $1/f$ and thermal components.

applied to the input of the oscillator can be written as

$$S_\theta(\omega) = \frac{kTF}{P_0}\left(1 + \frac{K\omega_\alpha}{\Delta\omega}\right), \tag{13.48}$$

where K is a constant accounting for the strength of the $1/f$ noise, and $\omega_\alpha = 2\pi f_\alpha$ is the *corner frequency* of the $1/f$ noise. The corner frequency depends primarily on the type of transistor used in the oscillator. Silicon junction FETs, for example, typically have corner frequencies ranging from 50 to 100 Hz, while GaAs MESFETs have corner frequencies ranging from 2 to 10 MHz, or higher. Silicon bipolar junction transistors have corner frequencies that range from 5 to 50 kHz.

Using (13.48) in (13.47) gives the power spectral density of the output phase noise as

$$\begin{aligned}
S_\phi(\omega) &= \frac{kT_0F}{P_0}\left(\frac{K\omega_0^2\omega_\alpha}{4Q_0^2\Delta\omega^3} + \frac{\omega_0^2}{4Q_0^2\Delta\omega^2} + \frac{K\omega_\alpha}{\Delta\omega} + 1\right) \\
&= \frac{kT_0F}{P_0}\left(\frac{K\omega_\alpha\omega_h^2}{\Delta\omega^3} + \frac{\omega_h^2}{\Delta\omega^2} + \frac{K\omega_\alpha}{\Delta\omega} + 1\right).
\end{aligned} \tag{13.49}$$

This result is sketched in Figure 13.17. There are two cases, depending on which of the middle two terms of (13.49) is more significant. In either case, for frequencies close to the carrier at f_0, the noise power decreases as $1/f^3$, or -18 dB/octave. If the resonator has a relatively low Q, so that its 3 dB bandwidth $f_h > f_\alpha$, then for frequencies between f_α and f_h the noise power drops as $1/f^2$, or -12 dB/octave. If the resonator has a relatively high Q, so that $f_h < f_\alpha$, then for frequencies between f_h and f_α the noise power drops as $1/f$, or -6 dB/octave.

At higher frequencies the noise is predominantly thermal, constant with frequency, and proportional to the noise figure of the amplifier. A noiseless amplifier with $F = 1$ (0 dB) would produce the minimum noise floor of $kT_0 = -174$ dBm/Hz. In accordance with Figure 13.13, the noise power is greatest at frequencies closest to the carrier frequency, but (13.49) shows that the $1/f^3$ component is proportional to $1/Q_0^2$, so that

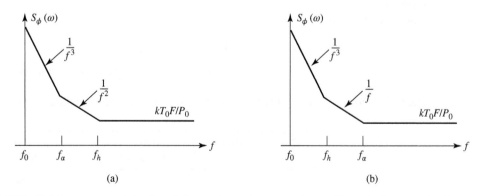

FIGURE 13.17 | Power spectral density of phase noise at the output of an oscillator. (a) Response for $f_h > f_\alpha$ (low Q). (b) Response for $f_\alpha > f_h$ (high Q).

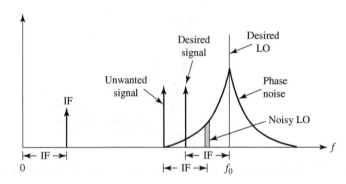

FIGURE 13.18 | Illustrating how local oscillator phase noise can lead to the reception of undesired signals adjacent to the desired signal.

better phase noise characteristics close to the carrier are achieved with a high-Q resonator. Finally, recall from (13.43) that the single-sideband phase noise will be one-half of the power spectral density of (13.49). These results give a reasonably good model for oscillator phase noise, and quantitatively explain the roll-off of noise power with frequency offset from the carrier.

The effect of phase noise in a receiver is to degrade both the signal-to-noise ratio (or bit error rate) and the selectivity [9]. Of these, the impact on selectivity is usually the most severe. Phase noise degrades receiver selectivity by causing down conversion of signals located nearby the desired signal frequency. The process is shown in Figure 13.18. A local oscillator at frequency f_0 is used to down convert a desired signal to an intermediate frequency (IF). Due to phase noise, however, an adjacent undesired signal can be down converted to the same IF frequency due to the phase noise spectrum of the local oscillator. The phase noise that leads to this conversion is located at an offset from the carrier equal to the IF frequency from the undesired signal. This process is called *reciprocal mixing*. From this diagram, it is easy to see that the maximum allowable phase noise in order to achieve an adjacent channel rejection (or selectivity) of S dB ($S \geq 0$) is given by

$$\mathscr{L}(f_m) = C \text{ (dBm)} - S \text{ (dB)} - I \text{ (dBm)} - 10\log(B), \quad \text{(dBc/Hz)}, \tag{13.50}$$

where C is the desired signal level (in dBm), I is the undesired (interference) signal level (in dBm), and B is the bandwidth of the IF filter (in Hz).

EXAMPLE 13.5 GSM RECEIVER PHASE NOISE REQUIREMENTS

The GSM cellular telephone standard requires a minimum of 9 dB rejection of interfering signal levels of -23 dBm at 3 MHz from the carrier, -33 dBm at 1.6 MHz from the carrier, and -43 dBm at 0.6 MHz from the carrier, for a carrier level of -99 dBm. Determine the required local oscillator phase noise at these carrier frequency offsets. The channel bandwidth is 200 kHz.

Solution
From (13.50) we have

$$\mathscr{L}(f_m) = C \text{ (dBm)} - S \text{ (dB)} - I \text{ (dBm)} - 10\log(B)$$

$$= -99 \text{ dBm} - 9 \text{ dB} - I \text{ (dBm)} - 10\log(2 \times 10^5).$$

The table below lists the required LO phase noise as computed from the above expression:

Frequency Offset f_m (MHz)	Interfering Signal Level (dBm)	$\mathscr{L}(f_m)$ (dBc/Hz)
3.0	−23	−138
1.6	−33	−128
0.6	−43	−118

This level of phase noise requires a phase-locked synthesizer. Bit errors in GSM systems are usually dominated by the reciprocal mixing effect, while errors due to thermal antenna and receiver noise are generally negligible.

13.4   FREQUENCY MULTIPLIERS

As frequency increases into the millimeter wave range it becomes increasingly difficult to build fundamental frequency oscillators with good power, stability, and noise characteristics. An alternative approach is to produce a harmonic of a lower frequency oscillator through the use of a *frequency multiplier*. As we saw in Section 10.3, a nonlinear element may generate many harmonics of an input sinusoidal signal, so frequency multiplication is a natural occurrence in circuits containing diodes and transistors. Designing a good-quality frequency multiplier, however, is a difficult task that generally requires nonlinear analysis, matching at multiple frequencies, stability analysis, and thermal considerations. We will discuss some of the general operational principles and properties of diode and transistor frequency multipliers, and refer the reader to the literature for more practical details [5].

Frequency multiplier circuits can be categorized as *reactive diode multipliers, resistive diode multipliers*, or *transistor multipliers*. A reactive diode multiplier uses either a varactor or a step-recovery diode biased to present a nonlinear junction capacitance. Since losses in such diodes are small, conversion efficiencies (the fraction of RF input power that is converted to the desired harmonic) can be relatively high. In fact, as we will show, ideal (lossless) reactive multipliers can achieve a theoretical conversion efficiency of 100%. Varactor multipliers are most useful for low harmonic conversion (multiplier factors of 2–4), while step-recovery diodes are able to generate more power at higher harmonics. Resistive multipliers exploit the nonlinear *I–V* characteristic of a forward-biased Schottky barrier diode. We will show that resistive multipliers have conversion efficiencies that decrease as the square of the harmonic number, and so these multipliers are only useful for low multiplication factors. Transistor multipliers can use both bipolar junction and FET devices, and can provide conversion gains. Transistor multipliers are limited by their cutoff frequency, however, and therefore are generally not useful at very high frequencies.

A disadvantage of frequency multipliers is that noise levels are increased by the multiplication factor. This is because frequency multiplication is effectively a phase multiplication process as well, so phase noise variations get multiplied in the same way that frequency is multiplied. The increase in noise power is given by $20 \log n$, where n is the multiplication factor. Thus a frequency doubler will increase the fundamental oscillator noise level by at least 6 dB, while a frequency tripler will lead to an increase of at least 9.5 dB. Reactive diode multipliers typically add little additional noise of their own since varactors and step-recovery diodes have very low series resistances, but resistive diode multipliers can generate significant additional noise power.

Reactive Diode Multipliers (Manley–Rowe Relations)

We begin our discussion with the *Manley–Rowe relations*, which result from a very general analysis of power conservation associated with frequency conversion in a nonlinear reactive element [10]. Consider the circuit of Figure 13.19, where two sources at frequencies ω_1 and ω_2 drive a nonlinear capacitor, C. The circuit also shows ideal bandpass filters to conceptually isolate powers in all harmonics of the form $n\omega_1 + m\omega_2$. Since the capacitor is nonlinear, its charge Q can be expressed as a power series in terms of the capacitor voltage, v:

$$Q = a_0 + a_1 v + a_2 v^2 + a_3 v^3 + \cdots$$

As in Section 10.3, this nonlinear relationship implies the generation of all frequency products of the form $n\omega_1 + m\omega_2$. Thus we can write the capacitor voltage as a Fourier series of the form

$$v(t) = \sum_{n=-\infty}^{\infty} \sum_{m=-\infty}^{\infty} V_{nm} e^{j(n\omega_1 + m\omega_2)t}. \tag{13.51}$$

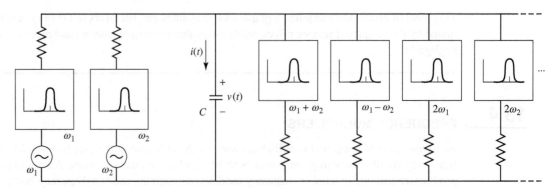

FIGURE 13.19 | Conceptual circuit for the derivation of the Manley–Rowe relations.

Similarly, the capacitor charge and current can be written as

$$Q(t) = \sum_{n=-\infty}^{\infty} \sum_{m=-\infty}^{\infty} Q_{nm} e^{j(n\omega_1 + m\omega_2)t}, \tag{13.52}$$

$$i(t) = \frac{dQ}{dt} = \sum_{n=-\infty}^{\infty} \sum_{m=-\infty}^{\infty} j(n\omega_1 + m\omega_2) Q_{nm} e^{j(n\omega_1 + m\omega_2)t} = \sum_{n=-\infty}^{\infty} \sum_{m=-\infty}^{\infty} I_{nm} e^{j(n\omega_1 + m\omega_2)t}. \tag{13.53}$$

Since $v(t)$ and $i(t)$ are real functions, we must have that $V_{-n,-m} = V_{nm}^*$ and $Q_{-n,-m} = Q_{nm}^*$.

No real power can be dissipated in the lossless capacitor. If ω_1 and ω_2 are not multiples of each other, there is no average power due to interacting harmonics. Then the average power (ignoring a factor of 4) at frequency $\pm |n\omega_1 + m\omega_2|$ is given as

$$P_{nm} = 2\,\mathrm{Re}\left\{ V_{nm} I_{nm}^* \right\} = V_{nm} I_{nm}^* + V_{nm}^* I_{nm} = V_{nm} I_{nm}^* + V_{-n,-m} I_{-n,-m}^* = P_{-n,-m}. \tag{13.54}$$

Conservation of power can then be expressed as

$$\sum_{n=-\infty}^{\infty} \sum_{m=-\infty}^{\infty} P_{nm} = 0. \tag{13.55}$$

Now multiply (13.55) by $\dfrac{n\omega_1 + m\omega_2}{n\omega_1 + m\omega_2}$ to obtain

$$\omega_1 \sum_{n=-\infty}^{\infty} \sum_{m=-\infty}^{\infty} \frac{n P_{nm}}{n\omega_1 + m\omega_2} + \omega_2 \sum_{n=-\infty}^{\infty} \sum_{m=-\infty}^{\infty} \frac{m P_{nm}}{n\omega_1 + m\omega_2} = 0. \tag{13.56}$$

Using (13.54) and the fact that $I_{nm} = j(n\omega_1 + m\omega_2)Q_{nm}$ gives

$$\omega_1 \sum_{n=-\infty}^{\infty} \sum_{m=-\infty}^{\infty} n\left(-j V_{nm} Q_{nm}^* - j V_{-n,-m} Q_{-n,-m}^* \right)$$

$$+ \omega_2 \sum_{n=-\infty}^{\infty} \sum_{m=-\infty}^{\infty} m\left(-j V_{nm} Q_{nm}^* - j V_{-n,-m} Q_{-n,-m}^* \right) = 0 \tag{13.57}$$

The double summation terms in (13.57) do not depend on ω_1 or ω_2 since we can always adjust the external circuitry so that all V_{nm} remain constant, and the Q_{nm} will remain constant as well since the capacitor charge depends directly on the voltage. Thus each summation in (13.56) must be identically zero:

$$\sum_{n=-\infty}^{\infty} \sum_{m=-\infty}^{\infty} \frac{n P_{nm}}{n\omega_1 + m\omega_2} = 0, \tag{13.58a}$$

$$\sum_{n=-\infty}^{\infty} \sum_{m=-\infty}^{\infty} \frac{m P_{nm}}{n\omega_1 + m\omega_2} = 0. \tag{13.58b}$$

Some simplification can be carried out by eliminating the negative indices of one summation by using the fact that $P_{-n,-m} = P_{nm}$. For example, from (13.58a),

$$\sum_{n=-\infty}^{\infty}\sum_{m=-\infty}^{\infty}\frac{nP_{nm}}{n\omega_1 + m\omega_2} = \sum_{n=0}^{\infty}\sum_{m=-\infty}^{\infty}\frac{nP_{nm}}{n\omega_1 + m\omega_2} + \sum_{n=0}^{\infty}\sum_{m=-\infty}^{\infty}\frac{-nP_{-n,-m}}{-n\omega_1 - m\omega_2}$$

$$= 2\sum_{n=0}^{\infty}\sum_{m=-\infty}^{\infty}\frac{nP_{nm}}{n\omega_1 + m\omega_2} = 0.$$

This results in the usual form for the Manley–Rowe relations:

$$\sum_{n=0}^{\infty}\sum_{m=-\infty}^{\infty}\frac{nP_{nm}}{n\omega_1 + m\omega_2} = 0, \tag{13.59a}$$

$$\sum_{n=-\infty}^{\infty}\sum_{m=0}^{\infty}\frac{mP_{nm}}{n\omega_1 + m\omega_2} = 0. \tag{13.59b}$$

The Manley–Rowe relations express power conservation for any lossless nonlinear reactance, and can be useful for harmonic generation, parametric amplifiers, and frequency converters at RF, microwave, and optical frequencies to predict the maximum possible power gain and conversion efficiency.

Reactive frequency multipliers involve a special case of the Manley–Rowe relations since only a single source is used. If we assume a source at frequency ω_1, then setting $m = 0$ in (13.59a) gives

$$\sum_{n=1}^{\infty} P_{n0} = 0,$$

or

$$\sum_{n=2}^{\infty} P_{n0} = -P_{10}, \tag{13.60}$$

where P_{n0} represents the power associated with the nth harmonic (the DC term for $n = 0$ is zero). In practice, $P_{10} > 0$ because this represents power delivered by the source, while the summation in (13.60) represents the total power contained in all the harmonics of the input signal, as generated by the nonlinear capacitor. If all harmonics but the nth are terminated with lossless reactive loads, the power balance of (13.60) reduces to

$$\left|\frac{P_{n0}}{P_{10}}\right| = 1, \tag{13.61}$$

indicating that it is theoretically possible to achieve 100% conversion efficiency for any harmonic. Of course, in practice, losses in the diode and matching circuitry serve to reduce the achievable efficiency substantially.

A block diagram of a diode frequency multiplier is shown in Figure 13.20. An input signal of frequency f_0 is applied to the diode, which is terminated with reactive loads at all frequencies except nf_0, the desired harmonic. If the diode junction capacitance has a square-law I–V characteristic, it is often necessary to terminate unwanted harmonics with short circuits if harmonics higher than the second are to be generated. This is because voltages at higher harmonics may not be generated unless lower harmonic currents are allowed to flow. These currents are commonly referred to as *idler currents*. For example, a varactor tripler will generally require terminations to allow idler currents at $2f_0$. Typical conversion efficiencies for varactor

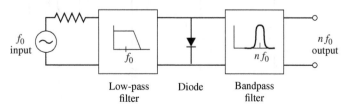

FIGURE 13.20 | Block diagram of a diode frequency multiplier.

multipliers range from 50 to 80% for doublers and triplers at 50 GHz. The upper frequency limit is controlled mainly by f_c, the cutoff frequency of the diode, which depends on the series resistance and dynamic junction capacitance. Typical varactor cutoff frequencies can exceed 1000 GHz, but efficient frequency multiplication requires that $nf_0 \ll f_c$.

Resistive Diode Multipliers

Resistive multipliers generally use forward-biased Schottky-barrier diodes to provide a nonlinear I–V characteristic. Resistive multipliers are less popular than reactive multipliers because their efficiencies are lower, especially for higher harmonic numbers. However, resistive multipliers offer better bandwidths, and more stable operation, than reactive multipliers. In addition, at high millimeter wave frequencies even the best varactor diodes begin to exhibit resistive properties. Since a resistive frequency multiplier is not lossless, the Manley–Rowe relations do not strictly apply. However, we can derive a similar set of relations for a nonlinear resistor, and demonstrate an important result for frequency conversion using nonlinear resistors.

Consider the resistive multiplier circuit shown in Figure 13.21. We have simplified the analysis by specializing to the frequency multiplier case by considering only a single source frequency—the more general case of two frequency sources is treated in reference [11]. For a source frequency ω, the nonlinear resistor generates harmonics of the form $n\omega$, so the resistor voltage and current can be written as a Fourier series:

$$v(t) = \sum_{m=-\infty}^{\infty} V_m e^{jm\omega t}, \tag{13.62a}$$

$$i(t) = \sum_{m=-\infty}^{\infty} I_m e^{jm\omega t}. \tag{13.62b}$$

The Fourier coefficients are determined as

$$V_m = \frac{1}{T} \int_{t=0}^{T} v(t) e^{-jm\omega t}\, dt, \tag{13.63a}$$

$$I_m = \frac{1}{T} \int_{t=0}^{T} i(t) e^{-jm\omega t}\, dt. \tag{13.63b}$$

Since $v(t)$ and $i(t)$ are real functions, we must have $V_m = V_{-m}^*$ and $I_m = I_{-m}^*$. The power associated with the mth harmonic is (ignoring a factor of 4)

$$P_m = 2\,\text{Re}\left\{V_m I_m^*\right\} = V_m I_m^* + V_m^* I_m. \tag{13.64}$$

Multiplying V_m of (13.63a) by $-m^2 I_m^*$ and summing gives

$$-\sum_{m=-\infty}^{\infty} m^2 V_m I_m^* = \frac{-1}{T} \int_{t=0}^{T} v(t) \sum_{m=-\infty}^{\infty} m^2 I_m^* e^{-jm\omega t}\, dt. \tag{13.65}$$

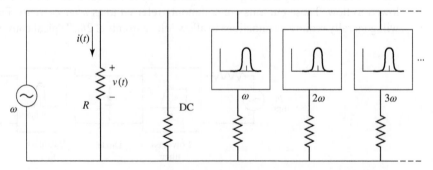

FIGURE 13.21 | Conceptual circuit for the derivation of power relations in a resistive frequency multiplier.

Next, use the result that

$$\frac{\partial^2 i(t)}{\partial t^2} = -\sum_{m=-\infty}^{\infty} m^2 \omega^2 I_m e^{jm\omega t} = -\sum_{m=-\infty}^{\infty} m^2 \omega^2 I_m^* e^{-jm\omega t}$$

to write (13.65) as

$$-\sum_{m=-\infty}^{\infty} m^2 V_m I_m^* = \frac{1}{\omega^2 T} \int_{t=0}^{T} v(t) \frac{\partial^2 i(t)}{\partial t^2} dt$$

$$= \frac{1}{2\pi\omega} v(t) \frac{\partial i(t)}{\partial t} \bigg|_{t=0}^{T} - \frac{1}{2\pi\omega} \int_{t=0}^{T} \frac{\partial v(t)}{\partial t} \frac{\partial i(t)}{\partial t} dt. \tag{13.66}$$

Since $v(t)$ and $i(t)$ are periodic functions (period T), we have $v(0) = v(T)$ and $i(0) = i(T)$. Derivatives of $i(t)$ have the same periodicity, so the second to last term in (13.66) vanishes. In addition, we can write

$$\frac{\partial v(t)}{\partial t} \frac{\partial i(t)}{\partial t} = \frac{\partial v(t)}{\partial t} \frac{\partial i}{\partial v} \frac{\partial v(t)}{\partial t} = \frac{\partial i}{\partial v} \left(\frac{\partial v(t)}{\partial t}\right)^2.$$

Equation (13.66) then reduces to

$$\sum_{m=-\infty}^{\infty} m^2 V_m I_m^* = \frac{1}{2\pi\omega} \int_{t=0}^{T} \frac{\partial i}{\partial v} \left(\frac{\partial v(t)}{\partial t}\right)^2 dt = \sum_{m=0}^{\infty} m^2 \left(V_m I_m^* + V_m^* I_m\right) = \sum_{m=0}^{\infty} m^2 P_m,$$

or

$$\sum_{m=0}^{\infty} m^2 P_m = \frac{1}{2\pi\omega} \int_{t=0}^{T} \frac{\partial i}{\partial v} \left(\frac{\partial v(t)}{\partial t}\right)^2 dt. \tag{13.67}$$

For positive nonlinear resistors (defined as having an I–V curve whose slope is always positive), the integrand of (13.67) will always be positive. Thus (13.67) can be reduced to

$$\sum_{m=0}^{\infty} m^2 P_m \geq 0. \tag{13.68}$$

If all harmonics are terminated in reactive loads except for ω (the fundamental) and $m\omega$ (the desired harmonic), then (13.68) reduces to $P_1 + m^2 P_m > 0$. The power $P_1 > 0$ is delivered by the source, while $P_m < 0$ represents harmonic power supplied by the device. The maximum theoretical conversion efficiency is then given as

$$\left|\frac{P_m}{P_1}\right| \leq \frac{1}{m^2}. \tag{13.69}$$

This result indicates that the efficiency of a resistive frequency multiplier drops as the square of the multiplication factor.

The performance of diode frequency multipliers can often be improved by using two diodes in a balanced configuration. This can lead to increased output power, improved input impedance characteristics, and the rejection of certain (all even or all odd) harmonics. Two diodes can be fed using a quadrature hybrid, or two diodes can be configured in an *antiparallel* arrangement (back-to-back with reversed polarities). The antiparallel configuration will reject all even harmonics of the input frequency.

Transistor Multipliers

Compared to diode frequency multipliers, transistor multipliers offer better bandwidth and the possibility of conversion efficiencies greater than 100% (conversion gain). FET multipliers also require less input and DC power than diode multipliers. In the past, before solid-state amplifiers were available at millimeter wave

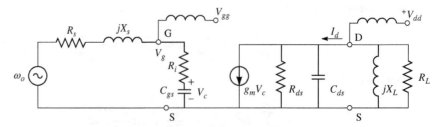

FIGURE 13.22 | Circuit diagram of a FET frequency multiplier. The transistor is modeled using a unilateral equivalent circuit.

frequencies, high-power diode multipliers were one of the few ways of generating millimeter wave power. Today, however, it is possible to generate the required frequency at low power, then amplify that signal to the desired power level using transistor amplifiers. This approach results in better efficiency and lower DC power requirements, and it allows the separate optimization of signal generation and amplification functions. Transistor multipliers are well suited for this application.

There are several nonlinearities that exist in a FET device that can be used for harmonic generation: the transconductance near pinch-off, the output conductance near pinch-off, the rectifying properties of the Schottky gate, and the varactor-like capacitances at the gate and drain. For frequency doubler operation, the most useful of these is the rectification property, where the FET is biased to conduct only during the positive half of the input signal waveform. This results in operation similar to that of a class B amplifier, and provides a multiplier circuit that is useful for low-power output (typically less than 10 dBm) at frequencies up to 60–100 GHz. Bipolar transistors can also be used for frequency multiplication, with the capacitance of the collector-base junction providing the necessary nonlinearity.

The basic circuit of a class B FET frequency multiplier is shown in Figure 13.22. A unilateral device is assumed here to simplify the analysis. The source is a generator of frequency ω_0, with period $T = 2\pi/\omega_0$, and matched to the FET with the source impedance $R_s + jX_s$. The drain of the FET is terminated with a load impedance $R_L + jX_L$, which is chosen to form a parallel RLC resonator with C_{ds} at the desired harmonic frequency, $n\omega_0$. The gate is biased at a DC voltage of $V_{gg} < 0$, while the drain is biased at $V_{dd} > 0$.

The operation of the FET multiplier can be understood with the help of the waveforms shown in Figure 13.23. As seen in Figure 13.23a, the FET is biased below the turn-on voltage, V_t, so the transistor does not conduct until the gate voltage exceeds V_t. The resulting drain current is shown in Figure 13.23b, and is seen to be similar in form to a half-wave rectified version of the gate voltage. This waveform is rich in harmonics, so the drain resonator can be designed to present a short circuit at the fundamental and all undesired harmonics, and an open circuit at the desired harmonic frequency. The resulting drain voltage for $n = 2$ is shown in Figure 13.23c.

We can make an approximate analysis of the FET multiplier by representing the drain current in terms of a Fourier series. If we assume that the drain current waveform is a half-cosine function of the form

$$i_d(t) = \begin{cases} I_{\max} \cos \dfrac{\pi t}{\tau} & \text{for } |t| < \tau/2 \\ 0 & \text{for } \tau/2 < |t| < T/2, \end{cases} \tag{13.70}$$

where τ is the duration of the drain current pulse, we can find the Fourier series as

$$i_d(t) = \sum_{n=0}^{\infty} I_n \cos \frac{2\pi n t}{T}, \tag{13.71}$$

with the Fourier coefficients given by

$$I_0 = I_{\max} \frac{2\tau}{\pi T}, \tag{13.72a}$$

$$I_n = I_{\max} \frac{4\tau}{\pi T} \frac{\cos(n\pi\tau/T)}{1 - (2n\tau/T)^2} \quad \text{for } n > 0. \tag{13.72b}$$

The coefficient I_n represents the drain current of harmonic frequency $n\omega_0$, so maximizing multiplier efficiency involves maximizing I_n. Since (13.72b) clearly shows that the maximum value of I_n decreases

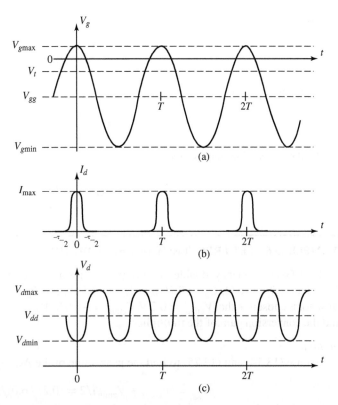

FIGURE 13.23 | Voltage and currents in the FET multiplier (doubler) circuit of Figure 13.22. (a) Gate voltage when the transistor is biased just below pinch-off. (b) Drain current, which conducts when the gate voltage is above the threshold voltage. (c) Drain voltage when the load resonator is tuned to the second harmonic.

with n, circuits of this type are generally limited to frequency doublers or triplers. For a given value of n, the maximum value of $I_n/I_{\max}$ depends on the ratio τ/T: for $n = 2$ the optimum occurs at $\tau/T = 0.35$, while for $n = 3$ the optimum occurs at $\tau/T = 0.22$. Because of device and biasing constraints, however, the designer usually has very little control of the pulse width τ, and practical values of τ/T are usually greater than optimum. Examination of Figure 13.23a shows that the normalized pulse duration is related to the gate voltages V_t, $V_{g\min}$, and $V_{g\max}$ as

$$\cos \frac{\pi \tau}{T} = \frac{2V_t - V_{g\max} - V_{g\min}}{V_{g\max} - V_{g\min}}. \tag{13.73}$$

The gate bias voltage satisfies the relation that

$$V_{gg} = (V_{g\max} + V_{g\min})/2, \tag{13.74}$$

and the peak value of the AC component of the gate voltage (frequency ω_0) is given by

$$V_g = V_{g\max} - V_{gg}. \tag{13.75}$$

Then the input power delivered to the FET can be expressed as

$$P_{\text{in}} = \frac{1}{2}|I_g|^2 R_i = \frac{|V_g|^2 R_i}{2|R_i - j/\omega_0 C_{gs}|^2}. \tag{13.76}$$

If the source is conjugately matched to the transistor, the input power will be equal to the available power, P_{avail}.

On the load side, the peak value of the AC component of the drain voltage (frequency $n\omega_0$) is given by

$$V_L = I_n R_L = (V_{d\max} - V_{d\min})/2, \tag{13.77}$$

assuming resonance of X_L and C_{ds}. This gives the optimal load resistance as

$$R_L = \frac{V_{dmax} - V_{dmin}}{2I_n}. \tag{13.78}$$

Then the output power at the harmonic $n\omega_0$ is

$$P_n = \frac{1}{2}|I_n|^2 R_L. \tag{13.79}$$

Finally, the conversion gain is given as

$$G_c = \frac{P_n}{P_{avail}}. \tag{13.80}$$

EXAMPLE 13.6 FET FREQUENCY DOUBLER DESIGN

A 12–24 GHz frequency doubler is designed using a GaAs MESFET with the following parameters: $V_t = -2.0$ V, $R_i = 10\,\Omega$, $C_{gs} = 0.20$ pF, $C_{ds} = 0.15$ pF, and $R_{ds} = 40\,\Omega$. Assume the operating point of the transistor is chosen so that $V_{gmax} = 0.2$ V, $V_{gmin} = -6.0$ V, $V_{dmax} = 5.0$ V, $V_{dmin} = 1.0$ V, and $I_{max} = 80$ mA. Find the conversion gain of the multiplier.

Solution
We first use (13.74) and (13.75) to find the peak value of the AC input voltage. The gate bias voltage is

$$V_{gg} = (V_{gmax} + V_{gmin})/2 = (0.2 - 6.0)/2 = -2.9 \text{ V},$$

and the peak AC input voltage is

$$V_g = V_{gmax} - V_{gg} = 0.2 + 2.9 = 3.1 \text{ V}.$$

Then the input power is given by (13.76):

$$P_{in} = \frac{|V_g|^2 R_i}{2|R_i - j/\omega_0 C_{gs}|^2} = \frac{(3.1)^2(10)}{2[(10)^2 + (1/2\pi(12 \times 10^9)(0.2 \times 10^{-12}))^2]}$$

$$= 10.7 \text{ mW}.$$

The pulse width is found from (13.73) as

$$\cos\frac{\pi\tau}{T} = \frac{2V_t - V_{gmax} - V_{gmin}}{V_{gmax} - V_{gmin}} = \frac{2(-2.0) - 0.2 + 6.0}{0.2 + 6.0} = 0.29,$$

for

$$\frac{\tau}{T} = 0.406.$$

Then the load current for the second harmonic is given by (13.72b):

$$I_2 = I_{max}\frac{4\tau}{\pi T}\frac{\cos(2\pi\tau/T)}{1 - (4\tau/T)^2} = 0.262 I_{max} = 21.0 \text{ mA}.$$

The load resistance required to match the transistor is found from (13.78):

$$R_L = \frac{V_{dmax} - V_{dmin}}{2I_2} = \frac{5-1}{2(0.021)} = 95.2\,\Omega.$$

The output power at 24 GHz is given by (13.79):

$$P_2 = \frac{1}{2}|I_2|^2 R_L = \frac{1}{2}(0.021)^2(95.2) = 21.0 \text{ mW}.$$

Finally, the conversion gain is, assuming the input is conjugately matched,

$$G_c = \frac{P_2}{P_{\text{avail}}} = \frac{21.0}{10.7} = 2.9 \text{ dB}.$$

The load reactance required to resonate the second harmonic is $X_L = 1/2\omega_0 C_{ds} = 44.2 \ \Omega$, which corresponds to an inductance of 0.293 nH.

13.5 MIXERS

A mixer is a three-port device that uses a nonlinear or time-varying element to achieve frequency conversion. As introduced in Section 11.1, an ideal mixer produces an output consisting of the sum and difference frequencies of its two input signals. Operation of practical RF and microwave mixers is usually based on the nonlinearity provided by either a diode or a transistor. As we have seen, a nonlinear component can generate a wide variety of harmonics and other products of input frequencies, so filtering must be used to select the desired frequency components. Modern microwave systems typically use several mixers and filters to perform the functions of frequency up-conversion and down-conversion between baseband signal frequencies and RF carrier frequencies.

We begin by discussing some of the important characteristics of mixers, such as image frequency, conversion loss, noise effects, and intermodulation distortion. Next we discuss the operation of single-ended mixers, using either a single diode or a transistor as the nonlinear element. The balanced diode mixer circuit is then described, followed by a brief description of more specialized mixer circuits.

Mixer Characteristics

The symbol and functional diagram for a mixer are shown in Figure 13.24. The mixer symbol is intended to imply that the output is proportional to the product of the two input signals. We will see that this is an idealized view of mixer operation, which in actuality produces a large variety of harmonics and other undesired products of the input signals. Figure 13.24a illustrates the operation of *frequency up-conversion*, as occurs in a transmitter. A local oscillator (LO) signal at the relatively high frequency f_{LO} is connected to one of the input ports of the mixer. The LO signal can be represented as

$$v_{\text{LO}}(t) = \cos 2\pi f_{\text{LO}} t. \tag{13.81}$$

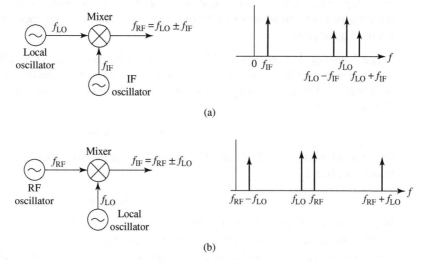

(a)

(b)

FIGURE 13.24 | Frequency conversion using a mixer. (a) Up-conversion. (b) Down-conversion.

A lower frequency baseband or intermediate frequency (IF) signal is applied to the other mixer input. This signal typically contains the information or data to be transmitted, and can be expressed for our purposes as

$$v_{IF}(t) = \cos 2\pi f_{IF} t. \tag{13.82}$$

The output of the idealized mixer is given by the product of the LO and IF signals:

$$v_{RF}(t) = K v_{LO}(t) v_{IF}(t) = K \cos 2\pi f_{LO} t \cos 2\pi f_{IF} t$$

$$= \frac{K}{2} [\cos 2\pi (f_{LO} - f_{IF}) t + \cos 2\pi (f_{LO} + f_{IF}) t], \tag{13.83}$$

where K is a constant accounting for the voltage conversion loss of the mixer. The RF output is seen to consist of the sum and differences of the input signal frequencies:

$$f_{RF} = f_{LO} \pm f_{IF}. \tag{13.84}$$

The spectra of the input and output signals are shown in Figure 13.24a, where we see that the mixer has the effect of modulating the LO signal with the IF signal. The sum and difference frequencies at $f_{LO} \pm f_{IF}$ are called the *sidebands* of the carrier frequency f_{LO}, with $f_{LO} + f_{IF}$ being the *upper sideband* (USB), and $f_{LO} - f_{IF}$ being the *lower sideband* (LSB). A *double-sideband* (DSB) signal contains both upper and lower sidebands, as in (13.83), while a *single-sideband* (SSB) signal can be produced by filtering or by using a single-sideband mixer.

Conversely, Figure 13.24b shows the process of *frequency down-conversion*, as used in a receiver. In this case an RF input signal of the form

$$v_{RF}(t) = \cos 2\pi f_{RF} t \tag{13.85}$$

is applied to the input of the mixer, along with the LO signal of (13.81). The output of the mixer is

$$v_{IF}(t) = K v_{RF}(t) v_{LO}(t) = K \cos 2\pi f_{RF} t \cos 2\pi f_{LO} t$$

$$= \frac{K}{2} [\cos 2\pi (f_{RF} - f_{LO}) t + \cos 2\pi (f_{RF} + f_{LO}) t]. \tag{13.86}$$

Thus the mixer output consists of the sum and difference of the input signal frequencies. The spectrum for these signals is shown in Figure 13.24b. In practice, the RF and LO frequencies are relatively close together, so the sum frequency is approximately twice the RF frequency, while the difference is much smaller than f_{RF}. The desired IF output in a receiver is the difference frequency, $f_{RF} - f_{LO}$, which is easily selected by low-pass filtering:

$$f_{IF} = f_{RF} - f_{LO}. \tag{13.87}$$

Note that the above discussion only considers the sum and difference outputs as generated by multiplication of the input signals, whereas in a realistic mixer many more products will be generated due to the more complicated nonlinear behavior of the diode or transistor. These products are usually undesirable and are removed by filtering.

Image frequency: In a receiver the RF input signal at frequency f_{RF} is typically delivered from the antenna, which may receive RF signals over a relatively wide band of frequencies. For a receiver with an LO frequency f_{LO} and IF frequency f_{IF}, (13.87) gives the RF input frequency that will be down-converted to the IF frequency as

$$f_{RF} = f_{LO} + f_{IF}, \tag{13.88a}$$

since the insertion of (13.88a) into (13.87) yields f_{IF} (after low-pass filtering). Now consider the RF input frequency given by

$$f_{IM} = f_{LO} - f_{IF}. \tag{13.88b}$$

Insertion of (13.88b) into (13.87) yields $-f_{IF}$ (after low-pass filtering). Mathematically, this frequency is identical to f_{IF} because the Fourier spectrum of any real signal is symmetric about zero frequency, and thus contains negative as well as positive frequencies. The RF frequency defined in (13.88b) is called the *image response*. The image response is important in receiver design because a received RF signal at the image frequency of (13.88b) is indistinguishable at the IF stage from the desired RF signal of frequency (13.88a) unless steps are taken in the RF stages of the receiver to preselect signals only within the desired RF frequency band.

The choice of which RF frequency in (13.88) is the desired and which is the image response is arbitrary, depending on whether the LO frequency is above or below the desired RF frequency. Another way of viewing this difference is to note that f_{IF} in (13.88) may be negative. Observe that the desired and image frequencies of (13.88a) and (13.88b) are separated by $2f_{IF}$.

Another implication of (13.87) and the fact that f_{IF} may be negative is that there are two LO frequencies that can be used for a given RF and IF frequency:

$$f_{LO} = f_{RF} \pm f_{IF}, \tag{13.89}$$

since taking the difference frequency of f_{RF} with these two LO frequencies gives $\pm f_{IF}$. These two frequencies correspond to the upper and lower sidebands when a mixer is operated as an up-converter. In practice, most receivers use a local oscillator set at the upper sideband, $f_{LO} = f_{RF} + f_{IF}$, because this requires a smaller LO tuning ratio when the receiver must select RF signals over a given band.

Conversion loss: Mixer design requires impedance matching at three ports, complicated by the fact that several frequencies and their harmonics are involved. Ideally, each mixer port would be matched at its particular frequency (RF, LO, or IF), and undesired frequency products would be absorbed with resistive loads, or blocked with reactive terminations. Resistive loads increase mixer losses, however, and reactive loads can be very frequency sensitive. In addition, there are inherent losses in the frequency conversion process because of the generation of undesired harmonics and other frequency products. An important figure of merit for a mixer is therefore the *conversion loss*, which is defined as the ratio of available RF input power to the available IF output power, expressed in dB:

$$L_c = 10 \log \frac{\text{available RF input power}}{\text{available IF output power}} \geq 0 \text{ dB}. \tag{13.90}$$

Conversion loss accounts for resistive losses in a mixer as well as loss in the frequency conversion process from RF to IF ports. Conversion loss applies to both up-conversion and down-conversion, even though the context of the above definition is for the latter case. Since the RF stages of receivers operate at much lower power levels than do transmitters, minimum conversion loss is more critical for receivers because of the importance of minimizing losses in the RF stages to maximize receiver noise figure.

Practical diode mixers typically have conversion losses between 4 and 7 dB in the 1–10 GHz range. Transistor mixers have lower conversion loss, and they may even have *conversion gain* of a few dB. One factor that strongly affects conversion loss is the LO power level; minimum conversion loss often occurs for LO powers between 0 and 10 dBm. This power level is large enough that the accurate characterization of mixer performance often requires nonlinear analysis.

Noise figure: Noise is generated in mixers by the diode or transistor elements, and by thermal sources due to resistive losses. Noise figures of practical mixers range from 1 to 5 dB, with diode mixers generally achieving lower noise figures than transistor mixers. The noise figure of a mixer depends on whether its input is a single-sideband signal or a double-sideband signal. This is because the mixer will down-convert noise at both sideband frequencies (since these have the same IF), but the power of an SSB signal is one-half that of a DSB signal (for the same amplitude). To derive the relation between the noise figure for these two cases, first consider a DSB input signal of the form

$$v_{DSB}(t) = A[\cos(\omega_{LO} - \omega_{IF})t + \cos(\omega_{LO} + \omega_{IF})t]. \tag{13.91}$$

Upon mixing with an LO signal $\cos \omega_{LO} t$ and low-pass filtering, the down-converted IF signal will be

$$v_{IF}(t) = \frac{AK}{2} \cos(\omega_{IF}t) + \frac{AK}{2} \cos(-\omega_{IF}t) = AK \cos \omega_{IF}t, \tag{13.92}$$

where K is a constant accounting for the conversion loss for each sideband. The average power of the DSB input signal of (13.91) is

$$S_i = \frac{A^2}{2} + \frac{A^2}{2} = A^2,$$

and the average power of the output IF signal is

$$S_o = \frac{A^2 K^2}{2}.$$

For noise figure, the input noise power is defined as $N_i = kT_0 B$, where $T_0 = 290$ K and B is the IF bandwidth. The total output noise power is equal to the input noise plus N_{added}, the noise power added by the mixer, divided by the conversion loss (assuming a reference at the mixer input):

$$N_o = \frac{(KT_0 B + N_{added})}{L_c}.$$

Then using the definition of noise figure gives the DSB noise figure of the mixer as

$$F_{DSB} = \frac{S_i N_o}{S_o N_i} = \frac{2}{K^2 L_c} \left(1 + \frac{N_{added}}{kT_0 B} \right). \tag{13.93}$$

The corresponding analysis for the SSB case begins with an SSB input signal of the form

$$v_{SSB}(t) = A \cos(\omega_{LO} - \omega_{IF})t. \tag{13.94}$$

Upon mixing with the LO signal $\cos \omega_{LO} t$ and low-pass filtering, the down-converted IF signal will be

$$v_{IF}(t) = \frac{AK}{2} \cos(\omega_{IF} t). \tag{13.95}$$

The average power of the SSB input signal of (13.94) is

$$S_i = \frac{A^2}{2},$$

and the average power of the output IF signal is

$$S_o = \frac{A^2 K^2}{8}.$$

The input and output noise powers are the same as for the DSB case, so the noise figure for an SSB input signal is

$$F_{SSB} = \frac{S_i N_o}{S_o N_i} = \frac{4}{K^2 L_c} \left(1 + \frac{N_{added}}{kT_0 B} \right). \tag{13.96}$$

Comparison with (13.93) shows that the noise figure of the SSB case is twice that of the DSB case:

$$F_{SSB} = 2F_{DSB}. \tag{13.97}$$

Other mixer characteristics: Since mixers involve nonlinearity, they will produce intermodulation products. Typical values of IIP_3 for mixers range from 15 to 30 dBm. Another important characteristic of a mixer is the isolation between the RF and LO ports. Ideally, the LO and RF ports would be decoupled, but internal impedance mismatches and limitations of coupler performance often result in some LO power being coupled out of the RF port. This is a potential problem for receivers that drive the RF port directly from the antenna because LO power coupled through the mixer to the RF port will be radiated by the antenna. Because such signals can interfere with other services or users, regulatory agencies often set stringent limits on the RF power radiated by receivers. This problem can be largely alleviated by using a bandpass filter between the antenna and mixer, or by using an RF amplifier ahead of the mixer. Isolation between the LO and RF ports is highly dependent on the type of coupler used for diplexing these two inputs, but typical values range from 20 to 40 dB.

EXAMPLE 13.7   IMAGE FREQUENCY

The E-GSM-900 digital cellular telephone system uses a receive frequency band of 925–960 MHz, with a first IF frequency of 87 MHz and a channel bandwidth of 30 kHz. What are the two possible ranges for the LO frequency? If the upper LO frequency range is used, determine the image frequency range. Does the image frequency fall within the receive passband?

Solution

By (13.89), the two possible LO frequency ranges are

$$f_{LO} = f_{RF} \pm f_{IF} = (925 \text{ to } 960) \pm 87 = \begin{cases} 1012 \text{ to } 1047 \text{ MHz} \\ 838 \text{ to } 873 \text{ MHz.} \end{cases}$$

Using the 1012–1047 MHz LO, we find that (13.87) gives the IF frequency as

$$f_{IF} = f_{RF} - f_{LO} = (925 \text{ to } 960) - (1012 \text{ to } 1047) = -87 \text{ MHz},$$

so from (13.88b) the RF image frequency range is

$$f_{IM} = f_{LO} - f_{IF} = (1012 \text{ to } 1047) + 87 = 1099 \text{ to } 1134 \text{ MHz},$$

which is well outside the receive passband.

The above treatment of mixers is idealized because of the assumption that the output was proportional to the product of the input signals, thus producing only sum and difference frequencies (for sinusoidal inputs). We now discuss more realistic mixers and show that the output does indeed contain a term proportional to the product of the inputs, but it also contains many higher order products as well.

Single-Ended Diode Mixer

A basic diode mixer circuit is shown in Figure 13.25a. This type of mixer is called a *single-ended mixer* because it uses a single diode element. The RF and LO inputs are combined in a *diplexer*, which superimposes the two input voltages to drive the diode. The diplexing function can be implemented using a directional coupler or hybrid junction to provide signal combining as well as isolation between the two inputs. The diode may be biased with a DC bias voltage, which must be decoupled from the RF signal paths. This is done by using DC blocking capacitors on either side of the diode, and an RF choke between the diode and the bias voltage source. The AC output of the diode is passed through a low-pass filter to provide the desired IF output voltage. This description is for application as a down-converter, but the same mixer can be used for up-conversion since each port may be used interchangeably as an input or output port.

The AC equivalent circuit of the mixer is shown in Figure 13.25b, where the RF and LO input voltages are represented as two series-connected voltage sources. Let the RF input voltage be a cosine wave of

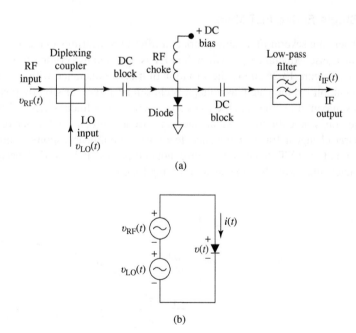

FIGURE 13.25 | (a) Circuit for a single-ended diode mixer. (b) Idealized equivalent circuit.

frequency ω_{RF}:

$$v_{RF}(t) = V_{RF} \cos \omega_{RF} t, \tag{13.98}$$

and let the LO input voltage be a cosine wave of frequency ω_{LO}:

$$v_{LO}(t) = V_{LO} \cos \omega_{LO} t. \tag{13.99}$$

Using the small-signal approximation of (11.6) gives the total diode current as

$$i(t) = I_0 + G_d[v_{RF}(t) + v_{LO}(t)] + \frac{G_d'}{2}[v_{RF}(t) + v_{LO}(t)]^2 + \cdots. \tag{13.100}$$

The first term in (13.100) is the DC bias current, which will be blocked from the IF output by the DC blocking capacitors. The second term is a replication of the RF and LO input signals, which will be filtered out by the low-pass IF filter. This leaves the third term, which can be rewritten using trigonometric identities as

$$\begin{aligned}
i(t) &= \frac{G_d'}{2}\left(V_{RF}\cos\omega_{RF}t + V_{LO}\cos\omega_{LO}t\right)^2 \\
&= \frac{G_d'}{2}\left(V_{RF}^2\cos^2\omega_{RF}t + 2V_{RF}V_{LO}\cos\omega_{RF}t\cos\omega_{LO}t + V_{LO}^2\cos^2\omega_{LO}t\right) \\
&= \frac{G_d'}{4}\Big[V_{RF}^2(1 + \cos 2\omega_{RF}t) + V_{LO}^2(1 + \cos 2\omega_{LO}t) + 2V_{RF}V_{LO}\cos(\omega_{RF}-\omega_{LO})t \\
&\quad + 2V_{RF}V_{LO}\cos(\omega_{RF}+\omega_{LO})t\Big].
\end{aligned}$$

This result is seen to contain several new signal components, only one of which produces the desired IF difference product. The two DC terms again will be blocked by the blocking capacitors, and the $2\omega_{RF}$, $2\omega_{LO}$, and $\omega_{RF} + \omega_{LO}$ terms will be blocked by the low-pass filter. This leaves the IF output current as

$$i_{IF}(t) = \frac{G_d'}{2}V_{RF}V_{LO}\cos\omega_{IF}t, \tag{13.101}$$

where $\omega_{IF} = \omega_{RF} - \omega_{LO}$ is the IF frequency. The spectrum of the down-converting single-ended mixer is thus identical to that of the idealized mixer shown in Figure 13.24b.

Single-Ended FET Mixer

There are several FET parameters that offer nonlinearities that can be used for mixing, but the strongest is the transconductance, g_m, when the FET is operated in a common source configuration with a negative gate bias. Figure 13.26 shows the variation of transconductance with gate bias for a typical FET. When used as an amplifier, the gate bias voltage is chosen near zero, or slightly positive, so the transconductance is near its maximum value, and the transistor operates as a linear device. When the gate bias is near the *pinch-off* region, where the transconductance approaches zero, a small positive variation of gate voltage can cause a large change in transconductance, leading to a nonlinear response. Thus the LO voltage can be applied to the gate of the FET to pump the transconductance to switch the FET between high- and low-transconductance states, thus providing the desired mixing function.

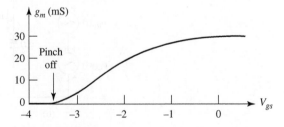

FIGURE 13.26 | Variation of FET transconductance versus gate-to-source voltage.

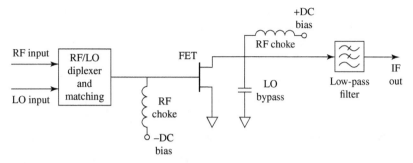

FIGURE 13.27 | Circuit for a single-ended FET mixer.

The circuit for a single-ended FET mixer is shown in Figure 13.27. A diplexing coupler is again used to combine the RF and LO signals at the gate of the FET. An impedance matching network is also usually required between the inputs and the FET, which typically presents a very low input impedance. RF chokes are used to bias the gate at a negative voltage near pinch-off, and to provide a positive bias for the drain of the FET. A bypass capacitor at the drain provides a return path for the LO signal, and a low-pass filter provides the final IF output signal.

Our analysis of the mixer of Figure 13.27 follows the original work described in reference [12]. The simplified equivalent circuit is shown in Figure 13.28, and is based on the unilateral equivalent circuit of a FET introduced in Section 11.3. The RF and LO input voltages are given in (13.98) and (13.99). Let $Z_g = R_g + jX_g$ be the Thevenin source impedance for the RF input port, and let $Z_L = R_L + jX_L$ be the Thevenin source impedance at the IF output port. These impedances are complex to allow a conjugate match at the input and output ports for maximum power transfer. The LO port has a real generator impedance of Z_0 since we are not concerned with maximum power transfer for the LO signal.

Since the FET transconductance is driven by the LO signal, its time variation can be expressed as a Fourier series in terms of harmonics of the LO:

$$g(t) = g_0 + 2 \sum_{n=1}^{\infty} g_n \cos n\omega_0 t. \tag{13.102}$$

Because we do not have an explicit formula for the transconductance, we cannot calculate directly the Fourier coefficients of (13.102), but must rely on measurements for these values. As we will see, the desired down-conversion result is due solely to the $n = 1$ term of the Fourier series, so we only need the g_1 coefficient. Measurements typically give a value in the range of 10 mS for g_1.

The conversion gain of the FET mixer can be found as

$$G_c = \frac{P_{\text{IF-avail}}}{P_{\text{RF-avail}}} = \frac{\dfrac{\left|V_D^{\text{IF}}\right|^2 R_L}{|Z_L|^2}}{\dfrac{|V_{\text{RF}}|^2}{4R_g}} = \frac{4R_g R_L}{|Z_L|^2} \left|\frac{V_D^{\text{IF}}}{V_{\text{RF}}}\right|^2, \tag{13.103}$$

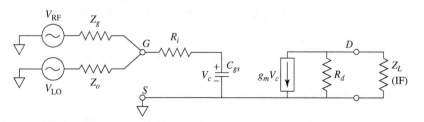

FIGURE 13.28 | Equivalent circuit for the FET mixer of Figure 13.27.

where V_D^{IF} is the IF drain voltage, and the impedances Z_g and Z_L are chosen for maximum power transfer at the RF and IF ports. The RF frequency component of the phasor voltage across the gate-to-source capacitance is given in terms of the voltage divider between Z_g, R_i, and C_{gs}:

$$V_c^{RF} = \frac{V_{RF}}{j\omega_{RF}C_{gs}\left[(R_i + Z_g) - \dfrac{j}{\omega_{RF}C_{gs}}\right]} = \frac{V_{RF}}{1 + j\omega_{RF}C_{gs}(R_i + Z_g)}. \tag{13.104}$$

Multiplying the transconductance of (13.102) by $v_c^{RF}(t) = V_c^{RF}\cos\omega_{RF}t$ gives terms of the form

$$g_m(t)v_c^{RF}(t) = g_0 V_c^{RF}\cos\omega_{RF}t + 2g_1 V_c^{RF}\cos\omega_{RF}t\cos\omega_{LO}t + \cdots. \tag{13.105}$$

The down-converted IF frequency component can be extracted from the second term of (13.105) using the usual trigonometric identity

$$g_m(t)v_c^{RF}(t)\Big|_{\omega_{IF}} = g_1 V_c^{RF}\cos\omega_{IF}t, \tag{13.106}$$

where $\omega_{IF} = \omega_{RF} - \omega_{LO}$. Then the IF component of the drain voltage is, in phasor form,

$$V_D^{IF} = -g_1 V_c^{RF}\left(\frac{R_d Z_L}{R_d + Z_L}\right) = \frac{-g_1 V_{RF}}{1 + j\omega_{RF}C_{gs}(R_i + Z_g)}\left(\frac{R_d Z_L}{R_d + Z_L}\right), \tag{13.107}$$

where (13.104) has been used. Using this result in (13.103) gives the conversion gain (before conjugate matching) as

$$G_c\Big|_{\substack{not \\ matched}} = \left(\frac{2g_1 R_d}{\omega_{RF}C_{gs}}\right)^2 \frac{R_g}{\left[(R_i + R_g)^2 + \left(X_g - \dfrac{1}{\omega_{RF}C_{gs}}\right)^2\right]}\frac{R_L}{\left[(R_d + R_L)^2 + X_L^2\right]}.$$

We now conjugately match the RF and IF ports to maximize the conversion gain. Thus we let $R_g = R_i, X_g = 1/\omega_{RF}C_{gs}, R_L = R_d$, and $X_L = 0$, which reduces the above result to

$$G_c = \frac{g_1^2 R_d}{4\omega_{RF}^2 C_{gs}^2 R_i}. \tag{13.108}$$

The quantities g_1, R_d, R_i, and C_{gs} are all parameters of the FET. Practical mixer circuits generally use matching circuits to transform the FET impedance to 50 Ω for the RF, LO, and IF ports.

EXAMPLE 13.8   MIXER CONVERSION GAIN

A single-ended FET mixer is to be designed for a wireless local area network receiver operating at 5 GHz. The parameters of the FET are $R_d = 200\ \Omega, R_i = 10\ \Omega, C_{gs} = 0.30$ pF, and $g_1 = 30$ mS. Calculate the maximum possible conversion gain.

Solution
This is a straightforward application of the formula for conversion gain given in (13.108):

$$G_c = \frac{g_1^2 R_d}{4\omega_{RF}^2 C_{gs}^2 R_i} = \frac{(30 \times 10^{-3})^2(200)}{4(2\pi \times 5 \times 10^9)^2(.30 \times 10^{-12})^2(10)} = 50.7 = 17.05\ \text{dB}.$$

Note that this value does not include losses due to the necessary impedance matching networks.

Balanced Mixer

RF input matching and RF-LO isolation can be improved through the use of a *balanced mixer*, which consists of two single-ended mixers combined with a hybrid junction. Figure 13.29 shows the basic

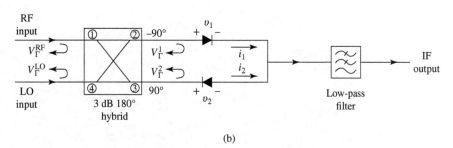

FIGURE 13.29 | Balanced mixer circuits. (a) Using a 90° hybrid. (b) Using a 180° hybrid.

configuration, with either a 90° hybrid (Figure 13.29a), or a 180° hybrid (Figure 13.29b). As we will see, a balanced mixer using a 90° hybrid junction will ideally lead to a perfect input match at the RF port over a wide frequency range, while the use of a 180° hybrid will ideally lead to perfect RF-LO isolation over a wide frequency range. In addition, both mixers will reject all even-order intermodulation products. Figure 13.30 shows a photograph of a microstrip circuit that contains several balanced mixers.

We can analyze the performance of a balanced mixer using the small-signal approach that was used for the single-ended diode mixer. Here we will concentrate on the balanced mixer with a 90° hybrid, shown in Figure 13.29a, and leave the 180° hybrid case as a problem. As usual, let the RF and LO voltages be defined as

$$v_{RF}(t) = V_{RF} \cos \omega_{RF} t, \tag{13.109}$$

and

$$v_{LO}(t) = V_{LO} \cos \omega_{LO} t. \tag{13.110}$$

From Section 7.5, the scattering matrix for the 90° hybrid junction is

$$[S] = \frac{-1}{\sqrt{2}} \begin{bmatrix} 0 & j & 1 & 0 \\ j & 0 & 0 & 1 \\ 1 & 0 & 0 & j \\ 0 & 1 & j & 0 \end{bmatrix}, \tag{13.111}$$

where the ports are numbered as shown in Figure 13.29a. The total RF and LO voltages applied to the two diodes can then be written as

$$v_1(t) = \frac{1}{\sqrt{2}} \left[V_{RF} \cos \left(\omega_{RF} t - 90° \right) + V_{LO} \cos \left(\omega_{LO} t - 180° \right) \right]$$

$$= \frac{1}{\sqrt{2}} \left(V_{RF} \sin \omega_{RF} t - V_{LO} \cos \omega_{LO} t \right), \tag{13.112a}$$

$$v_2(t) = \frac{1}{\sqrt{2}} \left[V_{RF} \cos \left(\omega_{RF} t - 180° \right) + V_{LO} \cos \left(\omega_{LO} t - 90° \right) \right]$$

$$= \frac{1}{\sqrt{2}} \left(-V_{RF} \cos \omega_{RF} t + V_{LO} \sin \omega_{LO} t \right). \tag{13.112b}$$

FIGURE 13.30 | Photograph of a 35 GHz microstrip monopulse radar receiver circuit. Three balanced mixers using ring hybrids can be seen, along with three stepped-impedance low-pass filters, and six quadrature hybrids. Eight feedlines are aperture coupled to microstrip antennas on the reverse side. The circuit also contains a Gunn diode source for the local oscillator.

Courtesy of Millitech Inc., Northampton, MA.

Using only the quadratic term from the small-signal diode approximation of (11.6) gives the diode currents as

$$i_1(t) = K v_1^2 = \frac{K}{2} \left(V_{RF}^2 \sin^2 \omega_{RF} t - 2 V_{RF} V_{LO} \sin \omega_{RF} t \cos \omega_{LO} t + V_{LO}^2 \cos^2 \omega_{LO} t \right),$$

$$(13.113a)$$

$$i_2(t) = -K v_2^2 = \frac{-K}{2} \left(V_{RF}^2 \cos^2 \omega_{RF} t - 2 V_{RF} V_{LO} \cos \omega_{RF} t \sin \omega_{LO} t + V_{LO}^2 \sin^2 \omega_{LO} t \right),$$

$$(13.113b)$$

where the negative sign on i_2 accounts for the reversed diode polarity, and K is a constant for the quadratic term of the diode response. Adding these two currents at the input to the low-pass filter gives

$$i_1(t) + i_2(t) = \frac{-K}{2} \left(V_{RF}^2 \cos 2\omega_{RF} t + 2 V_{RF} V_{LO} \sin \omega_{IF} t - V_{LO}^2 \cos 2\omega_{LO} t \right),$$

where the usual trigonometric identities have been used, and $\omega_{IF} = \omega_{RF} - \omega_{LO}$ is the IF frequency. Note that the DC components of the diode currents cancel upon combining. After low-pass filtering, the IF output is

$$i_{IF}(t) = -K V_{RF} V_{LO} \sin \omega_{IF} t, \qquad (13.114)$$

as desired.

We can also calculate the input match at the RF port and the coupling between the RF and LO ports. If we assume the diodes are matched and that each exhibits a voltage reflection coefficient Γ at the RF frequency, then the phasor expression for the reflected RF voltages at the diodes will be

$$V_{\Gamma_1} = \Gamma V_1 = \frac{-j\Gamma V_{RF}}{\sqrt{2}},$$
(13.115a)

$$V_{\Gamma_2} = \Gamma V_2 = \frac{-\Gamma V_{RF}}{\sqrt{2}}.$$
(13.115b)

These reflected voltages appear at ports 2 and 3 of the hybrid, respectively, and combine to form the following outputs at the RF and LO ports:

$$V_{\Gamma}^{RF} = \frac{-jV_{\Gamma_1}}{\sqrt{2}} - \frac{V_{\Gamma_2}}{\sqrt{2}} = -\frac{1}{2}\Gamma V_{RF} + \frac{1}{2}\Gamma V_{RF} = 0,$$
(13.116a)

$$V_{\Gamma}^{LO} = \frac{-V_{\Gamma_2}}{\sqrt{2}} - j\frac{V_{\Gamma_1}}{\sqrt{2}} = \frac{1}{2}j\Gamma V_{RF} + \frac{1}{2}j\Gamma V_{RF} = j\Gamma V_{RF}.$$
(13.116b)

Thus we see that the phase characteristics of the 90° hybrid lead to perfect cancellation of reflections at the RF port. The isolation between the RF and LO ports, however, is dependent on the matching of the diodes, which may be difficult to maintain over a reasonable frequency range.

Image Reject Mixer

We have already discussed the fact that two distinct RF input signals at frequencies $\omega_{RF} = \omega_{LO} \pm \omega_{IF}$ will down-convert to the same IF frequency when mixed with ω_{LO}. These two frequencies are the upper and lower sidebands of a double-sideband signal. The desired response can be arbitrarily selected as either the LSB ($\omega_{LO} - \omega_{IF}$) or the USB ($\omega_{LO} + \omega_{IF}$), assuming a positive IF frequency. The *image reject mixer*, shown in Figure 13.31, can be used to isolate these two responses into separate output signals. The same circuit can also be used for up-conversion, in which case it is usually called a *single-sideband modulator*. In this case, the IF input signal is delivered to either the LSB or the USB port of the IF hybrid, and the associated single-sideband signal is produced at the RF port of the mixer.

We can analyze the image reject mixer using the small-signal approximation. Let the RF input signal be expressed as

$$v_{RF}(t) = V_U \cos(\omega_{LO} + \omega_{IF})t + V_L \cos(\omega_{LO} - \omega_{IF})t,$$
(13.117)

where V_U and V_L represent the amplitudes of the upper and lower sidebands, respectively. Using the scattering matrix given in (13.111) for the 90° hybrid gives the RF voltages at the diodes as

$$v_A(t) = \frac{1}{\sqrt{2}}\left[V_U \cos\left(\omega_{LO}t + \omega_{RF}t - 90°\right) + V_L \cos\left(\omega_{LO}t - \omega_{IF}t - 90°\right)\right]$$

$$= \frac{1}{\sqrt{2}}[V_U \sin(\omega_{LO} + \omega_{IF})t + V_L \sin(\omega_{LO} - \omega_{IF})t],$$
(13.118a)

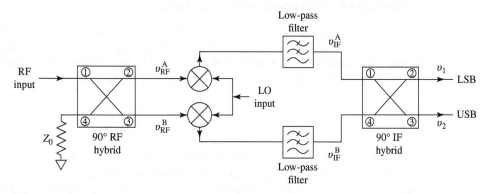

FIGURE 13.31 | Circuit for an image reject mixer.

$$v_B(t) = \frac{1}{\sqrt{2}} \left[V_U \cos\left(\omega_{\text{LO}}t + \omega_{\text{IF}}t - 180°\right) + V_L \cos\left(\omega_{\text{LO}}t - \omega_{\text{IF}}t - 180°\right) \right]$$

$$= \frac{-1}{\sqrt{2}} [V_U \cos(\omega_{\text{LO}} + \omega_{\text{IF}})t + V_L \cos(\omega_{\text{LO}} - \omega_{\text{IF}})t]. \tag{13.118b}$$

After mixing with the LO signal of (13.110) and low-pass filtering, the IF inputs to the IF hybrid are

$$v_{\text{IF}}^A(t) = \frac{KV_{\text{LO}}}{2\sqrt{2}}(V_U - V_L)\sin\omega_{\text{IF}}t, \tag{13.119a}$$

$$v_{\text{IF}}^B(t) = \frac{-KV_{\text{LO}}}{2\sqrt{2}}(V_U + V_L)\cos\omega_{\text{IF}}t, \tag{13.119b}$$

where K is the mixer constant for the squared term of the diode response. The phasor representation of the IF signals of (13.119) is

$$V_{\text{IF}}^A = \frac{-jKV_{\text{LO}}}{2\sqrt{2}}(V_U - V_L), \tag{13.120a}$$

$$V_{\text{IF}}^B = \frac{-KV_{\text{LO}}}{2\sqrt{2}}(V_U + V_L). \tag{13.120b}$$

Combining these voltages in the IF hybrid gives the following outputs:

$$V_1 = -j\frac{V_{\text{IF}}^A}{\sqrt{2}} - \frac{V_{\text{IF}}^B}{\sqrt{2}} = \frac{KV_{\text{LO}}V_L}{2} \quad \text{(LSB)}, \tag{13.121a}$$

$$V_2 = -\frac{V_{\text{IF}}^A}{\sqrt{2}} - j\frac{V_{\text{IF}}^B}{\sqrt{2}} = \frac{-jKV_{\text{LO}}V_U}{2} \quad \text{(USB)}, \tag{13.121b}$$

which we see are the separate sidebands of the down-converted input signal of (13.117). These outputs can be expressed in time domain form as

$$v_1(t) = \frac{KV_{\text{LO}}V_L}{2}\cos\omega_{\text{IF}}t, \tag{13.122a}$$

$$v_2(t) = \frac{KV_{\text{LO}}V_U}{2}\sin\omega_{\text{IF}}t, \tag{13.122b}$$

which clearly shows the presence of a 90° phase shift between the two sidebands. Also note that the image rejection mixer does not incur any additional losses beyond the usual conversion losses of the single rejection mixer. A practical difficulty with image rejection mixers is in fabricating a good hybrid at the relatively low IF frequency. Losses, and hence noise figure, are also usually greater than for a simpler mixer.

Differential FET Mixer and Gilbert Cell Mixer

The mixer shown in Figure 13.32a uses two FETs in a differential balanced configuration, similar to the differential amplifier discussed in Section 12.4. The LO input voltage and IF output voltage are balanced signals; baluns may be used at these ports to convert to single-ended signals. The RF input is single ended, and is applied to the bottom transistor. The RF and LO voltages can be written as

$$v_{\text{RF}}(t) = V_{\text{RF}}\cos\omega_{\text{RF}}t \tag{13.123a}$$

$$v_{\text{LO}}^{\pm}(t) = \pm V_{\text{LO}}\cos\omega_{\text{LO}}t \tag{13.123b}$$

Conceptually, the circuit operates as an alternating switch, with the LO turning the top two FETs on and off with alternate half-cycles of the LO voltage. As with the differential mode of the differential amplifier, the connection between the sources of the upper FETs is a virtual ground for the LO voltage. These transistors

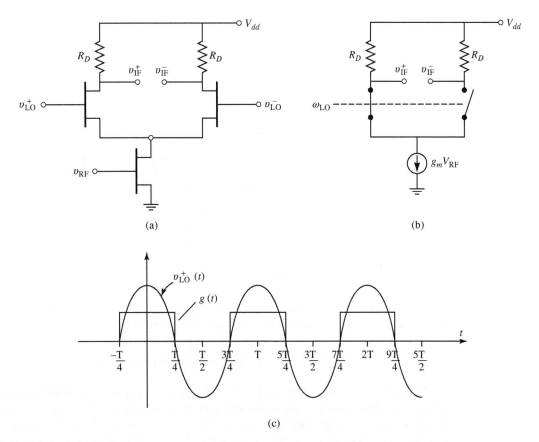

FIGURE 13.32 | (a) A singly balanced differential FET mixer. (b) Simplified equivalent circuit. (c) LO voltage wave-
form and idealized switching waveform of the top left FET.

are biased slightly above pinch-off, so each will be conducting for slightly more than half of each LO cycle.
During the positive half-cycle of v_{LO}^+ the top left FET will conduct with a low resistance, and it will turn
off during the negative half cycle. During the positive half cycle of v_{LO}^- (which occurs during the negative
half cycle of v_{LO}^+) the top right FET will turn on. Thus, one of the upper FETs is always conducting. The
lower FET is biased into saturation and operates as a normal RF amplifier, providing RF current through
the upper switches. The RF current at the drain of the bottom FET is approximately $I_{RF} = g_m V_{RF}$. The RF
and LO ports require impedance matching, and the IF output circuit should provide a return path to ground
for the LO signal. A current source, or inductive degeneration, is often used on the source of the lower FET.

A simplified equivalent circuit is shown in Figure 13.32b, with the upper FETs replaced with ideal
switches. The effect of the switches can be modeled by using a Fourier series for the idealized conductance
waveform shown in Figure 13.32c. The first few terms of this Fourier series can be found as

$$g(t) = \frac{1}{2} + \frac{2}{\pi} \cos \omega_{LO} t - \frac{2}{3\pi} \cos 3\omega_{LO} t + \cdots \tag{13.124}$$

Then the voltages at the IF terminals can be expressed as

$$v_{IF}^+(t) = -g(t) I_{RF} R_D = -g_m V_{RF} R_D \left(\frac{1}{2} + \frac{2}{\pi} \cos \omega_{LO} t \right) \cos \omega_{RF} t \tag{13.125a}$$

$$v_{IF}^-(t) = -[1 - g(t)] I_{RF} R_D = -g_m V_{RF} R_D \left(\frac{1}{2} - \frac{2}{\pi} \cos \omega_{LO} t \right) \cos \omega_{RF} t, \tag{13.125b}$$

where we have retained only the first two terms of (13.124). The net output IF voltage is

$$v_{IF}(t) = v_{IF}^+(t) - v_{IF}^-(t) = \frac{-4}{\pi} g_m V_{RF} R_D \cos \omega_{LO} t \cos \omega_{RF} t. \tag{13.126}$$

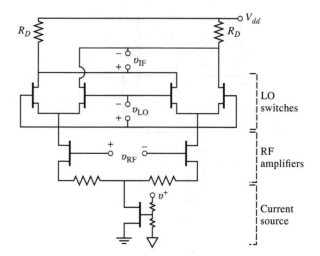

FIGURE 13.33 | A Gilbert cell mixer.

Note that voltages at the RF and LO frequencies are canceled (without filtering), leaving only terms at the frequencies $\omega_{LO} \pm \omega_{RF}$. After low-pass filtering, the IF output is

$$v_{IF}(t)\Big|_{LPF} = \frac{-2}{\pi} g_m V_{RF} R_D \cos \omega_{IF} t. \tag{13.127}$$

The *Gilbert cell mixer*, shown in Figure 13.33, uses two singly balanced FET mixers of the type shown in Figure 13.32a to form a *double-balanced mixer*. It has fully balanced (differential) ports for LO, RF, and IF signals. Due to the symmetry of the circuit and its excitations, the RF and LO signals are canceled at the IF output port. Operation is the same as the singly balanced FET mixer, with the four upper FETs operating as switches controlled by the LO voltage, and the lower two FETs operating as a balanced amplifier for the RF input signal. The circuit of Figure 13.33 includes a current source at the sources of the amplifier FETs. This mixer is frequently used in CMOS RFICs for wireless applications.

Other Mixers

There are a number of other mixer circuits that provide various advantages in terms of bandwidth, harmonic generation, and intermodulation products. The double-balanced mixer of Figure 13.34 uses two hybrid junctions or transformers, and provides good isolation between all three ports, as well as rejection of all even harmonics of the RF and LO signals. This leads to very good conversion loss, but less than ideal input matching at the RF port. The double-balanced mixer also provides a higher third-order intercept point than either a single-ended mixer or a balanced mixer.

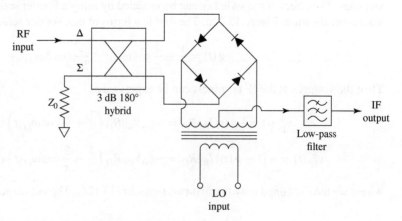

FIGURE 13.34 | Double-balanced mixer circuit.

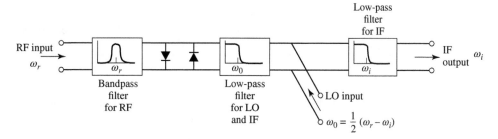

FIGURE 13.35 | Subharmonically pumped mixer using an antiparallel diode pair.

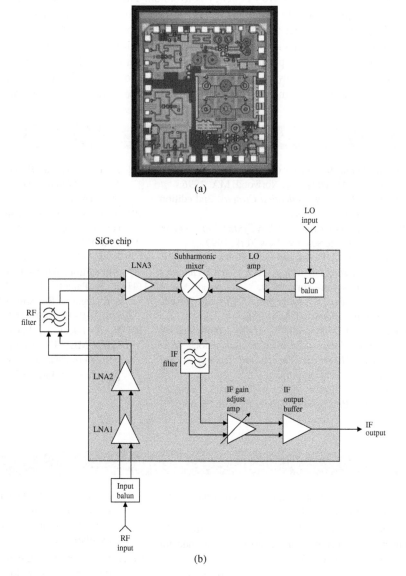

(a)

(b)

FIGURE 13.36 | (a) Photograph of a monolithic integrated millimeter wave down-converter using silicon germanium (SiGi). (b) Block diagram of the chip. The circuit operates from 43.5 to 45.5 GHz and includes differential LNA, LO, and IF amplifiers, a differential subharmonic mixer, and an off-chip RF filter. The noise figure is less than 6 dB.

Courtesy of Hittite Microwave Corporation.

TABLE 13.1 | Mixer Characteristics

Mixer Type	Number of Diodes	RF Input Match	RF-LO Isolation	Conversion Loss	Third-Order Intercept
Single ended	1	Poor	Fair	Good	Fair
Balanced (90°)	2	Good	Poor	Good	Fair
Balanced (180°)	2	Fair	Excellent	Good	Fair
Double balanced	4	Poor	Excellent	Excellent	Excellent
Image reject	2 or 4	Good	Good	Good	Good

Figure 13.35 shows the circuit for an *antiparallel diode mixer*, which is often used for subharmonically pumped millimeter wave frequency conversion. The back-to-back diodes function as a frequency doubler, thus requiring an LO frequency of one-half the usual value. The diode nonlinearity operates as a resistive frequency multiplier to generate the second harmonic of the LO to mix with the RF input to produce the desired output frequency. The antiparallel diode pair has a symmetric *I–V* characteristic that suppresses the fundamental mixing product of the RF and LO input signals, leading to better conversion loss. A photograph of a SiGe MMIC down-converter using a subharmonic mixer is shown in Figure 13.36.

Table 13.1 summarizes the characteristics of several of the mixers that we have discussed.

REFERENCES

[1] L. E. Larson, *RF and Microwave Circuit Design for Wireless Communications*, Artech House, Norwood, MA, 1996.

[2] J. R. Smith, *Modern Communication Circuits*, 2nd edition, McGraw-Hill, New York, 1998.

[3] U. L. Rohde, *Microwave and Wireless Synthesizers: Theory and Design*, Wiley-Interscience, New York, 1997.

[4] G. D. Vendelin, A. M. Pavio, and U. L. Rohde, *Microwave Circuit Design Using Linear and Nonlinear Techniques*, John Wiley & Sons, New York, 1990.

[5] S. A. Maas, *Nonlinear Microwave and RF Circuits*, 2nd edition, Artech House, Norwood, MA, 2003.

[6] Y. Komatsu and Y. Murakami, "Coupling Coefficient Between Microstrip Line and Dielectric Resonator," *IEEE Transactions on Microwave Theory and Techniques*, vol. MTT-31, pp. 34–40, January 1983.

[7] D. B. Leeson, "A Simple Model of Feedback Oscillator Noise Spectrum," *Proceedings of the IEEE*, vol. 54, pp. 329–330, 1966.

[8] A. Leon-Garcia, *Probability and Random Processes for Electrical Engineering,* 2nd edition, Addison-Wesley, Reading, MA, 1994.

[9] M. K. Nezami, "Evaluate the Impact of Phase Noise on Receiver Performance," *Microwaves & RF Magazine*, pp. 1–11, June 1998.

[10] R. E. Collin, *Foundations for Microwave Engineering*, 2nd edition, Wiley–IEEE Press, Hoboken, NJ, 2001.

[11] R. H. Pantell, "General Power Relationships for Positive and Negative Resistive Elements," *Proceedings of the IRE*, pp. 1910–1913, December 1958.

[12] R. A. Pucel, D. Masse, and R. Bera, "Performance of GaAs MESFET Mixers at X Band," *IEEE Transactions on Microwave Theory and Techniques*, vol. MTT-24, pp. 351–360, June 1976.

PROBLEMS

13.1 Derive the admittance matrix representation of the transistor oscillator circuit given in (13.3).

13.2 Derive the results in (13.20)–(13.22) for a Colpitts oscillator using a common emitter transistor with an inductor having a series resistance R.

13.3 Design a Colpitts oscillator operating at 50 MHz using a FET in a common gate configuration, including the effect of a lossy inductor. First derive equations for the resonant frequency and condition required for sustaining oscillation for an inductor with loss, corresponding to equations (13.20)–(13.22) for the BJT case. Use these results to find the required capacitances, assuming an inductor of 0.10 μH with a Q of 100, and a transistor with $g_m = 24$ mS and $R_o = 1/G_o = 1200$ Ω. Determine the minimum value of the inductor Q required to sustain oscillations.

13.4 Prove that the standard Smith chart can be used for negative resistances by plotting $1/\Gamma^*$ (instead of Γ). In this case, the resistance circle values are read as negative, while the reactance circles are unchanged.

13.5 Design a transistor oscillator at 2.4 GHz using a silicon BJT in a common emitter configuration driving a 50 Ω load on the drain side. The scattering parameters are as follows ($Z_0 = 50$ Ω): $S_{11} = 0.504\angle-125°$,

$S_{12} = 0.137\angle11.2°$, $S_{21} = 4.499\angle72°$, and $S_{22} = 0.471$ $\angle-87.6°$. Choose Γ_L for $|\Gamma_{in}| \gg 1$, and design appropriate load and terminating networks.

13.6 Repeat the oscillator design of Example 13.4 by replacing the dielectric resonator and microstrip feedline with a single-stub tuner to match Γ_S to a 50 Ω load. Find the Q of the tuner and 50 Ω load, then compute and plot $|\Gamma_{out}|$ versus $\Delta f/f_0$. Compare with the result in Figure 13.12b for the dielectric resonator case.

13.7 Repeat the dielectric oscillator design of Example 13.4 using a GaAs MESFET having the following scattering parameters at 1.5 GHz: $S_{11} = 1.7\angle135°$, $S_{12} = 0.3\angle50°$, $S_{21} = 3.5\angle40°$, and $S_{22} = 0.6\angle-60°$.

13.8 A HEMT device in the common gate configuration has the following scattering parameters at 8 GHz ($Z_0 = 50$ Ω): $S_{11} = 0.46\angle178°$, $S_{12} = 0.045\angle73°$, $S_{21} = 1.41\angle-19°$, $S_{22} = 1.02\angle-12°$. For application in an oscillator, a series inductor is added to the gate, as shown below, to increase instability. Compute and plot the μ-stability factor for L ranging from 0 to 20 nH, and determine the value that maximizes instability. (This can most easily be done with a microwave CAD package.)

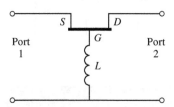

13.9 An oscillator uses an amplifier with a noise figure of 5.8 dB and a resonator having a Q of 425, and produces a 150 MHz output at a power level of 10 dBm. If the measured f_α is 50 kHz, plot the spectral density of the output noise power, and determine the phase noise (in dBc/Hz) at the following frequencies: (a) at 1 MHz from the carrier; (b) at 10 kHz from the carrier (assume $K = 1$).

13.10 Derive Equation (13.50) giving the required phase noise for a specified receiver selectivity.

13.11 Find the necessary LO phase noise specification if an 860 MHz cellular receiver with a 30 kHz channel spacing is required to have an adjacent channel rejection of 78 dB, assuming the interfering channel is at the same level as the desired channel. The final IF voice bandwidth is 15 kHz.

13.12 Apply the Manley–Rowe relations to an up-converting mixer. Assume a nonlinear reactance is excited at frequencies f_1 (RF) and f_2 (LO), and terminated with open circuits at all other frequencies except $f_3 = f_1 + f_2$. Show that the maximum possible conversion gain is given by $-P_{11}/P_{10} = 1 + \omega_2/\omega_1$.

13.13 Derive the relation between pulse duration and gate voltages given in (13.73) for the FET frequency multiplier.

13.14 A double-sideband signal of the form $v_{RF}(t) = V_{RF}[\cos(\omega_{LO} - \omega_{IF})t + \cos(\omega_{LO} + \omega_{IF})t]$ is applied to a mixer with an LO voltage given by $v_{LO}(t) = V_{LO} \cos \omega_{LO}t$. Derive the output of the mixer after low-pass filtering.

13.15 A diode has an I–V characteristic given by $i(t) = I_s(e^{3v(t)} - 1)$. Let $v(t) = 0.5 \cos \omega_1 t + 0.5 \cos \omega_2 t$, and expand $i(t)$ in a power series in v, retaining only the v, v^2, and v^3 terms. For $I_s = 2$ A, find the magnitudes of the current at each frequency.

13.16 An RF input signal for Personal Communications Service (PCS) at 1900 MHz is down-converted in a mixer to an IF frequency of 87 MHz. What are the two possible LO frequencies and the corresponding image frequencies?

13.17 Consider a diode mixer with a conversion loss of 6 dB and a noise figure of 5 dB, and a FET mixer with conversion gain of 4 dB and a noise figure of 8.5 dB. If each of these mixers is followed by an IF amplifier having a gain of 30 dB and a noise figure F_A, as shown below, calculate and plot the overall noise figure for both amplifier–mixer configurations for $F_A = 0$–12 dB.

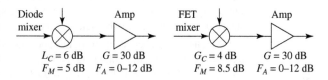

13.18 Let T_{SSB} be the equivalent noise temperature of a mixer receiving an SSB signal, and T_{DSB} be the temperature when it receives a DSB signal. Compute the output noise powers in each case, and show that $T_{SSB} = 2T_{DSB}$, and that therefore $F_{SSB} = 2F_{DSB}$. Assume that the conversion gains for the signal and its image are identical.

13.19 If the noise power $N_i = kTB$ is applied at the RF input port of a mixer having noise figure F (DSB) and conversion loss L_c, what is the available output noise power at the IF port? Assume the mixer is at a physical temperature T_0.

13.20 A *phase detector* produces an output signal proportional to the phase difference between two RF input signals. Let these input signals be expressed as

$$v_1 = v_0 \cos \omega t,$$
$$v_2 = v_0 \cos(\omega t + \theta).$$

If these two signals are applied to a single-balanced mixer using a 90° hybrid, show that the IF output signal, after low-pass filtering, is given by

$$i = kv_0^2 \sin \theta,$$

where k is a constant. If the mixer uses a 180° hybrid, show that the corresponding output signal is given by

$$i = kv_0^2 \cos \theta.$$

13.21 Analyze a balanced mixer using a 180° hybrid junction. Find the output IF current, and the input reflections at the RF and LO ports. Show that this mixer suppresses even harmonics of the LO. Assume that the RF signal is applied to the sum port of the hybrid, and that the LO signal is applied to the difference port.

13.22 For an image rejection mixer, let the RF hybrid have a dissipative insertion loss of L_R and the IF hybrid have a dissipative insertion loss of L_I. If the component single-ended mixers each have a conversion loss L_c and noise figure F, derive expressions for the overall conversion loss and noise figure of the image rejection mixer.

13.23 Shown in the figure below is the down-conversion chain for an RF receiver. The incoming signal of interest is centered around $f_c = 900$ MHz, and a low-side LO injection is used for down-conversion with $f_{LO} = 850$ MHz.

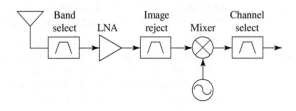

Determine intermediate frequency for this system. If the bandwidth of each channel is 2 MHz, what is the approximate Q required for the channel select filter? Also determine the image frequency for the signal of interest.

13.24 A 12–24 GHz frequency doubler is designed using a GaAs MOSFET with the following parameters: $V_t = -2.5$ V, $R_i = 15\ \Omega$, $C_{gs} = 0.29$ pF, $C_{ds} = 0.18$ pF, and $R_{ds} = 200\ \Omega$. Assume the operating point of the transistor is chosen so that $V_{gmax} = 0.25$ V, $V_{gmin} = -6.5$ V, $V_{dmax} = 5.0$ V, $V_{dmin} = 1.2$ V, and $I_{max} = 85$ mA. Find the conversion gain of the multiplier.

13.25 Find the voltage gain of a feedback network from the amplifier output to the input such that the circuit will oscillate at a frequency of 1 GHz. The voltage gain of the amplifier is

$$A(\omega) = \frac{2\angle 30°}{1 + \omega/(2\pi \times 10^9)}.$$

13.26 Consider the Colpitts oscillator shown below. The parameters are: $R_1 = 10$ kΩ, $R_2 = 5$ kΩ, $R_e = 8.6$ kΩ, $L = 2$ mH, $C_1 = 1.27$ nF, $C_2 = 0.127$ μF, $C_b = C_c = 10$ μF, and $V_{cc} = 15$ V. (a) Find the resonant frequency, and (b) determine R_L if this circuit oscillates at a frequency of 100 kHz.

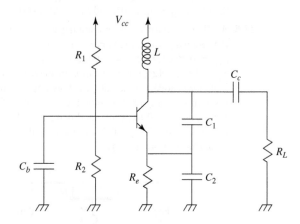

MULTIPLE CHOICE QUESTIONS

13.1 For an RF oscillator the transfer function is given by

(a) $A/(1 + AH(\omega))$ (c) $A/(-1 + AH(\omega))$
(b) $A/(1 - AH(\omega))$ (d) $1/(1 - AH(\omega))$

13.2 In an oscillator circuit, the oscillations are produced and the criterion is called

(a) Barkhausen criteria   (c) Colpitts criteria
(b) Shannon's criteria   (d) none of the above

13.3 In a Colpitts oscillator, the necessary condition for oscillation is

(a) $C_2/C_1 = g_m/G_i$ (c) $C_2/C_1 = g_m \times G_i$
(b) $C_1/C_2 = g_m/G_i$ (d) none of the above

13.4 Colpitts oscillator using a bipolar junction transistor operating at 50 MHz has an inductor in the feedback section of value 0.1 μH. The values of the capacitors in the feedback section are

(a) 150 pF, 50 pF   (c) 130 pF, 70 pF
(b) 200 pF, 200 pF   (d) 40 pF, 60 pF

13.5 In a Hartley oscillator, the necessary condition for oscillation is

(a) $L_1/L_2 = g_m/G_i$ (c) $L_2/L_1 = g_m/G_i$
(b) $L_1/L_2 = G_i/g_m$ (d) none of the above

13.6 Assuming the quality factor $Q_o = 100$, an inductor is operating at frequency of 50 MHz and its inductance is 0.2 μH. Then the series resistance associated with the inductor is

(a) 1.628 Ω   (c) 0.628 Ω
(b) 1.314 Ω   (d) 1.814 Ω

13.7 Hartley oscillator has inductance values of 15 mH and 5 mH in the feedback section and a capacitor of 3 nF. Then the resonant frequency of the circuit is

(a) 10.27 kHz   (c) 41.09 kHz
(b) 30.82 kHz   (d) 20.55 kHz

13.8 Colpitts oscillator in the feedback section has an inductance of 3 mH and capacitors of 15 nH and 5 nH. Then the resonant frequency of Colpitts oscillator is

(a) 35.59 kHz   (c) 47.45 kHz
(b) 23.73 kHz   (d) 71.18 kHz

13.9 As a feedback network, the quartz crystals are more efficient because

(a) circuit complexity is less
(b) they are cost effective
(c) they can operate at high voltage levels
(d) LC circuits have unloaded Q of about 10,000

13.10 The working principle of quartz crystal and tourmaline used in oscillators is

(a) piezoelectric effect   (c) Raman effect
(b) photoelectric effect   (d) black body radiation

13.11 If the input power for a frequency doubler is 11.2 mW and the output measured after the frequency doubling process is 28.5 mW, then the conversion gain for the frequency doubler is

(a) 2.9 dB
(b) 4.1 dB
(c) 8.1 dB
(d) 12.2 dB

13.12 The condition to be satisfied in terms of reflection coefficients for achieving steady-state oscillation is

(a) $\Gamma_{in} = -\Gamma_L$
(b) $\Gamma_{in} = \Gamma_L$
(c) $\Gamma_{in} = 1/\Gamma_L$
(d) none of the above

13.13 The load impedance used to achieve a stable oscillation if the input impedance of a diode used in the microwave oscillator is $35 - j18\ \Omega$ is given by

(a) $35 - j18\ \Omega$
(b) $-35 + j18\ \Omega$
(c) $50\ \Omega$
(d) $18 - j35\ \Omega$

13.14 In a transistor amplifier, if the input impedance is $-63 - j12.8\ \Omega$, then the terminating impedance required to create enough instability is

(a) $63 + j12.8\ \Omega$
(b) $21 + j12.8\ \Omega$
(c) $-(21 + j4.27)\ \Omega$
(d) $-(63 + j4.27)\ \Omega$

13.15 Dielectric resonators are preferred over waveguide resonators in oscillator tuning circuits because

(a) they have high Q factor
(b) they are compact in size
(c) they are easily integrated with microwave integrated circuits
(d) all of the above

13.16 Using either parallel or series arrangement, a dielectric resonator can be incorporated into a circuit to provide

(a) frequency stability
(b) high gain
(c) oscillations
(d) optimized reflection coefficient

13.17 For an oscillator or synthesizer, the phase variation is given by

(a) $\sin \omega_m t / \Delta f\, f_m$
(b) $\Delta f / f_m$
(c) $\sin \omega_m t / f_m$
(d) $\Delta f \sin \omega_m t / f_m$

13.18 The GSM 900 digital cellular telephone system uses a receive frequency band of 890 to 915 MHz, with a first IF frequency range of 45 MHz. One possible range of local oscillator frequency is

(a) 650 to 684 MHz
(b) 869 to 894 MHz
(c) 935 to 960 MHz
(d) 956 to 981 MHz

13.19 Determine the required local oscillator phase noise at 3 MHz carrier frequency offset if a GSM cellular telephone standard requires a minimum of 9 dB rejection of interfering signal levels of −43 dBm at 3 MHz from the carrier, −33 dBm at 1.6 MHz from the carrier, and −43 dBm at 0.6 MHz from the carrier, for a carrier level of −99 dBm.

(a) −138 dBc/Hz
(b) −128 dBc/Hz
(c) −118 dBc/Hz
(d) −108 dBc/Hz

13.20 If a frequency multiplier has a multiplication factor of 15, then the increase in noise level due to frequency multiplication is

(a) 11.76 dB
(b) 23.52 dB
(c) 35.28 dB
(d) 47.04 dB

ANSWER KEY

13.1 (b)	**13.5** (a)	**13.9** (d)	**13.13** (b)	**13.17** (d)
13.2 (a)	**13.6** (c)	**13.10** (a)	**13.14** (b)	**13.18** (c)
13.3 (a)	**13.7** (d)	**13.11** (b)	**13.15** (d)	**13.19** (c)
13.4 (b)	**13.8** (c)	**13.12** (c)	**13.16** (a)	**13.20** (b)

14

Introduction to Microwave Systems

Learning Objectives

After completing this chapter, you will be able to

- Learn about the system aspects of antennas
- Understand link budget and link margin, noise characterization of a receiver, and various aspects of wireless communication

- Describe different microwave systems such as radar systems and radiometer systems
- Learn how propagation phenomenon influences the operation of microwave systems

$\mathbf{A}$ microwave system consists of passive and active microwave components arranged to perform a useful function. Probably the two most important examples are microwave communication systems and microwave radar systems, but there are many others. In this chapter we will discuss the basic operation of several types of microwave systems to give a general overview of the application of microwave technology, and to show how the subjects of earlier chapters fit into the overall scheme of complete microwave systems.

An important component in any radar or wireless communication system is the antenna, so we will first discuss some of the basic properties of antennas. Then we treat wireless communication, radar, and radiometry systems as important applications of RF and microwave technology. We also briefly discuss propagation effects, biological effects, and other miscellaneous applications.

All of these topics are of sufficient depth that many books have been written for each. Our purpose here is to introduce these topics as a way of placing the earlier material in this book in the larger context of practical system applications. The interested reader is referred to the references at the end of the chapter for more complete treatments.

14.1   SYSTEM ASPECTS OF ANTENNAS

In this section we describe some of the basic characteristics of antennas that will be needed for our study of microwave communication, radar, and remote sensing systems. We are interested here not in the detailed electromagnetic theory of antenna operation, but rather in the systems aspect of the operation of an antenna in terms of its radiation patterns, directivity, gain, efficiency, and noise characteristics. References [1] and [2] can be reviewed for a more in-depth treatment of the fascinating subject of antenna theory and design. Figure 14.1 shows some of the different types of antennas that have been developed for commercial wireless systems.

A transmitting antenna can be viewed as a device that converts a guided electromagnetic wave on a transmission line into a plane wave propagating in free space. Thus, one side of an antenna appears as an electrical circuit element, while the other side provides an interface with a propagating plane wave. Antennas are inherently bidirectional, in that they can be used for both transmit and receive functions. Figure 14.2 illustrates the basic operation of transmitting and receiving antennas. The transmitter can be modeled as a Thevenin source consisting of a voltage generator and series impedance, delivering a power P_t to the transmitting antenna. A transmitting antenna radiates a spherical wave that, at large distances, approximates a plane wave over a localized area. A receiving antenna intercepts a portion of an incident plane wave, and delivers a receive power P_r to the receiver load impedance.

FIGURE 14.1 | Photograph of various millimeter wave antennas. Clockwise from top: a high-gain 38 GHz reflector antenna with radome, a prime-focus parabolic antenna, a corrugated conical horn antenna, a 38 GHz planar microstrip array, a pyramidal horn antenna with a Gunn diode module, and a multibeam reflector antenna.

A wide variety of antennas have been developed for different applications, as summarized in the following categories:

- *Wire antennas* include dipoles, monopoles, loops, sleeve dipoles, Yagi–Uda arrays, and related structures. Wire antennas generally have low gains, and are most often used at lower frequencies (HF to UHF). They have the advantages of light weight, low cost, and simple design.
- *Aperture antennas* include open-ended waveguides, rectangular or circular horns, reflectors, lenses, and reflectarrays. Aperture antennas are most commonly used at microwave and millimeter wave frequencies, and have moderate to high gains.

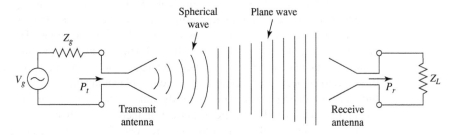

FIGURE 14.2 | Basic operation of transmitting and receiving antennas.

- *Printed antennas* include printed slots, printed dipoles, and microstrip patch antennas. These antennas can be made with photolithographic methods, with both radiating elements and associated feed circuitry fabricated on dielectric substrates. Printed antennas are most often used at microwave and millimeter wave frequencies, and can be easily arrayed for high gain.

- *Array antennas* consist of a regular arrangement of antenna elements with a feed network. Pattern characteristics such as beam pointing angle and sidelobe levels can be controlled by adjusting the amplitude and phase excitation of the array elements. An important type of array antenna is the *phased array*, in which variable-phase shifters are used to electronically scan the main beam of the antenna.

- *Ground-based antenna systems* consist of transportable, mobile, and fixed systems. For a transportable system, the mounting vehicle's physical characteristics should be examined carefully for parameters such as the resulting transmission of shock and natural frequency. When designing for Ka-Band operations, the trailer-mounted devices along with the antenna should form a part of the mechanical system analysis and should be designed accordingly. Ground-based antennas are used to establish ground-to-ground, ground-to-air, and point-to-point communication which could be used for defense, military, and various other industrial applications. To achieve a larger power in specific direction, ground-based antennas are directional in nature. Directional antennas are usually parabolic, or dish, type. A parabolic reflector can focus RF energy into a narrow beam, and it has got a very high directivity. Parabolic antennas are mostly used for long-distance communication links. Very large parabolic antennas can offer a gain of about 70 dBi and are used for radio astronomy. Permanent ground-based antennas are helpful in monitoring long-term terrestrial features. Large antennas located on the ground are useful in establishing links in order to receive as well as transmit data at low power levels for communication with satellites in deep space. These large antenna structures are costly to construct and maintain, and a failure in the systems can result in potential damage to critical information. Agencies, such as NASA, are exploring to develop alternative systems by using smaller aperture antennas joined in an array to create a much greater effective aperture structure. At advanced frequency bands, like Ka-band (26 GHz– 40 GHz), additional difficulties such as greater impact of atmospheric turbulence across the effective aperture of the antenna array are introduced, which induce phase scintillation errors.

- *Airborne antennas* are the nonground-based antennas. In an unmanned aerial vehicle (UAV) antennas are integrated as radio navigation and communication systems on a variety of airborne platforms. These antennas are omnidirectional and are used worldwide in military, commercial, and civilian aviation segments. All airborne antennas are designed to work for years in severe environments, with optimum efficiency and trouble-free operation. With the launch of Sputnik on October 4, 1957, the erstwhile USSR placed into orbit a small sphere with a wireless radio transmitter that was the first ever satellite launched into the Earth's orbit. Presently, more than hundred satellites are being launched every year. Space Shuttle, the low Earth orbital spacecraft system that was operated between 1981 and 2011, functioned as a remote sensing satellite and was used for a number of missions. Satellites can be classified by their orbital timing and geometry. Three orbits commonly used for satellites are sun synchronous, equatorial, and geostationary. Antennas are the most visible part of satellite communication systems, and these satellites are being used globally for such varied applications as DTH broadcasting, television broadcasting, Digital Satellite News Gathering (DSNG), and very small aperture terminal (VSAT). The frequency of operation of these antennas is in the range of 0.3 GHz to 30 GHz.

- *Planar antennas* include both printed circuit board (PCB) antennas and microstrip antennas. They place both the parasitic and active elements on a single plane, making them two dimensional. The antenna may be square, circular, or elliptical. Planar antennas can be made into a variety of shapes. They are low-cost and repeatedly used in wideband applications. The resonance of antenna for a particular band of frequency can be achieved with different shapes and with different feeding techniques. Planar antenna arrays can lead to high gain and by changing the phase of individual elements, the directional beam control can be achieved. A planar antenna on a flexible substrate which is highly thin does not distress the aircraft's aerodynamics, and thus it is mainly used on aircraft. If a planar antenna is constructed using PCB fabrication technology, then assimilating it with additional electronic components at the same time is more cost effective. Based on their design, planar antennas can support both linear and circular polarization.

Fields and Power Radiated by an Antenna

While we do not require detailed solutions to Maxwell's equations for our purposes, we do need to be familiar with the far-zone electromagnetic fields radiated by an antenna. Consider an antenna located at the origin of a spherical coordinate system. At large distances, where the localized near-zone fields are negligible, the radiated electric field of an arbitrary antenna can be expressed as

$$\bar{E}(r, \theta, \phi) = \left[\hat{\theta} F_\theta(\theta, \phi) + \hat{\phi} F_\phi(\theta, \phi)\right] \frac{e^{-jk_0 r}}{r} \text{ V/m}, \tag{14.1}$$

where $\bar{E}$ is the electric field vector, $\hat{\theta}$ and $\hat{\phi}$ are unit vectors in the spherical coordinate system, r is the radial distance from the origin, and $k_0 = 2\pi/\lambda$ is the free-space propagation constant, with wavelength $\lambda = c/f$. Also defined in (14.1) are the *pattern functions*, $F_\theta(\theta, \phi)$ and $F_\phi(\theta, \phi)$. The interpretation of (14.1) is that this electric field propagates in the radial direction with a phase variation of $e^{-jk_0 r}$ and an amplitude variation with distance of $1/r$. The electric field may be polarized in either the $\hat{\theta}$ or $\hat{\phi}$ direction, but not in the radial direction, since this is a TEM wave. The magnetic fields associated with the electric field of (14.1) can be found from (1.76) as

$$H_\phi = \frac{E_\theta}{\eta_0}, \tag{14.2a}$$

$$H_\theta = \frac{-E_\phi}{\eta_0}, \tag{14.2b}$$

where $\eta_0 = 377 \ \Omega$, the wave impedance of free-space. Note that the magnetic field vector is also polarized only in the transverse directions. The Poynting vector for this wave is given by (1.90) as

$$\bar{S} = \bar{E} \times \bar{H}^* \text{ W/m}^2, \tag{14.3}$$

and the time-average Poynting vector is

$$\bar{S}_{\text{avg}} = \frac{1}{2} \text{Re}\{\bar{S}\} = \frac{1}{2} \text{Re}\{\bar{E} \times \bar{H}^*\} \text{ W/m}^2. \tag{14.4}$$

We mentioned earlier that at large distances the near fields of an antenna are negligible, and that the radiated electric field can be written as in (14.1). We can give a more precise meaning to this concept by defining the *far-field distance* as the distance where the spherical wave front radiated by an antenna becomes a close approximation to the ideal planar phase front of a plane wave. This approximation applies over the radiating aperture of the antenna, and so it depends on the maximum dimension of the antenna. If we call this maximum dimension D, then the far-field distance is defined as

$$R_{ff} = \frac{2D^2}{\lambda} \text{m}. \tag{14.5}$$

This result is derived from the condition that the actual spherical wave front radiated by the antenna departs less than $\pi/8 = 22.5°$ from a true plane wave front over the maximum extent of the antenna. For electrically small antennas, such as short dipoles and small loops, this result may give a far-field distance that is too small; in this case, a minimum value of $R_{ff} = 2\lambda$ should be used.

EXAMPLE 14.1   FAR-FIELD DISTANCE OF AN ANTENNA

A parabolic reflector antenna used for reception with the direct broadcast system (DBS) is 20 inches in diameter and operates at 10 GHz. Find the far-field distance for this antenna.

Solution
The operating wavelength at 10 GHz is

$$\lambda = \frac{c}{f} = \frac{3 \times 10^8}{10 \times 10^9} = 3.0 \text{ cm}.$$

The far-field distance is found from (14.5), after converting 20 inches to 0.508 m:

$$R_{ff} = \frac{2D^2}{\lambda} = \frac{2(0.508)^2}{3.0 \times 10^{-2}} = 17.3 \text{ m}.$$

The actual distance from a DBS satellite to Earth is about 36,000 km, so it is safe to say that the receive antenna is in the far-field of the transmitting antenna.

Next, define the *radiation intensity* of the radiated electromagnetic field as

$$U(\theta, \phi) = r^2 |\bar{S}_{avg}| = \frac{r^2}{2} \operatorname{Re} \left\{ E_\theta \hat{\theta} \times H_\phi^* \hat{\phi} + E_\phi \hat{\phi} \times H_\theta^* \hat{\theta} \right\}$$

$$= \frac{r^2}{2\eta_0} \left[|E_\theta|^2 + |E_\phi|^2 \right] = \frac{1}{2\eta_0} \left[|F_\theta|^2 + |F_\phi|^2 \right] \text{W}, \tag{14.6}$$

where (14.1), (14.2), and (14.4) were used. The units of the radiation intensity are watts, or watts per unit solid angle, since the radial dependence has been removed. The radiation intensity gives the variation in radiated power versus position around the antenna. We can find the total power radiated by the antenna by integrating the Poynting vector over the surface of a sphere of radius r that encloses the antenna. This is equivalent to integrating the radiation intensity over a unit sphere:

$$P_{rad} = \int_{\phi=0}^{2\pi} \int_{\theta=0}^{\pi} \bar{S}_{avg} \cdot \hat{r} r^2 \sin \theta d\theta d\phi = \int_{\phi=0}^{2\pi} \int_{\theta=0}^{\pi} U(\theta, \phi) \sin \theta d\theta d\phi. \tag{14.7}$$

Antenna Pattern Characteristics

The *radiation pattern* of an antenna is a plot of the magnitude of the far-zone field strength versus position around the antenna, at a fixed distance from the antenna. Thus the radiation pattern can be plotted from the pattern function $F_\theta(\theta, \phi)$ or $F_\phi(\theta, \phi)$, versus either the angle θ (for an *elevation plane pattern*) or the angle ϕ (for an *azimuthal plane pattern*). The choice of plotting either F_θ or F_ϕ is dependent on the polarization of the antenna.

A typical antenna pattern is shown in Figure 14.3. This pattern is plotted in polar form, versus the elevation angle, θ, for a small horn antenna oriented in the vertical direction. The plot shows the relative variation

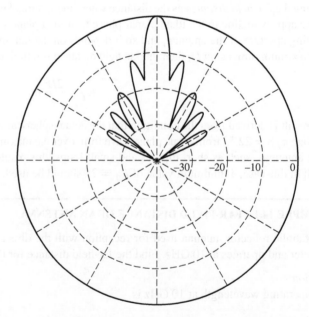

FIGURE 14.3 | The *E*-plane radiation pattern of a small horn antenna. The pattern is normalized to 0 dB at the beam maximum, with 10 dB per radial division.

of the radiated power of the antenna in dB, normalized to the maximum value. Since the pattern functions are proportional to voltage, the radial scale of the plot is computed as $20 \log |F(\theta, \phi)|$; alternatively, the plot could be computed in terms of the radiation intensity as $10 \log |U(\theta, \phi)|$. The pattern may exhibit several distinct lobes, with different maxima in different directions. The lobe having the maximum value is called the *main beam*, while those lobes at lower levels are called *sidelobes*. The pattern of Figure 14.3 has one main beam at $\theta = 0$ and several sidelobes, the largest of which are located at about $\theta = \pm 16°$. The level of these sidelobes is 13 dB below the level of the main beam. Radiation patterns may also be plotted in rectangular form; this is especially useful for antennas having a narrow main beam.

A fundamental property of an antenna is its ability to focus power in a given direction, to the exclusion of other directions. Thus an antenna with a broad main beam can transmit (or receive) power over a wide angular region, while an antenna having a narrow main beam will transmit (or receive) power over a small angular region. One measure of this focusing effect is the *3 dB beamwidth* of the antenna, defined as the angular width of the main beam at which the power level has dropped 3 dB from its maximum value (its half-power points). The 3 dB beamwidth of the pattern of Figure 14.3 is about 10°. Antennas having a constant pattern in the azimuthal plane are called *omnidirectional*, and are useful for applications such as broadcasting or for hand-held wireless devices, where it is desired to transmit or receive equally in all directions. Patterns that have relatively narrow main beams in both planes are known as *pencil beam* antennas, and are useful in applications such as radar and point-to-point radio links.

Another measure of the focusing ability of an antenna is the *directivity*, defined as the ratio of the maximum radiation intensity in the main beam to the average radiation intensity over all space:

$$D = \frac{U_{\text{max}}}{U_{\text{avg}}} = \frac{4\pi U_{\text{max}}}{P_{\text{rad}}} = \frac{4\pi U_{\text{max}}}{\displaystyle\int_{\theta=0}^{\pi} \int_{\phi=0}^{2\pi} U(\theta, \phi) \sin\theta \, d\theta \, d\phi}, \tag{14.8}$$

where (14.7) has been used for the radiated power. Directivity is a dimensionless ratio of power, and is usually expressed in dB as $D(\text{dB}) = 10 \log(D)$.

An antenna that radiates equally in all directions is called an *isotropic* antenna. Applying the integral identity that

$$\int_{\theta=0}^{\pi} \int_{\phi=0}^{2\pi} \sin\theta \, d\theta \, d\phi = 4\pi$$

to the denominator of (14.8) for $U(\theta, \phi) = 1$ shows that the directivity of an isotropic element is $D = 1$, or 0 dB. Since the minimum directivity of any antenna is unity, directivity is sometimes stated as relative to the directivity of an isotropic radiator, and written as *dBi*. Typical directivities for some common antennas are 2.2 dB for a wire dipole, 7.0 dB for a microstrip patch antenna, 23 dB for a waveguide horn antenna, and 35 dB for a parabolic reflector antenna.

Beamwidth and directivity are both measures of the focusing ability of an antenna: an antenna pattern with a narrow main beam will have a high directivity, while a pattern with a wide beam will have a lower directivity. We might therefore expect a direct relation between beamwidth and directivity, but in fact there is not an exact relationship between these two quantities. This is because beamwidth is only dependent on the size and shape of the main beam, whereas directivity involves integration of the entire radiation pattern. Thus it is possible for many different antenna patterns to have the same beamwidth but quite different directivities due to differences in sidelobes or the presence of more than one main beam. With this qualification in mind, however, it is possible to develop approximate relations between beamwidth and directivity that apply with reasonable accuracy to a large number of practical antennas. One such approximation that works well for antennas with pencil beam patterns is the following:

$$D \cong \frac{32,400}{\theta_1 \theta_2}, \tag{14.9}$$

where θ_1 and θ_2 are the beamwidths in two orthogonal planes of the main beam, in degrees. This approximation does not work well for omnidirectional patterns because there is a well-defined main beam in only one plane for such patterns.

EXAMPLE 14.2   PATTERN CHARACTERISTICS OF A DIPOLE ANTENNA

The far-zone electric field radiated by an electrically small wire dipole antenna oriented on the z-axis is given by

$$E_\theta(r, \theta, \phi) = V_0 \sin\theta \frac{e^{-jk_0 r}}{r} \text{ V/m},$$

$$E_\phi(r, \theta, \phi) = 0.$$

Find the main beam position of the dipole antenna, its beamwidth, and its directivity.

Solution
The radiation intensity for the above far-field is

$$U(\theta, \phi) = C \sin^2\theta,$$

where the constant $C = V_0^2/2\eta_0$. The radiation pattern is seen to be independent of the azimuthal angle ϕ, and so is omnidirectional in the azimuthal plane. The pattern has a "donut" shape, with nulls at $\theta = 0$ and $\theta = 180°$ (along the z-axis), and a beam maximum at $\theta = 90°$ (the horizontal plane). The angles where the radiation intensity has dropped by 3 dB are given by the solutions to

$$\sin^2\theta = 0.5;$$

thus the 3 dB, or half-power, beamwidth is $135° - 45° = 90°$.

The directivity is calculated using (14.8). The denominator of this expression is

$$\int_{\theta=0}^{\pi} \int_{\phi=0}^{2\pi} U(\theta, \phi) \sin\theta \, d\theta \, d\phi = 2\pi C \int_{\theta=0}^{\pi} \sin^3\theta \, d\theta = 2\pi C \left(\frac{4}{3}\right) = \frac{8\pi C}{3},$$

where the required integral identity is listed in Appendix E. Since $U_{max} = C$, the directivity reduces to

$$D = \frac{3}{2} = 1.76 \text{ dB}.$$

Antenna Gain and Efficiency

Resistive losses, due to nonperfect metals and dielectric materials, exist in all practical antennas. Such losses result in a difference between the power delivered to the input of an antenna and the power radiated by that antenna. As with many other electrical components, we can define the *radiation efficiency* of an antenna as the ratio of the desired output power to the supplied input power:

$$\eta_{rad} = \frac{P_{rad}}{P_{in}} = \frac{P_{in} - P_{loss}}{P_{in}} = 1 - \frac{P_{loss}}{P_{in}}, \tag{14.10}$$

where P_{rad} is the power radiated by the antenna, P_{in} is the power supplied to the input of the antenna, and P_{loss} is the power lost in the antenna. Note that there are other factors that can contribute to the effective loss of transmit power, such as impedance mismatch at the input to the antenna, or polarization mismatch with the receive antenna. However, these losses are external to the antenna and could be eliminated by the proper use of matching networks, or the proper choice and positioning of the receive antenna. Therefore losses of this type are usually not attributed to the antenna itself, as are dissipative losses due to metal conductivity or dielectric loss within the antenna.

Recall that antenna directivity is a function only of the shape of the radiation pattern (the radiated fields) of an antenna, and is not affected by losses in the antenna itself. To account for the fact that an antenna having a radiation efficiency less than unity will not radiate all of its input power, we define *antenna gain* as the product of directivity and efficiency:

$$G = \eta_{rad} D. \tag{14.11}$$

Thus, gain is always less than or equal to directivity. Gain can also be computed directly, by replacing P_{rad} in the denominator of (14.8) with P_{in}, since by the definition of radiation efficiency in (14.10) we have $P_{\text{rad}} = \eta_{\text{rad}} P_{\text{in}}$. Gain is usually expressed in dB, as $G(\text{dB}) = 10 \log(G)$. Sometimes the effect of impedance mismatch loss is included in the gain of an antenna; this is referred to as the *realized gain* [1].

Aperture Efficiency and Effective Area

Many types of antennas can be classified as *aperture antennas*, meaning that the antenna has a well-defined aperture area from which radiation occurs. Examples include reflector antennas, horn antennas, lens antennas, and array antennas. For such antennas, it can be shown that the maximum directivity that can be obtained from an electrically large aperture of area A is given as

$$D_{\text{max}} = \frac{4\pi A}{\lambda^2}. \tag{14.12}$$

For example, a rectangular horn antenna having an aperture $2\lambda \times 3\lambda$ has a maximum directivity of 24π, or about 19 dB. In practice, there are several factors that can serve to reduce the directivity of an antenna from its maximum possible value, such as nonideal amplitude or phase characteristics of the aperture field, aperture blockage, or, in the case of reflector antennas, spillover of the feed pattern. For this reason, we define an *aperture efficiency* as the ratio of the actual directivity of an aperture antenna to the maximum directivity given by (14.12). Then we can write the directivity of an aperture antenna as

$$D = \eta_{ap} \frac{4\pi A}{\lambda^2}. \tag{14.13}$$

Aperture efficiency is always less than or equal to unity.

The above definitions of antenna directivity, efficiency, and gain were stated in terms of transmitting antennas, but they apply to receiving antennas as well. For a receiving antenna it is also of interest to determine the received power for a given incident plane wave field. This is the converse problem of finding the power density radiated by a transmitting antenna, as given in (14.4). Determining received power is important for the derivation of the Friis radio system link equation, to be discussed in the following section. We expect that received power will be proportional to the power density, or Poynting vector, of the incident wave. Since the Poynting vector has dimensions of W/m^2, and the received power, P_r, has dimensions of W, the proportionality constant must have units of area. Thus we write

$$P_r = A_e S_{\text{avg}}, \tag{14.14}$$

where A_e is defined as the *effective aperture area* of the receive antenna. The effective aperture area has dimensions of m^2, and can be interpreted as the "capture area" of a receive antenna, intercepting part of the incident power density radiated toward the receive antenna. The quantity P_r in (14.14) is the power available at the terminals of the receive antenna, as delivered to a conjugately matched load.

The maximum effective aperture area of an antenna can be shown to be related to the directivity of the antenna as [1, 2]

$$A_e = \frac{D\lambda^2}{4\pi}, \tag{14.15}$$

where λ is the operating wavelength of the antenna. For electrically large aperture antennas the effective aperture area is often close to the actual physical aperture area. However, for many other types of antennas, such as dipoles and loops, there is no simple relation between the physical cross-sectional area of the antenna and its effective aperture area. The maximum effective aperture area as defined above does not include the effect of losses in the antenna, which can be accounted for by replacing D in (14.15) with G, the gain, of the antenna.

Background and Brightness Temperature

We have seen how noise power is generated by lossy components and active devices, but noise can also be delivered to the input of a receiver by the antenna. Antenna noise power may be received from the

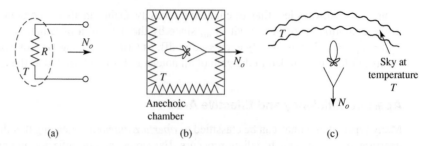

(a)   (b)   (c)

FIGURE 14.4 | Illustrating the concept of background temperature. (a) A resistor at temperature T. (b) An antenna in an anechoic chamber at temperature T. (c) An antenna viewing a uniform sky background at temperature T.

external environment, or generated internally as thermal noise due to losses in the antenna itself. While noise produced within a receiver is controllable to some extent (by judicious design and component selection), the noise received from the environment by a receiving antenna is generally not controllable, and may exceed the noise level of the receiver itself. Thus it is important to characterize the noise power delivered to a receiver by its antenna.

Consider the three situations shown in Figure 14.4. In Figure 14.4a we have the simple case of a resistor at temperature T, producing an available output noise power

$$N_o = kTB, \tag{14.16}$$

where B is the system bandwidth and k is Boltzmann's constant. In Figure 14.4b we have an antenna enclosed by an anechoic chamber at temperature T. The anechoic chamber appears as a perfectly absorbing enclosure, and is in thermal equilibrium with the antenna. Thus the terminals of the antenna are indistinguishable from the resistor terminals of Figure 14.4a (assuming an impedance-matched antenna), and therefore it produces the same output noise power as the resistor of Figure 14.4a. Figure 14.4c shows the same antenna directed at the sky. If the main beam of the antenna is narrow enough so that it sees a uniform region at physical temperature T, then the antenna again appears as a resistor at temperature T and produces the output noise power given in (14.16). This is true regardless of the radiation efficiency of the antenna, as long as the physical temperature of the antenna is also T.

In actuality an antenna typically sees a much more complex environment than the cases depicted in Figure 14.4. A general scenario of both naturally occurring and man-made noise sources is shown in Figure 14.5, where we see that an antenna with a relatively broad main beam may pick up noise power from a variety of origins. In addition, noise may be received through the sidelobes of the antenna pattern or via

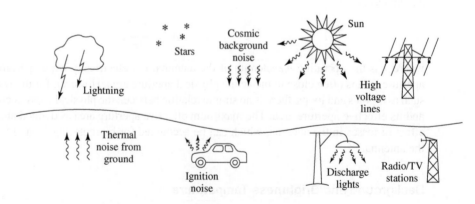

FIGURE 14.5 | Natural and man-made sources of background noise.

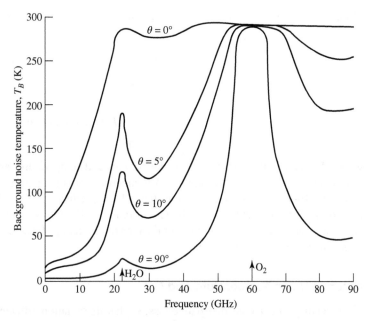

FIGURE 14.6 | Background noise temperature of sky versus frequency. θ is elevation angle measured from the horizon. Data are for sea level, with surface temperature of 15°C and surface water vapor density of 7.5 gm/m^3.

reflections from the ground or other large objects. As in Chapter 10, where the noise power from an arbitrary white noise source was represented as an equivalent noise temperature, we define the *background noise temperature*, T_B, as the equivalent temperature of a resistor required to produce the same noise power as the actual environment seen by the antenna. Some typical background noise temperatures that are relevant at low microwave frequencies are as follows:

- Sky (toward zenith) 3–5 K
- Sky (toward horizon) 50–100 K
- Ground 290–300 K

The overhead sky background temperature of 3–5 K is the cosmic background radiation believed to be a remnant of the big bang at the creation of the universe. This would be the noise temperature seen by an antenna with a narrow beam and high radiation efficiency pointed overhead, away from "hot" sources such as the Sun or stellar radio objects. The background noise temperature increases as the antenna is pointed toward the horizon because of the greater thickness of the atmosphere, so that the antenna sees an effective background closer to that of the anechoic chamber of Figure 14.4b. Pointing the antenna toward the ground further increases the effective loss, and hence the noise temperature.

Figure 14.6 gives a more complete picture of the background noise temperature, showing the variation of T_B versus frequency and for several elevation angles [3]. Note that the noise temperature shown in the graph follows the trends listed above, in that it is lowest for the overhead sky ($\theta = 90°$), and greatest for angles near the horizon ($\theta = 0°$). Also note the sharp peaks in noise temperature that occur at 22 and 60 GHz. The first is due to the resonance of molecular water, while the second is caused by resonance of molecular oxygen. Both of these resonances lead to increased atmospheric loss and hence increased noise temperature. The loss is great enough at 60 GHz that a high-gain antenna pointing through the atmosphere effectively appears as a matched load at 290 K. While loss in general is undesirable, these particular resonances can be useful for remote sensing applications, or for using the inherent attenuation of the atmosphere to limit propagation distances for radio communication over small distances.

When the antenna beamwidth is broad enough that different parts of the antenna pattern see different background temperatures, the effective *brightness temperature* seen by the antenna can be found by weighting the spatial distribution of background temperature by the pattern function of the antenna.

Mathematically we can write the *brightness temperature* T_b seen by the antenna as

$$T_b = \frac{\displaystyle\int_{\phi=0}^{2\pi}\int_{\theta=0}^{\pi} T_B(\theta,\phi)D(\theta,\phi)\sin\theta\, d\theta\, d\phi}{\displaystyle\int_{\phi=0}^{2\pi}\int_{\theta=0}^{\pi} D(\theta,\phi)\sin\theta\, d\theta\, d\phi}, \tag{14.17}$$

where $T_B(\theta,\phi)$ is the distribution of the background temperature, and $D(\theta,\phi)$ is the directivity (or the power pattern function) of the antenna. Antenna brightness temperature is referenced at the terminals of the antenna. Observe that when T_B is a constant, (14.17) reduces to $T_b = T_B$, which is essentially the case of a uniform background temperature shown in Figure 14.4b or 14.4c. Also note that this definition of antenna brightness temperature does not involve the gain or efficiency of the antenna, and so does not include thermal noise due to losses in the antenna.

Antenna Noise Temperature and G/T

If a receiving antenna has dissipative loss, so that its radiation efficiency η_{rad} is less than unity, the power available at the terminals of the antenna is reduced by the factor η_{rad} from that intercepted by the antenna (the definition of radiation efficiency is the ratio of output to input power). This reduction applies to received noise power, as well as received signal power, so the noise temperature of the antenna will be reduced from the brightness temperature given in (14.17) by the factor η_{rad}. In addition, thermal noise will be generated internally by resistive losses in the antenna, and this will increase the noise temperature of the antenna. In terms of noise power, a lossy antenna can be modeled as a lossless antenna and an attenuator having a power loss factor of $L = 1/\eta_{\text{rad}}$. Then, using (10.15) for the equivalent noise temperature of an attenuator, we can find the resulting noise temperature seen at the antenna terminals as

$$T_A = \frac{T_b}{L} + \frac{(L-1)}{L}T_p = \eta_{\text{rad}}T_b + (1-\eta_{\text{rad}})T_p. \tag{14.18}$$

The equivalent temperature T_A is called the *antenna noise temperature*, and is a combination of the external brightness temperature seen by the antenna and the thermal noise generated by the antenna. As with other equivalent noise temperatures, the proper interpretation of T_A is that a matched load at this temperature will produce the same available noise power as does the antenna. Note that this temperature is referenced at the output terminals of the antenna; since an antenna is not a two-port circuit element, it does not make sense to refer the equivalent noise temperature to its "input."

Observe that (14.18) reduces to $T_A = T_b$ for a lossless antenna with $\eta_{\text{rad}} = 1$. If the radiation efficiency is zero, meaning that the antenna appears as a matched load and does not see any external background noise, then (14.18) reduces to $T_A = T_p$, due to the thermal noise generated by the losses. If an antenna is pointed toward a known background temperature different than T_0, then (14.18) can be used to determine its radiation efficiency.

EXAMPLE 14.3   ANTENNA NOISE TEMPERATURE

A high-gain antenna has the idealized hemispherical elevation plane pattern shown in Figure 14.7, and is rotationally symmetric in the azimuthal plane. If the antenna is facing a region having a background

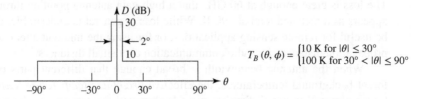

FIGURE 14.7 | Idealized antenna pattern and background noise temperature for Example 14.3.

temperature T_B approximated as given in Figure 14.7, find the antenna noise temperature. Assume the radiation efficiency of the antenna is 100%.

Solution
Since $\eta_{\text{rad}} = 1$, (14.18) reduces to $T_A = T_b$. The brightness temperature can be computed from (14.17), after normalizing the directivity to a maximum value of unity:

$$T_b = \frac{\int\limits_{\phi=0}^{2\pi}\int\limits_{\theta=0}^{\pi} T_B(\theta,\phi)D(\theta,\phi)\sin\theta\,d\theta\,d\phi}{\int\limits_{\phi=0}^{2\pi}\int\limits_{\theta=0}^{\pi} D(\theta,\phi)\sin\theta\,d\theta\,d\phi} = \frac{\int\limits_{\theta=0}^{1°} 10\sin\theta\,d\theta + \int\limits_{\theta=1°}^{30°} 0.1\sin\theta\,d\theta + \int\limits_{\theta=30°}^{90°}\sin\theta\,d\theta}{\int\limits_{\theta=0}^{1°}\sin\theta\,d\theta + \int\limits_{\theta=1°}^{90°} 0.01\sin\theta\,d\theta}$$

$$= \frac{-10\cos\theta\big|_0^{1°} - 0.1\cos\theta\big|_{1°}^{30°} - \cos\theta\big|_{30°}^{90°}}{-\cos\theta\big|_0^{1°} - 0.01\cos\theta\big|_{1°}^{90°}} = \frac{0.00152 + 0.0134 + 0.866}{0.0102} = 86.4 \text{ K}.$$

In this example most of the noise power is collected through the sidelobe region of the antenna.

The more general problem of a receiver connected through a lossy transmission line to an antenna viewing a background noise temperature distribution T_B can be represented by the system shown in Figure 14.8. The antenna is assumed to have a radiation efficiency η_{rad}, and the connecting transmission line has a power loss factor of $L \geq 1$, with both at physical temperature T_p. We also include the effect of an impedance mismatch between the antenna and the transmission line, represented by the reflection coefficient Γ. The equivalent noise temperature seen at the output terminals of the transmission line consists of three contributions: noise power from the antenna due to internal noise and the background brightness temperature, noise power generated from the lossy line in the forward direction, and noise power generated by the lossy line in the backward direction and reflected from the antenna mismatch toward the receiver. The noise due to the antenna is given by (14.18), but reduced by the loss factor of the line, $1/L$, and the reflection mismatch factor, $(1 - |\Gamma|^2)$. The forward noise power from the lossy line is given by (10.15), after reduction by the loss factor, $1/L$. The contribution from the lossy line reflected from the mismatched antenna is given by (10.15), after reduction by the power reflection coefficient, $|\Gamma|^2$, and the loss factor, $1/L^2$ (since the reference point for the back-directed noise power from the lossy line given by (10.15) is at the output terminals of the line). Thus the overall system noise temperature seen at the input to the receiver is given by

$$T_S = \frac{T_A}{L}(1 - |\Gamma|^2) + (L-1)\frac{T_p}{L} + (L-1)\frac{T_p}{L^2}|\Gamma|^2$$

$$= \frac{(1 - |\Gamma|^2)}{L}[\eta_{\text{rad}}T_b + (1 - \eta_{\text{rad}})T_p] + \frac{(L-1)}{L}\left(1 + \frac{|\Gamma|^2}{L}\right)T_p. \tag{14.19}$$

Observe that for a lossless line ($L = 1$) the effect of an antenna mismatch is to reduce the system noise temperature by the factor $(1 - |\Gamma|^2)$. Of course, the received signal power will be reduced by the same

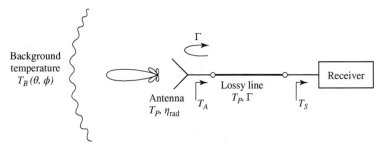

FIGURE 14.8 | A receiving antenna connected to a receiver through a lossy transmission line. An impedance mismatch exists between the antenna and the line.

amount. Also note that for the case of a matched antenna ($\Gamma = 0$), (14.19) reduces to

$$T_S = \frac{1}{L}[\eta_{\text{rad}}T_b + (1 - \eta_{\text{rad}})T_p] + \frac{L-1}{L}T_p, \tag{14.20}$$

as expected for a cascade of two noisy components.

Finally, it is important to realize the difference between radiation efficiency and aperture efficiency, and their effects on antenna noise temperature. While radiation efficiency accounts for resistive losses, and thus involves the generation of thermal noise, aperture efficiency does not. Aperture efficiency applies to the loss of directivity in aperture antennas, such as reflectors, lenses, or horns, due to feed spillover or suboptimum aperture excitation (e.g., a nonuniform amplitude or phase distribution), and by itself does not lead to any additional effect on noise temperature that would not be included through the pattern of the antenna.

The antenna noise temperature defined above is a useful figure of merit for a receive antenna because it characterizes the total noise power delivered by the antenna to the input of a receiver. Another useful figure of merit for receive antennas is the *G/T ratio*, defined as

$$G/T(\text{dB}) = 10\log\frac{G}{T_A} \text{ dB/K}, \tag{14.21}$$

where G is the gain of the antenna, and T_A is the antenna noise temperature. This quantity is important because, as we will see in Section 14.2, the signal-to-noise ratio (SNR) at the input to a receiver is proportional to G/T_A. The ratio G/T can often be maximized by increasing the gain of the antenna, since this increases the numerator and usually minimizes reception of noise from hot sources at low elevation angles. Of course, higher gain requires a larger and more expensive antenna, and high gain may not be desirable for applications requiring omnidirectional coverage (e.g., cellular telephones or mobile data networks), so often a compromise must be made. Finally, note that the dimensions given in (14.21) for $10\log(G/T)$ are not actually decibels per degree kelvin, but this is the nomenclature that is commonly used for this quantity.

14.2 WIRELESS COMMUNICATION

Wireless communication involves the transfer of information between two points without direct connection. While this may be accomplished using sound, infrared, optical, or radio frequency energy, most modern wireless systems rely on RF or microwave signals, usually in the UHF to millimeter wave frequency range. Because of spectrum crowding and the need for higher data rates, the trend is toward higher frequencies, so the majority of wireless systems today operate at frequencies ranging from about 800 MHz to a few gigahertz. RF and microwave signals offer wide bandwidths, and have the added advantage of being able to penetrate fog, dust, foliage, and even buildings and vehicles to some extent. Historically, wireless communication using RF energy has its foundations in the theoretical work of Maxwell, followed by the experimental verification of electromagnetic wave propagation by Hertz, and the practical development of radio techniques and systems by Tesla, Marconi, and others in the early part of the 20th century. Today, wireless systems include broadcast radio and television, cellular telephone and networking systems, direct broadcast satellite (DBS) television service, wireless local area networks (WLANs), paging systems, Global Positioning System (GPS) service, and radio frequency identification (RFID) systems [4]. These systems are beginning to provide, for the first time in history, worldwide connectivity for voice, video, and network communication.

One way to categorize wireless systems is according to the nature and placement of the users. In a *point-to-point* radio system a single transmitter communicates with a single receiver. Such systems generally use high-gain antennas in fixed positions to maximize received power and minimize interference with other radios that may be operating nearby in the same frequency range. Point-to-point radios are typically used for satellite communication, dedicated data communication by utility companies, and backhaul connection of cellular base stations to a central switching office. *Point-to-multipoint* systems connect a central station to a large number of possible receivers. The most common examples are commercial AM and FM radio and broadcast television, where a central transmitter uses an antenna with a broad azimuthal beam to reach many listeners and viewers. *Multipoint-to-multipoint* systems allow simultaneous communication between individual users (who may not be in fixed locations). Such systems generally do not connect two users directly, but instead rely on a grid of base stations to provide the desired interconnections between users. Cellular telephone systems and some types of WLANs are examples of this type of application.

Another way to characterize wireless systems is in terms of the directionality of communication. In a *simplex* system, communication occurs only in one direction—from the transmitter to the receiver. Examples of simplex systems include broadcast radio, television, and paging systems. In a *half-duplex* system, communication may occur in two directions, but not simultaneously. Early mobile radios and citizens band radio are examples of duplex systems, and generally rely on a "push-to-talk" function so that a single channel can be used for both transmitting and receiving at different times. *Full-duplex* systems allow simultaneous two-way transmission and reception. Examples include cellular telephone and point-to-point radio systems. Full-duplex transmission clearly requires a *duplexing* technique to avoid interference between transmitted and received signals. This can be done by using separate frequency bands for transmit and receive (*frequency division duplexing*), or by allowing users to transmit and receive only in certain predefined time intervals (*time division duplexing*).

While most wireless systems are ground based, it is also possible to use satellite systems for voice, video, and data communication [5]. Satellites offer the possibility of communication with a large number of users over wide areas, perhaps including the entire planet. Satellites in a *geosynchronous earth orbit* (GEO) are positioned approximately 36,000 km above Earth, and have a 24-hour orbital period. When a GEO satellite is positioned above the equator, it becomes *geostationary*, and will remain in a fixed position relative to Earth. Such satellites are useful for point-to-point radio links between widely separated stations, and are commonly used for television and data communication throughout the world. At one time transcontinental telephone service relied on such satellites, but undersea fiber optics cables have largely replaced satellites for transoceanic connections as being more economical and avoiding the annoying delay caused by the very long round-trip path between the satellite and Earth. Another drawback of GEO satellites is that their high altitude greatly reduces the received signal strength, making it difficult for two-way communication with handheld transceivers. *Low Earth orbit* (LEO) satellites orbit much closer to Earth, typically in the range of 500–2000 km. The shorter path length may allow line-of-sight communication between LEO satellites and hand-held radios, but satellites in LEO orbits are visible from a given point on the ground for only a short time, typically between a few minutes and about 20 minutes. Effective coverage therefore requires a large number of LEO satellites in different orbital planes. The ill-fated Iridium system is probably the best-known example of a LEO satellite communication system.

The Friis Formula

A general radio system link is shown in Figure 14.9, where the transmit power is P_t, the transmit antenna gain is G_t, the receive antenna gain is G_r, and the received power (delivered to a matched load) is P_r. The transmit and receive antennas are separated by the distance R.

From (14.6)–(14.7), the power density radiated by an isotropic antenna ($D = 1 = 0$ dB) at a distance R is given by

$$S_{\text{avg}} = \frac{P_t}{4\pi R^2} \ \text{W/m}^2. \tag{14.22}$$

This result reflects the fact that we must be able to recover all of the radiated power by integrating over a sphere of radius R surrounding the antenna; since the power is distributed isotropically, and the area of a sphere is $4\pi R^2$, (14.22) follows. If the transmit antenna has a directivity greater than 0 dB, we can find the radiated power density by multiplying by the directivity, since directivity is defined as the ratio of the actual radiation intensity to the equivalent isotropic radiation intensity. In addition, if the transmit antenna

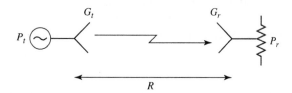

FIGURE 14.9 | A basic radio system.

has losses, we can include the radiation efficiency factor, which has the effect of converting directivity to gain. Thus, the general expression for the power density radiated by an arbitrary transmit antenna is

$$S_{\text{avg}} = \frac{G_t P_t}{4\pi R^2} \text{ W/m}^2. \tag{14.23}$$

If this power density is incident on the receive antenna, we can use the concept of effective aperture area, as defined in (14.14), to find the received power:

$$P_r = A_e S_{\text{avg}} = \frac{G_t P_t A_e}{4\pi R^2} \text{ W}.$$

Next, (14.15) can be used to relate the effective area to the directivity of the receive antenna. Again, the possibility of losses in the receive antenna can be accounted for by using the gain (rather than the directivity) of the receive antenna. Then the final result for the received power is

$$P_r = \frac{G_t G_r \lambda^2}{(4\pi R)^2} P_t \text{ W}. \tag{14.24}$$

This result is known as the *Friis radio link formula*, and it addresses the fundamental question of how much power is received by a radio antenna. In practice, the value given by (14.24) should be interpreted as the maximum possible received power, as there are a number of factors that can serve to reduce the received power in an actual radio system. These include impedance mismatch at either antenna, polarization mismatch between the antennas, propagation effects leading to attenuation or depolarization, and multipath effects that may cause partial cancellation of the received field.

Observe in (14.24) that the received power decreases as $1/R^2$ as the separation between transmitter and receiver increases. This dependence is a result of conservation of energy. While it may seem to be prohibitively large for large distances, in fact the space decay of $1/R^2$ is usually much better than the exponential decrease in power due to losses in a wired communication link. This is because the attenuation of power on a transmission line varies as $e^{-2\alpha z}$ (where α is the attenuation constant of the line), and at large distances the exponential function decreases faster than an algebraic dependence like $1/R^2$. Thus for long-distance communication, radio links will perform better than wired links. This conclusion applies to any type of transmission line, including coaxial lines, waveguides, and even fiber optic lines. (It may not apply, however, if the communication link is land or sea based, so that repeaters can be inserted along the link to recover lost signal power.)

As can be seen from the Friis formula, received power is proportional to the product $P_t G_t$. These two factors—the transmit power and transmit antenna gain—characterize the transmitter, and in the main beam of the antenna the product $P_t G_t$ can be interpreted equivalently as the power radiated by an isotropic antenna with input power $P_t G_t$. Thus, this product is defined as the *effective isotropic radiated power* (EIRP):

$$\text{EIRP} = P_t G_t \text{ W}. \tag{14.25}$$

For a given frequency, range, and receiver antenna gain, the received power is proportional to the EIRP of the transmitter and can only be increased by increasing the EIRP. This can be done by increasing the transmit power, or the transmit antenna gain, or both.

Link Budget and Link Margin

The various terms in the Friis formula of (14.24) are often tabulated separately in a *link budget*, where each of the factors can be individually considered in terms of its net effect on the received power. Additional loss factors, such as line losses or impedance mismatch at the antennas, atmospheric attenuation (see Section 14.5), and polarization mismatch can also be added to the link budget. One of the terms in a link budget is the *path loss*, accounting for the free-space reduction in signal strength with distance between the transmitter and receiver. From (14.24), path loss is defined (in dB) as

$$L_0(\text{dB}) = 20 \log \left(\frac{4\pi R}{\lambda} \right) > 0. \tag{14.26}$$

Note that path loss depends on wavelength (frequency), which serves to provide a normalization for the units of distance.

With the above definition of path loss, we can write the remaining terms of the Friis formula as shown in the following link budget:

Transmit power	P_t
Transmit antenna line loss	$(-)L_t$
Transmit antenna gain	G_t
Path loss (free-space)	$(-)L_0$
Atmospheric attenuation	$(-)L_A$
Receive antenna gain	G_r
Receive antenna line loss	$(-)L_r$
Receive power	P_r

We have also included loss terms for atmospheric attenuation and line attenuation. Assuming that all of the above quantities are expressed in dB (or dBm, in the case of P_t), we can write the receive power as

$$P_r(\text{dBm}) = P_t - L_t + G_t - L_0 - L_A + G_r - L_r. \tag{14.27}$$

If the transmit and/or receive antenna is not impedance matched to the transmitter/receiver (or to their connecting lines), impedance mismatch will reduce the received power by the factor $(1 - |\Gamma|^2)$, where Γ is the appropriate reflection coefficient. The resulting impedance mismatch loss,

$$L_{\text{imp}}(\text{dB}) = -10\log(1 - |\Gamma|^2) \geq 0, \tag{14.28}$$

can be included in the link budget to account for the reduction in received power.

Another possible entry in the link budget relates to the polarization matching of the transmit and receive antennas, as maximum power transmission between transmitter and receiver requires both antennas to be polarized in the same manner. If a transmit antenna is vertically polarized, for example, maximum power will only be delivered to a vertically polarized receiving antenna, while zero power would be delivered to a horizontally polarized receive antenna, and half the available power would be delivered to a circularly polarized antenna. Determination of the *polarization loss factor* is explained in references [1], [2], and [4].

In practical communication systems it is usually desired to have the received power level greater than the threshold level required for the minimum acceptable quality of service (usually expressed as the minimum carrier-to-noise ratio (CNR), or minimum SNR). This design allowance for received power is referred to as the *link margin*, and can be expressed as the difference between the design value of received power and the minimum threshold value of receive power:

$$\text{Link margin (dB)} = \text{LM} = P_r - P_{r(\min)} > 0, \tag{14.29}$$

where all quantities are in dB. Link margin should be a positive number; typical values may range from 3 to 20 dB. Having a reasonable link margin provides a level of robustness to the system to account for variables such as signal fading due to weather, movement of a mobile user, multipath propagation problems, and other unpredictable effects that can degrade system performance and quality of service. Link margin that is used to account for fading effects is sometimes referred to as *fade margin*. Satellite links operating at frequencies above 10 GHz, for example, often require fade margins of 20 dB or more to account for attenuation during heavy rain.

As seen from (14.29) and the link budget, link margin for a given communication system can be improved by increasing the received power (by increasing transmit power or antenna gains), or by reducing the minimum threshold power (by improving the design of the receiver, changing the modulation method, or by other means). Increasing link margin therefore usually involves an increase in cost and complexity, so excessive increases in link margin are usually avoided.

EXAMPLE 14.4   LINK ANALYSIS OF DBS TELEVISION SYSTEM

The direct broadcast system in North America operates at 12.2–12.7 GHz, with a transmit carrier power of 120 W, a transmit antenna gain of 34 dB, an IF bandwidth of 20 MHz, and a worst-case slant angle (30°) distance from the geostationary satellite to Earth of 39,000 km. The 18-inch receiving dish antenna has a gain of 33.5 dB and sees an average background brightness temperature of $T_b = 50$ K, with a receiver

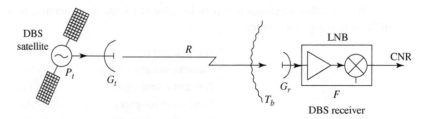

FIGURE 14.10 | Diagram of the DBS system for Example 14.4.

low-noise block (LNB) having a noise figure of 0.7 dB. The required minimum CNR is 15 dB. The overall system is shown in Figure 14.10. Find (a) the link budget for the received carrier power at the antenna terminals, (b) G/T for the receive antenna and LNB system, (c) the CNR at the output of the LNB, and (d) the link margin of the system.

Solution
We will take the operating frequency to be 12.45 GHz, so the wavelength is 0.0241 m. From (14.26) the path loss is

$$L_0 = 20 \log \left(\frac{4\pi R}{\lambda} \right) = 20 \log \left(\frac{(4\pi)(39 \times 10^6)}{0.0241} \right) = 206.2 \text{ dB}$$

(a) The link budget for the received power is

$$
\begin{array}{l}
P_t = 120 \text{ W} = 50.8 \text{ dBm} \\
G_t = 34.0 \text{ dB} \\
L_0 = (-)206.2 \text{ dB} \\
\underline{G_r = 33.5 \text{ dB}} \\
P_r = -87.9 \text{ dBm} = 1.63 \times 10^{-12} \text{ W.}
\end{array}
$$

(b) To find G/T we first find the noise temperature of the antenna and LNB cascade, referenced at the input of the LNB:

$$T_e = T_A + T_{\text{LNB}} = T_b + (F - 1)T_0 = 50 + (1.175 - 1)(290) = 100.8 \text{ K.}$$

Then G/T for the antenna and LNB is

$$G/T(\text{dB}) = 10 \log \frac{2239}{100.8} = 13.5 \text{ dB/K.}$$

(c) The CNR at the output of the LNB is

$$\text{CNR} = \frac{P_r G_{\text{LNB}}}{k T_e B G_{\text{LNB}}} = \frac{1.63 \times 10^{-12}}{(1.38 \times 10^{-23})(100.8)(20 \times 10^6)} = 58.6 = 17.7 \text{ dB.}$$

Note that G_{LNB}, the gain of the LNB module, cancels in the ratio for the output CNR.

(d) If the minimum required CNR is 15 dB, the system link margin is 2.7 dB.

Radio Receiver Architectures

The receiver is usually the most critical component of a wireless system, having the overall purpose of reliably recovering the desired signal from a wide spectrum of transmitting sources, interference, and noise. In this section we will describe some of the critical requirements for radio receiver design and summarize some of the most common types of receiver architectures.

A well-designed radio receiver must provide several different functions:

- *High gain* (~100 dB) to restore the low power of the received signal to a level near its original baseband value

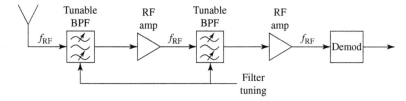

FIGURE 14.11 | Block diagram of a tuned radio frequency receiver.

- *Selectivity*, in order to receive the desired signal while rejecting adjacent channels, image frequencies, and interference
- *Down-conversion* from the received RF frequency to a lower IF frequency for processing
- *Detection* of the received analog or digital information
- *Isolation* from the transmitter to avoid saturation of the receiver

Because the typical signal power level from the receive antenna may be as low as -100 to -120 dBm, the receiver may be required to provide gain as high as 100 to 120 dB. This much gain should be spread over the RF, IF, and baseband stages to avoid instabilities and possible oscillation; it is generally good practice to avoid more than about 50–60 dB of gain at any one frequency band. The fact that amplifier cost generally increases with frequency is a further reason to spread gain over different frequency stages.

In principle, selectivity can be obtained by using a narrow bandpass filter at the RF stage of the receiver, but the bandwidth and cutoff requirements for such a filter are usually impractical to realize at RF frequencies. It is more effective to achieve selectivity by down-converting a relatively wide RF bandwidth around the desired signal, and using a sharp-cutoff bandpass filter at the IF stage to select only the desired frequency band. In addition, many wireless systems use a number of narrow but closely spaced channels, which must be selected using a tuned local oscillator, while the IF passband is fixed. The alternative of using an extremely narrowband, electronically tunable RF filter is not practical.

Tuned radio frequency receiver: One of the earliest types of receiving circuits to be developed was the *tuned radio frequency* (TRF) receiver. As shown in Figure 14.11, a TRF receiver employs several stages of RF amplification along with tunable bandpass filters to provide high gain and selectivity. Alternatively, filtering and amplification may be combined by using amplifiers with a tunable bandpass response. At relatively low broadcast radio frequencies, such filters and amplifiers have historically been tuned using mechanically variable capacitors or inductors. However, such tuning is problematic because of the need to tune several stages in parallel, and selectivity is poor because the passband of such filters is fairly broad. In addition, all the gain of the TRF receiver is achieved at the RF frequency, limiting the amount of gain that can be obtained before oscillation occurs, and increasing the cost and complexity of the receiver. Because of these drawbacks TRF receivers are seldom used today, and are an especially bad choice for higher RF or microwave frequencies.

Direct-conversion receiver: The *direct-conversion* receiver, shown in Figure 14.12, uses a mixer and local oscillator to perform frequency down-conversion with a zero IF frequency. The local oscillator is set to the same frequency as the desired RF signal, which is then converted directly to baseband. For this reason, the direct-conversion receiver is sometimes called a *homodyne* receiver. For AM reception the received baseband signal would not require any further detection. The direct-conversion receiver offers several advantages over the TRF receiver, as selectivity can be controlled with a simple low-pass baseband filter,

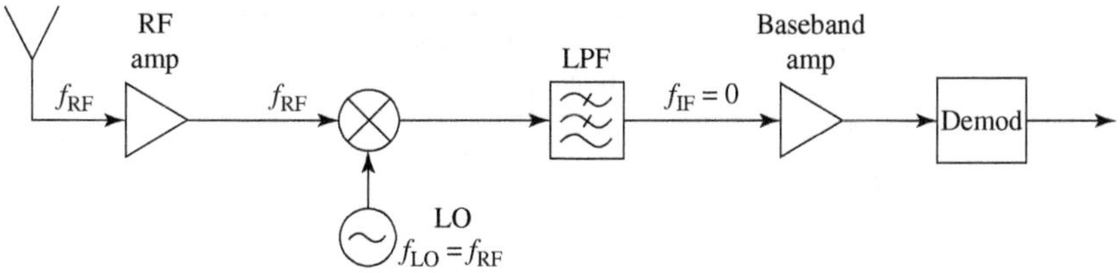

FIGURE 14.12 | Block diagram of a direct-conversion receiver.

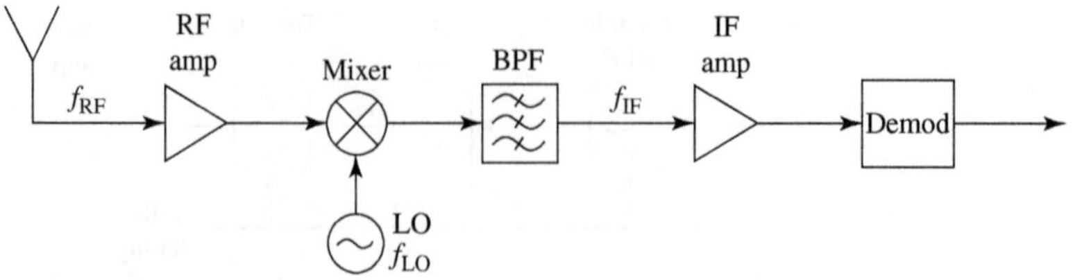

FIGURE 14.13 | Block diagram of a single-conversion superheterodyne receiver.

and gain may be spread through the RF and baseband stages (although it is difficult to obtain stable high gain at very low frequencies). Direct-conversion receivers are simpler and less costly than superheterodyne receivers since there is no IF amplifier, IF bandpass filter, or IF local oscillator required for final down-conversion. Another important advantage of direct-conversion is that there is no image frequency, since the mixer difference frequency is effectively zero, and the sum frequency is twice the LO and easily filtered. However, a serious disadvantage is that the LO must have a very high degree of precision and stability, especially for high RF frequencies, to avoid drift of the received signal frequency. This type of receiver is often used with Doppler radars, where the exact LO can be obtained from the transmitter, but a number of newer wireless systems are being designed with direct-conversion receivers.

Superheterodyne receiver: By far the most popular type of receiver in use today is the *superheterodyne* circuit, shown in Figure 14.13. The block diagram is similar to that of the direct-conversion receiver, but the IF frequency is now nonzero, and is generally selected to be between the RF frequency and baseband. A midrange IF allows the use of sharper cutoff filters for improved selectivity, and higher IF gain through the use of an IF amplifier. Tuning is conveniently accomplished by varying the frequency of the local oscillator so that the IF frequency remains constant. The superheterodyne receiver represents the culmination of over 50 years of receiver development, and is used in the majority of broadcast radios and televisions, radar systems, cellular telephone systems, and data communication systems.

At microwave and millimeter wave frequencies it is often necessary to use two stages of down-conversion to avoid problems due to LO stability. Such a *dual-conversion* superheterodyne receiver employs two local oscillators, two mixers, and two IF frequencies to achieve down-conversion to baseband.

Noise Characterization of a Receiver

We can now analyze the noise characteristics of a complete antenna–transmission line–receiver front end, as shown in Figure 14.14. In this system the total noise power at the output of the receiver, N_o, will be due to contributions from the antenna pattern, the loss in the antenna, the loss in the transmission line, and the receiver components. This noise power will determine the minimum detectable signal level for the receiver and, for a given transmitter power, the maximum range of the communication link.

The receiver components in Figure 14.14 consist of an RF amplifier with gain G_{RF} and noise temperature T_{RF}, a mixer with an RF-to-IF conversion loss factor L_M and noise temperature T_M, and an IF amplifier with gain G_{IF} and noise temperature T_{IF}. The noise effects of later stages can usually be ignored since the

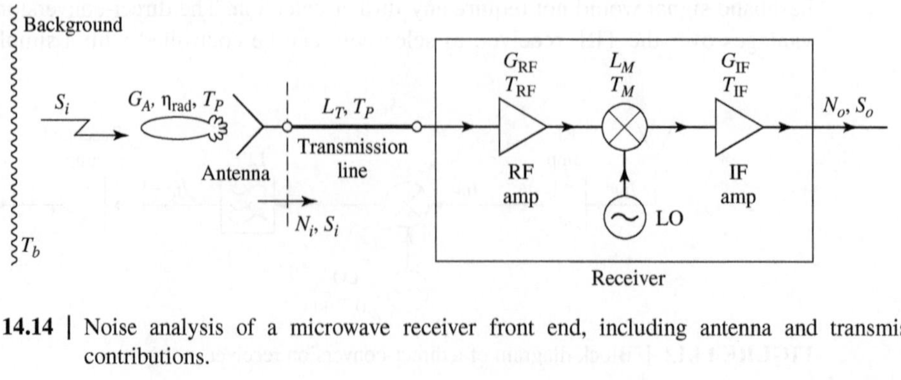

FIGURE 14.14 | Noise analysis of a microwave receiver front end, including antenna and transmission line contributions.

overall noise figure is dominated by the characteristics of the first few stages. The component noise temperatures can be related to noise figures as $T = (F - 1)T_0$. From (10.22) the equivalent noise temperature of the receiver can be found as

$$T_{\text{REC}} = T_{\text{RF}} + \frac{T_M}{G_{\text{RF}}} + \frac{T_{\text{IF}}L_M}{G_{\text{RF}}}. \tag{14.30}$$

The transmission line connecting the antenna to the receiver has a loss L_T, and is at a physical temperature T_p. So from (10.15) its equivalent noise temperature is

$$T_{\text{TL}} = (L_T - 1)T_p. \tag{14.31}$$

Again using (10.22), we find that the noise temperature of the transmission line (TL) and receiver (REC) cascade is

$$T_{\text{TL+REC}} = T_{\text{TL}} + L_T T_{\text{REC}}$$
$$= (L_T - 1)T_p + L_T T_{\text{REC}}. \tag{14.32}$$

This noise temperature is defined at the antenna terminals (the input to the transmission line).

As discussed in Section 14.1, the entire antenna pattern can collect noise power. If the antenna has a reasonably high gain with relatively low sidelobes, we can assume that all noise power comes via the main beam, so that the noise temperature of the antenna is given by (14.18):

$$T_A = \eta_{\text{rad}}T_b + (1 - \eta_{\text{rad}})T_p, \tag{14.33}$$

where η_{rad} is the efficiency of the antenna, T_p is its physical temperature, and T_b is the equivalent brightness temperature of the background seen by the main beam. (One must be careful with this approximation, as it is quite possible for the noise power collected by the sidelobes to exceed the noise power collected by the main beam, if the sidelobes are aimed at a hot background. See Example 14.3.) The noise power at the antenna terminals, which is also the noise power delivered to the transmission line, is

$$N_i = kBT_A = kB[\eta_{\text{rad}}T_b + (1 - \eta_{\text{rad}})T_p], \tag{14.34}$$

where B is the system bandwidth. If S_i is the received power at the antenna terminals, then the input SNR at the antenna terminals is S_i/N_i. The output signal power is

$$S_o = \frac{S_i G_{\text{RF}} G_{\text{IF}}}{L_T L_M} = S_i G_{\text{SYS}}, \tag{14.35}$$

where G_{SYS} has been defined as a system power gain. The output noise power is

$$N_o = \left(N_i + kBT_{\text{TL+REC}}\right) G_{\text{SYS}}$$
$$= kB(T_A + T_{\text{TL+REC}})G_{\text{SYS}}$$
$$= kB[\eta_{\text{rad}}T_b + (1 - \eta_{\text{rad}})T_p + (L_T - 1)T_p + L_T T_{\text{REC}}]G_{\text{SYS}}$$
$$= kBT_{\text{SYS}}G_{\text{SYS}}, \tag{14.36}$$

where T_{SYS} has been defined as the overall system noise temperature. The output SNR is

$$\frac{S_o}{N_o} = \frac{S_i}{kBT_{\text{SYS}}} = \frac{S_i}{kB[\eta_{\text{rad}}T_b + (1 - \eta_{\text{rad}})T_p + (L_T - 1)T_p + L_T T_{\text{REC}}]}. \tag{14.37}$$

It may be possible to improve this SNR by various signal processing techniques. Note that it may appear to be convenient to use an overall system noise figure to calculate the degradation in SNR from input to output for the above system, but one must be very careful with such an approach because noise figure is defined only for $N_i = kT_0 B$, which is not the case here. It is often less confusing to work directly with noise temperatures and powers, as we did above.

EXAMPLE 14.5   SIGNAL-TO-NOISE RATIO OF A MICROWAVE RECEIVER

A microwave receiver like that of Figure 14.14 has the following parameters:

$$
\begin{aligned}
f &= 4.0 \text{ GHz}, & G_{RF} &= 20 \text{ dB}, \\
B &= 1 \text{ MHz}, & F_{RF} &= 3.0 \text{ dB}, \\
G_A &= 26 \text{ dB}, & L_M &= 6.0 \text{ dB}, \\
\eta_{rad} &= 0.90, & F_M &= 7.0 \text{ dB}, \\
T_p &= 300 \text{ K}, & G_{IF} &= 30 \text{ dB}, \\
T_b &= 200 \text{ K}, & F_{IF} &= 1.1 \text{ dB}. \\
L_T &= 1.5 \text{ dB},
\end{aligned}
$$

If the received power at the antenna terminals is $S_i = -80$ dBm, calculate the input and output SNRs.

Solution

We first convert the above dB quantities to numerical values, and noise figures to noise temperatures:

$$
\begin{aligned}
G_{RF} &= 10^{20/10} = 100, \\
G_{IF} &= 10^{30/10} = 1000, \\
L_T &= 10^{1.5/10} = 1.41, \\
L_M &= 10^{6/10} = 4.0, \\
T_M &= (F_M - 1)T_0 = (10^{7/10} - 1)(290) = 1163 \text{ K}, \\
T_{RF} &= (F_{RF} - 1)T_0 = (10^{3/10} - 1)(290) = 289 \text{ K}, \\
T_{IF} &= (F_{IF} - 1)T_0 = (10^{1.1/10} - 1)(290) = 84 \text{ K}.
\end{aligned}
$$

Then from (14.30), (14.31), and (14.33) the noise temperatures of the receiver, transmission line, and antenna are

$$
\begin{aligned}
T_{REC} &= T_{RF} + \frac{T_M}{G_{RF}} + \frac{T_{IF}L_M}{G_{RF}} = 289 + \frac{1163}{100} + \frac{84(4.0)}{100} = 304 \text{ K}, \\
T_{TL} &= (L_T - 1)T_p = (1.41 - 1)300 = 123 \text{ K}, \\
T_A &= \eta_{rad}T_b + (1 - \eta_{rad})T_p = 0.9(200) + (1 - 0.9)(300) = 210 \text{ K}.
\end{aligned}
$$

The input noise power, from (14.34), is

$$
N_i = kBT_A = 1.38 \times 10^{-23}(10^6)(210) = 2.9 \times 10^{-15} \text{W} = -115 \text{ dBm}.
$$

Then the input SNR is

$$
\frac{S_i}{N_i} = -80 + 115 = 35 \text{ dB}.
$$

From (14.36) the total system noise temperature is

$$
T_{SYS} = T_A + T_{TL} + L_T T_{REC} = 210 + 123 + (1.41)(304) = 762 \text{ K}.
$$

This result clearly shows the noise contributions of the various components. The output SNR is found from (14.37) as

$$
\frac{S_o}{N_o} = \frac{S_i}{kBT_{SYS}},
$$

$$
kBT_{SYS} = 1.38 \times 10^{-23}(10^6)(762) = 1.05 \times 10^{-14} \text{ W} = -110 \text{ dBm},
$$

so

$$
\frac{S_o}{N_o} = -80 + 110 = 30 \text{ dB}.
$$

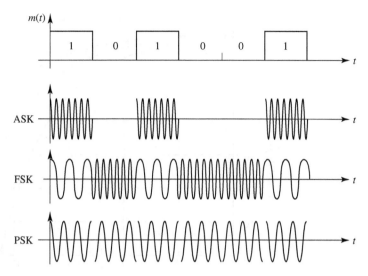

FIGURE 14.15 | Binary data and the resulting modulated carrier waveforms for amplitude shift keying, frequency shift keying, and phase shift keying.

Digital Modulation and Bit Error Rate

Information may be impressed upon a sinusoidal carrier using amplitude, frequency, or phase modulation. If the modulating signal is analog, as in the case of AM or FM radio, the amplitude, frequency, or phase of the carrier will undergo a continuous variation. If the modulating signal represents digital data in binary form, the variation in the amplitude, frequency, or phase of the carrier will be limited to two values. These types of modulations are usually referred to as *amplitude shift keying, frequency shift keying*, and *phase shift keying*, and abbreviated as ASK, FSK, and PSK, respectively. For example, ASK may involve a carrier that is turned on for a binary "1," and off for a binary "0." Frequency shift keying involves switching between two different carrier frequencies, while phase shift keying involves a 180° phase shift of the carrier, depending on the binary data. Binary phase shift keying is also referred to as BPSK. Figure 14.15 shows the carrier waveforms that result from binary digital modulation with ASK, FSK, and PSK methods.

The majority of modern wireless systems rely on digital modulation methods due to their superior performance in the presence of noise and signal fading, lower power requirements, and better suitability for the transmission of data with error-correcting codes or encryption. Besides the basic binary modulation schemes described above, there are a number of other digital modulation methods. One popular method is *quadrature phase shift keying* (QPSK), where two data bits are used to select one of four possible phase states (0°, 90°, 180°, or 270°). More generally, one can use *m-ary phase shift keying*, where one of 2^m phase states is selected on the basis of m data bits. It is also possible to modulate both amplitude and phase simultaneously, resulting in *quadrature amplitude modulation*, or QAM. Such higher order modulation methods allow higher data rates for a given channel bandwidth, but involve more system and processing complexity.

In an ideal situation a receiver will detect the same binary digit that was transmitted, but the presence of noise in the communication channel introduces the possibility that errors will be made during the detection process. The likelihood of an error in the detection of a single bit is quantified by the *bit error probability*, P_b, also known as the *bit error rate* (BER). The probability of error is dependent on the ratio of bit energy to noise power density, E_b/n_0, where E_b is the energy received during each bit interval, and n_0 is the power spectral density of the noise on the channel. The probability of error decreases as bit energy increases, or as noise density decreases. If S is the received signal (carrier) power (watts), with T_b being the bit period (seconds), and R_b the bit rate (bits per second), the bit energy can be written as

$$E_b = ST_b = S/R_b, \quad \text{(W-sec)} \tag{14.38}$$

Then the ratio E_b/n_0 is

$$\frac{E_b}{n_0} = \frac{ST_b}{n_0} = \frac{S}{n_0 R_b}. \tag{14.39}$$

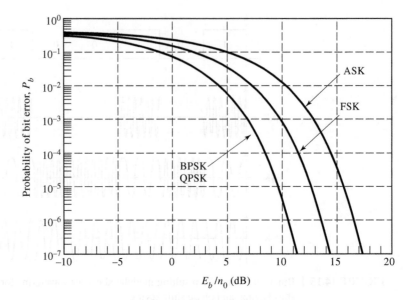

FIGURE 14.16 | Comparison of bit error rates for ASK, FSK, BPSK, and QPSK modulation methods versus E_b/n_0. (Coherent demodulation is assumed, with Gray coding for QPSK.)

Since the noise power is $N = n_0 B$, where B is the bandwidth of the receiver, the ratio of bit energy to noise power density can be expressed in terms of the SNR as

$$\frac{E_b}{n_0} = \frac{S}{N} B T_b = \frac{S}{N} \frac{B}{R_b}. \tag{14.40}$$

Note that this result indicates that, for a given SNR, the ratio of bit energy to noise power density will decrease (and the BER will increase) as the data rate increases. Depending on the type of modulation, the required receiver bandwidth may range from one to several times the bit rate.

Figure 14.16 shows bit error probability for four types of digital modulation (ASK, FSK, BPSK, and QPSK) versus the E_b/n_0 ratio. The bit error rate for QPSK is the same as for BPSK, but note that QPSK involves the transmission of two bits for every one bit sent by BPSK.

Each of the binary modulation methods transmits one bit during each bit period, and they are therefore said to have a *bandwidth efficiency* of 1 bps/Hz. Higher level modulation methods can achieve higher bandwidth efficiencies. For example, QPSK transmits two bits per period, and therefore has a bandwidth efficiency of 2 bps/Hz. Table 14.1 lists the bandwidth efficiency and the required E_b/n_0 ratio for a bit error rate of 10^{-5} for various digital modulation methods.

TABLE 14.1 | **Summary of Performance of Various Digital Modulation Methods**

Modulation Type	E_b/n_0(dB) for $P_b = 10^{-5}$	Bandwidth Efficiency
Binary ASK	15.6	1
Binary FSK	12.6	1
Binary PSK	9.6	1
QPSK	9.6	2
8-PSK	13.0	3
16-PSK	18.7	4
16-QAM	13.4	4
64-QAM	17.8	6

EXAMPLE 14.6   LINK ANALYSIS FOR LEO SATELLITE DOWNLINK

A LEO satellite at an orbital distance of 2000 km uses QPSK to communicate with a handset on Earth at a frequency of 16 GHz. The satellite has a transmit power of 100 W and an antenna gain of 25 dB, while the handset has an antenna gain of 1 dB and a system temperature of 750 K. If atmospheric attenuation is 2 dB, and the required link margin is 10 dB, what is the maximum data rate for a bit error probability of 0.01?

Solution
The wavelength is 1.875 cm, so from (14.26) the path loss is

$$L_0 = 20 \log\left(\frac{4\pi R}{\lambda}\right) = 20 \log\left(\frac{(4\pi)(2000 \times 10^3)}{0.01875}\right) = 182.5 \text{ dB}$$

The received power is

$$P_r = P_t + G_t - L_0 - L_A + G_r = 50 + 25 - 182.5 - 2 + 1 = -108.5 \text{ dBm.}$$

For a link margin of 10 dB, this received power level should be 10 dB above the threshold level. Thus, the threshold received signal level is

$$S_{\min} = P_r - \text{LM} = -108.5 - 10 = -118.5 \text{ dBm} = 1.41 \times 10^{-15} \text{ W.}$$

From Figure 14.16, the required E_b/n_0 for a bit error rate of 0.01 for QPSK is about 5 dB = 3.16. Solving (14.39) for the maximum bit rate gives

$$R_b = \left(\frac{E_b}{n_0}\right)^{-1}\frac{S_{\min}}{n_0} = \left(\frac{E_b}{n_0}\right)^{-1}\frac{S_{\min}}{kT_{\text{SYS}}} = \left(\frac{1}{3.16}\right)\frac{1.41 \times 10^{-15}}{(1.38 \times 10^{-23})(750)} = 43.2 \text{ kbps}$$

Wireless Communication Systems

We conclude this section with a summary of some of the most prevalent wireless communication systems in current use. Table 14.2 lists some of the commonly used frequency bands for wireless systems.

Cellular telephone and data systems: Cellular voice and networking systems are in constant evolution, involving the use of old and new technology, existing and newly available carrier frequencies, sophisticated multiple-access techniques, international agreements, and the special interests of commercial service providers, governments, and regulatory agencies. The objective is to provide mobile users with voice and data service (including Internet access and video), with high data rates and compatibility across systems. Much progress has been made, but there are still technical and organizational challenges that remain.

Cellular telephone systems were first proposed in the 1970s in response to the problem of providing mobile radio service to a large number of users in urban areas. Early mobile radio systems could handle only a very limited number of users due to inefficient use of the radio spectrum and interference between users. The cellular radio concept solved this problem by dividing a geographic area into nonoverlapping cells in which each cell has its own transmitter and receiver (the base station) to communicate with mobile users operating in that cell. Each cell site may allow as many as several hundred users to simultaneously communicate over voice and/or data channels. Frequency bands assigned to a particular cell can be reused in other, nonadjacent cells.

The first cellular telephone systems were built in Japan and Europe in 1979 and 1981, and in the United States (AMPS) in 1983. These systems used analog FM modulation and divided their allocated frequency bands into several hundred channels, each of which could support an individual telephone conversation. These early services grew slowly at first, because of the initial costs of developing an infrastructure of base stations and the initial expense of handsets, but by the 1990s growth became phenomenal.

Because of the rapidly growing business and consumer demand for wireless services, as well as advances in wireless technology, several second-generation standards were implemented in the United States, Europe, and Asia. These standards all employed digital modulation methods to provide better quality service and more efficient use of the radio spectrum, as well as multiple-access methods that could be categorized as either *time division multiple access* (TDMA), or *code division multiple access* (CDMA). Since then,

TABLE 14.2 | **Wireless System Frequencies**

Wireless System (Country)	Frequency
Advanced Mobile Phone System (AMPS, United States; obsolete)	U: 824–849 MHz
	D: 869–894 MHz
GSM 850 (Americas)	U: 824–849 MHz
	D: 869–894 MHz
GSM 900 (worldwide)	U: 890–915 MHz
	D: 935–960 MHz
GSM 1800 (worldwide)	U: 1710–1785 MHz
	D: 1805–1880 MHz
GSM 1900 (Americas)	U: 1850–1910 MHz
	D: 1930–1990 MHz
Universal Mobile Telecommunications System (UMTS),	U: 1920–1980 MHz
band 1 (most countries)	D: 2110–2170 MHz
UMTS, band 2 (most countries)	U: 1850–1910 MHz
	D: 1930–1990 MHz
UMTS, band 8 (most countries)	U: 880–916 MHz
	D: 925–960 MHz
Wireless local area networks (WiFi)	902–928 MHz
	2.400–2.484 GHz
	5.725–5.850 GHz
Global Positioning System (GPS)	L1: 1575.42 MHz
	L2: 1227.60 MHz
Direct Broadcast Satellite (DBS) (Europe, Russia)	10.7–12.75 GHz
(Americas)	12.2–12.7 GHz
(Asia, Australia)	11.7–12.2 GHz
Industrial, medical, and scientific bands (most countries)	902–928 MHz
	2.400–2.484 GHz
	5.725–5.850 GHz

U, uplink (mobile-to-base); D, downlink (base-to-mobile).

most countries have made more radio spectrum available, usually as a result of freeing up frequency bands that had been used by VHF broadcast television.

Today, most wireless cellular and smartphone systems have migrated to fifth generation (5G) standards, or are in the process of being upgraded to 5G standards. The International Mobile Telecommunications (IMT)-2020 project of the International Telecommunications Union (ITU) forms the basis for 5G standards, most of which rely on 3GPP and its variations, 5G NR (new radio access technology), LTE-M (LTE-MTC [Machine Type Communication], which includes enhanced Machine Type Communication [eMTC]), Narrowband Internet of Things (NB-IoT), New Radio-Industrial Internet of Things (NR-IIoT), and Digital Enhanced Cordless Telecommunications (DECT-5G). At the present time, IMT-2020 supports data rates of 10 Gbps and 20 Gbps for uplink and downlink, respectively. A related effort is the 3rd Generation Partnership Project (3GPP) and its family variations, which is based on a collaboration of various telecommunication groups to form a migration path from existing 4G infrastructure to 5G, and then toward a 10 Gbps data rate and 1 msec latency. Many countries have adopted the Universal Mobile Telecommunications System standard, which is based on 3GPP family and its variations NR, NR-IIoT, LTE-M, NB-IoT, and DECT-5G. At present, there are many proposed standards for interim use for capitalizing on existing infrastructure, as well as for new standards that will evolve into 5G systems. For digital cellular network 5G began wide deployment in 2019. As per earlier standards, the enclosed areas are divided into cells with a separate antenna. The frequency band of 5G is divided into low-band, mid-band, and millimeter waves.

Satellite systems for wireless voice and data: The conceptual advantage of satellite systems is that a relatively small number of satellites can provide coverage to users at any location in the world, including the oceans, deserts, and mountains—areas for which it is difficult or impossible to provide cellular service. In principle, as few as three geosynchronous satellites can provide complete global coverage, but the very high altitude of the geosynchronous orbit makes it difficult to communicate with hand-held terminals because the large path loss results in very low signal strength. Satellites in lower orbits can provide usable levels of signal power, but many more satellites are then needed to provide global coverage.

The Iridium project, originally financed by a consortium of companies headed by Motorola, was the first commercial satellite system to offer worldwide hand-held wireless telephone service. It consists of 66 LEO satellites in near-polar orbits, and connects mobile phone and paging subscribers to the public telephone system through a series of intersatellite relay links and land-based gateway terminals. Figure 14.17 shows a photo of one of the Iridium phased array antennas. The Iridium system cost was approximately $5 billion; it began service in November 1998, and filed for bankruptcy in August 1999. Iridium was acquired by the U.S. Defense Department in 2001 and is still operating at this time.

One drawback of using satellites for telephone service is that weak signal levels require a line-of-sight path from the mobile user to the satellite, meaning that satellite telephones generally cannot be used in buildings, automobiles, or even in many wooded or urban areas. This places satellite phone service at a definite performance disadvantage relative to land-based cellular services. Other commercial LEO satellite communication systems, such as Globalstar, have also ended in financial failure.

Most successful satellite communication systems rely on geostationary satellites. These include the INMARSAT systems, originally used to provide communication to maritime shipping, but also used in remote areas. Many financial services and businesses use *very small aperture terminals* (VSATs), which provide relatively low-rate data communication to geostationary satellites with 12- to 18-inch antennas. An example of a geostationary satellite telephone service is the Thuraya system, which provides coverage to parts of Africa, Europe, India, and the Middle East. The subscriber link operates at L band, with a fairly compact handset. There is a noticeable conversational delay with the Thuraya system due to the propagation time to and from the satellite.

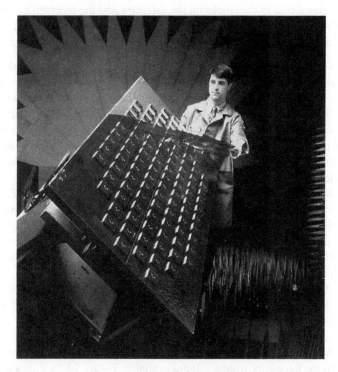

FIGURE 14.17 | Photograph of one of the three L-band antenna arrays for an Iridium communication satellite. The Iridium system consists of 66 satellites in low Earth orbit to provide global personal satellite TDMA communication services, including voice, fax, and paging.

Courtesy of Raytheon Company, Waltham, MA.

Global Positioning: The Global Positioning System (GPS) uses 24 satellites in medium Earth orbits to provide accurate position information (latitude, longitude, and elevation) to users on land, air, or sea. Originally developed as the NAVSTAR system by the U.S. Department of Defense, GPS has become one of the most pervasive applications of wireless technology for consumers and businesses throughout the world. GPS receivers are used on airplanes, ships, trucks, trains, and automobiles. Advances in technology have led to substantial reductions in size and cost, so that small GPS receivers can be integrated into cellular telephones and smart phones, and hand-held GPS devices are used by hikers and sportsmen. With differential GPS, accuracies on the order of 1 cm can be achieved, a capability that has revolutionized the surveying industry. An entirely new field of study, known as *geographic information systems* (GIS), is based on the relation of data to location, usually obtained in conjunction with GPS.

GPS positioning operates by using triangulation with a minimum of four satellites. GPS satellites are in orbits 20,200 km above Earth, with orbital periods of 12 hours. Distances from the user's GPS receiver to these satellites are found by timing the propagation delay between the satellites and the receiver. The orbital positions of the satellites (*ephemeris*) are known to very high accuracy, and each satellite contains an extremely accurate clock to provide a unique set of timing pulses. A GPS receiver decodes this timing information and performs the necessary calculations to find the position and velocity of the receiver. The GPS receiver usually must have a line-of-sight view to at least four satellites in the GPS constellation, although three satellites are adequate if altitude position is known (as in the case of ships at sea). Because of the low-gain antennas required for operation, the received signal level from a GPS satellite is very low—typically on the order of −130 dBm (for a receiver antenna gain of 0 dB). This signal level is usually below the noise power at the receiver, but spread-spectrum techniques are used to improve the received SNR.

GPS operates at two frequency bands: L1, at 1575.42 MHz, and L2, at 1227.60 MHz, transmitting spread-spectrum signals with BPSK modulation. The L1 frequency is used to transmit ephemeris data for each satellite, as well as timing codes, which are available to any commercial or public user. This mode of operation is referred to as the *Course/Acquisition* (C/A) code. In contrast, the L2 frequency is reserved for military use, and uses an encrypted timing code referred to as the *Protected* (P) code (there is also a P code signal transmitted at the L1 frequency). The P code offers much higher accuracy than the C/A code. The typical accuracy that can be achieved with an L1 GPS receiver is about 100 feet. Accuracy is limited by timing errors in the clocks on the satellites and the receiver, as well as error in the assumed position of the GPS satellites. The most significant error is generally caused by atmospheric and ionospheric effects, which introduce small but variable delays in signal propagation from the satellite to the receiver.

Wireless local area networks: Wireless local area networks provide connections between computers and peripherals over short distances. Wireless networks can be found in airports, coffee shops, office buildings, college campuses, and even on commercial airliners, busses, and cruise ships. Indoor coverage is usually less than a few hundred feet. Outdoors, in the absence of obstructions and with the use of high-gain antennas, much longer ranges can be obtained. Wireless networks are especially useful when it is impossible or prohibitively expensive to place network wiring in or between buildings, or when only temporary network access is needed. Mobile users, of course, can only be connected to a computer network through a wireless link.

Most commercial WLAN products are based on the IEEE 802.11 standards (Wi-Fi). These operate at either 2.4 or 5.7 GHz (in the industrial, scientific, and medical frequency bands), and use either frequency-hopping or direct-sequence spread-spectrum techniques. Standards 802.11a, 802.11b, and 802.11g can provide data rates up to 54 Mbps, while 802.11n (which uses multiple antennas) can achieve data rates of up to 150 Mbps. Actual data rates are often significantly lower due to nonideal propagation conditions and loading from other users.

Another wireless networking standard is *Bluetooth*, which is intended for short-range networking of portable devices, such as cameras, printers, headsets, games, and similar applications, to resident computers or routers. Bluetooth devices operate at 2.4 GHz, with RF power in the range of 1–100 mW and corresponding operating ranges of 1–100 m. Data rates range from 1 to 24 Mbps.

Millimeter wave frequencies are increasingly being considered for high-speed local area networking due to the large bandwidths that are available. Figure 14.18 shows a developmental model of a high-speed 60 GHz wireless networking transmitter.

Direct broadcast satellites: DBS systems provide television service with continental coverage from geosynchronous satellites directly to home users with a relatively small 18 inch diameter antenna. Prior to DBS

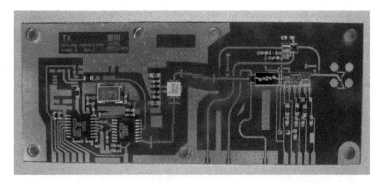

FIGURE 14.18 | Photograph of a high-speed 60 GHz wireless local area network transmitter. This WLAN operates at 59–62 GHz, with a data rate of 2.8 Gbps. It uses GaAs chips and a built-in four-element, circularly polarized microstrip antenna.

Courtesy of Newlans, Inc., Acton, MA.

technology, satellite TV service relied on analog signals that required an unsightly dish antenna as large as 6 feet in diameter to achieve the necessary SNR. The smaller DBS antenna became possible through the use of digital modulation techniques, which reduce the required received signal levels as compared to an analog system. DBS systems operate with carrier frequencies in the 10–12 GHz range (see Table 14.2), and typically use QPSK with digital multiplexing and error correction to deliver digital data at a rate of 40 Mbps. Several DBS satellites are used throughout the world to provide subscriber television service, sometimes with more than one satellite per coverage area. For North America, two satellites, DBS-1 and DBS-2, are in geostationary orbit at 101.2° and 100.8° longitude, and each provides 16 channels with 120 W of radiated power per channel. These satellites use opposite circular polarizations to minimize loss due to precipitation, and to avoid interference with each other (polarization duplexing).

Point-to-point radio systems: Point-to-point radios are used to provide dedicated data connections between two fixed points. Electric utility companies use point-to-point radios for transmission of telemetry information for the generation, transmission, and distribution of electric power between generating stations and substations. Point-to-point radios are also used to connect cellular base stations to the public switched telephone network, and are attractive because they are generally much cheaper than running high-bandwidth fiber-optic lines below ground level. Point-to-point radios usually operate in the 18, 24, or 38 GHz bands, and use a variety of digital modulation methods to provide data rates in excess of 50 Mbps. High-gain antennas are typically used to minimize power requirements and to avoid interference with other users.

Other wireless systems: Many other applications of wireless technology are being developed, and we can only briefly mention some of these. One of the most pervasive may turn out to be *Radio Frequency Identification* (RFID) systems, which rely on small, low-cost tags that can receive an interrogatory RF signal and reply with a signal containing preprogrammed data. RFID tags can be used for retail products, inventory control, industrial materials, security applications, or any application that requires identification or tracking. An interesting feature of RFID tags is that they can be passively powered, whereby they store the power required for signaling by rectifying the interrogatory signal and charging a small capacitor. This is then used to drive very low-power CMOS circuitry to transmit data back to the interrogating receiver.

Another area where wireless technology is beginning to experience growth is in motor vehicle and highway applications. These include toll collection, intelligent cruise control, collision avoidance radar, blind spot radar, traffic information, emergency messaging, and vehicle identification. Automatic toll collection is already in service in many parts of the United States and Europe. A number of automobile models are available with blind spot and collision sensors as optional equipment.

14.3   RADAR SYSTEMS

Radar, or *radio detection and ranging*, is the oldest application of microwave technology, dating back to World War II. In its basic operation, a transmitter sends out a signal, which is partly reflected by a distant target, and then detected by a sensitive receiver. If a narrow-beam antenna is used, the target's direction can be accurately given by the angular position of the antenna. The distance to the target is determined by the time required for a pulsed signal to travel to the target and back, and the radial velocity of the target is related to the Doppler shift of the return signal. Below are listed some of the typical applications of radar systems.

Civilian applications

- Airport surveillance
- Marine navigation
- Weather radar
- Altimetry
- Aircraft landing
- Security alarms
- Speed measurement (police radar)
- Geographic mapping

Military applications

- Air and marine navigation
- Detection and tracking of aircraft, missiles, and spacecraft
- Missile guidance
- Fire control for missiles and artillery
- Weapon fuses
- Reconnaissance

Scientific applications

- Astronomy
- Mapping and imaging
- Precision distance measurement
- Remote sensing of the environment

Early radar work in the United States and the United Kingdom began in the 1930s using very high frequency (VHF) sources. A major breakthrough occurred in the early 1940s with the British invention of the magnetron tube as a reliable source of high-power microwaves. Higher frequencies allowed the use of reasonably sized antennas with high gain, allowing mechanical tracking of targets with good angular resolution. Radar was quickly developed in the United Kingdom and the United States, and played an important role in World War II. Figure 14.19 shows a photograph of the phased array radar for the Patriot missile system.

We will now derive the radar equation, which governs the basic operation of most radars, and then describe some of the more common types of radar systems.

The Radar Equation

Two basic radar systems are illustrated in Figure 14.20; in a *monostatic radar* the same antenna is used for both transmit and receive, while a *bistatic radar* uses two separate antennas for these functions. Most radars are of the monostatic type, but in some applications (such as missile fire control) the target may be illuminated by a separate transmit antenna. Separate antennas are also sometimes used to achieve the necessary signal isolation between transmitter and receiver.

Here we will consider the monostatic case, but the bistatic case is very similar. If the transmitter radiates a power P_t through an antenna of gain G, the power density incident on the target is, from (14.23),

$$S_t = \frac{P_t G}{4\pi R^2}, \tag{14.41}$$

where R is the distance to the target. It is assumed that the target is in the main beam direction of the antenna. The target will scatter the incident power in various directions; the ratio of the scattered power in a given

FIGURE 14.19 | Photograph of the Patriot phased array radar. This is a C-band multifunction radar that provides tactical air defense, including target search and tracking, and missile fire control. The phased array antenna uses 5000 ferrite phase shifters to electronically scan the antenna beam.

Courtesy of Raytheon Company, Waltham, MA.

direction to the incident power density is defined as the *radar cross section*, σ, of the target. Mathematically,

$$\sigma = \frac{P_s}{S_t} \text{ m}^2, \tag{14.42}$$

where P_s is the total power scattered by the target, and S_t is the power density incident on the target. The radar cross section thus has the dimensions of area, and is a property of the target itself. It depends on the incident and reflection angles, as well as on the polarizations of the incident and reflected waves.

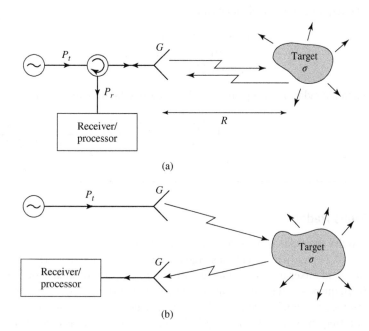

FIGURE 14.20 | Basic monostatic and bistatic radar systems. (a) Monostatic radar system. (b) Bistatic radar system.

Since the target scatters as a source of finite size, the power density of the reradiated field must decay as $1/4\pi R^2$ away from the target. Thus the power density of the scattered field back at the receive antenna must be

$$S_r = \frac{P_t G \sigma}{(4\pi R^2)^2}. \tag{14.43}$$

Using (14.15) for the effective area of the antenna gives the received power as

$$P_r = \frac{P_t G^2 \lambda^2 \sigma}{(4\pi)^3 R^4}. \tag{14.44}$$

This is the *radar equation*. Note that the received power varies as $1/R^4$, which implies that a high-power transmitter and a sensitive low-noise receiver are needed to detect targets at long ranges.

Because of noise received by the antenna and generated in the receiver, there will be some minimum detectable power that can be discriminated by the receiver. If this power is $P_{\min}$, then (14.44) can be rewritten to give the maximum range as

$$R_{\max} = \left[\frac{P_t G^2 \sigma \lambda^2}{(4\pi)^3 P_{\min}} \right]^{1/4}. \tag{14.45}$$

Signal processing can effectively reduce the minimum detectable signal, and so increase the usable range. One very common processing technique used with pulse radars is pulse integration, in which a sequence of N received pulses is integrated over time. The effect is to reduce the noise level, which has a zero mean, relative to the returned pulse level, resulting in an improvement factor of approximately N [6].

Of course, the above results seldom describe the performance of an actual radar system. Factors such as propagation effects, the statistical nature of the detection process, and external interference often reduce the usable range of a radar system.

EXAMPLE 14.7   APPLICATION OF THE RADAR RANGE EQUATION

A pulse radar operating at 12 GHz has an antenna with a gain of 27 dB and a transmitter power of 2 kW (pulse power). If it is desired to detect a target with a cross section of 15 m^2, and the minimum detectable signal is -90 dBm, what is the maximum range of the radar?

Solution
The required numerical values are

$$G = 10^{27/10} = 501.2,$$
$$P_{\min} = 10^{-90/10} \text{ mW} = 10^{-12} \text{ W},$$
$$\lambda = 0.025 \text{ m}.$$

Then the radar range equation of (14.45) gives the maximum range as

$$R_{\max} = \left[\frac{P_t G^2 \sigma \lambda^2}{(4\pi)^3 (P_{\min})} \right]^{1/4} = \left[\frac{(2 \times 10^3)(501.2)^2(15)(0.025)^2}{(4\pi)^3 (10^{-12})} \right]^{1/4}$$
$$= 6980 \text{ m}.$$

Pulse Radar

A pulse radar determines target range by measuring the round-trip time of a pulsed microwave signal. Figure 14.21 shows a typical pulse radar system block diagram. The transmitter portion consists of a single-sideband mixer used to frequency offset a microwave oscillator of frequency f_0 by an amount equal to the IF frequency. After power amplification, pulses of this signal are transmitted by the antenna. The transmit/receive switch is controlled by the pulse generator to give a transmit pulse width τ, with a pulse repetition frequency (PRF) of $f_r = 1/T_r$. The transmit pulse thus consists of a short burst of a microwave signal at the frequency $f_0 + f_{\text{IF}}$. Typical pulse durations range from 100 msec to 50 nsec; shorter pulses give better

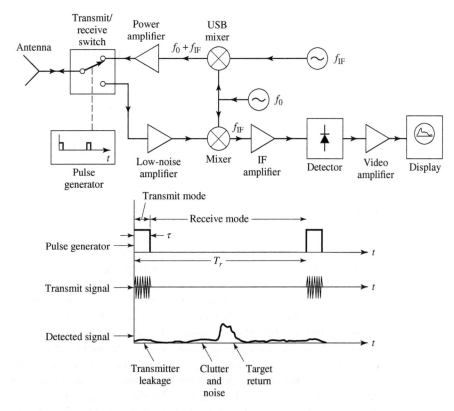

FIGURE 14.21 | A pulse radar system and timing diagram.

range resolution, but longer pulses result in a better SNR after receiver processing. Typical pulse repetition frequencies range from 100 Hz to 100 kHz; higher PRFs give more returned pulses per unit time, which improves performance, but lower PRFs avoid range ambiguities that can occur when $R > cT_r/2$.

In the receive mode, the returned signal is amplified and mixed with the local oscillator of frequency f_0 to produce the desired IF signal. The local oscillator is used for both up-conversion in the transmitter and down-conversion in the receiver; this simplifies the system and avoids the problem of frequency drift, which would be a consideration if separate oscillators were used. The IF signal is amplified, detected, and fed to a video amplifier/ display. Search radars often use a continuously rotating antenna for 360° azimuthal coverage; in this case the display shows a polar plot of target range versus angle. Modern radars use a computer for the processing of the detected signal and display of target information.

The transmit/receive (T/R) switch in the pulse radar actually performs two functions: forming the transmit pulse train, and switching the antenna between the transmitter and receiver. This latter function is also known as *duplexing*. In principle, the duplexing function could be achieved with a circulator, but an important requirement is that a high degree of isolation (about 80–100 dB) be provided between the transmitter and receiver to avoid transmitter leakage into the receiver, which would drown the target return (or possibly damage the receiver). As circulators typically achieve only 20–30 dB of isolation, some type of switch, with high isolation, is required. If necessary, further isolation can be obtained by using additional switches along the path of the transmitter circuit.

Doppler Radar

If the target has a velocity component along the line of sight of the radar, the returned signal will be shifted in frequency relative to the transmitted frequency due to the Doppler effect. If the transmitted frequency is f_0, and the radial target velocity is v, then the shift in frequency, or the *Doppler frequency*, will be

$$f_d = \frac{2vf_0}{c},$$

$$(14.46)$$

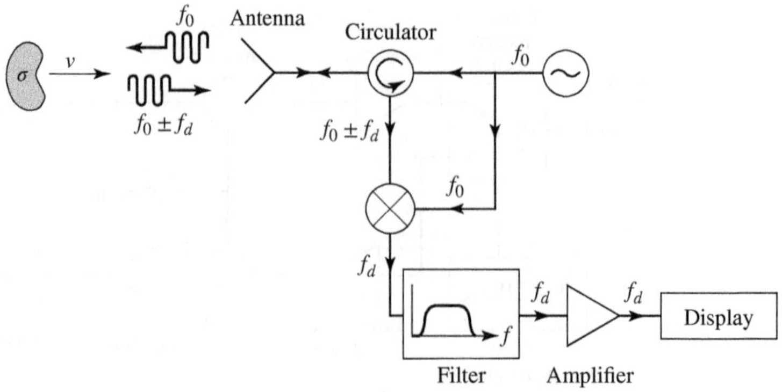

FIGURE 14.22 | Doppler radar system.

where c is the velocity of light. The received frequency is then $f_0 \pm f_d$, where the plus sign corresponds to an approaching target and the minus sign corresponds to a receding target.

Figure 14.22 shows a basic Doppler radar system. Observe that it is much simpler than a pulse radar since a continuous wave signal is used, and the transmit oscillator can also be used as a local oscillator for the receive mixer because the received signal is frequency offset by the Doppler frequency. The filter following the mixer should have a passband corresponding to the expected minimum and maximum target velocities. It is important that the filter have high attenuation at zero frequency, to eliminate the effect of clutter return and transmitter leakage at the frequency f_0, as these signals would down-convert to zero frequency. Then a high degree of isolation is not necessary between transmitter and receiver, and a circulator can be used. This type of filter response also helps to reduce the effect of $1/f$ noise.

The above radar cannot distinguish between approaching and receding targets, as the sign of f_d is lost in the detection process. Such information can be recovered, however, by using a mixer that produces separately the upper and lower sideband products.

Since the return of a pulse radar from a moving target will contain a Doppler shift, it is possible to determine both the range and velocity (and position, if a narrow-beam antenna is used) of a target with a single radar. Such a radar is known as a *pulse-Doppler radar*, and it offers several advantages over pulse or Doppler radars. One problem with a pulse radar is that it is impossible to distinguish between a true target and clutter returns from the ground, trees, buildings, etc. Such clutter returns may be picked up from the antenna sidelobes. However, if the target is moving (e.g., as in an airport surveillance radar application), the Doppler shift can be used to separate its return from clutter, which is stationary relative to the radar.

Radar Cross Section

A radar target is characterized by its radar cross section, as defined in (14.39), which gives the ratio of scattered power to incident power density. The cross section of a target depends on the frequency and polarizations of the incident and scattered waves, and on the incident and reflected angles relative to the target. Thus we can define a monostatic cross section (incident and reflected angles identical), and a bistatic cross section (incident and reflected angles different).

For simple shapes the radar cross section can be calculated as an electromagnetic boundary value problem; more complex targets require numerical techniques or measurement to find the cross section. The radar cross section of a conducting sphere can be calculated exactly; the monostatic result is shown in Figure 14.23, normalized to πa^2, the physical cross-sectional area of the sphere. Note that the cross section increases very quickly with size for electrically small spheres ($a \ll \lambda$). This region is called the *Rayleigh region*, and it can be shown that σ varies as $(a/\lambda)^4$ in this region. (This strong dependence on frequency explains why the sky is blue, as the blue component of sunlight scatters more strongly from atmospheric particles than do the lower frequency red components.)

For electrically large spheres, where $a \gg \lambda$, the radar cross section of the sphere is equal to its physical cross section, πa^2. This is the *optical region*, where geometrical optics is valid. Many other shapes, such as flat plates at normal incidence, also have cross sections that approach the physical area for electrically large sizes.

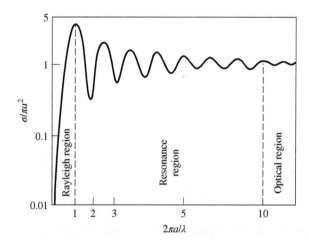

FIGURE 14.23 | Monostatic radar cross section of a conducting sphere.

Between the Rayleigh region and the optical region is the *resonance region*, where the electrical size of the sphere is on the order of a wavelength. Here the cross section is oscillating with frequency due to phase addition and cancellation of various scattered field components. Of particular note is the fact that the cross section may reach quite high values in this region.

Complex targets such as aircraft or ships generally have cross sections that vary rapidly with frequency and aspect angle. In military applications it is often desirable to minimize the radar cross section of vehicles to reduce detectability. This can be accomplished by using radar-absorbing materials (lossy dielectrics) in the construction of the vehicle. Table 14.3 lists the approximate radar cross sections of a variety of different targets.

14.4 RADIOMETER SYSTEMS

A radar system obtains information about a target by transmitting a signal and receiving the echo from the target, and thus can be described as an *active* remote sensing system. Radiometry, however, is a *passive* technique, which develops information about a target solely from the microwave portion of the blackbody radiation (noise) that it either emits directly or reflects from surrounding bodies. A *radiometer* is a sensitive receiver specially designed to measure this noise power.

Theory and Applications of Radiometry

As discussed in Section 10.1, a body in thermodynamic equilibrium at a temperature T radiates energy according to Planck's radiation law. In the microwave region this result reduces to $P = kTB$, where k is

TABLE 14.3 | **Typical Radar Cross Sections**

Target	$\sigma(m^2)$
Bird	0.01
Missile	0.5
Person	1
Small plane	1–2
Bicycle	2
Small boat	2
Fighter plane	3–8
Bomber	30–40
Large airliner	100
Truck	200

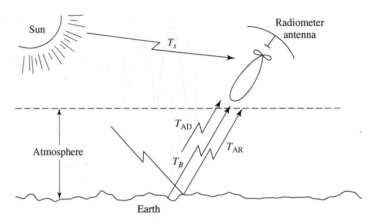

FIGURE 14.24 | Noise power sources in a typical radiometer application.

Boltzmann's constant, B is the system bandwidth, and P is the radiated power. This result strictly applies only to a *blackbody*, which is defined as an idealized material that absorbs all incident energy and reflects none; a blackbody also radiates energy at the same rate as it absorbs energy, thus maintaining thermal equilibrium. A nonideal body will partially reflect incident energy, and so it does not radiate as much power as would a blackbody at the same temperature. A measure of the power radiated by a body relative to that radiated by an ideal blackbody at the same temperature is the *emissivity*, e, defined as

$$e = \frac{P}{kTB}, \tag{14.47}$$

where P is the power radiated by the nonideal body, and kTB is the power that would be emitted by a perfect blackbody. Thus, $0 \leq e \leq 1$, and $e = 1$ for a perfect blackbody; emissivity may be thought of as the "efficiency" of blackbody radiation.

As we saw in Section 10.1, noise power can also be quantified in terms of equivalent temperature. Thus, for radiometric purposes, we can define a brightness temperature, T_B, as

$$T_B = eT, \tag{14.48}$$

where T is the physical temperature of the body. This shows that, radiometrically, a body never looks hotter than its actual temperature, since $0 \leq e \leq 1$.

Consider Figure 14.24, which shows the antenna of a microwave radiometer receiving noise powers from various sources. The antenna is pointed at a region of Earth that has an apparent brightness temperature T_B. The atmosphere emits radiation in all directions; the component radiated directly toward the antenna is T_{AD}, while the component reflected from Earth to the antenna is T_{AR}. There may also be noise powers that enter the sidelobes of the antennas from the Sun or other sources. Thus, we can see that the total brightness temperature seen by the radiometer is a function of the scene under observation, as well as the observation angle, frequency, polarization, attenuation of the atmosphere, and the antenna pattern. The objective of radiometry is to infer information about the scene from the measured brightness temperature and an analysis of the radiometric mechanisms that relate brightness temperature to physical conditions of the scene. For example, the power reflected from a uniform layer of snow over soil can be treated as plane wave reflection from a multilayer dielectric region, leading to the development of an algorithm that gives the thickness of the snow in terms of measured brightness temperature at various frequencies. Figure 14.25 shows a commercial multifrequency airborne radiometer for weather applications.

Microwave radiometry has developed over the last 20 years into a mature technology, one that is strongly interdisciplinary, drawing on results from fields such as electrical engineering, oceanography, geophysics, and atmospheric and space sciences, to name a few. Some of the more important applications of microwave radiometry are listed below.

Environmental applications

- Measurement of soil moisture
- Flood mapping

- Snow cover/ice cover mapping
- Ocean surface wind speed
- Atmospheric temperature profile
- Atmospheric humidity profile

Military applications

- Target detection
- Target recognition
- Surveillance
- Mapping

Astronomy applications

- Planetary mapping
- Solar emission mapping
- Mapping of galactic objects
- Measurement of cosmological background radiation

Total Power Radiometer

The aspect of radiometry that is of most interest to the microwave engineer is the design of the radiometer itself. The basic problem is to build a receiver that can distinguish between the desired external radiometric noise and the inherent noise of the receiver, even though the radiometric power is usually less than the receiver noise power. Although it is not a very practical instrument, we will first consider the total power radiometer because it represents a simple and direct approach to the problem and serves to illustrate the difficulties involved in radiometer design.

The block diagram of a typical total power radiometer is shown in Figure 14.26. The front end of the receiver is a standard superheterodyne circuit consisting of an RF amplifier, a mixer/local oscillator, and an IF stage. The IF filter determines the system bandwidth, B. The detector is generally a square-law device, so that its output voltage is proportional to the input power. The integrator is essentially a low-pass filter with a cutoff frequency of $1/\tau$, and serves to smooth out short-term variations in the noise power. For simplicity, we assume that the antenna is lossless, although in practice antenna loss will affect the apparent temperature of the antenna, as given in (14.18).

FIGURE 14.25 | Photograph of a stepped frequency microwave radiometer, operating at 4.7–7.2 GHz. This instrument is flown on aircraft to measure brightness temperature and infer ocean surface wind speed and rain rate estimation in hurricanes.

Courtesy of ProSensing, Inc., Amherst, MA.

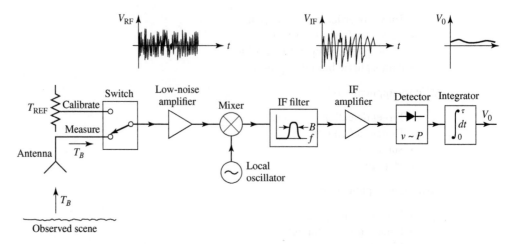

FIGURE 14.26 | Total power radiometer block diagram.

If the antenna is pointed at a background scene with a brightness temperature T_B, the antenna power will be $P_A = kT_BB$; this is the desired signal. The receiver contributes noise that can be characterized as a power $P_R = kT_RB$ at the receiver input, where T_R is the overall noise temperature of the receiver. Thus the output voltage of the radiometer is

$$V_o = G(T_B + T_R)kB, \tag{14.49}$$

where G is the overall gain constant of the radiometer. Conceptually, the system is calibrated by replacing the antenna input with two calibrated noise sources, from which the system constants GkB and GT_RkB can be determined. (This is similar to the Y-factor method for measuring noise temperature.) Then the desired brightness temperature, T_B, can be determined.

Two types of errors occur with this radiometer. First is an error, ΔT_N, in the measured brightness temperature due to noise fluctuations. Since noise is a random process, the measured noise power may vary from one integration period to the next. The integrator (or low-pass filter) acts to smooth out ripples in V_o with frequency components greater than $1/\tau$. It can be shown that the remaining error is [7]

$$\Delta T_N = \frac{T_B + T_R}{\sqrt{B\tau}}. \tag{14.50}$$

This result shows that if a longer measurement time, τ, can be tolerated, the error due to noise fluctuation can be reduced to a negligible value.

A more serious error is due to random variations in the system gain, G. Such variations generally occur in the RF amplifier, mixer, or IF amplifier, over a period of 1 s or longer. If the system is calibrated with a certain value of G, which changes by the time a measurement is made, an error will occur, as given in reference [7] as

$$\Delta T_G = (T_B + T_R)\frac{\Delta G}{G}, \tag{14.51}$$

where ΔG is the rms change in the system gain, G.

It will be useful to consider some typical numbers. For example, a 10 GHz total power radiometer may have a bandwidth of 100 MHz, a receiver temperature of $T_R = 500$ K, an integrator time constant of $\tau = 0.01$ s, and a system gain variation of $\Delta G/G = 0.01$. If the antenna temperature is $T_B = 300$ K, (14.50) gives the error due to noise fluctuations as $\Delta T_N = 0.8$ K, while (14.51) gives the error due to gain variations as $\Delta T_G = 8$ K. These results, which are based on reasonably realistic data, show that gain variation is the most detrimental factor affecting the accuracy of the total power radiometer.

The Dicke Radiometer

We have seen that the dominant factor affecting the accuracy of the total power radiometer is the variation of gain of the overall system. Since such gain variations have a relatively long time constant (>1 s), it is

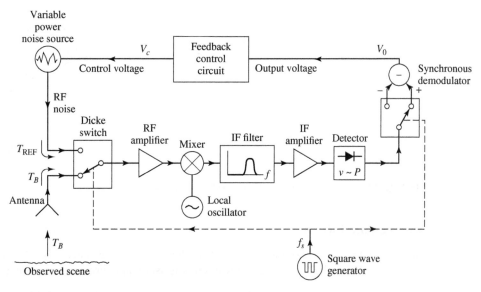

FIGURE 14.27 | Balanced Dicke radiometer block diagram.

conceptually possible to eliminate this error by repeatedly calibrating the radiometer at rapid rate. This is the principle behind the operation of the Dicke null-balancing radiometer.

A system diagram is shown in Figure 14.27. The superheterodyne receiver is identical to the total power radiometer, but the input is periodically switched between the antenna and a variable power noise source; this switch is called the *Dicke switch*. The output of the square-law detector drives a synchronous demodulator, which consists of a switch and a difference circuit. The demodulator switch operates in synchronism with the Dicke switch, so that the output of the subtractor is proportional to the difference between the noise powers from the antenna, T_B, and the reference noise source, T_{REF}. The output of the subtractor is then used as an error signal to a feedback control circuit, which controls the power level of the reference noise source so that V_o approaches zero. In this balanced state, $T_B = T_{REF}$, and T_B can be determined from the control voltage, V_c. The square-wave sampling frequency, f_s, is chosen to be much faster than the drift time of the system gain, so that this effect is virtually eliminated. Typical sampling frequencies range from 10 to 1000 Hz.

A typical radiometer would measure brightness temperature T_B over a range of about 50–300 K; this then implies that the reference noise source would have to cover this same range, which is difficult to do in practice. Thus, there are several variations on the above design, differing essentially in the way that the reference noise power is controlled or added to the system. One possible method is to use a constant T_{REF} that is somewhat hotter than the maximum T_B to be measured. The amount of reference noise power delivered to the system is then controlled by varying the pulse width of the sampling waveform. Another approach is to use a constant reference noise power, and vary the gain of the IF stage during the reference sample time to achieve a null output. Other possibilities, including alternatives to the Dicke radiometer, are discussed in the literature [7].

14.5 MICROWAVE PROPAGATION

In free-space, electromagnetic waves propagate in straight lines without attenuation or other adverse effects. Free-space, however, is an idealization that is only approximated when RF or microwave energy propagates through the atmosphere or in the presence of Earth. In practice, the performance of a communication, radar, or radiometry system may be seriously affected by propagation effects such as reflection, refraction, attenuation, or diffraction. Below we discuss some specific propagation phenomenon that can influence the operation of microwave systems. It is important to realize that propagation effects generally cannot be quantified in any exact or rigorous sense, but can only be described in terms of their statistics.

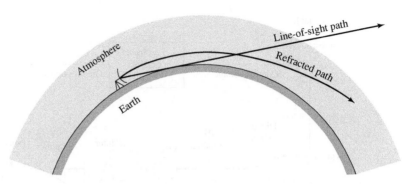

FIGURE 14.28 | Refraction of radio waves by the atmosphere.

Atmospheric Effects

The relative permittivity of the atmosphere is close to unity, but is actually a function of air pressure, temperature, and humidity. An empirical result that is useful at microwave frequencies is given by [6]

$$\epsilon_r = \left[1 + 10^{-6} \left(\frac{79P}{T} - \frac{11V}{T} + \frac{3.8 \times 10^5 V}{T^2} \right) \right]^2, \tag{14.52}$$

where P is the barometric pressure in millibars, T is the temperature in kelvins, and V is the water vapor pressure in millibars. This result shows that permittivity generally decreases (approaches unity) as altitude increases since pressure and humidity decrease with height faster than does temperature. This change in permittivity with altitude causes radio waves to bend toward Earth, as depicted in Figure 14.28. Such refraction of radio waves can sometimes be useful since it may extend the range of radar and communication systems beyond the limit imposed by the presence of Earth's horizon.

If an antenna is at a height, h, above Earth, simple geometry gives the line-of-sight distance to the horizon as

$$d = \sqrt{2Rh}, \tag{14.53}$$

where R is the radius of Earth. From Figure 14.28 we see that the effect of refraction on range can be accounted for by using an effective Earth radius kR, where $k > 1$. A value commonly used [6] is $k = 4/3$, but this is only an average value, which changes with weather conditions. In a radar system, refraction effects can lead to errors when determining the elevation of a target close to the horizon.

Weather conditions can sometimes produce a localized temperature inversion, where the temperature increases with altitude. Equation (14.52) then shows that the atmospheric permittivity will decrease much faster than normal with increasing altitude. This condition can sometimes lead to *ducting* (also called trapping, or anomalous propagation), where a radio wave can propagate long distances parallel to Earth's surface via the duct created by the layer of air along the temperature inversion. The situation is very similar to propagation in a dielectric waveguide. Such ducts can range in height from 50 to 500 feet, and may be near Earth's surface or higher in altitude.

Another atmospheric effect is attenuation, caused primarily by the absorption of microwave energy by water vapor or molecular oxygen. Maximum absorption occurs when the frequency coincides with one of the molecular resonances of water or oxygen, and thus atmospheric attenuation has distinct peaks at these frequencies. Figure 14.29 shows the atmospheric attenuation versus frequency. At frequencies below 10 GHz the atmosphere has very little effect on the strength of a signal. At 22.2 and 183.3 GHz, resonance peaks occur due to water vapor resonances, while resonances of molecular oxygen cause peaks at 60 and 120 GHz. Thus there are "windows" in the millimeter wave band near 35, 94, and 135 GHz where radar and communication systems can operate with minimum loss. Precipitation such as rain, snow, or fog will increase the attenuation, especially at higher frequencies. The effect of atmospheric attenuation can be included in system design when using the Friis transmission equation or the radar equation.

In some instances the system frequency may be chosen at a point of maximum atmospheric attenuation. Remote sensing of the atmosphere (temperature, water vapor, rain rate) is often done with radiometers operating near 20 or 55 GHz to maximize the sensing of atmospheric conditions. Another interesting example is spacecraft-to-spacecraft communication at 60 GHz. This millimeter wave frequency band has

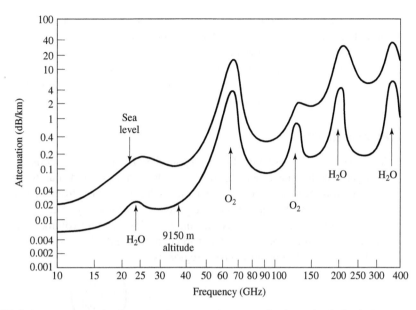

FIGURE 14.29 | Average atmospheric attenuation versus frequency (horizontal polarization).

the advantages of a large bandwidth and small antennas with high gains, and, since the atmosphere is very lossy at this frequency, the possibilities of interference, jamming, and eavesdropping from Earth are greatly reduced.

Ground Effects

The most obvious effect of the presence of the ground on RF and microwave propagation is reflection from Earth's surface (land or sea). As shown in Figure 14.30, a radar target (or receiver antenna) may be illuminated by both a direct wave from the transmitter and a wave reflected from the ground. The reflected wave is generally smaller in amplitude than the direct wave because of the larger distance it travels, the fact that it usually radiates from the sidelobe region of the transmit antenna, and the fact that the ground is not a perfect reflector. Nevertheless, the received signal at the target or receiver will be the vector sum of the two wave components and, depending on the relative phases of the two waves, may be greater or less than the direct wave alone. Because the distances involved are usually very large in terms of the electrical wavelength, even a small variation in the permittivity of the atmosphere can cause *fading* (long-term fluctuations) or *scintillation* (short-term fluctuations) in the signal strength. These effects can also be caused by reflections from inhomogeneities in the atmosphere.

In communication systems fading can sometimes be reduced by making use of the fact that the fading of two communication channels having different frequencies, polarizations, or physical locations is essentially independent. Thus a communication link can reduce fading effects by combining the outputs of two (or more) such channels; this is called a *diversity system*.

Another ground effect is *diffraction*, whereby a radio wave scatters energy in the vicinity of the line-of-sight boundary at the horizon, thus giving a range slightly beyond the horizon. This effect is usually very

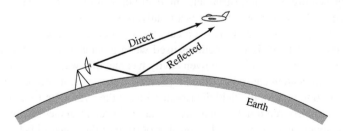

FIGURE 14.30 | Direct and reflected waves over Earth's surface.

small at microwave frequencies. Of course, when obstacles such as hills, mountains, or buildings are in the path of propagation, diffraction effects can be stronger.

In a radar system, unwanted reflections often occur from terrain, vegetation, trees, buildings, and the surface of the sea. Such clutter echoes generally degrade or mask the return of a true target, or show up as a false target, in the context of a surveillance or tracking radar. In mapping or remote sensing applications such clutter returns may actually constitute the desired signal.

Plasma Effects

A *plasma* is a gas consisting of ionized particles. The ionosphere consists of spherical layers of atmosphere with particles that have been ionized by solar radiation, and thus forms a plasma region. A very dense plasma is formed on a spacecraft as it reenters the atmosphere from outer space, due to the high temperatures produced by friction. Plasmas are also produced by lightning, meteor showers, and nuclear explosions.

A plasma is characterized by the number of ions per unit volume; depending on this density and the frequency, a wave might be reflected, absorbed, or transmitted by the plasma medium. An effective permittivity can be defined for a uniform plasma region as

$$\epsilon_e = \epsilon_0 \left(1 - \frac{\omega_p^2}{\omega^2} \right), \tag{14.54}$$

where

$$\omega_p = \sqrt{\frac{Nq^2}{m\epsilon_0}} \tag{14.55}$$

is the plasma frequency. In (14.55), q is the charge of the electron, m is the mass of the electron, and N is the number of ionized particles per unit volume. By studying the solution of Maxwell's equations for plane wave propagation in such a medium, it can be shown that wave propagation through a plasma is only possible for $\omega > \omega_p$. Lower frequency waves will be totally reflected. If a magnetic field is present, the plasma becomes anisotropic, and the analysis is more complicated. Earth's magnetic field may be strong enough to produce such an anisotropy in some cases.

The ionosphere consists of several different layers with varying ion densities; in order of increasing ion density, these layers are referred to as D, E, F_1, and F_2. The characteristics of these layers depend on seasonal weather and solar cycles, but the average plasma frequency is about 8 MHz. Thus, signals at frequencies less than 8 MHz (e.g., short-wave radio) can reflect off the ionosphere to travel distances well beyond the horizon. Higher frequency signals, however, will pass through the ionosphere.

In the case of a spacecraft entering the atmosphere, the high velocity produces a very dense plasma around the vehicle. The electron density is high enough that, from (14.55), the plasma frequency is very high, thus inhibiting communication with the spacecraft until its velocity has decreased. Besides this blackout effect, the plasma layer may also cause a large impedance mismatch between the antenna and its feed line.

14.6   OTHER APPLICATIONS AND TOPICS

Microwave Heating

To the average consumer the term "microwave" connotes a microwave oven, used in many households for heating food; industrial and medical applications also exist for microwave heating. As shown in Figure 14.31, a microwave oven is a relatively simple system consisting of a high-power microwave source, a waveguide feed, and the oven cavity. The source is generally a magnetron tube operating at 2.45 GHz, although 915 MHz is sometimes used when greater penetration is desired. Power output is usually in the range of 500–1500 W. The oven cavity has metallic walls, and is electrically large. To reduce the effect of uneven heating caused by standing waves in the oven, a "mode stirrer," which is just a metallic fan blade, is used to perturb the field distribution inside the oven. The food is also rotated with a motorized platter.

In a conventional oven a gas or charcoal fire, or an electric heating element, generates heat external to the material to be heated. The outside portion of the material is heated by convection, and the inside of the

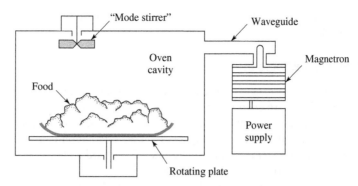

FIGURE 14.31 | A microwave oven.

material is warmed by conduction from the outer portion. In microwave heating, by contrast, the inside of the material is heated first. The process through which this occurs primarily involves the conduction losses in food materials having large loss tangents [8, 9]. An interesting fact is that the loss tangents of many foods decrease with increasing temperature, so that microwave heating is to some extent self-regulating. The result is that microwave cooking generally gives faster and more uniform heating of food as compared with conventional cooking. The efficiency of a microwave oven, when defined as the ratio of power converted to heat (in the food) to the power supplied to the oven, is generally less than 50%, but this is usually greater than the cooking efficiency of a conventional oven.

The most critical issue in the design of a microwave oven is safety. Since a very high power source is used, leakage must be very small to avoid exposing the user to harmful radiation. Thus the magnetron, feed waveguide, and oven cavity must all be carefully shielded. The door of the oven requires particular attention; besides close mechanical tolerances, the joint around the door usually employs RF-absorbing material and a $\lambda/4$ choke flange to reduce power leakage to an acceptable level.

Power Transfer

Electrical power transmission lines are a very efficient and convenient way to transfer energy from one point to another, as they have relatively low loss and initial costs, and can be easily routed. There are applications, however, where it is inconvenient or impossible to use such power lines. In such cases it is conceivable that electrical power can be transmitted without wires by a well-focused microwave beam [10].

One example is a solar satellite power station, where it has been proposed that electricity be generated in space by a large orbiting array of solar cells and transmitted to a receiving station on Earth by a microwave beam. We would thus be provided with a virtually inexhaustible source of electricity. Placing the solar arrays in space has the advantage of power delivery uninterrupted by darkness, clouds, or precipitation, which are problems encountered with Earth-based photovoltaic arrays.

To be economically competitive with other sources, a solar power satellite station would have to be very large. One proposal involves a solar array about 5×10 km in size feeding a 1 km diameter phased array antenna. The power output on Earth would be on the order of 5 GW. Such a project is extremely large in terms of cost and complexity. Also of legitimate concern is the operational safety of such a scheme, in terms of both the radiation hazards associated with the system when it is operating as designed, and the risks involved with a malfunction of the system. These considerations, as well as the political and philosophical ramifications of such a large, centralized power system, have made the future of the solar power satellite station doubtful.

Similar in concept, but on a much smaller scale, is the transmission of electrical power from Earth to a vehicle such as a small drone helicopter or airplane. The advantages are that such an aircraft could run indefinitely, and very quietly, at least over a limited area. Battlefield surveillance and weather prediction would be some possible applications. The concept has been demonstrated with several projects involving small pilotless aircraft.

On an even smaller scale is the wireless transmission of power to RFID tags, which is feasible primarily because of the very low DC power required for appropriately designed CMOS circuitry. A related idea is the collection of ambient RF power to charge batteries of portable devices. This sounds attractive, and may be possible in principle, but it is probably not feasible in most situations, especially when other power sources are available.

Biological Effects and Safety

The proven dangers of exposure to RF and microwave radiation are due to thermal effects. The body absorbs RF and microwave energy and converts it to heat; as in the case of a microwave oven, this heating occurs within the body and may not be felt at low levels. Such heating is most dangerous in the brain, the eye, the genitals, and the stomach organs. Excessive radiation can lead to cataracts, sterility, or cancer. It is important to determine a safe radiation level standard so that users of RF and microwave equipment will not be exposed to harmful power levels.

At the time of this writing, the most recent IEEE safety standard for human exposure to electromagnetic fields is given by IEEE Standard C95.1-2019. In the RF-microwave frequency range of 0 Hz to 300 GHz, exposure limits are specified for the peak external magnetic field as a function of frequency, as shown in Figure 14.32. These exposure limits are intended to apply generally to persons permitted in restricted environments and to the general public in unrestricted environments. The recommended safe power density limits are generally lower at lower frequencies because fields penetrate the body more deeply at these frequencies. At higher frequencies most of the power absorption occurs near the skin surface, so the safe limits can be higher. At frequencies below 100 MHz electric and magnetic fields interact with the body differently than higher frequency electromagnetic fields, and so separate limits are given for field components at these lower frequencies.

In the United States, the Federal Communications Commission (FCC) sets a separate exposure limit for hand-held wireless devices (cell phones, PDAs, smartphones, etc). These limits are given in terms of the *Specific Absorption Rate* (SAR), which measures how much power is dissipated as heat in a unit of tissue mass. Specific Absorption Rate is defined as

$$\text{SAR} = \frac{\sigma}{2\rho} |\bar{E}|^2 \text{ W/kg}, \tag{14.56}$$

where σ is the conductivity of the tissue (S/m), ρ is the density of the tissue (kg/m^3), and $\bar{E}$ is the electric field in the tissue sample. For partial body exposure (typically the head or hand), the FCC limit on SAR is 1.6 W/kg, averaged over 1 g of tissue. All wireless devices sold in the United States must meet this standard. Other countries have standards that are similar in nature and scope. The European Union, for example, requires hand-held wireless devices to have SAR exposure of less than 2 W/kg, averaged over 10 g of tissue. In India, as per the Department of Telecommunications, Ministry of Communications & IT, specifications, all mobile phones need to comply with the SAR value of 1.6 W/kg (averaged over 1 g of tissue) with effect from September 2012.

A separate standard applies to microwave ovens sold in the United States, requiring that all ovens be tested to ensure that the power level at 5 cm from any point on the oven does not exceed 1 mW/cm^2.

Most experts feel that the above limits represent safe levels, with a reasonable margin. Some researchers, however, feel that health hazards may occur due to nonthermal effects of long-term exposure to even low levels of microwave radiation.

EXAMPLE 14.8   POWER DENSITY IN THE VICINITY OF A MICROWAVE RADIO LINK

An 18 GHz common-carrier microwave communication link uses a tower-mounted antenna with a gain of 40 dB and a transmitter power of 20 W. To evaluate the radiation hazard of this system, calculate the power density at a distance of 20 m from the antenna. Do this for a position in the main beam of the antenna, and for a position in the sidelobe region of the antenna. Assume a worst-case sidelobe level of −10 dB.

Solution
The numerical gain of the antenna is

$$G_t = 10^{40/10} = 4000.$$

From (14.23), the power density in the main beam of the antenna at a distance of $R = 20$ m is

$$S_{\text{avg}} = \frac{P_t G_t}{4\pi R^2} = \frac{(20)(4000)}{4\pi(20)^2} = 15.92 \text{ W/m}^2.$$

The worst-case power density in the sidelobe region is 10 dB below this value, or 1.52 W/m^2.

Thus, the power density in the main beam at 20 m is below the U.S. standard for radiation hazard for the general population, while the power density in the sidelobe region is well below this limit. These power levels will diminish rapidly with increasing distance due to the $1/r^2$ dependence.

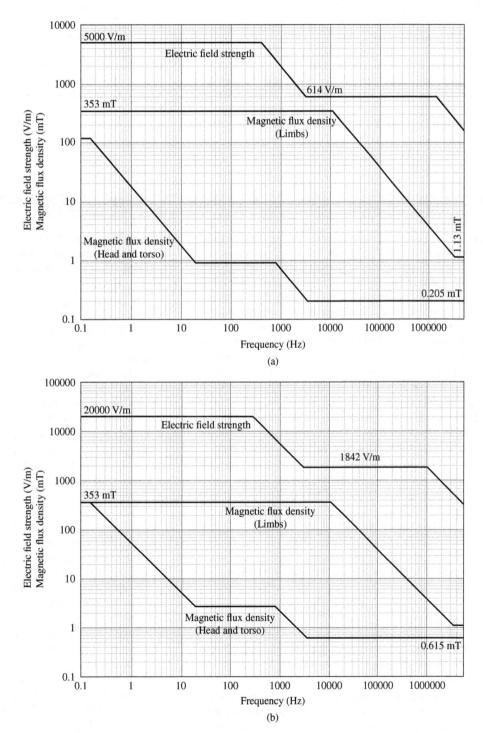

FIGURE 14.32 | IEEE Standard C95.1-2019 recommended peak external magnetic field for human exposure to RF and microwave electromagnetic fields. (a) Persons in unrestricted environments. (b) Persons permitted in restricted environments.

REFERENCES

[1] C. A. Balanis, *Antenna Theory: Analysis and Design*, 3rd edition, John Wiley & Sons, New York, 2005.

[2] W. L. Stutzman and G. A. Thiele, *Antenna Theory and Design*, 2nd edition, John Wiley & Sons, New York, 1998.

[3] L. J. Ippolito, R. D. Kaul, and R. G. Wallace, *Propagation Effects: Handbook for Satellite Systems Design*, 3rd edition, NASA, Washington, DC, 1983.

[4] D. M. Pozar, *Microwave and RF Design of Wireless Systems*, John Wiley & Sons, New York, 2001.

[5] E. Lutz, M. Werner, and A. Jahn, *Satellite Systems for Personal and Broadband Communications*, Springer-Verlag, Berlin, 2000.

[6] M. I. Skolnik, *Introduction to Radar Systems*, McGraw-Hill, New York, 1962.

[7] F. T. Ulaby, R. K. Moore, and A. K. Fung, *Microwave Remote Sensing: Active and Passive, Volume I, Microwave Remote Sensing, Fundamentals and Radiometry*. Addison-Wesley, Reading, MA, 1981.

[8] F. E. Gardiol, *Introduction to Microwaves*, Artech House, Dedham, MA, 1984.

[9] E. C. Okress, *Microwave Power Engineering*, Academic Press, New York, 1968.

[10] W. C. Brown, "The History of Power Transmission by Radio Waves," *IEEE Transactions on Microwave Theory and Techniques*, vol. MTT-32, pp. 1230–1242, September 1984.

PROBLEMS

14.1 The Iridium satellite communication system was designed with a link margin of 16 dB, and was originally advertised as being capable of providing service to users with hand-held phones in vehicles, buildings, and urban areas. Today, after bankruptcy and restructuring of the company, it is recommended that Iridium phones be used outdoors, with a line of sight to the satellites. Find some estimates of the link margins (due to fading) required for L-band communication into vehicles and buildings. Do you think the Iridium system would have operated reliably in these environments? If not, why was the system designed with a 16 dB link margin?

14.2 An antenna has a radiation pattern function given by $F_\theta(\theta, \phi) = A \sin^2 \theta \cos \phi$. Find the main beam position, the 3 dB beamwidths in the principal planes, and the directivity (in dB) for this antenna. What is the polarization of this antenna?

14.3 A monopole antenna on a large ground plane has a far-field pattern function given by $F_\theta(\theta, \phi) = A \sin \theta$ for $0 \leq \theta \leq 90°$. The radiated field is zero for $90° \leq \theta \leq 180°$. Find the directivity (in dB) of this antenna.

14.4 A DBS reflector antenna operating at 15 GHz has a diameter of 20 inches. If the aperture efficiency is 62%, find the directivity.

14.5 A reflector antenna used for a cellular base station backhaul radio link operates at 30 GHz, with a gain of 40 dB, a radiation efficiency of 90%, and a diameter of 18 inches. (a) Find the aperture efficiency of this antenna. (b) Find the half-power beamwidth, assuming the beamwidths are identical in the two principal planes.

14.6 A high-gain antenna array operating at 3.5 GHz is pointed toward a region of the sky for which the background can be assumed to be at a uniform temperature of 10 K. A noise temperature of 125 K is measured for the antenna temperature. If the physical temperature of the antenna is 290 K, what is its radiation efficiency?

14.7 Derive equation (14.20) by treating the antenna and lossy line as a cascade of two networks whose equivalent noise temperatures are given by (14.18) and (10.15).

14.8 Consider the replacement of a DBS dish antenna with a microstrip array antenna. A microstrip array offers an aesthetically pleasing flat profile, but suffers from relatively high dissipative loss in its feed network, which leads to a high noise temperature. If the background noise temperature is $T_B = 50$ K, with an antenna gain of 38 dB and a receiver LNB noise figure of 2.6 dB, find the overall G/T for the microstrip array antenna and the LNB if the array has a total loss of 3.5 dB. Assume the antenna is at a physical temperature of 290 K.

14.9 At a distance of 350 m from an antenna operating at 8 GHz, the radiated power density in the main beam is measured to be 8.5×10^{-3} W/m^2. If the input power to the antenna is known to be 80 W, find the gain of the antenna.

14.10 A cellular base station is to be connected to its Mobile Telephone Switching Office located 5 km away. Two possibilities are to be evaluated: (1) a radio link operating at 25 GHz, with $G_t = G_r = 28$ dB, and (2) a wired link using coaxial line having an attenuation of 0.05 dB/m, with four 30 dB repeater amplifiers along the line. If the minimum required received power level for both cases is the same, which option will require the smallest transmit power?

14.11 A GSM cellular telephone system operates at a downlink frequency of 935–960 MHz, with a channel bandwidth of 200 kHz, and a base station that transmits with an EIRP of 20 W. The mobile receiver has an antenna with a gain of 0 dBi and a noise temperature of 450 K, and the receiver has a noise figure of 8 dB. Find the maximum operating range if the required minimum SNR at the output of the receiver is 10 dB, and a link margin of 30 dB is required to account for propagation into vehicles, buildings, and urban areas.

14.12 Consider the GPS receiver system shown below. The guaranteed minimum L1 (1575 MHz) carrier power received by an antenna on Earth having a gain of 0 dBi is $S_i = -160$ dBW. A GPS receiver is usually specified as requiring a minimum carrier-to-noise ratio, relative to a 1 Hz bandwidth, of C/N (Hz). If the receiver antenna actually has a gain G_A and a noise temperature T_A, derive an expression for the maximum allowable amplifier noise figure F, assuming an amplifier gain G and a connecting line loss L. Evaluate this expression for $C/N = 35$ dB-Hz, $G_A = 8$ dB, $T_A = 320$ K, $G = 20$ dB, and $L = 35$ dB.

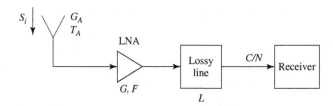

14.13 A key premise in many science fiction stories is the idea that radio and TV signals from Earth can travel through space and be received by listeners in another star system. Show that this is a fallacy by calculating the maximum distance from Earth where a signal could be received with a SNR of 0 dB. Specifically, assume TV channel 4, broadcasting at 65 MHz, with a 5 MHz bandwidth, a transmitter power of 1500 W, transmit and receive antenna gains of 5 dB, a cosmic background noise temperature of 5 K, and a perfectly noiseless receiver. How much would this distance decrease if an SNR of 30 dB is required at the receiver? (30 dB is a typical value for good reception of an analog video signal.) Relate these distances to the nearest planet in our solar system.

14.14 The Mariner 10 spacecraft used to explore the planet Mercury in 1974 used BPSK with $P_b = 0.05$ ($E_b/n_0 = 1.4$ dB) to transmit image data back to Earth (a distance of about 1.6×10^8 km). The spacecraft transmitter operated at 2.295 GHz, with an antenna gain of 27.6 dB and a carrier power of 16.8 W. The ground station had an antenna with a gain of 61.3 dB and an overall system noise temperature of 13.5 K. Find the maximum possible data rate.

14.15 Derive the radar equation for the bistatic case where the transmit and receive antennas have gains of G_t and G_r, and are at distances R_t and R_r from the target, respectively.

14.16 A pulse radar has a pulse repetition frequency $f_r = 1/T_r$. Determine the maximum unambiguous range of the radar. (Range ambiguity occurs when the round-trip time of a return pulse is greater than the pulse repetition time, so it becomes unclear as to whether a given return pulse belongs to the last transmitted pulse or some earlier transmitted pulse.)

14.17 A Doppler radar operating at 15 GHz is intended to detect target velocities ranging from 1 to 25 m/sec. What is the required passband of the Doppler filter?

14.18 A pulse radar operates at 2.4 GHz and has a per-pulse power of 1.5 kW. If it is to be used to detect a target with

$\sigma = 22$ m^2 at a range of 10 km, what should be the minimum isolation between the transmitter and receiver so that the leakage signal from the transmitter is at least 10 dB below the received signal? Assume an antenna gain of 30 dB.

14.19 An antenna having a gain G is shorted at its terminals. What is the minimum monostatic radar cross section in the direction of the main beam?

14.20 Consider the radiometer antenna shown below, where the antenna is at a physical temperature T_p and has a radiation efficiency η_{rad}, and an impedance mismatch Γ at its terminals. If T_S is the apparent temperature seen by the radiometer, show that $\Delta T_S/\Delta T_{true}$ is equal to the product of radiation efficiency and mismatch loss, by applying two background temperatures, $T_B = T_p$ and $T_B = T_2 \neq T_p$.

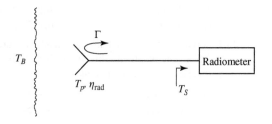

14.21 The atmosphere does not have a definite thickness since it gradually thins with altitude, with a consequent decrease in attenuation. However, if we use a simplified "orange peel" model and assume that the atmosphere can be approximated by a uniform layer of fixed thickness, we can estimate the background noise temperature seen through the atmosphere. Thus, let the thickness of the atmosphere be 4000 m, and find the maximum distance ℓ to the edge of the atmosphere along the horizon, as shown in the figure below (the radius of Earth is 6400 km). Now assume an average atmospheric attenuation of 0.005 dB/km, with a background noise temperature beyond the atmosphere of 4 K, and find the noise temperature seen on Earth by treating the cascade of the background noise with the attenuation of the atmosphere. Do this for an ideal antenna pointing toward the zenith, and toward the horizon.

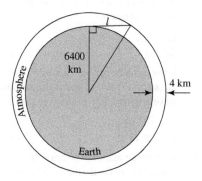

14.22 A 26 GHz radio link uses a tower-mounted reflector antenna with a gain of 35 dB and a transmitter power of 6 W.

(a) Find the minimum distance within the main beam of the antenna for which the U.S-recommended safe power density limit of 12 mW/cm^2 is not exceeded. (b) How does this distance change for a position within the side-lobe region of the antenna if we assume a worst-case sidelobe level of 10 dB below the main beam? (c) Are these distances in the far-field region of the antenna?

(Assume a circular reflector, with an aperture efficiency of 60%.)

14.23 On a clear day, with the sun directly overhead, the received power density from sunlight is about 1400 W/m^2. If we make the simplifying assumption that this power is transmitted via a single-frequency plane wave, find the resulting amplitude of the incident electric and magnetic fields.

MULTIPLE CHOICE QUESTIONS

14.1 A parabolic reflector antenna used for reception with the direct broadcast system (DBS) is operating at 15 GHz and is 20 inches in diameter. Find the far-field distance for this antenna.

(a) 12.9 m
(b) 25.8 m
(c) 17.3 m
(d) 34.6 m

14.2 A rectangular horn antenna has an aperture area of 3.5 λ × 2.5 λ. The maximum directivity that can be achieved by this rectangular horn antenna is

(a) 20.41 dB
(b) 10.21 dB
(c) 109.95 dB
(d) 54.98 dB

14.3 A rectangular horn antenna has an aperture area of 3.5 λ × 2.5 λ. If the aperture efficiency of the antenna is 85%, then the antenna directivity is

(a) 14.71 dB
(b) 17.17 dB
(c) 19.71 dB
(d) 23.17 dB

14.4 If an antenna has a directivity of 12.5 dB and is operating at a wavelength of λ, then the maximum effective aperture efficiency is

(a) 1.42 λ^2
(b) 2.83 λ^2
(c) 0.94 λ^2
(d) none of the above

14.5 If the beamwidth of an antenna in two orthogonal planes is 25° and 54°, respectively, then the directivity of the antenna is

(a) 12
(b) 18
(c) 24
(d) 30

14.6 If the power input to an antenna is 81 mW and if the radiated power is measured to be 69 mW, then the efficiency of the antenna is

(a) 75%
(b) 85%
(c) 90%
(d) insufficient data

14.7 If the distance between a transmitting station and a receiving station is 800 m and if the antennas are operating at a wavelength of 12 cm, then the path loss is

(a) 196.92 dB
(b) 98.46 dB
(c) 49.23 dB
(d) 24.62 dB

14.8 To provide accurate position information, a global positioning system in the medium Earth orbit uses _____ number of satellites.

(a) 48
(b) 52
(c) 34
(d) 24

14.9 The frequency of operation of Bluetooth devices is

(a) 3.5 GHz
(b) 4.4 GHz
(c) 2.4 GHz
(d) none of the above

14.10 The equivalent noise temperature of an antenna to the receiver connecting transmission line is

(a) $T_P(L_P + 1)$
(b) $T_P(L_P - 1)$
(c) $T_P/(L_P - 1)$
(d) $T_P/(L_P + 1)$

14.11 In a receiver, if the noise figure of the mixer stage is 8 dB, then the equivalent noise temperature, given that the receiver is operating at 290 K, is

(a) 1154 K
(b) 1445 K
(c) 1540 K
(d) 1735 K

14.12 If a transmission line connecting an antenna to a receiver has a loss of 1.8 dB, given the physical temperature is 26°C, the equivalent noise temperature is

(a) 56.4 K
(b) 112.7 K
(c) 153.5 K
(d) 307.1 K

14.13 Given that the antenna efficiency is 0.85, equivalent brightness temperature is 210 K, physical temperature is 27°C, the noise temperature of the antenna is

(a) 223.5 K
(b) 201.1 K
(c) 182.6 K
(d) none of the above

14.14 If a receiver is operating at a bandwidth of 2 MHz and has a noise temperature of 250 K, then the input noise power is

(a) −197 dBm
(b) −111.6 dBm
(c) −55.8 dBm
(d) −83.7 dBm

14.15 If the received power at antenna terminals is −86 dBm, and if the input noise power is −121 dBm, then the input SNR is

(a) 70 dB
(b) −17.5 dB
(c) −35 dB
(d) 35 dB

14.16 If the received power at the antenna terminals is −78 dBm and the output noise power is −118 dBm, then the output signal-to-noise ratio is given by

(a) 40 dB
(b) −20 dB
(c) 20 dB
(d) −40 dB

14.17 A pulse radar operating at 12 GHz frequency has an antenna with a gain of 28 dB and a transmitted power of 2 kW. If it is desired to detect a target of cross section 12 m^2, and the minimum detectable signal is −90 dBm, the maximum range of the radar is

(a) 3703 m
(b) 4057 m
(c) 8114 m
(d) 7407 m

14.18 The radar in which both reception and transmission is carried out using the same antenna is called

(a) monostatic radar
(b) bistatic radar
(c) monopole radar
(d) dipole radar

14.19 In the designing of a radiometer, the major challenge is

(a) design complexity
(b) high cost
(c) requirement of highly sensitive receivers
(d) none of the above

14.20 The total power radiometer system bandwidth is determined by

(a) RF amplifier
(b) IF filter
(c) local oscillator
(d) IF amplifier

ANSWER KEY

14.1 (b)	**14.5** (c)	**14.9** (c)	**14.13** (a)	**14.17** (d)
14.2 (a)	**14.6** (b)	**14.10** (b)	**14.14** (b)	**14.18** (a)
14.3 (c)	**14.7** (b)	**14.11** (c)	**14.15** (d)	**14.19** (c)
14.4 (a)	**14.8** (d)	**14.12** (c)	**14.16** (a)	**14.20** (b)

Appendices

APPENDIX A PREFIXES

Multiplying Factor	Prefix	Symbol
10^{12}	tera	T
10^{9}	giga	G
10^{6}	mega	M
10^{3}	kilo	k
10^{2}	hecto	h
10^{1}	deka	da
10^{-1}	deci	d
10^{-2}	centi	c
10^{-3}	milli	m
10^{-6}	micro	μ
10^{-9}	nano	n
10^{-12}	pico	p
10^{-15}	femto	f

APPENDIX B VECTOR ANALYSIS

Coordinate Transformations

Rectangular to cylindrical:

	$\hat{x}$	$\hat{y}$	$\hat{z}$
$\hat{\rho}$	$\cos\phi$	$\sin\phi$	0
$\hat{\phi}$	$-\sin\phi$	$\cos\phi$	0
$\hat{z}$	0	0	1

Rectangular to spherical:

	$\hat{x}$	$\hat{y}$	$\hat{z}$
$\hat{r}$	$\sin\theta\cos\phi$	$\sin\theta\sin\phi$	$\cos\theta$
$\hat{\theta}$	$\cos\theta\cos\phi$	$\cos\theta\sin\phi$	$-\sin\theta$
$\hat{\phi}$	$-\sin\phi$	$\cos\phi$	0

Cylindrical to spherical:

	$\hat{\rho}$	$\hat{\phi}$	$\hat{z}$
$\hat{r}$	$\sin\theta$	0	$\cos\theta$
$\hat{\theta}$	$\cos\theta$	0	$-\sin\theta$
$\hat{\phi}$	0	1	0

These tables can be used to transform unit vectors as well as vector components; e.g.,

$$\hat{\rho} = \hat{x}\cos\phi + \hat{y}\sin\phi$$
$$A_\rho = A_x\cos\phi + A_y\sin\phi$$

Vector Differential Operators

Rectangular coordinates:

$$\nabla f = \hat{x}\frac{\partial f}{\partial x} + \hat{y}\frac{\partial f}{\partial y} + \hat{z}\frac{\partial f}{\partial z}$$

$$\nabla \cdot \bar{A} = \frac{\partial A_x}{\partial x} + \frac{\partial A_y}{\partial y} + \frac{\partial A_z}{\partial z}$$

$$\nabla \times \bar{A} = \hat{x}\left(\frac{\partial A_z}{\partial y} - \frac{\partial A_y}{\partial z}\right) + \hat{y}\left(\frac{\partial A_x}{\partial z} - \frac{\partial A_z}{\partial x}\right) + \hat{z}\left(\frac{\partial A_y}{\partial x} - \frac{\partial A_x}{\partial y}\right)$$

$$\nabla^2 f = \frac{\partial^2 f}{\partial x^2} + \frac{\partial^2 f}{\partial y^2} + \frac{\partial^2 f}{\partial z^2}$$

$$\nabla^2 \bar{A} = \hat{x}\nabla^2 A_x + \hat{y}\nabla^2 A_y + \hat{z}\nabla^2 A_z$$

Cylindrical coordinates:

$$\nabla f = \hat{\rho}\frac{\partial f}{\partial \rho} + \hat{\phi}\frac{1}{\rho}\frac{\partial f}{\partial \phi} + \hat{z}\frac{\partial f}{\partial z}$$

$$\nabla \cdot \bar{A} = \frac{1}{\rho}\frac{\partial}{\partial \rho}(\rho A_\rho) + \frac{1}{\rho}\frac{\partial A_\phi}{\partial \phi} + \frac{\partial A_z}{\partial z}$$

$$\nabla \times \bar{A} = \hat{\rho}\left(\frac{1}{\rho}\frac{\partial A_z}{\partial \phi} - \frac{\partial A_\phi}{\partial z}\right) + \hat{\phi}\left(\frac{\partial A_\rho}{\partial z} - \frac{\partial A_z}{\partial \rho}\right) + \hat{z}\frac{1}{\rho}\left[\frac{\partial(\rho A_\phi)}{\partial \rho} - \frac{\partial A_\rho}{\partial \phi}\right]$$

$$\nabla^2 f = \frac{1}{\rho}\frac{\partial}{\partial \rho}\left(\rho\frac{\partial f}{\partial \rho}\right) + \frac{1}{\rho^2}\frac{\partial^2 f}{\partial \phi^2} + \frac{\partial^2 f}{\partial z^2}$$

$$\nabla^2 \bar{A} = \nabla(\nabla \cdot \bar{A}) - \nabla \times \nabla \times \bar{A}$$

Spherical coordinates:

$$\nabla f = \hat{r}\frac{\partial f}{\partial r} + \hat{\theta}\frac{1}{r}\frac{\partial f}{\partial \theta} + \frac{\hat{\phi}}{r\sin\theta}\frac{\partial f}{\partial \phi}$$

$$\nabla \cdot \bar{A} = \frac{1}{r^2}\frac{\partial}{\partial r}(r^2 A_r) + \frac{1}{r\sin\theta}\frac{\partial}{\partial \theta}(\sin\theta A_\theta) + \frac{1}{r\sin\theta}\frac{\partial A_\phi}{\partial \phi}$$

$$\nabla \times \bar{A} = \frac{\hat{r}}{r\sin\theta}\left[\frac{\partial}{\partial \theta}(A_\phi \sin\theta) - \frac{\partial A_\theta}{\partial \phi}\right] + \frac{\hat{\theta}}{r}\left[\frac{1}{\sin\theta}\frac{\partial A_r}{\partial \phi} - \frac{\partial}{\partial r}(rA_\phi)\right]$$
$$+ \frac{\hat{\phi}}{r}\left[\frac{\partial}{\partial r}(rA_\theta) - \frac{\partial A_r}{\partial \theta}\right]$$

$$\nabla^2 f = \frac{1}{r^2}\frac{\partial}{\partial r}\left(r^2\frac{\partial f}{\partial r}\right) + \frac{1}{r^2\sin\theta}\frac{\partial}{\partial \theta}\left(\sin\theta\frac{\partial f}{\partial \theta}\right) + \frac{1}{r^2\sin^2\theta}\frac{\partial^2 f}{\partial \phi^2}$$
$$\nabla^2 \bar{A} = \nabla\nabla \cdot \bar{A} - \nabla \times \nabla \times \bar{A}$$

Vector identities:

$$\bar{A} \cdot \bar{B} = |A||B|\cos\theta, \qquad \text{where } \theta \text{ is the angle between } \bar{A} \text{ and } \bar{B} \tag{B.1}$$

$$|\bar{A} \times \bar{B}| = |A||B|\sin\theta, \qquad \text{where } \theta \text{ is the angle between } \bar{A} \text{ and } \bar{B}. \tag{B.2}$$

$$\bar{A} \cdot \bar{B} \times \bar{C} = \bar{A} \times \bar{B} \cdot \bar{C} = \bar{C} \times \bar{A} \cdot \bar{B} \tag{B.3}$$

$$\bar{A} \times \bar{B} = -\bar{B} \times \bar{A} \tag{B.4}$$

$$\bar{A} \times (\bar{B} \times \bar{C}) = (\bar{A} \cdot \bar{C})\bar{B} - (\bar{A} \cdot \bar{B})\bar{C} \tag{B.5}$$

$$\nabla(fg) = g\nabla f + f\nabla g \tag{B.6}$$

$$\nabla \cdot (f\bar{A}) = \bar{A} \cdot \nabla f + f\nabla \cdot \bar{A} \tag{B.7}$$

$$\nabla \cdot (\bar{A} \times \bar{B}) = (\nabla \times \bar{A}) \cdot \bar{B} - (\nabla \times \bar{B}) \cdot \bar{A} \tag{B.8}$$

$$\nabla \times (f\bar{A}) = (\nabla f) \times \bar{A} + f\nabla \times \bar{A} \tag{B.9}$$

$$\nabla \times (\bar{A} \times \bar{B}) = \bar{A}\nabla \cdot \bar{B} - \bar{B}\nabla \cdot \bar{A} + (\bar{B} \cdot \nabla)\bar{A} - (\bar{A} \cdot \nabla)\bar{B} \tag{B.10}$$

$$\nabla \cdot (\bar{A} \cdot \bar{B}) = (\bar{A} \cdot \nabla)\bar{B} + (\bar{B} \cdot \nabla)\bar{A} + A \times (\nabla \times \bar{B}) + \bar{B} \times (\nabla \times \bar{A}) \tag{B.11}$$

$$\nabla \cdot \nabla \times \bar{A} = 0 \tag{B.12}$$

$$\nabla \times (\nabla f) = 0 \tag{B.13}$$

$$\nabla \times \nabla \times \bar{A} = \nabla\nabla \cdot \bar{A} - \nabla^2\bar{A} \tag{B.14}$$

Note: the term $\nabla^2\bar{A}$ has meaning only for rectangular components of $\bar{A}$.

$$\int_V \nabla \cdot \bar{A} \, dv = \oint_S \bar{A} \cdot d\bar{s} \qquad \text{(divergence theorem)} \tag{B.15}$$

$$\int_S (\nabla \times \bar{A}) \cdot d\bar{s} = \oint_C \bar{A} \cdot d\bar{\ell} \qquad \text{(Stokes' theorem)} \tag{B.16}$$

APPENDIX C BESSEL FUNCTIONS

Bessel functions are solutions to the differential equation,

$$\frac{1}{\rho}\frac{d}{d\rho}\left(\rho\frac{df}{d\rho}\right) + \left(k^2 - \frac{n^2}{\rho^2}\right)f = 0 \tag{C.1}$$

where k^2 is real and n is an integer. The two independent solutions to this equation are called ordinary Bessel functions of the first and second kind, written as $J_n(k\rho)$ and $Y_n(k\rho)$, and so the general solution to (C.1) is

$$f(\rho) = AJ_n(k\rho) + BY_n(k\rho) \tag{C.2}$$

where A and B are arbitrary constants to be determined from boundary conditions.

These functions can be written in series form as

$$J_n(x) = \sum_{m=0}^{\infty} \frac{(-1)^m (x/2)^{n+2m}}{m!(n+m)!} \tag{C.3}$$

$$Y_n(x) = \frac{2}{\pi}\left(\gamma + \ln\frac{x}{2}\right)J_n(x) - \frac{1}{\pi}\sum_{m=0}^{n-1}\frac{(n-m-1)!}{m!}\left(\frac{2}{x}\right)^{n-2m}$$

$$- \frac{1}{\pi}\sum_{m=0}^{\infty}\frac{(-1)^m(x/2)^{n+2m}}{m!(n+m)!}\left(1 + \frac{1}{2} + \frac{1}{3} + \cdots + \frac{1}{m} + 1 + \frac{1}{2} + \cdots + \frac{1}{n+m}\right) \tag{C.4}$$

where $\gamma = 0.5772\ldots$ is Euler's constant, and $x = k\rho$. Note that Y_n becomes infinite at $x = 0$, due to the ln term. From these series expressions, small argument formulas can be obtained as

$$J_n(x) \sim \frac{1}{n!}\left(\frac{x}{2}\right)^n \tag{C.5}$$

$$Y_0(x) \sim \frac{2}{\pi}\ln x \tag{C.6}$$

$$Y_n(x) \sim \frac{-1}{\pi}(n-1)!\left(\frac{x}{2}\right)^n, \qquad n > 0 \tag{C.7}$$

Large argument formulas can be derived as

$$J_n(x) \sim \sqrt{\frac{2}{\pi x}}\cos\left(x - \frac{\pi}{4} - \frac{n\pi}{2}\right) \tag{C.8}$$

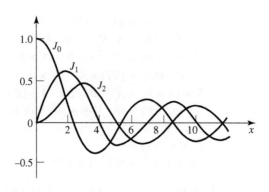

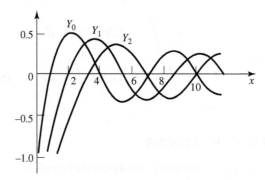

FIGURE C.1 | Bessel functions of the first and second kind.

$$Y_n(x) \sim \sqrt{\frac{2}{\pi x}} \sin\left(x - \frac{\pi}{4} - \frac{n\pi}{2}\right) \tag{C.9}$$

Figure C.1 shows graphs of a few of the lowest order Bessel functions of each type.

Recurrence formulas relate Bessel functions of different orders:

$$Z_{n+1}(x) = \frac{2n}{x} Z_n(x) - Z_{n-1}(x) \tag{C.10}$$

$$Z_n'(x) = \frac{-n}{x} Z_n(x) + Z_{n-1}(x) \tag{C.11}$$

$$Z_n'(x) = \frac{n}{x} Z_n(x) - Z_{n+1}(x) \tag{C.12}$$

$$Z_n'(x) = \frac{1}{2}[Z_{n-1}(x) - Z_{n+1}(x)] \tag{C.13}$$

where $Z_n = J_n$ or Y_n. The following integral relations involving Bessel functions are useful:

$$\int_0^x Z_m^2(kx)x\,dx = \frac{x^2}{2}\left[Z_n'^2(kx) + \left(1 - \frac{n^2}{k^2 x^2}\right) Z_n^2(kx)\right] \tag{C.14}$$

$$\int_0^x Z_n(kx)Z_n(\ell x)x\,dx = \frac{x}{k^2 - \ell^2}[kZ_n(\ell x)Z_{n+1}(kx) - \ell Z_n(kx)Z_{n+1}(\ell x)] \tag{C.15}$$

$$\int_0^{p_{nm}}\left[J_n'^2(x) + \frac{n^2}{x^2}J_n^2(x)\right]x\,dx = \frac{p_{nm}^2}{2}J_n'^2(p_{nm}) \tag{C.16}$$

$$\int_0^{p_{nm}'}\left[J_n'^2(x) + \frac{n^2}{x^2}J_n^2(x)\right]x\,dx = \frac{(p_{nm}')^2}{2}\left(1 - \frac{n^2}{(p_{nm}')^2}\right)J_n^2(p_{nm}') \tag{C.17}$$

where $J_n(p_{nm}) = 0$, and $J_n'(p_{nm}') = 0$. The zeros of $J_n(x)$ and $J_n'(x)$ are on the following two pages.

Zeros of Bessel Functions of First Kind: $J_n(x) = 0$ for $0 < x < 12$

n	1	2	3	4
0	2.4048	5.5201	8.6537	11.7915
1	3.8317	7.0156	10.1735	
2	5.1356	8.4172	11.6198	
3	6.3802	9.7610		
4	7.5883	11.0647		
5	8.7715			
6	9.9361			
7	11.0864			

Extrema of Bessel Functions of First Kind: $dJ_n(x)/dx = 0$ for $0 < x < 12$

n	1	2	3	4
0	3.8317	7.0156	10.1735	13.3237
1	1.8412	5.3314	8.5363	11.7060
2	3.0542	6.7061	9.9695	
3	4.2012	8.0152	11.3459	
4	5.3175	9.2824		
5	6.4156	10.5199		
6	7.5013	11.7349		
7	8.5778			
8	9.6474			
9	10.7114			
10	11.7709			

APPENDIX D USEFUL RESULTS

Maxwell's equations:

$$\nabla \times \bar{E} = -j\omega\mu\bar{H} - \bar{M} \qquad \nabla \cdot \bar{D} = \rho$$
$$\nabla \times \bar{H} = j\omega\epsilon\bar{E} + \bar{J} \qquad \nabla \cdot \bar{B} = 0$$

Surface resistance and skin depth:

$$R_s = \sqrt{\frac{\omega\mu}{2\sigma}} \qquad \delta_s = \sqrt{\frac{2}{\omega\mu\sigma}}$$

Input impedance of terminated lossless transmission lines:

$$Z_{\text{in}} = Z_0 \frac{Z_L + jZ_0 \tan \beta\ell}{Z_0 + jZ_L \tan \beta\ell} \qquad \text{(arbitrary load)}$$

$$Z_{\text{in}} = jZ_0 \tan \beta\ell \qquad \text{(short-circuited line)}$$

$$Z_{\text{in}} = -jZ_0 \cot \beta\ell \qquad \text{(open-circuited line)}$$

Relations between load impedance and reflection coefficient:

$$\Gamma = \frac{Z_L - Z_0}{Z_L + Z_0} \qquad Z_L = Z_0 \frac{1 + \Gamma}{1 - \Gamma}$$

Definitions of return loss, insertion loss and SWR:

$$\text{RL} = -20 \log |\Gamma|, \quad \text{IL} = -20 \log |T|, \quad \text{SWR} = \frac{1 + |\Gamma|}{1 - |\Gamma|}$$

Conversion between dB and nepers:

$$1 \text{ neper} = 8.686 \text{ dB}$$

Elements of the ferrite permeability tensor:

$$\mu = \mu_0 \left(1 + \frac{\omega_0 \omega_m}{\omega_0^2 - \omega^2} \right) \qquad \omega_0 = \mu_0 \gamma H_0$$

$$\omega_m = \mu_0 \gamma M_s$$

$$\kappa = \mu_0 \frac{\omega \omega_m}{\omega_0^2 - \omega^2} \qquad \text{(or 2.8 MHz/Oersted)}$$

Conversion between some values of reflection coefficient, SWR, and return loss:

$|\Gamma|$	0.024	0.032	0.048	0.050	0.056	0.10	0.178	0.200	0.316	0.33
SWR	1.05	1.07	1.10	1.11	1.12	1.22	1.43	1.50	1.92	2.00
RL (dB)	32.3	30.0	26.4	26.0	25.0	20.0	15.0	14.0	10.0	9.6

The *ABCD* Parameters of Some Useful Two-Port Circuits.

Circuit	*ABCD* Parameters	
Z	$A = 1$ $C = 0$	$B = Z$ $D = 1$
Y	$A = 1$ $C = Y$	$B = 0$ $D = 1$
Z_0, β, ℓ	$A = \cos \beta \ell$ $C = jY_0 \sin \beta \ell$	$B = jZ_0 \sin \beta \ell$ $D = \cos \beta \ell$
$N:1$	$A = N$ $C = 0$	$B = 0$ $D = \dfrac{1}{N}$
Y_3, Y_1, Y_2	$A = 1 + \dfrac{Y_2}{Y_3}$ $C = Y_1 + Y_2 + \dfrac{Y_1 Y_2}{Y_3}$	$B = \dfrac{1}{Y_3}$ $D = 1 + \dfrac{Y_1}{Y_3}$
Z_1, Z_2, Z_3	$A = 1 + \dfrac{Z_1}{Z_3}$ $C = \dfrac{1}{Z_3}$	$B = Z_1 + Z_2 + \dfrac{Z_1 Z_2}{Z_3}$ $D = 1 + \dfrac{Z_2}{Z_3}$

APPENDIX E OTHER MATHEMATICAL RESULTS

Useful Integrals

$$\int_0^a \cos^2 \frac{n\pi x}{a} dx = \int_0^a \sin^2 \frac{n\pi x}{a} dx = \frac{a}{2}, \qquad \text{for } n \geq 1 \tag{D.1}$$

$$\int_0^a \cos \frac{m\pi x}{a} \cos \frac{n\pi x}{a} dx = \int_0^a \sin \frac{m\pi x}{a} \sin \frac{n\pi x}{a} dx = 0, \qquad \text{for } m \neq n \tag{D.2}$$

$$\int_0^a \cos \frac{m\pi x}{a} \sin \frac{n\pi x}{a} dx = 0 \tag{D.3}$$

$$\int_0^\pi \sin^3 \theta d\theta = \frac{4}{3} \tag{D.4}$$

Taylor Series

$$f(x) = f(x_0) + (x - x_0) \frac{df}{dx}\bigg|_{x=x_0} + \frac{(x - x_0)^2}{2!} \frac{d^2f}{dx^2}\bigg|_{x=x_0} + \cdots \tag{D.5}$$

$$e^x = 1 + x + \frac{x^2}{2!} + \frac{x^3}{3!} + \cdots \tag{D.6}$$

$$\frac{1}{1 - x} = 1 + x + x^2 + x^3 + \cdots, \qquad \text{for } |x| < 1 \tag{D.7}$$

$$\sqrt{1 + x} = 1 + \frac{x}{2} - \frac{x^2}{8} + \cdots, \qquad \text{for } |x| < 1 \tag{D.8}$$

$$\ln x = 2 \left(\frac{x - 1}{x + 1} \right) + \frac{2}{3} \left(\frac{x - 1}{x + 1} \right)^3 + \cdots, \qquad \text{for } x > 0 \tag{D.9}$$

$$\sin x = x - \frac{x^3}{3!} + \frac{x^5}{5!} + \cdots \tag{D.10}$$

$$\cos x = 1 - \frac{x^2}{2!} + \frac{x^4}{4!} + \cdots \tag{D.11}$$

APPENDIX F PHYSICAL CONSTANTS

- Permittivity of free-space = $\epsilon_0 = 8.854 \times 10^{-12}$ F/m
- Permeability of free-space = $\mu_0 = 4\pi \times 10^{-7}$ H/m
- Impedance of free-space = $\eta_0 = 376.7$ Ω
- Velocity of light in free-space = $c = 2.998 \times 10^8$ m/sec
- Charge of electron = $q = 1.602 \times 10^{-19}$ C
- Mass of electron = $m = 9.107 \times 10^{-31}$ kg
- Boltzmann's constant = $k = 1.380 \times 10^{-23}$ J/°K
- Planck's constant = $h = 6.626 \times 10^{-34}$ J-sec
- Gyromagnetic ratio = $\gamma = 1.759 \times 10^{11}$ C/Kg   (for $g = 2$)

APPENDIX G CONDUCTIVITIES FOR SOME MATERIALS

Material	Conductivity S/m (20°C)	Material	Conductivity S/m (20°C)
Aluminum	3.816×10^7	Nichrome	1.0×10^6
Brass	2.564×10^7	Nickel	1.449×10^7
Bronze	1.00×10^7	Platinum	9.52×10^6
Chromium	3.846×10^7	Sea water	3–5
Copper	5.813×10^7	Silicon	4.4×10^{-4}
Distilled water	2×10^{-4}	Silver	6.173×10^7
Germanium	2.2×10^6	Steel (silicon)	2×10^6
Gold	4.098×10^7	Steel (stainless)	1.1×10^6
Graphite	7.0×10^4	Solder	7.0×10^6
Iron	1.03×10^7	Tungsten	1.825×10^7
Mercury	1.04×10^6	Zinc	1.67×10^7
Lead	4.56×10^6		

APPENDIX H DIELECTRIC CONSTANTS AND LOSS TANGENTS FOR SOME MATERIALS

Material	Frequency	ϵ_r	$\tan \delta$ (25°C)
Alumina (99.5%)	10 GHz	9.5–10.	0.0003
Barium tetratitanate	6 GHz	$37 \pm 5\%$	0.0005
Beeswax	10 GHz	2.35	0.005
Beryllia	10 GHz	6.4	0.0003
Ceramic (A-35)	3 GHz	5.60	0.0041
Fused quartz	10 GHz	3.78	0.0001
Gallium arsenide	10 GHz	13.0	0.006
Glass (pyrex)	3 GHz	4.82	0.0054
Glazed ceramic	10 GHz	7.2	0.008
Lucite	10 GHz	2.56	0.005
Nylon (610)	3 GHz	2.84	0.012
Parafin	10 GHz	2.24	0.0002
Plexiglass	3 GHz	2.60	0.0057
Polyethylene	10 GHz	2.25	0.0004
Polystyrene	10 GHz	2.54	0.00033
Porcelain (dry process)	100 MHz	5.04	0.0078
Rexolite (1422)	3 GHz	2.54	0.00048
Silicon	10 GHz	11.9	0.004
Styrofoam (103.7)	3 GHz	1.03	0.0001
Teflon	10 GHz	2.08	0.0004
Titania (D-100)	6 GHz	$96 \pm 5\%$	0.001
Vaseline	10 GHz	2.16	0.001
Water (distilled)	3 GHz	76.7	0.157

APPENDIX I PROPERTIES OF SOME MICROWAVE FERRITE MATERIALS

Material	Trans-Tech Number	$4\pi Ms$ G	ΔH Oe	ϵ_r	$\tan\delta$	T_c °C	$4\pi Mr$ G
Magnesium ferrite	TT1-105	1750	225	12.2	0.00025	225	1220
Magnesium ferrite	TT1-390	2150	540	12.7	0.00025	320	1288
Magnesium ferrite	TT1-3000	3000	190	12.9	0.0005	240	2000
Nickel ferrite	TT2-101	3000	350	12.8	0.0025	585	1853
Nickel ferrite	TT2-113	500	150	9.0	0.0008	120	140
Nickel ferrite	TT2-125	2100	460	12.6	0.001	560	1426
Lithium ferrite	TT73-1700	1700	<400	16.1	0.0025	460	1139
Lithium ferrite	TT73-2200	2200	<450	15.8	0.0025	520	1474
Yttrium garnet	G-113	1780	45	15.0	0.0002	280	1277
Aluminum garnet	G-610	680	40	14.5	0.0002	185	515

APPENDIX J STANDARD RECTANGULAR WAVEGUIDE DATA

Band*	Recommended Frequency Range (GHz)	TE_{10} Cutoff Frequency (GHz)	EIA Designation WR-XX	Inside Dimensions [Inches (cm)]	Outside Dimensions [Inches (cm)]
L	1.12–1.70	0.908	WR-650	6.500 × 3.250 (16.51 × 8.255)	6.660 × 3.410 (16.916 × 8.661)
R	1.70–2.60	1.372	WR-430	4.300 × 2.150 (10.922 × 5.461)	4.460 × 2.310 (11.328 × 5.867)
S	2.60–3.95	2.078	WR-284	2.840 × 1.340 (7.214 × 3.404)	3.000 × 1.500 (7.620 × 3.810)
H (G)	3.95–5.85	3.152	WR-187	1.872 × 0.872 (4.755 × 2.215)	2.000 × 1.000 (5.080 × 2.540)
C (J)	5.85–8.20	4.301	WR-137	1.372 × 0.622 (3.485 × 1.580)	1.500 × 0.750 (3.810 × 1.905)
W (H)	7.05–10.0	5.259	WR-112	1.122 × 0.497 (2.850 × 1.262)	1.250 × 0.625 (3.175 × 1.587)
X	8.20–12.4	6.557	WR-90	0.900 × 0.400 (2.286 × 1.016)	1.000 × 0.500 (2.540 × 1.270)
Ku (P)	12.4–18.0	9.486	WR-62	0.622 × 0.311 (1.580 × 0.790)	0.702 × 0.391 (1.783 × 0.993)
K	18.0–26.5	14.047	WR-42	0.420 × 0.170 (1.07 × 0.43)	0.500 × 0.250 (1.27 × 0.635)
Ka (R)	26.5–40.0	21.081	WR-28	0.280 × 0.140 (0.711 × 0.356)	0.360 × 0.220 (0.914 × 0.559)
Q	33.0–50.5	26.342	WR-22	0.224 × 0.112 (0.57 × 0.28)	0.304 × 0.192 (0.772 × 0.488)
U	40.0–60.0	31.357	WR-19	0.188 × 0.094 (0.48 × 0.24)	0.268 × 0.174 (0.681 × 0.442)
V	50.0–75.0	39.863	WR-15	0.148 × 0.074 (0.38 × 0.19)	0.228 × 0.154 (0.579 × 0.391)
E	60.0–90.0	48.350	WR-12	0.122 × 0.061 (0.31 × 0.015)	0.202 × 0.141 (0.513 × 0.356)
W	75.0–110.0	59.010	WR-10	0.100 × 0.050 (0.254 × 0.127)	0.180 × 0.130 (0.458 × 0.330)
F	90.0–140.0	73.840	WR-8	0.080 × 0.040 (0.203 × 0.102)	0.160 × 0.120 (0.406 × 0.305)
D	110.0–170.0	90.854	WR-6	0.065 × 0.0325 (0.170 × 0.083)	0.145 × 0.1125 (0.368 × 0.2858)
G	140.0–220.0	115.750	WR-5	0.051 × 0.0255 (0.130 × 0.0648)	0.131 × 0.1055 (0.333 × .2680)

* Letters in parentheses denote alternative designations.

APPENDIX K STANDARD COAXIAL CABLE DATA

RG/U Type	Impedance (Ω)	Inner cond. Diam. (in.)	Dielectric Material	Dielectric Diam. (in.)	Cable Type	Overall Diam. (in.)	Capacitance (pF/ft)	Max. Oper. Voltage	Loss at 1 GHz (dB/100 ft)
RG-8A/U	52	0.0855	P	0.285	braided	0.405	29.5	5000	9.0
RG-9B/U	50	0.0855	P	0.280	braided	0.420	30.8	5000	9.0
RG-55B/U	54	0.0320	P	0.116	braided	0.200	28.5	1900	16.5
RG-58B/U	54	0.0320	P	0.116	braided	0.195	28.5	1900	17.5
RG-59B/U	75	0.0230	P	0.146	braided	0.242	20.6	2300	11.5
RG-141A/U	50	0.0390	T	0.116	braided	0.190	29.4	1900	13.0
RG-142A/U	50	0.0390	T	0.116	braided	0.195	29.4	1900	13.0
RG-174/U	50	0.0189	P	0.060	braided	0.100	30.8	1500	31.0
RG-178B/U	50	0.0120	T	0.034	braided	0.072	29.4	1000	45.0
RG-179B/U	75	0.0120	T	0.063	braided	0.100	19.5	1200	25.0
RG-180B/U	95	0.0120	T	0.102	braided	0.140	15.4	1500	16.5
RG-187/U	75	0.0120	T	0.060	braided	0.105	19.5	1200	25.0
RG-188/U	50	0.0201	T	0.060	braided	0.105	29.4	1200	30.0
RG-195/U	95	0.0120	T	0.102	braided	0.145	15.4	1500	16.5
RG-213/U	50	0.0888	P	0.285	braided	0.405	30.8	5000	9.0
RG-214/U	50	0.0888	P	0.285	braided	0.425	30.8	5000	9.0
RG-223/U	50	0.0350	P	0.116	braided	0.211	30.8	1900	16.5
RG-316/U	50	0.0201	T	0.060	braided	0.102	29.4	1200	30.0
RG-401/U	50	0.0645	T	0.215	semi-rigid	0.250	29.3	3000	—
RG-402/U	50	0.0360	T	0.119	semi-rigid	0.141	29.3	2500	13.0
RG-405/U	50	0.0201	T	0.066	semi-rigid	0.0865	29.4	1500	—

Answers to Selected Problems

1.2 **(a)** $\vec{H} = 0.0424\cos(\omega t - kx)\hat{a}_z$, **(b)** $v_p = 1.875 \times 10^8$ m/sec, **(c)** $\lambda = 0.075$ m, **(d)** $\Delta\varphi = 83.78$ rad or $120°$

1.8 **(a)** $1 - |\Gamma|^2 = 4.911 \times 10^{-4}$ or -33.09 dB, **(b)** $t = 0.16$ mm

1.9 **(a)** $S_i = 46.0$ W/m^2, $S_r = 0.595$ W/m^2, **(b)** $S_{in} = 45.584$ W/m^2

2.1 **(a)** $f = 2.94$ GHz, **(b)** $v_p = 2.31 \times 10^8$ m/sec, **(c)** $\lambda = 0.0785$ m, **(d)** $\varepsilon_r = 1.68$, **(e)** $I(z) = 1.5e^{-j\beta z}$ mA, **(f)** $v(z,t) = 0.1125\cos(1.85 \times 10^{10}t - 80z)$ V

2.3 $L = \frac{\mu_0}{2\pi}\ln b/a = 2.28 \times 10^{-7}$, $C = 2\pi\epsilon_0\epsilon_r/\ln(b/a) = 9.83 \times 10^{-11}$ F/m, $R = \frac{R_s}{2\pi}(1/a + 1/b) = 3.68\,\Omega$/m, $G = 2\pi\omega\epsilon_0\epsilon_r/\ln(b/a)\tan\delta = 1.358 \times 10^{-4}$ S/m, $Z_0 = \sqrt{L/C} = 48.2\,\Omega$, $\alpha = 1/2\left(\frac{R}{Z_0} + GZ_0\right) = 0.0402$ Np/m $= 0.35$ dB/m, $\alpha_{\text{Total}} = \alpha_c + \alpha_d = 0.06697$ Np/m $= 0.582$ dB/m

2.8 $\Gamma_L = 0.359\angle255°$, SWR $= 2.12$, $\Gamma_{in} = 0.359\angle327°$, $\delta_{in} = 1.65 - j0.74$, $Z_{in} = 124 - j55.6\,\Omega$

2.9 $\delta_{in} = 0.42 - j0.85$, $\Gamma_L = 0.62\angle83°$, $\Gamma_{in} = 0.62\angle267°$, SWR $= 4.27$

2.11 $\ell = 2.147$ cm, $\ell = 3.324$ cm

2.12 $Z_0 = 66.7\,\Omega$, $600\,\Omega$

2.18 $P_{\text{INC}} - P_{\text{REF}} = 1.0 - 0.04 = 0.96$ W $= P_{\text{TRANS}}$, $P_{\text{DISS}} + P_{\text{TRANS}} = 0.64 + 0.96 = 1.6 = P_{\text{SOURCE}}$

2.20 **(a)** SWR $= 2.14$, **(b)** $\Gamma = 0.364\angle86.2°$, **(c)** $y_L = (0.73 - j0.62)/75 = 9.84 - j8.2$ mS, **(d)** $\Gamma_{in} = 0.364\angle158°$, $\delta_{in} = 0.48 + j0.15$, $Z_{in} = 35.97 + j11.21\,\Omega$, **(e)** $\ell_{\min} = 0.325\lambda$, **(f)** $\ell_{\max} = 0.075\lambda$

2.29 $\dfrac{(Z_L+Z_0^+)e^{j\beta^+\ell} + (Z_L-Z_0^-)e^{-j\beta^-\ell}}{\frac{1}{z_0^+}(Z_L+Z_0^+)e^{j\beta^+\ell} - \frac{1}{Z_0^-}(Z_L-Z_0^-)e^{-j\beta^-\ell}}$

3.5 loss $= 0.14688$ dB, $\Delta\phi = 1596.26°$

3.6 $\ell \simeq 5.8$ cm

3.9 $f_c = 4.82$ GHz

3.15 $k_c a = 3.12$

3.19 $\lambda_g = 8$ cm, $\alpha_d = 0.068$ dB/cm, $\alpha_c = 0.024$ dB/cm

3.20 $\lambda_g = 8.43$ cm, $\alpha_d = 0.0017$ dB/cm, $\alpha_c = 0.0154$ dB/cm

3.21 $\ell = 3.2$ cm, $Z_{in} = j75.4\,\Omega$

3.27 $v_p = 2.34 \times 10^8$ m/sec, $v_g = 1.85 \times 10^8$ m/sec

3.28 $b/a = 0.5$

4.4 $V_1^+ = 20\angle90°$, $V_1^- = 0$, $Z_{in}^{(2)} = 50\angle90°$

4.14 **(a)** lossy, **(b)** reciprocal, **(c)** return loss $= 1.94$ dB, **(d)** IL $= 4.43$ dB, delay $= 45°$, **(e)** $0.89\angle90°$

4.18 IL $= 12.51$ dB, delay $= -7.97°$

4.20 $P_L = 4.5$ W

4.24 $V_L = 2\angle-90°$

4.30 $\Delta = 0.071$ cm

5.1 **(a)** $C = 1.3263$ pF, $L = 8.62$ nH or $L_1 = 3.315$ nH, $L_2 = 3.315$ nH

5.3 $d = 0.204\lambda$, $\ell = 0.117\lambda$ or $d = 0.374\lambda$, $\ell = 0.382\lambda$

5.5 $\ell = 0.3\lambda$, $Z_1 = 107\,\Omega$

5.11 error $= 4\%$

5.14 $Z_1 = 1.1067Z_0$, $Z_2 = 1.3554Z_0$

5.21 RL < 4.56 dB

6.1 $f_0 = 900$ MHz, $Q_0 = 11.31$, $Q_L = 5.66$

6.5 **(a)** $C = 0.551$ pF, **(b)** $Q = 216.38$

6.9 $f_{101} = 8.249$ GHz, $f_{102} = 11.96$ GHz, $Q_{101} = 6255$, $Q_{102} = 7828$

6.14 $a = 1.305$ cm, $d = 1.535$ cm, $Q_0 = 972.13$

6.18 $f_0 = 9.88$ GHz

6.21 **(c)** $f_0 = 105$ GHz, $Q_c = 121,493$

7.2 through power $= 29.5$ dBm, coupled power $= 5$ dBm, isolated power $= -33$ dBm

7.8 change in output power $= 1.584$ dB

7.13 $S = 6.67$ mm, $r_0 = 2.47$ mm

7.18 $S = 0.3$ mm, $W = 0.6$ mm

7.21 $S = 0.863$ mm, $W = 2.83$ mm, $\ell = 5.06$ mm

7.32 $S_{11} = -1$

8.7 $N = 4$, $C_1 = 0.974$ pF, $L_2 = 5.88$ nH, $C_3 = 2.352$ pF, $L_4 = 2.436$ nH

8.8 $N = 3$, $L_1 = 0.891$ nH, $C_2 = 0.745$ pF, $L_3 = 0.891$ nH

8.10 attenuation $= 10$ dB

8.12

8.16 $\ell_1 = \ell_5 = 0.19$ cm, $\ell_2 = \ell_4 = 0.22$ cm, $\ell_3 = 0.28$ cm

8.18 attenuation $= 30$ dB

8.19 bandwidth about 1.9:1

9.1 (b) $\mu = 13.91\mu_0$, $\kappa = 12.29\mu_0$

9.4 $H_a = 442.86$ Oe

9.6 $L = 1.52$ cm

9.8 229 Oe $< H_0 < 950$ Oe

9.12 (a) $H_0 = 3021$ Oe, (b) $H_0 = 3571$ Oe

9.15 $L = 15.66$ mm

9.17 $L = 49.34$ cm

9.18 $L = 26.16$ cm

10.1 $F = 9.65$ dB

10.4 $F_{cas} = 3.56$ dB

10.7 (a) $F = 6$ dB, (b) $F = 1.76$ dB, (c) $F = 3$ dB

10.14 ratio $= 7$ dB

10.15 $OIP_3 = 6.95$ dBm (coherent)

10.17 LDR $= 73.2$ dB

10.18 LDR $= 86.7$ dB, SFDR $= 55.13$ dB

11.2 ON: IL $= 0.395$ dB, OFF: IL $= 21.47$ dB

11.3 ON: IL $= 0.061$ dB, OFF: IL $= 13.31$ dB

11.7 $R_i = 12.96$ Ω, $C_{gs} = 0.797$ pF, $R_{ds} = 275.86$ Ω, $C_{ds} = 0.52$ pF, $g_m = 47.8$ mS

12.1 (b) $G_A = 0.5$, $G_T = 0.444$, $G = 0.457$

12.4 $C_L = 1.36\angle 46.7°$, $R_L = 0.5$, $K = 0.607$

12.6 A and C are unconditionally stable, B is potentially stable

12.9 $G_T = 7.15$ dB

12.13 -2.9 dB $< G_T - G_{TU} < 4.3$ dB

12.15 $G_T = 3.5$ dB

12.21 $N_{opt} \approx 9$

13.3 $Q_{min} = 5.5$

13.8 $L = 2.5$ nH results in $\mu = -0.931$

13.11 $\mathscr{L}(30$ kHz$) = -120$ dBC/Hz

13.16 $f_{IM} = 1726$ MHz

13.25 $H(\omega) = 1\angle -30°$

14.2 $D = 5.7$ dB

14.4 $D = 35.96$ dB

14.6 $\eta_{rad} = 59\%$

14.8 $G/T = 11.77$ dB/K

14.11 $R = 15.2$ km

14.13 $R = 7.6 \times 10^7$ m (for SNR $= 0$ dB)

14.17 100–2500 Hz

14.23 $|E| = 990$ V/m

Index